Seed Biology: Basic and Translational Research

Seed Biology: Basic and Translational Research

Editor: Nash Belle

STATES
ACADEMIC PRESS
www.statesacademicpress.com

States Academic Press,
109 South 5th Street,
Brooklyn, NY 11249, USA

Visit us on the World Wide Web at:
www.statesacademicpress.com

ISBN: 978-1-63989-479-6 (Hardback)

Cataloging-in-Publication Data

Seed biology : basic and translational research / edited by Nash Belle.
p. cm.
Includes bibliographical references and index.
ISBN 978-1-63989-479-6
1. Seeds. 2. Seeds--Development. 3. Seed pathology. 4. Seed technology. I. Belle, Nash.
SB113.7 .S44 2022
631.521--dc23

Table of Contents

Preface

In my initial years as a student, I used to run to the library at every possible instance to grab a book and learn something new. Books were my primary source of knowledge and I would not have come such a long way without all that I learnt from them. Thus, when I was approached to edit this book; I became understandably nostalgic. It was an absolute honor to be considered worthy of guiding the current generation as well as those to come. I put all my knowledge and hard work into making this book most beneficial for its readers.

Structures and processes related to the evolution and formation of seeds are studied under seed biology. Seed producing plants are also known as spermatophytes. Fundamentally, seeds are means of reproduction. These are further classified into angiosperms and gymnosperms. Angiosperm seeds are produced in a hard protective structure called fruit. Gymnosperm seeds are unenclosed and develop either on the surface of leaves or scales. Seeds occur in various shapes and sizes. They perform several functions such as dispersal to new locations, nourishment to embryo and dormancy during unfavourable conditions. Various factors like weather, insects and diseases can affect the seed production. This book covers in detail some existent theories and innovative concepts revolving around seed biology. From theories to research to practical applications, case studies related to all contemporary topics of relevance to this field have been included herein. This book is a vital tool for all researching or studying seed biology as it gives incredible insights into emerging trends and concepts.

I wish to thank my publisher for supporting me at every step. I would also like to thank all the authors who have contributed their researches in this book. I hope this book will be a valuable contribution to the progress of the field.

Editor

1

Interploidy hybridization barrier of endosperm as a dosage interaction

James A. Birchler*

Division of Biological Science, University of Missouri-Columbia, Columbia, MO, USA

Edited by:
Paolo Sabelli, University of Arizona, USA

Reviewed by:
Claudia Köhler, Swedish Agricultural University, Sweden
Luca Comai, University of California Davis, USA

***Correspondence:**
James A. Birchler, Division of Biological Science, University of Missouri-Columbia, 311 Tucker Hall, Columbia, MO 65211-7400, USA
e-mail: birchlerj@missouri.edu

Crosses between plants at different ploidy levels will often result in failure of endosperm development. The basis of this phenomenon has been attributed to parental gene imprinting of genes involved with endosperm development but a review of the data from maize indicates a dosage interaction between the contributions of the female gametophyte and the primary endosperm nucleus to early endosperm development. However, it is noted that parental imprinting is a non-mutational means that can alter dosage sensitive factors and therefore can contribute to this effect. Operationally, the genes determining ploidy hybridization barrier would qualify for Dobzhansky-Muller incompatibilities that prevent gene flow between species.

Keywords: endosperm, dosage, imprinting, gene balance hypothesis, Dobzhansky-Muller incompatibility, small kernel effect, ploidy hybridization barrier, polyploidy

The endosperm is a nutritive tissue in angiosperms that results from the fusion of one of the two sperm involved in double fertilization with the central cell of the megagametophyte (Birchler, 1993). In many species it is consumed before seed maturation but is persistent in the grains and hence its value for human nutrition. It has long been recognized that interploidy crosses would result in failure of endosperm development and hence seed abortion (**Figure 1A**). The basis of this phenomenon has been a matter of debate but operationally it serves as a hybridization barrier between any polyploid and its diploid progenitor(s), at least in many plants.

One idea to explain the evolution of interploidy hybridization barrier involves parental conflict with regard to resource allocation to the progeny (Haig and Westoby, 1989). The concept is that select genes involved with resource development will be imprinted from one parent or the other in such a manner that a maternal parent will optimize resources to all progeny but any one paternal parent, when there are different ones, will optimize the resources for his own progeny over other potential fathers.

The phenomenon of imprinting of individual genes was discovered by Kermicle in his analysis of the maize anthocyanin gene, *r1* (Kermicle, 1970). It exhibits full color when transmitted through the female but a mottled expression across cells when transmitted through the male parent regardless of dose. Imprinting has also been attributed to endosperm size factors that are found when chromosomal segments are missing from the sperm (Lin, 1982); however, we revisit this interpretation below.

Endosperm size factors refer to the situation that occurs with some translocations between the supernumerary B chromosome and normal A chromosomal segments in maize (Birchler and Hart, 1987). The B chromosome is basically inert and is neither required nor detrimental to plants possessing them unless their numbers exceed about 15 copies. It is maintained in populations by an accumulation mechanism that consists of nondisjunction at the second pollen mitosis, which makes the two sperm, and then preferential fertilization of the egg by the sperm that has the two B chromosomes. Thus, translocations between the B chromosome and the A chromosomes will have nondisjunction of the chromosomal segment attached to the B centromere. When some B-A translocations are used as a male parent, the progeny missing a paternal contribution to the endosperm are smaller than normal siblings (Roman, 1947; Lin, 1982; Birchler and Hart, 1987).

Several regions of the genome will produce this effect to a greater or lesser degree including 1S, 1L, 4S, 5S, 7L, and 10L being the most prominent in most backgrounds (Birchler and Hart, 1987). At least for the effect of 1L and 10L, there is evidence that the effect is cumulative from several regions that contribute to the whole arm impact (Lin, 1982; Birchler and Hart, 1987).

The argument that this small kernel effect is a reflection of imprinting was that for 10L, introduction of extra copies through the female parent was not observed to have any effect nor could it rescue the absence of the paternal copy (Lin, 1982). While this is the case for 10L in some backgrounds, extra copies of other chromosome arms introduced through the female parent does not rescue but enhances the paternal small kernel effect (Birchler and Hart, 1987). Indeed, the specific region of 10L that itself is responsible for a small kernel effect will enhance the analogous effect of 1S when transmitted in extra dosage through the female parent. This observation suggests that these genes function when passed through the female (including in the zygote) or at least in the female gametophyte.

Moreover, by crossing B-A translocations among themselves, it was realized that the same arms that produce the paternal effect would enhance that response of other arms when present in extra copy through the female parent (Birchler and Hart, 1987). Indeed, in the author's materials, self pollination of the 10L translocations produce an additional class of further reduced

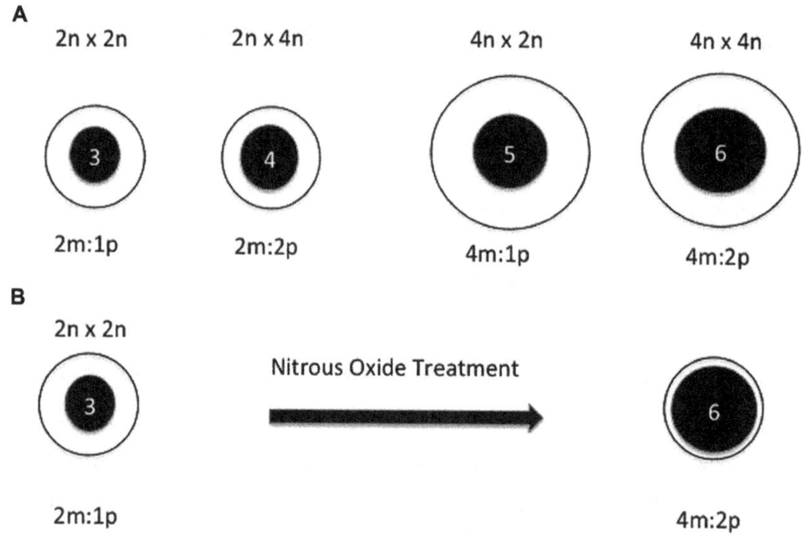

FIGURE 1 | (A) Depictions of the primary endosperm cell after fertilization of normal and interploidy crosses. From left to right: diploid, interploidy diploid by tetraploid, interploidy tetraploid by diploid, tetraploid. The crosses are noted across the top. The cell size is depicted by the clear circles and the nucleus size is depicted by the inner filled circles. The ploidy of each nucleus is noted within and the maternal to paternal genomic relationship is shown below each image. The interploidy endosperms both result in developmental failure. In the diploid and tetraploid crosses the maternal to paternal relationship is the same as is the maternal gametophytic relationship to the genomic dosage after fertilization. In both of the interploidy crosses, the maternal to paternal contributions are changed as is the maternal gametophytic relationship to the genomic dosage after fertilization. **(B)** Depiction of the primary endosperm cell before and after chromosome doubling via nitrous oxide treatment. At the left is depicted a primary endosperm cell from a normal diploid plant. At the right is shown a depiction when the nucleus is doubled in chromosome number from triploid to hexaploid in the progenitor smaller cell. In the latter, even though the maternal to paternal relationship is the same as a tetraploid, the maternal gametophytic contributions would be typical of a diploid but with a doubled genome with which to interact. Based on a figure from Bauer (2006).

sized kernels compared with crossing the translocation to normal females—just as occurs with all other regions of the genome that produce the small kernel effect. This result suggests that the responsible loci are in fact expressed when transmitted through the female parent and that they are involved in the same developmental process. The reason why this experience differs from Lin's results is unknown but would seem to reflect background effects. Furthermore, the enhancement results indicate that these factors operate in a dosage sensitive manner and also that the stoichiometry of one region relative to others would magnify the effect. It was postulated that the developmental program established in the female gametophyte has a quantitative component that interacts with the dosage of the primary endosperm nucleus following fertilization (Birchler and Hart, 1987; **Figure 1A**).

Indeed, if one postulates that the small kernel effects are due to the absence of a gene that is normally expressed from only the paternal allele, then no such hypothetical gene could be vital because no known region of the maize genome is lethal to the endosperm when missing in the sperm. The maternal enhancement seems unlikely to result from imprinting because it does not involve an all or none contribution from the female parent but rather the number of copies transmitted. Moreover, there is no apparent impact on endosperm growth unless the same or other regions of the genome are paternally absent. Also, it is important to note that there are many mutations that when homozygous recessive are highly defective to the endosperm but no paternal absence in the genome has any such effect suggesting that none of these genes are expressed exclusively from the paternally derived allele. These considerations point to a quantitative explanation for the small kernel effects.

Many years later, studies of tetraploid formation suggested a related explanation of the interploidy disruption of endosperm development. Kato and Birchler (2006) produced tetraploid derivatives of several inbred lines of maize by treating self-pollinated diploid plants with nitrous oxide gas, which causes chromosome doubling, at about the time of the first mitotic divisions after fertilizations. Interestingly, the kernels on the ears of such treated plants have many defective endosperms. If the endosperms were doubled in ploidy, there would be no change in maternal/paternal genomic relationship but rather a change of the quantitative relationship of the maternal gametophytic gene products to the copy number of genomes in the now doubled primary endosperm nucleus (**Figure 1B**).

Bauer (2006) examined this phenomenon in more detail. Treatment of pollinated diploid plants near the timing of the first endosperm mitosis produced numerous defective endosperms. Cytological analysis indicated doubling of chromosome numbers to 60 chromosomes in the defective kernel class although some defective kernels were doubled twice or three times to yield ~120 or ~240 chromosomes, respectively. A time course of nitrous oxide treatment revealed very little effect at 12–17 hours after pollination (HAP), which immediately precedes the first division in the endosperm. The percentage of defective endosperms with elevated ploidy increased progressively with treatments at 14–19

HAP, 16–21 HAP and 20–15 HAP and then sharply decreased with treatment at 24–29 HAP. The latter result is not due to a failure of the action of nitrous oxide because at this timepoint, there is a sharp increase in the number of normal-sized hexaploid endosperms at the expense of normal-sized triploid endosperms. The results are consistent with the interpretation that genome doubling to produce defective endosperms has an early developmental window.

It should be noted that interploidy crosses have multiple variables that complicate their interpretation. They vary the maternal to paternal genomic ratios within the endosperm itself but they also vary the contribution of female gametophyte contributions to the number of genomic targets after fertilization. The chromosome doubling experiment separates these variables. The maternal to paternal ratio within the endosperm is maintained but the relationship of the maternal gametophytic contributions to the genomic targets is altered suggesting this possibility as the basis of the endosperm failure.

The timing of this relationship can be further deduced based on the behavior of B-A translocations after fertilization (Birchler, 1980). At a low frequency, B-A chromosomes are lost during development of the endosperm. If they carry an anthocyanin pigment marker, such loss can be readily recognized as a mosaic kernel. For the regions of the genome that produce the small kernel effect when there is no paternal contribution to the endosperm at all, there is no detectable effect in these mosaic kernels even for those with loss at early divisions (Birchler, 1980). These observations indicate that the maternal/primary endosperm relationship is critical at the initiation of endosperm development but not shortly thereafter. This nonautonomy could potentially be a reflection of the syncytial nature of early endosperm development but this consideration does not rule out the noted critical relationship. It should be noted that the syncytial nature of early endosperm development does not obscure the mosaic pattern of the anthocyanin marker or other mosaicism that results from transposable element action or chemical or irradiation mutagenesis of pollen.

Given that the interploidy endosperm failure can be mimicked by changing the maternal/zygotic dosage without changing the maternal/paternal relationship and the small kernel effect of paternal absence of regions of the genome is enhanced by maternal increase might suggest that segmental or genomic relative dosage might be responsible. These effects might not necessarily be the result of parental imprinting. However, some factors involved with early endosperm development, which have been identified in Arabidopsis, are in fact expressed only from alleles originating from one parent but not the other (Dilkes and Comai, 2004; Kradolfer et al., 2013). But imprinting results in a type of dosage effect and so an entanglement of interpretations is potentially possible. Dilkes and Comai (2004) have discussed development of the endosperm and how differential resource allocation among different paternal parents seems unlikely. It is potentially the case then that imprinting is a non-mutational process of manipulating the dosage of genes (Beaudet and Jiang, 2002) involved with endosperm development (Kradolfer et al., 2013). Those that exhibit imprinting that are involved with endosperm development are likely a subset of the whole group of genes that affect this process. Clearly, genes can be imprinted in the endosperm that have no impact on kernel size (e.g., the *r1* locus), which fact might also be taken to suggest that imprinting and resource allocation are not necessarily connected. Indeed, endosperm size is basically determined maternally; there is no perceptible difference in size when pollen parents from lines with very different endosperm sizes are used onto a common female line. Thus, there is no evidence in maize to suggest that the paternal parent has any influence on endosperm resource allocation.

What then is the driving evolutionary force for ploidy hybridization barrier in the endosperm? There may be none: it might simply be a neutral reflection of developmental and gene regulatory processes that have dosage components. However, operationally, it might serve to prevent the widespread occurrence of triploids in populations if allotetraploids and the diploid progenitors were to hybridize. While such hybridization in itself would not be productive, the widespread occurrence of any resulting triploids that would hybridize with both tetraploids and diploids to produce many aneuploid progeny would likely disrupt the population fitness to a much greater degree. Individuals with an inability to produce triploids would have a higher reproductive fitness. Mechanisms that prevent triploid production would potentially be selected and were apparently established early in angiosperm evolution but, as noted, this might be a neutral reflection of developmental mechanism.

There is considerable evidence from many experimental avenues that genomic balance when upset can have detrimental effects on the phenotype (Birchler and Veitia, 2010, 2012). As described above, the small kernel effect and the endosperm doubling results are consistent with this type of stoichiometric relationship. Thus, the endosperm interploidy hybridization barrier is likely an extension of the genomic balance phenomena. Indeed, all other tissues show impacts of aneuploidy so it would be unusual if the endosperm did not: the small kernel effects are the likely manifestation.

It is known from work primarily in potato that the genes for interploidy barrier are multigenic and can be overcome with manipulation of genome dosage (Johnston et al., 1980; Johnston and Hanneman, 1982). With divergence of the quantitative expression of these genes, a hybridization barrier occurs without a change in ploidy. These cases might qualify as classical Muller-Dobzhansky species incompatibility genes (Dobzhansky, 1937; Muller, 1942) that are compatible within species but not between species (Kradolfer et al., 2013). Consequently, while the responsible genes might diverge in a neutral fashion, they serve de facto as an isolating mechanism between polyploids and related diploids.

Based on the evidence described above, we argue that the ploidy hybridization barrier and the small kernel effect from segmental paternal absence represent a type of dosage interaction between the maternal contributions from the female gametophyte and the genome or segmental dosage in the primary endosperm nucleus. Because this interaction has a stoichiometric character to its behavior, it is likely related to the basis of standard aneuploidy syndromes (Birchler and Veitia, 2012).

REFERENCES

Bauer, M. J. (2006). *The interploidy hybridization barrier in Zea mays* L. Ph.D. Dissertation, University of Missouri-Columbia, Columbia, MO

Beaudet, A. L., and Jiang, Y.-H. (2002). A rheostat model for a rapid and reversible form of imprinting-dependent evolution. *Am. J. Hum. Genet.* 70, 1389–1397. doi: 10.1086/340969

Birchler, J. A. (1980). On the nonautonomy of the small kernel phenotype produced by B-A translocations in maize. *Genet. Res.* 36, 111–116. doi: 10.1017/S0016672300019716

Birchler, J. A. (1993). Dosage analysis of maize endosperm development. *Annu. Rev. Genet.* 27, 181–204. doi: 10.1146/annurev.ge.27.120193.001145

Birchler, J. A., and Hart, J. R. (1987). Interaction of endosperm size factors in maize. *Genetics* 117, 309–317.

Birchler, J. A., and Veitia, R. A. (2010). The gene balance hypothesis: implications for gene regulation, quantitative traits and evolution. *New Phytol.* 186, 54–62. doi: 10.1111/j.1469-8137.2009.03087.x

Birchler, J. A., and Veitia, R. A. (2012). Gene balance hypothesis: connecting issues of dosage sensitivity across biological disciplines. *Proc. Natl. Acad. Sci. U.S.A.* 109, 14746–14753. doi: 10.1073/pnas.1207726109

Dilkes, B. P., and Comai, L. (2004). A differential dosage hypothesis for parental effects in seed development. *Plant Cell* 16, 3174–3180. doi: 10.1105/tpc.104.161230

Dobzhansky, T. (1937). Genetics and the Origin of Species. New York: Columbia University Press.

Haig, D., and Westoby, M. (1989). Parent specific gene expression and the triploid endosperm. *Am. Nat.* 134, 147–155. doi: 10.1086/284971

Johnston, S. A., den Nijs, T. P. M., Peloquin, S. J., and Hanneman, R. E. Jr. (1980). The significance of genic balance to endosperm development in interspecific crosses. *Theor. Appl. Genet.* 57, 5–9. doi: 10.1007/BF00276002

Johnston, S. A., and Hanneman, R. E. Jr. (1982). Manipulations of endosperm balance number overcome crossing barriers between diploid *Solanum* species. *Science* 217, 446–448. doi: 10.1126/science.217.4558.446

Kato, A., and Birchler, J. A. (2006). Induction of tetraploid derivatives of maize inbred lines by nitrous oxide gas treatment. *J. Hered.* 97, 39–44. doi: 10.1093/jhered/esj007

Kermicle, J. L. (1970). Dependence of the R-mottled phenotype in maize on mode of sexual transmission. *Genetics* 66, 69–85.

Kradolfer, D., Wolff, P., Jiang, H., Siretskiy, A., and Kohler, C. (2013). An imprinted gene underlies postzyogtic reproductive isolation in *Arabidopsis thaliana*. *Dev. Cell* 26, 1–11. doi: 10.1016/j.devcel.2013.08.006

Lin, B.-Y. (1982). Association of endosperm reduction with parental imprinting in maize. *Genetics* 100, 475–486.

Muller, H. J. (1942). Isolating mechanisms, evolution and temperature. *Biol. Symp.* 6, 71–125.

Roman, H. (1947). Mitotic nondisjunction in the case of interchanges involving the B-type chromosome in maize. *Genetics* 32, 391–409.

Conflict of Interest Statement: The author declares that the research was conducted in the absence of any commercial or financial relationships that could be construed as a potential conflict of interest.

2

Proteome balancing of the maize seed for higher nutritional value

Yongrui Wu[1] and Joachim Messing[2]*

[1] *Institute of Plant Physiology and Ecology, Shanghai Institutes for Biological Sciences, Chinese Academy of Sciences, Shanghai, China*
[2] *Waksman Institute of Microbiology, Rutgers University, Piscataway, NJ, USA*

Edited by:
Brian A. Larkins, University of Nebraska-Lincoln, USA

Reviewed by:
A. Mark Settles, University of Florida, USA
Marcelo Carnier Dornelas, Universidade Estadual de Campinas, Brazil

****Correspondence:***
Joachim Messing, Waksman Institute of Microbiology, Rutgers University, 190 Frelinghuysen Road, Piscataway, NJ 08854, USA
e-mail: messing@waksman.rutgers.edu

Most flowering plant seeds are composed of the embryo and endosperm, which are surrounded by maternal tissue, in particular the seed coat. Whereas the embryo is the dormant progeny, the endosperm is a terminal organ for storage of sugars and amino acids in proteins and carbohydrates, respectively. Produced in maternal leaves during photosynthesis, sugars, and amino acids are transported to developing seeds after flowering, and during germination they nourish early seedlings growth. Maize endosperm usually contains around 10% protein and 70% starch, and their composition ratio is rather stable, because it is strictly regulated through a pre-set genetic program that is woven by networks of many interacting or counteracting genes and pathways. Endosperm protein, however, is of low nutritional value due mainly to the high expression of the α-zein gene family, which encodes lysine-free proteins. Reduced levels of these proteins in the *opaque 2* (*o2*) mutant and α-zein RNAi (RNA interference) transgenic seed is compensated by an increase of non-zein proteins, leading to the rebalancing of the nitrogen sink and producing more or less constant levels of total proteins in the seed. The same rebalancing of zeins and non-zeins has been observed for maize seeds bred for 30% protein. In contrast to the nitrogen sink, storage of sulfur is controlled through the accumulation of specialized sulfur-rich proteins in maize endosperm. Silencing the synthesis of α-zeins through RNAi fails to raise sulfur-rich proteins. Although overexpression of the methionine-rich δ-zein can increase the methionine level in seeds, it occurs at least in part at the expense of the cysteine-rich β- and γ-zeins, demonstrating a balance between cysteine and methionine in sulfur storage. Therefore, we propose that the throttle for the flow of sulfur is placed before the synthesis of sulfur amino acids when sulfur is taken up and reduced during photosynthesis.

Keywords: nitrogen, lysine, sulfur, storage proteins, RNAi

STORAGE PROTEINS IN MAIZE SEED

Angiosperm seeds result from double fertilization and are usually the primary mode of reproduction. Besides their vital biological function, seeds are the most frequently harvested organs in agriculture. Therefore, their production has tremendous economic importance for humans and livestock. For this reason, maize has become one of the most productive cereal crops in the world in respect to yield per acreage. Maize seeds mainly consist of endosperm and embryo, which account for 90 and 10%, respectively, of the whole dry seed weight (Flint-Garcia et al., 2009). Starch and protein are mainly stored in endosperm, whereas most of the oil accumulates in the embryo.

Maize seeds contain ~10% proteins and ~70% of them are classified as storage proteins (Flint-Garcia et al., 2009). Based on their solubility in different solvents, endosperm proteins are divided into four groups: albumins, globulins, glutamines, and prolamins. The latter, called zeins, make up > 60% of total proteins (**Figure 1**). Zeins can be divided into four subfamilies, α (19 and 22 kDa), γ (50, 27, and 16 kDa), β (15 kDa), and δ (18 and 10 kDa; Esen, 1987; Coleman and Larkins, 1998; **Figure 1**). A common feature of all prolamins is internal tandem variable repeats of blocks of amino acids with primarily proline and glutamine, as first observed in a maize α-zein (Gerghaty et al., 1981). Because of this feature, α-zeins lack essential amino acids like lysine, methionine, and tryptophan. Due to the high expression of α-zein genes in maize endosperm, the final levels of these three essential amino acids in total protein are very low (Osborne and Mendel, 1914). Therefore, maize cannot serve as a balanced dietary protein source for humans and monogastric animals and has to be supplemented with these amino acids, raising the cost of food supply worldwide (Mertz et al., 1964). Interestingly, the level of one amino acid, methionine, can reach sufficient levels in some cultivars, making supplements redundant (Messing and Fisher, 1991). The reason for this is that minor zeins, β, γ, and δ, have a high proportion of sulfur-rich amino acids and can vary in expression levels among maize cultivars. The δ-zeins are very rich in methionine, whereas the γ-zeins are abundant in cysteine; β-zein has high percentages of cysteine and methionine, while α-zeins lack both of them (Wu et al., 2012).

PROTEOME BALANCING IN MAIZE ENDOSPERM

Maize domestication from its wild ancestor, teosinte, can be traced back to the Tehuacan Valley of Mexico as early as 8,000 years ago.

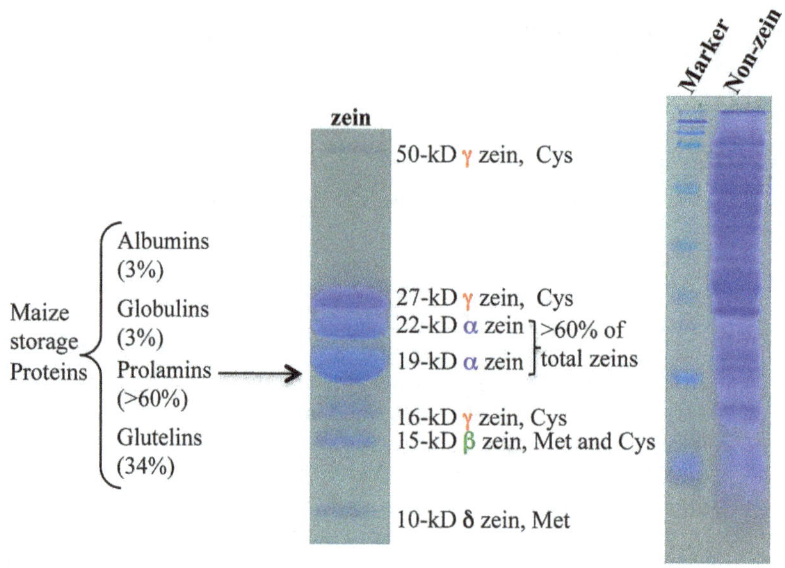

FIGURE 1 | Maize storage proteins. Prolamin proteins in maize are called zeins and the others are all classified as non-zeins. Adapted from Wu et al. (2012).

During this process, teosinte underwent dramatic changes, not only in plant morphology, but also in seed composition (Flint-Garcia et al., 2009). Teosinte contains ~30% protein and has a high level of the methionine-rich δ-zeins (Swarup et al., 1995), but modern maize has only ~10% total protein and a low level of methionine-rich proteins in cultivars grown for consumption. Although one can achieve among all crops the highest yields of grain with maize, its protein level is much lower than soybean, which contains ~35% protein with sufficient lysine levels.

To investigate whether artificial selection can significantly change seed compositions, a long-term selection experiment has been carried out for more than one century at the University of Illinois. This has yielded four strains with substantially different protein levels: Illinois high protein (IHP), Illinois low protein (ILP), Illinois high protein reverse (IHPR), and Illinois low protein reverse (ILPR) with protein levels of 30, 4, 7, and 15%, respectively (Hopkins, 1899; Dudley and Lambert, 2004; Moose et al., 2004; Dudley, 2007). However, elevated protein levels are mainly due to zeins, which of course lack lysine (Osborne and Mendel, 1914). Therefore, IHP contains even relatively lower lysine levels (Lys_{rel}) than those in normal maize (Wu and Messing, 2012).

An unexpected finding was when zein levels are lowered by the *opaque-2* mutation the relative lysine content was improved to a nearly sufficient level. This mutation affects an endosperm-specific transcription factor belonging to the bZIP family that is required for transactivation of several zein gene subfamilies (Schmidt et al., 1992; Cord Neto et al., 1995). The reduction in zein gene expression results in seeds with an opaque appearance. In the *o2* mutant, the main zein components, the α-zeins, are reduced by more than 60% in certain inbred lines. However, the total protein level remains almost unchanged by a compensatory increase of non-zein proteins with higher lysine levels (Holding and Larkins, 2009). As a consequence, the percentage of overall lysine is elevated. This compensation phenomenon indicates that nitrogen storage is controlled at the level of protein synthesis, leading to a more or less constant amount of total protein. However, as *o2* is recessive, pleiotropic and its penetration can vary in different α-zein haplotypes (Song and Messing, 2003), this trait requires two parental lines that are homozygous for *o2* and have additional QTLs for seed quality for hybrid seed production. Such QTLs, namely *o2* modifiers, are required to convert the starchy *o2* endosperm, which is unfavorable for storage and transport of large volumes of maize, to a hard kernel texture. This modified *o2* maize mutant is known as "Quality Protein Maize" or QPM (Vasal et al., 1980; Holding et al., 2008). Because of the loss of the opaque phenotype in QPM, it becomes difficult for breeders to maintain *o2* homozygosity through visual scoring. To simplify QPM breeding, high lysine maize lines can be created with RNA interference (RNAi) mutants, which reduce α-zein mRNA in a dominant and more targeted fashion. However, in the absence of *o2* modifiers, the resulting transgenic seeds also present an opaque phenotype (Segal et al., 2003; Huang et al., 2006; Wu and Messing, 2011).

Although QPM has a hard endosperm and contains higher lysine than normal (Vasal et al., 1980), the total protein levels are still lower than in soybeans (Prasanna et al., 2001). If one could create maize lines that rival the nutritional quality of soybeans being high-protein and high-lysine, while having a hard-endosperm texture, one could investigate whether total protein could also be rebalanced by the mechanism operating in IHP. Indeed, when an α-zein RNAi event was crossed with IHP, the total protein level was maintained, although zeins were substantially reduced. Consequently, the non-zein fraction was dramatically increased to compensate for the loss of zeins (**Figure 2**). Moreover, suppression of zeins with an α-zein RNAi is incomplete, leaving a

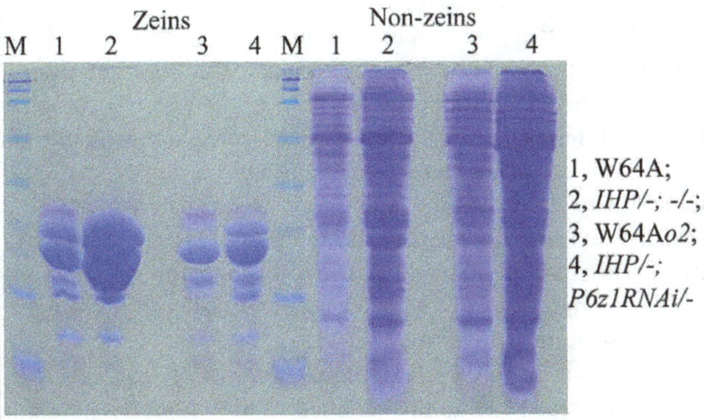

FIGURE 2 | Proteome rebalancing in *o2* mutant and IHP with suppressed α-zeins. Lane 1 and 4, W64A and W64Ao2; lane 2 and 4, two F1 progeny of IHP × α-zeinRNAi/− not inheriting and inheriting the RNAi. Adapted from Wu and Messing (2012).

considerable amount of residual zeins that provide a hard vitreous endosperm texture without the need for *o2* modifiers (Wu and Messing, 2012).

What could be the mechanism underlying rebalancing the seed proteome? Sugars and amino acids produced during photosynthesis are transported to seeds for deposition as starch and protein. It seems that developing maize seeds possess compensatory mechanisms that sense protein content when zein synthesis is interrupted, leading to translation of other mRNAs instead of zein mRNAs. This transfer of ribosomes to a different mRNA pool could be as simple as mass action, or involve an intracellular signal transduction to attain a predetermined protein level. Such signal transduction would likely occur, although not exclusively, at the transcriptional, posttranscriptional or translational levels (Frizzi et al., 2010; Jia et al., 2013). No matter how proteome rebalancing operates to alter seed composition, breeders had to take a long-term selection approach to accumulate QTLs to regulate this tightly controlled program (Dudley and Lambert, 2004). When synthesis of soybean's major storage proteins, glycinin and conglycinin, were suppressed in knockdown mutant lines, the seeds maintained nearly identical levels of total protein compared to the untransformed soybean cultivars, with similar seed size and weight (Schmidt et al., 2011), suggesting that proteome rebalancing in seeds might be a rather common event, providing a constant sink for reduced nitrogen during seed maturation. In addition, plant seeds seem to possess the ability to overcome a protein shortage by remodeling their protein composition for use during germination and early seeding growth. Profiling non-zein accumulation in *o2* and *α-zeinRNAi* mutants appears to follow two distinct patterns, with an overall slight increase of proteins in general and significant overexpression of several specific proteins (Holding and Larkins, 2009; Wu et al., 2012; Jia et al., 2013). Among the specifically enhanced expressed proteins, eIF2 and GAPDH have been identified as high-lysine containing proteins, which were thought to add a substantial contribution to the overall lysine elevation (Habben et al., 1993, 1995; Jia et al., 2013). However, what regulates their expression when zeins are suppressed remains unclear.

SULFUR REBALANCING IN MAIZE ENDOSPERM

Sulfur amino acid deficiency differs from lysine deficiency in several ways. The essential amino acid methionine is the only amino acid that is currently chemically synthesized for supplementation of animal feed, because even soybean proteins do not provide sufficient levels in a dietary ration. However, in contrast to lysine, maize produces β- and δ-zein proteins that are very high in methionine residues. But in most maize inbreds, they just do not accumulate at sufficient levels although it has been shown that increased levels of sulfur in the soil can increase synthesis of sulfur-rich proteins in peas (Beach et al., 1985). Screens of seeds from different maize genetic backgrounds by germinating in the presence of lysine and threonine (LT) resulted in the discovery of one LT-resistant line (Phillips, 1985), where the δ-zein gene, rich in methionine codons, was overexpressed (Kirihara et al., 1988). It also was shown that this differential expression was subject to parental imprinting in hybrid crosses (Chaudhuri and Messing, 1994). This high δ-zein line was sufficient to replace synthetic methionine in a regular feed for chickens, with a direct impact on weight and feather quality (Messing and Fisher, 1991). Interestingly, the high expression of the δ-zein gene is not due to transcription, but rather to post-transcriptional regulation of its mRNA (Schickler, 1993). In fact, it appears that the regulation occurs via the un-translated regions (UTRs) of the mRNA, which was shown in transgenic seeds when the δ-zein mRNA UTRs were replaced by other sequences (Lai and Messing, 2002).

The allele-specific regulation of the methionine content, however, would unlikely be a pathway for rebalancing protein composition in seeds. In fact, high-lysine maize lines, like the *o2* mutant, where lysine-free α-zein proteins are reduced with a compensatory increase of other proteins, have failed to show any increase but rather somewhat of a decreased methionine level (Mertz et al., 1964; Phillips, 1985; Wu et al., 2012). This could also be an effect on the transcription of δ-zeins by O_2 itself, as it also regulates another methionine-rich zein gene, the β-zein. Indeed, in knock-out mutants of δ-zeins in combination with a knock-down of β-zein, the accumulation of methionine is 40% less than that in normal maize lines (Wu et al., 2012). To eliminate the pleiotropic effects of O_2 and only reduce the expression of α-zeins, methionine levels were also evaluated in the presence of α-zein RNAi. In this case, the methionine level did not increase along with lysine, showing that non-zein proteins are not as rich in sulfur amino acids as in lysine. Indeed, only 8% of the proteins in the maize protein database have methionine residues above 4%, while about 57% have lysine residues above 4% (Wu et al., 2012). Therefore, it appears that the sink for reduced nitrogen and sulfur operate differently during seed development.

There is no apparent inferior kernel phenotype in high-methionine maize, in contrast to high-lysine α-zein RNAi maize, indicating that single gene manipulation could add a stable trait without agronomic compromises. Indeed, after several back-crosses of the chimeric δ-zein gene to a maize line that is low in methionine, overexpression of the 10-kDa zein gene remained stable (Lai and Messing, 2002). However, the transgenic line exhibited an interesting biochemical difference compared to normal maize (Wu et al., 2012). Unexpectedly, the accumulation of cysteine-rich γ- and β-zeins was dramatically suppressed (**Figure 3**). A recent study also found that selecting high-methionine variants from a maize population apparently resulted in low accumulation of the cysteine-rich 27-kDa γ-zein (Newell et al., 2014). These results suggest that increased methionine storage requires increased flow of reduced sulfur from cysteine to methionine, thereby reducing the translation of 27-kDa γ-zein mRNA. However, this shift could also be achieved with RNAi against γ- and β-zeins.

Perhaps, this could be explained as follows. Methionine and cysteine are the only two sulfur-containing amino acids among the twenty protein-containing L-amino acids. Sulfur is one of the essential elements for plant growth and is absorbed by root as sulfate (SO_4^{2-}) with an oxidation state +6. Sulfate has to be reduced to −2 through several enzymatic steps to form the intermediate product, cysteine. Because γ- and β-zeins are the cysteine-rich storage proteins, silencing their expression causes a significant reduction in cysteine level, indicating these proteins are the main sink for cysteine storage (Wu et al., 2012). If free cysteine is the first stable product with reduced sulfur, the majority of which is incorporated into cysteine-rich proteins, like γ-zeins and the excess would flow into methionine, its concentration would drive the translation of the δ- and β-zein (β-zein is rich both in cysteine and methionine) mRNA. Indeed, in RNAi against γ- and β-zeins, one can observe a boost in the accumulation of δ-zeins. Therefore, depriving the cysteine sink or increasing the methionine sink has the same result in that the flux of reduced sulfur flows through cysteine to methionine. This balance is made possible through the expression of different single/low copy number genes specialized for storage of these two amino acids (Wu et al., 2012).

Based on the above hypothesis, the major bottleneck for increasing seed methionine content is the capacity of sulfur absorption by roots and the efficiency by which sulfur can be reduced in the leaves during photosynthesis. Three enzymes ATP sulfurylase, APS reductase (APR) and sulfite reductase in this pathway combine coordinately to reduce sulfate with oxidation state +6 to sulfide with oxidation state −2 (**Figure 4**). Meanwhile, *O*-acetylserine (OAS), the other precursor for cysteine syhthesis, is formed from serine and acetyl-CoA catalyzed by serine acetyltransferase (SAT). And last, sulfide reacts with OAS, producing the end assimilation product of cysteine catalyzed by OAS thiol-lyase (Leustek et al., 2000). We propose such capacity of sulfur reduction could be enhanced by specific overexpression of the committing enzymes APR and SAT in leaf bundle sheath cells, where the sulfur reduction occurs and as a consequence improve cysteine and methionine sinks in seed.

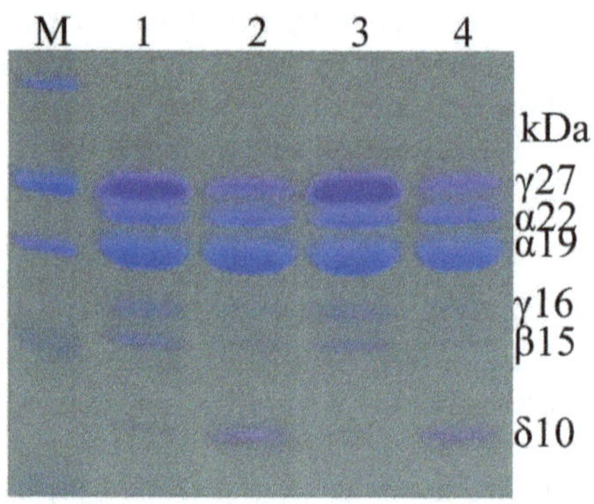

FIGURE 3 | Accumulation patterns of zeins in high-methionine transgenic and WT seeds. Hi-Met transgenic seeds in lane 2 and 4 express much higher levels of the 10-kDa δ-zein, but lower levels of β- and γ-zeins than WT in lane 1 and 4. Adapted from Wu et al. (2012).

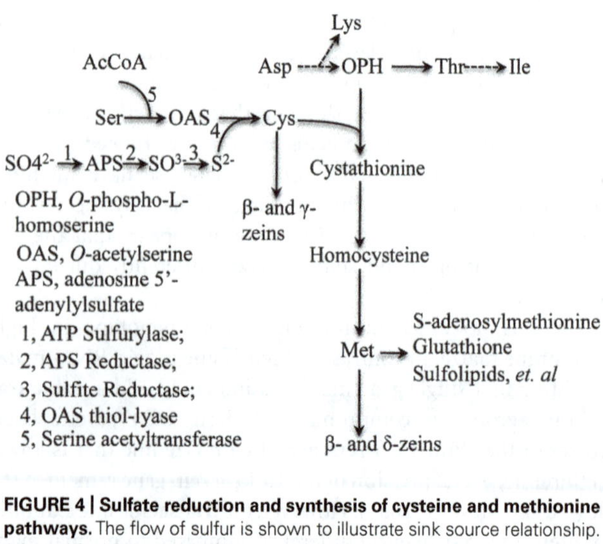

FIGURE 4 | Sulfate reduction and synthesis of cysteine and methionine pathways. The flow of sulfur is shown to illustrate sink source relationship. Adapted from Wu et al. (2012).

REFERENCES

Beach, L. R., Spencer, D., Randall, P. J., and Higgins, T. J. (1985). Transcriptional and post-transcriptional regulation of storage protein gene expression in sulfur-deficient pea seeds. *Nucleic Acids Res.* 13, 999–1013. doi: 10.1093/nar/13.3.999

Chaudhuri, S., and Messing, J. (1994). Allele-specific parental imprinting of dzr1, a posttranscriptional regulator of zein accumulation. *Proc. Natl. Acad. Sci. U.S.A.* 91, 4867–4871. doi: 10.1073/pnas.91.11.4867

Coleman, C. E., and Larkins, B. A. (1998). "The prolamins of maize," in *Seed Proteins*, eds P. R. Shewry and R. Casey (Dordrecht: Kluwer Academic Publishers), 109–139.

Cord Neto, G., Yunes, J. A., da Silva, M. J., Vettore, A. L., Arruda, P., and Leite, A. (1995). The involvement of Opaque 2 on beta-prolamin gene regulation in maize and Coix suggests a more general role for this transcriptional activator. *Plant Mol. Biol.* 27, 1015–1029. doi: 10.1007/BF00037028

Dudley, J. W. (2007). From means to QTL: the Illinois long-term selection experiment as a case study in quantitative genetics. *Crop Sci.* 47, S20–S31. doi: 10.2135/cropsci2007.04.0003IPBS

Dudley, J. W., and Lambert, R. J. (2004). 100 Generations of selection for oil and protein in corn. *Plant Breed. Rev.* 24, 79–110. doi: 10.1002/9780470650240.ch5

Esen, A. (1987). A proposed nomenclature for the alcohol-soluble proteins (zeins) of maize (*Zea Mays* L.). *J. Cereal Sci.* 5, 117–128. doi: 10.1016/S0733-5210(87)80015-2

Flint-Garcia, S. A., Bodnar, A. L., and Scott, M. P. (2009). Wide variability in kernel composition, seed characteristics, and zein profiles among diverse maize inbreds, landraces, and teosinte. *Theor. Appl. Genet.* 119, 1129–1142. doi: 10.1007/s00122-009-1115-1

Frizzi, A., Caldo, R. A., Morrell, J. A., Wang, M., Lutfiyya, L. L., Brown, W. E., et al. (2010). Compositional and transcriptional analyses of reduced zein kernels derived from the opaque2 mutation and RNAi suppression. *Plant Mol. Biol.* 73, 569–585. doi: 10.1007/s11103-010-9644-1

Geraghty, D., Peifer, M. A., Rubenstein, I., and Messing, J. (1981). The primary structure of a plant storage protein: zein. *Nucleic Acids Res.* 9, 5163–5174. doi: 10.1093/nar/9.19.5163

Habben, J. E., Kirleis, A. W., and Larkins, B. A. (1993). The origin of lysine-containing proteins in opaque-2 maize endosperm. *Plant Mol. Biol.* 23, 825–838. doi: 10.1007/BF00021537

Habben, J. E., Moro, G. L., Hunter, B. G., Hamaker, B. R., and Larkins, B. A. (1995). Elongation factor 1 alpha concentration is highly correlated with the lysine content of maize endosperm. *Proc. Natl. Acad. Sci. U.S.A.* 92, 8640–8644. doi: 10.1073/pnas.92.19.8640

Holding, D. R., Hunter, B. G., Chung, T., Gibbon, B. C., Ford, C. F., Bharti, A. K., et al. (2008). Genetic analysis of opaque2 modifier loci in quality protein maize. *Theor. Appl. Genet.* 117, 157–170. doi: 10.1007/s00122-008-0762-y

Holding, D. R., and Larkins, B. A. (2009) "Zein storage proteins," in *Molecular Genetic Approaches to Maize Improvement*, Vol. 63, eds A. L. Kriz and B. A. Larkins (Berlin: Springer), 269–286. doi: 10.1007/978-3-540-68922-5_19

Hopkins, C. G. (1899). Improvement in the chemical composition of the corn kernel. *Ill. Agric. Exp. Stn. Bull.* 55, 205–240.

Huang, S., Frizzi, A., Florida, C. A., Kruger, D. E., and Luethy, M. H. (2006). High lysine and high tryptophan transgenic maize resulting from the reduction of both 19- and 22-kD alpha-zeins. *Plant Mol. Biol.* 61, 525–535. doi: 10.1007/s11103-006-0027-6

Jia, M., Wu, H., Clay, K. L., Jung, R., Larkins, B. A., and Gibbon, B. C. (2013). Identification and characterization of lysine-rich proteins and starch biosynthesis genes in the opaque2 mutant by transcriptional and proteomic analysis. *BMC Plant Biol.* 13:60. doi: 10.1186/1471-2229-13-60

Kirihara, J. A., Hunsperger, J. P., Mahoney, W. C., and Messing, J. W. (1988). Differential expression of a gene for a methionine-rich storage protein in maize. *Mol. Gen. Genet.* 211, 477–484. doi: 10.1007/BF00425704

Lai, J., and Messing, J. (2002). Increasing maize seed methionine by mRNA stability. *Plant J.* 30, 395–402. doi: 10.1046/j.1365-313X.2001.01285.x

Leustek, T., Martin, M. N., Bick, J. A., and Davies, J. P. (2000). Pathways and regulation of sulfur metabolism revealed through molecular and genetic studies. *Annu. Rev. Plant Physiol. Plant Mol. Biol.* 51, 141–165. doi: 10.1146/annurev.arplant.51.1.141

Mertz, E. T., Bates, L. S., and Nelson, O. E. (1964). Mutant gene that changes protein composition and increases lysine content of maize endosperm. *Science* 145, 279–280. doi: 10.1126/science.145.3629.279

Messing, J., and Fisher, H. (1991). Maternal effect on high methionine levels in hybrid corn. *J. Biotechnol.* 21, 229–237. doi: 10.1016/0168-1656(91)90044-V

Moose, S. P., Dudley, J. W., and Rocheford, T. R. (2004). Maize selection passes the century mark: a unique resource for 21st century genomics. *Trends Plant Sci.* 9, 358–364. doi: 10.1016/j.tplants.2004.05.005

Newell, M. A., Vogel, K. E., Adams, M., Aydin, N., Bodnar, A. L., Ali, M., et al. (2014). Genetic and biochemical differences in populations bred for extremes in maize grain methionine concentration. *BMC Plant Biol.* 14:49. doi: 10.1186/1471-2229-14-49

Osborne, T. B., and Mendel, L. B. (1914). Amino-acids in nutrition and growth. *J. Biol. Chem.* 17, 325–349.

Osborne, T. B., and Mendel, L. B. (1914). Nutritional properties of proteins of the maize kernel. *J. Biol. Chem.* 18, 1–16.

Phillips, R. (1985). Elevated protein-bound methionine in seeds of a maize line resistant to lysine plus threonine. *Cereal Chem.* 62, 213–218.

Prasanna, B. M., Vasal, S. K., Kassahun, B., and Singh, N. N. (2001). Quality protein maize. *Curr. Sci.* 81, 1308–1319.

Schickler, H. (1993). Repression of the high-methionine zein gene in the maize inbred line Mo17. *Plant J.* 3, 221–229. doi: 10.1046/j.1365-313X.1993.t01-14-00999.x

Schmidt, M. A., Barbazuk, W. B., Sandford, M., May, G., Song, Z., Zhou, W., et al. (2011). Silencing of soybean seed storage proteins results in a rebalanced protein composition preserving seed protein content without major collateral changes in the metabolome and transcriptome. *Plant Physiol.* 156, 330–345. doi: 10.1104/pp.111.173807

Schmidt, R. J., Ketudat, M., Aukerman, M. J., and Hoschek, G. (1992). Opaque-2 is a transcriptional activator that recognizes a specific target site in 22-kD zein genes. *Plant Cell* 4, 689–700. doi: 10.1105/tpc.4.6.689

Segal, G., Song, R., and Messing, J. (2003). A new opaque variant of maize by a single dominant RNA-interference-inducing transgene. *Genetics* 165, 387–397.

Song, R., and Messing, J. (2003). Gene expression of a gene family in maize based on noncollinear haplotypes. *Proc. Natl. Acad. Sci. U.S.A.* 100, 9055–9060. doi: 10.1073/pnas.1032999100

Swarup, S., Timmermans, M. C., Chaudhuri, S., and Messing, J. (1995). Determinants of the high-methionine trait in wild and exotic germplasm may have escaped selection during early cultivation of maize. *Plant J.* 8, 359–368. doi: 10.1046/j.1365-313X.1995.08030359.x

Vasal, S. K., Villegas, E., Bjarnason, M., Gelaw, B., and Goertz, P. (1980). "Genetic modifiers and breeding strategies in developing hard endosperm opaque 2 materials," in *Improvement of Quality Traits of Maize for Grain and Silage Use*, eds W. G. Pollmer and R. H. Phipps (London: Martinus Nijhoff), 37–73.

Wu, Y., and Messing, J. (2011). Novel genetic selection system for quantitative trait loci of quality protein maize. *Genetics* 188, 1019–1022. doi: 10.1534/genetics.111.131078

Wu, Y., and Messing, J. (2012). RNA interference can rebalance the nitrogen sink of maize seeds without losing hard endosperm. *PLoS ONE* 7:e32850. doi: 10.1371/journal.pone.0032850

Wu, Y., Wang, W., and Messing, J. (2012) Balancing of sulfur storage in maize seed. *BMC Plant Biol.* 12:77. doi: 10.1186/1471-2229-12-77

Conflict of Interest Statement: The authors declare that the research was conducted in the absence of any commercial or financial relationships that could be construed as a potential conflict of interest.

3

Ubiquitin-mediated control of seed size in plants

Na Li and Yunhai Li*

State Key Laboratory of Plant Cell and Chromosome Engineering, Institute of Genetics and Developmental Biology, Chinese Academy of Sciences, Beijing, China

Edited by:
Paolo Sabelli, University of Arizona, USA

Reviewed by:
Toshiro Ito, Temasek Life Sciences Laboratory, Singapore
Rita Crinelli, University of Urbino, Italy

****Correspondence:***
Yunhai Li, Institute of Genetics and Developmental Biology, No.1 West Beichen Road, Chaoyang District, Beijing 100101, China
e-mail: yhli@genetics.ac.cn

Seed size in higher plants is an important agronomic trait, and is also crucial for evolutionary fitness. In flowering plants, the seed comprises three major anatomical components, the embryo, the endosperm and the seed coat, each with different genetic compositions. Therefore, seed size is coordinately determined by the growth of the embryo, endosperm and maternal tissue. Recent studies have revealed multiple pathways that influence seed size in plants. Several factors involved in ubiquitin-related activities have been recently known to determine seed size in *Arabidopsis* and rice. In this review, we summarize current knowledge of ubiquitin-mediated control of seed size and discuss the role of the ubiquitin pathway in seed size control.

Keywords: seed size, seed development, ubiquitin, ubiquitin receptor, E3 ubiquitin ligase

INTRODUCTION

In angiosperms, seed development is an important process in the life cycle. The seed contains the basic architecture of the plant and accumulates nutrients for germination and early seedling growth. The size of seeds is important for evolutionary fitness and stress responses. In addition, seed size is one of the most important components of seed yield. Crop plants have undergone selection for large seed size during domestication (Sundaresan, 2005; Song et al., 2007; Shomura et al., 2008; Fan et al., 2009).

Seed development begins with double fertilization in which one sperm cell fuses with the egg cell to form the diploid embryo, and the other sperm cell fuses with the central cell to give rise to the triploid endosperm (Chaudhury et al., 2001; Sundaresan, 2005). In monocots, the endosperm constitutes the major part of the mature seed. In most dicots, the endosperm grows rapidly in the beginning and is eventually consumed, and the embryo occupies most of the mature seed. Maternal integuments surrounding the developing embryo and endosperm develop into seed coat after fertilization (Chaudhury and Berger, 2001; Chaudhury et al., 2001). Therefore, the size of a seed is determined by the coordinated growth of the diploid embryo, the triploid endosperm and the maternal sporophytic integuments. However, it is only in recent decades that we have begun to identify some genes involved in seed size control (reviewed in Kesavan et al., 2013).

The ubiquitin pathway has been recently known to play an important part in plant seed size determination (Song et al., 2007; Li et al., 2008; Xia et al., 2013; Du et al., 2014). Ubiquitin is a conserved 76-amino-acid protein that is covalently attached to target proteins through the sequential action of three enzymes (Hershko and Ciechanover, 1998; Moon et al., 2004). Firstly, the ubiquitin activating enzyme (E1) forms a thioester bond with the C-terminal glycine of ubiquitin in an ATP-dependent manner and transfers the activated ubiquitin to a cysteinyl residue on the ubiquitin conjugating enzyme (E2). The E2 can either bind with the ubiquitin protein ligase (E3) to transfer ubiquitin directly to substrate proteins, or transfer ubiquitin to E3 in the case of HECT (homology to E6-AP C terminus) E3s, which then transfers it to the substrates (Pickart, 2001). In both cases, E3 defines the substrate specificity. Conjugation of a single ubiquitin to a substrate protein can modify its activity (Mukhopadhyay and Riezman, 2007); however, the ubiquitination process can repeat several times to attach new ubiquitin to the lysine residue of the conjugated ubiquitin on the substrate to form a polyubiquitin chain. The number and the location of ubiquitin molecules that are attached define the fates of the target. One of the famous forms of ubiquitylation with Lys-48 linked polyubiquitin chains often leads the substrate protein to the 26S proteasome for degradation (Vierstra, 2009).

The 26S proteasome is a multi-subunit protease that consists of a cylindrical 20S core particle (CP), capped on each end by a 19S regulatory particle (RP) (Finley, 2009). The 19S RP contains lid and base components, which recognize the ubiquitinated substrates, remove and recycle the ubiquitin moieties, unfold the target proteins and transport them into the central chamber of CP. The CP is a core protease in which proteolysis takes place and unfolded proteins are broken into peptides (Vierstra, 2009). During the degradation of polyubiquitinated proteins, ubiquitin chains linked to the substrates can be cleaved and recycled by deubiquitinating enzymes (DUBs) (Sadanandom et al., 2012). DUBs also generate free ubiquitin moieties from their initial translation products, or reverse the effects of ubiquitination by removing ubiquitin from the targets (Smalle and Vierstra, 2004). Thus, the ubiquitylation process in the cell is dynamic and highly controlled (Sadanandom et al., 2012).

Genomic analysis revealed that more than 1400 genes in *Arabidopsis thaliana* encode components of ubiquitin-26S proteasome pathway (Smalle and Vierstra, 2004). Ubiquitin-mediated signaling is involved in diverse aspects of plant life cycle, such as hormone signaling, circadian rhythm, pathogen responses, and abiotic stress responses (Sadanandom et al., 2012). Recently, several components of the ubiquitin pathway have been found to play critical roles in the regulation of seed and organ size

(**Table 1**). In this review, we aim to summarize current knowledge on ubiquitin-mediated control of seed size and discuss the role of the ubiquitin pathway in seed growth.

REGULATION OF SEED SIZE BY THE UBIQUITIN RECEPTORS DA1 AND DAR1

The *Arabidopsis da1-1* (DA means "large" in Chinese) mutant was isolated from a genetic screen for mutations that increase seed and organ size (Li et al., 2008). The *da1-1* mutant produced larger and heavier seeds than the wild type (Li et al., 2008). The increased seed size in *da1-1* was a result of enlargement of sporophytic integuments. In addition, *da1-1* plants formed large flowers, siliques, leaves and increased biomass compared with wild-type plants. *DA1* controls seed and organ growth by restricting cell proliferation. The *da1-1* mutation causes an arginine-to -lysine mutation in the position 358 of the DA1 protein ($DA1^{R358K}$). In *Arabidopsis*, seven DA1-related (DAR) proteins share extensive amino acid similarity with DA1. DA1 homologs were also found in other plant species but not in animals, indicating a plant-specific mechanism to control seed and organ growth. Interestingly, the disruption of *DA1* or its closest family member *DAR1* with T-DNA insertions did not cause obvious seed and organ size phenotypes, while the simultaneous disruption of both *DA1* and *DAR1* resulted in large seeds and organs, indicating that *DA1* and *DAR1* act redundantly to restrict seed and organ growth. This genetic analysis also suggests that the mutant protein encoded by *da1-1* may have negative effects on DA1 and DAR1. Consistent with this notion, overexpression of a *da1-1* cDNA dramatically increased seed and organ size of wild-type plants.

DA1 encodes a ubiquitin receptor containing two ubiquitin interacting motifs (UIMs) and a single zinc-binding LIM domain defined by its conservation with the canonical Lin-11, Isl-1, and Mec-3 domains (Li et al., 2008). UIM-containing proteins are characterized by coupled ubiquitin binding and ubiquitylation, which generally bring about monoubiquitylation of the ubiquitin receptor proteins. This, in turn, promotes the conformation change of the receptors, regulates their activity or binding capacity with other proteins, and initiates a signal cascade (Hicke et al., 2005). Considering that UIM domains of DA1 have the ubiquitin-binding activity, DA1 may be involved in ubiquitin-mediated signaling processes by coupled ubiquitin binding and ubiquitylation. On the other hand, ubiquitin receptors could bind polyubiquitinated proteins and mediate their degradation by the 26S proteasome (Verma et al., 2004). Thus, it is also possible that DA1 may interact with its polyubiquitinated substrates via UIM domains and facilitate their degradation.

REGULATION OF SEED SIZE BY THE E3 UBIQUITIN LIGASES BB/EOD1, DA2, AND GW2

There are two E1s, at least 37 E2s and more than 1300 E3s in *Arabidopsis* (Smalle and Vierstra, 2004). E3s function at the last step of the ubiquitylation cascade and recognize the specific substrates. E3s fall into two groups according to their conserved domains: HECT or RING (Really Interesting New Gene)/U-box type. The RING-type E3 ubiquitin ligases can act independently or as components of multi-subunit E3 complexes including SCF (SKP1-CULLIN-F-box), CUL3 (CULLIN 3)- BTB/POZ (Bric a brac, Tramtrack and Broad complex/Pox virus and Zinc finger), CUL4-DDB1 (UV-Damaged DNA Binding Protein 1) and APC (Anaphase Promoting Complex) (Mazzucotelli et al., 2006). Currently, several RING-type E3 ubiquitin ligases have been identified as key factors of seed size control in dicot and monocot plants.

Two RING-type E3 ubiquitin ligases, DA2 and Big Brother (BB)/Enhancer of DA1 (EOD1), were identified as negative regulators of seed size in *Arabidopsis* (Li et al., 2008; Xia et al., 2013). Loss-of-function *da2-1* and *eod1/bb* mutants shared similar phenotypes, such as large organs and increased biomass. Overexpression of either *DA2* or *BB/EOD1* resulted in a reduction in organ size (Disch et al., 2006; Xia et al., 2013). In addition, both EOD1 and DA2 act maternally to regulate seed size by restricting cell proliferation in the integuments of ovules and developing seeds (Li et al., 2008; Xia et al., 2013), suggesting that these two E3 ubiquitin ligases may share similar mechanisms in seed size control. Importantly, both the *eod1*and *da2-1* mutations

Table 1 | List of ubiquitin-related proteins involved in seed size control.

Categories	Protein name	Species	Accession number	Reference(s)
Ubiquitin receptors	DA1	*Arabidopsis*	AT1G19270	Li et al., 2008
	DAR1	*Arabidopsis*	AT4G36860	Li et al., 2008
E3 ligases	DA2	*Arabidopsis*	AT1G78420	Xia et al., 2013
	EOD1/BB	*Arabidopsis*	AT3G63530	Disch et al., 2006; Li et al., 2008
	DA2L	*Arabidopsis*	AT1G17145	Xia et al., 2013
	GW2	Rice	EF447275	Song et al., 2007
	TaGW2	Wheat	JN896622, JN896623	Su et al., 2011; Bednarek et al., 2012; Yang et al., 2012
	ZmGW2	Maize	EU968771, FJ573211, EU962093	Li et al., 2010
Other regulators	UBP15	*Arabidopsis*	AT1G17110	Liu et al., 2008; Du et al., 2014
	Rpt2a	*Arabidopsis*	AT4G29040	Kurepa et al., 2009
	SAMBA	*Arabidopsis*	AT1G32310	Eloy et al., 2012
	GW5	Rice	AB433345	Shomura et al., 2008; Weng et al., 2008

synergistically enhance the seed size and weight phenotypes of *da1-1*, suggesting that both EOD1 and DA2 may function with DA1 to control seed size by modulating the activity of common downstream targets. However, genetic analyses show that *DA2* and *EOD1* function independently to control seed size (**Figure 1**) (Xia et al., 2013), suggesting that DA2 and EOD1 may target distinct growth stimulators for degradation, with common regulation via DA1. The synergistic effects could result from the simultaneous disruption of two components of a protein complex (Perez-Perez et al., 2009; Lanctot et al., 2013). It has been demonstrated that the ubiquitin receptor DA1 interacts with the E3 ligase DA2 through its C-terminal region (Xia et al., 2013), and the UIM domains of DA1 can bind ubiquitin (Li et al., 2008). Thus, it is likely that the interaction between DA1 and DA2 helps DA1 to bind the ubiquitinated substrates of DA2 and facilitate their degradation by the proteasome.

In rice (*Oryza sativa*), a quantitative trait locus (QTL) for *GRAIN WIDTH AND WEIGHT2* (*GW2*) encodes a RING-type E3 ubiquitin ligase (Song et al., 2007). Loss-of-function *GW2* allele caused wide spikelet hulls and accelerated grain milk-filling rates, resulting in increased grain width, weight and yield. The naturally occurring WY3 allele of *GW2* encoding a truncated version of the protein with a 310-amino acid deletion produced wide and heavy grains due to increased cell proliferation in spikelet hulls. In contrast, transgenic rice plants overexpressing *GW2* formed smaller and lighter grains than wild-type plants. Thus, GW2 might negatively affect the level or the activity of factors promoting cell proliferation. Interestingly, GW2 shares significant

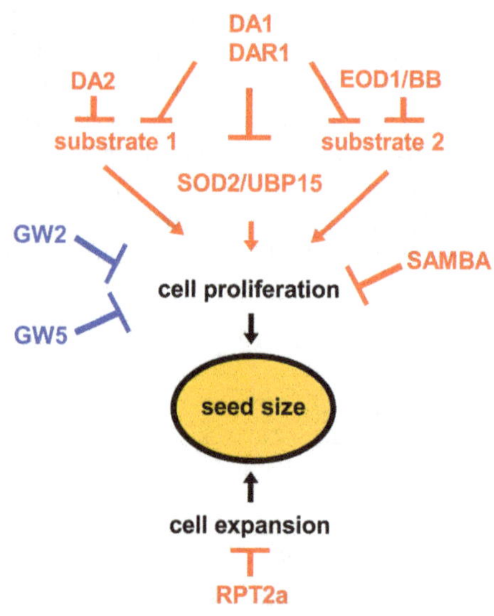

FIGURE 1 | A genetic and molecular framework for ubiquitin-mediated control of seed size. DA1 and DAR1 act redundantly to restrict cell proliferation. DA1 and DA2 may control cell proliferation by suppressing a common substrate (substrate 1). Similarly, DA1 and EOD1 may regulate cell division through a common target (substrate 2). DA1 acts upstream of UBP15 and modulates its stability to control cell proliferation. GW2, GW5 and SAMBA control seed size by restricting cell proliferation, respectively. RPT2a regulates seed size by limiting cell expansion. The seed size regulators in *Arabidopsis* and rice are shown as red and blue, respectively.

similarity with *Arabidopsis* DA2 and DA2-like protein (DA2L) (Xia et al., 2013). Overexpression of *GW2* in *Arabidopsis* resulted in small seeds and organs, as it has been observed in *35S:DA2* and *35S:DAL2* transgenic plants (Xia et al., 2013), indicating a possible conserved function in *Arabidopsis* and rice. The RING domain of GW2 is characterized by a Cys at metal ligand position 5 and a His at metal ligand position 6 (C5HC2) (Song et al., 2007). This feature is shared by the RING domain of maize, wheat, yeast and fungal homologs. Although the spacing of the Cys residues in the RING domain of DA2 is similar to that in the RING domain of GW2, the RING domain of DA2 or its dicot homologs lacks a conserved His residue that is replaced by Asn (Asn-91) (Xia et al., 2013). Biochemical and genetic analyses showed that this amino acid (Asn-91) is not required for DA2 E3 ligase activity and the roles of DA2 in seed size control, suggesting that the RING domain of DA2 might be a variant of that found in GW2.

In wheat (*Triticum aestivum*), there are three *GW2* homologs (originating from A, B, and D genomes, respectively) (Su et al., 2011). Analysis of modern varieties showed that *TaGW2-6A* Hap-6A-A is a superior allele for grain size. Varieties with *TaGW2-6A* Hap-6A-A allele had higher mean grain width than those with Hap-6A-G. This effect was due to the higher expression level of *TaGW2* in the varieties with Hap-6A-G allele, indicating that the expression level of *TaGW2* was negatively correlated with grain width. Meanwhile, a single base (T) insertion in the eighth exon of *TaGW2-6A* was detected in a large-kernel wheat variety, Lankaodali. This mutation produced a truncated protein, indicating that TaGW2-6A has a negative effect on grain size. In contrast, another report showed that overall down-regulation of *TaGW2* by RNA interference resulted in decreased grain size and weight, suggesting that *TaGW2* may positively regulate grain size (Bednarek et al., 2012). Further studies are needed to elucidate the role of TaGW2 in grain size control.

ZmGW2-CHR4 and *ZmGW2-CHR5*, two homologs of the rice *GW2*, have been found in maize (*Zea mays*) (Li et al., 2010). These two loci were located on duplicated maize chromosomal regions that have co-orthologous relationships with the rice region containing *GW2*. Single nucleotide polymorphism (SNP) in the promoter region of *ZmGW2-CHR4* was significantly associated with kernel width and one-hundred kernel weight, and the expression level of *ZmGW2-CHR4* was negatively correlated with kernel width. Similarly, *ZmGW2- CHR5* also affected kernel width (Li et al., 2010).

REGULATION OF SEED SIZE BY THE UBIQUITIN-SPECIFIC PROTEASE UBP15

SUPPRESSOR2 OF DA1 (*SOD2*) encodes UBIQUITIN-SPECIFIC PROTEASE15 (UBP15), which is a deubiquitinating enzyme (Liu et al., 2008; Du et al., 2014). UBP15 contains a ubiquitin-specific protease (UBP) domain that is required for deubiquitination activity, and a signature MYND-type zinc finger domain (Zf-MYND) that is supposed to function in protein-protein interaction. *sod2/ubp15* mutants were identified as suppressors of *da1-1*. *sod2/ubp15* plants produced small leaves, flowers and seeds, whereas plants overexpressing *UBP15* formed large seeds and organs, indicating that UBP15 is a positive regulator of seed and organ growth. UBP15 functions to regulate seed size by promoting cell proliferation in maternal integuments of ovules and developing seeds. Genetic analyses show that *ubp15* is epistatic to *da1-1* with respect to seed size, suggesting that UBP15 acts downstream of DA1 to promote seed growth (**Figure 1**). UBP15 protein is stabilized by adding proteasome inhibitor MG132, suggesting that UBP15 is degraded by the 26S proteasome. Furthermore, DA1 physically interacts with UBP15 and modulates its stability. It is likely that the ubiquitin receptor DA1 targets UBP15 and mediates its degradation by the proteasome. However, UBP15 acts independently of the E3 ubiquitin ligases BB/EOD1 and DA2 to control seed size (**Figure 1**), indicating that UBP15 is not the substrate of the E3 ubiquitin ligases DA2 or EOD1, and also suggesting that other E3 ligase(s) might be involved in proteasome-dependent degradation of UBP15.

The *Arabidopsis* genome encodes 27 UBPs, which were clustered into 14 subfamilies (Yan et al., 2000). The *UBP15* subfamily contains five genes (*UBP15-19*). Although loss of function in the *UBP16* gene had no obvious growth defects, the *ubp16* mutation enhanced the organ growth phenotypes of *ubp15*, indicating that UBP15 and UBP16 function redundantly to control organ size (Liu et al., 2008). It would be interesting to investigate whether UBP16 is involved in seed size control. It is also a worthwhile challenge to know if DA1 could interact genetically and physically with UBP16 and target it for degradation.

REGULATION OF SEED SIZE BY RPT2a, A SUBUNIT OF THE 26S PROTEASOME

The RP of the 26S proteasome is composed of a lid containing non-ATPase subunits (RPN3, 5–9, and 11–12) and a base consisting of six related AAA-ATPases (RPT1-6) and three non-ATPase subunits (RPN1, 2, and 10) (Smalle and Vierstra, 2004). One of the base subunits, the regulatory particle RPT2, has been found to affect seed size (Kurepa et al., 2009). There are two RPT2 homologs (RPT2a and RPT2b) in *Arabidopsis*, which share 98.8% identity in amino acid sequences. Loss-of-function of RPT2a caused a weak defect in 26S proteasome activity and led to enlargement of most organs including seeds. The size of cells in *rpt2a* mutants was increased compared with that in the wild type, while the number of cells in *rpt2a* mutants was reduced, suggesting a possible compensation mechanism between cell proliferation and cell expansion. It is plausible that the RP of 26S is required for the degradation of the positive regulators of cell expansion, thereby influencing seed size.

REGULATION OF SEED SIZE BY SAMBA, A PLANT- SPECIFIC APC/C REGULATOR

Plant organ growth is coordinated by cell division and cell expansion. Cell cycle progression is controlled by the degradation of essential cell cycle regulators such as securin or cyclins (De Veylder et al., 2003). In plants, A- and B-type cyclins are specifically recognized by a multi-subunit E3 ubiquitin ligase complex called anaphase-promoting complex/cyclosome (APC/C) (Heyman and De Veylder, 2012). The cyclins are then subjected to proteolysis by the 26S proteasome, and this promotes the mitotic progression. The activities of plant APC/C are regulated by different activating proteins or inhibitors including CELL DIVISION CYCLE 20 (CDC20), CDC20 HOMOLOGY1/CELL CYCLE SWITCH 52 (CDH1/CCS52), ULTRAVIOLENT-B-INSINSITIVE4 (UVI4), UVI4-like/OMISSION OF SECOND DIVISION1/GIGAS CELL1 (UVI4/OSD1/GIG1), and SAMBA. SAMBA is a plant-specific APC/C regulator that plays a role in seed size control (Eloy et al., 2012). In *Arabidopsis thaliana*, *SAMBA* is expressed in developing seeds and during early plant development stages. Loss of function of *SAMBA* stabilizes the A-type cyclin CYCA2;3 and promotes cell proliferation and endoreduplication, resulting in large seeds and organs. The yeast two-hybrid assay showed that SAMBA specifically interacts with A-type cyclins. These results indicate that SAMBA targets A-type cyclins for APC/C- mediated degradation and acts as a negative regulator of seed growth.

REGULATION OF SEED SIZE BY GW5

Rice *GW5* is a major QTL that controls rice grain width and weight (Shomura et al., 2008; Wan et al., 2008; Weng et al., 2008). Fine mapping of this locus uncovered that a 1212-bp deletion including the *GW5* gene is correlated with increased grain width (Shomura et al., 2008; Weng et al., 2008). Genotyping analysis of rice cultivars revealed that an intact *GW5* was detected in the slender-grain rice, whereas the 1212-bp deletion was observed in the wide-grain lines, suggesting a strong artificial selection during breeding (Shomura et al., 2008). *GW5* encodes a nucleus-localized protein of 144 amino acids with a predicted nuclear localization signal and an arginine-rich domain. GW5 interacted with polyubiquitin in a yeast two-hybrid assay (Weng et al., 2008), suggesting that GW5 might be involved in the ubiquitin-proteasome pathway to regulate cell division during seed development. As the E3 ubiquitin ligase GW2 also controls glume cell division and grain width, it has been hypothesized that GW5 and GW2 might act in the same pathway. However, genetic analyses showed that plants pyramiding *gw2* and *gw5* exhibited an enhanced phenotype of grain width compared with those carrying one of the two major QTLs (Ying et al., 2012), suggesting that they may act in different pathways or function in a same complex to regulate rice grain size.

CHALLENGES AND FUTURE PERSPECTIVES

During the past decade, several factors involved in ubiquitin-related activities have been identified to influence seed size in plants, indicating that the ubiquitin pathway plays an important role in seed size control. Interestingly, most of these factors affect not only seed size but also organ growth. For example, *da1* mutant showed large seeds, leaves, and flowers (Li et al., 2008), whereas *sod2* mutants produced small seeds and organs (Du et al., 2014), suggesting a possible link between seed size control and organ growth. By contrast, several other mutants with large organs formed normal-sized seeds (Horiguchi et al., 2005; White, 2006; Xu and Li, 2011), implying that seed and organ size is not always positively related. These results suggest that seeds and organs may possess both common and distinct pathways to regulate their respective size.

Our current knowledge of ubiquitin-mediated control of seed size is rather fragmented, relying on several seemingly independent pathways full of gaps (**Figure 1**). One of the major challenges in the future is to define the molecular function of the known factors in seed size control. For example, what are the specific targets of the ubiquitin receptors and the E3 ubiquitin ligases? How are the activities of these receptors and E3 ligases regulated? Thus, identification of their interacting proteins and downstream targets by biochemical and genetic approaches will help fill up the major gaps in each pathway and understand the molecular mechanisms of these factors in seed size control. To identify novel ubiquitin-related factors in seed size control, both forward and reverse genetic approaches could be used. Genetic screens for modifiers of the known genes will help identify downstream targets of the ubiquitin receptors or the E3 ligases. The use of the newly developed genome editing technology (reviewed in Gaj et al., 2013) will greatly facilitate the functional characterization of candidate ubiquitin-related genes involved in seed size regulation. On the other hand, systems biology approaches, such as transcriptomic, proteomic, and metabolomics analysis, should yield novel insights into the molecular networks of ubiquitin-mediated control of seed size.

ACKNOWLEDGMENTS

This work was supported by the grants from the National Natural Science Foundation of China (91017014, 31221063, and 31300242) and National Basic Research Program of China (2013CBA01401). We apologize to the colleagues whose work is not covered in this review due to limited space.

REFERENCES

Bednarek, J., Boulaflous, A., Girousse, C., Ravel, C., Tassy, C., Barret, P., et al. (2012). Down-regulation of the TaGW2 gene by RNA interference results in decreased grain size and weight in wheat. *J. Exp. Bot.* 63, 5945–5955. doi: 10.1093/jxb/ers249

Chaudhury, A. M., and Berger, F. (2001). Maternal control of seed development. *Semin. Cell Dev. Biol.* 12, 381–386. doi: 10.1006/scdb.2001.0267

Chaudhury, A. M., Koltunow, A., Payne, T., Luo, M., Tucker, M. R., Dennis, E. S., et al. (2001). Control of early seed development. *Annu. Rev. Cell Dev. Biol.* 17, 677–699. doi: 10.1146/annurev.cellbio.17.1.677

De Veylder, L., Joubes, J., and Inze, D. (2003). Plant cell cycle transitions. *Curr. Opin. Plant Biol.* 6, 536–543. doi: 10.1016/j.pbi.2003.09.001

Disch, S., Anastasiou, E., Sharma, V. K., Laux, T., Fletcher, J. C., and Lenhard, M. (2006). The E3 ubiquitin ligase BIG BROTHER controls Arabidopsis organ size in a dosage-dependent manner. *Curr. Biol.* 16, 272–279. doi: 10.1016/j.cub.2005.12.026

Du, L., Li, N., Chen, L., Xu, Y., Li, Y., Zhang, Y., et al. (2014). The ubiquitin receptor DA1 regulates seed and organ size by modulating the stability of the ubiquitin-specific protease UBP15/SOD2 in Arabidopsis. *Plant Cell* 26, 665–677. doi: 10.1105/tpc.114.122663

Eloy, N. B., Gonzalez, N., Van Leene, J., Maleux, K., Vanhaeren, H., De Milde, L., et al. (2012). SAMBA, a plant-specific anaphase-promoting complex/cyclosome regulator is involved in early development and A-type cyclin stabilization. *Proc. Natl. Acad. Sci. U.S.A.* 109, 13853–13858. doi: 10.1073/pnas.1211418109

Fan, C., Yu, S., Wang, C., and Xing, Y. (2009). A causal C-A mutation in the second exon of GS3 highly associated with rice grain length and validated as a functional marker. *Theor. Appl. Genet.* 118, 465–472. doi: 10.1007/s00122-008-0913-1

Finley, D. (2009). Recognition and processing of ubiquitin-protein conjugates by the proteasome. *Annu. Rev. Biochem.* 78, 477–513. doi: 10.1146/annurev.biochem.78.081507.101607

Gaj, T., Gersbach, C. A., and Barbas, C. F. 3rd (2013). ZFN, TALEN, and CRISPR/Cas-based methods for genome engineering. *Trends Biotechnol.* 31, 397–405. doi: 10.1016/j.tibtech.2013.04.004

Hershko, A., and Ciechanover, A. (1998). The ubiquitin system. *Annu. Rev. Biochem.* 67, 425–479. doi: 10.1146/annurev.biochem.67.1.425

Heyman, J., and De Veylder, L. (2012). The anaphase-promoting complex/cyclosome in control of plant development. *Mol. Plant* 5, 1182–1194. doi: 10.1093/mp/sss094

Hicke, L., Schubert, H. L., and Hill, C. P. (2005). Ubiquitin-binding domains. *Nat. Rev. Mol. Cell Biol.* 6, 610–621. doi: 10.1038/nrm1701

Horiguchi, G., Kim, G. T., and Tsukaya, H. (2005). The transcription factor AtGRF5 and the transcription coactivator AN3 regulate cell proliferation in leaf primordia of *Arabidopsis thaliana*. *Plant J.* 43, 68–78. doi: 10.1111/j.1365-313X.2005.02429.x

Kesavan, M., Song, J. T., and Seo, H. S. (2013). Seed size: a priority trait in cereal crops. *Physiol. Plant* 147, 113–120. doi: 10.1111/j.1399-3054.2012.01664.x

Kurepa, J., Wang, S., Li, Y., Zaitlin, D., Pierce, A. J., and Smalle, J. A. (2009). Loss of 26S proteasome function leads to increased cell size and decreased cell number in Arabidopsis shoot organs. *Plant Physiol.* 150, 178–189. doi: 10.1104/pp.109.135970

Lanctot, A. A., Peng, C. Y., Pawlisz, A. S., Joksimovic, M., and Feng, Y. (2013). Spatially dependent dynamic MAPK modulation by the Nde1-Lis1-Brap complex patterns mammalian CNS. *Dev. Cell* 25, 241–255. doi: 10.1016/j.devcel.2013.04.006

Li, Q., Li, L., Yang, X., Warburton, M. L., Bai, G., Dai, J., et al. (2010). Relationship, evolutionary fate and function of two maize co-orthologs of rice GW2 associated with kernel size and weight. *BMC Plant Biol.* 10:143. doi: 10.1186/1471-2229-10-143

Li, Y., Zheng, L., Corke, F., Smith, C., and Bevan, M. W. (2008). Control of final seed and organ size by the DA1 gene family in *Arabidopsis thaliana*. *Genes Dev.* 22, 1331–1336. doi: 10.1101/gad.463608

Liu, Y., Wang, F., Zhang, H., He, H., Ma, L., and Deng, X. W. (2008). Functional characterization of the Arabidopsis ubiquitin-specific protease gene family reveals specific role and redundancy of individual members in development. *Plant J.* 55, 844–856. doi: 10.1111/j.1365-313X.2008.03557.x

Mazzucotelli, E., Belloni, S., Marone, D., De Leonardis, A., Guerra, D., Di Fonzo, N., et al. (2006). The e3 ubiquitin ligase gene family in plants: regulation by degradation. *Curr. Genomics* 7, 509–522. doi: 10.2174/138920206779315728

Moon, J., Parry, G., and Estelle, M. (2004). The ubiquitin-proteasome pathway and plant development. *Plant Cell* 16, 3181–3195. doi: 10.1105/tpc.104.161220

Mukhopadhyay, D., and Riezman, H. (2007). Proteasome-independent functions of ubiquitin in endocytosis and signaling. *Science* 315, 201–205. doi: 10.1126/science.1127085

Perez-Perez, J. M., Candela, H., and Micol, J. L. (2009). Understanding synergy in genetic interactions. *Trends Genet.* 25, 368–376. doi: 10.1016/j.tig.2009.06.004

Pickart, C. M. (2001). Mechanisms underlying ubiquitination. *Annu. Rev. Biochem.* 70, 503–533. doi: 10.1146/annurev.biochem.70.1.503

Sadanandom, A., Bailey, M., Ewan, R., Lee, J., and Nelis, S. (2012). The ubiquitin-proteasome system: central modifier of plant signalling. *New Phytol.* 196, 13–28. doi: 10.1111/j.1469-8137.2012.04266.x

Shomura, A., Izawa, T., Ebana, K., Ebitani, T., Kanegae, H., Konishi, S., et al. (2008). Deletion in a gene associated with grain size increased yields during rice domestication. *Nat. Genet.* 40, 1023–1028. doi: 10.1038/ng.169

Smalle, J., and Vierstra, R. D. (2004). The ubiquitin 26S proteasome proteolytic pathway. *Annu. Rev. Plant Biol.* 55, 555–590. doi: 10.1146/annurev.arplant.55.031903.141801

Song, X. J., Huang, W., Shi, M., Zhu, M. Z., and Lin, H. X. (2007). A QTL for rice grain width and weight encodes a previously unknown RING-type E3 ubiquitin ligase. *Nat. Genet.* 39, 623–630. doi: 10.1038/ng2014

Su, Z., Hao, C., Wang, L., Dong, Y., and Zhang, X. (2011). Identification and development of a functional marker of TaGW2 associated with grain weight in bread wheat (*Triticum aestivum* L.). *Theor. Appl. Genet.* 122, 211–223. doi: 10.1007/s00122-010-1437-z

Sundaresan, V. (2005). Control of seed size in plants. *Proc. Natl. Acad. Sci. U.S.A.* 102, 17887–17888. doi: 10.1073/pnas.0509021102

Verma, R., Oania, R., Graumann, J., and Deshaies, R. J. (2004). Multiubiquitin chain receptors define a layer of substrate selectivity in the ubiquitin-proteasome system. *Cell* 118, 99–110. doi: 10.1016/j.cell.2004.06.014

Vierstra, R. D. (2009). The ubiquitin-26S proteasome system at the nexus of plant biology. *Nat. Rev. Mol. Cell Biol.* 10, 385–397. doi: 10.1038/nrm2688

Wan, X., Weng, J., Zhai, H., Wang, J., Lei, C., Liu, X., et al. (2008). Quantitative trait loci (QTL) analysis for rice grain width and fine mapping of an identified QTL allele gw-5 in a recombination hotspot region on chromosome 5. *Genetics* 179, 2239–2252. doi: 10.1534/genetics.108.089862

Weng, J., Gu, S., Wan, X., Gao, H., Guo, T., Su, N., et al. (2008). Isolation and initial characterization of GW5, a major QTL associated with rice grain width and weight. *Cell Res.* 18, 1199–1209. doi: 10.1038/cr.2008.307

White, D. W. (2006). PEAPOD regulates lamina size and curvature in Arabidopsis. *Proc. Natl. Acad. Sci. U.S.A.* 103, 13238–13243. doi: 10.1073/pnas.0604349103

Xia, T., Li, N., Dumenil, J., Li, J., Kamenski, A., Bevan, M. W., et al. (2013). The ubiquitin receptor DA1 interacts with the E3 ubiquitin ligase DA2 to regulate seed and organ size in Arabidopsis. *Plant Cell* 25, 3347–3359. doi: 10.1105/tpc.113.115063

Xu, R., and Li, Y. (2011). Control of final organ size by mediator complex subunit 25 in *Arabidopsis thaliana*. *Development* 138, 4545–4554. doi: 10.1242/dev.071423

Yan, N., Doelling, J. H., Falbel, T. G., Durski, A. M., and Vierstra, R. D. (2000). The ubiquitin-specific protease family from Arabidopsis. AtUBP1 and 2 are required for the resistance to the amino acid analog canavanine. *Plant Physiol.* 124, 1828–1843. doi: 10.1104/pp.124.4.1828

Yang, Z., Bai, Z., Li, X., Wang, P., Wu, Q., Yang, L., et al. (2012). SNP identification and allelic-specific PCR markers development for TaGW2, a gene linked to wheat kernel weight. *Theor. Appl. Genet.* 125, 1057–1068. doi: 10.1007/s00122-012-1895-6

Ying, J. Z., Gao, J. P., Shan, J. X., Zhu, M. Z., Shi, M., and Lin, H. X. (2012). Dissecting the genetic basis of extremely large grain shape in rice cultivar 'JZ1560'. *J. Genet. Genomics* 39, 325–333. doi: 10.1016/j.jgg.2012.03.001

Conflict of Interest Statement: The authors declare that the research was conducted in the absence of any commercial or financial relationships that could be construed as a potential conflict of interest.

4

The role of photosynthesis and amino acid metabolism in the energy status during seed development

Gad Galili[1]*, Tamar Avin-Wittenberg[2], Ruthie Angelovici[3] and Alisdair R. Fernie[2]

[1] Department of Plant Sciences, The Weizmann Institute of Science, Rehovot, Israel
[2] Max-Planck-Institut für Molekulare Pflanzenphysiologie, Potsdam-Golm, Germany
[3] Department of Biochemistry and Molecular Biology, Michigan State University, East Lansing, MI, USA

Edited by:
Brian A. Larkins, University of Nebraska–Lincoln, USA

Reviewed by:
Alessandro Vitale, Consiglio Nazionale delle Ricerche – National Research Council of Italy, Italy
Simona Masiero, Universitá degli Studi di Milano, Italy

***Correspondence:**
Gad Galili, Department of Plant Sciences, The Weizmann Institute of Science, 234 Herzl Street, Rehovot 7610001, Israel
e-mail: gad.galili@weizmann.ac.il

Seeds are the major organs responsible for the evolutionary upkeep of angiosperm plants. Seeds accumulate significant amounts of storage compounds used as nutrients and energy reserves during the initial stages of seed germination. The accumulation of storage compounds requires significant amounts of energy, the generation of which can be limited due to reduced penetration of oxygen and light particularly into the inner parts of seeds. In this review, we discuss the adjustment of seed metabolism to limited energy production resulting from the suboptimal penetration of oxygen into the seed tissues. We also discuss the role of photosynthesis during seed development and its contribution to the energy status of developing seeds. Finally, we describe the contribution of amino acid metabolism to the seed energy status, focusing on the Asp-family pathway that leads to the synthesis and catabolism of Lys, Thr, Met, and Ile.

Keywords: seed development, metabolism and bioenergetics, photosynthesis, branched chain amino acids, TCA cycle

INTRODUCTION

Seeds are the major organ responsible for the evolutionary upkeep of the plant lineage. They store the genetic material of the plant within the embryo, thus guarantying continuation of the plant's life cycle in the next generation. Seed development is highly regulated, with distinct transcript, protein and metabolite switches occurring in a concerted manner throughout its progression (Fait et al., 2006; Sreenivasulu et al., 2008; Xu et al., 2008; Angelovici et al., 2010). Seeds produce a great variety of storage compounds, among them a myriad of carbohydrates (especially starch), storage proteins, and storage lipids. These provide – both directly in the form of food and indirectly in the form of feed ~70% of the world's human caloric intake (Sreenivasulu and Wobus, 2013). Storage metabolites are subsequently used as nutrient and energy sources to support early stages of seed germination, while the seedling is still heterotrophic. Production of storage compounds whose generation by respiration becomes limiting due to a limitation of oxygen penetration into the dense seed tissues requires a considerable amount of energy. The requirement for oxygen and hence energy production becomes particularly critical during reserve metabolites accumulation and seed desiccation, in which the dense seed tissues limit the oxygen penetration and thus limit mitochondrial energy production (Weber et al., 2005; Angelovici et al., 2010). Photosynthesis in green seeds, and other energy conservation mechanisms, such as a decrease of overall respiration rates during embryo development, somewhat relieve the major drop in energy demand, but do not sufficiently compensate for the energy requirement (Weber et al., 2005; Angelovici et al., 2010). In seeds of the model plant *Arabidopsis thaliana*, amino acid metabolism is modulated so that many amino acids are synthesized and accumulated during seed desiccation (Fait et al., 2006). In recent years, several lines of evidence have implied that amino acids are not only used for the synthesis of storage proteins, but also, upon energy demand, can be catabolized and their catabolic products fed into the TCA cycle to generate energy (Zhu and Galili, 2004; Angelovici et al., 2010; Galili, 2011). This was also shown for plants exposed to extended darkness, another condition characterized by energy deprivation (Ishizaki et al., 2005, 2006; Araujo et al., 2010). In this mini-review, we describe lines of evidence demonstrating the contribution of photosynthesis and amino acid metabolism to the energy production in developing seeds and the contribution of amino acid catabolism to the energy status of seedlings exposed to extended darkness.

DEVELOPING SEEDS FACE AN EXTREME SHORTAGE OF ENERGY DUE TO LIMITED PENETRATION OF OXYGEN, PARTICULARLY INTO THE INNER SEED PARTS

Since developing seeds are quite dense, they suffer from limited penetration of oxygen, particularly into the inner seed tissues. The shortage of oxygen extensively limits energy-requiring metabolic processes crucial for seed development and embryo maturation (Vigeolas et al., 2003, 2011; van Dongen et al., 2004). In developing seeds of dicotyledonous (dicot) plant species, photosynthesis in the developing embryo contributes a considerable amount of oxygen to the seed tissue, which further fuels energy-generating biochemical pathways, such as glycolysis and respiration (Rolletschek et al., 2003; Ruuska et al., 2004; Borisjuk et al., 2005; Tschiersch et al., 2011, 2012). This is in contrast to developing non-green seeds of monocotyledonous (monocot) plants that do not generate energy via photosynthesis, and hence must depend on alternative energy-generating metabolic processes in addition to glycolysis and respiration in order to fuel seed development. The chemical donors for energy-generation are hypothetically produced in the vegetative parts of the plant, using particularly energy derived from photosynthesis, and are transported to the developing seeds (van Dongen et al., 2004). A number of different metabolites produced in the vegetative tissues are transported to the seeds; however, at a quantitative level, the amino acid Asn, which accumulates to relatively high levels during senescence, is particularly notable as a major energy source (Credali et al., 2013). Its role as an energy donor occurs via the conversion of Asn into Asp, the direct precursor of the branched Asp-family pathway (**Figure 1**). The Asp-family pathway synthesizes the amino acids Lys, Thr, Met, and Ile, which, under conditions of energy shortage, are further catabolized by the TCA cycle to generate energy (Galili, 2011). An energy shortage does not occur only in seeds, but also in vegetative tissues that lack photosynthesis, and upon exposure of plants to extended darkness. Under these energy shortage conditions, Lys is directly catabolized into the TCA cycle, while Thr and Met are converted into Ile, which is then directly catabolized into the TCA cycle (Joshi et al., 2006; Araujo et al., 2010). We focus in this review on the catabolism of Lys and Ile in respect to the contribution of amino acid catabolism to the energy status either during seed development or upon exposure to extended darkness.

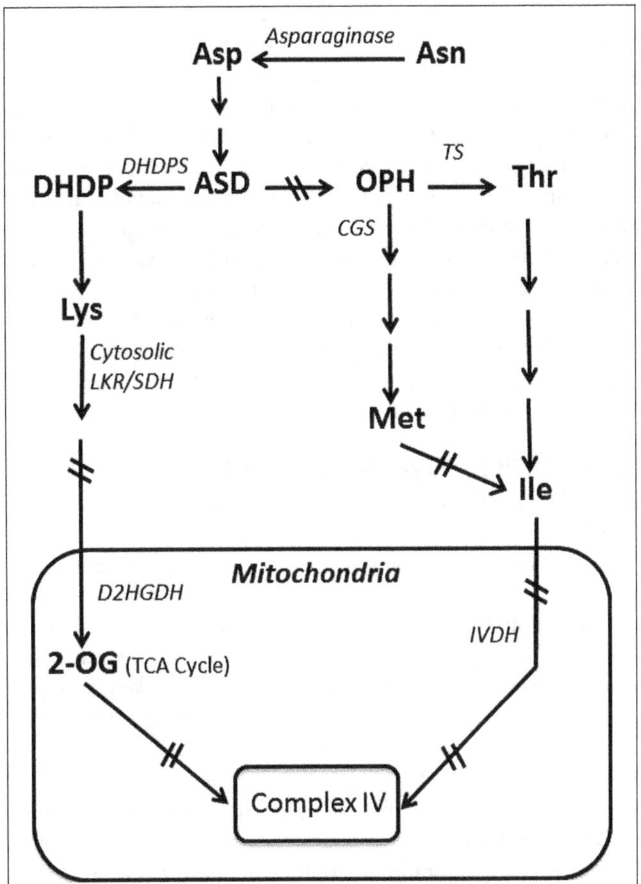

FIGURE 1 | Schematic diagram showing the synthesis of the amino acids of the Asp-family pathway and their catabolism by the TCA cycle. The mitochondrial enzymatic complex is highlighted by a black box. Abbreviations: 2-OG, 2-oxoglutarate; ASD, aspartic semialdehyde dehydrogenase; CGS, cystathionine gamma synthase; D2HGDH, 2-D-hydroxyglutarate dehydrogenase; DHDP, dihydrodipicolinate; DHDPS, dihydrodipicolinate synthase; IVDH, isovaleryl-CoA dehydrogenase; LKR/SDH, lysine ketoglutarate reductase/saccharopine dehydrogenase; OPH, *O*- phospho_L_homoserine; TS, Thr synthase. Arrows with two crossed lines represent several enzymatic reactions.

CONTRIBUTION OF SEED PHOTOSYNTHESIS TO OXYGEN AVAILABILITY IN DEVELOPING SEEDS

A significant amount of information on the role of seed photosynthesis for the synthesis of seed reserve compounds has been gathered at the IPK Institute for Plant Genetics in Gatersleben Germany (http://www.ipk-gatersleben.de/en/), where they analyzed developing soybean seeds, which possess active seed photosynthesis (Rolletschek et al., 2003; Ruuska et al., 2004; Borisjuk et al., 2005; Tschiersch et al., 2011, 2012). Light intensity and oxygen level progressively decrease with a progressive increase in seed depth (Vigeolas et al., 2003). Oxygen level is highest in light illuminated seeds, progressively lower in seeds that are not illuminated and lowest in seeds exposed to darkness. The ATP/ADP ratio is markedly higher in developing seeds exposed to light, indicating that photosynthesis in illuminated seeds contributes to the production of energy (Borisjuk et al., 2005). These results indicate that photosynthesis, which produces oxygen, significantly contributes to the energy status of developing seeds. Interestingly, in an independent study, overexpression of hemoglobin 2 in *Arabidopsis* seeds improved the energy status of seeds under low oxygen conditions and led to an increase in the fatty acid content of developing and mature seeds, further emphasizing the importance of oxygen availability during seeds development (Vigeolas et al., 2011). Photosynthesis could also be important in relatively large seeds, such as soybean seeds, whose inner cell layers receive significantly lower oxygen levels that the outer seed tissues. Borisjuk et al. (2005) demonstrated that developing dicot seeds with green embryos display a reduction in photosynthetic activity during development, starting from the interior of the embryo. This indicates the contribution of this photosynthetic activity to seed energy status.

Photosynthesis encompasses two different reactions, namely: (i) the light reactions that lead to reduction of NADP to NADPH and create a proton gradient across the chloroplast membrane that is used for ATP synthesis; and (ii) the light-independent reactions in which RuBisCO fixes CO_2 from the atmosphere in a NADPH-requiring process, and the Calvin–Benson cycle produces sugars (Buchanan et al., 2000). Interestingly, developing oilseed rape (*Brassica napus* L) embryos possess active photosynthesis, and RuBisCO acts in a special metabolic context without the Calvin–Benson cycle in order to improve the efficiency of carbon utilization during the synthesis of oil storage (Schwender et al., 2004a). This special pathway generates 20% more acetyl-CoA, a precursor for fatty acid biosynthesis, in comparison to glycolysis and saves 40% of the carbon that would otherwise be lost as CO_2.

CONTRIBUTION OF AMINO ACID METABOLISM TO THE ENERGY STATUS OF DEVELOPING SEEDS

Amino acids are constituents of proteins and hence essential components for the life of all organisms. Yet, in response to specific developmental and stress-associated conditions, amino acids also serve as energy donors through their catabolism in the TCA cycle (Araujo et al., 2010; Angelovici et al., 2011; Kirma et al., 2012). In vegetative tissues, photosynthesis is the major source of energy during daytime, and thus during daytime amino acids are used for the synthesis of proteins and, as precursors of multiple secondary metabolites (Less and Galili, 2008). Yet, during night-time or in response to stresses that cause major energy shortages due to the lack of or reduction in photosynthesis, amino acids also serve as important energy sources through their catabolism via the TCA cycle (Araujo et al., 2011). Therefore, the contributions of amino acid catabolism to the energy status requirements of developing seeds appears even more critical than in vegetative tissues, due to the limits of oxygen diffusion.

CONTRIBUTION OF Lys SYNTHESIS AND CATABOLISM IN THE TCA CYCLE TO SEED ENERGY STATUS

Insight into the potential importance of Lys catabolism to the energy status of developing seeds was obtained in research aiming to enhance the nutritional quality of seeds by increasing Lys accumulation (Zhu and Galili, 2003). Lys is an essential amino acid that serves a vital role in human food and livestock feed, since humans and some livestock (such as chicken and pigs, i.e., mono-gastric livestock) are unable to synthesize Lys (Kirma et al., 2012). The synthesis of Lys in plants (**Figure 1**) is subject to post-translational regulation, where increasing Lys feedback inhibits the activity of dihydrodipicolinate synthase (DHDPS), which catalyzies the first committed step in the Lys biosynthetic branch of the Asp-family pathway. Plants possess two DHDPS isozymes (DHDPS1 and DHDPS2), with DHDPS2 accounting for the majority of the total DHDPS activity (Jones-Held et al., 2012). Because Lys is an essential amino acid, an attempt was made to increase Lys in seeds of the model plant *Arabidopsis*, using a recombinant gene encoding in a seed-specific manner a mutant bacterial DHDPS2 enzyme insensitive to Lys feedback inhibition (Zhu and Galili, 2003). This approach yielded only a relatively mild increase in seed Lys, leading to the hypothesis that this amino acid is not an end product metabolite, but rather might be catabolized in the TCA cycle to generate energy. To test this hypothesis, the same bacterial feedback-insensitive DHDPS was expressed in seeds of a transgenic *Arabidopsis* genotype that lacks the capacity for Lys catabolism due to a knockout mutation in the gene encoding the bi-functional enzyme lysine-ketoglutarate reductase/saccharopine dehydrogenase (LKR/SDH), which contains the first two enzymes of Lys catabolism linked as a single polypeptide. In this combined genotype, seed Lys content was nearly 1000-fold higher than wild type (WT), indicating that developing seeds have a strong flux of Lys synthesis as well as catabolism into the TCA cycle (Zhu and Galili, 2003).

Developing seeds possessing significantly enhanced Lys levels due to increased Lys synthesis and suppression of Lys catabolism also exhibited notable differences in their transcriptomes and primary metabolomes, indicating that Lys metabolism (synthesis and catabolism) is well connected to the TCA cycle (Angelovici et al., 2009, 2011). For example, compared to WT, the levels of the TCA cycle metabolites fumarate, citrate, and ketoglutarate are lower in the genotype possessing enhanced Lys synthesis and a block of Lys catabolism, compared to the wild-type, genotype (Angelovici et al., 2009, 2011). Lys catabolism into the TCA cycle is apparently used to generate energy essential for seed development. As shown in **Figure 1**, Lys is not the only amino acid of the Asp-family pathway that feeds into the TCA cycle. A second branch of the Asp-family pathway leads to the synthesis of Thr and Met, which are further metabolized into Ile, which also feeds into the TCA cycle and serves as a major substrate for energy generation (Joshi et al., 2006; Kochevenko and Fernie, 2011).

In the developing seed, the upstream substrate that feeds the Asp-family pathway on route to the TCA cycle is Asn. Asn is transported from vegetative tissues to developing seeds where it is converted by asparaginase into Asp, the starting point metabolite of the Asp-family pathway (**Figure 1**; Credali et al., 2013). Indeed, the asparaginase level is generally stimulated during early seed development (Dickson et al., 1992; Credali et al., 2013).

AMINO ACID CATABOLISM FACILITATES RESPIRATION UNDER STRESS CONDITIONS

Under normal conditions, respiration depends on the oxidation of carbohydrates. However, during situations in which carbohydrate supply is limited, the plant cell can modify its metabolism to utilize alternative respiratory substrates. Among these substrates are proteins. Protein degradation is a highly regulated process, involving a multitude of cellular reactions, such as ubiquitinylation and degradation *via* the proteasome, the autophagy machinery and nutrient sensing by the TOR pathway. The product of protein degradation, free amino acids, can be further catabolize to generate energy (ATP; Zhu and Galili, 2003; Joshi et al., 2006; Araujo et al., 2011; Kochevenko and Fernie, 2011). Indeed, it has been shown that under abiotic stress there is increased transcription of amino acid catabolic genes (Less and Galili, 2008).

In mammals, the mitochondrial protein, electron-transfer flavoprotein:ubiquinone oxireductase (ETFQO), accepts electrons from electron transfer flavoprotein (ETF) to reduce ubiquinone. ETF serves as an obligatory electron acceptor for nine mitochondrial flavoprotein dehydrogenases. The ETF/ETFQO system facilitates electron transfer from these flavoprotein dehydrogenases to the main respiratory chain (Ishizaki et al., 2005, 2006). Homologs of ETF and ETFQO have been characterized in plants, as well as two flavoprotein dehydrogenases: isovaleryl-CoA dehydrogenase (IVDH) and 2-D-hydroxyglutarate dehydrogenase (D2HGDH; Ishizaki et al., 2005, 2006; Araujo et al., 2010). T-DNA insertion mutants of each of these genes display increased sensitivity to prolonged darkness, a condition that leads to carbon starvation. In addition, metabolic analysis of these mutants under prolonged darkness revealed an accumulation of several amino acids and an intermediate metabolite of Leu catabolism in comparison to WT. These results suggest a role for amino acid catabolism in supplying respiratory intermediates during carbon starvation (**Figure 2**; Ishizaki et al., 2005, 2006). It has been demonstrated that IVDH is apparently predominantly responsible for the transport of electrons from breakdown products of phytol and branched-chain amino acids (BCAA: Ile, Leu, and Val), while D2HGDH

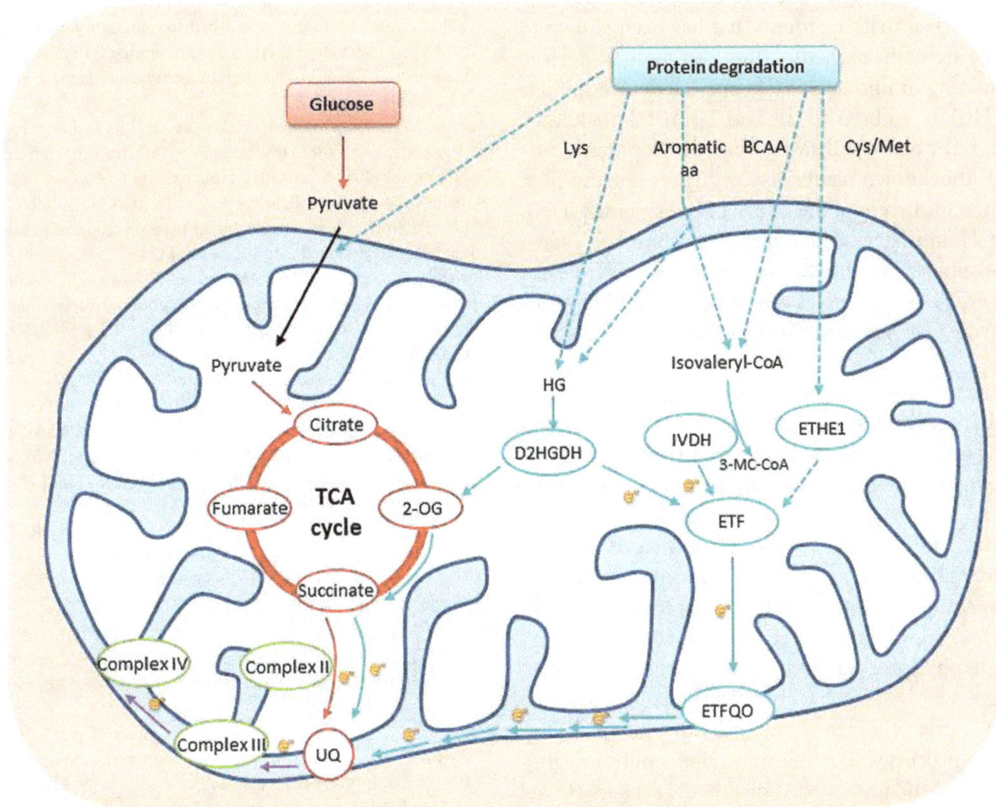

FIGURE 2 | Amino acids can be used as electron donors to the mitochondrial electron transport chain. Protein degradation yields a range of amino acids that are further metabolized either into isovaleryl-CoA or HG. Isovaleryl-CoA can be produced by catabolism of the branched chain and aromatic amino acids and by both phytol and Lys degradation, whereas HG can be produced by aromatic amino acid degradation or from the Lys derivative L-pipecolate. The electrons generated are transferred to the respiratory chain through to the ubiquinol pool via an ETF/ETFQO system. Possible involvement of sulfur containing amino acids has also been implicated by the phenotype of ETHE1 knockdown plants. Some amino acids can facilitate energy production *via* the TCA cycle, either by conversion to pyruvic acid or acetyl-CoA or by direct conversion to TCA cycle intermediates, such as 2-OG, and direct electron supply to the ubiquinone pool of the mitochondrial electron transport chain in plants. Dotted arrows represent possible transport processes and multi enzymatic reactions. Abbreviations: BCAA, branched chain amino acids; D2HGDH, 2-D-hydroxyglutarate dehydrogenase; e–, electron; ETF, electron transfer flavoprotein; ETFQO, ETF:ubiquinone oxidoreductase; ETHE1, ethylmalonic encephalopathy protein1; HG, hydroxyglutarate; IVDH, isovaleryl-CoA dehydrogenase; 3-MC-CoA, 3-methylcrotonyl-CoA; 2-OG, 2-oxoglutarate; TCA cycle, tricarboxylic acid cycle; UQ, ubiquinone (Adapted from Araujo et al., 2011; Krüssel et al., 2014).

is responsible for the transport of electrons from breakdown products of Lys (**Figure 2**). This last finding is interesting, as it demonstrates the ability of Lys degradation products to contribute electrons directly to the mitochondrial electron transport chain, as well as supply intermediates to the TCA cycle (Araujo et al., 2010).

Although a direct role for the ETF/ETFQO system in seed development has not yet been demonstrated, considerable experimental evidence has accumulated to support this hypothesis. First, the existence of hypoxic conditions and lack of photosynthesis in some seeds point to the need for alternative electron donors in seeds for the mitochondrial electron transport chain. Secondly, analysis of a mutant with high levels of 12 out of the 20 amino acids in seeds was shown to be IVDH defective, further strengthening the role of this enzyme in the regulation of free amino acid homeostasis during seed development (Gu et al., 2010). Thirdly, a recent publication highlighted the role of BCAT2, a BCAA catabolic enzyme, in establishing the variation of BCAA levels in different *Arabidopsis* accessions (Angelovici et al., 2013). BCAT2 was shown to localize in the mitochondria, and its expression was shown to increase during seed development, suggesting the involvement of BCAA catabolism in seed development, possibly via the ETF/ETFQO system (Angelovici et al., 2013). Additional evidence suggesting the involvement of the ETF/ETFQO system in seed development was obtained using genetic manipulation of amino acid delivery to pea (*Pisum sativum*) embryos. When an increase of amino acid delivery was induced, transcripts of the ETF complex were most highly induced (Weigelt et al., 2008). It is important to note that *etf* and *etfqo* mutants display lower seed set and shorter siliques when grown under long day conditions (Ishizaki et al., 2005, 2006). This phenotype is attributed to the maternal tissue, as demonstrated by reciprocal crosses of etfqo mutants and WT plants (Ishizaki et al., 2005). Another finding strengthening the role of the ETF/ETFQO system in seed development is a recent publication describing ETHE1, a mitochondrial sulfur dioxygenase. ETHE1 is a matrix protein that has been shown to be involved in the detoxification of sulfur compounds derived from sulfur-containing amino acids (Cys and Met). A complete knockout of ETHE1 is embryo lethal and ETHE1 knockdown plants are viable, but present a delay in seed development. Surprisingly, ETHE1 knockdown plants display higher sensitivity to carbon starvation, reminiscent of the ETF/ETFQO system T-DNA mutants (**Figure 2**) and have been shown to accumulate BCAA under these conditions. These data suggest a possible role for ETHE1 in the ETF/ETFQO system as well as hinting at the importance of the enzyme, *per se*, during seed development (Krüssel et al., 2014).

In summary, in this review we discussed a number of studies that have enhanced our understanding of the parallel roles of photosynthesis and amino acid-derived respiratory catabolic processes in meeting the demands for normal energy status during seed development. In respect to amino acids, we focused on Lys degradation and highlighted the involvement of the ETF pathway. Technical advances, including improved understanding of sub-cellular compartmentization of metabolism (Sweetlove and Fernie, 2013), alongside mathematical approaches to understand whole plant physiology as well as metabolite-metabolite associations (Toubiana et al., 2012; Grafahrend-Belau et al., 2013) are now available. These will greatly enhance our understanding of these enzymatic processes. Furthermore, metabolite imaging techniques, such as genetically encoded metabolite sensors (Okumoto et al., 2012), nuclear magnetic imaging (Borisjuk et al., 2013), and flux profiling techniques (Schwender et al., 2004b), alongside miniaturization of respiration measurements (Sew et al., 2013), will additionally enhance our understanding of the shifts in plant energy metabolism that occur during the process of seed development. Our potential to metabolically engineer seeds in a highly tailored manner will be radically improved once such information is available and integrated with our current knowledge.

ACKNOWLEDGMENTS

Our research was supported by grants from Israel Science Foundation; Bi-national Agriculture Research and Development (BARD); Israel Science Foundation; Ministry of Agriculture; Alternative Energy Research Initiation (AERI) Program and ICORE Biofuels Program (GG); Minerva, Alexander von Humboldt and EMBO fellowships (TAW) and by the Max-Planck Society (ARF).

REFERENCES

Angelovici, R., Fait, A., Fernie, A. R., and Galili, G. (2011). A seed high-lysine trait is negatively associated with the TCA cycle and slows down *Arabidopsis* seed germination. *New Phytol.* 189, 148–159. doi: 10.1111/j.1469-8137.2010.03478.x

Angelovici, R., Fait, A., Zhu, X., Szymanski, J., Feldmesser, E., Fernie, A. R., et al. (2009). Deciphering transcriptional and metabolic networks associated with lysine metabolism during *Arabidopsis* seed development. *Plant Physiol.* 151, 2058–2072. doi: 10.1104/pp.109.145631

Angelovici, R., Galili, G., Fernie, A. R., and Fait, A. (2010). Seed desiccation: a bridge between maturation and germination. *Trends Plant Sci.* 15, 211–2118. doi: 10.1016/j.tplants.2010.01.003

Angelovici, R., Lipka, A. E., Deason, N., Gonzalez-Jorge, S., Lin, H., Cepela, J., et al. (2013). Genome-wide analysis of branched-chain amino acid levels in *Arabidopsis* seeds. *Plant Cell* 25, 4827–4843. doi: 10.1105/tpc.113.119370

Araujo, W. L., Ishizaki, K., Nunes-Nesi, A., Larson, T. R., Tohge, T., Krahnert, I., et al. (2010). Identification of the 2-hydroxyglutarate and isovaleryl-CoA dehydrogenases as alternative electron donors linking lysine catabolism to the electron transport chain of *Arabidopsis* mitochondria. *Plant Cell* 22, 1549–1563. doi: 10.1105/tpc.110.075630

Araujo, W. L., Tohge, T., Ishizaki, K., Leaver, C. J., and Fernie, A. R. (2011). Protein degradation – an alternative respiratory substrate for stressed plants. *Trends Plant Sci.* 16, 489–498. doi: 10.1016/j.tplants.2011.05.008

Borisjuk, L., Neuberger, T., Schwender, J., Heinzel, N., Sunderhaus, S., Fuchs, J., et al. (2013). Seed architecture shapes embryo metabolism in oilseed rape. *Plant Cell* 25, 1625–1640. doi: 10.1105/tpc.113.111740

Borisjuk, L., Nguyen, T. H., Neuberger, T., Rutten, T., Tschiersch, H., Claus, B., et al. (2005). Gradients of lipid storage, photosynthesis and plastid differentiation in developing soybean seeds. *New Phytol.* 167, 761–776. doi: 10.1111/j.1469-8137.2005.01474.x

Buchanan, B. B., Gruissem, W., and Jones, R. L. (2000). *Biochemistry and Molecular Biology of Plants.* Rockville, MD: American Society of Plant Physiologists.

Credali, A., Garcia-Calderon, M., Dam, S., Perry, J., Diaz-Quintana, A., Parniske, M., et al. (2013). The K^+-dependent asparaginase, NSE1, is crucial for plant growth and seed production in *Lotus japonicus. Plant Cell Physiol.* 54, 107–118. doi: 10.1093/pcp/pcs156

Dickson, J. M., Vincze, E., Grant, M. R., Smith, L. A., Rodber, K. A., Farnden, K. J., et al. (1992). Molecular cloning of the gene encoding developing seed L-asparaginase from *Lupinus angustifolius. Plant Mol. Biol.* 20, 333–336. doi: 10.1007/BF00014503

Fait, A., Angelovici, R., Less, H., Ohad, I., Urbanczyk-Wochniak, E., Fernie, A. R., et al. (2006). *Arabidopsis* seed development and germination is associated with temporally distinct metabolic switches. *Plant Physiol.* 142, 839–854. doi: 10.1104/pp.106.086694

Galili, G. (2011). The aspartate-family pathway of plants: linking production of essential amino acids with energy and stress regulation. *Plant Signal. Behav.* 6, 192–195. doi: 10.4161/psb.6.2.14425

Grafahrend-Belau, E., Junker, A., Eschenroder, A., Muller, J., Schreiber, F., and Junker, B. H. (2013). Multiscale metabolic modeling: dynamic flux balance analysis on a whole-plant scale. *Plant Physiol.* 163, 637–647. doi: 10.1104/pp.113.224006

Gu, L., Jones, A. D., and Last, R. L. (2010). Broad connections in the *Arabidopsis* seed metabolic network revealed by metabolite profiling of an amino acid catabolism mutant. *Plant J.* 61, 579–590. doi: 10.1111/j.1365-313X.2009.04083.x

Ishizaki, K., Larson, T. R., Schauer, N., Fernie, A. R., Graham, I. A., and Leaver, C. J. (2005). The critical role of *Arabidopsis* electron-transfer flavoprotein:ubiquinone oxidoreductase during dark-induced starvation. *Plant Cell* 17, 2587–2600. doi: 10.1105/tpc.105.035162

Ishizaki, K., Schauer, N., Larson, T. R., Graham, I. A., Fernie, A. R., and Leaver, C. J. (2006). The mitochondrial electron transfer flavoprotein complex is essential for survival of *Arabidopsis* in extended darkness. *Plant J.* 47, 751–760. doi: 10.1111/j.1365-313X.2006.02826.x

Jones-Held, S., Ambrozevicius, L. P., Campbell, M., Drumheller, B., Harrington, E., and Leustek, T. (2012). Two *Arabidopsis thaliana* dihydrodipicolinate synthases, DHDPS1 and DHDPS2, are unequally redundant. *Funct. Plant Biol.* 39, 1058–1067. doi: 10.1071/Fp12169

Joshi, V., Laubengayer, K. M., Schauer, N., Fernie, A. R., and Jander, G. (2006). Two *Arabidopsis* threonine aldolases are nonredundant and compete with threonine deaminase for a common substrate pool. *Plant Cell* 18, 3564–3575. doi: 10.1105/tpc.106.044958

Kirma, M., Araujo, W. L., Fernie, A. R., and Galili, G. (2012). The multifaceted role of aspartate-family amino acids in plant metabolism. *J. Exp. Bot.* 63, 4995–5001. doi: 10.1093/jxb/ers119

Kochevenko, A., and Fernie, A. R. (2011). The genetic architecture of branched-chain amino acid accumulation in tomato fruits. *J. Exp. Bot.* 62, 3895–3906. doi: 10.1093/jxb/err091

Krüssel, L., Junemann, J., Wirtz, M., Birke, H., Thornton, J. D., Browning, L. W., et al. (2014). The mitochondrial sulfur dioxygenase ETHE1 is required for amino acid catabolism during carbohydrate starvation and embryo development in *Arabidopsis thaliana. Plant Physiol.* 165, 92–104. doi: 10.1104/pp.114.239764

Less, H., and Galili, G. (2008). Principal transcriptional programs regulating plant amino acid metabolism in response to abiotic stresses. *Plant Physiol.* 147, 316–330. doi: 10.1104/pp.108.115733

Okumoto, S., Jones, A., and Frommer, W. B. (2012). Quantitative imaging with fluorescent biosensors. *Annu. Rev. Plant Biol.* 63, 663–706. doi: 10.1146/annurev-arplant-042110-103745

Rolletschek, H., Weber, H., and Borisjuk, L. (2003). Energy status and its control on embryogenesis of legumes. Embryo photosynthesis contributes to oxygen supply and is coupled to biosynthetic fluxes. *Plant Physiol.* 132, 1196–1206. doi: 10.1104/pp.102.017376

Ruuska, S. A., Schwender, J., and Ohlrogge, J. B. (2004). The capacity of green oilseeds to utilize photosynthesis to drive biosynthetic processes. *Plant Physiol.* 136, 2700–2709. doi: 10.1104/pp.104.047977

Schwender, J., Goffman, F., Ohlrogge, J. B., and Shachar-Hill, Y. (2004a). Rubisco without the Calvin cycle improves the carbon efficiency of developing green seeds. *Nature* 432, 779–782. doi: 10.1038/nature03145

Schwender, J., Ohlrogge, J., and Shachar-Hill, Y. (2004b). Understanding flux in plant metabolic networks. *Curr. Opin. Plant Biol.* 7, 309–317. doi: 10.1016/j.pbi.2004.03.016

Sew, Y. S., Stroher, E., Holzmann, C., Huang, S., Taylor, N. L., Jordana, X., et al. (2013). Multiplex micro-respiratory measurements of *Arabidopsis* tissues. *New Phytol.* 200, 922–932. doi: 10.1111/nph.12394

Sreenivasulu, N., Usadel, B., Winter, A., Radchuk, V., Scholz, U., Stein, N., et al. (2008). Barley grain maturation and germination: metabolic pathway and regulatory network commonalities and differences highlighted by new MapMan/PageMan profiling tools. *Plant Physiol.* 146, 1738–1758. doi: 10.1104/pp.107.111781

Sreenivasulu, N., and Wobus, U. (2013). Seed-development programs: a systems biology-based comparison between dicots and monocots. *Annu. Rev. Plant Biol.* 64, 189–217. doi: 10.1146/annurev-arplant-050312-120215

Sweetlove, L. J., and Fernie, A. R. (2013). The spatial organization of metabolism within the plant cell. *Annu. Rev. Plant Biol.* 64, 723–746. doi: 10.1146/annurev-arplant-050312-120233

Toubiana, D., Semel, Y., Tohge, T., Beleggia, R., Cattivelli, L., Rosental, L., et al. (2012). Metabolic profiling of a mapping population exposes new insights in the regulation of seed metabolism and seed, fruit, and plant relations. *PLoS Genet.* 8:e1002612. doi: 10.1371/journal.pgen.1002612

Tschiersch, H., Borisjuk, L., Rutten, T., and Rolletschek, H. (2011). Gradients of seed photosynthesis and its role for oxygen balancing. *Biosystems* 103, 302–308. doi: 10.1016/j.biosystems.2010.08.007

Tschiersch, H., Liebsch, G., Borisjuk, L., Stangelmayer, A., and Rolletschek, H. (2012). An imaging method for oxygen distribution, respiration and photosynthesis at a microscopic level of resolution. *New Phytol.* 196, 926–936. doi: 10.1111/j.1469-8137.2012.04295.x

van Dongen, J. T., Roeb, G. W., Dautzenberg, M., Froehlich, A., Vigeolas, H., Minchin, P. E., et al. (2004). Phloem import and storage metabolism are highly coordinated by the low oxygen concentrations within developing wheat seeds. *Plant Physiol.* 135, 1809–1821. doi: 10.1104/pp.104.040980

Vigeolas, H., Huhn, D., and Geigenberger, P. (2011). Nonsymbiotic hemoglobin-2 leads to an elevated energy state and to a combined increase in polyunsaturated fatty acids and total oil content when overexpressed in developing seeds of transgenic *Arabidopsis* plants. *Plant Physiol.* 155, 1435–1444. doi: 10.1104/pp.110.166462

Vigeolas, H., Van Dongen, J. T., Waldeck, P., Huhn, D., and Geigenberger, P. (2003). Lipid storage metabolism is limited by the prevailing low oxygen concentrations within developing seeds of oilseed rape. *Plant Physiol.* 133, 2048–2060. doi: 10.1104/pp.103.031963

Weber, H., Borisjuk, L., and Wobus, U. (2005). Molecular physiology of legume seed development. *Annu. Rev. Plant Biol.* 56, 253–279. doi: 10.1146/annurev.arplant.56.032604.144201

Weigelt, K., Kuster, H., Radchuk, R., Muller, M., Weichert, H., Fait, A., et al. (2008). Increasing amino acid supply in pea embryos reveals specific interactions of N and C metabolism, and highlights the importance of mitochondrial metabolism. *Plant J.* 55, 909–926. doi: 10.1111/j.1365-313X.2008.03560.x

Xu, S. B., Li, T., Deng, Z. Y., Chong, K., Xue, Y., and Wang, T. (2008). Dynamic proteomic analysis reveals a switch between central carbon metabolism and alcoholic fermentation in rice filling grains. *Plant Physiol.* 148, 908–925. doi: 10.1104/pp.108.125633

Zhu, X., and Galili, G. (2003). Increased lysine synthesis coupled with a knockout of its catabolism synergistically boosts lysine content and also transregulates the metabolism of other amino acids in *Arabidopsis* seeds. *Plant Cell* 15, 845–853. doi: 10.1105/tpc.009647

Zhu, X., and Galili, G. (2004). Lysine metabolism is concurrently regulated by synthesis and catabolism in both reproductive and vegetative tissues. *Plant Physiol.* 135, 129–136. doi: 10.1104/pp.103.037168

Conflict of Interest Statement: The authors declare that the research was conducted in the absence of any commercial or financial relationships that could be construed as a potential conflict of interest.

5

Programmed cell death (PCD): an essential process of cereal seed development and germination

Fernando Domínguez and Francisco J. Cejudo*

Instituto de Bioquímica Vegetal y Fotosíntesis, Universidad de Sevilla – Consejo Superior de Investigaciones Científicas, Sevilla, Spain

Edited by:
Paolo Sabelli, University of Arizona, USA

Reviewed by:
Daniel Hofius, Swedish University of Agricultural Sciences, Sweden
Patrick Gallois, University of Manchester, UK

***Correspondence:**
Francisco J. Cejudo, Instituto de Bioquímica Vegetal y Fotosíntesis, Universidad de Sevilla – Consejo Superior de Investigaciones Científicas, Avda Américo Vespucio 49, Sevilla 41092, Spain
e-mail: fjcejudo@us.es

The life cycle of cereal seeds can be divided into two phases, development and germination, separated by a quiescent period. Seed development and germination require the growth and differentiation of new tissues, but also the ordered disappearance of cells, which takes place by a process of programmed cell death (PCD). For this reason, cereal seeds have become excellent model systems for the study of developmental PCD in plants. At early stages of seed development, maternal tissues such as the nucellus, the pericarp, and the nucellar projections undergo a progressive degeneration by PCD, which allows the remobilization of their cellular contents for nourishing new filial tissues such as the embryo and the endosperm. At a later stage, during seed maturation, the endosperm undergoes PCD, but these cells remain intact in the mature grain and their contents will not be remobilized until germination. Thus, the only tissues that remain alive when seed development is completed are the embryo axis, the scutellum and the aleurone layer. In germinating seeds, both the scutellum and the aleurone layer play essential roles in producing the hydrolytic enzymes for the mobilization of the storage compounds of the starchy endosperm, which serve to support early seedling growth. Once this function is completed, scutellum and aleurone cells undergo PCD; their contents being used to support the growth of the germinated embryo. PCD occurs with tightly controlled spatial-temporal patterns allowing coordinated fluxes of nutrients between the different seed tissues. In this review, we will summarize the current knowledge of the tissues undergoing PCD in developing and germinating cereal seeds, focussing on the biochemical features of the process. The effect of hormones and redox regulation on PCD control will be discussed.

Keywords: cereal, development, germination, plant, programmed cell death, seed

INTRODUCTION

Two phases may be distinguished in the life cycle of the cereal seeds, development and germination. Seed development is initiated by the fertilization events and culminates with the formation of a mature seed, which has a low content of water and remains in a quiescent status. Upon imbibition, the quiescent seed initiates the phase of germination in which the reserves stored in the starchy endosperm are remobilized to support the initial stages of seedling growth. The phase of development may be subdivided in three stages (**Figure 1**): (I), early development, which includes double fertilization, syncytium formation and endosperm cellularization; (II), differentiation, which comprises the formation of the different cell types of the seed (embryo-surrounding cells, transfer cells, starchy endosperm and aleurone), endoreduplication and accumulation of storage reserves in the endosperm; and (III), maturation, which includes desiccation and dormancy (Sabelli and Larkins, 2009). At the morphological level, early seed development is characterized by a large increase in seed size, which reaches about 80% of its final length at this stage (**Figure 1**; Bosnes et al., 1992; Domínguez and Cejudo, 1996). Seed length increases predominantly due to the elongation of the pericarp cells in the longitudinal direction; this growth making room to accommodate the growing endosperm (Radchuk et al., 2011). Prior to the stage of storage accumulation in the growing endosperm, a remobilization of nutrients from maternal tissues takes place, which involves the participation of a complex set of hydrolytic enzymes, including proteases (Domínguez and Cejudo, 1996) and starch-degrading enzymes (Radchuk et al., 2009). Following germination, a large process of remobilization of the storage compounds of the starchy endosperm occurs. This process, which supports early seedling growth until the new plant becomes autonomous, requires the synthesis and secretion of hydrolytic enzymes initially by the scutellum epithelium cells. Then, the aleurone cells, which surround the starchy endosperm, become activated and secrete a large amount of hydrolases (Fincher, 1989), in a process regulated by gibberellins synthesized by the scutellum and then released into the starchy endosperm (Appleford and Lenton, 1997).

Although the formation of a new seed, as well as the generation of a new plant upon germination, requires the generation and differentiation of a large number of cells, the ordered disappearance of cells is also of high relevance. Cell degeneration takes place by programmed cell death (PCD), which may be considered as an ordered process of selective removal of cells. There are several reasons that make PCD an important process for the successful completion of seed development and germination. Both seed development and germination rely on a continuous remobilization of nutrients, which is supported by cell degeneration. This

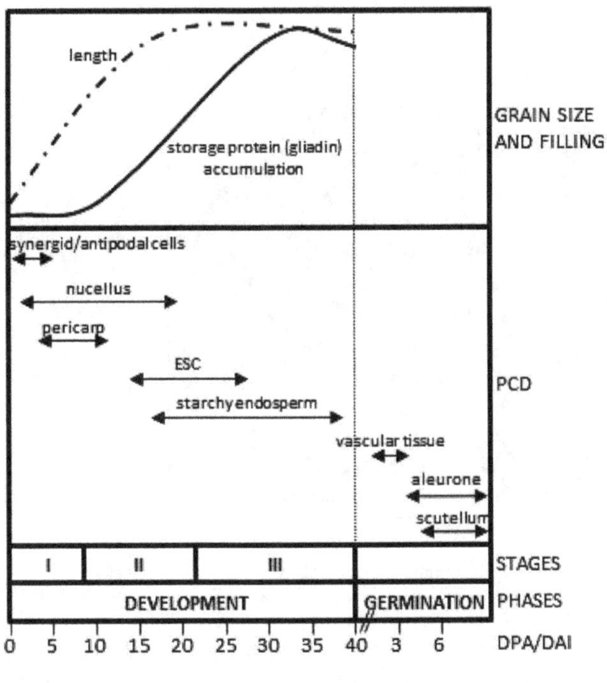

FIGURE 1 | Tissues undergoing PCD in developing and germinating cereal seeds. The upper panel represents the increase in seed length and storage accumulation taking as models the wheat and barley seeds. The lower panel indicates the periods of seed development in which the tissues undergo PCD. Stages of development: I, early development; II, differentiation; III, maturation. ESC, embryo-surrounding cells; DPA, days post anthesis (development); DAI, days after imbibition (germination).

is the case of antipodal, nucellar and pericarp cells, which degenerate at early stages of seed development (Domínguez et al., 2001; An and You, 2004; Radchuk et al., 2011). The starchy endosperm undergoes PCD during the final stages of seed development; however, this tissue remains intact after cell death in the mature seed and the reserves stored therein will be mobilized only after germination. PCD also has the important function of creating new structures, such as nucellar projection cells at middle stages of seed development (Domínguez et al., 2001), and the vascular tissue of the scutellum, the differentiation of which is completed immediately after imbibition (Domínguez et al., 2002). Finally, PCD of pericarp cells is associated with the enlargement of the seed, facilitating the growth of the endosperm (Radchuk et al., 2011) and the formation of the seed cuticle, which has a protective function.

In summary, PCD is an essential process for the successful completion of cereal seed development and germination and, thus, cereal seeds have become an important model system for the study of cell death in plants. Based on morphological characteristics, two categories of plant PCD were recently proposed: vacuolar PCD, which includes autophagy, and necrosis, in conjunction with some forms of PCD that present mixed features and are not clearly ascribed to these categories (van Doorn et al., 2011). Most of the tissues that undergo PCD in cereal seeds show features of vacuolar PCD in agreement with the notion that cell death is part of developmental programs that characterize the formation and germination of the seed. In this review we will summarize the present knowledge of the tissues undergoing PCD in developing and germinating cereal seeds, emphasizing their morphological and biochemical characteristics. In addition, we will focus on the function of the process of PCD for the successful completion of these developmental programs.

PCD OF MATERNAL TISSUES IS AN IMPORTANT PROCESS OF CEREAL SEED DEVELOPMENT

In monocots, fertilization is a double event that results in a diploid embryo and a triploid endosperm (Olsen, 2004). The formation of the seed involves the generation, growth and differentiation of new tissues; however, the ordered disappearance of cells plays as well an important function in this complex developmental process. In particular, several maternal tissues undergo PCD to help the formation of the seed; among them, it is worth mentioning the deaths of cells of the embryo sac, the nucellus, the pericarp, and the nucellar projections, which take place sequentially, as outlined in **Figure 1**. It follows below a brief description of how these cells undergo PCD.

SYNERGID AND ANTIPODAL CELLS

The embryo sac is composed of an egg cell, which is accompanied by two synergid cells at one pole and three antipodal cells at the opposite pole. Synergid cells are the first to undergo PCD at this initial stage of seed development (**Figure 1**). This death process occurs shortly before the pollen tube discharges and seems to be important to guide the growth of the pollen tube (An and You, 2004). In *Arabidopsis*, it was shown that the signaling cascade leading to the death of one of the two synergids is initiated by the contact with the pollen tube (Sandaklie-Nikolova et al., 2007). Synergids control sperm delivery through the *FERONIA* signaling pathway to initiate and modulate their distinct calcium signatures in response to calcium dynamics and growth behavior of the pollen tube (Ngo et al., 2014). PCD of antipodal cells occurs later, at 2–3 days post-anthesis (DPA), and contributes to the development of the adjacent free-nuclear endosperm. Nuclear materials from the dying antipodal cells support the nuclear divisions in the growing coenocyte (Engell, 1994; An and You, 2004).

NUCELLUS

At early stages of cereal seed development, the nucellus is among the first tissues to degenerate; nucellar cells undergoing a process of PCD, which has been well characterized at the morphological and biochemical levels (Domínguez et al., 2001; Radchuk et al., 2011). After the double fertilization event, the endosperm nucleus suffers several rounds of divisions to form a multinucleate syncytium surrounding the characteristic central vacuole (**Figure 2**). The technique of terminal deoxynucleotidil transferase dUTP end labeling (TUNEL), which allows the direct staining of fragmented DNA and, thus, the visualization of nuclei from cells undergoing PCD, has been of great aid in characterizing the pattern of PCD in early developing seeds. The TUNEL assay allowed the identification of degenerating nuclei of the inner cell layers of the nucellus very early after anthesis; the degenerative process spreading to the outer nucellar layers at 2 DPA (Radchuk et al., 2011). It has been proposed that PCD of the nucellus serves for the remobilization

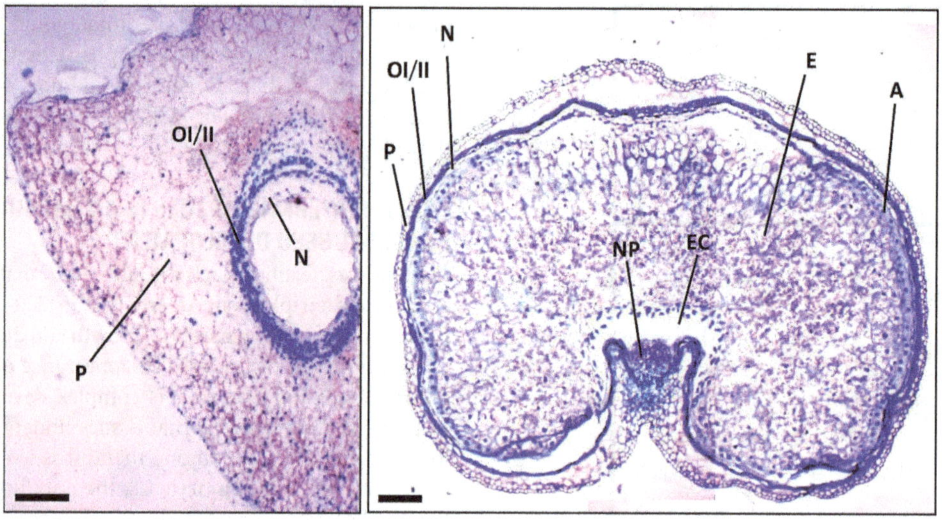

FIGURE 2 | Sections of wheat seeds at 3 DPA (left) and 16 DPA (right) stained with toluidine blue. Note the different tissues and how they change between these stages of seed development. P, pericarp; OI, outer integument; II, inner integument; N, nucellus; NP, nucellar projections; A, aleurone; E, starchy endosperm; EC, endosperm cavity. Bars, 100 μm.

of its cellular contents, which are needed for the nourishment of the growing coenocyte and the cellularization process. Additional markers of cell degeneration are the expression of different hydrolytic enzymes, such as the aspartic protease nucellin (Chen and Foolad, 1997), a cathepsin B-like protease (Domínguez and Cejudo, 1998), the vacuolar processing enzyme nucellain (Linnestad et al., 1998), and the α-amylase AMY 4 (Radchuk et al., 2009). A gradient from internal to external layers is observed in the degeneration of the nucellus in developing wheat grains that culminates when only the nucellar epidermis remains. At 5 DPA, the nucellar parenchyma is completely disorganized, and TUNEL-labeled nuclei of the nucellar epidermis and the two-cell layer inner integuments are observed (**Figure 3**; Domínguez et al., 2001). At 15 DPA, the nucellus is reduced to a single-cell layer, which shows high level of expression of genes encoding cathepsin B-like thiol protease and serine carboxypeptidase III, suggesting a high hydrolytic activity in this tissue (Domínguez and Cejudo, 1998). It is noteworthy that besides that of genes involved in the hydrolytic activity, these nucellar cells also show the expression of genes encoding enzymes involved in biosynthetic metabolism. This is the case of phosphoenolpyruvate carboxylase (PEPC), which is expressed at high level at early stages of seed development (5 DPA) in the nucellus, the multinucleate syncytium and the vascular tissue. This activity may generate carbon skeletons to support the demand of amino acid biosynthesis in the growing endosperm (González et al., 1998).

PERICARP

In the case of wheat and barley seeds, the pericarp is a tissue of maternal origin which, at early stages of development, is formed by several layers of parenchymatic cells, a two cell layer chlorenchyma and the inner epidermis (**Figure 2**). During the pre-storage phase (0–10 DPA), this tissue shows a decrease in cell divisions, which is accompanied by twofold to threefold increase in cell elongation, while the rows of parenchymatic cells localized between the inner integument and the outer epidermis are reduced twofold to fourfold (Radchuk et al., 2011). The first symptoms of pericarp cell degeneration appear at 4 DPA (Domínguez et al., 2001; Radchuk et al., 2011). Then, during the period of 6–10 DPA, PCD is extended to the whole tissue (Radchuk et al., 2011), so that by 15 DPA the pericarp is reduced to several layers of cuticle, as shown for the case of the wheat seed (**Figure 3**). Biochemical analyses showed the presence of proteolytic activities in the pericarp of wheat seeds at early stages of development (Domínguez and Cejudo, 1996). Furthermore, transcriptomic analyses performed in barley seeds revealed the expression of different genes encoding proteolytic enzymes in this tissue (Sreenivasulu et al., 2006), including the vacuolar processing enzyme VPE4 (Radchuk et al., 2011), suggesting that the degeneration of these cells occurs by a process of PCD. In addition, the expression of α-amylases suggests the existence of starch degradation in the pericarp (Radchuk et al., 2009). Indeed, pericarp PCD is accompanied by dynamic changes in starch accumulation patterns (**Figure 3**); starch granules being synthesized, deposited and degraded in plastids temporarily until their utilization by the growing endosperm (Zhou et al., 2009). PCD of the maternal tissues at these stages of seed development may have two purposes: to make room for the expanding endosperm and to support the nourishment of this tissue.

NUCELLAR PROJECTIONS

The nucellar projections cells form a complex tissue of great relevance during seed development. These cells differentiate into transfer cells, thus enabling the transfer of nutrients from the pericarp to the endosperm cavity (**Figure 2**). Based on morphological features, several stages of differentiation through the radial axis can be established: roundish meristematic cells close to the pigment strand and adjacent to the vascular tissue (stage I); middle

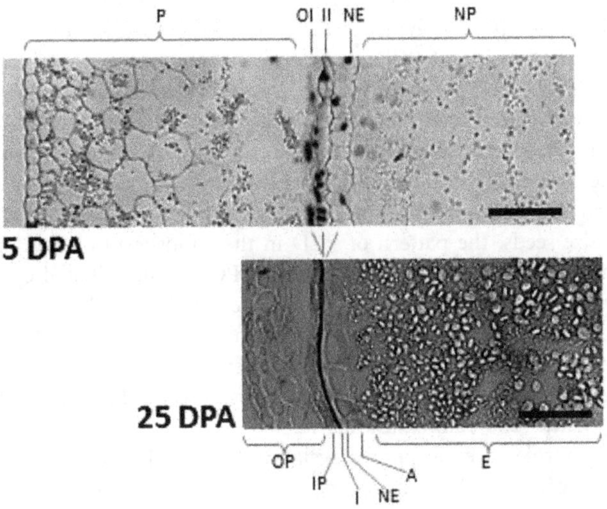

FIGURE 3 | TUNEL assay showing maternal tissues undergoing PCD at early and late stages of wheat seed development. In seeds at 5 DPA, nuclei from the nucellus and the integuments showed TUNEL-positive signal. Note the reduction of pericarp width and the formation of seed cuticles as a consequence of PCD (compare 5 DPA vs 25 DPA). Starch granules of different sizes are also observed in the starchy endosperm. P, pericarp; OI, outer integument; II, inner integument; NE, nucellar epidermis; NP, nucellar parenchyma; OP, outer pericarp; IP, inner pericarp; I, pigmented cuticle derived from outer and inner integuments; A, aleurone; E, starchy endosperm. Bars, 50 μm.

zone with elongated cells (stage II); transfer cells with peculiar cell wall invaginations (stage III); and autolysing cells adjacent to the endosperm cavity (stage IV). PCD plays a relevant function in this process of differentiation (Domínguez et al., 2001). While nucellar projection cells show no symptoms of PCD at early stages of development (5 DPA), most cells show TUNEL-stained nuclei as seed development progresses (13 DPA). At 18 DPA, TUNEL staining is restricted to nucellar projection cells adjacent to the pigment strand, because cells located near the endosperm cavity seem to be completely degraded (Domínguez et al., 2001). The high level of expression of cathepsin B-like thiol protease (Domínguez and Cejudo, 1998) and the vacuolar processing enzyme nucellain (Linnestad et al., 1998) in nucellar projection cells at 9–16 DPA is in agreement with the known role of these proteases in the process of cell degeneration. Another marker of the differentiation and PCD of the nucellar projections is the JEKILL protein (Radchuk et al., 2006). The increase in the level of this protein has been associated with structural changes in the nucellar projections, so that a gradient is generated from the crease region to the autolysing cells close to the endosperm cavity. Repression of the gene encoding JEKILL impairs the differentiation of the nucellar projections, which affects the exchange of nutrients between the pericarp and the endosperm (Radchuk et al., 2006). The gradient observed in the nucellar projections, from the crease region to the endosperm cavity, has also been observed by large-scale *in situ* hybridization expression analyses (Drea et al., 2005) and by laser micro-dissection pressure catapulting-based transcriptome analyses (Thiel et al., 2008). At 8 DPA, the meristematic zone (stage I of differentiation) shows the expression of genes characteristic of tissues with high mitotic activity. Genes involved in cell wall biosynthesis and expansin/extensin genes are expressed along the elongation zone (stage II); finally, genes involved in PCD-related proteolysis and nitrogen remobilization are expressed in the disintegration zone (stages III and IV). Several genes associated with the degeneration of the nucellar projection cells have been identified (Domínguez and Cejudo, 1998; Linnestad et al., 1998; Radchuk et al., 2006), however, little is known about upstream regulatory factors controlling their patterns of expression. The MADS29 transcription factor was proposed to play a relevant function in the control of nucellus and nucellar projection cells degeneration (Yin and Xue, 2012; Yang et al., 2012). The *MADS29* gene is expressed at high levels in the nucellus at early stages of seed development, up to 3 DPA. At later stages, 3–10 DPA, this gene shows a high level of expression in nucellar projection cells but not in the pericarp, integuments and endosperm (Yin and Xue, 2012). Under-expression of the *MADS29* gene provokes abnormal development and formation of shrunken seeds with reduced rate of grain-filling and altered starch granules. Histological analysis of these seeds revealed that the degeneration of the nucellar projection cells and other maternal tissues is blocked (Yang et al., 2012; Yin and Xue, 2012), suggesting that MADS29 directly regulates the expression of several PCD-related genes (Yin and Xue, 2012). Nutrient transport in temperate cereals, such as wheat and barley, occurs along the entire length of the grain through a single vascular band embedded in the maternal pericarp. This vascular band distributes nutrients to the endosperm cavity through the nucellar projection. However, in tropical crops, such as maize and sorghum, the transfer of nutrients occurs through a placento-chalazal layer localized in the basal region of the grain. Interestingly, like the nucellar projections, the placento-chalazal layer also degenerates during maize seed development (Kladnik et al., 2004).

FILIAL TISSUES UNDERGO PCD IN DEVELOPING AND GERMINATING CEREAL SEEDS

With regard to the processes of PCD of filial tissues, these may be classified in those that occur during seed development, such as in the suspensor, embryo-surrounding layers and the starchy endosperm, and those taking place after germination, which include PCD in the parenchymal, epithelial, and vascular tissue of the scutellum, and the aleurone layer (**Figure 1**).

SUSPENSOR AND EMBRYO-SURROUNDING LAYERS

In cereals such as maize, the first asymmetric division of the diploid zygote produces an apical cell, which develops into the embryo proper, and a basal cell, which generates the suspensor. The analysis of the pattern of expression of different genes suggests a marked regionalization in the differentiation of the embryo (Okamoto et al., 2005), which might also be affected by auxin transport (Forestan et al., 2010). Several tissues undergo PCD during embryogenesis. This is the case of the suspensor, which participates in the transfer of nutrients from maternal tissues to the developing embryo proper. The suspensor undergoes PCD in conjunction with PCD in the embryo-surrounding tissues. In maize seeds at 14 DPA, the scutellum cell layers that surround the shoot primordium, as well as the coleoptile and the root cap, show TUNEL-stained nuclei. Nuclei of the shoot primordium-surrounding layers appear completely degraded by 17 DPA (Giuliani et al., 2002). The suspensor undergoes a process of PCD, which is extended between 14 and 27 DPA defining a top-to-bottom gradient of DNA fragmentation, chromatin condensation and nuclei degeneration (Giuliani et al., 2002). In the so-called *emb* (embryo-specific) mutants of maize, which show arrested embryo development but a normal endosperm, the process of PCD in the suspensor is impaired (Consonni et al., 2003). More in-depth analyses at the molecular level have been carried out in non-cereal crops such as in the gymnosperm *Norway spruce* (Filonova et al., 2000). In this system it was identified a metacaspase, termed mcII-Pa, which has autoprocessing activity and is translocated from the cytoplasm to the nucleus in embryo cells undergoing PCD. Cell death thus relies on the proteolytic activity of metacaspase mcII-Pa, which acts as an executioner of PCD (Bozhkov et al., 2005). The death of the embryo suspensor requires the activation of autophagy-related components downstream of metacaspase mcII-Pa, as shown by the fact that the genetic suppression of the metacaspase-autophagy pathway promotes a switch from vacuolar PCD to necrosis. This suppression results in failure of suspensor differentiation and embryonic arrest (Minina et al., 2013). In addition, VEIDase, a caspase-6-like activity, was also identified in *Norway spruce* embryogenesis. The activity of this protease increases at early stages of embryo development and it has been proposed to participate in embryo pattern formation. When VEIDase activity is inhibited, the differentiation of the embryo-suspensor is blocked and the development of the embryo arrested (Bozhkov et al., 2004). VEIDase activity was also detected in barley developing embryo (Boren et al., 2006), suggesting common PCD mechanisms in gymnosperms and monocots. In tobacco, it was shown that the molecular mechanism triggering suspensor PCD is based on the antagonistic actions of two proteins: the cystatin NtCYS, a protease inhibitor, and its target, the cathepsin H-like protease NtCP14. NtCYS prevents precocious PCD in the basal cell of the proembryo by inhibiting NtCP14 protease. Transcriptional down-regulation of NtCYS leads to an increase in NtCP14 activity, which promotes PCD (Zhao et al., 2013). Silencing of the NtCYS inhibitor or overexpression of NtCP14 protease genes provoke precocious cell death with consequent embryonic arrest and grain abortion (Zhao et al., 2013).

ENDOSPERM

In wheat seeds, the expansion of the endosperm is preceded by PCD in cells adjacent to the nucellar projections. The endosperm cavity is thus formed, allowing the transfer of nutrients from the vascular bundle embedded in the pericarp to the growing endosperm. The development of the endosperm in cereal seeds encompasses different processes, such as endoreduplication, accumulation of storage materials and PCD. During endoreduplication, DNA replication is not followed by cytokinesis, resulting in an altered cell cycle and polyploidy (Sabelli, 2012). In maize seeds, endoreduplication is initiated in the central area of the endosperm around 8–10 DPA; the process being extended toward the periphery and producing a high level of polyploidy in endosperm cells at 20 DPA (Sabelli, 2012). It should be noted that endoreduplication is not a homogeneous process since central endosperm cells have higher levels of polyploidy than external cells (Sabelli, 2012). The progression of PCD in the endosperm of developing maize seeds follows a two-wave pattern: the first wave begins around 16 DPA in central cells, coincident with an increase in DNA content; the second starts at the upper crown and progresses toward the base of the seed between 24 and 40 DPA, paralleling the pattern of starch accumulation (Young et al., 1997). In contrast with maize seeds, the pattern of PCD in the endosperm of developing wheat seeds progresses randomly; PCD being initiated by 16 DPA and extended in a random manner until 30 DPA, when the entire endosperm is affected (Young et al., 1997). Retinoblastoma-related (RBR) proteins and cyclin-dependent kinase (CDK) have been identified as fundamental players in cereal endosperm development. The retinoblastoma-related pathway seems to play a major role in endosperm development of maize seeds since it is involved in the control of processes such as endoreduplication, cell proliferation, cell size and cell death (Sabelli et al., 2013).

EMBRYO VASCULAR TISSUE

In cereals, the differentiation of the scutellum vascular tissue proceeds only to the provascular stage of seed development, and is completed following germination. This pattern of differentiation has the purpose of avoiding translocation of nutrients from the scutellum to the embryonic axis before grain maturation (Swift and O'Brien, 1970, 1971). However, a fully functional vascular system is needed immediately after seed germination. Therefore, differentiation of the scutellar tracheary elements is completed in seeds at 2–3 days after imbibition (DAI), forming the characteristic annular thickenings, in a process that involves PCD (Domínguez et al., 2002). Differentiation of the tracheary elements implies the participation of endo- and exopeptidases. Up to six genes encoding carboxypeptidases have been identified in cereals (Dal Degan et al., 1994; Washio and Ishikawa, 1994); some of them being expressed in the embryo of germinating grains. Only the GA_3-induced serine carboxypeptidase III, which was isolated from wheat aleurone cells (Baulcombe et al., 1987), has been shown to participate in vessel formation. The pattern of expression of this gene, as determined by in situ hybridization, shows a clear coincidence with TUNEL-stained nuclei in the tracheary elements of the scutellum, which suggests a role for this cartboxypeptidase as executioner in this PCD process (Domínguez et al., 2002). The participation of the serine carboxypeptidase III in the differentiation of vascular tissues of other organs, such as shoots and roots, has also been suggested (Domínguez et al., 2002).

ALEURONE

The differentiation of the aleurone layer, the outermost cell layer of the endosperm, initiates in developing seeds around 8 DPA (Bosnes et al., 1992; **Figure 2**). In contrast with the endosperm, which undergoes a process of PCD at later stages of seed development (Young et al., 1997; Young and Gallie, 1999), the aleurone layer and the embryo remain alive in the mature seed. In wheat seeds, both tissues show the expression of protease inhibitors (Corre-Menguy et al., 2002), which may have a protective function since the surrounding pericarp shows intense proteolytic activity during seed development (Domínguez and Cejudo, 1996). Following germination, the aleurone layer displays a high metabolic activity to synthesize and secrete hydrolytic enzymes, which are released into the starchy endosperm and promote the degradation of storage reserves (Cejudo et al., 1992, 1995; Domínguez and Cejudo, 1995, 1999). Once this function is completed, the aleurone undergoes a process of PCD (Domínguez et al., 2004). DNA fragmentation in aleurone layer nuclei was observed in seeds after 4 DAI, and then increased progressively. TUNEL assays revealed a very characteristic spatial pattern of aleurone layer PCD; the process being initiated in cells proximal to the embryo and extended to distal cells, both in wheat and barley (Wang et al., 1998; Domínguez et al., 2004). In contrast to wheat and barley, the process of aleurone PCD was delayed in maize seeds; DNA laddering being detected only after 12 DAI (Domínguez et al., 2004). This delay may reflect differences in the germination strategies of both types of grains. In this regard, it should be noted that the scutellum is larger in maize than in wheat or barley seeds and, thus, may have a more relevant function supporting the initial stages of seedling growth. The progression of PCD in the aleurone layer is a tightly regulated process, which takes place only when these cells have completed the synthesis and secretion of hydrolytic enzymes. Indeed, this function of the aleurone cells is essential for germination as shown by the fact that the aleurone-deficient maize mutant seeds, *dek1*, are unable to germinate (Domínguez et al., 2004).

SCUTELLUM

Although, as mentioned above, some of the tissues surrounding the embryo undergo PCD during embryogenesis in developing seeds, the bulk of scutellum PCD occurs after germination (Domínguez et al., 2012). This process was analyzed in wheat seeds in which the first symptoms of PCD appear at 4 DAI and increase progressively up to 7 DAI, affecting both the epidermal and parenchymal cells. The spatial progression of scutellum PCD in wheat grains occurs with an apical-to-basal gradient. PCD is initiated in the apical zone once the adjacent aleurone cells have completed PCD (Domínguez et al., 2012). This pattern of PCD progression suggests that one of the major functions of the scutellum in germinated seeds, which is the transfer of nutrients from the starchy endosperm to support the initial seedling growth (West et al., 1998; Aoki et al., 2006), does not cease abruptly (Domínguez et al., 2012). In fact, the degeneration of the scutellum seems to be coordinated with the sequential production of hydrolases. In wheat seeds at early stages of germination (1 DAI), the *AmyI* gene, which encodes the α-amylase I isoform, is expressed exclusively in the scutellar epithelium; its expression being transient and independent of GA_3. At a later stage (2 DAI), *AmyI* expression in the scutellar epithelium decreases while it increases in the aleurone layer (Cejudo et al., 1995). In contrast, the gene encoding a cathepsin B-like is expressed in scutellum parenchymal cells but not in the epithelium of wheat seeds at 2 DAI (Cejudo et al., 1992), thus suggesting a function for this protease other than mobilization of the starchy endosperm. Whether cathepsin B-like is involved in a lysosomal-like function, in the final degradation of peptide products of the proteases secreted into the endosperm, or in pro-death roles, is not yet known. As mentioned above, the gene encoding the serine carboxipeptidase III is expressed during the differentiation of the scutellum vascular tissue (Domínguez et al., 2002). Therefore, PCD progression in scutellar parenchyma and epithelium appears as the last events in nutrients remobilization before the autonomous growth of the embryo (Domínguez et al., 2012).

DIFFERENT HALLMARKS CHARACTERIZE THE PROCESS OF PCD IN CEREAL SEEDS

The well-defined patterns of PCD in developing and germinating cereal seeds, as well as the variety of cells undergoing PCD, has favored the use of these systems to study PCD at the biochemical level. One of the more characteristic hallmarks of PCD in most developmental processes is the fragmentation of DNA, which is highly dependent of the level of DNA packaging (Domínguez and Cejudo, 2012). Nuclear DNA is packed into chromatin loops of ca. 50 kb, six of which are grouped in a rosette-like structure. In cereal seeds at initial stages of endosperm development (4–6 DPA), the identification of DNA fragments of 50–300 kb suggests the participation of proteases that cleave chromatin folding points of the rosette-like structure as an early event in DNA fragmentation and PCD (Young and Gallie, 2000). The second stage of DNA degradation corresponds to the internucleosomal fragmentation, which results in the typical ladder of multimers of 180–200 bp. The analysis of endosperm PCD in developing cereal seeds identified internucleosomal laddering only at late stages of seed development (20 DPA to the end of maturation); this process being thus an irreversible phase in cell death (Young et al., 1997; Young and Gallie, 1999, 2000). Finally, a third stage of DNA fragmentation of the starchy endosperm cells occurs in germinated seeds, which yields completely digested nuclear DNA. Internucleosomal fragmentation of nuclear DNA, which is a hallmark of animal apoptosis, is a clear feature of PCD in cereal seeds as shown in maternal tissues (Domínguez et al., 2001; Domínguez and Cejudo, 2006) and starchy endosperm cells (Young et al., 1997; Young and Gallie, 1999, 2000) during development, as well as in aleurone layer (Wang et al., 1996; Domínguez et al., 2004) and epithelial and parenchymal cells of the scutellum (Domínguez et al., 2012) following germination.

The fact that DNA fragmentation is central to cell death implies the participation of nucleolytic enzymes in this process. Biochemical analyses of cells undergoing PCD in cereal seeds have allowed the identification of nuclear- and cytoplasmic-localized endonucleases. While nuclear-localized nucleases promote the cleavage of nuclear DNA into high- and low-molecular weight fragments, cytoplasmic-localized endonucleases participate in the degradation of naked, double- or single-stranded DNA fragments as the final step that culminates the complete degradation of the cellular DNA. The action of nuclear-localized endonucleases seems to be a key event in pre-mortem nuclear dismantling, whereas cytoplasmic-localized endonucleases seem to carry out the completion of DNA degradation after vacuolar tonoplast disruption during post-mortem nuclear dismantling (Domínguez and Cejudo, 2012). Wheat grains have been a model system to identify nuclear-localized factors involved in internucleosomal DNA fragmentation. Two nuclear-localized neutral Ca^{2+}/Mg^{2+} endonucleases of ca. 30 and 50 kDa were identified, respectively, in aleurone (Domínguez et al., 2004) and nucellus cells undergoing PCD (Domínguez and Cejudo, 2006). An acid Zn^{2+}-dependent endonuclease of ca. 70 kDa was also identified in the nucleus of wheat scutellum cells undergoing PCD (Domínguez et al., 2012). The differences in cation requirement, electrophoretic mobility and optimal pH reveal that internucleosomal DNA fragmentation is performed by different nucleases in the different tissues of the wheat grain (Domínguez et al., 2004, 2012; Domínguez and Cejudo, 2006). Among the so-called waste-management endonucleases responsible for a third level of DNA fragmentation, it is worth mentioning the nucleases acting in the starchy endosperm cells of germinating barley seeds (Brown and Ho, 1986, 1987).

Finally, caspases are very well characterized proteases that participate as initiators and executioners in the process of apoptosis in animals. Because of this central role in the execution of apoptosis, the search for caspase counterparts has been a central focus of PCD studies in plants. Different approaches have revealed the complex set of endoproteolytic activities that participate in cell death in cereal seeds. These include serine-endoproteases in maternal tissues at early stages of development (Domínguez and Cejudo, 1996), and thiol-proteases in aleurone layer, scutellum, and starchy endosperm following germination (Domínguez and Cejudo, 1995). A caspase 6-like proteolytic activity, which acts at the sequence VEID, was identified in starchy endosperm and embryo cells from developing barley seeds (Boren et al., 2006). This VEIDase activity has been localized to autophagosome-like vesicles in randomly distributed cells of the starchy endosperm of barley seeds (Boren et al., 2006), in parallel with the progression of PCD in wheat (Young and Gallie, 1999). Therefore, VEIDase activity might be considered as an executioner with caspase-like activity in cereals. Nuclear-localized proteases have also been identified in cells undergoing PCD from developing wheat seeds. This is the case of a serine endoprotease of ca. 60 kDa identified in nuclear extracts from maternal tissues, which might be potentially responsible for the cleavage of structural proteins in the nucleus (Domínguez and Cejudo, 2006).

HORMONAL REGULATION OF PCD IN DEVELOPING AND GERMINATING CEREAL SEEDS

The spatial-temporal patterns of PCD affecting different tissues of developing and germinating cereal grains, as described above, suggest the participation of mechanisms able to orchestrate these complex patterns of cell death. Although it is presumed that different factors are involved in the regulation of PCD in cereal seeds, hormonal action seems to be an obvious candidate. Yet the information concerning their participation in controlling the patterns of PCD in cereal seeds is scarce. Here, we summarize results showing the involvement of hormones in the control of PCD (**Figure 4**).

At early stages of *Arabidopsis* seed development, synergid cell death has been associated with the activation of the ethylene signaling pathway that takes place during the process of fertilization, in which EIN3 and EIN2 have been identified as critical factors (Völz et al., 2013). Ethylene appears also to be a crucial hormone controlling endosperm development in cereal seeds. During cereal seed development, two peaks of ethylene production occur; the first one being coincident with the onset of PCD in the central region of the endosperm, whereas the second one is associated with the increase in endonucleolytic activity (Young et al., 1997). Exogenously added ethylene results in the acceleration of cell death and DNA fragmentation in maize and wheat developing seeds, while treatments with inhibitors of ethylene biosynthesis or perception have the opposite effect (Young et al., 1997; Young and Gallie, 1999). The positive effect of ethylene on PCD induction was confirmed with the *shrunken2* mutant of maize, a starch-deficient

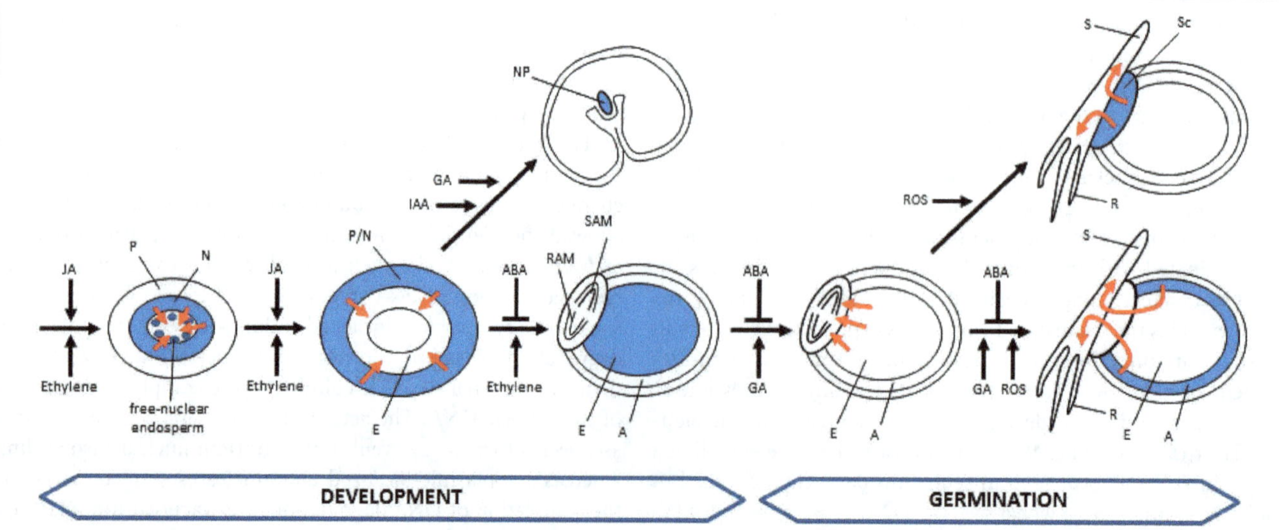

FIGURE 4 | Hormonal control of PCD in developing and germinating cereal seeds. Tissues undergoing PCD are represented in blue. Fluxes of nutrients promoted by PCD events are indicated by red arrows. Cells of the starchy endosperm undergo PCD during development but remain intact until germination. A, aleurone; E, endosperm; N, nucellus; NP, nucellar projections; P, pericarp; R, root; RAM, root apical meristem; S, shoot; SAM, shoot apical meristem; Sc, scutellum; ABA, abscisic acid; IAA, indol acetic acid; JA, jasmonic acid; GA, gibberellic acid; ROS, reactive oxygen species.

mutant that accumulates sugars, thereby producing ethylene levels threefold to fivefold higher than in wild-type seeds, which shows accelerated PCD (Young et al., 1997).

Two hormones, jasmonic acid (JA) and ethylene, have been proposed to participate in the control of pericarp PCD, based on the high level of expression of genes involved in the biosynthesis and signaling of both hormones during PCD of maternal tissues (Sreenivasulu et al., 2006). Lipases, lipid transfer proteins and lipo-oxygenases, which are involved in biosynthesis of JA precursors, were found among the genes with increased expression in barley seeds at 6–12 DPA. In addition, genes of the ethylene signal transduction pathway are also induced at these stages of seed development. The genes showing a higher level of expression include the ethylene receptor (ETR3), the raf-like protein kinase (CTR1), a MAP kinase, the ethylene-insensitive (EIN2) and the ethylene response factor ERF2 (Sreenivasulu et al., 2006).

Gibberellins and auxins have also been proposed to participate in the control of PCD of maternal tissues, such as the nucellar projections. The gradient of differentiation observed in nucellar projection cells in wheat seeds at 8 DPA is coincident with the expression patterns of genes involved in GA biosynthesis and signaling (Thiel et al., 2008). In addition, the auxin-dependent MADS29 transcription factor appears as a key regulator of early seed development by stimulating the expression of a Cys-protease and other PCD-related proteins participating in the degradation of the nucellus and the nucellar projections (Yin and Xue, 2012).

The important function of ABA in the control of cereal seed maturation is well known. During seed development, ABA biosynthesis takes place both in the developing embryo and the endosperm with the participation of different members of the aldehyde oxidase gene family (Sreenivasulu et al., 2006). Furthermore, ABA plays an essential role in the acquisition of desiccation tolerance of the embryo by stimulating stress-responsive genes, such as those encoding late-embryogenesis abundant proteins and dehydrins (Sreenivasulu et al., 2006). The implication of ABA in cereal endosperm PCD was demonstrated with two maize *viviparous (vp)* mutants, the ABA-insensitive *vp1* and the ABA-deficient *vp9* mutants. It should be noted that ethylene levels in the developing endosperm of these mutants are twofold to fourfold higher than in wild-type seeds, and thus ethylene might promote an acceleration of cell death in these mutants. The treatment of wild-type seeds with fluridone, an inhibitor of ABA biosynthesis, also promoted DNA fragmentation and cell death. Based on these findings, ABA was proposed as a negative regulator of ethylene biosynthesis and/or action during maize development (Young and Gallie, 2000). Thus a balance between ABA and ethylene regulates the onset and progression of PCD during cereal endosperm development (Young et al., 1997; Young and Gallie, 1999, 2000).

While ABA has a key function in seed maturation by inhibiting precocious germination, gibberellins have the opposite function and are the most relevant hormones promoting seed germination. The aleurone layer, the only endosperm tissue that remains alive in mature cereal seeds, shows a high sensitivity to gibberellins. The response of aleurone cells to gibberellins may be subdivided in two phases. In the short term, gibberellins stimulate the metabolic activation of these quiescent cells upon seed imbibition. In response to gibberellins, aleurone cells induce the expression of genes encoding hydrolytic enzymes, such as amylases and proteases, which are secreted into the starchy endosperm (Cejudo et al., 1992, 1995; Domínguez and Cejudo, 1995, 1999). In addition, the hormone triggers the acidification of the starchy endosperm, thus facilitating the mobilization of storage compounds (Domínguez and Cejudo, 1999). Once the metabolic activation is achieved, aleurone cells show a long-term response to gibberellins, which involves the induction of PCD, a process that is counteracted by ABA (Wang et al., 1996, 1998; Bethke et al., 1999; Domínguez et al., 2004). Treatments of wheat grains with paclobutrazol, an inhibitor of gibberellin synthesis, delayed germination and avoided DNA fragmentation (Domínguez et al., 2004). Similarly, aleurone protoplasts treated with LY83583, which has an antagonistic effect on gibberellin signaling, showed no symptoms of PCD (Bethke et al., 1999). The activating role of gibberellins in PCD was corroborated with the analysis of the wheat gibberellin-insensitive mutant *Rht,* which shows altered post-germination, including delayed aleurone PCD (Domínguez et al., 2004). Taken into account the spatial–temporal gradients observed in the wheat aleurone layer, which affect the expression of genes encoding hydrolytic enzymes (Cejudo et al., 1992, 1995; Domínguez and Cejudo, 1999), the acidification of the starchy endosperm (Domínguez and Cejudo, 1999), and the process of DNA fragmentation and cell death (Domínguez et al., 2004), a model was generated to explain the action of gibberellins during the different PCD phases of a single aleurone cell: (1) quiescent phase, which is the status of aleurone cells in mature grains prior to gibberellin perception; (2) active phase, in which gibberellin perception induces the synthesis of hydrolytic enzymes and the acidification of the starchy endosperm; and (3) death phase, in which aleurone cells undergo PCD (Cejudo et al., 2001). According to this model, the aleurone layer of germinating grains is composed by a heterogeneous group of cells: those located adjacent to the embryo are the first to perceive and respond to gibberellins and, thus, to enter the death phase, whereas those located far from the embryo may be still in the quiescent phase (Cejudo et al., 2001).

REDOX REGULATION OF PCD IN DEVELOPING AND GERMINATING CEREAL SEEDS

In addition to hormone control, redox regulation has been considered to play an important role in establishing the patterns of PCD in cereal seeds. Aerobic metabolism inevitably generates reactive oxygen species (ROS), including hydrogen peroxide, superoxide anion and hydroxyl radicals. These compounds show reactivity with almost all cellular macromolecules and, hence, their accumulation above certain levels may have a harmful effect causing oxidative stress (Bailly, 2004). At the end of the phase of development, the cereal seed enters a phase of desiccation, which involves a massive loss of water that provokes oxidative stress in tissues that remain alive in the mature seed, such as the embryo and the aleurone layer. Moreover, these tissues also suffer oxidative stress in germinating seeds due to resumption of respiration (Serrato and Cejudo, 2003). ROS, in conjunction with hormones, has been proposed to exert an important regulatory role in the regulation of PCD of cereal aleurone cells (Bethke and Jones, 2001; Fath et al., 2002). The GA-induced progression of PCD in these cells is accelerated in the presence of internally generated or exogenously applied hydrogen peroxide, whereas antioxidant agents, such as ascorbic acid or dithiothreitol, have the opposite effect (Bethke and Jones, 2001). Nitric oxide (NO) has the effect of delaying PCD of aleurone cells, most probably by counteracting the accelerating effect of ROS (Beligni et al., 2002). To avoid the harmful effect of ROS and prevent precocious PCD, aleurone cells are equipped with detoxification systems based on the scavenger activity of catalase, ascorbate peroxidase and superoxide dismutase (Fath et al., 2001), as well as haem oxygenase (Wu et al., 2011), some of which are down-regulated before the process of PCD is initiated (Fath et al., 2001). A nuclear detoxification system has also been described in wheat seed cells suffering oxidative stress, which is composed by an NADPH thioredoxin reductase (NTR) and a 1-Cys-peroxiredoxin (1-Cys Prx; Stacy et al., 1996, 1999; Pulido et al., 2009). This nuclear-localized redox system may have the function of preventing oxidative damage to DNA and nuclear structures, as suggested by DNA protection assays *in vitro*. In addition, this redox system may control the level of hydrogen peroxide in the nucleus, thus having a potential signaling function. Besides nuclear-localized detoxification systems, wheat aleurone cells also display a set of gibberellin-induced glycosylases participating in the base excision repair pathway to remove non-bulky DNA base lesions generated by ROS (Bissenbaev et al., 2011). Under the oxidant conditions generated in germinating seeds, the 1-Cys Prx undergoes a progressive inactivation by overoxidation of its catalytic cysteine residue (Pulido et al., 2009). This inactivation would promote a further accumulation of hydrogen peroxide; the nuclear environment becoming more oxidant and promoting cell death.

CONCLUDING REMARKS AND PERSPECTIVES

The cereal seed constitutes an excellent example illustrating the essential function of cell death for developmental programs in plants. Different techniques, such as TUNEL staining of nuclei of dying cell, have been a valuable aid in identifying tissues undergoing PCD and also in studying the spatial-temporal patterns of PCD in developing and germinating cereal seeds. Moreover, although most of the tissues undergoing PCD in cereal seeds present autophagic-like morphology, other tissues such as the starchy endosperm show peculiar characteristics. The identification of proteolytic and nucleolytic activities associated with PCD in the different tissues reveal the biochemical complexity of cell death in this plant model system. PCD in cereal seeds occurs with very well-defined spatial-temporal patterns, which are established by different factors; hormones playing a relevant function. So far, most of the knowledge of PCD in cereal seeds has been obtained from descriptive approaches. However, the last decade has brought impressive advance in our understanding of the cereal genomes that have been sequenced. In cereal models such as rice and *Brachypodium*, larger collections of mutants are available; in addition, genetic transformation of cereals is becoming a routine technology. All these molecular and genetic tools now available for cereals are expected to facilitate functional approaches applied to the study of PCD, which may allow a more precise dissection of the process in this plant system and, eventually, the identification of genes acting as key regulators of cell death in plants.

ACKNOWLEDGMENTS

Work in our lab is supported by European Regional Development Fund-cofinanced grants from the Spanish Ministry of Science and Innovation (BIO2010-15430) and Junta de Andalucía (BIO-182 and CVI-5919).

REFERENCES

An, L.-H., and You, R.-L. (2004). Studies on nuclear degeneration during programmed cell death of synergid and antipodal cells in *Triticum aestivum*. *Sex. Plant Reprod.* 17, 195–201. doi: 10.1007/s00497-004-0220-1

Aoki, N., Scofiel, G. N., Wang, X.-D., Offler, C. E., Patrick, J. W., and Furbank, R. T. (2006). Pathway of sugar transport in germinating wheat seeds. *Plant Physiol.* 141, 1255–1263. doi: 10.1104/pp.106.082719

Appleford, N. E. J., and Lenton, J. R. (1997). Hormonal regulation of α-amylase gene expression in germinating wheat (*Triticum aestivum*) grains. *Physiol. Plant.* 100, 534–542. doi: 10.1111/j.1399-3054.1997.tb03058.x

Bailly, C. (2004). Active oxygen species and antioxidants in seed biology. *Seed Sci. Res.* 14, 93–107. doi: 10.1079/SSR2004159

Baulcombe, D. C., Barker, R. F., and Jarvis, M. G. (1987). A gibberellin responsive wheat gene has homology to yeast carboxypeptidase Y. *J. Biol. Chem.* 262, 13726–13735.

Beligni, M. V., Fath, A., Bethke, P. C., Lamattina, L., and Jones, R. L. (2002). Nitric oxide acts as an anti-oxidant and delays programmed cell death in barley aleurone layers. *Plant Physiol.* 129, 1642–1650. doi: 10.1104/pp.002337

Bethke, P. C., and Jones, R. L. (2001). Cell death of barley aleurone protoplasts is mediated by reactive oxygen species. *Plant J.* 25, 19–29. doi: 10.1111/j.1365-313X.2001.00930.x

Bethke, P. C., Lonsdale, J. E., Fath, A., and Jones, R. L. (1999). Hormonally regulated programmed cell death in barley aleurone cells. *Plant Cell* 11, 1033–1045. doi: 10.1105/tpc.11.6.1033

Bissenbaev, A. K., Ishchenko, A. A., Taipakova, S. M., and Saparbaev, M. K. (2011). Presence of base excision repair enzymes in the wheat aleurone and their activation in cells undergoing programmed cell death. *Plant Physiol. Biochem.* 49, 1155–1164. doi: 10.1016/j.plaphy.2011.07.017

Boren, M., Höglund, A., Bozhkov, P., and Jansson, C. (2006). Developmental regulation of a VEIDase caspase-like proteolytic activity in barley caryopsis. *J. Exp. Bot.* 57, 3747–3753. doi: 10.1093/jxb/erl136

Bosnes, M., Weideman, F., and Olsen, O.-A. (1992). Endosperm differentiation in barley wild-type and sex mutants. *Plant J.* 2, 661–674. doi: 10.1111/j.1365-313X.1992.tb00135.x

Bozhkov, P., Filonova, L. H., Suarez, M. F., Helmersson, A., Smertenko, A. P., Zhivotovsky, B., et al. (2004). VEIDase is a principal caspase-like activity involved in plant programmed cell death and essential for embryonic pattern formation. *Cell Death Diff.* 11, 175–182. doi: 10.1038/sj.cdd.4401330

Bozhkov, P. V., Suárez, M. F., Filonova, L. H., Daniel, G., Zamyatnin, A. A., Rodríguez-Nieto, S., et al. (2005). Cysteine protease mcII-Pa executes programmed cell death during plant embryogenesis. *Proc. Natl. Acad. Sci. U.S.A.* 102, 14463–14468. doi: 10.1073/pnas.0506948102

Brown, P. H., and Ho, T.-H. D. (1986). Barley aleurone layers secrete a nuclease in response to gibberellic acid. *Plant Physiol.* 82, 801–806. doi: 10.1104/pp.82.3.801

Brown, P. H., and Ho, T.-H. D. (1987). Biochemical properties and hormonal regulation of barley nuclease. *Eur. J. Biochem.* 168, 357–364. doi: 10.1111/j.1432-1033.1987.tb13427.x

Cejudo, F. J., Cubo, M. T., and Baulcombe, D. C. (1995). Amy I expression during wheat seed germination. *Plant Sci.* 106, 207–213. doi: 10.1016/0168-9452(95)04077-8

Cejudo, F. J., González, M. C., Serrato, A. J., Sánchez, R., and Domínguez, F. (2001). Function and hormonal control of proteases in cereal grains. *Recent Res. Dev. Plant Physiol.* 2, 57–65.

Cejudo, F. J., Murphy, G., Chinoy, C., and Baulcombe, D. C. (1992). A gibberellin-regulated gene from wheat with sequence homology to cathepsin B of mammalian cells. *Plant J.* 2, 937–948. doi: 10.1046/j.1365-313X.1992.t01-7-00999.x

Chen, F., and Foolad, M. R. (1997). Molecular organization of a gene in barley which encodes a protein similar to aspartic protease and its specific expression in nucellar cells during degeneration. *Plant Mol. Biol.* 35, 821–831. doi: 10.1023/A:1005833207707

Consonni, G., Aspesi, C., Barbante, A., Dolfini, S., Giuliani, C., Giulini, A., et al. (2003). Analysis of four maize mutants arrested in early embryogenesis reveals an irregular pattern of cell division. *Sex. Plant Reprod.* 15, 281–290. doi: 10.1007/s00497-003-0165-9

Corre-Menguy, F., Cejudo, F. J., Mazubert, C., Vidal, J., Lelandais-Briere, C., Torres, G., et al. (2002). Chracterization of the expression of a wheat cystatin gene during caryopsis development. *Plant Mol. Biol.* 50, 687–698. doi: 10.1023/A:1019906031305

Dal Degan, F., Rocher, A., Cameron-Mills, V., and Von Wettstein, D. (1994). The expression of serine carboxypeptidases during maturation and germination of the barley grain. *Proc. Natl. Acad. Sci. U.S.A.* 91, 8209–8213. doi: 10.1073/pnas.91.17.8209

Domínguez, F., and Cejudo, F. J. (1995). Pattern of endoproteolysis following wheat grain germination. *Physiol. Plant.* 95, 253–259. doi: 10.1111/j.1399-3054.1995.tb00835.x

Domínguez, F., and Cejudo, F. J. (1996). Characterization of endoproteases appearing during wheat grain development. *Plant Physiol.* 112, 1211–1217.

Domínguez, F., and Cejudo, F. J. (1998). Germination-related genes encoding proteolytic enzymes are expressed in the nucellus of developing wheat grains. *Plant J.* 15, 569–574. doi: 10.1046/j.1365-313X.1998.00233.x

Domínguez, F., and Cejudo, F. J. (1999). Patterns of starchy endosperm acidification and protease gene expression in wheat grains following germination. *Plant Physiol.* 119, 81–87. doi: 10.1104/pp.119.1.81

Domínguez, F., and Cejudo, F. J. (2006). Identification of a nuclear-localized nuclease from wheat cells undergoing programmed cell death that is able to trigger DNA fragmentation and apoptotic morphology on nuclei from human cells. *Biochem. J.* 307, 529–536. doi: 10.1042/BJ20051809

Domínguez, F., and Cejudo, F. J. (2012). A comparison between nuclear dismantling during plant and animal programmed cell death. *Plant Sci.* 197, 114–121. doi: 10.1016/j.plantsci.2012.09.009

Domínguez, F., González, M. C., and Cejudo, F. J. (2002). A germination-related gene encoding a serine carboxypeptidase is expressed during the differentiation of the vascular tissue in wheat grains and seedlings. *Planta* 215, 727–734. doi: 10.1007/s00425-002-0809-2

Domínguez, F., Moreno, J., and Cejudo, F. J. (2001). The nucellus degenerates by a process of programmed cell death during the early stages of wheat grain development. *Planta* 213, 352–360. doi: 10.1007/s004250000517

Domínguez, F., Moreno, J., and Cejudo, F. J. (2004). A gibberellin-induced nuclease is localized in the nucleus of wheat aleurone cells undergoing programmed cell death. *J. Biol. Chem.* 279, 11530–11536. doi: 10.1074/jbc.M308082200

Domínguez, F., Moreno, J., and Cejudo, F. J. (2012). The scutellum of germinated wheat grains undergoes programmed cell death: identification of an acidic nuclease involved in nucleus dismantling. *J. Exp. Bot.* 63, 5475–5485. doi: 10.1093/jxb/ers199

Drea, S., Leader, D. J., Arnold, B. C., Shaw, P., Dolan, L., and Doonan J. H. (2005). Systematic spatial analysis of gene expression during wheat caryopsis development. *Plant Cell* 17, 2172–2185. doi: 10.1105/tpc.105.034058

Engell, K. (1994). Embryology of barley. IV. Ultrastructure of the antipodal cells of Hordeum vulgare L. cv. Bomi before and after fertilization of the egg cell. *Sex. Plant Reprod.* 7, 333–346. doi: 10.1007/BF00230512

Fath, A., Bethke, P. C., Beligni, V., and Jones, R. L. (2002). Active oxygen and cell death in cereal aleurone cells. *J. Exp. Bot.* 53, 1273–1282. doi: 10.1093/jexbot/53.372.1273

Fath, A., Bethke, P. C., and Jones, R. L. (2001). Enzymes that scavenge reactive oxygen species are down-regulated prior to gibberellic acid-induced programmed cell death in barley aleurone. *Plant Physiol.* 126, 156–166. doi: 10.1104/pp.126.1.156

Filonova, L. H., Bozhkov, P. V., Brukhin, V. B., Daniel, G., Zhivotovsky, B., and von Arnold, S. (2000). Two waves of programmed cell death occur during formation and development of somatic embryos in the gymnosperm, Norway spruce. *J. Cell Sci.* 113, 4399–4411.

Fincher, G. B. (1989). Molecular and cellular biology associated with endosperm mobilization in germinating cereals grains. *Annu. Rev. Plant Physiol. Plant Mol. Biol.* 40, 305–346. doi: 10.1146/annurev.pp.40.060189.001513

Forestan, C., Meda, S., and Varotto, S. (2010). ZmPIN1-mediated auxin transport is related to cellular differentiation during maize embryogenesis and endosperm development. *Plant Physiol.* 152, 1373–1390. doi: 10.1104/pp.109.150193

Giuliani, C., Consonni, G., Gavazzi, G., Colombo, M., and Dolfini, S. (2002). Programmed cell death during embryogenesis in maize. *Ann. Bot.* 90, 287–292. doi: 10.1093/aob/mcf173

González, M. C., Osuna, L., Echevarría, C., Vidal, J., and Cejudo, F. J. (1998). Expression and localization of phosphoenolpyruvate carboxylase in developing and germinating wheat grains. *Plant Physiol.* 116, 1249–1258. doi: 10.1104/pp.116.4.1249

Kladnik, A., Chamusco, K., Dermastia, M., and Chourey, P. (2004). Evidence of programmed cell death in post-phloem transport cells of the maternal pedicel tissue in developing caryopsis of maize. *Plant Physiol.* 136, 3572–3581. doi: 10.1104/pp.104.045195

Linnestad, C., Doan, D. N. P., Brown, R. C., Lemmon, B. E., Meyer, D. J., Jung, R., et al. (1998). Nucellain, a barley homolog of the dicot vacuolar-processing protease, is localized in nucellar cell walls. *Plant Physiol.* 118, 1169–1180. doi: 10.1104/pp.118.4.1169

Minina, E. A., Filonova, L. H., Fukada, K., Savenkov, E. I., Gogvadze, V., Clapham, D., et al. (2013). Autophagy and metacaspase determine the mode of cell death in plants. *J. Cell Biol.* 203, 917–927. doi: 10.1083/jcb.201307082

Ngo, Q. A., Vogler, H., Lituiev, D. S., Nestorova, A., and Grossniklaus, U. (2014). A calcium dialog mediated by the FERONIA signal transduction pathway controls plant sperm delivery. *Dev. Cell* 29, 491–500. doi: 10.1016/j.devcel.2014.04.008

Okamoto, T., Scholten, S., Lörz, H., and Kranz, E. (2005). Identification of genes that are up- or down-regulated in the apical or basal cell of maize two-celled embryos and monitoring their expression during zygote development by a cell manipulation- and PCR-based approach. *Plant Cell Physiol.* 46, 332–338. doi: 10.1093/pcp/pci032

Olsen, O.-A. (2004). Nuclear endosperm development in cereals and *Arabidopsis thaliana*. *Plant Cell* 16, S214–S227. doi: 10.1105/tpc.017111

Pulido, P., Cazalis, R., and Cejudo, F. J. (2009). An antioxidant redox system in the nucleus of wheat seed cells suffering oxidative stress. *Plant J.* 57, 132–145. doi: 10.1111/j.1365-313X.2008.03675.x

Radchuk, V., Borisjuk, L., Radchuk, R., Steinbiss, H.-H., Rolletschek, H., Broeders, S., et al. (2006). Jekyll encodes a novel protein involved in the sexual reproduction of barley. *Plant Cell* 18, 1652–1666. doi: 10.1105/tpc.106.041335

Radchuk, V., Borisjuk, L., Sreenivasulu, N., Merx, K., Mock, H.-P., Rolletschek, H., et al. (2009). Spatio-temporal profiling of starch biosynthesis and degradation in the developing barley grain. *Plant Physiol.* 150, 190–204. doi: 10.1104/pp.108.133520

Radchuk, V., Weier, D., Radchuk, R., Weschke, W., and Weber, H. (2011). Development of maternal seed tissue in barley is mediated by regulated cell expansion and cell disintegration and coordinated with endosperm growth. *J. Exp. Bot.* 62, 1217–1227. doi: 10.1093/jxb/erq348

Sabelli, P. A. (2012). Replicate and die for your own good: endoreduplication and cell death in the cereal endosperm. *J. Cereal Sci.* 56, 9–20. doi: 10.1016/j.jcs.2011.09.006

Sabelli, P. A., and Larkins, B. A. (2009). The development of endosperm in grasses. *Plant Physiol.* 149, 14–26. doi: 10.1104/pp.108.129437

Sabelli, P. A., Liu, Y., Dante, R. A., Lizarraga, L. E., Nguyen, H. N., Brown, S. W., et al. (2013). Control of cell proliferation, endoreduplication, cell size and cell death by the retinoblastoma-related pathway in maize endosperm. *Proc. Natl. Acad. Sci. U.S.A.* 110, E1827–E1836. doi: 10.1073/pnas.1304903110

Sandaklie-Nikolova, L., Palanivelu, R., King, E. J., Copenhaver, G. P., and Drews, G. N. (2007). Synergid cell death in *Arabidopsis* is triggered following direct interaction with the pollen tube. *Plant Physiol.* 144, 1753–1762. doi: 10.1104/pp.107.098236

Serrato, A. J., and Cejudo, F. J. (2003). Type-h thioredoxins accumulate in the nucleus of developing wheat seed tissues suffering oxidative stress. *Planta* 217, 392–399. doi: 10.1007/s00425-003-1009-4

Sreenivasulu, N., Radchuk, V., Strickert, M., Miersch, O., Weschke, W., and Wobus, U. (2006). Gene expression patterns reveal tissue-specific signalling networks controlling programmed cell death and ABA-regulated maturation in developing barley grains. *Plant J.* 47, 310–327. doi: 10.1111/j.1365-313X.2006.02789.x

Stacy, R. A. P., Nordeng, T. W., Culiáñez-Maciá, F. A., and Aalen, R. (1999). The dormancy-related peroxiredoxin anti-oxidant, PER 1, is localized to the nucleus of barley embryo and aleurone cells. *Plant J.* 19, 1–8. doi: 10.1046/j.1365-313X.1999.00488.x

Stacy, R. A. P., Munthe, E., Steinum, T., Sharma, B., and Aalen, R. B. (1996). A peroxiredoxin antioxidant is encoded by a dormancy-related gene, Per1, expressed during late development in the aleurone and embryo of barley grains. *Plant Mol. Biol.* 31, 1205–1216. doi: 10.1007/BF00040837

Swift, J. G., and O'Brien, T. P. (1970). Vascularization of the scutellum of wheat. *Aust. J. Bot.* 18, 45–53. doi: 10.1071/BT9700045

Swift, J. G., and O'Brien, T. P. (1971). Vascular differentiation in the wheat embryo. *Aust. J. Bot.* 19, 63–71 doi: 10.1071/BT9710063

Thiel, J., Weier, D., Sreenivasulu, N., Strickert, M., Weichert, N., Melzer, M., et al. (2008). Different hormonal regulation of cellular differentiation and function in nucellar projection and endosperm transfer cells: a microdissection-based transcriptome study of young barley grains. *Plant Physiol.* 148, 1436–1452. doi: 10.1104/pp.108.127001

van Doorn, W. G., Beers, E. P., Dangl, J. L., Franklin-Tong, V. E., Gallois, P., Hara-Nishimura, I., et al. (2011). Morphological classification of plant cell deaths. *Cell Death. Differ* 18, 1241–1246. doi: 10.1038/cdd.2011.36

Völz, R., Heydlauff, J., Ripper, D., Von Lyncker, L., and GroB-Hardt, R. (2013). Ethylene signalling is required for synergid degeneration and the establishment of a pollen tube block. *Dev. Cell* 25, 310–316. doi: 10.1016/j.devcel.2013.04.001

Wang, M., Oppedijk, B. J., Lu, X., Van Duijn, B., and Schilperoort, R. A. (1996). Apoptosis in barley aleurone during germination and its inhibition by abscisic acid. *Plant Mol. Biol.* 32, 1125–1134. doi: 10.1007/BF00041396

Wang, M., Oppedijk, B. J., Caspers, M. P. M., Lamers, G. E. M., Boot, M. J., Geerlings D. N. G., et al. (1998). Spatial and temporal regulation of DNA fragmentation in the aleurone of germinating barley. *J. Exp. Bot.* 49, 1293–1301. doi: 10.1093/jxb/49.325.1293

Washio, K., and Ishikawa, K. (1994). Organ-specific and hormone-dependent expression of genes for serine carboxypeptidase during development and following germination of rice grains. *Plant Physiol.* 105, 1275–1280. doi: 10.1104/pp.105.4.1275

West, C. E., Waterworth, W. M., Stephens, S. M., Smith, C. P., and Bray, C. M. (1998). Cloning and functional characterization of a peptide transporter expressed in the scutellum of barley grain during the early stages of germination. *Plant J.* 15, 221–229. doi: 10.1046/j.1365-313X.1998.00199.x

Wu, M., Huang, J., Xu, S., Ling, T., Xie, Y., and Shen, W. (2011). Haem oxygenase delays programmed cell death in wheat aleurone layers by modulation of hydrogen peroxide metabolism. *J. Exp. Bot.* 62, 235–248. doi: 10.1093/jxb/erq261

Yang, X., Wu, F., Lin, X., Du, X., Chong, K., Gramzow, L., et al. (2012). Live and let die – The Bsister MADS-box gene OsMADS29 controls the degeneration of cells in maternal tissues during seed development of rice (*Oryza sativa*). *PLoS ONE* 7:e51435. doi: 10.1371/journal.pone.0051435

Yin, L.-L., and Xue, H.-W. (2012). The MADS29 transcription factor regulates the degradation of the nucellus and the nucellar projection during rice seed development. *Plant Cell* 24, 1049–1065. doi: 10.1105/tpc.111.094854

Young, T. E., Gallie, D. R., and DeMason, D. A. (1997). Ethylene mediated programmed cell death during maize endosperm development of wild type and shrunken2 genotypes. *Plant Physiol.* 115, 737–751.

Young, T. E., and Gallie, D. R. (1999). Analysis of programmed cell death in wheat endosperm reveals differences in endosperm development between cereals. *Plant Mol. Biol.* 39, 915–926. doi: 10.1023/A:1006134027834

Young, T. E., and Gallie, D. R. (2000). Regulation of programmed cell death in maize endosperm by abscisic acid. *Plant Mol. Biol.* 42, 397–414. doi: 10.1023/A:1006333103342

Zhao, P., Zhou, X.-M., Zhang, L.-Y., Wang, W., Ma, L.-G., Yang, L.-B., et al. (2013). A bipartite molecular module controls cell death activation in the basal cell lineage of plant embryos. *PLoS Biol.* 11:e1001655. doi: 10.1371/journal.pbio.1001655

Zhou, Z., Wang, L., Li, J., Song, X., and Yang, C. (2009). Study on programmed cell death and dynamic changes of starch accumulation in pericarp cells of *Triticum aestivum* L. *Protoplasma* 236, 49–58. doi: 10.1007/s00709-009-0046-7

Conflict of Interest Statement: The authors declare that the research was conducted in the absence of any commercial or financial relationships that could be construed as a potential conflict of interest.

6

The dynamic behavior of storage organelles in developing cereal seeds and its impact on the production of recombinant proteins

Elsa Arcalis*†, *Verena Ibl*†, *Jenny Peters, Stanislav Melnik and Eva Stoger*

Department of Applied Genetics and Cell Biology, University of Natural Resources and Life Sciences, Vienna, Austria

Edited by:
Paolo Sabelli, University of Arizona, USA

Reviewed by:
Martin Hajduch, Slovak Academy of Sciences, Slovakia
Fumio Takaiwa, National Institute of Agrobiological Sciences, Japan

Correspondence:
Eva Stoger, Department of Applied Genetics and Cell Biology, University of Natural Resources and Life Sciences, Muthgasse 18, 1190 Vienna, Austria
e-mail: eva.stoger@boku.ac.at

† Elsa Arcalis and Verena Ibl have contributed equally to this work.

Cereal endosperm is a highly differentiated tissue containing specialized organelles for the accumulation of storage proteins, which are ultimately deposited either within protein bodies derived from the endoplasmic reticulum, or in protein storage vacuoles (PSVs). During seed maturation endosperm cells undergo a rapid sequence of developmental changes, including extensive reorganization and rearrangement of the endomembrane system and protein transport via several developmentally regulated trafficking routes. Storage organelles have been characterized in great detail by the histochemical analysis of fixed immature tissue samples. More recently, *in vivo* imaging and the use of tonoplast markers and fluorescent organelle tracers have provided further insight into the dynamic morphology of PSVs in different cell layers of the developing endosperm. This is relevant for biotechnological applications in the area of molecular farming because seed storage organelles in different cereal crops offer alternative subcellular destinations for the deposition of recombinant proteins that can reduce proteolytic degradation, allow control over glycan structures and increase the efficacy of oral delivery. We discuss how the specialized architecture and developmental changes of the endomembrane system in endosperm cells may influence the subcellular fate and post-translational modification of recombinant glycoproteins in different cereal species.

Keywords: endosperm, molecular farming, recombinant glycoproteins, cereal biotechnology, storage organelles

INTRODUCTION

The endosperm is a short-lived tissue adapted for nutrient storage, which is found in both dicotyledonous and monocotyledonous seeds. However, its relative contribution to the mass of the mature seed varies greatly in different species. In cereals, the endosperm occupies the major part of the seed and consists of starchy inner endosperm cells, which are dead at the time of full maturation, and one or several outer layers of living aleurone cells, which secrete the enzymes needed to break down reserves following germination (Young and Gallie, 2000).

The endomembrane system of the developing cereal endosperm is clearly influenced by its functional specialization. Storage protein synthesis requires a highly active and well-developed endoplasmic reticulum (ER), and different types of storage organelles are rapidly formed during endosperm development to accommodate various types of storage proteins (Muntz, 1998; Shewry and Halford, 2002). ER-derived protein bodies mostly contain prolamin aggregates, whereas protein storage vacuoles (PSVs) tend to accumulate different storage protein classes, typically in separate phases, and often incorporate the content of ER-derived storage bodies (Shewry et al., 1995; Galili, 2004; Tosi et al., 2009). At the peak of storage protein production, programmed cell death occurs in the inner endosperm to generate a starchy endosperm core. In the surrounding aleurone cells, PSVs are converted into lytic vacuoles as germination commences (Bethke et al., 1998) and programmed cell death is delayed until a few days later.

To accommodate this sequence of intracellular changes in developing and germinating seeds, the endomembrane system must be extremely flexible and capable of rapid morphological and functional adaptation. Seed development and germination are therefore ideal platforms to study endomembrane plasticity and reshaping, and to follow the transition from PSV to lytic vacuole, representing a shift between resource accumulation and redistribution.

Techniques that allow the real-time visualization of membranes and organelle markers within the endosperm tissue are particularly suitable for this type of analysis because they provide insight into membrane dynamics during seed development and germination. For example, several transgenic cereal lines are now available expressing fluorescent subcellular markers and tagged or mutated storage proteins (Mohanty et al., 2009; Onda et al., 2009; Tosi et al., 2009; Ibl et al., 2014; Oszvald et al., 2014) and other fluorescent probes have been developed that can track organelles in living cells and report the subcellular organization and dynamics *in vivo* (Landrum et al., 2010). These techniques can be combined with transcriptional analysis and proteomics, which help to identify the molecular regulators for storage organelle development in cereals (Walley et al., 2013).

The mechanisms of protein trafficking and deposition in cereal endosperm can also be investigated using recombinant proteins, particularly glycoproteins (Stoger et al., 2005a; He et al., 2012; Wakasa and Takaiwa, 2013). Seeds are an attractive platform for the production of high-value recombinant pharmaceutical proteins, an approach known as molecular farming, because the specialized storage organelles allow such proteins to accumulate to high concentrations in small-volume compartments that contain molecular chaperones and disulfide isomerases to facilitate protein folding and maintain stability (De Jaeger et al., 2002; Stoger et al., 2005a; Ramessar et al., 2008a). This supportive intracellular environment means that the recombinant proteins are inert, and can be stored for long periods of time in the form of dry seeds without losing stability or activity. The accumulation of inert proteins in seeds also allows the expression of toxic proteins because these are inactive and are not expressed in vegetative tissues and therefore they do not impact on plant growth and development. The accumulation of proteins is also boosted by the triploid genome and the several rounds of endoreduplication that occur in the endosperm, thus increasing the effective number of transgene copies (Sabelli and Larkins, 2009). On this basis, many recombinant proteins have been produced successfully in cereal seeds, including commercial products such as lactoferrin, lysozyme, and human serum albumin (HSA) produced in rice by Ventria Bioscience (Fort Collins, CO, USA; Ramessar et al., 2008b; He et al., 2011b) and human growth factor for cosmetic use produced in barley by ORF Genetics (Iceland; Erlendsson et al., 2010).

Storage organelles in cereal seeds provide a clear advantage for the production of recombinant proteins, but the plasticity and remodeling of endomembrane organelles affect the accumulation and modification of proteins and so a better understanding of species-dependent and development-specific factors is needed to optimize cereal seeds for molecular farming applications. In this review article, we summarize the key features of the endomembrane organelles in major cereal species, and address the issues of cargo trafficking to these organelles in the developing endosperm of cereal species, endomembrane reshaping, and vacuolar transition during development and germination. Finally, we discuss how these endosperm-specific features affect the production of recombinant glycoproteins in plants.

STORAGE ORGANELLES AND ENDOMEMBRANE TRAFFIC IN DIFFERENT CEREAL SPECIES

The endomembrane system in the developing cereal endosperm is highly specialized to accommodate different storage proteins, and differences are observed between cereal species. A representative endosperm cell from each of the four major cereals is shown in **Figure 1**. The vacuolar storage compartment can be identified easily in wheat (**Figure 1A**) and barley (**Figure 1B**) because large prolamin deposits are found in a central vacuole. In contrast, rice and maize endosperm cells do not have a single prominent protein storage site. Rice endosperm cells uniquely separate globulins and prolamins into separate compartments: globulins are sorted into numerous small PSVs whereas prolamins accumulate in ER-derived protein bodies (**Figure 1C**). Both compartments are evenly distributed within the endosperm cell cytoplasm. Maize, like wheat and barley, predominantly stores prolamins, which in this species are known as zeins. These form highly conserved protein bodies derived from the ER (Lending et al., 1989) and fill up the cytoplasm of endosperm cells together with a small number of vacuolar compartments that accumulate globulins (Woo et al., 2001; Arcalis et al., 2010; **Figure 1D**).

SECRETION VS. STORAGE IN ENDOSPERM CELLS

The endosperm tissue of many plants is composed of uniform, living reserve cells that secrete hydrolytic enzymes to degrade the endosperm cell walls, allowing the radicle to emerge during germination. In contrast, cereal endosperm contains a core of mature starchy endosperm cells that are already dead before germination, and the enzymes that break down reserves are secreted by the cells in the aleurone layer (Leubner-Metzger et al., 1995; Nonogaki and

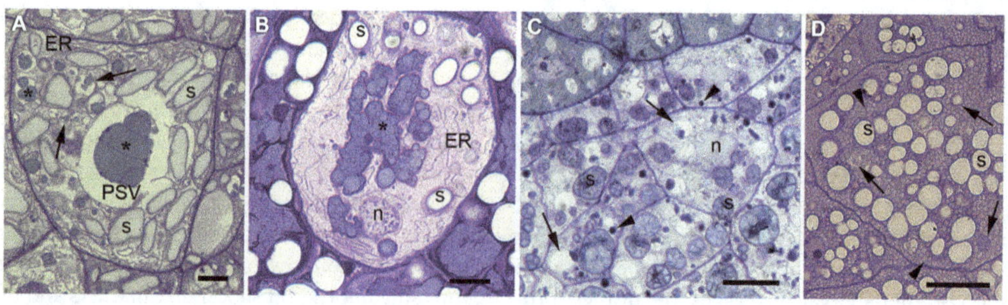

FIGURE 1 | Comparison of endosperm cells from the subaleurone layer at mid-maturation stage in wheat (A), barley (B), rice (C), and maize (D). Semithin sections (1 μm in thickness), stained with toluidine blue for light microscopy (Arcalis et al., 2004). **(A)** Wheat endosperm cells show a predominant, large central PSV. Note the huge prolamin aggregates stained in light blue (*) and the scarce globulin (triticin) bodies stained in dark blue at the periphery of the prolamins. Several smaller vacuolar compartments containing prolamins can also be observed (arrows), and well developed endoplasmic reticulum (ER) and abundant spindle like starch grains (s). **(B)** Barley endosperm cells show abundant protein deposits forming a multiphasic protein body (see the different shades of blue), habitually within a PSV. Note the abundant ER, the spheroidal starch grains (s) and an apoptotic nucleus (n). **(C)** Rice endosperm cells are the smallest depicted in this figure. Rice endosperm accumulates mainly glutelins in protein storage vacuoles (arrows). Note the abundant PSVs containing a blue stained inclusion, normally close to the tonoplast and the spherical, densely packed prolamin bodies or PB-I (arrowheads). Starch (s). **(D)** Maize endosperm stores mainly prolamins in ER-derived protein bodies (zein bodies). Zein bodies (arrowheads) reach a diameter of 1 μm, are very abundant and appear evenly spread within the cytoplasm. Several storage vacuoles can be also observed (arrows) either completely filled or with peripherical protein deposits. Starch grains (s) are spheroidal and similar in size to the PSVs. Scale bar equals 10 μm.

Morohashi, 1996; Chen and Bradford, 2000; Brown and Lemmon, 2007). Secretory activity is therefore less likely in the starchy cereal endosperm, as has indeed been reported when cereal seeds are used for the expression of recombinant proteins. Recombinant proteins such as fungal phytase and the HIV-neutralizing antibody 2G12, carrying a signal peptide directing them to the endomembrane system but no further targeting signal, were shown to be transported to the PSV via the Golgi apparatus as evidenced by immunoelectron microscopy and glycan analysis in wheat, rice, and maize (Arcalis et al., 2004, 2010; Drakakaki et al., 2006; Peters et al., 2013). The same phytase protein, carrying the same targeting information, was secreted to the apoplast in leaves of transgenic rice plants (Drakakaki et al., 2006). These results suggest that the final destination for otherwise secreted proteins in the cereal endosperm is the PSV, although it is also possible that signal sequences are recognized and interpreted differently by cereal leaf mesophyll and seed endosperm cells.

TRANSPORT BETWEEN THE ER AND PSV

The tracking of proteins in different storage compartments has shown that the PSVs in cereal endosperm cells are not simply post-Golgi compartments. Prolamins such as wheat glutenins, barley hordeins and oat avenins are initially deposited in ER-derived protein bodies, but later they are also found in the PSVs (Cameron-Mills and Wettstein, 1980; Lending et al., 1989; Rechinger et al., 1993; Shewry et al., 1995; Galili, 2004; Tosi et al., 2009; Ibl et al., 2014). How they get there is a matter of controversy, but it appears that both Golgi-dependent and Golgi-independent routes are involved. Levanony et al. (1992) initially suggested, based on a series of electron micrographs, that ER-derived prolamin bodies in wheat endosperm are eventually incorporated in the PSV through an autophagy-like process that bypasses the Golgi apparatus. This is supported by the presence of internal membranes in barley endosperm PSVs, including ER-derived membranes associated with individual and clustered protein bodies, suggesting that also hordein transport to the PSV at least partially bypasses the Golgi (Ibl et al., 2014). Ectopic zein bodies are also sequestered into PSVs in tobacco endosperm cells (Coleman et al., 2004). Reyes et al. (2011) proposed a Golgi-independent, autophagy-like route for the vacuolar delivery of zeins based on the presence of pre-vacuolar compartments that would sequester zein bodies and deliver them to the PSV in aleurone cells. Although the lipidation of ATG8 was observed, which is consistent with the macroautophagic mechanism (Klionsky et al., 2008; Rubinsztein et al., 2009; Chung et al., 2010), there was no subsequent vacuolar degradation of lipidated ATG8, indicating that the vacuolar delivery of zeins to the PSVs of maize aleurone cells does not use the typical ATG8-dependent macroautophagic process, but rather an atypical autophagy-like sequence of events (Aamodt et al., 2011). As an alternative to this model, Rogers (2011) proposed a mechanism by which membrane-containing vesicles bud from the ER as pre-vacuolar organelles that subsequently deliver materials into the PSV, which has been documented in dicots (Oufattole et al., 2005).

The direct route from the ER to PSVs in seeds involves a lot of traffic. Torres et al. (2001) reported the presence of the ER-resident molecular chaperone calreticulin in the rice endosperm PSV, providing evidence for a Golgi-independent pathway. Accordingly, KDEL-tagged recombinant proteins are often found at least partially in PSVs, e.g., a single chain antibody fragment (scFv) in rice (Torres et al., 2001) and HSA in wheat (Arcalis et al., 2004). The direct ER-to-PSV transport of recombinant proteins has also been reported in dicot seeds (Petruccelli et al., 2006; Floss et al., 2009; Loos et al., 2011a; Morandini et al., 2011; Arcalis et al., 2013). The endogenous KDEL-tagged proteinase sulfhydryl-endopeptidase (SH-EP) behaves in a similar manner in the cotyledons of germinating mung bean seeds (Toyooka et al., 2000; Okamoto et al., 2003), and storage protein precursors in pumpkin seeds are delivered to PSVs in precursor-accumulating (PAC) vesicles (Hara-Nishimura et al., 1998). PAC-like vesicles have also been described in rice endosperm by Takahashi et al. (2005). They are larger than the PAC vesicles in pumpkin seeds but they also arise directly from the ER and contain storage proteins destined for the PSV, such as glutelins and globulins (Takahashi et al., 2005). Viotti (2014) recently speculated that PAC vesicles may be the precursors of PSVs that slowly acquire their final size and identity via Golgi-mediated and post-Golgi-mediated transport, similar to the transition from pro-vacuoles to lytic vacuoles (Viotti et al., 2013; Viotti, 2014). Indeed, the main vacuolar proton pumps (H^+-ATPase, VHA-a3 and H^+-PPase AVP1) appear to follow a Golgi-independent route from the ER to the PSV in *Arabidopsis* seeds (Viotti et al., 2013), and alpha-TIP may follow the same, brefeldin A (BFA)-insensitive route. Considering this possibility, the trafficking of storage proteins and the relationships between the different storage compartments in cereal seeds could be reinterpreted, well in line with a model proposed earlier for barley aleurone PSVs (Bethke et al., 1998). In this model, it has been proposed that the ER membrane producing the neutral lipid may also serve as the site of storage protein accumulation and could mature into a PSV, supported by the observation that oleosomes forming on the surface of the ER are later found surrounding PSVs.

DEVELOPMENTAL CHANGES OF PROTEIN TRAFFICKING ROUTES

A number of studies have suggested that the intracellular trafficking and distribution of storage proteins may change during seed development (Shy et al., 2001; Vitale and Hinz, 2005; Tosi et al., 2009). Accordingly, the fates of the recombinant glycoprotein phytase, and the endogenous vacuolar corn proteins legumin-1 (CL-1) and α-globulin (CAG), were found to change during maize endosperm development (Arcalis et al., 2010). All three proteins were found in the PSVs during early development as expected, but closer to maturity they were found at the periphery of the ER-derived zein bodies. In the case of recombinant phytase, the switch in localization was accompanied by a switch in the glycan profile, consistent with a stage-dependent protein trafficking behavior comprising a Golgi-dependent pathway to the PSVs in younger cells switching to the deposition in ER-associated compartments later. There was also a significant reduction in the number of PSVs during development (Arcalis et al., 2010), suggesting a net decrease in number or a change in appearance reflecting the different content. There was no significant decline in the number of Golgi organelles in developing maize, although in wheat the number of transcripts representing Golgi-associated proteins was shown to decline during seed maturation (Shy et al., 2001).

The endomembrane system of specialized and transient tissues such as endosperm should therefore not be considered as a static and rigid structure, but rather as a plastic and dynamic system that involves interactions between organelles and changes in protein trafficking to fulfill distinct developmental functions.

SPATIOTEMPORAL MORPHOLOGICAL CHANGES OF STORAGE ORGANELLES IN CEREAL SEEDS

ARCHITECTURAL CHANGES IN THE ENDOMEMBRANE SYSTEM DURING DEVELOPMENT AND GERMINATION

The massive reorganization of the endomembrane system during seed development involves significant morphological changes to the endosomal and storage organelles (Hoh et al., 1995; Wang et al., 2010; Ibl et al., 2014). The observation of *in vitro* transgenic maize endosperm cultures has shown that the ER changes from its typical reticulated distribution pattern into a more punctate architecture during the development of both starchy endosperm and aleurone cells, probably representing the formation of ER-derived protein bodies (Aamodt et al., 2011). In barley, the aleurone ER remains predominantly reticulated, whereas prominent protein bodies appear in the subaleurone and central starchy endosperm ERs (Ibl et al., 2014). In the germinating grain, the synthesis of secretory proteins in the living aleurone cells is accompanied by the proliferation of the ER to form stacks, an increase in the size and complexity of the Golgi apparatus, a reduction in the number of oleosomes and an increase in the number of glyoxysomes and mitochondria (reviewed by Fath et al., 2000).

Conventional studies of subcellular structures, intracellular pathways and protein storage organelles in cereal seeds involve the acquisition of static images based on immunofluorescence and electron microscopy (Cameron-Mills and Wettstein, 1980; Arcalis et al., 2004, 2010; Gubatz et al., 2007; Holding et al., 2007; Fukuda et al., 2013; Tian et al., 2013). This reveals little about the dynamic restructuring events during endosperm development, which have to be inferred from samples at different developmental stages. However, fluorescent membrane markers can be used to follow the dynamic morphological changes *in situ* by live cell imaging. This was reported in the context of barley endosperm development, comparing PSVs in the aleurone, subaleurone and central starchy endosperm layers (Ibl et al., 2014). Whereas the spherical PSVs in the aleurone remain constant, those in the subaleurone and central starchy endosperm cells undergo substantial but cell type-specific morphological changes. In both the subaleurone and central starchy endosperm cells, the PSVs initially appear as large compartments. In the subaleurone, they subsequently go through cycles of fusion and rupture, first allowing the protein bodies to form larger, composite aggregates within the vacuole, and finally producing unconfined protein bodies and ER-derived as well as TIP3-labeled membrane fragments (**Figures 2A,B**). In contrast, PSVs in the starchy endosperm become smaller so that the protein bodies are tightly enclosed, and later in development some of the PSV membranes lose their integrity. This membrane degeneration may be induced by the desiccation process. The aleurone is protected against desiccation-induced injury (Stacy et al., 1999) but membrane dehydration may contribute to the dynamic behavior of membranes in the starchy endosperm due to the physical impact on

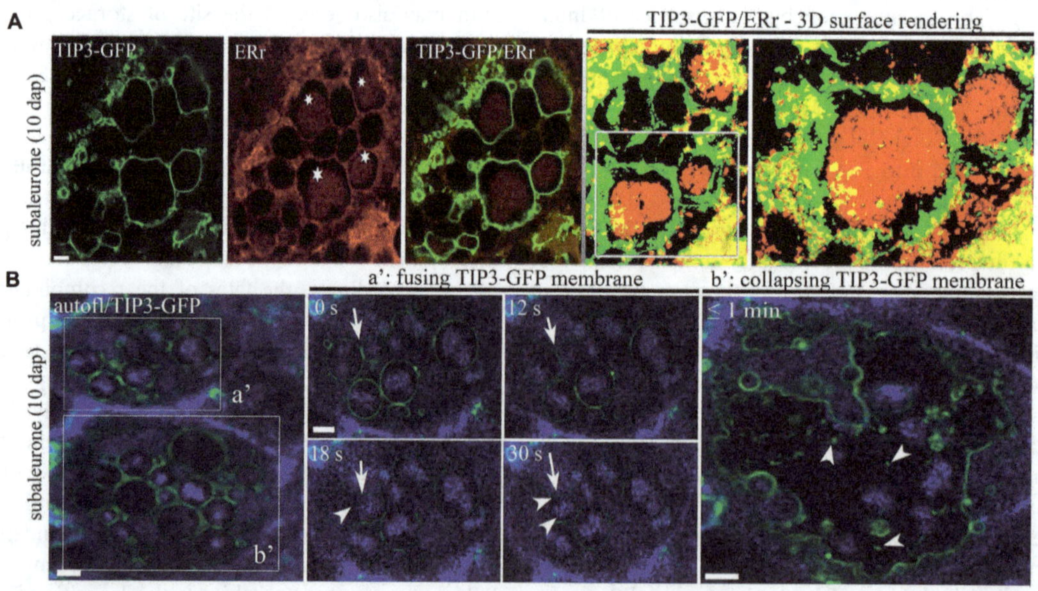

FIGURE 2 | TIP3-GFP-labeled PSVs contain protein bodies and are involved in fusion and rupture in barley endosperm. (A) TIP3-GFP-labeled PSVs comprise large protein bodies stained with ER-Tracker^{TM} red (asterisks). Note the three-dimensional surface rendering (and the magnified inset) of 16 sections with a step size of 0.5 μm. **(B)** Live cell imaging of TIP3-GFP-labeled PSVs show fusion (a′) and collapse (b′) processes in the subaleurone at 10 DAP. Note the presence of TIP3-GFP-labeled vesicles (a′, arrowheads) after PSV fusion (a′, arrows). Images were acquired every 6 s. Scale equals 5 μm. Figure partially reproduced from Ibl et al. (2014).

membrane lipids, including demixing, fluid-to-gel phase transition, increasing cytosolic viscosity, protein denaturation, and membrane fusion (Bryant et al., 2001; Hoekstra et al., 2001). In rice endosperm, the synthesis of large amounts of disulfide-rich storage proteins during seed development is accompanied by the production of H_2O_2, which causes the peroxidation of membrane lipids (Onda et al., 2009). This promotes lipid chain fracture, which alters intrinsic membrane properties such as fluidity and permeability, finally leading to the leakage of small molecules and electrolytes (Sattler et al., 2006; Onda et al., 2009; Sharma et al., 2012). Notably, mutant rice seeds that cannot produce normal amounts of sulfhydryl groups generate less H_2O_2 and desiccate more slowly than wild-type seeds (Onda et al., 2009). The production of H_2O_2 in the ER of endosperm cells may also induce programmed cell death during seed desiccation and maturation (Onda, 2013).

THE ALEURONE AS A STORAGE TISSUE

In addition to providing hydrolases for the breakdown of reserves in the starchy endosperm, the aleurone is also a storage tissue in its own right (Ritchie et al., 2000). Aleurone cells are filled with spherical PSVs that retain a constant appearance during endosperm development (Cameron-Mills and Wettstein, 1980; Bethke et al., 1996; Ibl et al., 2014). PSVs in the barley aleurone contain storage proteins, phytate (which is rich in phosphorus, potassium and magnesium) as well as storage carbohydrates (Jacobsen et al., 1971). *In situ* hybridization, RT-PCR and immunoblot analyses have shown that zein transcripts and proteins are present in the aleurone (Woo et al., 2001; Aamodt et al., 2011). Subcellular localization studies have revealed the presence of zeins, α-globulin and legumin-1 in maize aleurone cells at 18 and 22 days after pollination (DAP), predominantly as large inclusions within PSVs and small ER-derived protein bodies (Aamodt et al., 2011). As reported for the barley aleurone, PSVs and lipid bodies occupy most of the aleurone cellular volume in maize, along with multi-vesicular bodies (MVBs) and double-membrane autophagosome-like structures. Electron tomography has shown that aleurone cells at 22 DAP also contain one or more globoids (crystals of phytic acid salts) and a large system of intravacuolar membranes, possibly derived from the ER (Aamodt et al., 2011). Moreover, typical ER-resident proteins were detected in the PSVs (Aamodt et al., 2011). The dynamic analysis of the intravacuolar membrane system and the zein-rich inclusions in the maize aleurone revealed developmental changes in the PSV size, the total membrane surface area and the percentage corresponding to the intravacuolar membranes (Aamodt et al., 2011). The PSVs were smaller and more homogenous at 14 DAP compared to 22 DAP. The total membrane surface area declined by ~50% during this interval, whereas zein-rich inclusions occupied almost 10-fold more of the PSV lumen at 22 DAP (Aamodt et al., 2011). The content and intravacuolar membrane system of maize aleurone PSVs therefore appears exquisitely dependent on the developmental stage.

MORPHOLOGICAL CHANGES WITHIN THE ALEURONE DURING GERMINATION

Following imbibition, the embryo releases gibberellic acid (GA_3), which diffuses into the aleurone layer and induces the highly differentiated and specialized aleurone cells to synthesize and secrete digestive enzymes that subsequently mobilize the insoluble reserves in the starchy endosperm (Ritchie et al., 2000). During this process, the aleurone cells break down their storage proteins and use the resulting amino acids to synthesize a spectrum of hydrolases (Jones and Jacobsen, 1991; Bethke et al., 1998). After a few days, the storage reserves of the endosperm are depleted and the aleurone cells die.

The process is accompanied by profound morphological changes in the endomembrane system of aleurone cells (Jacobsen et al., 1985). De-embryonated half-grains, isolated aleurone layers and aleurone protoplasts were used to study the responses of cereal aleurone cells to hormones and to understand molecular and cellular aspects of signaling and regulation in the aleurone (reviewed by Ritchie et al., 2000; Bethke and Jones, 2001).

Barley aleurone protoplasts treated with GA_3 undergo dramatic morphological changes. Freshly isolated aleurone protoplasts contain many small PSVs, but 4 days of hormone treatment causes them to coalesce into one large central PSV (Jacobsen et al., 1985; Bush et al., 1986; Swanson and Jones, 1996; Fath et al., 1999). This GA_3-dependent vacuolation process correlates strongly with the duration of GA_3 treatment and is completed after 5 days of incubation (Bethke and Jones, 2001; Hwang et al., 2005). Granules within the vacuoles accumulate in larger numbers as the vacuoles increase in size. These morphological changes reflect a functional transition from nutrient-storing compartments to lytic organelles. Noninvasive measurements of the vacuolar pH in barley aleurone protoplasts showed that protoplasts respond to GA_3, acidifying the vacuole lumen in the course of a few hours from pH 6.6 to 5.8 or below (Swanson and Jones, 1996; Swanson et al., 1998). The cells also showed an abrupt loss of membrane integrity (Fath et al., 2000). The prolonged incubation of aleurone protoplasts with GA_3 was lethal.

Interestingly, gene expression profiling studies show that GA_3 and abscisic acid (ABA) are both synthesized during seed maturation and germination and build a possible interaction network during seed germination (Sreenivasulu et al., 2006, 2008). GA_3 and ABA signals also appear to influence the activity of a subset of stored proteases and the expression of further proteases and cell wall-loosening enzymes during grain maturation and germination. ABA antagonizes the events induced by GA_3 in the aleurone at many levels, indicating a sensitive balance between these hormones involving a spatiotemporal interaction network of key regulators involved in storage organelle dynamics. Isolated aleurone protoplasts and layers have been used to study aleurone characteristics, especially the roles of GA_3 and ABA in the regulation of vacuolar dynamics (Swanson and Jones, 1996; Swanson et al., 1998). Even if the same cellular events occur in imbibed whole grains and GA_3-treated aleurone layers or protoplasts, the timing of events is faster in isolated aleurone layers and protoplasts and the model is simplified by the absence of interacting tissues and the application of GA or ABA, rather than their spatiotemporal ratios found *in vivo* (Bethke and Jones, 2001). The characteristics of aleurone protoplasts may also vary according to culture conditions (Jacobsen et al., 1985). These aspects emphasize the importance of confirming that results obtained using isolated cell cultures are also relevant in intact seeds.

IDENTIFYING MOLECULAR REGULATORS OF STORAGE ORGANELLE BEHAVIOR

As discussed above, the dynamic behavior of storage organelles in cereal seeds strongly indicates the presence of a spatiotemporal regulatory network (**Figure 3**). Transcriptomics and proteomics have been instrumental in many areas of cereal research (reviewed by Pechanova et al., 2013). One opportunity to identify the molecular regulators of storage organelle-reshaping events is the transcriptional and/or proteomic analysis of developing and germinating seeds, to build an atlas of proteotypes. This has recently been reported for maize seeds, resulting in an atlas comprising ~14,000 proteins (Walley et al., 2013). Total protein was extracted from dissected maize seeds at seven stages of development, including separated embryo, endosperm and aleurone/pericarp tissues. Normalized and averaged data showed that peptide homologs of *Arabidopsis* TIP3 were minimally expressed in the endosperm after 12 DAP, but increased in the endosperm and in the aleurone/pericarp tissue by 27 DAP. Similarly, the comparative transcriptional analysis of two tissue fractions during barley grain maturation, desiccation and germination (starchy endosperm/aleurone vs. embryo/scutellum) revealed that TIP3 expression increases steadily during maturation and declines steadily throughout germination (Sreenivasulu et al., 2008). Because the storage endosperm dies during late maturation, any increase in transcript abundance in this fraction during germination must depend on RNA that is newly synthesized in the aleurone (Sreenivasulu et al., 2008).

Precise sampling for the more specific spatial analysis of expressed proteins is challenging and a system that can differentiate between aleurone and starchy endosperm cells is necessary. Laser microdissection (LMD) overcomes the heterogeneous distribution of different tissues and cells by allowing spatially resolved sampling of individual cells, cell populations or tissues (Fang and Schneider, 2013). This approach was recently used for the tissue-specific transcriptomic analysis of barley seeds, allowing the detection of transcriptional networks in the context of endosperm development (Thiel et al., 2011). In combination with studies using transgenic lines with fluorescent marker proteins for cellular compartments, LMD-based transcriptomics provides a promising strategy to investigate the spatiotemporal regulation of storage organelle dynamics.

Recent developments in high-resolution, two-dimensional gel electrophoresis (2DE), multidimensional liquid chromatography and mass spectrometry (MS) have also helped to characterize the dynamic cereal seed proteome, e.g., in maize (Mechin et al., 2004) and barley (Finnie et al., 2004; Finnie and Svensson, 2009). Specialized extraction procedures allow the enrichment of enzymes and other proteins that control seed metabolism and storage product mobilization, in some cases allowing selective extraction from specific tissues and compartments such as the aleurone plasma membrane and starch granule proteomes in barley (Boren et al., 2004; Hynek et al., 2006). The resulting spatiotemporal profiling of the barley seed proteome during grain filling and germination allowed proteins to be assigned to distinct functional categories (Finnie and Svensson, 2003; Bonsager et al., 2007; Finnie et al., 2011). In maize, the availability of different mutants has facilitated proteomic analysis (Damerval and Devienne, 1993).

More insight into the dynamic nature of the seed proteome has been gained using comparative proteomics, and particularly quantitative proteomics. For example, comparative 2DE has been used to study the time course of protein mobilization during rice seed germination, revealing that germination causes the radical reprogramming of the proteome, not only by consuming protein reserves but also rebuilding defense pathways and inducing morphological changes (Yang et al., 2007; He et al., 2011a). Isobaric tags for relative and absolute quantification (iTRAQ) has been used to determine precise quantitative changes in the proteome during seed development (Lan et al., 2011; Owiti et al., 2011), and the same technique has been combined with multiple reaction monitoring (MRM) and gene ontology (GO) classification to profile the rice embryogenesis-dependent proteins, thus providing a dataset that could be used to investigate the molecular basis of rice embryogenesis (Zi et al., 2013).

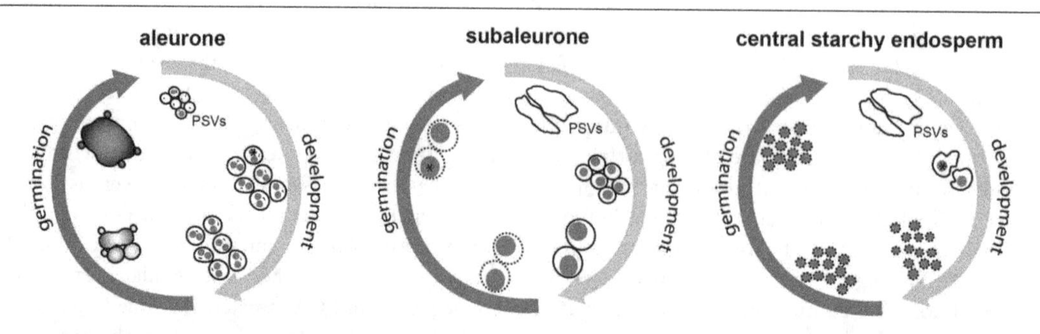

FIGURE 3 | Schematic overview of the spatiotemporally regulated dynamics of storage organelles in cereal seeds during development and germination. The spherical PSVs in the aleurone remain constant throughout seed maturation. Note the presence of zein-rich protein inclusions (asterisk) and intravacuolar membranes (indicated as short black lines). Upon germination, GA_3-dependent vacuolation results in the formation of one large central vacuole and secondary lytic vacuoles. The dramatic change from nutrient-storing compartments to lytic organelles is indicated by different shadings. In the subaleurone, large PSVs turn into smaller, spherical structures enclosing protein bodies (gray), followed by vacuolar fusion resulting in the formation of larger, composite protein bodies. The surrounding membranes gradually lose their integrity (indicated by broken lines). In the central starchy endosperm, the large PSVs containing protein bodies (gray) are shrinking. At later developmental stages, the protein bodies are tightly enclosed by PSV membranes, which degenerate during development, leaving protein bodies with no continuous PSV membrane or no membrane at all.

RECOMBINANT PROTEIN PRODUCTION IN CEREAL ENDOSPERM

UNUSUAL TRAFFICKING BEHAVIOR OF RECOMBINANT PROTEINS

Protein trafficking and organelle reshaping can also be investigated by looking at the behavior of recombinant proteins, including storage proteins fused to fluorescent tags (Tosi et al., 2009; Saito et al., 2012; Shigemitsu et al., 2013; Wakasa and Takaiwa, 2013) or foreign proteins that have been produced in the context of molecular farming. Endogenous glycoproteins are rarely found in the endosperm (Woo et al., 2001) and are difficult to follow due to the lack of appropriate detection antibodies, so recombinant glycoproteins are particularly useful because they can be traced through intracellular pathways by immunolocalization and by characterizing their glycan modifications (Drakakaki et al., 2006; Arcalis et al., 2010).

A better understanding of the peculiar rules of protein trafficking in the cereal endosperm is also desirable for the production of valuable recombinant proteins such as pharmaceuticals, which need to be produced in a predictable and homogeneous form (Stoger et al., 2014). Targeting sequences that have been shown to work in vegetative plant organs do not always guarantee the predictable trafficking of recombinant proteins to a given compartment in cereal endosperm cells (Arcalis et al., 2004; Takaiwa et al., 2009; Wakasa and Takaiwa, 2013). For example, the same protein may be secreted from rice leaf cells, but sequestered in PSVs within the endosperm (Drakakaki et al., 2006). The addition of a KDEL signal usually targets proteins to the ER lumen in vegetative tissues, but in the endosperm the final destination appears to differ by species and protein, such as in the case of endogenous prolamins accumulating in the ER and ER-derived protein bodies in rice (Wakasa and Takaiwa, 2013) and maize (Rademacher et al., 2008), or in the PSVs of wheat endosperm (Arcalis et al., 2004). Storage protein fusions usually accumulate in predictable compartments, i.e., protein bodies or PSVs (Sugita et al., 2005; Wakasa and Takaiwa, 2013). Storage organelles are generally a desirable target site for recombinant proteins because they have evolved to facilitate stable protein accumulation and thus offer a protective environment for recombinant proteins, which can remain stable in dry cereal seeds at ambient temperatures for several years (Stoger et al., 2005b). Secondly, the accumulation of recombinant proteins into storage organelles can protect the protein not only from degradation within the plant cell but also beyond harvest. For example, prolamin bodies are highly resistant to gastrointestinal digestion *in vitro* and *in vivo*, resulting in the excretion of significant numbers of prolamin bodies, making them suitable for bioencapsulation (Tanaka et al., 1975; Ogawa et al., 1987). It should be borne in mind that the trafficking of recombinant proteins can change during seed development and affect the glycan profile as discussed above, so the harvesting time can significantly affect the characteristics of the purified protein (Arcalis et al., 2010).

Another factor that may impact not only the expression, but also the subcellular localisation and modification of recombinant proteins in the endosperm is the choice of promoter to drive transgene expression. The structure and arrangement of storage organelles varies according to the tissue layer within the endosperm (Tosi et al., 2009; Ibl et al., 2014), and endosperm-specific promoters also vary significantly in their spatiotemporal activity (Qu and Takaiwa, 2004), which should make possible to favor recombinant protein expression in cell types with a specific organelle architecture. Endosperm-specific promoters often have minimal or no activity in the aleurone (Lamacchia et al., 2001), so their use can be contrasted with constitutive or even aleurone-specific promoters such as the amylase promoter, which has been used to drive the expression of a recombinant enzyme during malting in barley (Nuutila et al., 1999). The secretory profile and *N*-glycoproteome differ markedly between the aleurone and the inner maturing endosperm (Barba-Espin et al., 2014). However, recombinant protein production during malting/germination is challenging because of the higher intrinsic proteolytic activity in the tissue at this stage, which affects the stability of the product.

Finally, the fate of a recombinant protein in developing seeds may occasionally be affected by interactions with endogenous storage proteins. Proteins with free cysteine residues and multimeric proteins that assemble via disulfide bonds (such as antibodies) are particularly prone to covalent interactions with prolamins. For example, free antibody chains interact with zeins (Peters et al., 2013), and a cysteine-rich peptide was shown to mediate protein retention in rice prolamin bodies (Takaiwa et al., 2009).

THE BIOENCAPSULATION IN STORAGE ORGANELLES

Plant tissue can provide some protection from proteolytic enzymes in the gut, and the sequestration of recombinant proteins in rice protein bodies appears to extend the protection from digestive proteolysis following oral administration in an animal model. This was shown by directly comparing the digestibility of a model oral tolerogen from three sources: ER-derived protein bodies in rice endosperm, glutelin bodies (PSVs) in rice endosperm and a chemically synthesized peptide (Takagi et al., 2010). Both endosperm-derived peptides were more resistant to *in vitro* digestion with pepsin than the soluble form, but the prolamin bodies were more resistant than that from PSVs. Similar results were reported in feeding studies, indicating that bioencapsulated forms of the tolerogen significantly enhanced its immunological efficacy. The cholera toxin B subunit (CTB) also accumulated within protein bodies and the PSV in rice endosperm, and was similarly protected from *in vitro* pepsin degradation when fed to mice, allowing the induction of CTB-specific serum IgG and mucosal IgA antibodies (Nochi et al., 2007).

These studies suggest that protein storage organelles offer natural bioencapsulation strategies compatible with oral vaccines or prophylactic mucosal antibodies for passive immunization against gastro-intestinal disease. Oral vaccines must be protected from digestion to ensure sufficient amounts of the antigen reach the Peyer's patches (lymphoid tissue in the ileum), which is necessary for the induction of an oral immune response. Storage proteins such as zeins have already been considered as *in vitro* protein encapsulation reagents for slow-release pharmaceuticals (Lai and Guo, 2011; Lau et al., 2013), thus naturally occurring protein bodies may provide the ideal encapsulation platform for recombinant pharmaceuticals produced in plants.

ER STRESS AND NOVEL ER-DERIVED COMPARTMENTS

Some recombinant proteins have been shown to induce the formation of additional, prolamin-free ER-derived protein bodies, e.g., a modified house dust allergen and a modified birch pollen allergen expressed in rice seeds (Yang et al., 2012a; Wang et al., 2013). Similarly, human IL-10 expressed in rice seeds was found to be localized in ER-derived prolamin bodies and aberrant ER-derived compartments (Fujiwara et al., 2010; Yang et al., 2012b). The aberrant deposition of recombinant proteins in newly formed ER-derived structures has also been observed in other plant seeds, including a scFv-Fc antibody that was partially localized in ER-derived compartments delimited by ribosome-associated membranes in *Arabidopsis* seeds (Van Droogenbroeck et al., 2007). Similar observations have been reported for other scFv-Fc molecules with or without KDEL signals (Loos et al., 2011b). These novel ER-derived compartments may play a protective role in the host cell, and may be induced preferentially by aggregation-prone protein candidates. Although additional factors such as protein expression levels may also come into play, seeds could be more prone to the induction of ER-derived compartments because the mechanism that generates ER-derived protein bodies is already prevalent, especially in the endosperm.

Recombinant protein expression can sometimes elicit an ER stress response reflected by the coordinated expression of ER-resident molecular chaperones (Oono et al., 2010; Wakasa et al., 2012). This correlates with the appearance of distorted protein bodies at the microscopic level (Yang et al., 2012b; Wang et al., 2013). In some cases, there is also a specific grain phenotype at the macroscopic level, such as the opaque phenotype with floury and shrunken features reported in transgenic rice seeds (Oono et al., 2010), phenotypically similar to the maize *floury2* mutant that is also characterized by an ER stress response (Hunter et al., 2002). It is unclear whether seeds are more prone to ER stress induced by recombinant protein expression than vegetative tissues, perhaps due to the temporally confined high protein expression levels, or whether the stress symptoms are more obvious and therefore more frequently noticed in seeds due to the opaque phenotype.

UNUSUAL *N*-GLYCAN STRUCTURES

N-glycan structures on recombinant proteins vary according to the final subcellular location and the trafficking route (Lerouge et al., 1998). The lack of secretion in cereal endosperm often results in a distinct lack of typical apoplast *N*-glycan structures such as GnGnXF, which are abundant in proteins expressed in leaves (reviewed in Samyn-Petit et al., 2003; Arcalis et al., 2013). Glycoproteins that have not passed through the Golgi apparatus typically bear oligomannosidic structures (high-mannose glycans) whereas those in post-Golgi locations tend to bear complex, xylosylated and fucosylated *N*-glycans. The ratio between these glycan classes sometimes correlates with the distribution of a protein between ER-derived protein bodies and PSVs, but this is not necessarily the case because the direct ER-to-PSV transport route discussed above may allow a proportion of proteins deposited in the PSV to bypass the Golgi apparatus completely. Interestingly, a number of recent studies involving the analysis of glycopeptides, rather than *N*-glycans digested off glycoproteins prior to analysis, have revealed significant proportions of aglycosylated seed-derived recombinant glycoproteins (Van Droogenbroeck et al., 2007). This may reflect the rapid production of high levels of the recombinant protein based on temporally confined extreme transcriptional and translational activities, but it may also indicate the limited capacity of the glycosylation machinery in maturing seeds. Under-glycosylation may also be linked to ER-stress and result in the formation of small, ER-derived organelles as discussed above.

Another, perhaps more puzzling phenomenon, is the abundance of trimmed *N*-glycan structures consisting of only a single GlcNAc residue, which may be useful for specialty applications but may not support the functionality of some recombinant proteins such as antibodies with effector functions (Umana et al., 1999; Schuster et al., 2007). These structures are often found on KDEL-tagged proteins, and most likely reflect the activity of Endo-β-*N*-acetylglucosaminidase (ENGase) on oligomannosidic *N*-glycan substrates (Fischl et al., 2011). Single GlcNAc residues have been detected on glycoproteins produced in maize, and may even represent the predominant glycoform suggesting a high level of ENGase) activity in cereal seeds (Rademacher et al., 2008; Ramessar et al., 2008a; Arcalis et al., 2010). Indeed, ENGase activity has been observed in cereal seeds (Vuylsteker et al., 2000), and free oligomannosidic glycan structures, which are released by ENGase activity, have been identified in seeds of diverse plant species (Kimura and Kitahara, 2000; Kimura et al., 2002) and a physiological role for them in plant development, fruit and seed maturation has been proposed (Priem and Gross, 1992; Kimura and Kitahara, 2000). However, *Arabidopsis* plants lacking ENGase activity showed no obvious morphological phenotype (Fischl et al., 2011; Kimura et al., 2011).

It is still unclear why ENGase, an enzyme with a presumed cytosolic localization (Suzuki et al., 2002; Fischl et al., 2011), can access substrates that accumulate within endomembrane organelles. Although we cannot completely exclude the possibility that glycan trimming occurs during extraction, our preliminary data instead support the *in vivo* action of the enzyme. In addition, based on the fact that endomembrane structures break down and compartments rapture, possibly in response to oxidative stress and desiccation (Ibl et al., 2014), combined with the incorporation of cytosolic material into the interior of vacuoles via atypical pre-vacuolar compartments during autophagy (Reyes et al., 2011), it could be speculated that these events might also allow ENGases to come into contact with glycoprotein substrates. Further investigations are required to determine whether this is the case, perhaps providing a better understanding of the physiological role of ENGases in plant seeds and increasing the spectrum of its endogenous substrates.

CONCLUSION

Endosperm is a short-lived tissue that undergoes a rapid sequence of developmental changes, including extensive endomembrane system reorganization and rearrangement. The development and maturation of storage organelles is linked with storage protein synthesis and protein transport via several developmentally regulated trafficking routes. The investigation of spatially and temporally regulated gene expression will provide insight into the underlying molecular mechanisms, and transgenic cereal plants expressing fluorescent marker proteins combined with live cell imaging techniques allow the investigation of dynamic membrane reorganization events. The analysis of recombinant protein transport, deposition and glycosylation provides additional information about cargo trafficking to storage organelles. By understanding these processes, we will be able to control the production of recombinant proteins and design optimized production strategies for molecular farming that exploit the unique bioencapsulation properties of storage organelles.

ACKNOWLEDGMENTS

The authors would like to acknowledge financial support by the Austrian Science Fund FWF (P25736-B20 and I1461-B16).

REFERENCES

Aamodt, K., Abrahantes Quintana, A., Adamova, D., Adare, A. M., Aggarwal, M. M., Aglieri Rinella, G., et al. (2011). Centrality dependence of the charged-particle multiplicity density at midrapidity in Pb-Pb collisions at $\sqrt{S}$NN = 2.76 TeV. *Phys. Rev. Lett.* 106, 032301. doi: 10.1103/PhysRevLett.106.032301

Arcalis, E., Marcel, S., Altmann, F., Kolarich, D., Drakakaki, G., Fischer, R., et al. (2004). Unexpected deposition patterns of recombinant proteins in post-endoplasmic reticulum compartments of wheat endosperm. *Plant Physiol.* 136, 3457–3466. doi: 10.1104/pp.104.050153

Arcalis, E., Stadlmann, J., Marcel, S., Drakakaki, G., Winter, V., Rodriguez, J., et al. (2010). The changing fate of a secretory glycoprotein in developing maize endosperm. *Plant Physiol.* 153, 693–702. doi: 10.1104/pp.109.152363

Arcalis, E., Stadlmann, J., Rademacher, T., Marcel, S., Sack, M., Altmann, F., et al. (2013). Plant species and organ influence the structure and subcellular localization of recombinant glycoproteins. *Plant Mol. Biol.* 83, 105–117. doi: 10.1007/s11103-013-0049-9

Barba-Espin, G., Dedvisitsakul, P., Hagglund, P., Svensson, B., and Finnie, C. (2014). Gibberellic acid-induced aleurone layers responding to heat shock or tunicamycin provide insight into the *N*-glycoproteome, protein secretion, and endoplasmic reticulum stress. *Plant Physiol.* 164, 951–965. doi: 10.1104/pp.113.233163

Bethke, P. C., Hillmer, S., and Jones, R. L. (1996). Isolation of intact protein storage vacuoles from barley aleurone (identification of aspartic and cysteine proteases). *Plant Physiol.* 110, 521–529.

Bethke, P. C., and Jones, R. L. (2001). Cell death of barley aleurone protoplasts is mediated by reactive oxygen species. *Plant J.* 25, 19–29. doi: 10.1046/j.1365-313x.2001.00930.x

Bethke, P. C., Swanson, S. J., Hillmer, S., and Jones, R. L. (1998). From storage compartment to lytic organelle: the metamorphosis of the aleurone protein storage vacuole. *Ann. Bot.* 82, 399–412. doi: 10.1006/anbo.1998.0702

Bonsager, B. C., Finnie, C., Roepstorff, P., and Svensson, B. (2007). Spatio-temporal changes in germination and radical elongation of barley seeds tracked by proteome analysis of dissected embryo, aleurone layer, and endosperm tissues. *Proteomics* 7, 4528–4540. doi: 10.1002/pmic.200700766

Boren, M., Larsson, H., Falk, A., and Jansson, C. (2004). The barley starch granule proteome—internalized granule polypeptides of the mature endosperm. *Plant Sci.* 166, 617–626. doi: 10.1016/j.plantsci.2003.10.028

Brown, R. C., and Lemmon, B. E. (2007). The developmental biology of cereal endosperm. *Plant Cell Monogr.* 8, 1–20. doi: 10.1007/7089_2007_106

Bryant, G., Koster, K. L., and Wolfe, J. (2001). Membrane behaviour in seeds and other systems at low water content: the various effects of solutes. *Seed Sci. Res.* 11, 17–25. doi: 10.1079/SSR200056

Bush, D. S., Cornejo, M. J., Huang, C. N., and Jones, R. L. (1986). Ca-stimulated secretion of alpha-amylase during development in barley aleurone protoplasts. *Plant Physiol.* 82, 566–574. doi: 10.1104/pp.82.2.566

Cameron-Mills, V., and Wettstein, D. V. (1980). Protein body formation in the developing barley endosperm. *Carlsberg Res. Commun.* 45, 577–594. doi: 10.1007/BF02932924

Chen, F., and Bradford, K. J. (2000). Expression of an expansin is associated with endosperm weakening during tomato seed germination. *Plant Physiol.* 124, 1265–1274. doi: 10.1104/pp.124.3.1265

Chung, T., Phillips, A. R., and Vierstra, R. D. (2010). ATG8 lipidation and ATG8-mediated autophagy in *Arabidopsis* require ATG12 expressed from the differentially controlled ATG12A AND ATG12B loci. *Plant J.* 62, 483–493. doi: 10.1111/j.1365-313X.2010.04166.x

Coleman, C. E., Yoho, P. R., Escobar, S., and Ogawa, M. (2004). The accumulation of α-zein in transgenic tobacco endosperm is stabilized by co-expression of β-zein. *Plant Cell Physiol.* 45, 864–871. doi: 10.1093/pcp/pch104

Damerval, C., and Devienne, D. (1993). Quantification of dominance for proteins pleiotropically affected by *Opaque*-2 in maize. *Heredity* 70, 38–51. doi: 10.1038/Hdy.1993.6

De Jaeger, G., Scheffer, S., Jacobs, A., Zambre, M., Zobell, O., Goossens, A., et al. (2002). Boosting heterologous protein production in transgenic dicotyledonous seeds using *Phaseolus vulgaris* regulatory sequences. *Nat. Biotechnol.* 20, 1265–1268. doi: 10.1038/nbt755

Drakakaki, G., Marcel, S., Arcalis, E., Altmann, F., Gonzalez-Melendi, P., Fischer, R., et al. (2006). The intracellular fate of a recombinant protein is tissue dependent. *Plant Physiol.* 141, 578–586. doi: 10.1104/pp.106.076661

Erlendsson, L. S., Muench, M. O., Hellman, U., Hrafnkelsdottir, S. M., Jonsson, A., Balmer, Y., et al. (2010). Barley as a green factory for the production of functional Flt3 ligand. *Biotechnol. J.* 5, 163–171. doi: 10.1002/biot.200900111

Fang, J., and Schneider, B. (2013). Laser microdissection: a sample preparation technique for plant micrometabolic profiling. *Phytochem. Anal.* 25, 307–313. doi: 10.1002/pca.2477

Fath, A., Bethke, P., Lonsdale, J., Meza-Romero, R., and Jones, R. (2000). Programmed cell death in cereal aleurone. *Plant Mol. Biol.* 44, 255–266. doi: 10.1023/A:1026584207243

Fath, A., Bethke, P. C., and Jones, R. L. (1999). Barley aleurone cell death is not apoptotic: characterization of nuclease activities and DNA degradation. *Plant J.* 20, 305–315. doi: 10.1046/j.1365-313X.1999.t01-2-00605.x

Finnie, C., Andersen, B., Shahpiri, A., and Svensson, B. (2011). Proteomes of the barley aleurone layer: a model system for plant signalling and protein secretion. *Proteomics* 11, 1595–1605. doi: 10.1002/pmic.201000656

Finnie, C., Maeda, K., ØStergaard, O., Bak-Jensen, K. S., Larsen, J., and Svensson, B. (2004). Aspects of the barley seed proteome during development and germination. *Biochem. Soc. Trans.* 32, 517–519. doi: 10.1042/BST0320517

Finnie, C., and Svensson, B. (2003). Feasibility study of a tissue-specific approach to barley proteome analysis: aleurone layer, endosperm, embryo and single seeds. *J. Cereal Sci.* 38, 217–227. doi: 10.1016/S0733-5210(03)00033-X

Finnie, C., and Svensson, B. (2009). Barley seed proteomics from spots to structures. *J. Proteomics* 72, 315–324. doi: 10.1016/j.jprot.2008.12.001

Fischl, R. M., Stadlmann, J., Grass, J., Altmann, F., and Leonard, R. (2011). The two endo-β-*N*-acetylglucosaminidase genes from *Arabidopsis thaliana* encode cytoplasmic enzymes controlling free *N*-glycan levels. *Plant Mol. Biol.* 77, 275–284. doi: 10.1007/s11103-011-9808-7

Floss, D. M., Sack, M., Arcalis, E., Stadlmann, J., Quendler, H., Rademacher, T., et al. (2009). Influence of elastin-like peptide fusions on the quantity and quality of a tobacco-derived human immunodeficiency virus-neutralizing antibody. *Plant Biotechnol. J.* 7, 899–913. doi: 10.1111/j.1467-7652.2009.00452.x

Fujiwara, Y., Aiki, Y., Yang, L., Takaiwa, F., Kosaka, A., Tsuji, N. M., et al. (2010). Extraction and purification of human interleukin-10 from transgenic rice seeds. *Protein Expr. Purif.* 72, 125–130. doi: 10.1016/j.pep.2010.02.008

Fukuda, M., Wen, L., Satoh-Cruz, M., Kawagoe, Y., Nagamura, Y., Okita, T. W., et al. (2013). A guanine nucleotide exchange factor for Rab5 proteins is essential for intracellular transport of the proglutelin from the Golgi apparatus to the protein storage vacuole in rice endosperm. *Plant Physiol.* 162, 663–674. doi: 10.1104/pp.113.217869

Galili, G. (2004). ER-derived compartments are formed by highly regulated processes and have special functions in plants. *Plant Physiol.* 136, 3411–3413. doi: 10.1104/pp.104.900125

Gubatz, S., Dercksen, V. J., Bruss, C., Weschke, W., and Wobus, U. (2007). Analysis of barley (*Hordeum vulgare*) grain development using three-dimensional digital models. *Plant J.* 52, 779–790. doi: 10.1111/j.1365-313X.2007.03260.x

Hara-Nishimura, I., Shimada, T., Hatano, K., Takeuchi, Y., and Nishimura, M. (1998). Transport of storage proteins to protein storage vacuoles is mediated by large precursor-accumulating vesicles. *Plant Cell* 10, 825–836. doi: 10.1105/tpc.10.5.825

He, D., Han, C., Yao, J., Shen, S., and Yang, P. (2011a). Constructing the metabolic and regulatory pathways in germinating rice seeds through proteomic approach. *Proteomics* 11, 2693–2713. doi: 10.1002/pmic.201000598

He, Y., Ning, T., Xie, T., Qiu, Q., Zhang, L., Sun, Y., et al. (2011b). Large-scale production of functional human serum albumin from transgenic rice seeds. *Proc. Natl. Acad. Sci. U.S.A.* 108, 19078–19083. doi: 10.1073/pnas.1109736108

He, X., Haselhorst, T., Von Itzstein, M., Kolarich, D., Packer, N. H., Gloster, T. M., et al. (2012). Production of α-*L*-iduronidase in maize for the potential treatment of a human lysosomal storage disease. *Nat. Commun.* 3, 1062. doi: 10.1038/ncomms2070

Hoekstra, F. A., Golovina, E. A., and Buitink, J. (2001). Mechanisms of plant desiccation tolerance. *Trends Plant Sci.* 6, 431–438. doi: 10.1016/S1360-1385(01)02052-0

Hoh, B., Hinz, G., Jeong, B. K., and Robinson, D. G. (1995). Protein storage vacuoles form de-novo during pea cotyledon development. *J. Cell Sci.* 108, 299–310.

Holding, D. R., Otegui, M. S., Li, B., Meeley, R. B., Dam, T., Hunter, B. G., et al. (2007). The maize floury1 gene encodes a novel endoplasmic reticulum protein involved in zein protein body formation. *Plant Cell* 19, 2569–2582. doi: 10.1105/tpc.107.053538

Hunter, B. G., Beatty, M. K., Singletary, G. W., Hamaker, B. R., Dilkes, B. P., Larkins, B. A., et al. (2002). Maize opaque endosperm mutations create extensive changes in patterns of gene expression. *Plant Cell* 14, 2591–2612. doi: 10.1105/tpc.003905

Hwang, Y. S., Bethke, P. C., Cheong, Y. H., Chang, H. S., Zhu, T., and Jones, R. L. (2005). A gibberellin-regulated calcineurin B in rice localizes to the tonoplast and is implicated in vacuole function. *Plant Physiol.* 138, 1347–1358. doi: 10.1104/pp.105.062703

Hynek, R., Svensson, B., Jensen, O. N., Barkholt, V., and Finnie, C. (2006). Enrichment and identification of integral membrane proteins from barley aleurone layers by reversed-phase chromatography, SDS-PAGE, and LC-MS/MS. *J. Proteome Res.* 5, 3105–3113. doi: 10.1021/pr0602850

Ibl, V., Kapusi, E., Arcalis, E., Kawagoe, Y., and Stoger, E. (2014). Fusion, rupture and degeneration: the fate of in vivo-labeled PSVs in developing barley endosperm. *J. Exp. Bot.* 65, 3249–3261. doi: 10.1093/jxb/eru175

Jacobsen, J. V., Knox, R. B., and Pyliotis, N. A. (1971). Structure and composition of aleurone grains in barley aleurone layer. *Planta* 101, 189–209. doi: 10.1007/Bf00386828

Jacobsen, J. V., Zwar, J. A., and Chandler, P. M. (1985). Gibberellic-acid-responsive protoplasts from mature aleurone of Himalaya barley. *Planta* 163, 430–438. doi: 10.1007/BF00395154

Jones, R. L., and Jacobsen, J. V. (1991). Regulation of synthesis and transport of secreted proteins in cereal aleurone. *Int. Rev. Cytol.* 126, 49–88. doi: 10.1016/S0074-7696(08)60682-8

Kimura, Y., and Kitahara, E. (2000). Structural analysis of free *N*-glycans occurring in soybean seedlings and dry seeds. *Biosci. Biotechnol. Biochem.* 64, 1847–1855. doi: 10.1271/bbb.64.1847

Kimura, Y., Matsuo, S., Tsurusaki, S., Kimura, M., Hara-Nishimura, I., and Nishimura, M. (2002). Subcellular localization of endo-β-*N*-acetylglucosaminidase and high-mannose type free *N*-glycans in plant cell. *Biochim. Biophys. Acta* 1570, 38–46. doi: 10.1016/S0304-4165(02)00149-6

Kimura, Y., Takeoka, Y., Inoue, M., Maeda, M., and Fujiyama, K. (2011). Double-knockout of putative endo-β-*N*-acetylglucosaminidase (ENGase) genes in *Arabidopsis thaliana*: loss of ENGase activity induced accumulation of high-mannose type free *N*-glycans bearing N,N′-acetylchitobiosyl unit. *Biosci. Biotechnol. Biochem.* 75, 1019–1021. doi: 10.1271/bbb.110148

Klionsky, D. J., Abeliovich, H., Agostinis, P., Agrawal, D. K., Aliev, G., Askew, D. S., et al. (2008). Guidelines for the use and interpretation of assays for monitoring autophagy in higher eukaryotes. *Autophagy* 4, 151–175. doi: 10.4161/auto.5338

Lai, L. F., and Guo, H. X. (2011). Preparation of new 5-fluorouracil-loaded zein nanoparticles for liver targeting. *Int. J. Pharm.* 404, 317–323. doi: 10.1016/j.ijpharm.2010.11.025

Lamacchia, C., Shewry, P. R., Di Fonzo, N., Forsyth, J. L., Harris, N., Lazzeri, P. A., et al. (2001). Endosperm-specific activity of a storage protein gene promoter in transgenic wheat seed. *J. Exp. Bot.* 52, 243–250. doi: 10.1093/jexbot/52.355.243

Lan, P., Li, W., Wen, T. N., Shiau, J. Y., Wu, Y. C., Lin, W., et al. (2011). iTRAQ protein profile analysis of *Arabidopsis* roots reveals new aspects critical for iron homeostasis. *Plant Physiol.* 155, 821–834. doi: 10.1104/pp.110.169508

Landrum, M., Smertenko, A., Edwards, R., Hussey, P. J., and Steel, P. G. (2010). BODIPY probes to study peroxisome dynamics *in vivo*. *Plant J.* 62, 529–538. doi: 10.1111/j.1365-313X.2010.04153.x

Lau, E. T., Giddings, S. J., Mohammed, S. G., Dubois, P., Johnson, S. K., Stanley, R. A., et al. (2013). Encapsulation of hydrocortisone and mesalazine in zein microparticles. *Pharmaceutics* 5, 277–293. doi: 10.3390/pharmaceutics5020277

Lending, C. R., Chesnut, R. S., Shaw, K. L., and Larkins, B. A. (1989). Immunolocalization of avenin and globulin storage proteins in developing endosperm of *Avena sativa* L. *Planta* 178, 315–324. doi: 10.1007/BF00391859

Lerouge, P., Cabanes-Macheteau, M., Rayon, C., Fischette-Laine, A. C., Gomord, V., and Faye, L. (1998). *N*-glycoprotein biosynthesis in plants: recent developments and future trends. *Plant Mol. Biol.* 38, 31–48. doi: 10.1023/A:1006012005654

Leubner-Metzger, G., Frundt, C., Vogeli-Lange, R., and Meins, F. Jr. (1995). Class I β-1,3-glucanases in the endosperm of tobacco during germination. *Plant Physiol.* 109, 751–759.

Levanony, H., Rubin, R., Altschuler, Y., and Galili, G. (1992). Evidence for a novel route of wheat storage proteins to vacuoles. *J. Cell Biol.* 119, 1117–1128. doi: 10.1083/jcb.119.5.1117

Loos, A., Van Droogenbroeck, B., Hillmer, S., Grass, J., Kunert, R., Cao, J., et al. (2011a). Production of monoclonal antibodies with a controlled *N*-glycosylation pattern in seeds of *Arabidopsis thaliana*. *Plant Biotechnol. J.* 9, 179–192. doi: 10.1111/j.1467-7652.2010.00540.x

Loos, A., Van Droogenbroeck, B., Hillmer, S., Grass, J., Pabst, M., Castilho, A., et al. (2011b). Expression of antibody fragments with a controlled *N*-glycosylation pattern and induction of endoplasmic reticulum-derived vesicles in seeds of *Arabidopsis*. *Plant Physiol.* 155, 2036–2048. doi: 10.1104/pp.110.171330

Mechin, V., Balliau, T., Chateau-Joubert, S., Davanture, M., Langella, O., Negroni, L., et al. (2004). A two-dimensional proteome map of maize endosperm. *Phytochemistry* 65, 1609–1618. doi: 10.1016/j.phytochem.2004.04.035

Mohanty, A., Luo, A., Deblasio, S., Ling, X., Yang, Y., Tuthill, D. E., et al. (2009). Advancing cell biology and functional genomics in maize using fluorescent protein-tagged lines. *Plant Physiol.* 149, 601–605. doi: 10.1104/pp.108.130146

Morandini, F., Avesani, L., Bortesi, L., Van Droogenbroeck, B., De Wilde, K., Arcalis, E., et al. (2011). Non-food/feed seeds as biofactories for the high-yield production of recombinant pharmaceuticals. *Plant Biotechnol. J.* 9, 911–921. doi: 10.1111/j.1467-7652.2011.00605.x

Muntz, K. (1998). Deposition of storage proteins. *Plant Mol. Biol.* 38, 77–99. doi: 10.1023/A:1006020208380

Nochi, T., Takagi, H., Yuki, Y., Yang, L., Masumura, T., Mejima, M., et al. (2007). Rice-based mucosal vaccine as a global strategy for cold-chain- and needle-free vaccination. *Proc. Natl. Acad. Sci. U.S.A.* 104, 10986–10991. doi: 10.1073/pnas.0703766104

Nonogaki, H., and Morohashi, Y. (1996). An Endo-β-mannanase develops exclusively in the micropylar endosperm of tomato seeds prior to radicle emergence. *Plant Physiol.* 110, 555–559. doi: 10.1104/pp.110.2.555

Nuutila, A. M., Ritala, A., Skadsen, R. W., Mannonen, L., and Kauppinen, V. (1999). Expression of fungal thermotolerant endo-1,4-β-glucanase in transgenic barley seeds during germination. *Plant Mol. Biol.* 41, 777–783. doi: 10.1023/A:1006318206471

Ogawa, M., Kumamaru, T., Satoh, H., Iwata, N., Omura, T., Kasai, Z., et al. (1987). Purification of protein body-I of rice seed and its polypeptide composition. *Plant Cell Physiol.* 28, 1517–1527.

Okamoto, T., Shimada, T., Hara-Nishimura, I., Nishimura, M., and Minamikawa, T. (2003). C-terminal KDEL sequence of a KDEL-tailed cysteine proteinase (sulfhydryl-endopeptidase) is involved in formation of KDEL vesicle and in efficient vacuolar transport of sulfhydryl-endopeptidase. *Plant Physiol.* 132, 1892–1900. doi: 10.1104/pp.103.021147

Onda, Y. (2013). Oxidative protein-folding systems in plant cells. *Int. J. Cell Biol.* 2013, 585431. doi: 10.1155/2013/585431

Onda, Y., Kumamaru, T., and Kawagoe, Y. (2009). ER membrane-localized oxidoreductase ero1 is required for disulfide bond formation in the rice endosperm. *Proc. Natl. Acad. Sci. U.S.A.* 106, 14156–14161. doi: 10.1073/pnas.0904429106

Oono, Y., Wakasa, Y., Hirose, S., Yang, L., Sakuta, C., and Takaiwa, F. (2010). Analysis of ER stress in developing rice endosperm accumulating β-amyloid peptide. *Plant Biotechnol. J.* 8, 691–718. doi: 10.1111/j.1467-7652.2010.00502.x

Oszvald, M., Tamas, L., Shewry, P. R., and Tosi, P. (2014). The trafficking pathway of a wheat storage protein in transgenic rice endosperm. *Ann. Bot.* 113, 807–815. doi: 10.1093/aob/mcu008

Oufattole, M., Park, J. H., Poxleitner, M., Jiang, L., and Rogers, J. C. (2005). Selective membrane protein internalization accompanies movement from the endoplasmic reticulum to the protein storage vacuole pathway in *Arabidopsis*. *Plant Cell* 17, 3066–3080. doi: 10.1105/tpc.105.035212

Owiti, J., Grossmann, J., Gehrig, P., Dessimoz, C., Laloi, C., Hansen, M. B., et al. (2011). iTRAQ-based analysis of changes in the cassava root proteome reveals pathways associated with post-harvest physiological deterioration. *Plant J.* 67, 145–156. doi: 10.1111/j.1365-313X.2011.04582.x

Pechanova, O., Takac, T., Samaj, J., and Pechan, T. (2013). Maize proteomics: an insight into the biology of an important cereal crop. *Proteomics* 13, 637–662. doi: 10.1002/pmic.201200275

Peters, J., Sabalza, M., Ramessar, K., Christou, P., Capell, T., Stoger, E., et al. (2013). Efficient recovery of recombinant proteins from cereal endosperm is affected by interaction with endogenous storage proteins. *Biotechnol. J.* 8, 1203–1212. doi: 10.1002/biot.201300068

Petruccelli, S., Otegui, M. S., Lareu, F., Tran Dinh, O., Fitchette, A. C., Circosta, A., et al. (2006). A KDEL-tagged monoclonal antibody is efficiently retained in the endoplasmic reticulum in leaves, but is both partially secreted and sorted to protein storage vacuoles in seeds. *Plant Biotechnol. J.* 4, 511–527. doi: 10.1111/j.1467-7652.2006.00200.x

Priem, B., and Gross, K. C. (1992). Mannosyl-and Xylosyl-containing glycans promote tomato (*Lycopersicon esculentum* Mill.) fruit ripening. *Plant Physiol.* 98, 399–401. doi: 10.1104/pp.98.1.399

Qu, L. Q., and Takaiwa, F. (2004). Evaluation of tissue specificity and expression strength of rice seed component gene promoters in transgenic rice. *Plant Biotechnol. J.* 2, 113–125. doi: 10.1049/j.1467-7653.2003.00055.x

Rademacher, T., Sack, M., Arcalis, E., Stadlmann, J., Balzer, S., Altmann, F., et al. (2008). Recombinant antibody 2G12 produced in maize endosperm efficiently neutralizes HIV-1 and contains predominantly single-GlcNAc *N*-glycans. *Plant Biotechnol. J.* 6, 189–201. doi: 10.1111/j.1467-7652.2007.00306.x

Ramessar, K., Rademacher, T., Sack, M., Stadlmann, J., Platis, D., Stiegler, G., et al. (2008a). Cost-effective production of a vaginal protein microbicide to prevent HIV transmission. *Proc. Natl. Acad. Sci. U.S.A.* 105, 3727–3732. doi: 10.1073/pnas.0708841104

Ramessar, K., Sabalza, M., Capell, T., and Christou, P. (2008b). Maize plants: an ideal production platform for effective and safe molecular pharming. *Plant Sci.* 174, 409–419. doi: 10.1016/j.plantsci.2008.02.002

Rechinger, K. B., Simpson, D. J., Svendsen, I., and Cameron-Mills, V. (1993). A role for gamma 3 hordein in the transport and targeting of prolamin polypeptides to the vacuole of developing barley endosperm. *Plant J.* 4, 841–853. doi: 10.1046/j.1365-313X.1993.04050841.x

Reyes, F. C., Chung, T., Holding, D., Jung, R., Vierstra, R., and Otegui, M. S. (2011). Delivery of prolamins to the protein storage vacuole in maize aleurone cells. *Plant Cell* 23, 769–784. doi: 10.1105/tpc.110.082156

Ritchie, S., Swanson, S. J., and Gilroy, S. (2000). Physiology of the aleurone layer and starchy endosperm during grain development and early seedling growth: new insights from cell and molecular biology. *Seed Sci. Res.* 10, 193–212. doi: 10.1017/S0960258500000234

Rogers, J. C. (2011). Internal membranes in maize aleurone protein storage vacuoles: beyond autophagy. *Plant Cell* 23, 4168–4171. doi: 10.1105/tpc.111.092551

Rubinsztein, D. C., Cuervo, A. M., Ravikumar, B., Sarkar, S., Korolchuk, V., Kaushik, S., et al. (2009). In search of an "autophagomometer." *Autophagy* 5, 585–589. doi: 10.4161/auto.5.5.8823

Sabelli, P. A., and Larkins, B. A. (2009). The development of endosperm in grasses. *Plant Physiol.* 149, 14–26. doi: 10.1104/pp.108.129437

Saito, Y., Shigemitsu, T., Yamasaki, R., Sasou, A., Goto, F., Kishida, K., et al. (2012). Formation mechanism of the internal structure of type I protein bodies in rice endosperm: relationship between the localization of prolamin species and the expression of individual genes. *Plant J.* 70, 1043–1055. doi: 10.1111/j.1365-313X.2012.04947.x

Samyn-Petit, B., Wajda Dubos, J. P., Chirat, F., Coddeville, B., Demaizieres, G., Farrer, S., et al. (2003). Comparative analysis of the site-specific *N*-glycosylation of human lactoferrin produced in maize and tobacco plants. *Eur. J. Biochem.* 270, 3235–3242. doi: 10.1046/j.1432-1033.2003.03706.x

Sattler, S. E., Mene-Saffrane, L., Farmer, E. E., Krischke, M., Mueller, M. J., and Dellapenna, D. (2006). Nonenzymatic lipid peroxidation reprograms gene expression and activates defense markers in *Arabidopsis* tocopherol-deficient mutants. *Plant Cell* 18, 3706–3720. doi: 10.1105/tpc.106.044065

Schuster, M., Jost, W., Mudde, G. C., Wiederkum, S., Schwager, C., Janzek, E., et al. (2007). *In vivo* glyco-engineered antibody with improved lytic potential produced by an innovative non-mammalian expression system. *Biotechnol. J.* 2, 700–708. doi: 10.1002/biot.200600255

Sharma, P., Jha, A. B., Dubey, R. S., and Pessarakli, M. (2012). Reactive oxygen species, oxidative damage, and antioxidative defense mechanism in plants under stressful conditions. *J. Bot.* 2012, 217037. doi: 10.1155/2012/217037

Shewry, P. R., and Halford, N. G. (2002). Cereal seed storage proteins: structures, properties and role in grain utilization. *J. Exp. Bot.* 53, 947–958. doi: 10.1093/jexbot/53.370.947

Shewry, P. R., Napier, J. A., and Tatham, A. S. (1995). Seed storage proteins: structures and biosynthesis. *Plant Cell* 7, 945–956. doi: 10.1105/tpc.7.7.945

Shigemitsu, T., Masumura, T., Morita, S., and Satoh, S. (2013). Accumulation of rice prolamin-GFP fusion proteins induces ER-derived protein bodies in transgenic rice calli. *Plant Cell Rep.* 32, 389–399. doi: 10.1007/s00299-012-1372-3

Shy, G., Ehler, L., Herman, E., and Galili, G. (2001). Expression patterns of genes encoding endomembrane proteins support a reduced function of the Golgi in wheat endosperm during the onset of storage protein deposition. *J. Exp. Bot.* 52, 2387–2388. doi: 10.1093/jexbot/52.365.2387

Sreenivasulu, N., Radchuk, V., Strickert, M., Miersch, O., Weschke, W., and Wobus, U. (2006). Gene expression patterns reveal tissue-specific signaling networks controlling programmed cell death and ABA- regulated maturation in developing barley seeds. *Plant J.* 47, 310–327. doi: 10.1111/j.1365-313X.2006.02789.x

Sreenivasulu, N., Usadel, B., Winter, A., Radchuk, V., Scholz, U., Stein, N., et al. (2008). Barley grain maturation and germination: metabolic pathway and regulatory network commonalities and differences highlighted by new MapMan/PageMan profiling tools. *Plant Physiol.* 146, 1738–1758. doi: 10.1104/pp.107.111781

Stacy, R. A. P., Nordeng, T. W., Culianez-Macia, F. A., and Aalen, R. B. (1999). The dormancy-related peroxiredoxin anti-oxidant, PER1, is localized to the nucleus of barley embryo and aleurone cells. *Plant J.* 19, 1–8. doi: 10.1046/j.1365-313X.1999.00488.x

Stoger, E., Fischer, R., Moloney, M., and Ma, J. K. (2014). Plant molecular pharming for the treatment of chronic and infectious diseases. *Annu. Rev. Plant Biol.* 65, 743–768. doi: 10.1146/annurev-arplant-050213-035850

Stoger, E., Ma, J. K., Fischer, R., and Christou, P. (2005a). Sowing the seeds of success: pharmaceutical proteins from plants. *Curr. Opin. Biotechnol.* 16, 167–173. doi: 10.1016/j.copbio.2005.01.005

Stoger, E., Sack, M., Nicholson, L., Fischer, R., and Christou, P. (2005b). Recent progress in plantibody technology. *Curr. Pharm. Des.* 11, 2439–2457. doi: 10.2174/1381612054367535

Sugita, K., Endo-Kasahara, S., Tada, Y., Lijun, Y., Yasuda, H., Hayashi, Y., et al. (2005). Genetically modified rice seeds accumulating GLP-1 analogue stimulate insulin secretion from a mouse pancreatic β-cell line. *FEBS Lett.* 579, 1085–1088. doi: 10.1016/j.febslet.2004.12.082

Suzuki, T., Yano, K., Sugimoto, S., Kitajima, K., Lennarz, W. J., Inoue, S., et al. (2002). Endo-β-*N*-acetylglucosaminidase, an enzyme involved in processing of free oligosaccharides in the cytosol. *Proc. Natl. Acad. Sci. U.S.A.* 99, 9691–9696. doi: 10.1073/pnas.152333599

Swanson, S. J., Bethke, P. C., and Jones, R. L. (1998). Barley aleurone cells contain two types of vacuoles: characterization of lytic organelles by use of fluorescent probes. *Plant Cell* 10, 685–698. doi: 10.1105/tpc.10.5.685

Swanson, S. J., and Jones, R. L. (1996). Gibberellic acid induces vacuolar acidification in barley aleurone. *Plant Cell* 8, 2211–2221. doi: 10.1105/tpc.8.12.2211

Takagi, H., Hiroi, T., Hirose, S., Yang, L., and Takaiwa, F. (2010). Rice seed ER-derived protein body as an efficient delivery vehicle for oral tolerogenic peptides. *Peptides* 31, 1421–1425. doi: 10.1016/j.peptides.2010.04.032

Takahashi, H., Saito, Y., Kitagawa, T., Morita, S., Masumura, T., and Tanaka, K. (2005). A novel vesicle derived directly from endoplasmic reticulum is involved in the transport of vacuolar storage proteins in rice endosperm. *Plant Cell Physiol.* 46, 245–249. doi: 10.1093/pcp/pci019

Takaiwa, F., Hirose, S., Takagi, H., Yang, L., and Wakasa, Y. (2009). Deposition of a recombinant peptide in ER-derived protein bodies by retention with cysteine-rich prolamins in transgenic rice seed. *Planta* 229, 1147–1158. doi: 10.1007/s00425-009-0905-7

Tanaka, Y., Hayashida, S., and Hongo, M. (1975). The relationship of the feces protein particles to rice protein bodies. *Agric. Biol. Chem.* 39, 515–518. doi: 10.1271/bbb1961.39.515

Thiel, J., Weier, D., and Weschke, W. (2011). Laser-capture microdissection of developing barley seeds and cDNA array analysis of selected tissues. *Methods Mol. Biol.* 755, 461–475. doi: 10.1007/978-1-61779-163-5_39

Tian, L., Dai, L. L., Yin, Z. J., Fukuda, M., Kumamaru, T., Dong, X. B., et al. (2013). Small GTPase Sar1 is crucial for proglutelin and α-globulin export from the endoplasmic reticulum in rice endosperm. *J. Exp. Bot.* 64, 2831–2845. doi: 10.1093/jxb/ert128

Torres, E., Gonzalez-Melendi, P., Stoger, E., Shaw, P., Twyman, R. M., Nicholson, L., et al. (2001). Native and artificial reticuloplasmins co-accumulate in distinct domains of the endoplasmic reticulum and in post-endoplasmic reticulum compartments. *Plant Physiol.* 127, 1212–1223. doi: 10.1104/Pp.010260

Tosi, P., Parker, M., Gritsch, C. S., Carzaniga, R., Martin, B., and Shewry, P. R. (2009). Trafficking of storage proteins in developing grain of wheat. *J. Exp. Bot.* 60, 979–991. doi: 10.1093/jxb/ern346

Toyooka, K., Okamoto, T., and Minamikawa, T. (2000). Mass transport of proform of a KDEL-tailed cysteine proteinase (SH-EP) to protein storage vacuoles by endoplasmic reticulum-derived vesicle is involved in protein mobilization in germinating seeds. *J. Cell Biol.* 148, 453–464. doi: 10.1083/jcb.148.3.453

Umana, P., Jean-Mairet, J., Moudry, R., Amstutz, H., and Bailey, J. E. (1999). Engineered glycoforms of an antineuroblastoma IgG1 with optimized antibody-dependent cellular cytotoxic activity. *Nat. Biotechnol.* 17, 176–180. doi: 10.1038/6179

Van Droogenbroeck, B., Cao, J., Stadlmann, J., Altmann, F., Colanesi, S., Hillmer, S., et al. (2007). Aberrant localization and underglycosylation of highly accumulating single-chain Fv-Fc antibodies in transgenic *Arabidopsis* seeds. *Proc. Natl. Acad. Sci. U.S.A.* 104, 1430–1435. doi: 10.1073/pnas.0609997104

Viotti, C. (2014). ER and vacuoles: never been closer. *Front. Plant Sci.* 5:20. doi: 10.3389/fpls.2014.00020

Viotti, C., Kruger, F., Krebs, M., Neubert, C., Fink, F., Lupanga, U., et al. (2013). The endoplasmic reticulum is the main membrane source for biogenesis of the lytic vacuole in *Arabidopsis*. *Plant Cell* 25, 3434–3449. doi: 10.1105/tpc.113.114827

Vitale, A., and Hinz, G. (2005). Sorting of proteins to storage vacuoles: how many mechanisms? *Trends Plant Sci.* 10, 316–323. doi: 10.1016/j.tplants.2005.05.001

Vuylsteker, C., Cuvellier, G., Berger, S., Faugeron, C., and Karamanos, Y. (2000). Evidence of two enzymes performing the de-*N*-glycosylation of proteins in barley: expression during germination, localization within the grain and set-up during grain formation. *J. Exp. Bot.* 51, 839–845. doi: 10.1093/jexbot/51.346.839

Wakasa, Y., Hayashi, S., and Takaiwa, F. (2012). Expression of OsBiP4 and OsBiP5 is highly correlated with the endoplasmic reticulum stress response in rice. *Planta* 236, 1519–1527. doi: 10.1007/s00425-012-1714-y

Wakasa, Y., and Takaiwa, F. (2013). The use of rice seeds to produce human pharmaceuticals for oral therapy. *Biotechnol. J.* 8, 1133–1143. doi: 10.1002/biot.201300065

Walley, J. W., Shen, Z., Sartor, R., Wu, K. J., Osborn, J., Smith, L. G., et al. (2013). Reconstruction of protein networks from an atlas of maize seed proteotypes. *Proc. Natl. Acad. Sci. U.S.A.* 110, E4808–E4817. doi: 10.1073/pnas.1319113110

Wang, S., Takahashi, H., Kajiura, H., Kawakatsu, T., Fujiyama, K., and Takaiwa, F. (2013). Transgenic rice seeds accumulating recombinant hypoallergenic birch pollen allergen Bet v 1 generate giant protein bodies. *Plant Cell Physiol.* 54, 917–933. doi: 10.1093/pcp/pct043

Wang, Y., Ren, Y., Liu, X., Jiang, L., Chen, L., Han, X., et al. (2010). OsRab5a regulates endomembrane organization and storage protein trafficking in rice endosperm cells. *Plant J.* 64, 812–824. doi: 10.1111/j.1365-313X.2010.04370.x

Woo, Y. M., Hu, D. W., Larkins, B. A., and Jung, R. (2001). Genomics analysis of genes expressed in maize endosperm identifies novel seed proteins and clarifies patterns of zein gene expression. *Plant Cell* 13, 2297–2317. doi: 10.1105/tpc.13.10.2297

Yang, L., Hirose, S., Suzuki, K., Hiroi, T., and Takaiwa, F. (2012a). Expression of hypoallergenic Der f 2 derivatives with altered intramolecular disulphide bonds induces the formation of novel ER-derived protein bodies in transgenic rice seeds. *J. Exp. Bot.* 63, 2947–2959. doi: 10.1093/jxb/ers006

Yang, L., Hirose, S., Takahashi, H., Kawakatsu, T., and Takaiwa, F. (2012b). Recombinant protein yield in rice seed is enhanced by specific suppression of endogenous seed proteins at the same deposit site. *Plant Biotechnol. J.* 10, 1035–1045. doi: 10.1111/j.1467-7652.2012.00731.x

Yang, P., Li, X., Wang, X., Chen, H., Chen, F., and Shen, S. (2007). Proteomic analysis of rice (*Oryza sativa*) seeds during germination. *Proteomics* 7, 3358–3368. doi: 10.1002/pmic.200700207

Young, T. E., and Gallie, D. R. (2000). Programmed cell death during endosperm development. *Plant Mol. Biol.* 44, 283–301. doi: 10.1023/A:1026588408152

Zi, J., Zhang, J., Wang, Q., Zhou, B., Zhong, J., Zhang, C., et al. (2013). Stress responsive proteins are actively regulated during rice (*Oryza sativa*) embryogenesis as indicated by quantitative proteomics analysis. *PLoS ONE* 8:e74229. doi: 10.1371/journal.pone.0074229

Conflict of Interest Statement: The authors declare that the research was conducted in the absence of any commercial or financial relationships that could be construed as a potential conflict of interest.

7

Imprinting in plants as a mechanism to generate seed phenotypic diversity

Fang Bai and A. M. Settles*

Horticultural Sciences Department and Plant Molecular and Cellular Biology Program, University of Florida, Gainesville, FL, USA

Edited by:
Paolo A. Sabelli, University of Arizona, USA

Reviewed by:
Claudia Köhler, Swedish Agricultural University, Sweden
Matthew Mount Stuart Evans, Carnegie Institution for Science, USA

****Correspondence:***
A. M. Settles, Horticultural Sciences Department and Plant Molecular and Cellular Biology Program, University of Florida, P. O. Box 110690, Gainesville, FL 32611-0690, USA
e-mail: settles@ufl.edu

Normal plant development requires epigenetic regulation to enforce changes in developmental fate. Genomic imprinting is a type of epigenetic regulation in which identical alleles of genes are expressed in a parent-of-origin dependent manner. Deep sequencing of transcriptomes has identified hundreds of imprinted genes with scarce evidence for the developmental importance of individual imprinted loci. Imprinting is regulated through global DNA demethylation in the central cell prior to fertilization and directed repression of individual loci with the Polycomb Repressive Complex 2 (PRC2). There is significant evidence for transposable elements and repeat sequences near genes acting as cis-elements to determine imprinting status of a gene, implying that imprinted gene expression patterns may evolve randomly and at high frequency. Detailed genetic analysis of a few imprinted loci suggests an imprinted pattern of gene expression is often dispensable for seed development. Few genes show conserved imprinted expression within or between plant species. These data are not fully explained by current models for the evolution of imprinting in plant seeds. We suggest that imprinting may have evolved to provide a mechanism for rapid neofunctionalization of genes during seed development to increase phenotypic diversity of seeds.

Keywords: epigenetics, DNA methylation, histone modification, imprinting, genomics, seed development, maize endosperm, *Arabidopsis* endosperm

OVERVIEW OF ANGIOSPERM SEED DEVELOPMENT

In this review, we focus on the developmental role of epigenetic regulation, specifically genomic imprinting, in maize and *Arabidopsis* seeds. Imprinting, or parent-of-origin specific gene expression, has evolved convergently in mammals and angiosperms (Pires and Grossniklaus, 2014). Imprinted gene expression in angiosperms is found in developing seeds. Angiosperm seeds initiate with double fertilization of the megagametophyte (Peris et al., 2010). The pollen tube delivers two haploid sperm cells to the embryo sac. One sperm cell fuses with the haploid egg to generate a diploid embryo, and the other sperm cell fuses with the diploid central cell to form the triploid endosperm. The resulting embryo and endosperm are genetically identical except for their ploidy level with the endosperm having two maternal doses of the genome and one paternal dose. Although the endosperm and embryo have essentially the same genotype, they have markedly different developmental programs (**Figure 1**; Kiesselbach, 1949; Brown et al., 1999; Chandler et al., 2008; Peris et al., 2010).

The endosperm starts development by dividing nuclei without completing cytokinesis (reviewed in Olsen, 2004; Sabelli and Larkins, 2009b). This syncytial development transitions to cellularization in which the nuclei become enclosed in cell walls (**Figure 1**). As the endosperm cellularizes, the cells begin to take on differentiated fates with internal endosperm cells accumulating nutrient storage reserves (Kiesselbach, 1949; Brown et al., 1994, 1996, 1999; Stangeland et al., 2003). In many eudicots, like *Arabidopsis*, the embryo consumes the endosperm reserves as it develops resulting in most of the endosperm degenerating by seed maturity. By contrast, the internal storage cells in the maize endosperm persist through seed development and the storage reserves are used during seedling growth (Kiesselbach, 1949). Epidermal endosperm cells take on different fates from the internal storage cell types. In *Arabidopsis*, there are distinct endosperm cell morphologies at the micropyllar and chalazal ends of the embryo sac (Brown et al., 1999). In maize, epidermal endosperm cells differentiate into basal transfer cells, embryo surrounding region, and aleurone (Kiesselbach, 1949). All maize endosperm cells, except the aleurone, undergo programmed cell death prior to seed maturation (**Figure 1**; Young et al., 1997; Young and Gallie, 2000).

Embryo development starts with asymmetric cell division of the zygote to form an apical-basal axis (Chandler et al., 2008; Peris et al., 2010). Basal cells divide to develop the suspensor and contribute to the root meristem. Apical cells initially develop a globular embryo, which transitions to form the shoot and root apical meristems along with cotyledons in *Arabidopsis* or a scutellum, coleoptile, and embryonic leaves in maize (i.e., transition stage). The genetic programs controlling meristem specification and lateral organ initiation have been extensively reviewed (Chandler et al., 2008; De Smet et al., 2010; Wendrich and Weijers, 2013).

Imprinted genes primarily show parent-of-origin expression patterns in the endosperm although there are imprinted genes also in the developing embryo (Jahnke and Scholten, 2009; Raissig et al., 2013). Endosperm growth has a significant impact

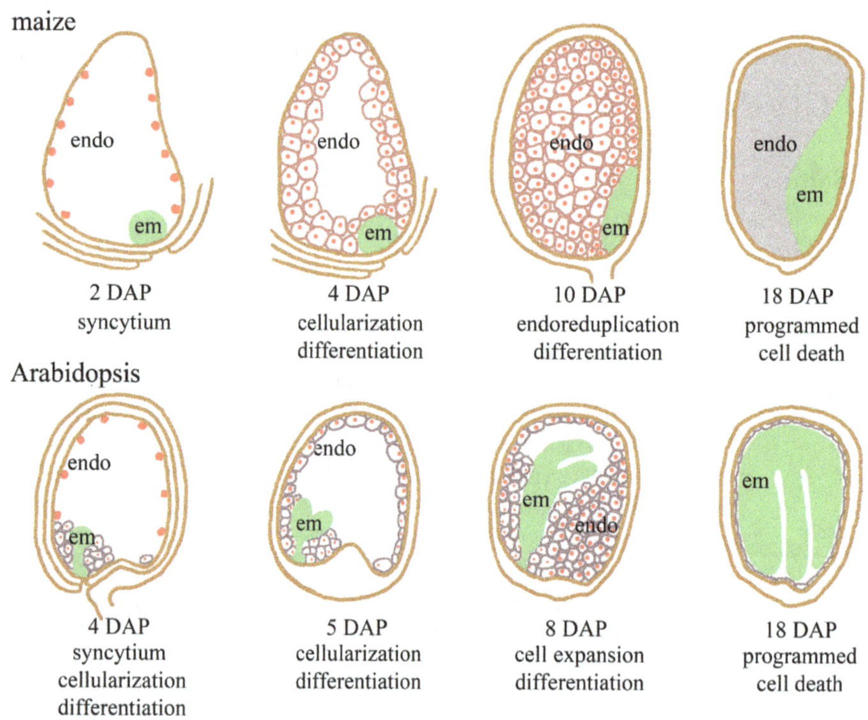

FIGURE 1 | Comparison of seed development in maize and *Arabidopsis thaliana*. The endosperm proliferates initially as a multi-nucleate syncytium, while the globular embryo develops. Endosperm cellularization precedes embryo transition. Endosperm cells expand and accumulate storage molecules once cellularization is complete. In maize, the endosperm is persistent and undergoes programmed cell death starting around 18 days after pollination (DAP). The *Arabidopsis* embryo consumes most of the endosperm prior to seed maturation. Endosperm (endo) nuclei are indicated in red. The embryo (em) is in green.

on final seed size, and imprinting has been hypothesized to regulate seed size (Arnaud and Feil, 2006; Xiao et al., 2006; Li and Berger, 2012; Fatihi et al., 2013). However, there is significant data arguing that the endosperm has developmental functions beyond providing nutrition for the developing embryo. Embryo transition occurs soon after endosperm cell differentiation, and recent evidence indicates differentiated endosperm is important for embryo developmental programs. For example, the embryo surrounding endosperm in *Arabidopsis* secretes the ESF1 signaling peptide to promote normal basal embryo development (Costa et al., 2014). Failure to differentiate the embryo surrounding endosperm in maize causes an embryo developmental block at the transition stage suggesting a similar function for this cell type in maize (Fouquet et al., 2011). Later in *Arabidopsis* seed development, the ZHOUPI basic-helix-loop-helix transcription factor is expressed exclusively in the embryo surrounding region and activates a signaling pathway required for normal epidermal differentiation in the embryo (Yang et al., 2008; Xing et al., 2013). These data show that the endosperm plays an active role in promoting embryo development and argue that epigenetic regulation of endosperm gene expression could have consequences for seed size as well as embryo developmental programs.

WHAT IS IMPRINTING?

Genomic imprinting in plants is an epigenetic phenomenon by which genetically identical alleles are differentially expressed in a parent-of-origin dependent manner. Imprinted gene expression primarily occurs in the endosperm and there is strong data for imprinted genes controlling early endosperm cell divisions as well as regulating the transfer of nutrients to the seed (Gutierrez-Marcos et al., 2004; Day et al., 2008; Sabelli and Larkins, 2009a; Tiwari et al., 2010; Shirzadi et al., 2011; Costa et al., 2012). Imprinting is an exception from Mendel's Laws on the expression and inheritance of the two parental alleles in which dominant alleles express phenotypes over recessive alleles irrespective of the parental source of the allele. Instead, imprinted genes will express either the maternal or paternal allele even though the primary sequences of these alleles may be identical.

It is easiest to understand imprinted inheritance through an example. The *A1* locus of maize encodes a structural gene for anthocyanin biosynthesis (O'Reilly et al., 1985), while the *R* locus encodes a transcription factor that induces anthocyanin biosynthesis (Ludwig et al., 1989; Perrot and Cone, 1989). The *A1* locus shows Mendelian inheritance, while certain haplotypes of the *R* locus are imprinted (Kermicle, 1969). Indeed, *R* was the first imprinted locus described in plants. The R^r allele shows altered expression when R^r is inherited from pollen. The expression pattern of paternally inherited R^r can be seen by contrasting self-pollinations of heterozygous individuals for *A1/a1* or R^r/r (**Figure 2**). When a plant heterozygous for *A1/a1* is self-pollinated, the seeds segregate in a 3:1 ratio for full color to yellow kernels (**Figure 2**). Self-pollination of R^r/r yields three kernel color types

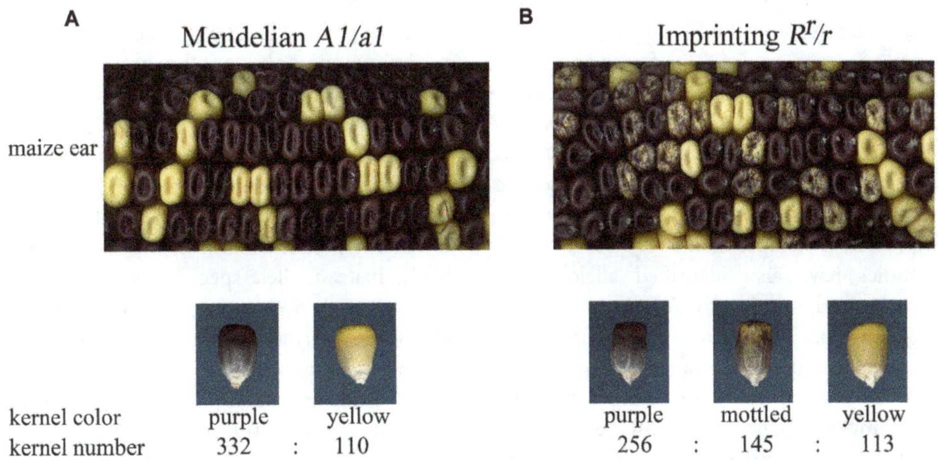

FIGURE 2 | Comparison of Mendelian genetic inheritance and imprinted inheritance. (A) The *a1* locus shows Mendelian inheritance in self-pollinations of *A1/a1* individuals. Full purple color kernels are dominant over yellow kernels and the progeny segregate in a 3:1 ratio. **(B)** The R^r allele shows imprinted inheritance. Self-pollination of R^r/r yields three kernel color types in a 2:1:1 ratio of purple to mottled to yellow kernels. Purple kernels inherited the R^r allele from the megagametophyte. Mottled kernels are heterozygous individuals that inherited the R^r allele from the pollen. Kernel counts are given for the ears shown in the upper panels.

in a 2:1:1 ratio of purple to yellow to mottled purple kernels. These mottled purple kernels have an endosperm genotype of $r\ r/R^r$ where the dominant allele inherited from the male is repressed in a stochastic pattern in the seed. The same paternal R^r allele is not affected in the embryo and plants from mottled kernels will yield full color kernels if crossed as female to an *r/r* plant and mottled kernels if crossed as a male to the *r/r* genotype.

One interpretation of the mottled $r\ r/R^r$ kernel phenotype is that it is due to an insufficient dosage as a consequence of the endosperm fusing a diploid maternal central cell with a haploid paternal sperm cell. However, introducing multiple copies of R^r with translocation and trisomic stocks does not alter the anthocyanin phenotypes. When more than two doses are inherited maternally, the kernel is always full color, and when multiple R^r alleles are inherited paternally, the mottled phenotype always results (Kermicle, 1970). These and more recent data indicate that imprinting is an independent phenomenon from dosage effects.

The *r* locus is an example of maternal bias of gene expression. Imprinted genes can show bias for either the maternal or paternal allele and consequently are classified into maternally expressed genes (MEGs) and paternally expressed genes (PEGs). MEGs and PEGs can be identified molecularly by examining allele-specific expression in reciprocal crosses. Single nucleotide polymorphisms and small insertion-deletions within transcripts from diverse parents are used to identify the expression level of both the maternal and paternal allele. By carrying out the same gene expression analysis on reciprocal crosses, it is possible to identify genes that express only the maternal or only the paternal allele irrespective of the polymorphisms found within the alleles of the genes. A variety of molecular strategies have been employed to identify individual imprinted genes such as *Maternally expressed gene1* (*Meg1*) and *Maternally expressed in embryo1* (*Mee1*) in maize (Gutierrez-Marcos et al., 2004; Jahnke and Scholten, 2009).

RNA-seq transcriptomics allows global analysis of imprinted gene expression at a much larger scale. Maize is particularly well suited for allele identification in RNA-seq experiments, because maize inbred lines show high levels of polymorphism allowing for a large number of genes to be assayed for imprinting in a single experiment (Chia et al., 2012; Jiao et al., 2012). Initial experiments in maize examined reciprocal crosses between the reference genome inbred line, B73, and Mo17 (Waters et al., 2011; Zhang et al., 2011). These studies identified hundreds of MEGs and PEGs with relatively little overlap between them. By examining additional time-points during seed development and additional inbred crosses more than 500 genes show statistically significant bias for gene expression based on the parent of origin (Waters et al., 2013; Zhang et al., 2014). Many imprinted genes only show parent-of-origin bias transiently. For example, *Meg1* is maternally expressed early during seed development and is biallelic, expressed from both maternal and paternal alleles, by mid-seed development (Gutierrez-Marcos et al., 2004). Attempts to apply transcriptomics to early stages of developing maize seeds did not effectively isolate endosperm or embryo tissue from maternal tissue, so most identified maize MEGs and PEGs show imprinted expression patterns after embryo transition (Xin et al., 2013). The overlap of imprinted genes between all data sets is low. For example, Waters et al. (2013) found that only 5–10% of the imprinted genes in a survey of four inbred lines showed imprinting in all genotype combinations. These results have raised questions about whether sequencing depth, statistical approaches, allele-specific effects, or environmental factors have significant impact on the genes detected as imprinted.

Similar transcriptomic approaches have been applied to identify imprinted genes in *Arabidopsis*. Endosperm and embryo transcripts from reciprocal crosses between *Col-0* and *Ler* accessions identified over 200 imprinted genes (Gehring et al., 2011; Hsieh et al., 2011). Expanding these studies to additional accessions found that, like in maize, only a small number of genes consistently show imprinting in all accessions (Wolff et al., 2011; Pignatta et al., 2014). For example, Pignatta et al. (2014) found about 10% of MEGs and 5% of PEGs are shared between the three accessions they surveyed. It seems surprising that both maize and *Arabidopsis* transcriptome surveys have found only a few conserved imprinted genes within each species. Considering both the maize and *Arabidopsis* observations, the data suggest that relatively few loci are stably selected for imprinted gene expression.

Transcriptomic studies have also identified allele-specific imprinting in both maize and *Arabidopsis* (Waters et al., 2013; Pignatta et al., 2014). Allele-specific imprinted genes have MEG or PEG expression for a single allele from a single accession or inbred line, much like the R^r allele of maize. Using kernel phenotypes, allele-specific imprinting has been observed in other maize loci including the *dzr1*^{Mo17} and *B-Boliva* alleles that control zein and anthocyanin accumulation, respectively (Chaudhuri and Messing, 1994; Selinger and Chandler, 2001). These older examples indicate that allele-specific imprinting can have significant effects on kernel phenotypes. However, phenotypes have not been associated with the more recently discovered allele-specific imprints.

With hundreds of imprinted genes identified, annotation of these genes suggests that imprinted loci do function in processes proximate to developmental programs. Imprinted genes include proteins involved in chromatin modification, transcription factors, hormone signaling, ubiquitin-targeted protein degradation, and RNA processing (Gehring et al., 2011; Hsieh et al., 2011; Wolff et al., 2011; Zhang et al., 2011; Xin et al., 2013; Pignatta et al., 2014). MEGs show some enrichment for transcription factors, such as MYB family genes (Hsieh et al., 2011; Pignatta et al., 2014), while PEGs show enrichment for chromatin and transcriptional modifiers (Waters et al., 2013; Pignatta et al., 2014). However, only two genes show conserved imprinted expression between *Arabidopsis*, rice, and maize (Waters et al., 2013). A loss of function allele of one of these PEGs, *ZmYuc1*, is tightly linked to the recessive *defective endosperm18* locus of maize, suggesting that residual maternal expression is sufficient to confer normal seed development (Bernardi et al., 2012). Both allele-specific imprinting and the low conservation of imprinted gene expression across angiosperms suggest that deeply conserved developmental circuits have not been selected for this type of epigenetic regulation in angiosperms. Based on these and additional arguments below, we suggest that imprinting is primarily a form of regulation to enable rapid diversifying selection of seed phenotypes.

MOLECULAR MECHANISMS OF IMPRINTING

Altering the expression state of an allele depending upon the parent-of-origin requires epigenetic modification of the alleles inherited by the male and female gametes. The mechanisms by which MEGs and PEGs are identified and programmed have been extensively reviewed (Köhler et al., 2012; Gehring, 2013; Zhang et al., 2013). As a brief overview, both histone modification and DNA methylation have essential roles in setting imprinted patterns of gene expression. The *Arabidopsis* model for establishing contrasting epigenetic states in the male and female gametes starts with differential demethylation of the genome. The DNA glycosylase gene, *DEMETER (DME)*, is expressed in the central cell of the megagametophyte but not the sperm cells of the pollen (Choi et al., 2002; Schoft et al., 2011). *DME* activity removes 5-methylcytosine predominantly from transposable element and repeat sequences leading to most repetitive sequences having reduced methylation in the developing endosperm (Gehring et al., 2009; Hsieh et al., 2009). Surprisingly, maize does not show these global patterns of DNA hypomethylation in the endosperm (Zhang et al., 2011, 2014). Instead, allele-specific bisulfite sequencing of endosperm DNA revealed a pattern of DNA hypomethylation at maternal alleles with corresponding hypermethylation at paternal alleles for specific sites within the genome (Zhang et al., 2014). These maize results are consistent with DNA demethylation specifically occurring in the central cell.

The differential loss of DNA methylation sets-up contrasting chromatin marks in *Arabidopsis* repeat sequences near the paternal and maternal alleles. Methylation marks can then be interpreted by the genome with a variety of molecular mechanisms. For example, methylation of the paternal allele can lead to a transcriptionally silent state, while the demethylated maternal allele would become transcriptionally active (Kinoshita et al., 2004; Jullien et al., 2006a; Hermon et al., 2007; Tiwari et al., 2008). There are also a few examples where RNA-directed DNA methylation (RdDM) is critical in the male parent to ensure silencing of the paternal allele at MEG loci, suggesting that small RNAs can have a significant role in setting MEG expression patterns (Bratzel et al., 2012; Vu et al., 2013). Although these models can explain MEG patterns of expression, PEGs can also be hypermethylated at the paternal allele and hypomethylated at the maternal allele (Gehring et al., 2009; Hsieh et al., 2009; Zhang et al., 2014). This maternal hypomethylation is essential for silencing of the maternal allele for many PEGs (Hsieh et al., 2011; Wolff et al., 2011).

How can the same epigenetic mark of reduced DNA methylation in the maternal allele result in opposite MEG and PEG expression patterns? Trimethylation of lysine 27 on histone H3 (H3K27me3) is another chromatin mark that is required for imprinted gene expression (Schuettengruber and Cavalli, 2009; Köhler et al., 2012). H3K27me3 marks are catalyzed by the Polycomb Repressive Complex2 (PRC2). In the *Arabidopsis* endosperm, the PRC2 complex is referred to as the FERTILIZATION INDEPENDENT SEED (FIS) complex and is composed of four core subunits: the MEDEA (MEA) Enhancer of zeste homolog, the FERTILIZATION INDEPENDENT SEED2 (FIS2) Suppressor of zeste homolog, the FERTILIZATION INDEPENDENT ENDOSPERM (FIE) Extra sex combs homolog, and the MULTICOPY SUPPRESSOR OF IRA1 (MSI1), which is a WD-40 repeat protein that is homologus to Drosophila p55 (Grossniklaus et al., 1998; Kiyosue et al., 1999; Luo et al., 1999; Köhler et al., 2003). The H3K27me3 post-translational modification is a repressive chromatin mark, and FIS-PRC2 is known to be required to repress paternal alleles of MEGs as well as maternal alleles of PEGs (Köhler et al., 2003, 2005; Baroux et al., 2006; Makarevich et al., 2006; Fitz Gerald et al., 2009; Weinhofer et al., 2010).

It is not inherently obvious how PRC2 would differentially target hypo- or hypermethylated DNA. PRC2 is recruited to cis-elements at repressed loci. These PRC2 recruitment elements have been identified in multiple organisms and can include repeat sequences (Kinoshita et al., 2007; Makarevich et al., 2008), small segments of CG-rich sequence (Jermann et al., 2014), or transcription factor binding sites (Berger et al., 2011; Liu et al., 2011; Lodha et al., 2013). In addition, non-coding RNA has been shown to interact with PRC2 and target it to specific loci in plants (Heo and Sung, 2011). DNA methylation interferes with PRC2 function and prevents H3K27me3 modification (Weinhofer et al., 2010; Deleris et al., 2012; Jermann et al., 2014). Thus, hypermethylation of paternal alleles can interfere with PRC2 recruitment sites allowing expression of the paternal allele, while PRC2 activity at the maternal, hypomethylated allele would result in transcriptional silencing to give a PEG pattern of expression. Global analysis of H3K27me3 sites in maize supports this model (Makarevitch et al., 2013). Indeed, Zhang et al. (2014) found that PEGs showed enrichment for maternal H3K27me3 marks concomitant with hypomethylation at the maternal allele and hypermethylation at the paternal allele.

PRC2 is also required to repress the paternal allele of some MEGs (Baroux et al., 2006; Gehring et al., 2006; Jullien et al., 2006b). However, there is no molecular mechanism proposed for how PRC2 would preferentially target the hypermethylated, paternal allele at a MEG locus. Maternal specific expression is also observed for the *Arabidopsis ZIX* locus, but the MEG pattern of expression does not dependent upon DME or FIS-PRC2 (Ngo et al., 2012). Moreover, imprinted gene expression is documented within the embryo of both maize and *Arabidopsis* (Jahnke and Scholten, 2009; Raissig et al., 2013). The DME DNA glycosylase is not expressed significantly in the *Arabidopsis* egg cell (Choi et al., 2002), and there is no evidence for global DNA demethylation in the embryo (Hsieh et al., 2009). The FIS-PRC2 complex is required for some embryo imprinted gene expression suggesting that H3K27me3 does have a functional role in setting-up MEG and PEG expression in the embryo (Raissig et al., 2013). These observations indicate that we are far from completely understanding the molecular mechanisms guiding imprinted expression patterns during seed development.

IS IMPRINTING NECESSARY FOR SEED DEVELOPMENT?

Genetic analysis of imprinted genes suggests a similar spectrum of developmental functions as for biallelic-expressed genes. As mentioned earlier, imprinting in maize can affect non-essential genes regulating anthocyanin biosynthesis or storage protein accumulation. There are also numerous imprinted genes that have been shown to have critical roles in seed development. Superficially, it is simple to conclude that imprinted expression patterns are therefore critical to seed development. However, we argue that most of these examples fail to provide conclusive evidence that the imprinted pattern is indispensable as opposed to a minimum expression level of the gene being critical.

The *MEA, FIE,* and *FIS2* genes encode subunits of PRC2 and are MEGs in *Arabidopsis* (Luo et al., 2000). Loss-of-function mutations in these genes have profound effects on seed development with 50% seed abortion, delays in endosperm and embryo development, and increased cell proliferation in the developmentally delayed endosperm and embryo (Ohad et al., 1996; Chaudhury et al., 1997; Grossniklaus et al., 1998; Kiyosue et al., 1999). Moreover, these mutants can begin central cell divisions even when the ovule is not fertilized. The FIS-PRC2 phenotypes have been interpreted as imprinted expression of PRC2 being a key repressor of seed growth. However, very similar phenotypes are observed in mutants of the non-imprinted subunit of the FIS-PRC2 complex, *MSI1*, suggesting that imprinted gene expression is not directly responsible for the seed phenotypes (Köhler et al., 2003; Guitton and Berger, 2005; Leroy et al., 2007). This conclusion is further supported by the maize *MEA/Enhancer of zeste* (*Mez1*) gene, which is an endosperm MEG (Haun et al., 2007). A transposon insertion in the promoter region of the locus causes biallelic expression of *Mez1* but no change in seed phenotype (Haun et al., 2009). These data show that FIS-PRC2 function needs to be expressed at sufficient levels in both the female gametophyte and developing seed. However, direct evidence is lacking to support the hypothesis that imprinted expression of the PRC2 subunits is required.

The *PHERES1* (*PHE1*) gene is consistently up-regulated in *Arabidopsis mea, fie,* and *fis2* mutants (Köhler et al., 2003). The *PHE1* locus was the first PEG to be identified and encodes the AGAMOUS-LIKE37 (AGL37) MADS-domain protein that is a predicted transcription factor (Köhler et al., 2003, 2005). Knocking-down expression of *PHE1* in a *mea* mutant can partially rescue *mea* defective seed phenotypes suggesting that part of the FIS-PRC2 mutant phenotypes are due to increased *PHE1* expression (Köhler et al., 2003). However, insertion mutants in the 3′ regulatory region of *PHE1* can cause a loss of imprinting, switching to a biallelic pattern, with no effect on seed phenotype reported (Makarevich et al., 2008). *PHE1* is one of several *AGL* genes, including *AGL28, AGL36, AGL40, AGL62,* and *AGL90,* which are up-regulated when endosperm cellularization is delayed either by PRC2 mutants or genome dosage imbalances (Kradolfer et al., 2013a). Although *AGL28* and *AGL36* are MEGs (Shirzadi et al., 2011; Wolff et al., 2011), the other *AGL* genes are expressed from both parental alleles. Mutations in *AGL62* cause recessive seed defects illustrating that FIS-PRC2 complex influences biallelic genes as well as imprinted genes (Kang et al., 2008). Total expression levels of the *AGL* co-expression network correlates well with the timing of endosperm cellularization and embryo development (Walia et al., 2009; Tiwari et al., 2010; Kradolfer et al., 2013a). The divergent mechanisms of epigenetic control for these *AGL* genes and the lack of a requirement for paternal expression of *PHE1* suggest that imprinting *per se* is not likely the primary regulator of this developmental node.

An additional *Arabidopsis* PEG, *ADMETOS* (*ADM*), has been implicated in regulating the *AGL* gene node (Kradolfer et al., 2013b). *ADM* encodes a recently evolved J-domain protein that is only found in a few genera of the *Brassicaceae.* Consistent with its PEG expression, the *adm* locus was identified as a paternal-specific suppressor of seed abortion due to paternal genome excess (Kradolfer et al., 2013b). When mutated, *adm* reduces the over-expression of *PHE1* and other *AGL* genes toward normal both in interploidy crosses and in *mea* mutants. Although *ADM* is a PEG in multiple *Arabidopsis* accessions (Hsieh et al., 2011; Wolff et al., 2011), natural variation reducing *ADM* expression level in the L*er* accession is correlated with improved seed development and viability in paternal genome excess crosses (Kradolfer et al., 2013b). When *adm* is mutant in both maternal and paternal gametes, *adm* more effectively reduces *AGL* expression as well as more effectively suppresses seed abortion due to either paternal genome excess or *mea*, suggesting the maternal allele expresses at a developmentally significant level (Kradolfer et al., 2013b). Interestingly, homozygous *adm/adm* plants, overexpression of *ADM*, and biallelic expression of *ADM* have no seed phenotype in diploid crosses, suggesting that imprinting of this gene is not necessary for normal diploid seed development. The wild-type function of *ADM* is primarily to block interploidy and interspecific hybridizations.

The *Arabidopsis FORMIN HOMOLOGUES5* (*AtFH5*) gene is a MEG in which the paternal allele is repressed by PRC2 (Fitz Gerald et al., 2009). ATFH5 is an actin nucleator and is critical for cell plate formation and endosperm cellularization (Ingouff et al., 2005). Ectopic expression of the paternal allele of *AtFH5* does not impact *mea* mutant phenotypes (Fitz Gerald et al., 2009), suggesting paternal silencing of *AtFH5* may not be required for normal endosperm development. Moreover, double mutants of *mea* and *atfh5* show additive endosperm cellularization and morphogenic defects (Fitz Gerald et al., 2009). These genetic results typically would be interpreted as indicating *mea* and *atfh5* act in different genetic pathways. Although *AtFH5* is clearly an imprinted gene with a critical endosperm development function, it is unclear whether the imprinted gene expression pattern has a significant role in endosperm development.

The role of imprinting for the *Arabidopsis MATERNALLY EXPRESSED PAB C-TERMINAL (MPC)* gene is even less clear than for *AtFH5*. *MPC* encodes the C-terminal domain of poly(A) binding proteins (PABP) and is hypothesized to have a role in regulating translation of mRNA (Tiwari et al., 2008). The *MPC* gene is a MEG, and homozygous *mpc* RNAi lines show abnormal embryo and endosperm development. However, the role of imprinted gene expression for *MPC* function is difficult to address, since the gene body sequence is necessary to confer maternal specific expression (Tiwari et al., 2008).

In maize, only one imprinted gene has been functionally characterized in seed development. The *meg1* locus encodes a small, secreted peptide that is expressed specifically in the basal endosperm transfer cell layer of the developing endosperm (Gutierrez-Marcos et al., 2004). *Meg1* is initially expressed from the maternal allele and becomes biallelic around 12 days after pollination. RNAi of *meg1* results in reduced transfer cell differentiation and smaller seeds than non-transgenic controls (Costa et al., 2012). Ectopic expression of *Meg1* results in patchy, ectopic transfer cell differentiation throughout the epidermal endosperm, indicating that MEG1 protein is a positive regulator of transfer cell fate.

Meg1 is part of a gene family with six members expressing at significant levels in transfer cells during seed development (Xiong et al., 2014). Only *Meg1* shows imprinted expression with all other *Meg* family members expressing similarly when inherited through either parent. The developmental function of these other *Meg* family members has not yet been experimentally tested. However, the expression patterns of these genes are very similar to *Meg1* in both developmental timing and location, suggesting that these genes are also likely to be regulators of transfer cell differentiation (Xiong et al., 2014). To directly test the role of imprinting on *Meg1* function, Costa et al. (2012) developed a non-imprinted, synthetic *Meg1* gene with two different transfer cell specific promoters. These transgenics show a dosage-sensitive increase both in the number of transfer cells and in seed size suggesting that imprinting of *Meg1* serves to limit nutrient uptake and seed size. Thus, among more than a dozen seed developmental genes studied in detail, only *Meg1* has strong evidence indicating that imprinted gene expression has a significant impact on development and growth of the seed.

WHY DOES IMPRINTING EXIST IN ANGIOSPERMS?

Parent-of-origin specific gene expression is a fascinating pattern of molecular regulation of the genome, and its evolution has been the subject of extensive theoretical debate (Patten et al., 2014). The most accredited explanation for imprinting in plants is provided by the parental-conflict hypothesis, also known as the kinship theory of selection (Haig and Westoby, 1989; Haig and Westoby, 1991; Haig, 2013). This hypothesis argues that imprinting evolves when the maternal parent provides resources during offspring development. In angiosperms, seeds require nutrition from the maternal parent from fertilization until seed maturation. The parental-conflict hypothesis states that the paternal genome expression is selected to increase support for individual progeny, while the maternal genome expression is selected to limit resources to maximize seed set.

The parental-conflict hypothesis predicts that MEGs should reduce seed size and potentially reduce seed set in unfavorable conditions. Conversely, PEGs would increase seed size and promote seed set. Loss-of-function phenotypes of the FIS-PRC2 mutants *mea*, *fis2*, *fie*, and *msi1* have been interpreted to support parental-conflict theory, because these mutants extend cell proliferation in the endosperm and embryo at the cost of failing to complete development (Grossniklaus et al., 1998; Kiyosue et al., 1999; Guitton and Berger, 2005; Ingouff et al., 2005). However, parental-conflict predicts stable networks of imprinted genes with MEGs and PEGs balancing each other for normal seed development (Patten et al., 2014). As discussed above, most imprinted genes, except for *Meg1*, that have been studied in detail can lose imprinted gene expression without significant consequence to seed development, suggesting MEG or PEG expression does not undergo significant selection pressure. The case of the *Meg1* gene also argues against the parental-conflict hypothesis. *Meg1* is normally maternally expressed, and a non-imprinted *Meg1* transgene shows a positive dosage effect for increasing seed size (Costa et al., 2012). The parental-conflict hypothesis would predict that a maternally expressed peptide like MEG1 should be a repressor of transfer cell development or that *Meg1* should be a PEG.

An alternate hypothesis to explain imprinting is the maternal-offspring coadaptation model of gene expression, in which maternal alleles may be selected for imprinted expression to provide the greatest combined fitness for the mother and offspring (Wolf and Hager, 2006). This model is meant primarily to explain the larger number of MEGs over PEGs that have been identified in both mammals and angiosperms. For *Meg1*, the model correctly predicts maternal specific expression, but maternal-offspring coadaptation does not explain the relatively extensive number of

PEGs or the apparent mutability of most angiosperm imprinted genes to switch between biallelic and imprinted states (Patten et al., 2014).

More in-depth evolutionary analysis of the identified imprinted genes in *Arabidopsis* suggests that imprinting correlates with rapid evolution of gene duplicates. More than two-thirds of *Arabidopsis* imprinted genes derive from recent gene duplication events (Qiu et al., 2014). *Arabidopsis* imprinted genes also show reduced domains of expression and increased evolutionary rates over non-imprinted paralogs. This analysis argues strongly that imprinted genes are undergoing neofunctionalization. Neither the parental-conflict nor the coadaptation models predict that recent gene duplication events would be favored for imprinted expression, although the bias toward gene duplicates does not specifically argue against these evolutionary models (Patten et al., 2014).

The current understanding of cis-elements targeting genes for imprinting further suggests that imprinting is primarily a rapid form of evolution. Transposons and short repeats appear to be the targets of differential demethylation in the central cell versus sperm cells (Gehring et al., 2009; Hsieh et al., 2009). Transposon movement allows random conversion of genes to an imprinted pattern of expression. For example, differences in transposon insertions near genes are associated with allele-specific imprinting in *Arabidopsis* (Pignatta et al., 2014). Transposon and other insertions can also convert imprinted genes to biallelic expression patterns, providing a fast mechanism to revert alleles into Mendelian, diploid expression (Haun et al., 2009). Importantly, transposition is known to increase in plants exposed to abiotic and biotic stress, suggesting imprinted gene expression is expected to change more rapidly when plants are poorly adapted to an environment (reviewed in Chénais et al., 2012). Although transposon insertions are generally thought to reduce gene expression, genome-wide analysis of gene expression and DNA methylation of 140 *Arabidopsis* accessions suggests transposon insertion within genes is associated with increased expression levels specifically during seed and pollen development (Schmitz et al., 2013). This more permissive epigenetic state allows genes silenced in other tissues to be expressed during seed development.

Based on the recent genome-wide analyses of imprinting and epigenetic regulation, we suggest that imprinting is a form of epigenetic regulation that allows more rapid selection on recent gene duplicates. Imprinting uncovers individual alleles by converting genes into a pseudohaploid mode of expression during seed development. There is no evidence for prolonged, imprinted expression of genes after germination, and many imprinted genes are expressed later in plant development (Pignatta et al., 2014). Thus, imprinting of one copy of a gene duplicate enables the imprinted gene to accumulate mutations without compromising whole plant fitness. Monoallelic expression in the seed exposes an imprinted allele to more rapid selection acting primarily upon the seed phenotype. Imprinting of recessive, advantageous alleles can confer greater fitness if only expressed from one parent. By contrast, selection against deleterious imprinted alleles is not as strong as in true haploid inheritance. Deleterious imprinted alleles would only be selected against when inherited from the parent conferring expression. For example, a deleterious PEG would be neutral when inherited from the mother. If a deleterious PEG is linked to an advantageous MEG allele, it could be maintained in a population for a significant period, potentially allowing time for additional compensatory mutations. Thus, imprinted expression may allow plant genomes to explore a larger space of allelic and phenotypic variation in the seed while avoiding deleterious plant phenotypes. The mature seed phenotype is expected to be a major driver of species fitness, and we suggest imprinting is a form of gene expression that allows for more efficient diversifying selection on the seed phenotype.

An important consequence of hypothesizing imprinting as a form of diversifying selection is that most imprinted expression patterns would be expected to have neutral effects on the fitness of the seed. Gene networks that appear to fit parental-conflict or coadaptation models are expected to evolve under diversifying selection. However, the prediction is that the bulk of imprinted expression patterns could revert to biallelic expression with no consequence on seed phenotype. Similarly, allele specific imprinting and novel imprinted loci would be expected to evolve at high frequency. Additional functional data of imprinted genes in outcrossing species such as maize, would help resolve whether parental-conflict or other types of selection is the primary driving force for the evolution of imprinted genes.

ACKNOWLEDGMENTS

We thank Mary Daliberti for assistance in preparing **Figure 1** and the reviewers for their helpful comments on the manuscript. We apologize to the authors of relevant research articles that were not highlighted in this review due to space constraints. The authors' research on seed development is supported by grants from the National Science Foundation (awards IOS-1031416 and MCB-1412218), the National Institute of Food and Agriculture (awards 2010-04228 and 2011-67013-30032), and the Vasil-Monsanto Endowment.

REFERENCES

Arnaud, P., and Feil, R. (2006). MEDEA takes control of its own imprinting. *Cell* 124, 468–470. doi: 10.1016/j.cell.2006.01.020

Baroux, C., Gagliardini, V., Page, D. R., and Grossniklaus, U. (2006). Dynamic regulatory interactions of polycomb group genes: MEDEA autoregulation is required for imprinted gene expression in *Arabidopsis*. *Genes Dev.* 20, 1081–1086. doi: 10.1101/gad.378106

Berger, N., Dubreucq, B., Roudier, F., Dubos, C., and Lepiniec, L. (2011). Transcriptional regulation of *Arabidopsis* LEAFY COTYLEDON2 involves RLE, a cis-element that regulates trimethylation of histone H3 at lysine-27. *Plant Cell* 23, 4065–4078. doi: 10.1105/tpc.111.087866

Bernardi, J., Lanubile, A., Li, Q. B., Kumar, D., Kladnik, A., Cook, S. D., et al. (2012). Impaired auxin biosynthesis in the defective endosperm18 mutant is due to mutational loss of expression in the ZmYuc1 gene encoding endosperm-specific YUCCA1 protein in maize. *Plant Physiol.* 160, 1318–1328. doi: 10.1104/pp.112.204743

Bratzel, F., Yang, C., Angelova, A., López-Torrejón, G., Koch, M., del Pozo, J. C., et al. (2012). Regulation of the new *Arabidopsis* imprinted gene AtBMI1C requires the interplay of different epigenetic mechanisms. *Mol. Plant* 5, 260–269. doi: 10.1093/mp/ssr078

Brown, R., Lemmon, B. E., Nguyen, H., and Olsen, O.-A. (1999). Development of endosperm in *Arabidopsis thaliana*. *Sex. Plant Reprod.* 12, 32–42. doi: 10.1007/s004970050169

Brown, R. C., Lemmon, B. E., and Olsen, O.-A. (1994). Endosperm development in barley: microtubule involvement in the morphogenetic pathway. *Plant Cell* 6, 1241–1252. doi: 10.1105/tpc.6.9.1241

Brown, R. C., Lemmon, B. E., and Olsen, O.-A. (1996). Development of the endosperm in rice (*Oryza sativa* L.): cellularization. *J. Plant Res.* 109, 301–313. doi: 10.1007/BF02344477

Chandler, J., Nardmann, J., and Werr, W. (2008). Plant development revolves around axes. *Trends Plant Sci.* 13, 78–84. doi: 10.1016/j.tplants.2007.11.010

Chaudhuri, S., and Messing, J. (1994). Allele-specific parental imprinting of dzr1, a posttranscriptional regulator of zein accumulation. *Proc. Natl. Acad. Sci. U.S.A.* 91, 4867–4871. doi: 10.1073/pnas.91.11.4867

Chaudhury, A. M., Ming, L., Miller, C., Craig, S., Dennis, E. S., and Peacock, W. J. (1997). Fertilization-independent seed development in *Arabidopsis thaliana. Proc. Natl. Acad. Sci. U.S.A.* 94, 4223–4228. doi: 10.1073/pnas.94.8.4223

Chénais, B., Caruso, A., Hiard, S., and Casse, N. (2012). The impact of transposable elements on eukaryotic genomes: from genome size increase to genetic adaptation to stressful environments. *Gene* 509, 7–15. doi: 10.1016/j.gene.2012.07.042

Chia, J. M., Song, C., Bradbury, P. J., Costich, D., De Leon, N., Doebley, J., et al. (2012). Maize HapMap2 identifies extant variation from a genome in flux. *Nat. Genet.* 44, 803–807. doi: 10.1038/ng.2313

Choi, Y., Gehring, M., Johnson, L., Hannon, M., Harada, J. J., Goldberg, R. B., et al. (2002). DEMETER, a DNA glycosylase domain protein, is required for endosperm gene imprinting and seed viability in *Arabidopsis. Cell* 110, 33–42. doi: 10.1016/S0092-8674(02)00807-3

Costa, L. M., Marshall, E., Tesfaye, M., Silverstein, K. A., Mori, M., Umetsu, Y., et al. (2014). Central cell-derived peptides regulate early embryo patterning in flowering plants. *Science* 344, 168–172. doi: 10.1126/science.1243005

Costa, L. M., Yuan, J., Rouster, J., Paul, W., Dickinson, H., and Gutierrez-Marcos, J. F. (2012). Maternal control of nutrient allocation in plant seeds by genomic imprinting. *Curr. Biol.* 22, 160–165. doi: 10.1016/j.cub.2011.11.059

Day, R. C., Herridge, R. P., Ambrose, B. A., and Macknight, R. C. (2008). Transcriptome analysis of proliferating *Arabidopsis* endosperm reveals biological implications for the control of syncytial division, cytokinin signaling, and gene expression regulation. *Plant Physiol.* 148, 1964–1984. doi: 10.1104/pp.108.128108

Deleris, A., Stroud, H., Bernatavichute, Y., Johnson, E., Klein, G., Schubert, D., et al. (2012). Loss of the DNA methyltransferase MET1 induces H3K9 hypermethylation at PcG target genes and redistribution of H3K27 trimethylation to transposons in *Arabidopsis thaliana. PLoS Genet.* 8:e1003062. doi: 10.1371/journal.pgen.1003062

De Smet, I., Lau, S., Mayer, U., and Jürgens, G. (2010). Embryogenesis – the humble beginnings of plant life. *Plant J.* 61, 959–970. doi: 10.1111/j.1365-313X.2010.04143.x

Fatihi, A., Zbierzak, A. M., and Dormann, P. (2013). Alterations in seed development gene expression affect size and oil content of *Arabidopsis* seeds. *Plant Physiol.* 163, 973–985. doi: 10.1104/pp.113.226761

Fitz Gerald, J. N., Hui, P. S., and Berger, F. (2009). Polycomb group-dependent imprinting of the actin regulator AtFH5 regulates morphogenesis in *Arabidopsis thaliana. Development* 136, 3399–3404. doi: 10.1242/dev.036921

Fouquet, R., Martin, F., Fajardo, D. S., Gault, C. M., Gomez, E., Tseung, C. W., et al. (2011). Maize rough endosperm3 encodes an RNA splicing factor required for endosperm cell differentiation and has a nonautonomous effect on embryo development. *Plant Cell* 23, 4280–4297. doi: 10.1105/tpc.111.092163

Gehring, M. (2013). Genomic imprinting: insights from plants. *Annu. Rev. Genet.* 47, 187–208. doi: 10.1146/annurev-genet-110711-155527

Gehring, M., Bubb, K. L., and Henikoff, S. (2009). Extensive demethylation of repetitive elements during seed development underlies gene imprinting. *Science* 324, 1447–1451. doi: 10.1126/science.1171609

Gehring, M., Huh, J. H., Hsieh, T. F., Penterman, J., Choi, Y., Harada, J. J., et al. (2006). DEMETER DNA glycosylase establishes MEDEA polycomb gene self-imprinting by allele-specific demethylation. *Cell* 124, 495–506. doi: 10.1016/j.cell.2005.12.034

Gehring, M., Missirian, V., and Henikoff, S. (2011). Genomic analysis of parent-of-origin allelic expression in *Arabidopsis thaliana* seeds. *PLoS ONE* 6:e23687. doi: 10.1371/journal.pone.0023687

Grossniklaus, U., Vielle-Calzada, J. P., Hoeppner, M. A., and Gagliano, W. B. (1998). Maternal control of embryogenesis by MEDEA, a polycomb group gene in *Arabidopsis. Science* 280, 446–450. doi: 10.1126/science.280.5362.446

Guitton, A. E., and Berger, F. (2005). Loss of function of MULTICOPY SUPPRESSOR OF IRA 1 produces nonviable parthenogenetic embryos in *Arabidopsis. Curr. Biol.* 15, 750–754. doi: 10.1016/j.cub.2005.02.066

Gutierrez-Marcos, J. F., Costa, L. M., Biderre-Petit, C., Khbaya, B., O'sullivan, D. M., Wormald, M., et al. (2004). Maternally expressed gene1 is a novel maize endosperm transfer cell-specific gene with a maternal parent-of-origin pattern of expression. *Plant Cell* 16, 1288–1301. doi: 10.1105/tpc.019778

Haig, D. (2013). Kin conflict in seed development: an interdependent but fractious collective. *Annu. Rev. Cell Dev. Biol.* 29, 189–211. doi: 10.1146/Annurev-Cellbio-101512-122324

Haig, D., and Westoby, M. (1989). Parent-specific gene-expression and the triploid endosperm. *Am. Nat.* 134, 147–155. doi: 10.1086/284971

Haig, D., and Westoby, M. (1991). Genomic imprinting in endosperm - its effect on seed development in crosses between species, and between different ploidies of the same species, and its implications for the evolution of apomixis. *Philos. Trans. R. Soc. Lond. B Biol. Sci.* 333, 1–13. doi: 10.1098/Rstb.1991.0057

Haun, W. J., Danilevskaya, O. N., Meeley, R. B., and Springer, N. M. (2009). Disruption of imprinting by mutator transposon insertions in the 5' proximal regions of the Zea mays Mez1 locus. *Genetics* 181, 1229–1237. doi: 10.1534/genetics.108.093666

Haun, W. J., Laoueille-Duprat, S., O'Connell, M. J., Spillane, C., Grossniklaus, U., Phillips, A. R., et al. (2007). Genomic imprinting, methylation and molecular evolution of maize enhancer of zeste (Mez) homologs. *Plant J.* 49, 325–337. doi: 10.1111/j.1365-313X.2006.02965.x

Heo, J. B., and Sung, S. (2011). Vernalization-mediated epigenetic silencing by a long intronic noncoding RNA. *Science* 331, 76–79. doi: 10.1126/science.1197349

Hermon, P., Srilunchang, K. O., Zou, J., Dresselhaus, T., and Danilevskaya, O. N. (2007). Activation of the imprinted polycomb group fie1 gene in maize endosperm requires demethylation of the maternal allele. *Plant Mol. Biol.* 64, 387–395. doi: 10.1007/s11103-007-9160-0

Hsieh, T. F., Ibarra, C. A., Silva, P., Zemach, A., Eshed-Williams, L., Fischer, R. L., et al. (2009). Genome-wide demethylation of *Arabidopsis* endosperm. *Science* 324, 1451–1454. doi: 10.1126/science.1172417

Hsieh, T. F., Shin, J., Uzawa, R., Silva, P., Cohen, S., Bauer, M. J., et al. (2011). Regulation of imprinted gene expression in *Arabidopsis* endosperm. *Proc. Natl. Acad. Sci. U.S.A.* 108, 1755–1762. doi: 10.1073/pnas.1019273108

Ingouff, M., Fitz Gerald, J. N., Guerin, C., Robert, H., Sorensen, M. B., Van Damme, D., et al. (2005). Plant formin AtFH5 is an evolutionarily conserved actin nucleator involved in cytokinesis. *Nat. Cell Biol.* 7, 374–380. doi: 10.1038/ncb1238

Jahnke, S., and Scholten, S. (2009). Epigenetic resetting of a gene imprinted in plant embryos. *Curr. Biol.* 19, 1677–1681. doi: 10.1016/j.cub.2009.08.053

Jermann, P., Hoerner, L., Burger, L., and Schübeler, D. (2014). Short sequences can efficiently recruit histone H3 lysine 27 trimethylation in the absence of enhancer activity and DNA methylation. *Proc. Natl. Acad. Sci. U.S.A.* 111, E3415–E3421. doi: 10.1073/pnas.1400672111

Jiao, Y., Zhao, H., Ren, L., Song, W., Zeng, B., Guo, J., et al. (2012). Genome-wide genetic changes during modern breeding of maize. *Nat. Genet.* 44, 812–815. doi: 10.1038/ng.2312

Jullien, P. E., Kinoshita, T., Ohad, N., and Berger, F. (2006a). Maintenance of DNA methylation during the *Arabidopsis* life cycle is essential for parental imprinting. *Plant Cell* 18, 1360–1372. doi: 10.1105/tpc.106.041178

Jullien, P. E., Katz, A., Oliva, M., Ohad, N., and Berger, F. (2006b). Polycomb group complexes self-regulate imprinting of the polycomb group gene MEDEA in *Arabidopsis. Curr. Biol.* 16, 486–492. doi: 10.1016/j.cub.2006.01.020

Kang, I. H., Steffen, J. G., Portereiko, M. F., Lloyd, A., and Drews, G. N. (2008). The AGL62 MADS domain protein regulates cellularization during endosperm development in *Arabidopsis. Plant Cell* 20, 635–647. doi: 10.1105/tpc.107.055137

Kermicle, J. L. (1969). Androgenesis conditioned by a mutation in maize. *Science* 166, 1422–1424. doi: 10.1126/science.166.3911.1422

Kermicle, J. L. (1970). Dependence of the R-mottled aleurone phenotype in maize on mode of sexual transmission. *Genetics* 66, 69–85. doi: 10.3410/f.718486711.793497084

Kiesselbach, T. A. (1949). *The Structure and Reproduction of Corn. Research Bulletin 161.* Lincoln: Agricultural Experiment Station, University of Nebraska College of Agriculture.

Kinoshita, T., Miura, A., Choi, Y., Kinoshita, Y., Cao, X., Jacobsen, S. E., et al. (2004). One-way control of FWA imprinting in *Arabidopsis* endosperm by DNA methylation. *Science* 303, 521–523. doi: 10.1126/science.1089835

Kinoshita, Y., Saze, H., Kinoshita, T., Miura, A., Soppe, W. J., Koornneef, M., et al. (2007). Control of FWA gene silencing in *Arabidopsis thaliana* by SINE-related direct repeats. *Plant J.* 49, 38–45. doi: 10.1111/j.1365-313X.2006.02936.x

Kiyosue, T., Ohad, N., Yadegari, R., Hannon, M., Dinneny, J., Wells, D., et al. (1999). Control of fertilization-independent endosperm development by the MEDEA polycomb gene in *Arabidopsis. Proc. Natl. Acad. Sci. U.S.A.* 96, 4186–4191. doi: 10.1073/pnas.96.7.4186

Köhler, C., Hennig, L., Bouveret, R., Gheyselinck, J., Grossniklaus, U., and Gruissem, W. (2003). *Arabidopsis* MSI1 is a component of the MEA/FIE polycomb group complex and required for seed development. *EMBO J.* 22, 4804–4814. doi: 10.1093/emboj/cdg444

Köhler, C., Page, D. R., Gagliardini, V., and Grossniklaus, U. (2005). The *Arabidopsis thaliana* MEDEA polycomb group protein controls expression of PHERES1 by parental imprinting. *Nat. Genet.* 37, 28–30. doi: 10.1038/ng1495

Köhler, C., Wolff, P., and Spillane, C. (2012). Epigenetic mechanisms underlying genomic imprinting in plants. *Annu. Rev. Plant Biol.* 63, 331–352. doi: 10.1146/annurev-arplant-042811-105514

Kradolfer, D., Hennig, L., and Köhler, C. (2013a). Increased maternal genome dosage bypasses the requirement of the FIS polycomb repressive complex 2 in *Arabidopsis* seed development. *PLoS Genet.* 9:e1003163. doi: 10.1371/journal.pgen.1003163

Kradolfer, D., Wolff, P., Jiang, H., Siretskiy, A., and Köhler, C. (2013b). An imprinted gene underlies postzygotic reproductive isolation in *Arabidopsis thaliana. Dev. Cell* 26, 525–535. doi: 10.1016/j.devcel.2013.08.006

Leroy, O., Hennig, L., Breuninger, H., Laux, T., and Köhler, C. (2007). Polycomb group proteins function in the female gametophyte to determine seed development in plants. *Development* 134, 3639–3648. doi: 10.1242/dev.009027

Li, J., and Berger, F. (2012). Endosperm: food for humankind and fodder for scientific discoveries. *New Phytol.* 195, 290–305. doi: 10.1111/j.1469-8137.2012.04182.x

Liu, X., Kim, Y. J., Müller, R., Yumul, R. E., Liu, C., Pan, Y., et al. (2011). AGAMOUS terminates floral stem cell maintenance in *Arabidopsis* by directly repressing WUSCHEL through recruitment of polycomb group proteins. *Plant Cell* 23, 3654–3670. doi: 10.1105/tpc.111.091538

Lodha, M., Marco, C. F., and Timmermans, M. C. (2013). The ASYMMETRIC LEAVES complex maintains repression of KNOX homeobox genes via direct recruitment of polycomb-repressive complex2. *Genes Dev.* 27, 596–601. doi: 10.1101/gad.211425.112

Ludwig, S. R., Habera, L. F., Dellaporta, S. L., and Wessler, S. R. (1989). Lc, a member of the maize R gene family responsible for tissue-specific anthocyanin production, encodes a protein similar to transcriptional activators and contains the myc-homology region. *Proc. Natl. Acad. Sci. U.S.A.* 86, 7092–7096. doi: 10.1073/pnas.86.18.7092

Luo, M., Bilodeau, P., Dennis, E. S., Peacock, W. J., and Chaudhury, A. (2000). Expression and parent-of-origin effects for FIS2, MEA, and FIE in the endosperm and embryo of developing *Arabidopsis* seeds. *Proc. Natl. Acad. Sci. U.S.A.* 97, 10637–10642. doi: 10.1073/pnas.170292997

Luo, M., Bilodeau, P., Koltunow, A., Dennis, E. S., Peacock, W. J., and Chaudhury, A. M. (1999). Genes controlling fertilization-independent seed development in *Arabidopsis thaliana. Proc. Natl. Acad. Sci. U.S.A.* 96, 296–301. doi: 10.1073/pnas.96.1.296

Makarevich, G., Leroy, O., Akinci, U., Schubert, D., Clarenz, O., Goodrich, J., et al. (2006). Different polycomb group complexes regulate common target genes in *Arabidopsis. EMBO Rep.* 7, 947–952. doi: 10.1038/sj.embor.7400760

Makarevich, G., Villar, C. B., Erilova, A., and Köhler, C. (2008). Mechanism of PHERES1 imprinting in *Arabidopsis. J. Cell Sci.* 121, 906–912. doi: 10.1242/jcs.023077

Makarevitch, I., Eichten, S. R., Briskine, R., Waters, A. J., Danilevskaya, O. N., Meeley, R. B., et al. (2013). Genomic distribution of maize facultative heterochromatin marked by trimethylation of H3K27. *Plant Cell* 25, 780–793. doi: 10.1105/tpc.112.106427

Ngo, Q. A., Baroux, C., Guthörl, D., Mozerov, P., Collinge, M. A., Sundaresan, V., et al. (2012). The Armadillo repeat gene ZAK IXIK promotes *Arabidopsis* early embryo and endosperm development through a distinctive gametophytic maternal effect. *Plant Cell* 24, 4026–4043. doi: 10.1105/tpc.112.102384

Ohad, N., Margossian, L., Hsu, Y. C., Williams, C., Repetti, P., and Fischer, R. L. (1996). A mutation that allows endosperm development without fertilization. *Proc. Natl. Acad. Sci. U.S.A.* 93, 5319–5324. doi: 10.1073/pnas.93.11.5319

Olsen, O.-A. (2004). Nuclear endosperm development in cereals and *Arabidopsis thaliana. Plant Cell* 16, S214–S227. doi: 10.1105/tpc.017111

O'Reilly, C., Shepherd, N. S., Pereira, A., Schwarz-Sommer, Z., Bertram, I., Donald, S., et al. (1985). Molecular cloning of the a1 locus of Zea mays using the transposable elements En and Mu1. *EMBO J.* 4, 877–882.

Patten, M. M., Ross, L., Curley, J. P., Queller, D. C., Bonduriansky, R., and Wolf, J. B. (2014). The evolution of genomic imprinting: theories, predictions and empirical tests. *Heredity (Edinb.)* 113, 119–128. doi: 10.1038/hdy.2014.29

Peris, C. I., Rademacher, E. H., and Weijers, D. (2010). Green beginnings - pattern formation in the early plant embryo. *Curr. Top. Dev. Biol.* 91, 1–27. doi: 10.1016/S0070-2153(10)91001-6

Perrot, G. H., and Cone, K. C. (1989). Nucleotide sequence of the maize R-S gene. *Nucleic Acids Res.* 17:8003. doi: 10.1093/nar/17.19.8003

Pignatta, D., Erdmann, R. M., Scheer, E., Picard, C. L., Bell, G. W., and Gehring, M. (2014). Natural epigenetic polymorphisms lead to intraspecific variation in *Arabidopsis* gene imprinting. *Elife* 3:e03198. doi: 10.7554/eLife.03198

Pires, N. D., and Grossniklaus, U. (2014). Different yet similar: evolution of imprinting in flowering plants and mammals. *F1000Prime Rep.* 6, 63. doi: 10.12703/P6-63

Qiu, Y., Liu, S. L., and Adams, K. L. (2014). Frequent changes in expression profile and accelerated sequence evolution of duplicated imprinted genes in *Arabidopsis. Genome Biol. Evol.* 6, 1830–1842. doi: 10.1093/gbe/evu144

Raissig, M. T., Bemer, M., Baroux, C., and Grossniklaus, U. (2013). Genomic imprinting in the *Arabidopsis* embryo is partly regulated by PRC2. *PLoS Genet.* 9:e1003862. doi: 10.1371/journal.pgen.1003862

Sabelli, P. A., and Larkins, B. A. (2009a). The contribution of cell cycle regulation to endosperm development. *Sex. Plant Reprod.* 22, 207–219. doi: 10.1007/s00497-009-0105-4

Sabelli, P. A., and Larkins, B. A. (2009b). The development of endosperm in grasses. *Plant Physiol.* 149, 14–26. doi: 10.1104/pp.108.129437

Schmitz, R. J., Schultz, M. D., Urich, M. A., Nery, J. R., Pelizzola, M., Libiger, O., et al. (2013). Patterns of population epigenomic diversity. *Nature* 495, 193–198. doi: 10.1038/nature11968

Schoft, V. K., Chumak, N., Choi, Y., Hannon, M., Garcia-Aguilar, M., Machlicova, A., et al. (2011). Function of the DEMETER DNA glycosylase in the *Arabidopsis thaliana* male gametophyte. *Proc. Natl. Acad. Sci. U.S.A.* 108, 8042–8047. doi: 10.1073/pnas.1105117108

Schuettengruber, B., and Cavalli, G. (2009). Recruitment of polycomb group complexes and their role in the dynamic regulation of cell fate choice. *Development* 136, 3531–3542. doi: 10.1242/dev.033902

Selinger, D. A., and Chandler, V. L. (2001). B-Bolivia, an allele of the maize b1 gene with variable expression, contains a high copy retrotransposon-related sequence immediately upstream. *Plant Physiol.* 125, 1363–1379. doi: 10.1104/pp.125.3.1363

Shirzadi, R., Andersen, E. D., Bjerkan, K. N., Gloeckle, B. M., Heese, M., Ungru, A., et al. (2011). Genome-wide transcript profiling of endosperm without paternal contribution identifies parent-of-origin-dependent regulation of AGAMOUS-LIKE36. *PLoS Genet.* 7:e1001303. doi: 10.1371/journal.pgen.1001303

Stangeland, B., Salehian, Z., Aalen, R., Mandal, A., and Olsen, O.-A. (2003). Isolation of GUS marker lines for genes expressed in *Arabidopsis* endosperm, embryo and maternal tissues. *J. Exp. Bot.* 54, 279–290. doi: 10.1093/jxb/erg031

Tiwari, S., Schulz, R., Ikeda, Y., Dytham, L., Bravo, J., Mathers, L., et al. (2008). MATERNALLY EXPRESSED PAB C-TERMINAL, a novel imprinted gene in *Arabidopsis*, encodes the conserved C-terminal domain of polyadenylate binding proteins. *Plant Cell* 20, 2387–2398. doi: 10.1105/tpc.108.061929

Tiwari, S., Spielman, M., Schulz, R., Oakey, R. J., Kelsey, G., Salazar, A., et al. (2010). Transcriptional profiles underlying parent-of-origin effects in seeds of *Arabidopsis thaliana. BMC Plant Biol.* 10:72. doi: 10.1186/1471-2229-10-72

Vu, T. M., Nakamura, M., Calarco, J. P., Susaki, D., Lim, P. Q., Kinoshita, T., et al. (2013). RNA-directed DNA methylation regulates parental genomic imprinting at several loci in *Arabidopsis. Development* 140, 2953–2960. doi: 10.1242/dev.092981

Walia, H., Josefsson, C., Dilkes, B., Kirkbride, R., Harada, J., and Comai, L. (2009). Dosage-dependent deregulation of an AGAMOUS-LIKE gene cluster contributes to interspecific incompatibility. *Curr. Biol.* 19, 1128–1132. doi: 10.1016/j.cub.2009.05.068

Waters, A. J., Bilinski, P., Eichten, S. R., Vaughn, M. W., Ross-Ibarra, J., Gehring, M., et al. (2013). Comprehensive analysis of imprinted genes in maize reveals allelic variation for imprinting and limited conservation with other species. *Proc. Natl. Acad. Sci. U.S.A.* 110, 19639–19644. doi: 10.1073/pnas.1309182110

Waters, A. J., Makarevitch, I., Eichten, S. R., Swanson-Wagner, R. A., Yeh, C. T., Xu, W., et al. (2011). Parent-of-origin effects on gene expression and DNA methylation in the maize endosperm. *Plant Cell* 23, 4221–4233. doi: 10.1105/tpc.111.092668

Weinhofer, I., Hehenberger, E., Roszak, P., Hennig, L., and Köhler, C. (2010). H3K27me3 profiling of the endosperm implies exclusion of polycomb group protein targeting by DNA methylation. *PLoS Genet.* 6:e1001152. doi: 10.1371/journal.pgen.1001152

Wendrich, J. R., and Weijers, D. (2013). The *Arabidopsis* embryo as a miniature morphogenesis model. *New Phytol.* 199, 14–25. doi: 10.1111/nph.12267

Wolf, J. B., and Hager, R. (2006). A maternal-offspring coadaptation theory for the evolution of genomic imprinting. *PLoS Biol.* 4:e380. doi: 10.1371/journal.pbio.0040380

Wolff, P., Weinhofer, I., Seguin, J., Roszak, P., Beisel, C., Donoghue, M. T., et al. (2011). High-resolution analysis of parent-of-origin allelic expression in the *Arabidopsis* endosperm. *PLoS Genet.* 7:e1002126. doi: 10.1371/journal.pgen.1002126

Xiao, W., Brown, R. C., Lemmon, B. E., Harada, J. J., Goldberg, R. B., and Fischer, R. L. (2006). Regulation of seed size by hypomethylation of maternal and paternal genomes. *Plant Physiol.* 142, 1160–1168. doi: 10.1104/pp.106.088849

Xin, M., Yang, R., Li, G., Chen, H., Laurie, J., Ma, C., et al. (2013). Dynamic expression of imprinted genes associates with maternally controlled nutrient allocation during maize endosperm development. *Plant Cell* 25, 3212–3227. doi: 10.1105/tpc.113.115592

Xing, Q., Creff, A., Waters, A., Tanaka, H., Goodrich, J., and Ingram, G. C. (2013). ZHOUPI controls embryonic cuticle formation via a signaling pathway involving the subtilisin protease ABNORMAL LEAF-SHAPE1 and the receptor kinases GASSHO1 and GASSHO2. *Development* 140, 770–779. doi: 10.1242/dev.088898

Xiong, Y., Mei, W., Kim, E. D., Mukherjee, K., Hassanein, H., Barbazuk, W. B., et al. (2014). Adaptive expansion of the maize maternally expressed gene (Meg) family involves changes in expression patterns and protein secondary structures of its members. *BMC Plant Biol.* 14:204. doi: 10.1186/s12870-014-0204-8

Yang, S., Johnston, N., Talideh, E., Mitchell, S., Jeffree, C., Goodrich, J., et al. (2008). The endosperm-specific ZHOUPI gene of *Arabidopsis thaliana* regulates endosperm breakdown and embryonic epidermal development. *Development* 135, 3501–3509. doi: 10.1242/dev.026708

Young, T. E., and Gallie, D. R. (2000). Programmed cell death during endosperm development. *Plant Mol. Biol.* 44, 283–301. doi: 10.1023/A:1026588408152

Young, T. E., Gallie, D. R., and DeMason, D. A. (1997). Ethylene-mediated programmed cell death during maize endosperm development of wild type and shrunken2 genotypes. *Plant Physiol.* 115, 737–751. doi: 10.1104/pp.115.2.737

Zhang, H., Chaudhury, A., and Wu, X. (2013). Imprinting in plants and its underlying mechanisms. *J. Genet. Genomics* 40, 239–247. doi: 10.1016/j.jgg.2013.04.003

Zhang, M., Xie, S., Dong, X., Zhao, X., Zeng, B., Chen, J., et al. (2014). Genome-wide high resolution parental-specific DNA and histone methylation maps uncover patterns of imprinting regulation in maize. *Genome Res.* 24, 167–176. doi: 10.1101/gr.155879.113

Zhang, M., Zhao, H., Xie, S., Chen, J., Xu, Y., Wang, K., et al. (2011). Extensive, clustered parental imprinting of protein-coding and noncoding RNAs in developing maize endosperm. *Proc. Natl. Acad. Sci. U.S.A.* 108, 20042–20047. doi: 10.1073/pnas.1112186108

Conflict of Interest Statement: The authors declare that the research was conducted in the absence of any commercial or financial relationships that could be construed as a potential conflict of interest.

8

Protein body formation in the endoplasmic reticulum as an evolution of storage protein sorting to vacuoles: insights from maize γ-zein

Davide Mainieri[†], Francesca Morandini[†], Marie Maîtrejean, Andrea Saccani, Emanuela Pedrazzini and Alessandro Vitale*

Istituto di Biologia e Biotecnologia Agraria, Consiglio Nazionale delle Ricerche, Milano, Italy

Edited by:
Brian A. Larkins, University of Nebraska-Lincoln, USA

Reviewed by:
Eliot Herman, University of Arizona, USA
David Richard Holding, University of Nebraska, USA

***Correspondence:**
Alessandro Vitale, Istituto di Biologia e Biotecnologia Agraria, Consiglio Nazionale delle Ricerche, via Bassini 15, 20133 Milano, Italy
e-mail: vitale@ibba.cnr.it

[†]Davide Mainieri and Francesca Morandini have contributed equally to this work.

The albumin and globulin seed storage proteins present in all plants accumulate in storage vacuoles. Prolamins, which are the major proteins in cereal seeds and are present only there, instead accumulate within the endoplasmic reticulum (ER) lumen as very large insoluble polymers termed protein bodies. Inter-chain disulfide bonds play a major role in polymerization and insolubility of many prolamins. The N-terminal domain of the maize prolamin 27 kD γ-zein is able to promote protein body formation when fused to other proteins and contains seven cysteine residues involved in inter-chain bonds. We show that progressive substitution of these amino acids with serine residues in full length γ-zein leads to similarly progressive increase in solubility and availability to traffic from the ER along the secretory pathway. Total substitution results in very efficient secretion, whereas the presence of a single cysteine is sufficient to promote partial sorting to the vacuole via a wortmannin-sensitive pathway, similar to the traffic pathway of vacuolar storage proteins. We propose that the mechanism leading to accumulation of prolamins in the ER is a further evolutionary step of the one responsible for accumulation in storage vacuoles.

Keywords: disulfide bonds, endoplasmic reticulum, evolution of the secretory pathway, seed storage proteins, *Zea mays*

INTRODUCTION

Seed storage proteins, the major food proteins, have unique characteristics that allow very high accumulation during seed development, low digestibility by predators, but rapid hydrolysis during germination. Despite these shared features, as well as structures that strongly suggest an evolution through variable combinations of very few domain types (Shewry et al., 1995; Adachi et al., 2001; Xu and Messing, 2008; Gu et al., 2010), the different seed storage proteins have specific assembly, solubility, and intracellular localization properties and can thus be divided into three major classes: (i) 2S albumins (the number refers to the sedimentation constant) are monomeric and soluble in water, (ii) 7S and 11S globulins are homotrimers and homohexamers, respectively, and are soluble in salt water, (iii) prolamins, the most heterogeneous class, are soluble in strong alcohol/water solution or after reduction of disufide bonds. How the evolution of biochemical properties relates to that of subcellular localization is still not clear (Ibl and Stoger, 2012). All storage proteins are co-translationally inserted into the endoplasmic reticulum (ER), where 7S and 11S globulins rapidly trimerize. The ER is the port of entry of the secretory pathway: from here, proteins traffic through the Golgi complex and are secreted by default, unless they have structural features/signals that sort them to intracellular compartments of the pathway. Albumins and globulins follow traffic pathways shared by the other soluble vacuolar proteins and are sorted to storage vacuoles of cotyledonary cells, where they accumulate (Vitale and Hinz, 2005). Endoproteolytic cleavage of 11S globulin polypeptides occurring in the vacuole promotes further assembly of trimers into the final 11S hexamers. Most prolamins, instead, form large polymers that in many cases rapidly become insoluble and do not proceed along the secretory pathway, accumulating as protein bodies (PB) within the ER lumen of endosperm cells (Shewry et al., 1995; Herman, 2008). The developmentally programmed use of the ER as a protein storage compartment seems to be unique to plants, whereas the formation of insoluble protein accretions in the ER is associated to many human diseases (Anelli and Sitia, 2010). Storage albumins and globulins are present throughout land plant evolution, but prolamins have been found only in cereals and are therefore much less ancient (Xu and Messing, 2008). In all cereals, each PB contains many different prolamin polypeptides, indicating promiscuous interactions among the products of the different prolamin genes. When expressed individually in vegetative tissues of transgenic plants, some prolamins form homotypic PBs within the ER, indicating that they contain all the information sufficient to form an ER-retained polymer, whereas others remain soluble and are delivered to the vacuole. Examples of these two different destinies are maize 27 kD γ-zein and wheat γ-gliadin, respectively (Geli et al., 1994; Napier et al., 1997). Individual storage globulins can also be retained in the ER as large accretions, instead of being sorted to storage vacuoles, if the synthesis of other members of the family is suppressed (Kinney et al., 2001). Altogether, these data suggest a close relationship between the mechanisms of storage protein accumulation in vacuoles or the ER, and that the latter may have evolved directly from the former.

The 27 kD γ-zein (hereon termed γ-zein, for simplicity) is among the most ancient maize prolamins (Xu and Messing, 2008, 2009); it is therefore a good model to study the early events that may have caused a shift of accumulation from the vacuole to the ER. After co-translational removal of the signal peptide for entry into the ER, mature γ-zein consists of two major regions, each corresponding to about half of the polypeptide (Prat et al., 1985). As schematically illustrated in **Figure 1A**, the N-terminal region is characterized by eight repeats of the hexapeptide PPPVHL (the repeated hexapeptide is also VHLPPP, because the last of the eight PPPVHL sequences is followed by PPP) and seven Cys residues involved in inter-chain bonds. A synthetic $(VHLPPP)_8$ peptide forms *in vitro* an amphipathic structure that has affinity with lipids derived from plant ER (Kogan et al., 2004). The C-terminal region is homologous to 2S albumins (Shewry et al., 1995), which are characterized by three domains named A, B, and C, containing eight Cys that form four inter-chain disulfide bonds (**Figure 1A**). Consistently, when a fusion between thioredoxin and the C-terminal domain of γ-zein was expressed in *E. coli*, these eight Cys residues formed the intra-chain bonds typical of 2S albumins (Ems-McClung et al., 2002). The presence of a repeated sequence, as well as the 2S albumin domains, are common features of many prolamins (Shewry et al., 1995).

Early study established that γ-zein is soluble only in the presence of reducing agents (Vitale et al., 1982). The N-terminal region is retained in the ER if the 2S albumin-like region is deleted, whereas efficient secretion occurs when most of the N-terminal region is deleted (Geli et al., 1994). When a fragment (amino acids 24–112, from the translation start) that includes the eight Pro-rich repeats and the first six Cys residues was fused to the C-terminus of the vacuolar 7S storage globulin of common bean, phaseolin, the chimeric construct, termed zeolin, formed polymers with the main features of γ-zein: they are insoluble unless reduced and accumulate as homotypic PB in the ER (Mainieri et al., 2004). The Zera sequence, which consists of amino acids 1–112 of γ-zein, similarly promotes the formation of PB in the ER when fused to the N-terminus of a number of proteins (Torrent et al., 2009). Collectively, these experiments indicate that the N-terminal region of γ-zein forms inter-chain disulfide bonds and contains information sufficient for ER retention, independently of its position in fusion proteins.

When the six Cys residues of zeolin were mutated to Ser, efficient secretion occurred from transiently transfected protoplasts (Pompa and Vitale, 2006). Consistently, *in vivo* treatment with the reducing agent 2-mercaptoethanol caused secretion of zeolin from protoplasts of transgenic tobacco (Pompa and Vitale, 2006). Therefore, the inter-chain disulfide bonds are necessary for zeolin retention in the ER. This result was confirmed and extended by mutagenesis of the fluorescent chimera Zera-ECFP (Llop-Tous et al., 2010): when all the six Zera Cys residues were mutated or all eight repeats were deleted, PB formation was totally impaired, leading to efficient secretion, indicating cooperation between hydrophobic interactions and disulfide bonds in the formation of PB. Progressive mutagenesis of Cys residues or deletion of the Pro-rich repeats caused parallel gradual decrease in the size and abundance of Zera-ECFP PBs and increased secretion. The first two N-terminal Cys residues were identified

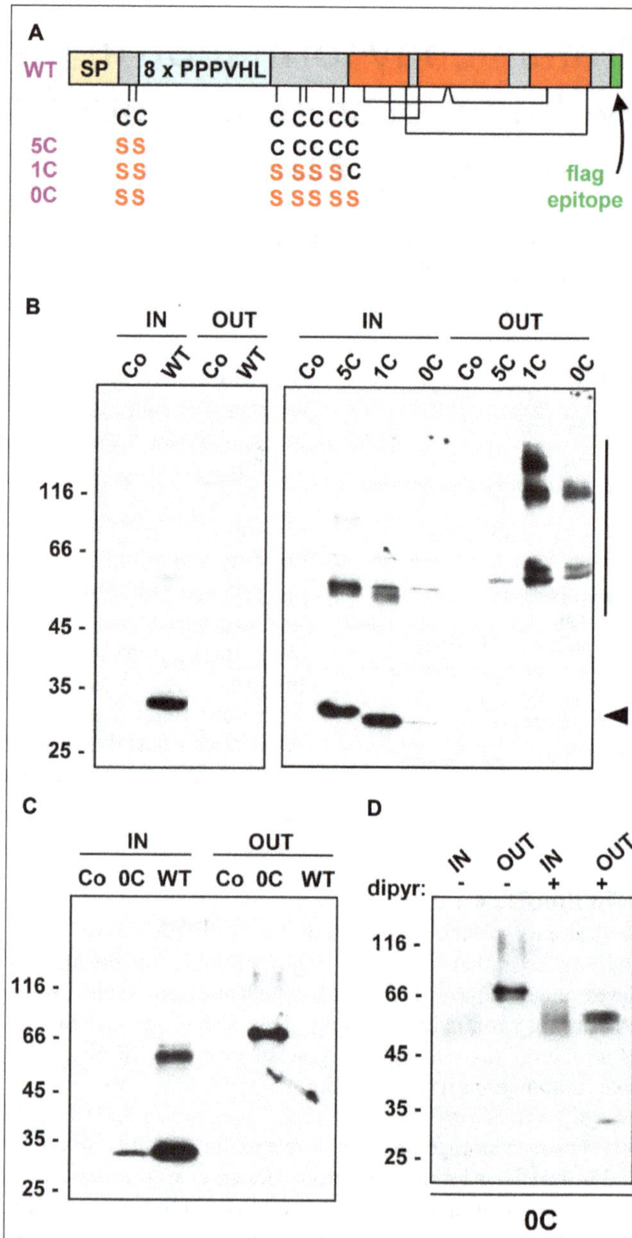

FIGURE 1 | Synthesis and secretion of γ-zein constructs. (A) Schematic drawing of the different protein constructs, in which the following features are indicated: signal peptide (orange, SP); Cys residues in the N-terminal region and their substitutions with Ser in the mutated forms; repeated domain (light blue, 8 × PPPVHL); 2S albumin homologous domains (red), with the four putative intrachain disufide bonds; added flag epitope (green). **(B,C)** Tobacco protoplasts were transfected with plasmid containing the indicated γ-zein constructs or with empty plasmid as control (Co) and incubated for 24 h. Protoplasts (IN) or incubation media (OUT) were then homogenated in the presence of 2ME. Aliquots corresponding to equal number of protoplasts or the corresponding incubation medium were analyzed by SDS-PAGE in reducing conditions and protein blot with anti-flag antiserum. **(D)** Protoplasts transfected with plasmid encoding 0C-γ-zein were incubated for 24 h in the presence (+) or absence (−) of 0.25 mM 2,2′-dipyridyl (dipyr). Proteins were then analyzed as in **(B)**. In **(B)**, the positions of monomers (arrowhead) and polymers (verical bar) are indicated at right. In **(B–D)**, numbers at left indicate the positions of molecular mass markers, in kD.

as critical, since their mutagenesis was sufficient for full secretion, as visualized by fluorescence microscopy (Llop-Tous et al., 2010).

Using the full length γ-zein as a model, we have studied here the relationships between prolamin solubility, assembly, and intracellular traffic. We found a progressive change in subcellular location from the ER, to the vacuole, to secretion, that casts light on the structural features that may have allowed the evolutionary shift of storage protein localization from the vacuole to the ER, thus giving rise to a new subcellular compartment.

MATERIALS AND METHODS

PLASMID CONSTRUCTION

To prepare WT γ-zein for transient expression and detection with anti-flag antiserum, plasmid pBSKS.G1L (Bellucci et al., 1997), which contains the coding sequence of 27 kD γ-zein, was PCR amplified with the following oligos: 5′-TGTAGTCGACATGAGGGTGTTGCTCGTTGCCCT-3′ (termed forward1, the SalI restriction site is underlined) and 5′-ACATGCATGCCTATCATTACTTGTCGTCGTCGTCCTTGTAGTCGTGGGGACACCGCCGGCAGCA-3′, (termed reverse1, the SphI restriction site is underlined, the reverse complement of the codons encoding the flag epitope DYKDDDDK is double underlined). The sequence was restricted with SalI and SphI and inserted into the transient expression vector pDHA (Tabe et al., 1995). 1C γ-zein was constructed starting from pDHAzeolin(Cys$^-$), see Pompa and Vitale (2006), first by amplifying its zein domain with the oligos 5′-TGTAGTCGAC*ATG* AGGGTGTTGCTCGTTGCCCTCGCTCTCCTGGCTCTCGCTGCGAGCGCCACCTCCACGCATACAAGCGGCGGTAGCGGATCTCAGCCA-3′ (forward2, the SalI site is underlined, the translation start ATG is in Italic, the codons that substitute the first two Cys with Ser are double underlined) and 5′-AAAACTGCAG TTGCGACGGACTAGGATGAGG-3′ (reverse2, the PstI site is underlined). The amplified sequence was restricted with SalI and PstI and used to substitute the corresponding fragment of WT γ-zein inserted into pDHA. To construct 5C γ-zein, WT γ-zein inserted into pDHA was amplified with the oligos forward2 and reverse1, restricted with SalI and SphI and reinserted into pDHA. 0C γ-zein was constructed using a QuickChange protocol, starting from 1C γ-zein inserted into pDHA, using the oligos 5′-CAACAGGGAACCTCCGGCGTTGGCAGC-3′ (forward3, the codon that substitutes the last Cys of the N-terminal domain of γ-zein with Ser is double underlined) and its reverse complement (reverse3).

TRANSIENT EXPRESSION IN PROTOPLASTS

Protoplasts were isolated from small (4–7 cm) leaves of tobacco (*Nicotiana tabacum*) SR1 plants grown in axenic conditions and subjected to polyethylene glycol-mediated transfection as described (Pedrazzini et al., 1997). Unless otherwise stated in the Results section, 40 μg of plasmid were used to transform 10^6 protoplasts. Transformed protoplasts were resuspended in K3 medium [Gamborg's B5 basal medium with minimal organics (Sigma), supplemented with 750 mg/L $CaCl_2$ $2H_2O$, 250 mg/L NH_4NO_3, 136.2 g/L sucrose, 250 mg/L xylose, 1 mg/L 6-benzylaminopurine, and 1 mg/L 1-naphthalenacetic acid, pH 5.5], supplemented with 150 μg/ml bovine serum albumin as a competing substrate for extracellular proteases, and were incubated in the dark at 25°C. After the desired time, the incubation medium containing secreted proteins was carefully collected with a 2 ml syringe and protoplasts were washed and concentrated by the addition of three volumes of ice-cold W5 medium (154 mM NaCl, 5 mM KCl, 125 mM $CaCl_2$ $2H_2O$, and 5 mM glucose) and centrifugation at 60 *g* for 10 min. The incubation medium and the protoplast pellet were frozen in liquid nitrogen and stored at −80°C, but freezing of protoplasts was avoided when subcellular fractionation was performed. When needed, inhibitors of protein trafficking brefeldin A (BFA: Roche; final concentration 10 μg/ml) or wortmannin (Sigma–Aldrich; final concentration 10 μM), were added to the protoplast incubation medium 1 h after transfection and maintained at the same concentration throughout the incubation. To inhibit prolyl hydroxylation, transfected protoplasts were incubated in the presence of 0.25 mM 2,2′-dipyridyl (dipyr), a chemical that inhibits *in vivo* activity of prolyl-4-hydroxylase (Moriguchi et al., 2011).

PROTEIN ANALYSIS

Protoplasts and incubation media were homogenated in homogenation buffer [150 mM Tris-Cl, pH 7.5, 150 mM NaCl, 1.5 mM EDTA, 1.5% Triton X-100, Complete protease inhibitor cocktail (Roche)], supplemented (reducing conditions) or not (oxidizing conditions) with 4% (v/v) 2-mercaptoethanol (2-ME). Separation of soluble and insoluble proteins was performed by centrifugation at 1,500 *g*, 10 min, 4°C. In the experiment shown in **Figure 2B**, the oxidizing homogenation buffer was supplemented with the alkylating agent iodoacetamide (Sigma–Aldrich, final concentration 70 mM). To analyze proteins, the homogenates were adjusted to 20 mM Tris-Cl pH 8.6, 1.0% SDS, 4% 2-ME, 8% glycerol (denaturation buffer), heated at 90°C for 5 min and separated by 15% acrylamide SDS-PAGE. For SDS-PAGE in non-reducing conditions, 2-ME was omitted from the denaturation buffer. Gels were blotted to Hybond-P membrane (GE Healthcare) and proteins revealed using with anti-flag rabbit polyclonal antibodies (Sigma–Aldrich; 1:2,000 dilution), anti-BiP rabbit antiserum (Pedrazzini et al., 1997, 1:10,000 dilution) or anti-endoplasmin rabbit antiserum (Klein et al., 2006, 1:2,500 dilution) and the Super-Signal West Pico Chemiluminescent Substrate (Pierce Chemical, Rockford, IL, USA). Protein Molecular Weight Markers (Fermentas, Vilnius, Lithuania) were used as molecular mass markers.

For velocity ultracentrifugation, the incubation medium of protoplasts expressing 0C γ-zein was homogenized with oxidizing homogenation buffer and centrifuged 1,500 *g*, 10 min, 4°C. The supernatant was loaded on top of a linear 5–25% (w/v) Suc gradient made in 150 mM NaCl, 1 mM EDTA, 0.1% Triton X-100, 50 mM Tris-Cl, pH 7.5. After centrifugation at 200,000 *g* average, 4°C for 20 h, an equal aliquot of each gradient fraction was analyzed by SDS-PAGE and protein blot. An identical gradient loaded with molecular mass markers was run in parallel.

SUBCELLULAR FRACTIONATION AND VACUOLE PURIFICATION

Separation of microsomes derived from the different subcellular compartments was performed by protoplast homogenation in

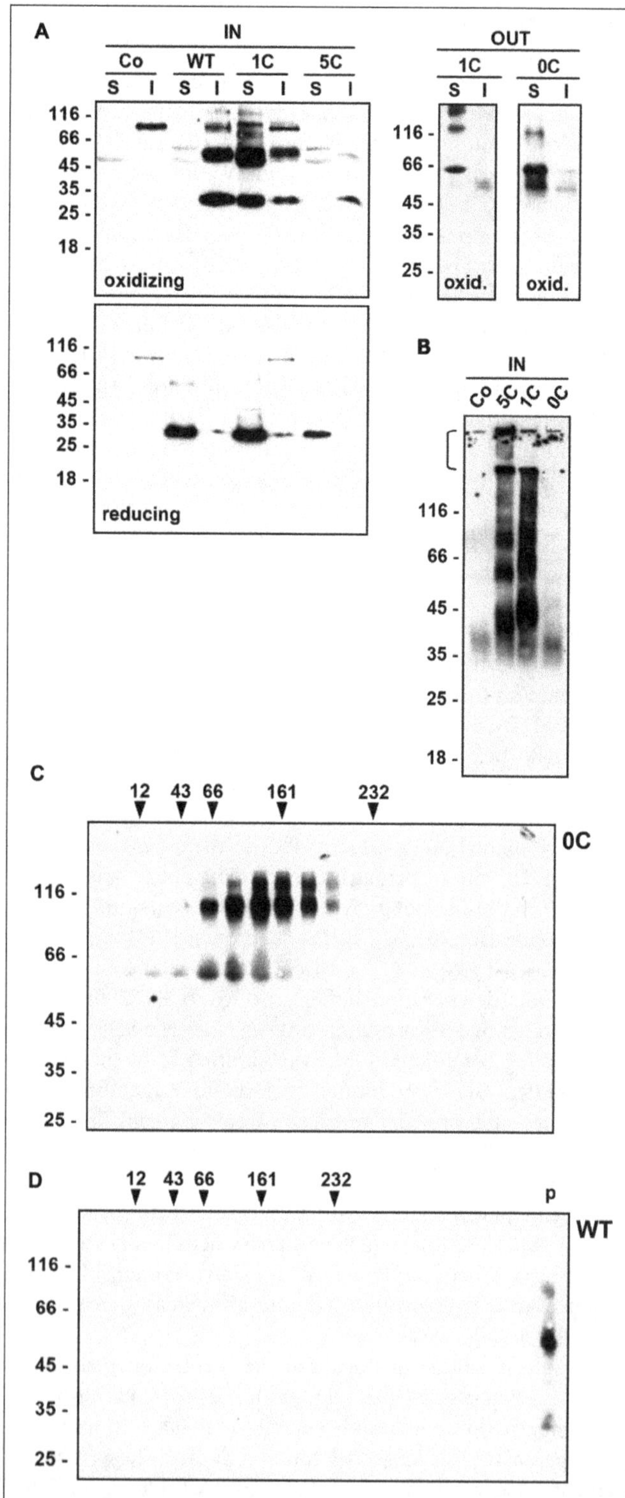

FIGURE 2 | Solubility and assembly of γ-zein constructs. (A) Tobacco protoplasts were transfected with plasmid containing the indicated γ-zein constructs or with empty plasmid as control (Co) and incubated for 24 h. Protoplasts (IN) or incubation medium (OUT) were homogenated in the presence (reducing) or absence (oxidizing) of 2ME. After centrifugation, soluble (S) and insoluble (I) proteins were analyzed by SDS-PAGE in reducing conditions and protein blot with anti-flag antiserum. **(B)** Protoplasts transfected as in **(A)** were homogenated in the absence *(Continued)*

FIGURE 2 | Continued
of 2ME. After centrifugation, soluble proteins were analyzed by SDS-PAGE in non-reducing conditions and protein blot with anti-flag antiserum. The stacking gel was not removed and is indicated by the bracket at left. **(C)** Protoplasts were transfected with plasmid encoding 0C-γ-zein. After 24 h the incubation medium was homogenated and loaded on top of a linear 5–25% (w/v) Suc gradient. After velocity separation of molecules by ultracentrifugation, an aliquot of each gradient fraction was analyzed by SDS-PAGE and protein blot with anti-flag antiserum. Top of the gradient is at left. Numbers on top indicate the position along the gradient and the molecular mass, in kD, of velocity centrifugation markers. **(D)** Protoplasts were transfected with plasmid encoding WT γ-zein. After 24 h protoplasts were homogenated and analyzed by velocity ultracentrifugation performed as in **(C)**. *p* indicates the material precipitated at the bottom of the tube. In all panels, numbers at left indicate the positions of molecular mass markers along the SDS-PAGE gel, in kD.

10 mM KCl, 100 mM Tris-Cl, pH 7.8, 2 mM $MgCl_2$, 12% (w/w) sucrose, followed by isopycnic centrifugation using linear 16–55% (w/w) sucrose gradient in the same buffer, as described (Mainieri et al., 2004). Isolation of vacuoles by floatation in Ficoll/betaine gradients (Dombrowski et al., 1994) and determination of the activity of the vacuolar marker α-mannosidase (Foresti et al., 2008) have also been described.

RESULTS

SOLUBILITY AND SECRETION OF γ-ZEIN ARE INVERSELY RELATED TO THE NUMBER OF Cys RESIDUES IN THE N-TERMINAL REGION AND THE EXTENT OF OLIGOMERIZATION

Three mutated constructs were produced, in which the first two, first six or all seven Cys residues of the N-terminal region were mutated to Ser (**Figure 1A**). The positions of these Cys residues, from the translation start, are 26, 28, 83, 101, 103, 111, and 117. To facilitate detection by protein blot, the flag epitope DYKDDDDK was added to the C-terminus of these constructs and of WT γ-zein. For brevity, here on the constructs will be called based on the number of remaining Cys residues of the N-terminal region: 5C indicates γ-zein$C^{26,28}$S, 1C indicates γ-zein$C^{26,28,83,101,103,111}$S, and 0C indicates γ-zein$C^{26,28,83,101,103,111,117}$S; WT indicates wild type γ-zein. All constructs were transiently expressed in tobacco leaf protoplasts under the control of the constitutive CaMV 35S promoter.

To first test whether the mutations promoted traffic of γ-zein, at 24 h after transfection proteins were extracted from protoplasts or the incubation media in the presence of non-ionic detergent and the reducing agent 2-mercaptoethanol (2-ME), separated by SDS-PAGE and detected with anti-flag antiserum. Polypeptides of the expected size, migrating between the 25 and 35 kD molecular mass markers (**Figure 1B**, arrowhead), as well as higher molecular mass forms that indicated oligomer formation or post-translational modifications, in part related to secretion (**Figure 1B**, vertical bar), were detected in protoplasts transfected with all γ-zein constructs, but not when transfection was performed with the empty plasmid. WT γ-zein was exclusively recovered in protoplasts, as expected, whereas a progressive decrease in the number of Cys residues caused a parallel, gradual increase in secretion (**Figure 1B**). In the experiment shown in **Figure 1B**, secretion of 0C-γ-zein was almost complete. This was not always the case in fully independent transfections, probably reflecting variability in intracellular redox balance among the different protoplast preparations (Lombardi et al., 2012), but the ratio of secreted/intracellular recombinant protein was always 0C > 1C > 5C ≥ WT γ-zein, where secreted WT γ-zein was always virtually 0% (**Table 1**). Even in experiments in which the total amount of WT γ-zein synthesized was clearly higher than that of 0C, the former was not secreted at all, indicating that the differences in the traffic properties are not dependent on expression levels (**Figure 1C**).

The γ-zein forms migrating around 50–60 kD could represent dimers particularly difficult to denature, whereas the larger forms, that are mainly detected extracellularly, could indicate further oligomerization or post-translational modifications related to traffic (**Figure 1B**). Higher molecular mass forms difficult to denature have also been detected to variable extents upon expression of γ-zein or protein fusions containing the N-terminal γ-zein domain in transgenic plants (Geli et al., 1994; Bellucci et al., 2000; Mainieri et al., 2004; Torrent et al., 2009; Virgili-López et al., 2013).

Some of the γ-zein oligomerizations/modifications are clearly dependent on the presence of Cys residues in the N-terminal domain, as can be appreciated comparing secreted 1C- and 0C-γ-zein, but others are not (**Figure 1B**). It has been suggested that hydroxylation of proline residues could occur on γ-zein (Geli et al., 1994). *In vivo* inhibition of prolyl-4-hydroxylase abolished most of the differences between intracellular and secreted 0C-γ-zein (**Figure 1D**), indicating that the increase in molecular mass beyond 50–60 kD was indeed dependent on this modification, possibly followed by O-glycosylation of hydroxyproline occurring only along the route to secretion or extracellularly.

The inverse relationship between the number of Cys residues and secretion suggested that the removal of Cys residues increases γ-zein solubility in the naturally oxidizing ER environment. Homogenation in the absence or presence of 2-ME was therefore performed (**Figure 2A**, oxidizing and reducing, respectively), followed by centrifugation to separate soluble and insoluble material (S and I samples in **Figure 2A**), denaturation of all samples and analysis by SDS-PAGE in reducing conditions. Putative dimers were much more evident if extraction was performed in oxidizing conditions. WT γ-zein was insoluble unless extracted in the presence of reducing agent, as expected (Vitale et al., 1982), whereas decreasing the number of Cys residues caused a parallel increase in solubility: a very small proportion of intracellular 5C, more than 50% of intracellular 1C, and more than 90% of secreted 0C- and 1C-γ-zein were solubilized also in the absence of reducing agent (**Figure 2A**, compare S and I for each sample).

Analysis by non-reducing SDS-PAGE indicated that soluble 1C-γ-zein and, even more, 5C-γ-zein undergo oligomerization beyond the dimer stage (**Figure 2B**; notice that a relevant proportion of 5C γ-zein does not even enter the separating gel). Oligomerization of the very small proportion of intracellular 0C-γ-zein was difficult to determine (notice that the diffused band around 40 kD is present also in the control and is thus irrelevant), but the native molecular mass of secreted 0C γ-zein polypeptides analyzed by velocity gradient ultracentrifugation corresponded roughly to that determined by SDS-PAGE, indicating that dimers particularly difficult to denature are formed by all constructs and that 0C-γ-zein does not form larger oligomers beyond dimers (**Figure 2C**). WT γ-zein could not be analyzed by non-reducing SDS-PAGE because of its insolubility, but upon velocity centrifugation it migrated at the bottom of the tube, indicating that it formed very large disulfide-bonded polymers, as expected (**Figure 2D**).

It can be concluded that γ-zein expressed in leaf protoplasts is insoluble unless reduced, as in maize endosperm, and that lowering the number of Cys residues progressively reduces the ability to form large disulfide-bonded polymers, resulting in increased solubility and ability to traffic along the secretory pathway.

Table 1 | Secretion levels of γ-zein constructs.

γ-zein construct	WT	5C	1C	0C
	Percentage of recombinant protein that was secreted			
Experiment 1[a]	0	0	5	79
Experiment 2	0	10	68	90
Experiment 3	0	0	32	51

[a] *Experiments 1, 2, and 3 are three fully independent protoplast preparation and transfection experiments.*

SOLUBLE γ-ZEIN TRAFFICS IN PART TO THE VACUOLE UNLESS ALL N-TERMINAL Cys RESIDUES ARE MUTATED

To determine whether the introduced mutations also changed the intracellular distribution of γ-zein polypeptides that were not secreted during the 24 h incubation, transfected protoplasts were homogenized in the absence of detergent and the homogenate subjected to isopycnic gradient centrifugation to separate subcellular compartments. Protein blots were analyzed using either anti-flag serum or serum against the chaperone BiP, a major soluble protein of the ER. WT γ-zein migrated as a clear peak in the region around density 1.20 g/L, which contained also BiP molecules (**Figures 3A,B**). This is an expected behavior for a protein forming PBs in the ER. An extremely low proportion of putative dimers migrated at top of the gradient (**Figure 3A**, first lane at left, faint band below the 66 kD marker). The chaperone BiP is commonly also found at top of isopycnic gradients, possibly due to partial *in vitro* release from broken ER membranes before they seal into microsomes (**Figure 3B**; see Gomord et al., 1997; Frigerio et al., 2001). Both 5C-γ-zein (**Figure 3C**) and 1C-γ-zein (**Figure 3D**) formed a peak at the same density of WT γ-zein, but they were also present at top of the gradients in much more relevant proportions compared to WT γ-zein, more markedly in the case of 1C- than 5C-γ-zein. This could indicate *in vitro* release from the ER; however, leaf vacuoles completely break during homogenation and release their soluble content, which remains on top of isopycnic gradients together with cytosolic soluble proteins (Pedrazzini et al., 1997; Frigerio et al., 2001). We therefore verified whether mutated, soluble γ-zein polypeptides

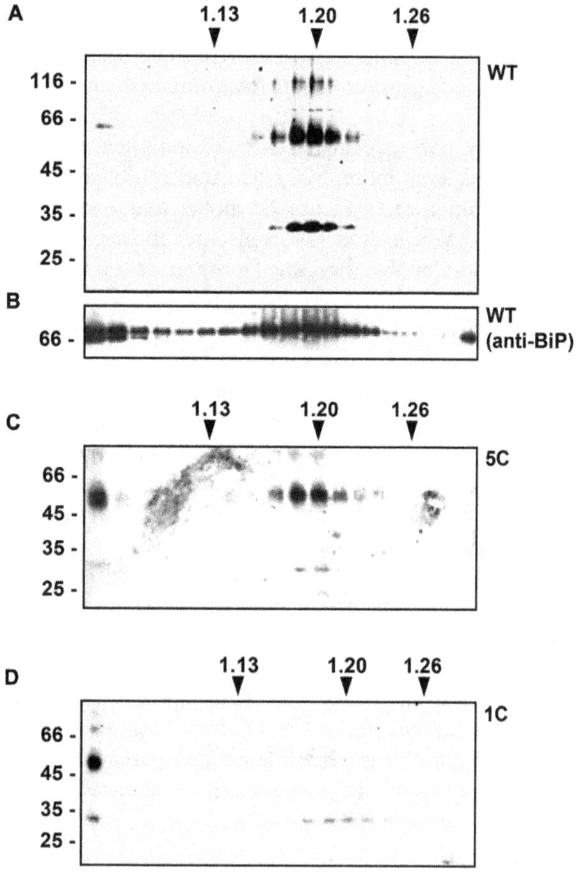

FIGURE 3 | Subcellular distribution of the different γ-zein constructs. Tobacco protoplasts were transfected with plasmid encoding WT γ-zein **(A,B)**, 5C-γ-zein **(C)** or 1C-γ-zein **(D)**. After 24 h incubation, protoplasts were homogenated in oxidizing conditions, in the absence of detergent and presence of sucrose. Homogenates were fractionated by isopycnic ultracentrifugation on sucrose density gradients. Equal amounts of each gradient fraction were analyzed by SDS-PAGE and protein blot using anti-flag **(A,C,D)** or anti-BiP **(B)** antiserum. Numbers on top indicate fraction density (g/ml). In each panel, numbers at left indicate the positions of molecular mass markers, in kD.

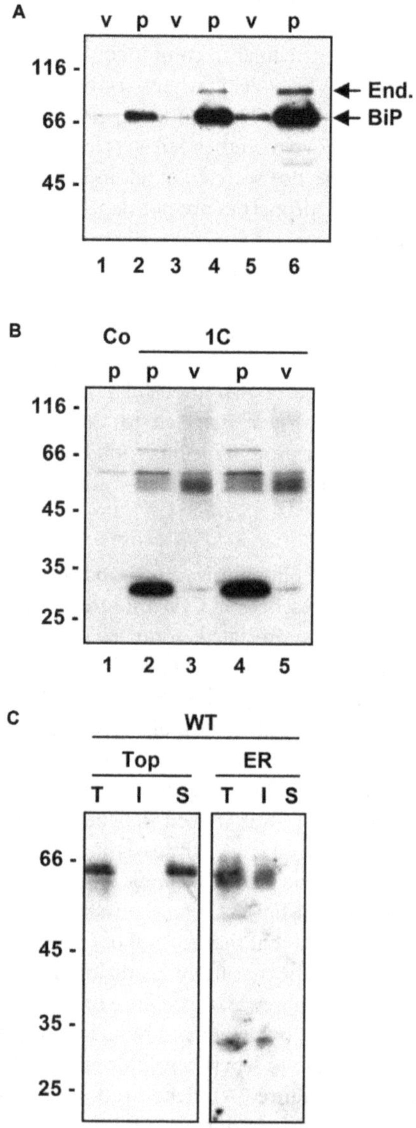

FIGURE 4 | Vacuole purification. (A) Tobacco protoplasts were transfected with empty plasmid. After 24 h incubation, either protoplast homogenates (p) or isolated vacuoles (v) were prepared and analyzed by SDS-PAGE and protein blot using a mixture of anti-BiP and anti-endoplasmin antisera. Lane 1 and 2: equal number of protoplasts and vacuoles, lanes 3 and 4 equal activity of the vacuolar enzyme α-mannosidase, lanes 5 and 6 fivefold amount of material with respect to lanes 3 and 4, respectively. **(B)** Tobacco protoplasts were transfected with empty plasmid (Co) or plasmid encoding 1C-γ-zein. After 24 h incubation, either protoplast homogenates (p) or isolated vacuoles (v) were prepared and analyzed by SDS-PAGE and protein blot using anti-flag antiserum. Lane 1-3: equal number of protoplasts and vacuoles, lanes 4 and 5 equal activity of the vacuolar enzyme α-mannosidase. **(C)** Tobacco protoplasts were transfected with plasmid encoding WT γ-zein. After 24 h incubation, protoplasts were homogenated in oxidizing conditions, in the absence of detergent and presence of sucrose. The homogenate was fractionated by isopycnic ultracentrifugation on sucrose density gradient. The fractions remaining on top of the gradient (top) or containing ER microsomes (ER) were diluted with homogenation buffer in oxidizing conditions and centrifuged. Equal amounts of total homogenate (T), soluble material (S) or insoluble precipitate (I) were analyzed by SDS-PAGE and protein blot using anti-flag antiserum. In each panel, numbers at left indicate the positions of molecular mass markers, in kD.

are in part delivered to the vacuole. Protoplasts were subjected to gentle lysis, followed by centrifugation in buffers that allow vacuole floatation. These preparations of intact vacuoles contain very low proportions of the major soluble ER chaperones BiP and endoplasmin, indicating almost no contamination by microsomes originating from other endomembranes (**Figure 4A**). Vacuoles prepared from protoplasts expressing 1C-γ-zein were highly enriched in γ-zein dimers, indicating that this mutated protein is in part delivered to the vacuole (**Figure 4B**) and suggesting that also 5C- and WT γ-zein dimers present at top of isopycnic gradients (**Figure 3**) are located in the vacuole. Consistently, the extremely low amount of WT γ-zein dimers at top of isopycnic gradients (see the first lane in **Figure 3A**) are fully soluble in the absence of reducing agent, whereas those in the ER fraction are insoluble (**Figure 4C**). This indicates that WT γ-zein polypeptides can, albeit extremely rarely, remain soluble and available for vacuolar delivery.

SECRETION OF SOLUBLE γ-ZEIN IS INHIBITED BY BREFELDIN A AND STIMULATED BY WORTMANNIN

Most soluble seed storage proteins are delivered to storage vacuoles following a traffic pathway that involves the Golgi apparatus and multivesicular bodies (MVB, also termed prevacuolar compartment). The chemicals BFA and wortmannin affect this pathway: BFA blocks the traffic step from the ER to the Golgi apparatus and therefore inhibits vacuolar delivery as well as Golgi-mediated protein secretion (Gomez and Chrispeels, 1993; Jones and Herman, 1993; Pedrazzini et al., 1997); wortmannin inhibits the recycling of vacuolar sorting receptors from MVB to the Golgi/*trans* Golgi network, thus inducing default secretion of soluble proteins that use these receptors to reach vacuoles (daSilva et al., 2005). However, highly condensed insoluble prolamins can be delivered to vacuoles by autophagy (Levanony et al., 1992; Coleman et al., 1996; Reyes et al., 2011); furthermore, vacuoles can also be reached by BFA-insensitive vesicular traffic pathways that seem to bypass the Golgi apparatus or MVB (Gomez and Chrispeels, 1993; Hara-Nishimura et al., 1998; Pedrazzini et al., 2013). We have therefore investigated the effects of BFA and wortmannin on γ-zein trafficking. Treatment of protoplasts with BFA fully inhibited the secretion of 1C- and 0C-γ-zein as well as the modifications that increase the molecular mass of dimers, indicating that secretion occurs via the Golgi apparatus and that the post-translational modifications require traffic (**Figure 5**). There was no evident increase in accumulation of intracellular polypeptides, most probably because the drug also partially inhibits protein synthesis (Mellor et al., 1994; de Virgilio et al., 2008); consistently, WT γ-zein, which is not secreted by untreated protoplasts, accumulated at lower amounts upon BFA treatment (**Figure 5**). Wortmannin markedly stimulated the secretion of 1C- and to a much less extent that of 0C-γ-zein, whereas no WT γ-zein could be detected in the protoplast incubation medium even upon treatment with this drug (**Figure 5**). This indicates that the vacuolar delivery of a relevant proportion of 1C-γ-zein polypeptides occurs through the same pathway followed by most vacuolar storage proteins. In agreement with the other data presented in this study, the results of wortmannin treatment also indicate that WT γ-zein does not enter this traffic pathway and that traffic of 0C-γ-zein leads mainly to secretion rather than vacuolar delivery.

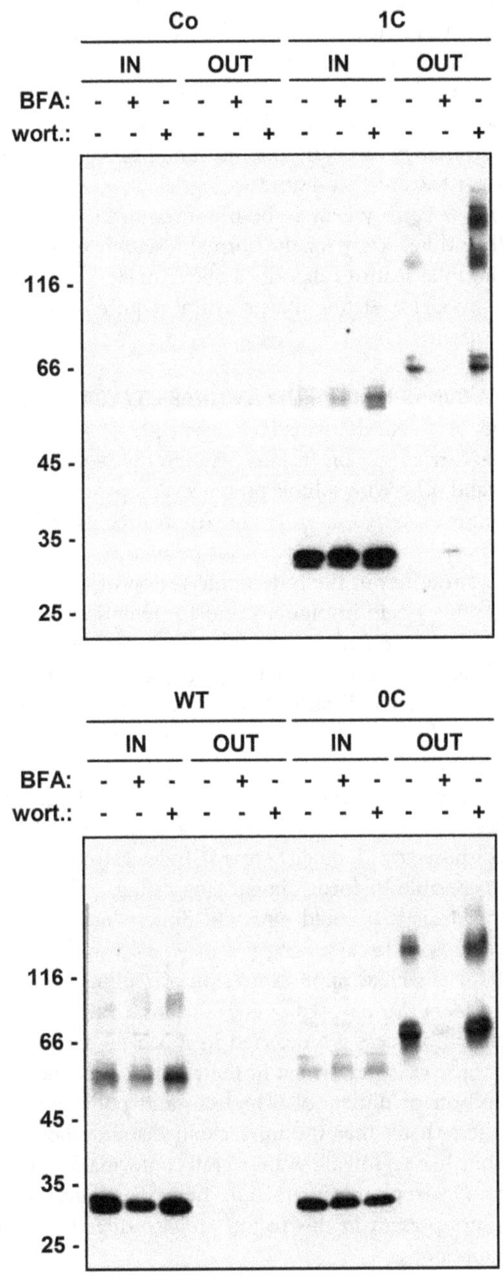

FIGURE 5 | Effect of intracellular traffic inhibitors on the secretion of γ-zein constructs. Tobacco protoplasts were transfected with plasmid containing the indicated γ-zein construct or with empty plasmid as control (Co). After 24 h incubation in the presence (+) or absence (–) of brefeldin A (BFA) or wortmannin (wort.), aliquots corresponding to equal number of protoplasts (IN) or the corresponding incubation medium (OUT) were analyzed by SDS-PAGE in reducing conditions, followed by protein blot using anti-flag antiserum. In each panel, numbers at left indicate the positions of molecular mass markers, in kD.

DISCUSSION

In all cereals, PBs contain the products of many different prolamin genes. Detailed characterization of maize seed development showed that zein genes have distinct temporal patterns of expression (Woo et al., 2001) and produce proteins with distinct ordered locations within individual PBs (Lending and Larkins, 1989). Characterization of the *floury1* (Holding et al., 2007) and *opaque1* (Wang et al., 2012) mutations indicated that specific proteins of the ER membrane and of the acto-myosin system are also involved in the shaping of PBs with normal size and morphology. These features indicate that a natural PB is assembled through a sequence of complex molecular interactions, whose details remain in large part unknown. The discovery that the ectopic expression of certain individual prolamins is sufficient to form large electron-dense accretions in the ER has, however, opened the way to a reductionist approach toward understanding prolamin assembly into polymers and their ER retention (Geli et al., 1994; Bagga et al., 1995). The 27 kD γ-zein, perhaps the best characterized of these proteins, is among the most ancient prolamins (Xu and Messing, 2008, 2009), is synthesized starting at very early stages of seed development and is specifically localized at the PB periphery, in proximity to the ER membrane, in mature PBs (Lending and Larkins, 1989). Elimination of 27 kD γ-zein by RNAi or induced gene deletion does not abolish maize PB formation, but strongly alters their morphology, size, and number, and has a general effect on endosperm texture (Wu and Messing, 2010; Yuan et al., 2014). Therefore, 27 kD γ-zein is both sufficient to form a homotypic PB and necessary for the normal assembly of a natural PB. Altogether, the features described above make 27 kD γ-zein an excellent model to define the structural requirements to initiate PB formation.

DISULFIDE BONDS, INSOLUBILITY AND INABILITY TO TRAFFIC

Deletions of γ-zein domains (Geli et al., 1994), amino acid substitutions carried out on fusions between the γ-zein N-terminal domain and otherwise soluble proteins (Pompa and Vitale, 2006; Llop-Tous et al., 2010), as well as *in vivo* treatments with reducing agent (Pompa and Vitale, 2006), all pointed to the importance of the Cys residues in the N-terminal region of the polypeptide in promoting γ-zein insolubility and ER retention, but this had not been demonstrated for the full-length protein. The results reported here demonstrate that 0C-γ-zein is fully soluble and very efficiently secreted, indicating that the presence of the C-terminal, 2S albumin-like domain and the N-terminal repeated region are not sufficient to form insoluble polymers and for ER retention in the absence of the seven N-terminal Cys residues. This efficient secretion also indicates that 0C-γ-zein is not a misfolded protein disposed by ER quality control. Indeed all constructs studied here are able to form dimers, suggesting a similar folding pathway. Forms that could represent dimers have also detected in transgenic plants expressing γ-zein (Geli et al., 1994; Bellucci et al., 2000) as well as upon expression of zeolin (Mainieri et al., 2004) and Zera (Torrent et al., 2009) fusions. Intracellular dimers and further polymers are detected in much higher proportions when extraction is performed in the absence of reducing agent. The detection of dimers of 0C-γ-zein and polymers of 1C-γ-zein suggest that either the intra-chain disulfide bonds of the 2S albumin-like region allow an overall conformation that favors inter-chain hydrophobic interactions or, perhaps surprisingly, the Cys residues present in this region are also directly involved in inter-chain bonds.

The results presented here confirm those obtained by Zera-ECFP mutagenesis regarding the direct relationship between number of Cys residues in the N-terminal domain and ability to form ER-located PB (Llop-Tous et al., 2010). However, the high importance of the first two N-terminal Cys in Zera-ECFP was not confirmed in full-length γ-zein: mutagenesis of these residues in Zera-ECFP resulted in full secretion (Llop-Tous et al., 2010), whereas 5C γ-zein is largely insoluble and ER-located, and it is almost unavailable for secretion. This indicates that the contribution of individual Cys residues of the N-terminal region to PB formation are less strict in full length γ-zein, as also suggested by domain deletion experiments (Geli et al., 1994).

As mentioned in the Introduction, synthetic $(VHLPPP)_8$ forms an amphipathic helix that interacts *in vitro* both with itself and with liposomes (Kogan et al., 2004). The inter-chain disulfide bonds may thus determine folding properties that stabilize and further promote similar interactions *in vivo*. The affinity with lipids may explain the natural position of γ-zein at the periphery of the PB, in close contact with the luminal face of the ER membrane, its ability form stable PBs also when expressed alone, and the altered PB shape in maize mutants that lack γ-zein. Thus, even if the very abundant α-zeins also assemble into disulfide bonded polymers and do not seem to interact covalently with γ-zein (Vitale et al., 1982), the latter seems fundamental to provide a scaffold that favors the stable architecture of a natural PB.

POLYMERIZATION AND VACUOLAR SORTING

As two, six or all seven Cys residues of the N-terminal domain were mutated, a parallel increase in BFA-sensitive secretion was observed. However, unlike WT γ-zein, intracellular 5C- and 1C-γ-zein were not exclusively located in the ER. The presence of a relevant proportion of protein at top of isopycnic gradients suggested a vacuolar localization, confirmed by vacuole isolation. The polypeptides delivered to the vacuole are soluble. Treatments with wortmannin indicated that this vacuolar delivery is due to traffic rather than autophagy and that the route is the same followed by most 2S albumins and 7/11S globulins. Very efficient secretion of 0C- and partial vacuolar sorting of 1C-γ-zein indicate that Cys^{117} acts as a determinant for vacuolar sorting. These results support a model of the evolution of PB formation from the mechanism of storage protein sorting to vacuoles and not directly from default secretion.

The γ-zein fragments used in zeolin and Zera constructs never included Cys^{117}, because they stopped a few amino acids ahead of it. We do not know whether any other Cys residue of the N-terminal region would be sufficient in promoting vacuolar sorting of full-length γ-zein, but the following observations indicate that this promotion occurs because the Cys residue stabilizes polymerization events that require other γ-zein domains. First, a γ-zein deletion construct that lacks most of the N-terminal region is very efficiently secreted but still contains Cys^{26}, Cys^{28}, and Cys^{117} (Geli et al., 1994), indicating that these residues are not sufficient *per se* to drive vacuolar sorting or ER retention. Second, experiments performed on the bean 7S globulin phaseolin indicate that the short C-terminal hydrophobic peptide that acts as a vacuolar sorting signal and promotes transient polymerization events (Frigerio et al., 1998; Holkeri and Vitale, 2001; Castelli and Vitale, 2005) can be at least in part replaced by a Cys residue that forms an inter-chain bond (Pompa et al., 2010). There is thus a striking similarity between the effect of the presence of a single additional Cys in 1C- compared to 0C-γ-zein and the artificial addition of a single Cys to a mutated, secreted phaseolin: in both cases vacuolar sorting is stimulated. It should be underlined that addition of Cys residues in phaseolin domains distant from the C-terminal end does not have any effect, indicating that disulfide bonds are formed when close interactions anyway exist, as expected (Pompa et al., 2010). In γ-zein, these interactions clearly require the Pro-rich amphipathic, repeated region.

AN EVOLUTIONARY MODEL

Prolamins have been divided into three groups, named I, II, and III starting from the most recently evolved (Xu and Messing, 2009).

The γ- and β-zeins have been assigned to group II. Their synthesis during seed development starts before that of the α- and δ-zeins, which belong to group I (Lending and Larkins, 1989). Maize does not have group III members. The 27 kD γ-zein is therefore among the most ancient maize prolamins (Xu and Messing, 2008, 2009). The A, B, and C domains that are common to 2S albumins and to other vacuolar seed proteins – such as trypsin inhibitor – are the most common feature of group II and III prolamins, indicating an evolution involving regions of vacuolar proteins (Xu and Messing, 2008). It has been hypothesized that conversion of Cys residues of the A, B, or C domains from an involvement in intra-chain disulfide bonds to inter-chain bonds may have been important events in this evolution (Kawagoe et al., 2005). The results shown here indicate a different model, in which at least one of the key mechanisms was the addition of new Cys residues that further increase and stabilize inter-chain interactions otherwise involved in vacuolar sorting (**Figure 5**). According to this model, the progressive addition reached a level that promoted the formation of large, insoluble polymers unable to traffic out of the ER (**Figure 6**).

The ER is the port of entry of the secretory pathway; intuitively, the development of a mechanism for protein accumulation in this compartment would be both more rudimentary and more energy-saving than using protein trafficking and sorting to vacuoles. However, the very complex array of players and molecular interactions involved in productive protein folding and quality control in the ER, the main functions of this compartment, may be negatively affected, at least in theory, by the presence of large amounts of stored material. Evolution of proteins that accumulate in the ER as insoluble polymers, apparently a very unusual event in any kingdom, may have been favored by the fact that the endosperm, unlike cotyledons, undergoes programmed cell death at the late stages of cereal seed development, making thus irrelevant the final location of stored proteins for what regards their mobilization during germination. It should, however, be underlined that the many experiments in which proteins that accumulate in the ER were ectopically expressed in vegetative tissues indicate that the plant ER is anyway very tolerant for what regards protein accumulation (Herman, 2008), another feature that may have allowed protein body formation in plants but not in members of other kingdoms.

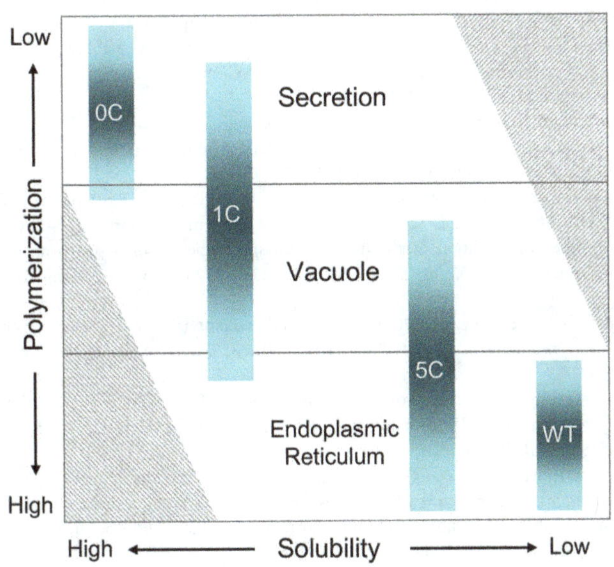

FIGURE 6 | A model for the relationships between solubility, polymerization and subcellular localization of seed storage proteins, based on the behaviors of WT, 5C-, 1C-, and 0C-γ-zein. The shaded columns indicate the subcellular localization of each construct, where the relative space occupied in each of the three locations is roughly proportional with the observed average distributions. The striped areas at lower left and upper right indicate situations predicted to be impossible. For simplicity, the model does not take into consideration other factors that determine or influence the subcellular localization, such as for example the short vacuolar sorting signals identified in several soluble 2S albumins and 7/11S globulins.

ACKNOWLEDGMENTS

We thank Angelo Viotti and Michele Bellucci for providing plasmid pBSKS.G1L. This work was supported by Programs "Risorse biologiche e tecnologie innovative per lo sviluppo sostenibile del sistema agroalimentare" and "Filagro" of CNR-Regione Lombardia.

REFERENCES

Adachi, M., Takenaka, Y., Gidamis, A. B., Mikami, B., and Utsumi, S. (2001). Crystal structure of soybean proglycinin alaB1b homotrimer. *J. Mol. Biol.* 305, 291–305. doi: 10.1006/jmbi.2000.4310

Anelli, T., and Sitia, R. (2010). Physiology and pathology of proteostasis in the early secretory compartment. *Semin. Cell Dev. Biol.* 21, 520–525. doi: 10.1016/j.semcdb.2010.02.006

Bagga, S., Adams, H., Kemp, J. D., and Sengupta-Gopalan, C. (1995). Accumulation of 15-kilodalton zein in novel protein bodies in transgenic tobacco. *Plant Physiol.* 107, 13–23.

Bellucci, M., Alpini, M., and Arcioni, S. (2000). Expression of maize γ-zein and β-zein genes in transgenic *Nicotiana tabacum* and *Lotus corniculatus*. *Plant Cell Tissue Organ Cult.* 62, 141–151. doi: 10.1023/A:1026735106843

Bellucci, M., Lazzari, B., Viotti, A., and Arcioni, S. (1997). Differential expression of a γ-zein gene in *Medicago sativa, Lotus corniculatus and Nicotiana tabacum*. *Plant Sci.* 127, 161–169. doi: 10.1016/S0168-9452(97)00092-7

Castelli, S., and Vitale, A. (2005). The phaseolin vacuolar sorting signal promotes transient, strong membrane association and aggregation of the bean storage protein in transgenic tobacco. *J. Exp. Bot.* 56, 1379–1387. doi: 10.1093/jxb/eri139

Coleman, C. E., Herman, E. M., Takasaki, K., and Larkins, B. A. (1996). The maize gamma-zein sequesters alpha-zein and stabilizes its accumulation in protein bodies of transgenic tobacco endosperm. *Plant Cell* 8, 2335–2345 doi: 10.1105/tpc.8.12.2335

daSilva, L. L., Taylor, J. P., Hadlington, J. L., Hanton, S. L., Snowden, C. J., Fox, S. J., et al. (2005). Receptor salvage from the prevacuolar compartment is essential for efficient vacuolar protein targeting. *Plant Cell* 17, 132–148. doi: 10.1105/tpc.104.026351

de Virgilio, M., De Marchis, F., Bellucci,M., Mainieri, D., Rossi, M., Benvenuto, E., et al. (2008). The human immunodeficiency virus antigen Nef forms protein bodies in leaves of transgenic tobacco when fused to zeolin. *J. Exp. Bot.* 59, 2815–2829. doi: 10.1093/jxb/ern143

Dombrowski, J. E., Gomez, L., Chrispeels, M. J., and Raikhel, N. V. (1994). "Targeting of proteins to the vacuole," in *Plant Molecular Biology Manual*, eds S. B. Gelvin, R. A. Schilperoort (Dordrecht: Kluwer Academic Publishers), 1–29.

Ems-McClung, S. C., Benmoussa, M., and Hainline, B. E. (2002). Mutational analysis of the maize gamma zein C-terminal cysteine residues. *Plant Sci.* 162, 131–141. doi: 10.1016/S0168-9452(01)00549-0

Foresti, O., De Marchis, F., de Virgilio M., Klein E. M., Arcioni S., Bellucci, M., et al. (2008). Protein domains involved in assembly in the endoplasmic reticulum promote vacuolar delivery when fused to secretory GFP, indicating a protein quality control pathway for degradation in the plant vacuole. *Mol. Plant* 1, 1067–1076. doi: 10.1093/mp/ssn066

Frigerio, L., de Virgilio, M., Prada, A., Faoro, F., and Vitale, A. (1998). Sorting of phaseolin to the vacuole is saturable and requires a short C-terminal peptide. *Plant Cell* 10, 1031–1042. doi: 10.1105/tpc.10.6.1031

Frigerio, L., Pastres, A., Prada, A., and Vitale, A. (2001). Influence of KDEL on the fate of Trimeric or assembly-defective phaseolin: selective use of an alternative route to vacuoles. *Plant Cell* 13, 1109–1126. doi: 10.1105/tpc.13.5.1109

Geli, M. I., Torrent, M., and Ludevid, D. (1994). Two structural domains mediate two sequential events in γ-zein targeting: protein endoplasmic reticulum retention and protein body formation. *Plant Cell* 6, 1911–1922.

Gomez, L., and Chrispeels, M. J. (1993). Tonoplast and soluble vacuolar proteins are targeted by different mechanisms. *Plant Cell* 5, 1113–1124. doi: 10.1105/tpc.5.9.1113

Gomord, V., Denmat, L. A., Fitchette-Lainé, A. C., Satiat Jeunemaitre, B., Hawes, C., and Faye, L. (1997). The C-terminal HDEL sequence is sufficient for retention of secretory proteins in the endoplasmic reticulum (ER) but promotes vacuolar targeting of proteins that escape the ER. *Plant J.* 11, 313–325. doi: 10.1046/j.1365-313X.1997.11020313.x

Gu, Y. Q., Wanjugi, H., Coleman-Derr, D., Kong, X., and Anderson, O. D. (2010). Conserved globulin gene across eight grass genomes identify fundamental units of the loci encoding seed storage proteins. *Funct. Integr. Genomics* 10, 111–122. doi: 10.1007/s10142-009-0135-x

Hara-Nishimura, I., Shimada, T., Hatano, K., Takeuchi, Y., and Nishimura M. (1998). Transport of storage proteins to protein storage vacuoles is mediated by large precursor-accumulating vesicles. *Plant Cell* 10, 825–836. doi: 10.1105/tpc.10.5.825

Herman, E. M. (2008). Endoplasmic reticulum bodies: solving the insoluble. *Curr. Opin. Plant Biol.* 11, 672–679. doi: 10.1016/j.pbi.2008.08.004

Holding, D. R., Otegui, M. S., Li, B., Meeley, R. B., Dam, T., Hunter, B. G., et al. (2007). The maize floury1 gene encodes a novel endoplasmic reticulum protein involved in zein protein body formation. *Plant Cell* 19, 2569–2582. doi: 10.1105/tpc.107.053538

Holkeri, H., and Vitale, A. (2001). Vacuolar sorting determinants within a plant storage protein trimer act cumulatively. *Traffic* 2, 737–741. doi: 10.1034/j.1600-0854.2001.21008.x

Ibl, V., and Stoger, E. (2012). The formation, function and fate of protein storage compartments in seeds. *Protoplasma* 249, 379–392. doi: 10.1007/s00709-011-0288-z

Jones, A. M., and Herman, E. M. (1993). KDEL-containing auxin binding protein is secreted to the plasma membrane and cell wall. *Plant Physiol.* 101, 595–606.

Kawagoe, Y., Suzuki, K., Tasaki, M., Yasuda, H., Akagi, K., Katoh, E., et al. (2005). The critical role of disulfide bond formation in protein sorting in the endosperm of rice. *Plant Cell* 17, 1141–1153. doi: 10.1105/tpc.105.030668

Kinney, A. J., Jung, R., and Herman, E. M. (2001). Cosuppression of the α subunits of β-conglycinin in transgenic soybean seeds induces the formation of endoplasmic reticulum-derived protein bodies. *Plant Cell* 13, 1165–1178.

Klein, E. M., Mascheroni, L., Pomp, A., Ragni, L., Weimar, T., Lilley, K. S., et al. (2006). Plant endoplasmin supports the protein secretory pathway and has a role in proliferating tissues. *Plant J.* 48, 657–673. doi: 10.1111/j.1365-313X.2006.02904.x

Kogan, M. J., Lopez, O., Cocera, M., Lopez-Iglesias, C., De La Maza, A., and Giralt, E. (2004). Exploring the interaction of the surfactant N-terminal domain of gamma-Zein with soybean phosphatidylcholine liposomes. *Biopolymers* 73, 258–268. doi: 10.1002/bip.10578

Lending, C. R., and Larkins, B. A. (1989). Changes in the zein composition of protein bodies during maize endosperm development. *Plant Cell* 1, 1011–1023. doi: 10.1105/tpc.1.10.1011

Levanony, H., Rubin, R., Altschuler, Y., and Galili, G. (1992). Evidence or a novel route of wheat storage proteins to vacuoles. *J. Cell Biol.* 119, 1117–1128. doi: 10.1083/jcb.119.5.1117

Llop-Tous, I., Madurga, S., Giralt, E., Marzabal, P., Torrent, M., and Ludevid, M. D. (2010). Relevant elements of a maize γ-zein domain involved in protein body biogenesis. *J. Biol. Chem.* 285, 35633–35644. doi: 10.1074/jbc.M110.116285

Lombardi, A., Marshall, R. S., Castellazzi, C. L., and Ceriotti, A. (2012). Redox regulation of glutenin subunit assembly in the plant endoplasmic reticulum. *Plant J.* 72, 1015–1026. doi: 10.1111/tpj.12020

Mainieri, D., Rossi, M., Archinti, M., Bellucci, M., De Marchis, F., Vavassori, S., et al. (2004). Zeolin: a new recombinant storage protein constructed using maize γ-zein and bean phaseolin. *Plant Physiol.* 136, 3447–3456. doi: 10.1104/pp.104.046409

Mellor, H., Kimball, S. R., and Jefferson, L. S. (1994). Brefeldin A inhibits protein synthesis through the phosphorylation of the alpha-subunit of eukaryotic initiation factor-2. *FEBS Lett.* 350, 143–146. doi: 10.1016/0014-5793(94)00756-X

Moriguchi, R., Matsuoka, C., Suyama, A., and Matsuoka, K. (2011). Reduction of plant-specific arabinogalactan-type O-glycosylation by treating tobacco plants with ferrous chelator 2,2 ′-dipyridyl. *Biosci. Biotechnol. Biochem.* 75, 994–996. doi: 10.1271/bbb.100884

Napier, J. A., Richard, G., Turner, M. F. P., and Shewry, P. R. (1997). Trafficking of wheat gluten proteins in transgenic tobacco plants: gamma-gliadin does not contain an endoplasmic reticulum retention signal. *Planta* 203, 488–494. doi: 10.1007/s004250050218

Pedrazzini, E., Giovinazzo, G., Bielli, A., de Virgilio, M., Frigerio, L., Pesca, M., et al. (1997). Protein quality control along the route to the plant vacuole. *Plant Cell* 9, 1869–1880. doi: 10.1105/tpc.9.10.1869

Pedrazzini, E., Komarova, N., Rentsch, D., and Vitale, A. (2013). Traffic routes and signals for the tonoplast. *Traffic* 14, 622–628. doi: 10.1111/tra.12051

Pompa, A., De Marchis, F., Vitale, A., Arcioni, S., and Bellucci, M. (2010). An engineered C-terminal disulfide bond can partially replace the phaseolin vacuolar sorting signal. *Plant J.* 61, 782–791. doi: 10.1111/j.1365-313X.2009.04113.x

Pompa, A., and Vitale, A. (2006). Retention of a bean phaseolin/maize gamma-zein fusion in the endoplasmic reticulum depends on disulfide bond formation. *Plant Cell* 18, 2608–2621. doi: 10.1105/tpc.106.042226

Prat, S., Cortadas, J., Puigdomenech, P., and Palau, J. (1985). Nucleic acid (cDNA) and amino acid sequences of the maize endosperm protein glutelin-2. *Nucleic Acids Res.* 13, 1493–1504. doi: 10.1093/nar/13.5.1493

Reyes, F. C., Chung, T., Holding. D., Jung, R., Vierstra, R., and Otegui, M. S. (2011). Delivery of prolamins to the protein storage vacuole in maize aleurone cells. *Plant Cell* 23, 769–784. doi: 10.1105/tpc.110.082156

Shewry, P. R., Napier, J. A., and Tatham, A. S. (1995). Seed storage proteins: structures and biosynthesis. *Plant Cell* 7, 945–956. doi: 10.1105/tpc.7.7.945

Tabe, L. M., Wardley-Richardson, T., Ceriotti, A., Aryan, A., McNabb, W., Moore, A., et al. (1995). A biotechnological approach to improving the nutritive value of alfalfa. *J. Anim. Sci.* 73, 2752–2759.

Torrent, M., Llompart, B., Lasserre-Ramassamy, S., Llop-Tous, I., Bastida, M., Marzabal P., et al. (2009). Eukaryotic protein production in designed storage organelles. *BMC Biol.* 7:5. doi: 10.1186/1741-7007-7-5

Virgili-López, G., Langhans, M., Bubeck, J., Pedrazzini, E., Gouzerh, G., Neuhaus, J.-M., et al. (2013). Comparison of membrane targeting strategies for the accumulation of the human immunodeficiency virus p24 protein in transgenic tobacco. *Int. J. Mol. Sci.* 14, 13241–13265. doi: 10.3390/ijms140713241

Vitale, A., and Hinz, G. (2005). Sorting of proteins to storage vacuoles: how many mechanisms? *Trends Plant Sci.* 10, 316–323. doi: 10.1016/j.tplants.2005.05.001

Vitale, A., Smaniotto, E., Longhi, R., and Galante, E. (1982). Reduced soluble proteins associated with maize endosperm protein bodies. *J. Exp. Bot.* 33, 439–448. doi: 10.1093/jxb/33.3.439

Wang, G., Wang, F., Wang, G., Wang, F., Zhang, X., Zhong, M., et al. (2012). Opaque1 encodes a myosin XI motor protein that is required for endoplasmic reticulum motility and protein body formation in maize endosperm. *Plant Cell* 24, 3447–3462. doi: 10.1105/tpc.112.101360

Woo, Y.-M., Hu, D. W.-N., Larkins, B. A., and Jung, R. (2001). Genomics analysis of genes expressed in maize endosperm identifies novel seed proteins and clarifies patterns of zein gene expression. *Plant Cell* 13, 2297–2317. doi: 10.1105/tpc.13.10.2297

Wu, Y., and Messing, J. (2010). RNA interference-mediated change in protein body morphology and seed opacity through loss of different zein proteins. *Plant Physiol.* 153, 337–347. doi: 10.1104/pp.110.154690

Xu, J. H., and Messing, J. (2008). Organization of the prolamin gene family provides insight into the evolution of the maize genome and gene duplications in grass species. *Proc. Natl. Acad. Sci. U.S.A.* 105, 14330–14335. doi: 10.1073/pnas.0807026105

Xu, J. H., and Messing, J. (2009). Amplification of prolamin storage protein genes in different subfamilies of the Poaceae. *Theor. Appl. Genet.* 119, 1397–1412. doi: 10.1007/s00122-009-1143-x

Yuan, L., Dou, Y., Kianian, S. F., Zhang, C., and Holding, D. R. (2014). Deletion mutagenesis identifies a haploinsufficient role for γ-zein in opaque2 endosperm modification. *Plant Physiol.* 164, 119–130. doi: 10.1104/pp.113.230961

Conflict of Interest Statement: The authors declare that the research was conducted in the absence of any commercial or financial relationships that could be construed as a potential conflict of interest.

9

The beginning of a seed: regulatory mechanisms of double fertilization

Andrea Bleckmann[1], Svenja Alter[2] and Thomas Dresselhaus[1*]

[1] _Cell Biology and Plant Biochemistry, Biochemie-Zentrum Regensburg, University of Regensburg, Regensburg, Germany_
[2] _Plant Breeding, Center of Life and Food Sciences Weihenstephan, Technische Universität München, Freising, Germany_

Edited by:
Paolo Sabelli, University of Arizona, USA

Reviewed by:
Yanhai Yin, Iowa State University, USA
Meng-Xiang Sun, Wuhan University, China

*Correspondence:
Thomas Dresselhaus, Cell Biology and Plant Biochemistry, Biochemie-Zentrum Regensburg, University of Regensburg, Universitätsstrasse 18, 93053 Regensburg, Germany
e-mail: thomas.dresselhaus@ur.de

The launch of seed development in flowering plants (angiosperms) is initiated by the process of double fertilization: two male gametes (sperm cells) fuse with two female gametes (egg and central cell) to form the precursor cells of the two major seed components, the embryo and endosperm, respectively. The immobile sperm cells are delivered by the pollen tube toward the ovule harboring the female gametophyte by species-specific pollen tube guidance and attraction mechanisms. After pollen tube burst inside the female gametophyte, the two sperm cells fuse with the egg and central cell initiating seed development. The fertilized central cell forms the endosperm while the fertilized egg cell, the zygote, will form the actual embryo and suspensor. The latter structure connects the embryo with the sporophytic maternal tissues of the developing seed. The underlying mechanisms of double fertilization are tightly regulated to ensure delivery of functional sperm cells and the formation of both, a functional zygote and endosperm. In this review we will discuss the current state of knowledge about the processes of directed pollen tube growth and its communication with the synergid cells resulting in pollen tube burst, the interaction of the four gametes leading to cell fusion and finally discuss mechanisms how flowering plants prevent multiple sperm cell entry (polyspermy) to maximize their reproductive success.

Keywords: pollen tube, ovule, gamete interaction, cell fusion, signaling, fertilization, polyspermy

INTRODUCTION

High crop yield strongly depends on efficient formation of numerous ovules, which after successful fertilization, develop into seeds comprising seed coat, embryo, and endosperm. In angiosperms, the haploid gametophytic generations produce the male and female gametes required to execute double fertilization. Both gametophytes are reduced to only a few cells. The female gametophyte is deeply embedded and thus protected by the maternal sporophytic tissues of the pistil (**Figure 1**). It harbors the female gametes (egg and central cell) and is surrounded by the nucellus tissue as well as the inner and outer integuments. After fertilization these different tissues form the seed coat. The female gametophyte arises from a megaspore mother cell though processes known as megasporogenesis and megagametogenesis (for review see Evans and Grossniklaus, 2009; Drews and Koltunow, 2011). In ~70% of all angiosperm species including *Arabidopsis* and maize the embryo sac develops according to the Polygonum type (Drews et al., 1998). The functional megaspore undergoes three mitotic divisions resulting in a syncytium containing eight nuclei. After nuclei migration and cellularization seven cells are differentiated: the haploid egg cell and its two adjoining synergid cells are located at the micropylar pole forming the egg apparatus. The homodiploid central cell containing two fused or attached nuclei is located more centrally, whereas three antipodal cells are found at the chalazal pole of the ovule opposite to the egg apparatus. While synergid cells are essential for pollen tube attraction, burst and sperm cell release (see below), the function of antipodal cells is so far unknown. During female gametophyte maturation antipodal cells are degenerating in the ovule of the eudicot model plant *Arabidopsis* (Mansfield et al., 1991), whereas they proliferate in other species including grasses and form a cluster of about 20–40 cells (Diboll and Larson, 1966).

The haploid male gametophyte (pollen grain) is formed during the processes of microsporogenesis and microgametogenesis from the microspore mother cell by meiosis and two successive mitotic divisions resulting in the formation of a tricellulate pollen grain. The vegetative cell encases the two sperm cells, which are connected with the vegetative cell nucleus by the generative cell plasma membrane, forming the male germ unit (MGU). MGU formation ensures the simultaneous delivery of both gametes to the ovule (for review see McCue et al., 2011). The major task of the vegetative cell is to deliver the sperm cells through the maternal tissues of the style and ovary to an unfertilized ovule. After pollen germination, the vegetative cell forms a tube and grows by tip-based-growth mechanism along papillae cells of the stigma into the style toward the transmitting tract. Inside the transmitting tract, pollen tubes are guided toward the ovules by mechanical and chemotactic cues involving numerous interactions with the sporophytic style tissues. In many eudicots pollen tubes exit the transmitting tract and grow along the septum, the funiculus and the outer integument

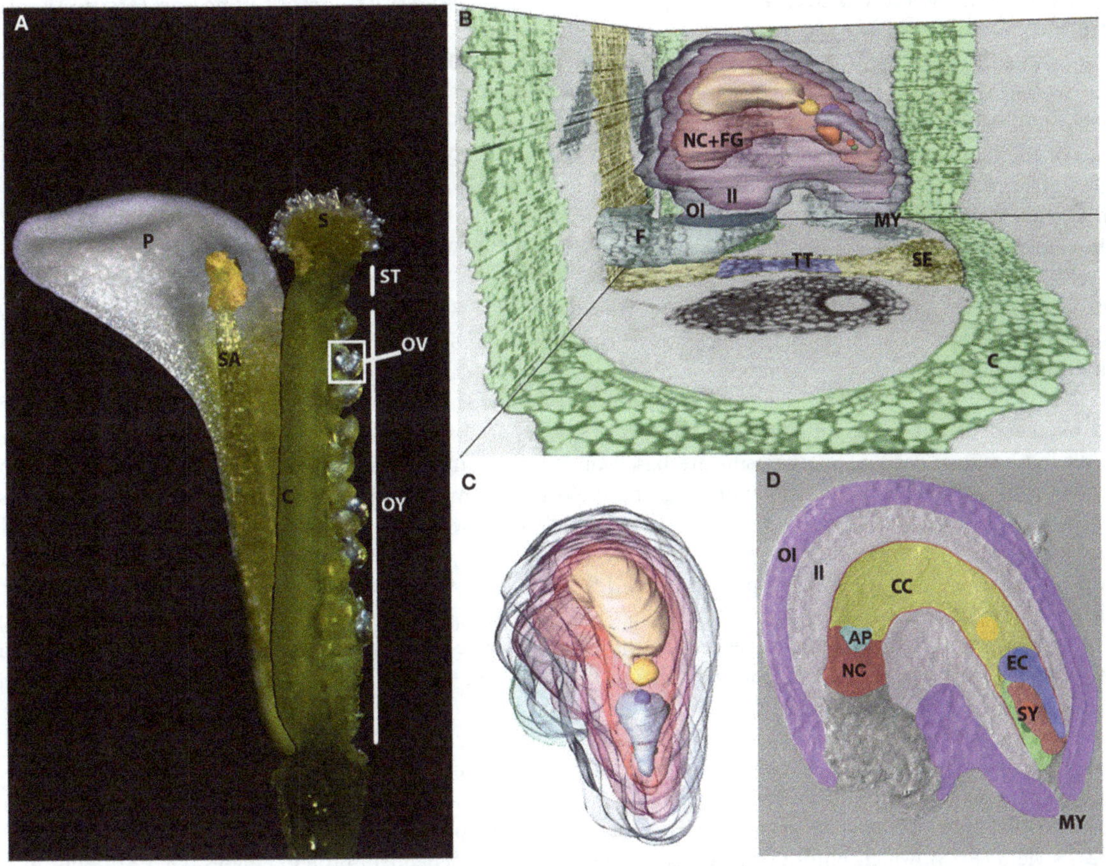

FIGURE 1 | The female gametophyte is deeply imbedded inside the female flower organs. (A) Dissected and reconstructed *Arabidopsis* flower. One of four petals (P) and one of six stamina (SA) are shown. They surround the pistil, which represents the female flower organ. It can be dissected into three parts. The upper part contains the papilla cells and forms the stigma (S), which is connected to the ovary (OY) by the style (ST). The ovary is formed by two fused carpels (C), which harbor two rows of ovules (OV). A side view **(B)** and front view **(C)** of a 3D-remodeled ovule reconstructed from toluidine blue stained single, successive ultra-thin sections of a dissected pistil. See Supplemental Movie 1 for whole series of sections. The ovule is connected to the septum (SE, yellow) containing the transmitting tract (TT, blue) by the funiculus (F, petrol) and surrounded by the carpel tissue (C) (green). A 3D-model of a dissected ovule shown from various angles is shown in Supplemental Movie 2. The mature female gametophyte cells (FG) and the nucellus tissue (NC) are surrounded by the outer (OI) and inner integuments (II) (OI, blue; II, purple). The vacuole and nucleus of the different female gametophyte cells showed highest contrast and are therefore shown individually. Near to the micropyle (MY), the two nuclei of the two synergid cells (SY) are shown in red and green. The egg cell, indicated by EC in **(D)**, has a comparably large vacuole (light blue) and its nucleus (blue) is located at its chalazal pole. The center of the female gametophyte is filled by the vacuole (light yellow) of the central cell, indicated by CC in **(D)**, and its homo-diploid nucleus (yellow). The three degenerating antipodal cells, indicated by AP in turquoise color in **(D)** at the chalazal pole are not highlighted. **(D)** DIC microscopic image of a mature female gametophyte surrounded by the maternal sporophytic tissues of the ovule. The cell types and tissues are artificially colored as shown in **(B,C)**. At full maturity the nucellus cell (NC) layer surrounding the developing embryo sac is flattened between inner integument (II) and female gametophyte cells.

toward the micropyle of unfertilized ovules. In grasses the ovary contains a single ovule and the pollen tube is directly guided toward its surface after leaving the blind ending transmitting tract. The pollen tube continues to grow along its surface toward the micropylar region (for review see Lausser and Dresselhaus, 2010). Finally, the pollen tube enters the micropyle, an opening between the inner and outer integuments, and grows toward the two synergid cells. The pollen tube bursts and sperm cells are released. This process is associated with the degeneration of the receptive synergid cell due to programmed cell death. Subsequently, both sperm cells arrive at the gamete fusion site and fertilize the egg and central cell (Hamamura et al., 2011). From the moment of germination until sperm discharge the pollen grain/tube communicates with at least five different sporophytic and three different gametophytic cell types to successfully accomplish fertilization (Palanivelu and Tsukamoto, 2012). Its extended growth inside the female flower tissue is regulated by many different guidance, attraction and support mechanisms. After sperm cell release all gametes are activated, followed by fusion of their membranes and nuclei by processes known as plasma- and karyogamy, respectively. After successful double fertilization further signaling events are activated to prevent polyspermy. In this review we will summarize and discuss the cell–cell communication processes, which are essential to successfully accomplish double fertilization and to initiate seed development in angiosperms.

POLLEN TUBE GROWTH AND ATTRACTION

POLLEN REJECTION

Pollen tube growth and guidance toward the female gametes are controlled at various stages by chemotactic signals and growth support molecules derived from the sporophytic and gametophytic tissues of the female flower organs. Pollen grains placed on the stigma (**Figure 1A**) by contact, wind or different pollinators stick to the papilla cells and start to hydrate followed by their germination. The efficient adhesion of the pollen grain to the papilla cell is regulated by interaction events between these cells and may activate thereby inter- and intra-species barriers to prevent unsuccessful pollination and fertilization events already at this early time point during reproduction. Angiosperms possess different strategies to recognize self from alien pollen and evolved independent self-incompatibility (SI) mechanisms to prevent self-fertilization. Early SI mechanisms are based on cell-cell communication events between the papilla cells and the pollen grains, whereas later SI mechanisms occur while the growing pollen tube interacts with the cells of the transmitting tract. Species of the *Solanaceae*, for example, use a pistil-expressed S-RNase, which penetrates the pollen tube (McClure et al., 1989; Luu et al., 2000). A compatible pollen tube expresses the S-locus F-box protein (SLF), which leads to the degradation of the S-RNase (Hua and Kao, 2006; Kubo et al., 2010), while in incompatible interactions intact S-RNase degrades RNAs resulting, for example, in the disruption of the actin cytoskeleton and other cellular processes (Liu et al., 2007; Roldán et al., 2012). In *Papaveraceae* SI depends on the small pistil secreted protein *Papaver rhoeas* style S (PrsS), which binds to the S-locus pollen tube membrane protein *P. r.* pollen S (PrpS) and activates a Ca^{2+}-dependent signaling cascade resulting in pollen inhibition and programmed cell death (Wheeler et al., 2009; Wu et al., 2011). SI is best understood in *Brassicaceae*, which use a surface-localized S-locus receptor kinase (SRK) in papilla cells (Takasaki et al., 2000) and a pollen coat localized cysteine-rich protein (SP11/SCR) (Schopfer, 1999; Shiba et al., 2001) to distinguish self from alien pollen. Their successful interaction leads to proteasome dependent degradation of Exo70A1, an essential component of the exocyst complex. It is thought to be involved in secretion of essential pollen germination factors necessary for pollen hydration (Synek et al., 2006; Samuel et al., 2009). Rejection of pollen in *Brassicaceae* thus occurs already during pollen hydration and germination at the surface of papilla cells. Little is known about SI in the economically important grasses (reviewed in Dresselhaus et al., 2011). Pollen hydration and germination appear not to be affected, although only grass pollen tubes are capable of penetrating the style and reach the transmitting tract. This indicates that SI in the grasses depends on successful interaction of the pollen tube with the sporophytic cells of the style and transmitting tract. The signaling events involved in this recognition process still await their discovery. More details about SI mechanisms can be found in Iwano and Takayama (2012), Watanabe et al. (2012) and Dresselhaus and Franklin-Tong (2013).

POLLEN TUBE GUIDANCE TOWARD AND THROUGH THE TRANSMITTING TRACT

After adhesion and hydration, compatible pollen germinates, penetrates the style and grows through the extracellular space of stylar cells toward the transmitting tract (**Figure 1B**). The growth direction of the pollen tube is regulated by the formation of different gradients including water, γ-amino butyric acid (GABA), calcium and other small molecules such as D-serine. The water flow during hydration forms an external gradient specifying the site of pollen tube outgrowth and was shown to be controlled by triacylglyceride (Lush et al., 1998; Wolters-Arts et al., 1998). Ca^{2+} influx into the pollen tube tip region is known to be essential for germination and tube growth (Brewbaker and Kwack, 1963; for a review see Steinhorst and Kudla, 2013a) and leads to the generation of an oscillating apex-based cytoplasmic Ca^{2+} (Ca^{2+}_{cyto}) gradient (Miller et al., 1992; Calder et al., 1997). Initially, papilla cells export Ca^{2+}_{cyto} by the auto-inhibited Ca^{2+}-ATPase13 (ACA13) at the pollen grain adhesion site (Iwano et al., 2004, 2014). Extracellular Ca^{2+} is then imported into the pollen tube by glutamate receptor-like channels (GLRs), which can be stimulated by D-serine (Michard et al., 2011). In animal systems it was shown that GLRs are non-selective cation channels catalyzing Na^{+} and/or Ca^{2+} influx into cells. Binding of the agonist D-serine to GLRs should thus lead to channel opening resulting in a Ca^{2+}_{cyto} increase (Gilliham et al., 2006). D-serine is produced by Serine-Racemase1 (SR1), which shows an expression peak in the style indicating D-serine availability. The induced changes in Ca^{2+}_{cyto} concentration in the pollen tube might thereafter regulate and coordinate many different signaling events like actin polymerization and thus influence pollen tube growth behavior and growth direction. Ca^{2+}_{cyto}-sensors, belonging to the protein families of calmodulin (CaM), calmodulin-like proteins (CMLs), calcium-dependent protein kinases (CDPKs), and calcineurin B-like proteins (CBLs) are expressed in the pollen tube and are thought to control different cell-cell communication events indicated by their localization and overexpression phenotypes. The presence of these different Ca^{2+}_{cyto} receptors around the sperm cells and at the pollen tube tip indicate an essential role of Ca^{2+} signals both during pollen tube growth and double fertilization (Zhou et al., 2009; Steinhorst and Kudla, 2013b).

During pollen tube growth the tip needs to modulate the surrounding cell wall of stylar cells enabling its penetration through the extra-cellular space, most likely by interaction with extensin-like proteins and arabinogalactan proteins as well as the secretion of cell wall softening enzymes and inhibitors such as polygalacturonases and pectin methylesterase inhibitors (Cosgrove et al., 1997; Grobe et al., 1999; Stratford et al., 2001; Ogawa et al., 2009; Nguema-Ona et al., 2012; Woriedh et al., 2013). The transmitting tract is composed of small cylindrical cells that are surrounded by an extracellular matrix (ECM), which contains a mixture of glycoproteins, glycolipids, and polysaccharides (Lennon et al., 1998). The ECM provides essential nutrients as well as components for an accelerated, extended and guided pollen tube growth (Palanivelu and Preuss, 2006). Without an intact transmitting tract like in the *NO TRANSMITTING TRACT* (*NTT*) mutant or its target *HALF FILLED* (*HAF*), pollen tube growth is severely affected and either slowed down or prematurely terminated. *NTT* encodes a C2H2/C2HC zinc finger transcription factor involved in ECM production and is essential for programmed cell death in the transmitting tract upon pollination (Crawford et al., 2007). *HAF* encodes a bHLH transcription factor and is involved in NTT dependent transmitting tract regulation (Crawford and Yanofsky,

2011). The transmitting tract-specific arabinogalactan glycoproteins TTS1 and TTS2 have a positive effect on *in vitro* grown tobacco pollen and show a gradient of increased glycosylation correlating with pollen tube growth direction inside the transmitting tract (Cheung et al., 1995; Wu et al., 1995). Another factor which has a positive effect on pollen tube growth and guidance is chemocyanin, a small secreted peptide in the style of lily (Kim et al., 2003). The different sporophyte-derived signals do not only guide or increase pollen tube growth rate, but rather lead to a change in the pollen transcriptome and thereby activate the pollen for female gametophyte-derived attraction signals (Higashiyama et al., 1998; Palanivelu and Preuss, 2006). Recently, *de novo* expression of closely related MYB transcription factors and other genes were reported to be induced during pollen tube growth through the style regulating themselves a number of downstream genes. Hence pollen tubes maturate during their growth through the sporophytic tissue and thereby become competent for fertilization (Leydon et al., 2013, 2014).

OVULAR POLLEN TUBE GUIDANCE

The signaling events that control pollen tube exit from the transmitting tract and guidance toward the ovule are not known. In *Arabidopsis* this process was shown to be tightly regulated and usually only a single pollen tube exits the transmitting tract in proximity of an unfertilized ovule. The pollen tube grows on the septum surface toward the funiculus, the tissue connecting the ovule with the septum (**Figures 1B–D**; Supplemental Movies 1, 2). At the funiculus the pollen tube is directed through the micropyle inside the ovule by a mechanism known as micropylar guidance (Shimizu and Okada, 2000). In *Arabidopsis* a gradient of GABA was reported in front of the ovule. The transaminase POLLEN ON PISTIL2 (POP2) forms this gradient through GABA degradation. At moderate concentrations GABA stimulates pollen tube growth and thus likely supports growth toward the ovule (Palanivelu et al., 2003). Another candidate involved in micropylar guidance is D-serine, which was already described above. Its synthesizing enzyme gene *SR1* is also expressed in the ovule indicating the presence of D-serine (Michard et al., 2011). Semi-*in vitro* fertilization experiments revealed an oscillation of Ca^{2+}_{cyto} levels in growing pollen tubes depending on their distance from an unfertilized ovule and especially from the synergid cells (Shi et al., 2009; Iwano et al., 2012). The connection between Ca^{2+}_{cyto} and D-serine by GLR channels in growing pollen tubes was already described above. The observed changes in the Ca^{2+}_{cyto} level depending on its distance from the synergid cells might again result from this interplay.

Recently, two pollen-expressed mitogen-activated protein kinases (MAPKs), MPK3 and MPK6, were identified in *Arabidopsis*, which are part of the ovular guidance network. *In vivo* pollination assays revealed that *mpk3/6* double mutant pollen tubes were not capable of growing along the funiculus after transmitting tract exit but micropylar guidance (see below) was not effected in the double mutants (Guan et al., 2014). MPK3/6 are two cytoplasmic protein kinases, which seem to be part of the signaling cascade mediating extracellular stimuli to changes in pollen tube growth direction.

In summary, our current understanding of ovular pollen tube guidance is very limited, but a whole orchestra of small molecules derived from the ovule seem to be involved in pollen tube growth support and attraction, and multiple signaling networks are required in pollen tubes to respond to the diverse set of signals and to direct their growth behavior.

MICROPYLAR POLLEN TUBE GUIDANCE

After arrival at the surface of the ovule, the pollen tube reaches the last phase of its journey, which is known as micropylar pollen tube guidance. It enters the micropyle, an opening between the two integuments, and directly grows toward the egg apparatus in species such as *Arabidopsis* (**Figure 2A**). In grasses the pollen tube first has to overcome a few layers of nucellus cells (Márton et al., 2005) before it also gets in contact with the filiform apparatus of the synergid cells, a thickened and elaborated cell wall at their micropylar pole, where the cell surface is extensively invaginated (Willemse and van Went, 1984; Huang and Russell, 1992). It was believed for a long time that the pollen tube grows through the filiform apparatus to enter one synergid cell, leading to pollen tube burst and cell death of the receptive synergid cell. Recently, it was shown that the pollen tube is repelled by the filiform apparatus and instead grows along the cell wall of the synergid cells until it reaches a certain point after the filiform apparatus where its growth is arrested and burst occurs explosively (Leshem et al., 2013). Pollen tube burst results in the discharge of its cytoplasmic contents including the two sperm cells. The synergid cells represent the main source for chemo-attractants required for micropylar pollen tube guidance. Moreover, laser ablation experiments in *Torenia fournieri* have demonstrated that a single synergid cell is sufficient and necessary to attract pollen tubes (Higashiyama et al., 2001). The major function of the filiform apparatus may thus be to considerably increase the micropylar surface of the synergid cells, which represent glandular cells of the egg apparatus. Many known components required for pollen tube growth and guidance are membrane-associated and accumulate at the filiform apparatus, which gives it the additional role of a signaling platform. It contains, for example, a high Ca^{2+} concentration, which is known to play a key role during the regulation of pollen tube growth (Brewbaker and Kwack, 1963; Chaubal and Reger, 1990; Iwano et al., 2004; Michard et al., 2011) and also seems to trigger pollen tube burst afterwards (see below). In *Arabidopsis* the formation of the filiform apparatus as well as the expression of different attractants in the synergid cells depend on the activity of the R2R3-type Myb transcription factor MYB98 (Kasahara et al., 2005; Punwani et al., 2007). Among other genes, *MYB98* regulates the expression of genes encoding cysteine-rich proteins (CRPs), including those representing a subgroup of defensin-like (DEFL) polypeptides (Punwani et al., 2008; Takeuchi and Higashiyama, 2012).

In *Torenia* it was shown that a DEFL subgroup of CRPs called LUREs are secreted from the synergid cells and accumulate at the filiform apparatus (Okuda et al., 2009; Kanaoka et al., 2011). LUREs attract pollen tubes in a species-preferential manner from a distance of about 100–150 μm and were recently shown to bind to the tip region of pollen tubes (Okuda et al., 2009, 2013). Due to their rapid molecular evolution it was difficult to

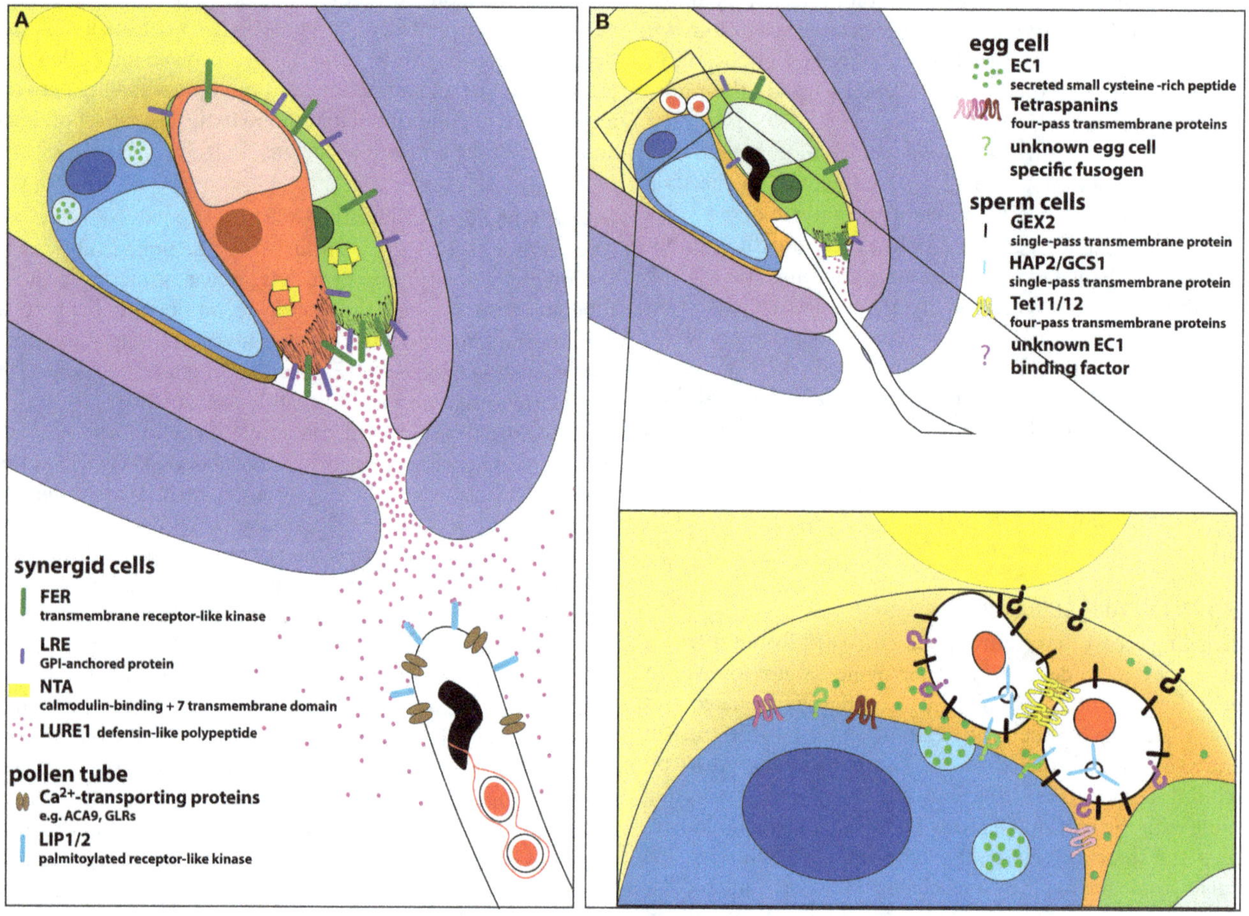

FIGURE 2 | Model of signaling events during micropylar pollen tube attraction and double fertilization in *Arabidopsis*. (A) The micropylar opening of the ovule is formed by the inner and outer integuments. The female gametophyte is "naked" at its micropylar pole containing one egg (colored in blue) and two synergid cells (colored in green and red) representing the egg apparatus. The central cell surrounds the egg apparatus. The synergid cells are the main sources of pollen tube attractants. Among other components, they secrete LURE peptides, which bind to pollen expressed LIP1/2 receptors thus directing pollen tube growth. Calcium transporters are involved in pollen tube growth control. The plasma membranes of synergid cells harbor a high concentration of receptors like FER and LRE, especially in the region of the filiform apparatus. Upon pollen tube perception NTA is relocated to the plasma membrane by FER activity likely regulated by Ca^{2+} oscillations inside the synergid cells. **(B)** The pollen tube bursts when it reaches a certain point beyond the filiform apparatus and releases its cytoplasmic contents including the two sperm cells. Pollen tube burst depends on the presence and activation of FER-, LER-, NTA-, and VDD-dependent signaling cascades culminating in the death of the receptive synergid cell (indicated by diffusing red color). Released sperm cells are located at the gamete fusion side, between the two female gametes. The two sperm cells are connected to each other, likely involving tetraspanins. The male gametes adhere to female gametes by GEX2 located at their surface. After activation, the egg cell secrets EC1 leading to sperm cell activation and HAP2/GCS1 localization to the plasma membrane. HAP2/GCS1 and tetraspanins at the surface of gametes may be involved in mediating membrane fusion. Unknown egg and central cell-specific fusogenic proteins as well as EC1 receptor are indicated by question marks in green, black, and purple, respectively.

identify orthologs in other plant species, but finally the DEFL subgroup CRP810/AtLURE1 of *Arabidopsis* was discovered to be involved in micropylar pollen tube guidance (Takeuchi and Higashiyama, 2012). In *Zea mays*, EGG APPARATUS1 (ZmEA1), a small hydrophobic precursor protein of 94 aa was reported as an egg apparatus-specific protein required for micropylar pollen tube guidance (Márton et al., 2005, 2012). ZmEA1 was shown to bind in a species-specific manner to the apical region of the pollen tube, where it is quickly internalized and degraded, likely keeping the pollen tube susceptible to pollen tube attractants while growing through the micropylar nucellus cell layers (Márton et al., 2012; Uebler et al., 2013).

More puzzling is the role of the central cell in micropylar guidance of the pollen tube. For example *magatama* (*maa*) mutants show defects in central cell maturation; both haploid nuclei are smaller and often fail to fuse. Pollen tubes grow in the direction of an unfertilized *maa* ovule but loose their way just before entering the micropyle. Moreover, mutant female gametophytes attracted two pollen tubes at a high frequency (Shimizu and Okada, 2000). *MAA3* was recently shown to encode a helicase required for general RNA metabolism, which could explain the central cell maturation defect but not the defect in pollen tube guidance (Shimizu et al., 2008). Another example of central cell-dependent defects in micropylar pollen tube guidance is the transcriptional regulator CENTRAL CELL GUIDANCE (CCG), which is expressed exclusively in the central cell of the female gametophyte (Chen et al., 2007). These guidance defects may be indirect and caused by non-functional or immature central cells influencing maturation of egg apparatus cells and thus the generation of guidance components in these cells. It might also be possible that molecules generated by the MAA3 and CCG pathways directly regulate the generation of guidance molecules in the neighboring cells. Also the egg cell seems to be involved in micropylar guidance. GAMETE EXPRESSED 3 (GEX3) is a plasma membrane-localized protein, which is expressed in the unfertilized egg cell. Down-regulation of *GEX3* by antisense RNA in the egg cell leads to defects in micropylar guidance by an unknown mechanism (Alandete-Saez et al., 2008).

Until recently, male factors and signaling pathways reacting to attractants secreted from the egg apparatus were unknown. The receptor-like kinases (RLKs) LOST IN POLLEN TUBE GUIDANCE1 (LIP1) and 2 (LIP2) have been identified, which are preferentially expressed in the pollen tube. Both proteins show membrane localization due to a palmitoylation site and are involved in the AtLURE1-dependent guidance mechanism. *lip1/2* double mutant pollen reach the funiculus but fail to grow through the micropyle inside the ovule, and the pollen tube shows a reduced attraction toward AtLURE1 (Liu et al., 2013). However, it is unclear whether LIP1/2 are directly involved in LURE perception.

POLLEN TUBE BURST AND SPERM CELL DISCHARGE

Pollen tube burst seems to be regulated by RLKs located at surfaces of both male and female interaction partners (**Figures 2A,B**). The RLK FERONIA/SIRENE (FER/SRN) is expressed in most tissues including the synergid cells, where it localizes predominately at their surface in the filiform apparatus region. Loss-of-function mutants display a pollen tube-overgrowth phenotype. Pollen tube growth arrest and sperm cell discharge fail in *fer* ovules (Huck et al., 2003; Rotman et al., 2003; Escobar-Restrepo et al., 2007). *FER* acts as a cell surface regulator for RAC/ROP GTPases. Recently it was shown that FER binds to the small secreted peptide Rapid Alkalinization Factor (RALF), which leads to the inhibition of a plasma membrane H^+-ATPase resulting in the suppression of cell elongation in the primary root (Haruta et al., 2014). Besides changes in the pH, RALF also induces the increase of Ca^{2+}_{cyto} (Pearce et al., 2001; Haruta and Constabel, 2003; Haruta et al., 2008) and thus may influence pollen tube growth arrest and eventually its burst. The *Arabidopsis* genome contains around 30 *RALF*-like genes indicating the possibility that a pollen secreted RALF-like peptide may indeed be involved in FER-dependent pollen tube perception (Olsen et al., 2002). Other proteins were identified whose loss-of-function resemble the *fer* phenotype such as the glycosylphosphatidylinositol (GPI)-anchored protein LORELEI (LRE) and the Mildew Resistance Locus O (MLO) family protein NORTIA (NTA). Both genes are expressed in synergid cells and show a similar pollen tube overgrowth phenotype (Capron et al., 2008; Kessler et al., 2010). *NTA* encodes a protein with multiple potential transmembrane domains as well as a calmodulin-binding site. The frequency of unfertilized ovules in *lre/lre* and *nta/nta* is less pronounced compared to *fer/fer* pistils. This finding indicates that FER activity is essential for pollen tube perception, while other, yet unknown factors act redundantly with LRE and NTA. These factors are all present at the synergid plasma membrane during pollen tube contact. However, while FER accumulates at the filiform apparatus already prior to pollen tube arrival, NTA relocalizes to the plasma membrane of the synergid cells at the filiform apparatus region upon pollen tube contact. In a transient expression system NTA is directly targeted to the plasma membrane. However, in *Arabidopsis* ovules under control of its endogenous promoter, NTA localizes to uncharacterized compartments within the cell, and becomes relocalized to the plasma membrane upon pollen tube arrival, indicating the presence of an active retention mechanism. This relocalization is FER-dependent and therefore connects FER to NTA in the same signaling network (Kessler et al., 2010). The presence of a calmodulin-binding domain in NTA supports the idea of a Ca^{2+}-dependent signaling network, which is activated upon pollen tube arrival. Due to its predicted signal peptide and GPI anchor, LRE is expected to localize to the extracellular side of the plasma membrane after passage through the secretory pathway. But so far plasma membrane localization could only be shown in a transient expression system and not in synergid cells themselves (Capron et al., 2008). Whether LRE localization also changes during the process of fertilization needs to be elucidated. Another factor required for successful pollen tube/synergid cell communication is VERDANDI (VDD), a member of the plant-specific B3 superfamily of transcription factors. *Vdd* mutants show defects in antipodal and synergid cell identity and result in the lack of pollen tube burst after reaching the synergid cells. In contrast to *fer, nta*, or *lre* mutants, an overgrowth phenotype was not reported indicating that VDD may act downstream of cell surface signaling components (Matias-Hernandez et al., 2010). Little is known about male factors involved in pollen tube/synergid cell communication. Two closely related homologs of FER, ANXUR1 (ANX1), and ANX2 were reported to be involved in the timing of pollen tube burst or more precisely in the inhibition of pollen tube burst. In an *in vitro* pollen tube growth assay, pollen of double mutants show spontaneous discharge already after pollen bulge formation, whereas *in vivo*-grown pollen tubes germinate normally on a stigma but rupture in the style before arriving at the egg apparatus. Both receptors localize mainly to the apical tip of the pollen tube as well as in small vesicles (Boisson-Dernier et al., 2009; Miyazaki et al., 2009). Their over-expression inhibits growth by over-acting exocytosis and over-accumulation of secreted cell wall material (Boisson-Dernier et al., 2013) suggesting that the main function is associated with coordination of growth through the style rather than sperm cell discharge. Other male factors, which are involved in pollen tube growth and reception, are the pollen-expressed transcription factors MYB97, MYB101 and MYB120. *myb97/101/120* triple mutants exhibited uncontrolled growth and failed to discharge their sperm cells after entering the embryo sac (Liang et al., 2013). It is thought that these factors are required to enable pollen tube to communicate with the pistil tissues and the female gametophyte (Leydon et al., 2014). As already mentioned, the level of Ca^{2+}_{cyto} alters during pollen tube elongation. Upon pollen tube arrival Ca^{2+}_{cyto} level starts to oscillate in the synergid cells, triggered by the contact of the pollen tube tip with the synergid cell. This oscillation can be observed until pollen tube burst, which leads to the degeneration of one synergid cell (Sandaklie-Nikolova et al., 2007; Iwano et al., 2012; Denninger et al., 2014; Ngo et al., 2014). These changes in Ca^{2+}_{cyto} level are essential for sperm delivery and are depending on FER and LRE activity, respectively. Downstream of FER, NTA localization to the synergid cell surface and its activity likely depend on sufficient Ca^{2+}_{cyto} level in the synergid cells (Ngo et al., 2014). The synergid cell, which is in contact with the pollen tube follows a regulated cell death program that is somehow associated and controlled by pollen tube burst and linked to oscillation of Ca^{2+}_{cyto} level (Higashiyama et al., 2000; Sandaklie-Nikolova et al., 2007; Denninger et al., 2014; Ngo et al., 2014). It cannot be explained by mechanical breakdown due to an invading pollen tube. In *fer* mutant, for example, pollen tube growth continues around the synergid cells, which must be associated with mechanical stress but does not induce synergid cell death (Escobar-Restrepo et al., 2007). However, the signaling events, which are responsible for programmed cell death in the synergid cell, are not understood yet. In maize pollen tube arrival is associated with the secretion of defensin-like ZmES proteins, inducing pollen tube burst by activating the K^+ -channel *Zea mays* 1 (KZM1) in the pollen tube membrane (Amien et al., 2010). Whether these "toxin"-like molecules are also capable of inducing synergid cell burst remains to be shown.

GAMETE INTERACTION AND PREVENTION OF POLYSPERMY

SPERM CELL DELIVERY, ACTIVATION, AND GAMETIC MEMBRANE INTERACTIONS

Once released, the two sperm cells are delivered to the so-called gamete fusion site between egg and central cell (**Figure 2B**). It is controversial whether this requires active transport or is solely based on cytoplasmic flow associated with burst of both pollen tube and receptive synergid cell and/or the architecture of the egg apparatus. Most flowering plants, like *Arabidopsis*, generate isomorphic sperm cells and therefore fusion of sperm cell appears to be random, either with the egg or the central cell (Berger et al., 2008; Ingouff et al., 2009; Liu et al., 2010). Some reports suggest that fertilization of the egg cell is preferred, which was demonstrated, for example, in mutants of *CYCLIN DEPENDENT KINASE A1* (*CDKA;1*), which generate only one sperm-like germ cell (Iwakawa et al., 2006; Nowack et al., 2006). Experiments with photo-labeled sperm cells have demonstrated that there is no preference for either female gamete (Hamamura et al., 2011). The differentiation into two equal sperm cells depends on the activity of the MYB transcription factor DUO POLLEN 1 (DUO1), which is required for correct male germ cell differentiation by regulating key genes essential for fertilization such as *GAMETE EXPRESSED 2* (*GEX2*) and *GENERATIVE CELL SPECIFIC 1* (*GCS1*), also known as *HAPLESS 2* (*HAP2*) (Brownfield et al., 2009). *GEX2* encodes a single-pass transmembrane protein with filamin repeats exposed to the extracellular space. GEX2 localizes to the sperm cell plasma membrane and contains extracellular immunoglobulin-like domains, similar to gamete interaction factors reported in algae and mammals (Misamore et al., 2003; Inoue et al., 2005). In the presence of GEX2 the two gametes adhere to the egg and central cell. *gex2* mutant sperm cells show reduced adhesion to female gametes, likely causing cell fusion failure (Mori et al., 2014). GCS1/HAP2 is another factor required for gamete interaction in *Arabidopsis*. After pollen tube burst, both sperm cells of *gcs1/hap2* loss of function mutants remain at the fusion site and fail to fuse with female gametes, leading to the attraction of additional pollen tubes (polytubey). It was further shown that in the absence of the potential fusogen GCS1/HAP2, attachment of male to the female gamete occurs but no membrane fusion is visible, implying that the protein mediates membrane fusion as a component of signaling events, or more likely that it is directly involved in the fusion event (Wong and Johnson, 2010; Mori et al., 2014). GCS1/HAP2 is a conserved protein and has been identified in genomes of all major eukaryotic taxa except fungi. *Gcs1/hap2* mutants in protozoan and algal gametes result in fusion failure, suggesting that this protein is required for a common mechanism of membrane fusion in eukaryotes (Mori et al., 2006; Hirai et al., 2008; Liu et al., 2008; Steele and Dana, 2009; Wong and Johnson, 2010). Upon sperm cell arrival at the gamete fusion site (**Figure 2B**) the egg cell starts to secrete small cysteine-rich proteins of the EGG CELL 1 (EC1) family. EC1 leads to the relocalization of HAP2/GCS1 from the endomembrane system to the sperm cell plasma membrane and thus activates sperm cells enabling them to fuse with the female gametes (Sprunck et al., 2012). The egg cell appears to require activation itself and calcium may play a key role in this process; this is indicated by a single strong Ca^{2+}_{cyto} transient in the egg cell associated with pollen tube burst and sperm delivery (Denninger et al., 2014), which thus precedes EC1 secretion.

The relocalization of a transmembrane fusogen was also described for the mammalian-specific fusogen IZUMO1 (Inoue et al., 2005). Female components, which are directly involved in gamete fusion are so far unknown in higher plants. In mammals, CD9-like membrane spanning proteins of the tetraspanin family are located at the plasma membrane of eggs and were shown to be required for gamete fusion (Kaji et al., 2000; Le Naour et al., 2000; Miyado et al., 2000). In *Arabidopsis* the conserved tetraspanin family consists of 17 members. While TET11 and especially TET12 are located at the surface of sperm cells and reach high concentrations in the membrane region connecting both sperm cells, TET9 appears at the surface of female gametophyte cells including the egg and central cell (Boavida et al., 2013). *Arabidopsis* tetraspanins were shown to form homo- and heterodimers, but so far functional studies are missing. However, their presence at the surface of plant gametes and structural homology to mammalian CD9-like proteins suggest that they may possess a similar role during gamete interaction.

DEGRADATION OF FUSOGENS AND PREVENTION OF POLYSPERMY

In general polyspermy blocks prevent multiple fertilization events that would otherwise lead to abnormal development or even embryo lethality, and thus reproductive failure. In *Chlamydomonas* FUS1, a single-pass transmembrane protein with a high similarity to prokaryotic invasion and adhesion molecules, mediates membrane fusion (Ferris et al., 1996; Misamore et al., 2003). In *Chlamydomonas* both GCS1/HAP2 and FUS1 are rapidly degraded after cell fusion, resulting in a fast membrane block to prevent polygamy (Misamore et al., 2003; Liu et al., 2010). In mammals, it was recently shown that IZUMO1 is recognized by the GPI-anchored protein JUNO on the egg cell surface. Rapid degradation of JUNO after fertilization suggests an additional mechanism for membrane block to prevent polyspermy (Bianchi et al., 2014).

In *Arabidopsis* it was shown that the polyspermy block only functions in the egg cell and not in the central cell, which is capable of fusing with more than one sperm cell, as demonstrated in the *tetraspore* (*tes*) mutant. This mutant produces more than one sperm pair, which is released simultaneously at the gamete fusion site. After fertilization, polyploidy resulting from multiple fertilization events was observed in the developing endosperm, but not in the embryo (Scott et al., 2008). The cause of the egg cell-specific fast block to polyspermy is unclear. *In vitro* fertilization experiments with maize egg and sperm cells have shown that cell wall material is detectable already within 30 sec after fusion (Kranz et al., 1995) and thus may prevent further gametic membrane interactions. Additionally, a quick block to polyspermy may also depend on the degradation of fusogens as described above. Calcium may play a role in immediate signaling of successful plasmogamy and release of cell wall material as an extended Ca^{2+}_{cyto} transient is observed in the egg cell associated with successful gamete fusion (Denninger et al., 2014). However, the precise cellular function of calcium signaling during gamete interaction is currently unclear and will require further experimentation.

Another way to prevent polyspermy is the deactivation of pollen tube guidance and the activation of repelling mechanisms. In *Arabidopsis* usually only a single pollen tube is guided inside the ovule to execute double fertilization. After unsuccessful fertilization events, for example by failure of cell-cell fusion in *gcs1/hap2*, *duo1*, *duo3*, *gex2*, *cdka;1*, or *ec1-RNAi* gametes (Beale et al., 2012; Kasahara et al., 2012; Sprunck et al., 2012; Maruyama et al., 2013; Mori et al., 2014), secondary pollen tubes are attracted by the remaining synergid cell by a process named as polytubey. This process is delayed by a couple of hours (Kasahara et al., 2012), suggesting that pollen tube repellents are released upon sperm cell discharge and require degradation until additional pollen tubes can be attracted by the remaining synergid cell. After successful fertilization this cell quickly disintegrates, but remains viable for significantly longer times upon fertilization failure (Beale et al., 2012). A recent report showed that both female gametes independently control successful fertilization thus maximizing reproductive success (Maruyama et al., 2013). The key to prevent polytubey is the quick degeneration of the 2nd synergid cell. In order to investigate its death, it was recently reported that an Ethylene-Insensitive (EIN3-EIN2)/Ethylene-Insensitive3-like2 (EIL2)-dependent, ethylene-response cascade is activated after fertilization. Its artificial activation results in premature synergid cell disintegration and thus a block to pollen tube attraction (Völz et al., 2013). The degeneration of the 2nd synergid cell thus leads to the stop of attractant secretion and ultimately prevents polyspermy.

ACTIVATION OF SEED DEVELOPMENT

Both female gametes need to be fertilized to produce viable progeny. Although equipped with the genetic repertoire to generate every cell type (in the case of the fertilized egg cell) or a number of highly specialized cell types (in the case of the fertilized central cell), both female gametes appear in an arrested state until activated through fertilization. In contrast, parthenogenetic egg cells do not arrest and initiate cell division without fertilization. The central cell, however, requires fertilization in most plant species, even in those containing parthenogenetic egg cells. Its activation is closely related to seed development as both parthenogenetic embryogenesis and seed development arrest at an early stage without central cell fertilization (Koltunow and Grossniklaus, 2003; Barcaccia and Albertini, 2013). Recent reports in *Arabidopsis* confirm intensive cross-talk between endosperm and embryo as well as between endosperm and seed coat shortly after fertilization (Costa et al., 2014; Figueiredo and Köhler, 2014). It was further shown that the zygotic genome is activated shortly after fertilization in this species and both maternal and paternal genomes contribute equally to the transcriptome of the early embryo (Nodine and Bartel, 2012). Research in the last two decades has discovered many differences in epigenetic modification between male and female genomes, which lead to variations in expression profiles between their genes before and after fertilization. Polycomb group genes and RNA silencing mechanisms play a major role in these processes, but will not be considered here in more detail as excellent reviews can be found elsewhere (e.g., Van Ex et al., 2011; Gehring, 2013).

It is well conceivable that sperm cells deliver factors, which activate female gametes after fusion. A transcript encoding the Interleukin-1 Receptor-Associated Kinase (IRAK)/Pelle-like kinase gene *SHORT SUSPENSOR* (*SSP*), which was shown to be delivered by sperm cells, becomes translated in the zygote and acts in the YODA (YDA) MAPK pathway during zygote elongation (Bayer et al., 2009). In the zygote, the regulatory network activated by SSP-YDA is yet unknown. Activation of the cell cycle might also represent a key mechanism for the activation and progression of seed development. However, *cdka;1* mutant single sperm-like germ cells, defective in a master cell cycle regulator are capable of fertilizing egg cells and activating the embryonic program (Iwakawa et al., 2006; Nowack et al., 2006). Although the mutant *cdka;1* central cell showed mitotic divisions upon egg cell fertilization, it appeared mostly unfertilized, and endosperm proliferation and thereby seed development stopped after a certain time point. This finding suggested a positive proliferation signal from the zygote leading to cell cycle activation in the central cell. However, occasionally two *cdka;1* sperm cells are delivered to the gamete fusion site leading to cell-cell fusion of both female gametes with one *cdka;1* mutant sperm cell each. It was further reported that fusion between nuclei of sperm and central cell fails. The failure of karyogamy in the central cell prevents incorporation of the paternal genome, impairs endosperm development and causes seed abortion. This and the above findings using pathenogenetic species imply that the paternal genome plays an essential role during early seed development and that sperm cell factors are also required to activate central cell development (Aw et al., 2010). In summary, very little is known about the molecular mechanisms activating both female gametes that lead to the initiation of seed development.

CONCLUSIONS

Efficient and successful fertilization of all developed ovules is a key to reproductive success and essential for high crop yield. Using the model plant *Arabidopsis* tremendous progress has been made in the past couple of years to understand the underlying cellular and molecular mechanisms that regulate pollen tube growth and guidance, sperm delivery and gamete interaction resulting in blocks to polytubey and polyspermy. Little is known about the activation of gametes and thus seed development immediately after fertilization. Many of the processes described above involve conserved mechanisms and proteins. Some of these proteins are highly polymorphic and species-specific, allowing female flower organs to discriminate self from alien pollen grains/pollen tubes to avoid reproductive failure after pollination and fertilization with incompatible gametophytes and gametes, respectively. In summary, we have learned that species-specific or even ecotype-specific molecules and plant family-specific mechanisms are required during compatible interactions. These are used by papilla cells to control pollen germination, by transmitting tract cells during pollen tube growth and by the ovule and female gametophyte cells during the last steps of pollen tube journey. Moreover, even sperm cell discharge is regulated in a species-preferential manner. Whether gamete interactions depend on species-specific molecules remains to be shown. Up to now the final processes of fertilization seem to involve partly conserved proteins even from lower to higher eukaryotes. The knowledge generated can now be used to investigate, for example, speciation mechanisms or can be applied to overcome hybridization barriers between species. Initial attempts enabling ovules to attract pollen tubes from unrelated plant families have been successful (Márton et al., 2012). However, as outlined above double fertilization mechanisms are very complex and regulated at multiple levels, and it will be a challenge to overcome all steps simultaneously allowing wide hybridization between plant species that presently cannot be crossed. A major challenge for the near future is to understand fertilization mechanisms also in crop plants, especially in the grasses, which represent the most economically important plant family. Maize was suggested as a grass and crop model to investigate these processes (Dresselhaus et al., 2011), but transformation difficulties, the low number of available insertion mutants, the requirement of sufficient greenhouse space and especially technical problems to visualize the fertilization process *in vivo* still limit its utilization for reproduction biologists. Concerted efforts are now required to understand the molecular mechanisms of double fertilization in crop plants, which significantly differ from *Arabidopsis* in both reproductive structures and genetic repertoire.

ACKNOWLEDGMENTS

We acknowledge Erhard Strohm for his support in 3D modeling of *Arabidopsis* ovules. Work in the Dresselhaus lab is supported by grants from the German Research Council (DFG) via the Collaborative Research Centers SFB924 und SFB960.

REFERENCES

Alandete-Saez, M., Ron, M., and McCormick, S. (2008). *GEX3*, expressed in the male gametophyte and in the egg cell of *Arabidopsis thaliana*, is essential for micropylar pollen tube guidance and plays a role during early embryogenesis. *Mol. Plant* 1, 586–598. doi: 10.1093/mp/ssn015

Amien, S., Kliwer, I., Márton, M. L., Debener, T., Geiger, D., Becker, D., et al. (2010). Defensin-like ZmES4 mediates pollen tube burst in maize via opening of the potassium channel KZM1. *PLoS Biol.* 8:e1000388. doi: 10.1371/journal.pbio.1000388

Aw, S. J., Hamamura, Y., Chen, Z., Schnittger, A., and Berger, F. (2010). Sperm entry is sufficient to trigger division of the central cell but the paternal genome is required for endosperm development in *Arabidopsis*. *Development* 137, 2683–2690. doi: 10.1242/dev.052928

Barcaccia, G., and Albertini, E. (2013). Apomixis in plant reproduction: a novel perspective on an old dilemma. *Plant Reprod.* 26, 159–179. doi: 10.1007/s00497-013-0222-y

Bayer, M., Nawy, T., Giglione, C., Galli, M., Meinnel, T., and Lukowitz, W. (2009). Paternal control of embryonic patterning in *Arabidopsis thaliana*. *Science* 323, 1485–1488. doi: 10.1126/science.1167784

Beale, K. M., Leydon, A. R., and Johnson, M. (2012). Gamete fusion is required to block multiple pollen tubes from entering an *Arabidopsis* ovule. *Curr. Biol.* 22, 1090–1094. doi: 10.1016/j.cub.2012.04.041

Berger, F., Hamamura, Y., Ingouff, M., and Higashiyama, T. (2008). Double fertilization - caught in the act. *Trends Plant Sci.* 13, 437–443. doi: 10.1016/j.tplants.2008.05.011

Bianchi, E., Doe, B., Goulding, D., and Wright, G. J. (2014). Juno is the egg Izumo receptor and is essential for mammalian fertilization. *Nature* 508, 483–487. doi: 10.1038/nature13203

Boavida, L. C., Qin, P., Broz, M., Becker, J. D., and McCormick, S. (2013). *Arabidopsis* tetraspanins are confined to discrete expression domains and cell types in reproductive tissues and form homo- and heterodimers when expressed in yeast. *Plant Physiol.* 163, 696–712. doi: 10.1104/pp.113.216598

Boisson-Dernier, A., Lituiev, D. S., Nestorova, A., Franck, C. M., Thirugnanarajah, S., and Grossniklaus, U. (2013). ANXUR receptor-like kinases coordinate cell wall integrity with growth at the pollen tube tip via NADPH oxidases. *PLoS Biol.* 11:e1001719. doi: 10.1371/journal.pbio.1001719

Boisson-Dernier, A., Roy, S., Kritsas, K., Grobei, M. A., Jaciubek, M., Schroeder, J. I., et al. (2009). Disruption of the pollen-expressed FERONIA homologs ANXUR1 and ANXUR2 triggers pollen tube discharge. *Development* 136, 3279–3288. doi: 10.1242/dev.040071

Brewbaker, J. L., and Kwack, B. H. (1963). The essential role of calcium ion in pollen germination and pollen tube growth. *Am. J. Bot.* 50, 859–865. doi: 10.2307/2439772

Brownfield, L., Hafidh, S., Borg, M., Sidorova, A., Mori, T., and Twell, D. (2009). A plant germline-specific integrator of sperm specification and cell cycle progression. *PLoS Genet.* 5:e1000430. doi: 10.1371/journal.pgen.1000430

Calder, G. M., Franklin-Tong, V. E., Shaw, P. J., and Drøbak, B. K. (1997). Ca2+ oscillations in plant cells: initiation by rapid elevation in cytosolic free Ca^{2+} levels. *Biochem. Biophys. Res. Commun.* 234, 690–694. doi: 10.1006/bbrc.1997.6514

Capron, A., Gourgues, M., Neiva, L. S., Faure, J. -E., Berger, F., Pagnussat, G., et al. (2008). Maternal control of male-gamete delivery in *Arabidopsis* involves a putative GPI-anchored protein encoded by the *LORELEI* gene. *Plant Cell* 20, 3038–3049. doi: 10.1105/tpc.108.061713

Chaubal, R., and Reger, B. (1990). Relatively high calcium is localized in synergid cells of wheat ovaries. *Sex. Plant Reprod.* 3, 98–102.

Chen, Y.-H., Li, H.-J., Shi, D.-Q., Yuan, L., Liu, J., Sreenivasan, R., et al. (2007). The central cell plays a critical role in pollen tube guidance in *Arabidopsis*. *Plant Cell* 19, 3563–3577. doi: 10.1105/tpc.107.053967

Cosgrove, D. J., Bedinger, P., and Durachko, D. M. (1997). Group I allergens of grass pollen as cell wall-loosening agents. *Proc. Natl. Acad. Sci. U.S.A.* 94, 6559–6564. doi: 10.1073/pnas.94.12.6559

Costa, L. M., Marshall, E., Tesfaye, M., Silverstein, K. A., Mori, M., Umetsu, Y., et al. (2014). Central cell-derived peptides regulate early embryo patterning in flowering plants. *Science* 344, 168–172. doi: 10.1126/science.1243005

Crawford, B. C. W., Ditta, G., and Yanofsky, M. F. (2007). The *NTT* gene is required for transmitting-tract development in carpels of *Arabidopsis thaliana*. *Curr. Biol.* 17, 1101–1108. doi: 10.1016/j.cub.2007.05.079

Crawford, B. C. W., and Yanofsky, M. F. (2011). HALF FILLED promotes reproductive tract development and fertilization efficiency in *Arabidopsis thaliana*. *Development* 138, 2999–3009. doi: 10.1242/dev.067793

Denninger, P., Bleckmann, A., Lausser, A., Vogler, F., Ott, T., Ehrhardt, D., et al. (2014). Male-female communication triggers calcium signatures during fertilization in *Arabidopsis*. *Nat. Commun.* 5:4645. doi: 10.1038/ncomms5645

Diboll, A. G., and Larson, D. A. (1966). An electron microscopic study of the mature megagametophyte in *Zea mays*. *Am. J. Bot.* 53, 391–402. doi: 10.2307/2439880

Dresselhaus, T., and Franklin-Tong, N. (2013). Male-female crosstalk during pollen germination, tube growth and guidance, and double fertilization. *Mol. Plant* 6, 1018–1036. doi: 10.1093/mp/sst061

Dresselhaus, T., Lausser, A., and Márton, M. L. (2011). Using maize as a model to study pollen tube growth and guidance, cross-incompatibility and sperm delivery in grasses. *Ann. Bot.* 108, 727–737. doi: 10.1093/aob/mcr017

Drews, G. N., and Koltunow, A. M. G. (2011). The female gametophyte. *Arabidopsis Book* 9:e0155. doi: 10.1199/tab.0155

Drews, G. N., Lee, D., and Christensen, C. A. (1998). Genetic analysis of female gametophyte development and function. *Plant Cell* 10, 5–17. doi: 10.1105/tpc.10.1.5

Escobar-Restrepo, J.-M., Huck, N., Kessler, S., Gagliardini, V., Gheyselinck, J., Yang, W.-C., et al. (2007). The FERONIA receptor-like kinase mediates male-female interactions during pollen tube reception. *Science* 317, 656–660. doi: 10.1126/science.1143562

Evans, M. M. S., and Grossniklaus, U. (2009). "The maize megagametophyte," in *Handbook of Maize: Its Biology*, eds J. L. Bennetzen and S. C. Hake (New York, NY: Springer New York), 79–104.

Ferris, P. J., Woessner, J. P., and Goodenough, U. W. (1996). A sex recognition glycoprotein is encoded by the plus mating-type gene *fus1* of *Chlamydomonas reinhardtii*. *Mol. Biol. Cell* 7, 1235–1248. doi: 10.1091/mbc.7.8.1235

Figueiredo, D. D., and Köhler, C. (2014). Signalling events regulating seed coat development. *Biochem. Soc. Trans.* 42, 358–363. doi: 10.1042/BST20130221

Gehring, M. (2013). Genomic imprinting: insights from plants. *Annu. Rev. Genet.* 47, 187–208. doi: 10.1146/annurev-genet-110711-155527

Gilliham, M., Campbell, M., Dubos, C., Becker, D., and Davenport, R. (2006). "The *Arabidopsis thaliana* Glutamate-like Receptor Family (AtGLR)," in *Communication in Plants*, eds F. Baluška, S. Mancuso, and D. Volkmann (Berlin, Heidelberg: Springer Berlin Heidelberg), 205–222.

Grobe, K., Becker, W. M., Schlaak, M., and Petersen, A. (1999). Grass group I allergens (beta-expansins) are novel, papain-related proteinases. *Eur. J. Biochem.* 263, 33–40. doi: 10.1046/j.1432-1327.1999.00462.x

Guan, Y., Lu, J., Xu, J., McClure, B., and Zhang, S. (2014). Two mitogen-activated protein kinases, MPK3 and MPK6, are required for funicular guidance of pollen tubes in *Arabidopsis*. *Plant Physiol.* 165, 528–533. doi: 10.1104/pp.113.231274

Hamamura, Y., Saito, C., Awai, C., Kurihara, D., Miyawaki, A., Nakagawa, T., et al. (2011). Live-cell imaging reveals the dynamics of two sperm cells during double fertilization in *Arabidopsis thaliana*. *Curr. Biol.* 21, 497–502. doi: 10.1016/j.cub.2011.02.013

Haruta, M., and Constabel, C. P. (2003). Rapid alkalinization factors in poplar cell cultures. Peptide isolation, cDNA cloning, and differential expression in leaves and methyl jasmonate-treated cells. *Plant Physiol.* 131, 814–823. doi: 10.1104/pp.014597

Haruta, M., Monshausen, G., Gilroy, S., and Sussman, M. R. (2008). A cytoplasmic Ca^{2+} functional assay for identifying and purifying endogenous cell signaling peptides in *Arabidopsis* seedlings: identification of AtRALF1 peptide. *Biochemistry* 47, 6311–6321. doi: 10.1021/bi8001488

Haruta, M., Sabat, G., Stecker, K., Minkoff, B. B., and Sussman, M. R. (2014). A peptide hormone and its receptor protein kinase regulate plant cell expansion. *Science* 343, 408–411. doi: 10.1126/science.1244454

Higashiyama, T., Kuroiwa, H., Kawano, S., and Kuroiwa, T. (1998). Guidance *in vitro* of the pollen tube to the naked embryo sac of *Torenia fournieri*. *Plant Cell* 10, 2019–2032. doi: 10.1105/tpc.10.12.2019

Higashiyama, T., Kuroiwa, H., Kawano, S., and Kuroiwa, T. (2000). Explosive discharge of pollen tube contents in *Torenia fournieri*. *Plant Physiol.* 122, 11–14. doi: 10.1104/pp.122.1.11

Higashiyama, T., Yabe, S., Sasaki, N., Nishimura, Y., Miyagishima, S, Kuroiwa, H., et al. (2001). Pollen tube attraction by the synergid cell. *Science* 293, 1480–1483. doi: 10.1126/science.1062429

Hirai, M., Arai, M., Mori, T., Miyagishima, S.-Y., Kawai, S., Kita, K., et al. (2008). Male fertility of malaria parasites is determined by GCS1, a plant-type reproduction factor. *Curr. Biol.* 18, 607–613. doi: 10.1016/j.cub.2008.03.045

Hua, Z., and Kao, T.-H. (2006). Identification and characterization of components of a putative petunia S-locus F-box-containing E3 ligase complex involved in S-RNase-based self-incompatibility. *Plant Cell* 18, 2531–2553. doi: 10.1105/tpc.106.041061

Huang, B.-Q., and Russell, S. D. (1992). Female germ unit: organization, isolation, and function. *Int. Rev. Cytol.* 140, 233–292. doi: 10.1016/s0074-7696(08)61099-2

Huck, N., Moore, J. M., Federer, M., and Grossniklaus, U. (2003). The *Arabidopsis* mutant *feronia* disrupts the female gametophytic control of pollen tube reception. *Development* 130, 2149–2159. doi: 10.1242/dev.00458

Ingouff, M., Sakata, T., Li, J., Sprunck, S., Dresselhaus, T., and Berger, F. (2009). The two male gametes share equal ability to fertilize the egg cell in *Arabidopsis thaliana*. *Curr. Biol.* 19, R19–R20. doi: 10.1016/j.cub.2008.11.025

Inoue, N., Ikawa, M., Isotani, A., and Okabe, M. (2005). The immunoglobulin superfamily protein Izumo is required for sperm to fuse with eggs. *Nature* 434, 234–438. doi: 10.1038/nature03362

Iwakawa, H., Shinmyo, A., and Sekine, M. (2006). *Arabidopsis* CDKA;1, a cdc2 homologue, controls proliferation of generative cells in male gametogenesis. *Plant J.* 45, 819–831. doi: 10.1111/j.1365-313X.2005.02643.x

Iwano, M., Igarashi, M., Tarutani, Y., Kaothien-Nakayama, P., Nakayama, H., Moriyama, H., et al. (2014). A pollen coat-inducible autoinhibited Ca^{2+}-ATPase expressed in stigmatic papilla cells is required for compatible pollination in the Brassicaceae. *Plant Cell* 26, 636–649. doi: 10.1105/tpc.113.121350

Iwano, M., Ngo, Q., Entani, T., Shiba, H., Nagai, T., Miyawaki, A., et al. (2012). Cytoplasmic Ca^{2+} changes dynamically during the interaction of the pollen tube with synergid cells. *Development* 139, 4202–4209. doi: 10.1242/dev.081208

Iwano, M., Shiba, H., Miwa, T., Che, F., Takayama, S., Nagai, T., et al. (2004). Ca^{2+} dynamics in a pollen grain and papilla cell during pollination of *Arabidopsis*. *Plant Physiol.* 136, 3562–3571. doi: 10.1104/pp.104.046961

Iwano, M., and Takayama, S. (2012). Self/non-self discrimination in angiosperm self-incompatibility. *Curr. Opin. Plant Biol.* 15, 78–83. doi: 10.1016/j.pbi.2011.09.003

Kaji, K., Oda, S., Shikano, T., Ohnuki, T., Uematsu, Y., Sakagami, J., et al. (2000). The gamete fusion process is defective in eggs of Cd9-deficient mice. *Nat. Genet.* 24, 279–282. doi: 10.1038/73502

Kanaoka, M. M., Kawano, N., Matsubara, Y., Susaki, D., Okuda, S., Sasaki, N., et al. (2011). Identification and characterization of TcCRP1, a pollen tube attractant from *Torenia concolor*. *Ann. Bot.* 108, 739–747. doi: 10.1093/aob/mcr111

Kasahara, R. D., Maruyama, D., Hamamura, Y., Sakakibara, T., Twell, D., and Higashiyama, T. (2012). Fertilization recovery after defective sperm cell release in *Arabidopsis*. *Curr. Biol.* 22, 1084–1089. doi: 10.1016/j.cub.2012.03.069

Kasahara, R. D., Portereiko, M. F., Sandaklie-Nikolova, L., Rabiger, D. S., and Drews, G. N. (2005). *MYB98* is required for pollen tube guidance and synergid cell differentiation in *Arabidopsis*. *Plant Cell* 17, 2981–2992. doi: 10.1105/tpc.105.034603

Kessler, S. A., Shimosato-Asano, H., Keinath, N. F., Wuest, S. E., Ingram, G., Panstruga, R., et al. (2010). Conserved molecular components for pollen tube reception and fungal invasion. *Science* 330, 968–971. doi: 10.1126/science.1195211

Kim, S., Mollet, J.-C., Dong, J., Zhang, K., Park, S.-Y., and Lord, E. M. (2003). Chemocyanin, a small basic protein from the lily stigma, induces pollen tube chemotropism. *Proc. Natl. Acad. Sci. U.S.A.* 100, 16125–16130. doi: 10.1073/pnas.2533800100

Koltunow, A. M., and Grossniklaus, U. (2003). Apomixis: a developmental perspective. *Annu. Rev. Plant Biol.* 54, 547–574. doi: 10.1146/annurev.arplant.54.110901.160842

Kranz, E., Wiegen, P., and Lörz, H. (1995). Early cytological events after induction of cell division in egg cells and zygote development following *in vitro* fertilization with angiosperm gametes. *Plant J.* 8, 9–23. doi: 10.1046/j.1365-313X.1995.08010009.x

Kubo, K., Entani, T., Takara, A., Wang, N., Fields, A. M., Hua, Z., et al. (2010). Collaborative non-self recognition system in S-RNase-based self-incompatibility. *Science* 330, 796–799. doi: 10.1126/science.1195243

Lausser, A., and Dresselhaus, T. (2010). Sporophytic control of pollen tube growth and guidance in grasses. *Biochem. Soc. Trans.* 38, 631–634. doi: 10.1042/BST0380631

Le Naour, F., Rubinstein, E., Jasmin, C., Prenant, M., and Boucheix, C. (2000). Severely reduced female fertility in CD9-deficient mice. *Science* 287, 319–321. doi: 10.1126/science.287.5451.319

Lennon, K. A., Roy, S., Hepler, P. K., and Lord, E. M. (1998). The structure of the transmitting tissue of *Arabidopsis thaliana* (L.) and the path of pollen tube growth. *Sex. Plant Reprod.* 11, 49–59. doi: 10.1007/s004970050120

Leshem, Y., Johnson, C., and Sundaresan, V. (2013). Pollen tube entry into the synergid cell of *Arabidopsis* is observed at a site distinct from the filiform apparatus. *Plant Reprod.* 26, 93–99. doi: 10.1007/s00497-013-0211-1

Leydon, A. R., Beale, K. M., Woroniecka, K., Castner, E., Chen, J., Horgan, C., et al. (2013). Three MYB transcription factors control pollen tube differentiation required for sperm release. *Curr. Biol.* 23, 1209–1214. doi: 10.1016/j.cub.2013.05.021

Leydon, A. R., Chaibang, A., and Johnson, M. (2014). Interactions between pollen tube and pistil control pollen tube identity and sperm release in the *Arabidopsis* female gametophyte. *Biochem. Soc. Trans.* 42, 340–345. doi: 10.1042/BST20130223

Liang, Y., Tan, Z.-M., Zhu, L., Niu, Q.-K., Zhou, J.-J., Li, M., et al. (2013). MYB97, MYB101 and MYB120 function as male factors that control pollen tube-synergid interaction in *Arabidopsis thaliana* fertilization. *PLoS Genet.* 9:e1003933. doi: 10.1371/journal.pgen.1003933

Liu, J., Zhong, S., Guo, X., Hao, L., Wei, X., Huang, Q., et al. (2013). Membrane-bound RLCKs LIP1 and LIP2 are essential male factors controlling male-female attraction in *Arabidopsis*. *Curr. Biol.* 23, 993–998. doi: 10.1016/j.cub.2013.04.043

Liu, Y., Misamore, M. J., and Snell, W. J. (2010). Membrane fusion triggers rapid degradation of two gamete-specific, fusion-essential proteins in a membrane block to polygamy in *Chlamydomonas*. *Development* 137, 1473–1481. doi: 10.1242/dev.044743

Liu, Y., Tewari, R., Ning, J., Blagborough, A. M., Garbom, S., Pei, J., et al. (2008). The conserved plant sterility gene *HAP2* functions after attachment of fusogenic membranes in *Chlamydomonas* and *Plasmodium* gametes. *Genes Dev.* 22, 1051–1068. doi: 10.1101/gad.1656508

Liu, Z.-Q., Xu, G.-H., and Zhang, S.-L. (2007). *Pyrus pyrifolia* stylar S-RNase induces alterations in the actin cytoskeleton in self-pollen and tubes *in vitro*. *Protoplasma* 232, 61–67. doi: 10.1007/s00709-007-0269-4

Lush, W., Grieser, F., and Wolters-Arts, M. (1998). Directional guidance of *Nicotiana alata* pollen tubes *in vitro* and on the stigma. *Plant Physiol.* 118, 733–741. doi: 10.1104/pp.118.3.733

Luu, D. T., Qin, X., Morse, D., and Cappadocia, M. (2000). S-RNase uptake by compatible pollen tubes in gametophytic self-incompatibility. *Nature* 407, 649–651. doi: 10.1038/35036623

Mansfield, S. G., Briarty, L. G., and Erni, S. (1991). Early embryogenesis in *Arabidopsis thaliana*. I. The mature embryo sac. *Can. J. Bot.* 69, 447–460. doi: 10.1139/b91-062

Márton, M. L., Cordts, S., Broadhvest, J., and Dresselhaus, T. (2005). Micropylar pollen tube guidance by egg apparatus 1 of maize. *Science* 307, 573–576. doi: 10.1126/science.1104954

Márton, M. L., Fastner, A., Uebler, S., and Dresselhaus, T. (2012). Overcoming hybridization barriers by the secretion of the maize pollen tube attractant ZmEA1 from *Arabidopsis* ovules. *Curr. Biol.* 22, 1194–1198. doi: 10.1016/j.cub.2012.04.061

Maruyama, D., Hamamura, Y., Takeuchi, H., Susaki, D., Nishimaki, M., Kurihara, D., et al. (2013). Independent control by each female gamete prevents the attraction of multiple pollen tubes. *Dev. Cell* 25, 317–323. doi: 10.1016/j.devcel.2013.03.013

Matias-Hernandez, L., Battaglia, R., Galbiati, F., Rubes, M., Eichenberger, C., Grossniklaus, U., et al. (2010). *VERDANDI* is a direct target of the MADS domain ovule identity complex and affects embryo sac differentiation in *Arabidopsis*. *Plant Cell* 22, 1702–1715. doi: 10.1105/tpc.109.068627

McClure, B. A., Haring, V., Ebert, P. R., Anderson, M. A., Simpson, R. J., Sakiyama, F., et al. (1989). Style self-incompatibility gene products of *Nicotiana alata* are ribonucleases. *Nature* 342, 955–957. doi: 10.1038/342955a0

McCue, A. D., Cresti, M., Feijó, J. A., and Slotkin, R. K. (2011). Cytoplasmic connection of sperm cells to the pollen vegetative cell nucleus: potential roles of the male germ unit revisited. *J. Exp. Bot.* 62, 1621–1631. doi: 10.1093/jxb/err032

Michard, E., Lima, P. T., Borges, F., Silva, A. C., Portes, M. T., Carvalho, J. E., et al. (2011). Glutamate receptor-like genes form Ca^{2+} channels in pollen tubes and are regulated by pistil D-serine. *Science* 332, 434–437. doi: 10.1126/science.1201101

Miller, D. D., Callaham, D. A., Gross, D. J., and Hepler, P. K. (1992). Free Ca^{2+} gradient in growing pollen tubes of *Lillium*. *J. Cell Sci.* 101, 7–12.

Misamore, M. J., Gupta, S., and Snell, W. J. (2003). The *Chlamydomonas* Fus1 protein is present on the mating type plus fusion organelle and required for a critical membrane adhesion event during fusion with minus gametes. *Mol. Biol. Cell* 14, 2530–2542. doi: 10.1091/mbc.E02-12-0790

Miyado, K., Yamada, G., Yamada, S., Hasuwa, H., Nakamura, Y., Ryu, F., et al. (2000). Requirement of CD9 on the egg plasma membrane for fertilization. *Science* 287, 321–324. doi: 10.1126/science.287.5451.321

Miyazaki, S., Murata, T., Sakurai-Ozato, N., Kubo, M., Demura, T., Fukuda, H., et al. (2009). *ANXUR1* and *2*, sister genes to *FERONIA/SIRENE*, are male factors for coordinated fertilization. *Curr. Biol.* 19, 1327–1331. doi: 10.1016/j.cub.2009.06.064

Mori, T., Igawa, T., Tamiya, G., Miyagishima, S.-Y., and Berger, F. (2014). Gamete attachment requires GEX2 for successful fertilization in *Arabidopsis*. *Curr. Biol.* 24, 170–175. doi: 10.1016/j.cub.2013.11.030

Mori, T., Kuroiwa, H., Higashiyama, T., and Kuroiwa, T. (2006). GENERATIVE CELL SPECIFIC 1 is essential for angiosperm fertilization. *Nat. Cell Biol.* 8, 64–71. doi: 10.1038/ncb1345

Ngo, Q. A., Vogler, H., Lituiev, D. S., Nestorova, A., and Grossniklaus, U. (2014). A calcium dialog mediated by the FERONIA signal transduction pathway controls plant sperm delivery. *Dev. Cell* 29, 491–500. doi: 10.1016/j.devcel.2014.04.008

Nguema-Ona, E., Coimbra, S., Vicré-Gibouin, M., Mollet, J.-C., and Driouich, A. (2012). Arabinogalactan proteins in root and pollen-tube cells: distribution and functional aspects. *Ann. Bot.* 110, 383–404. doi: 10.1093/aob/mcs143

Nodine, M. D., and Bartel, D. P. (2012). Maternal and paternal genomes contribute equally to the transcriptome of early plant embryos. *Nature* 482, 94–97. doi: 10.1038/nature10756

Nowack, M. K., Grini, P. E., Jakoby, M. J., Lafos, M., Koncz, C., and Schnittger, A. (2006). A positive signal from the fertilization of the egg cell sets off endosperm proliferation in angiosperm embryogenesis. *Nat. Genet.* 38, 63–67. doi: 10.1038/ng1694

Ogawa, M., Kay, P., Wilson, S., and Swain, S. M. (2009). ARABIDOPSIS DEHISCENCE ZONE POLYGALACTURONASE1 (ADPG1), ADPG2, and QUARTET2 are polygalacturonases required for cell separation during reproductive development in *Arabidopsis*. *Plant Cell* 21, 216–233. doi: 10.1105/tpc.108.063768

Okuda, S., Suzuki, T., Kanaoka, M. M., Mori, H., Sasaki, N., and Higashiyama, T. (2013). Acquisition of LURE-binding activity at the pollen tube tip of *Torenia fournieri*. *Mol. Plant* 6, 1074–1090. doi: 10.1093/mp/sst050

Okuda, S., Tsutsui, H., Shiina, K., Sprunck, S., Takeuchi, H., Yui, R., et al. (2009). Defensin-like polypeptide LUREs are pollen tube attractants secreted from synergid cells. *Nature* 458, 357–361. doi: 10.1038/nature07882

Olsen, A. N., Mundy, J., and Skriver, K. (2002). Peptomics, identification of novel cationic *Arabidopsis* peptides with conserved sequence motifs. *In Silico Biol.* 2, 441–451.

Palanivelu, R., Brass, L., Edlund, A. F., and Preuss, D. (2003). Pollen tube growth and guidance is regulated by *POP2*, an *Arabidopsis* gene that controls GABA levels. *Cell* 114, 47–59. doi: 10.1016/S0092-8674(03)00479-3

Palanivelu, R., and Preuss, D. (2006). Distinct short-range ovule signals attract or repel *Arabidopsis thaliana* pollen tubes *in vitro*. *BMC Plant Biol.* 6:7. doi: 10.1186/1471-2229-6-7

Palanivelu, R., and Tsukamoto, T. (2012). Pathfinding in angiosperm reproduction: pollen tube guidance by pistils ensures successful double fertilization. *Wiley Interdiscip. Rev. Dev. Biol.* 1, 96–113. doi: 10.1002/wdev.6

Pearce, G., Moura, D. S., Stratmann, J., and Ryan, C. A. (2001). RALF, a 5-kDa ubiquitous polypeptide in plants, arrests root growth and development. *Proc. Natl. Acad. Sci. U.S.A.* 98, 12843–12847. doi: 10.1073/pnas.201416998

Punwani, J. A., Rabiger, D. S., Lloyd, A., and Drews, G. N. (2008). The MYB98 subcircuit of the synergid gene regulatory network includes genes directly and indirectly regulated by MYB98. *Plant J.* 55, 406–414. doi: 10.1111/j.1365-313X.2008.03514.x

Punwani, J., Rabiger, D. S., and Drews, G. N. (2007). MYB98 positively regulates a battery of synergid-expressed genes encoding filiform apparatus localized proteins. *Plant Cell* 19, 2557–2568. doi: 10.1105/tpc.107.052076

Roldán, J. A., Rojas, H. J., and Goldraij, A. (2012). Disorganization of F-actin cytoskeleton precedes vacuolar disruption in pollen tubes during the *in vivo* self-incompatibility response in *Nicotiana alata*. *Ann. Bot.* 110, 787–795. doi: 10.1093/aob/mcs153

Rotman, N., Rozier, F., Boavida, L., Dumas, C., Berger, F., and Faure, J. (2003). Female control of male gamete delivery during fertilization in

Arabidopsis thaliana. Supplements. *Curr. Biol.* 13, 432–436. doi: 10.1016/S0960-9822(03)00093-9

Samuel, M. A., Chong, Y. T., Haasen, K. E., Aldea-Brydges, M. G., Stone, S. L., and Goring, D. R. (2009). Cellular pathways regulating responses to compatible and self-incompatible pollen in *Brassica* and *Arabidopsis* stigmas intersect at Exo70A1, a putative component of the exocyst complex. *Plant Cell* 21, 2655–2671. doi: 10.1105/tpc.109.069740

Sandaklie-Nikolova, L., Palanivelu, R., King, E. J., Copenhaver, G. P., and Drews, G. N. (2007). Synergid cell death in *Arabidopsis* is triggered following direct interaction with the pollen tube. *Plant Physiol.* 144, 1753–1762. doi: 10.1104/pp.107.098236

Schopfer, C. R. (1999). The male determinant of self-incompatibility in *Brassica*. *Science* 286, 1697–1700. doi: 10.1126/science.286.5445.1697

Scott, R. J., Armstrong, S. J., Doughty, J., and Spielman, M. (2008). Double fertilization in *Arabidopsis thaliana* involves a polyspermy block on the egg but not the central cell. *Mol. Plant* 1, 611–619. doi: 10.1093/mp/ssn016

Shi, Y.-Y., Tao, W.-J., Liang, S.-P., Lü, Y., and Zhang, L. (2009). Analysis of the tip-to-base gradient of CaM in pollen tube pulsant growth using *in vivo* CaM-GFP system. *Plant Cell Rep.* 28, 1253–1264. doi: 10.1007/s00299-009-0725-z

Shiba, H., Takayama, S., Iwano, M., Shimosato, H., Funato, M., Nakagawa, T., et al. (2001). A pollen coat protein, SP11/SCR, determines the pollen S-specificity in the self-incompatibility of *Brassica* species. *Plant Physiol.* 125, 2095–2103. doi: 10.1104/pp.125.4.2095

Shimizu, K. K., Ito, T., Ishiguro, S., and Okada, K. (2008). *MAA3* (*MAGATAMA3*) helicase gene is required for female gametophyte development and pollen tube guidance in *Arabidopsis thaliana*. *Plant Cell Physiol.* 49, 1478–1483. doi: 10.1093/pcp/pcn130

Shimizu, K. K., and Okada, K. (2000). Attractive and repulsive interactions between female and male gametophytes in *Arabidopsis* pollen tube guidance. *Development* 127, 4511–4518. doi: 10.5167/uzh-71801

Sprunck, S., Rademacher, S., Vogler, F., Gheyselinck, J., Grossniklaus, U., and Dresselhaus, T. (2012). Egg cell-secreted EC1 triggers sperm cell activation during double fertilization. *Science* 338, 1093–1097. doi: 10.1126/science.1223944

Steele, R. E., and Dana, C. E. (2009). Evolutionary history of the *HAP2/GCS1* gene and sexual reproduction in metazoans. *PLoS ONE* 4:e7680. doi: 10.1371/journal.pone.0007680

Steinhorst, L., and Kudla, J. (2013a). Calcium - a central regulator of pollen germination and tube growth. *Biochim. Biophys. Acta* 1833, 1573–1581. doi: 10.1016/j.bbamcr.2012.10.009

Steinhorst, L., and Kudla, J. (2013b). Calcium and reactive oxygen species rule the waves of signaling. *Plant Physiol.* 163, 471–485. doi: 10.1104/pp.113.222950

Stratford, S., Barne, W., Hohorst, D. L., Sagert, J. G., Cotter, R., Golubiewski, A., et al. (2001). A leucine-rich repeat region is conserved in pollen extensin-like (Pex) proteins in monocots and dicots. *Plant Mol. Biol.* 46, 43–56. doi: 10.1023/A:1010659425399

Synek, L., Schlager, N., Eliás, M., Quentin, M., Hauser, M., and Zárskı, V. (2006). AtEXO70A1, a member of a family of putative exocyst subunits specifically expanded in land plants, is important for polar growth and plant development. *Plant J.* 48, 54–72. doi: 10.1111/j.1365-313X.2006.02854.x

Takasaki, T., Hatakeyama, K., Suzuki, G., Watanabe, M., Isogai, A., and Hinata, K. (2000). The S receptor kinase determines self-incompatibility in *Brassica* stigma. *Nature* 403, 913–916. doi: 10.1038/35002628

Takeuchi, H., and Higashiyama, T. (2012). A species-specific cluster of defensin-like genes encodes diffusible pollen tube attractants in *Arabidopsis*. *PLoS Biol.* 10:e1001449. doi: 10.1371/journal.pbio.1001449

Uebler, S., Dresselhaus, T., and Márton, M.-L. (2013). Species-specific interaction of EA1 with the maize pollen tube apex. *Plant Signal. Behav.* 8, e25682. doi: 10.4161/psb.25682

Van Ex, F., Jacob, Y., and Martienssen, R. A. (2011). Multiple roles for small RNAs during plant reproduction. *Curr. Opin. Plant Biol.* 14, 588–593. doi: 10.1016/j.pbi.2011.07.003

Völz, R., Heydlauff, J., Ripper, D., von Lyncker, L., and Groß-Hardt, R. (2013). Ethylene signaling is required for synergid degeneration and the establishment of a pollen tube block. *Dev. Cell* 25, 310–316. doi: 10.1016/j.devcel.2013.04.001

Watanabe, M., Suwabe, K., and Suzuki, G. (2012). Molecular genetics, physiology and biology of self-incompatibility in Brassicaceae. *Proc. Jpn. Acad,. Ser. B,. Phys. Biol. Sci.* 88, 519–535. doi: 10.2183/pjab.88.519

Wheeler, M. J., de Graaf, B. H. J., Hadjiosif, N., Perry, R. M., Poulter, N. S., Osman, K., et al. (2009). Identification of the pollen self-incompatibility determinant in *Papaver rhoeas*. *Nature* 459, 992–995. doi: 10.1038/nature08027

Willemse, M. T. M., and van Went, J. L., (1984). "The female gametophyte," in *Embryology of Angiosperms*, ed B. M. Johri (Berlin, Heidelberg: Springer Berlin Heidelberg), 159–196. doi: 10.1007/978-3-642-69302-1_4

Wolters-Arts, M., Lush, W. M., and Mariani, C. (1998). Lipids are required for directional pollen-tube growth. *Nature* 392, 818–821. doi: 10.1038/33929

Wong, J. L., and Johnson, M. (2010). Is HAP2-GCS1 an ancestral gamete fusogen? *Trends Cell Biol.* 20, 134–141. doi: 10.1016/j.tcb.2009.12.007

Woriedh, M., Wolf, S., Márton, M. L., Hinze, A., Gahrtz, M., Becker, D., et al. (2013). External application of gametophyte-specific ZmPMEI1 induces pollen tube burst in maize. *Plant Reprod.* 26, 255–266. doi: 10.1007/s00497-013-0221-z

Wu, H. M., Wang, H., and Cheung, A. Y. (1995). A pollen tube growth stimulatory glycoprotein is deglycosylated by pollen tubes and displays a glycosylation gradient in the flower. *Cell* 82, 395–403. doi: 10.1016/0092-8674(95)90428-X

Wu, J., Wang, S., Gu, Y., Zhang, S., Publicover, S. J., and Franklin-Tong, V. E. (2011). Self-incompatibility in *Papaver rhoeas* activates nonspecific cation conductance permeable to Ca^{2+} and K^{+}. *Plant Physiol.* 155, 963–973. doi: 10.1104/pp.110.161927

Cheung, A, Y., Wang, H., and Wu, H. M. (1995). A floral transmitting tissue-specific glycoprotein attracts pollen tubes and stimulates their growth. *Cell* 82, 383–393. doi: 10.1016/0092-8674(95)90427-1

Zhou, L., Fu, Y., and Yang, Z. (2009). A genome-wide functional characterization of *Arabidopsis* regulatory calcium sensors in pollen tubes. *J. Integr. Plant Biol.* 51, 751–761. doi: 10.1111/j.1744-7909.2009.00847.x

Conflict of Interest Statement: The authors declare that the research was conducted in the absence of any commercial or financial relationships that could be construed as a potential conflict of interest.

10

The role of the testa during development and in establishment of dormancy of the legume seed

Petr Smýkal[1]*, Vanessa Vernoud[2], Matthew W. Blair[3], Aleš Soukup[4] and Richard D. Thompson[2]

[1] Department of Botany, Faculty of Sciences, Palacký University in Olomouc, Olomouc, Czech Republic
[2] INRA, UMR 1347 Agroécologie, Dijon, France
[3] Department of Agricultural and Environmental Sciences, Tennessee State University, Nashville, TN, USA
[4] Department of Experimental Plant Biology, Charles University, Prague, Czech Republic

Edited by:
Paolo Sabelli, University of Arizona, USA

Reviewed by:
Françoise CORBINEAU, Université Pierre et Marie Curie, France
Carol C. Baskin, University of Kentucky, USA

***Correspondence:**
Petr Smýkal, Department of Botany, Faculty of Sciences, Palacký University in Olomouc, Šlechtitelů 11, 783 71 Olomouc, Czech Republic
e-mail: petr.smykal@upol.cz

Timing of seed germination is one of the key steps in plant life cycles. It determines the beginning of plant growth in natural or agricultural ecosystems. In the wild, many seeds exhibit dormancy and will only germinate after exposure to certain environmental conditions. In contrast, crop seeds germinate as soon as they are imbibed usually at planting time. These domestication-triggered changes represent adaptations to cultivation and human harvesting. Germination is one of the common sets of traits recorded in different crops and termed the "domestication syndrome." Moreover, legume seed imbibition has a crucial role in cooking properties. Different seed dormancy classes exist among plant species. Physical dormancy (often called hardseededness), as found in legumes, involves the development of a water-impermeable seed coat, caused by the presence of phenolics- and suberin-impregnated layers of palisade cells. The dormancy release mechanism primarily involves seed responses to temperature changes in the habitat, resulting in testa permeability to water. The underlying genetic controls in legumes have not been identified yet. However, positive correlation was shown between phenolics content (e.g., pigmentation), the requirement for oxidation and the activity of catechol oxidase in relation to pea seed dormancy, while epicatechin levels showed a significant positive correlation with soybean hardseededness. myeloblastosis family of transcription factors, WD40 proteins and enzymes of the anthocyanin biosynthesis pathway were involved in seed testa color in soybean, pea and *Medicago*, but were not tested directly in relation to seed dormancy. These phenolic compounds play important roles in defense against pathogens, as well as affecting the nutritional quality of products, and because of their health benefits, they are of industrial and medicinal interest. In this review, we discuss the role of the testa in mediating legume seed germination, with a focus on structural and chemical aspects.

Keywords: domestication, dormancy, hardseededness, legumes, proanthocyanidins, seed coat, testa, water permeability

LEGUMES

Legume seeds are the second most important plant protein source, on a world basis, after cereals. While in cereals the major storage molecule is starch, which is deposited in the endosperm, in most of the grain legumes (pulses) the endosperm is transitory and consumed by the embryo during seed maturation, which contains a high proportion of proteins (20–40%), and either lipids (soybean, peanut) or starch (or both) as a further carbon source. Nutritionally, they are generally deficient in sulfur-containing amino acids (cysteine and methionine), but unlike cereal grains, their lysine content is relatively high. The major storage proteins are globulins, which account for up to 70% of the total seed nitrogen. The ability of legumes to fix atmospheric nitrogen allows them to colonize poor soils; however adequate nitrogen reserves in the seed are vital to allow the seedling to survive the heterotrophic growth phase before nitrogen fixation is established in root nodules. Fabaceae, the third largest family of flowering plants, are divided into three subfamilies: Caesalpinioideae, Mimosoideae, and Papilionoideae, all together with 800 genera and 20,000 species (Lewis et al., 2005). The latter subfamily contains most of the cultivated major food and feed crops (Smýkal et al., 2014). It is a diverse family with a worldwide distribution, encompassing a broad range of plant forms, from annual and perennial herbs to trees and lianas. This variation is also reflected by widely diverse seed shapes and sizes, ranging from 1-mm seeds of the native Australian legume species *Pycnospora lutescens* to 18-cm seeds of the coastal tree *Mora oleifera* (Gunn, 1981). Legume seeds develop either within pods (e.g., legume), or less frequently, samara (Lewis et al., 2005), which can have several functions; protective, dispersal, and nutritive, can comprise a significant source of remobilized nutrients during seed filling.

DORMANCY CONCEPTS

The function of a seed is to establish a new plant but it can do this only once, because the completion of germination essentially is an irreversible process. Plants have evolved several dormancy mechanisms to optimize the time of germination (Foley, 2001). Since seed dormancy is a physiological adaptation to environmental heterogeneity, it is a primary factor that influences natural population dynamics (reviewed in Bewley et al., 2013). Dormancy provides a strategy for seeds to spread germination in time in order to reduce the risk of plant death and possible species extinction in an unfavorable environment. Dormancy occurs in three ways: (1) Seeds are dispersed from the parent plant with different degrees of dormancy. Frequently, the variation in dormancy is reflected by the appearance of the seeds or dispersal units in terms of color, size, and thickness of the coat. (2) Through the dependence of dormancy breakage on environmental factors. (3) Through seed dispersal via animals, wind or water (reviewed in Bewley et al., 2013). A classical concept of seed dormancy was formulated by Harper (1957) who distinguished three types: (1) Seeds born as dormant (innate); (2) Those with achieved dormancy (induced); and (3) Seeds with dormancy thrust upon them (enforced). Moreover, Harper distinguished two categories of plants living in a community; those which are growing at present and those which are dormant (in the form of a soil seed bank). A problem in distinguishing dormancy-relieving factors from factors stimulating or initiating germination is that the actual state of dormancy cannot be measured directly (Thompson and Ooi, 2013). Vleeshouwers et al. (1995) defined dormancy as: "a seed characteristic, the degree of which defines what conditions should be met to make the seed germinate" (Vleeshouwers et al., 1995). Consequently, seeds of many species that form a persistent seed bank exhibit annual changes in dormancy (Karssen, 1982; Milberg and Andersson, 1997). This phenomenon of dormancy cycling (Baskin and Baskin, 1985) is regulated by various factors such as temperature. Moreover, the intensity of dormancy within a given species varies at several levels: among populations, within populations and between seeds collected in different years from the same population (Foley, 2001; Lacerda et al., 2004). There is also heterogeneity in dormancy among seeds at the level of the individual plant (reviewed in Matilla et al., 2005), depending on the age and the nutritional status of the mother plant during seed maturation, seed position on the mother plant, seed size and shape, the time since seed harvest, and the duration of seed storage (Probert, 2000). Despite of all this variation, seed dormancy has a clear genetic basis (Graeber et al., 2012). Several dormancy classes were defined by Nikolaeva (1969, 1977) and more recently reviewed by Finch-Savage and Leubner-Metzger (2006). Morphological dormancy refers to seeds that have an underdeveloped embryo and require longer time to grow and germinate. Physiological dormancy, the most prevalent form of dormancy, appears to broadly involve abscisic acid (ABA) and gibberellins (GAs) metabolism. In addition, there are morphophysiological and combinational dormancies. In contrast to hormone-mediated seed dormancy, extensively studied in *Arabidopsis* or cereals, we have still limited knowledge of the regulation of physical dormancy, which involves the development of a water-impermeable seed coat (Baskin et al., 2000). This type of dormancy is found in at least 17 plant families, including agronomically important families like the Fabaceae, Malvaceae, Cannaceae, Geraniaceae, and Convolvulaceae (Baskin et al., 2000) and is present in the wild progenitors of cultivated legumes (Dueberrn de Sousa and Marcos-Filho, 2001; Zohary et al., 2012; Abbo et al., 2014).

Germination begins with water uptake (imbibition) by the quiescent dry seed and is completed by radicle protrusion through the tissues surrounding the embryo. There are three phases of seed imbibition. Dry seeds have very low water potentials, which causes rapid water influx during phase I. As this process is driven by water potential, it also occurs in dead seeds. Similar phenomena can be observed in resurrection plants and pollen. The permeability of the testa, being the part of the seed that comes into contact with the ambient water, plays a central role in water uptake. Phase II encompasses the rupture of the testa, and during phase III endosperm rupture and radicle protrusion occur (Finch-Savage and Leubner-Metzger, 2006). Germination occurs when embryo growth overcomes the constraints imposed by the seed coat (Bewley et al., 2013).

In nature, exposure to high temperature or fluctuating temperatures is the most likely cause of release from seed dormancy. The interactions between seed dormancy mechanisms and accumulated and current environmental conditions determine whether and what fraction of seeds in a seed bank will germinate at a given time. Weather and soil physical characteristics largely determine the microclimate to which seeds are exposed. The most critical environmental factor is water. Seeds imbibe water from their surroundings, and the water potential of the soil determines the maximum water potential that the seeds can attain. Seed banks may be composed of seeds from different years, which experienced different after ripening- or dormancy-breaking regimes, resulting in multiple subpopulations with different dormancy characteristics. Temperature is the second most important environmental determinant of seed germination. In extreme cases this is associated with fire (Hanley and Fenner, 1998) or vegetation gaps that are hotter than the surrounding forest soils (Vázquez-Yanes and Orozco-Segovia, 1982). Moreover, water and temperature regimes interact and also light plays a role in the onset of germination via regulation of phytochrome activity. Other factors include oxygen and other gasses. Dependence on exogenous factors for the initiation of germination suggests that physically dormant seeds should be limited in their ability to spread germination risk over multiple time periods or "recruitment opportunities" (reviewed in Bewley et al., 2013). However, in a few taxa the responsiveness of seeds to dormancy-breaking cues varies seasonally (Karssen, 1982), which suggests temperature, rainfall and perhaps deciduous tree leaf-drop as key factors. In summary, seasonal germination patterns are largely controlled by the seeds' responses to prevailing environmental factors, such as moisture, temperature, light, and various chemicals in conjunction with seasonal environmental factors (e.g., chilling, after-ripening) that sensitize the seeds to the environment.

LEGUME SEED COAT (TESTA) DEVELOPMENT AND STRUCTURE

The angiosperm seed develops from the fertilized ovule and depending on the stage of development is usually composed of (1) the embryo, arising by fertilization of the egg cell by one of the pollen tube nuclei; (2) the nutritive tissue of the endosperm, generated by the fusion of two polar nuclei of the embryo sac with the other sperm nucleus; and (3) a protective seed coat (testa), derived from the inner, outer or both ovular integuments (Bradford and Nonogaki, 2009).

THE SEED COAT TISSUE COMPONENTS

The origin of the seed coat can be traced back to the L1 sporophyte layer of the ovule primordium (Schneitz et al., 1997) with an emerging network of regulatory pathways coordinating growth of the inner and/or outer integuments surrounding the ovule (Skinner et al., 2004; Galbiati et al., 2013; Kurdyukov et al., 2014). The number of ovule integuments varies depending on the species; legumes have two integuments (bitegmic ovules). The inner integument largely vanishes during development (Esau, 1965) while the outer one produces several distinct cell layers and establishes the "typical" seed coat structure. The chalazal region is an important part of the testa where connections of the vascular tissues of the maternal funiculus terminate. The scar where the funiculus was attached is called hilum (**Figure 1B**). Current seed identification criteria are based upon morphological characteristics including seed size, general shape, surface shape, color, pattern, hilum length and width. These are often used in taxonomical classifications (Lestern and Gunn, 1981; Chernoff et al., 1992; Güneş, 2013) and archeobotany (Zohary et al., 2012). Legume seed characters support the concept of one family (Fabaceae) as advocated already by de Candolle (1825). Although the seed coats of different species vary greatly in structure and composition, they undergo similar phases of development in relation to the embryo and endosperm (Butler, 1988). In legumes, the seed coat and endosperm develop first, followed by development of the embryo (Weber et al., 2005).

In spite of some known exceptions such as peanut (*Arachis hypogaea*) with lignified pod or *Archidendron* and *Pithecellobium*, which have a partly pulpy and edible testa (Gunn, 1981), there is a rather common blueprint of seed coat structure for the Fabaceae family (Lush and Evans, 1980). Interspecific variation comprises mainly the patterns of differentiation, dimensions, and modifications of cell walls of individual layers. There are a number of publications in this area (Pammel, 1899; Rolston, 1978; Lush and Evans, 1980) with considerable emphasis on economically important soybean.

THE OUTER INTEGUMENTS

The epidermis of the outer integument forms a single layer of tightly packed palisade of radially elongated sclereids (called Malpighian cells, macrosclereids, or palisade cells) with heavily and unevenly thickened cell walls (**Figure 1A**). The outer tangential cell walls are covered with cuticle and, because of specific cell wall thickening and modifications, they are commonly described as terminal caps. Their shape together with cuticle and waxy depositions determine the texture of the seed coat surface (Güneş, 2013). The architecture of this layer and the structure of its cuticle attract considerable attention as their properties are generally related to water-impermeability of hard seeds (White, 1908; Hamly, 1932, 1935; Riggio Bevilacqua et al., 1989; Argel and Paton, 1999). The cuticle forms a continuous layer covering the seed, except for the hilum, and is considered the outermost barrier to imbibition (White, 1908; Spurny, 1963; Ma et al., 2004; Shao et al., 2007). Interestingly, the chemical composition of soybean cuticle seems to differ in fatty acid composition from the shoot cuticle (Shao et al., 2007). Its continuity, being a crucial feature of hardseededness, might be compromised by the emergence of cracks during seed expansion and development (Ranathunge et al., 2010). In some legume species, such as soybean, the seed surface is also modified by secretory activity of the fruit (pod) wall (Yaklich et al., 1986) which might deposit hydrophobic proteins on the cuticle (Clarkson and Robards, 1975; Gijzen et al., 1999).

THE SCLEREID LAYERS

There is frequently a lucent region of macrosclereid cell walls separating macrosclereid terminal caps from their basal parts. This border line, which extends transversally across the macrosclereid layer is named light-line or *linea lucida* (**Figure 1A**) in many species. Its appearance derives from local variation in refractive indices and stainability attributed to the modifications in polysaccharide deposition and/or impregnation of this cell wall region (Hamly, 1935; Harris, 1983; Bhalla and Slattery, 1984; Bevilacqua et al., 1987). The strength of the light line was related to seed coat impermeability (Stevenson, 1937; Harris, 1987). The lumen of macrosclereids is usually irregular and tapered toward the seed coat surface due to prominent cell wall thickening. Macrosclereids differentiation was followed in cytological detail in pea (Spurny, 1963; Harris, 1983), soybean (Harris, 1987), and clover (Algan and Büyükkartal, 2000). The length of macrosclereids seems to be under environmental control in soybean (Noodén et al., 1985). Unevenly distributed pores (pits) in soybean seed coat develop during the desiccation phase (Yaklich et al., 1986; Vaughan et al., 1987) and the presence of such cracks, pits and other irregularities on the seed coat surface, seems to be related to its water permeability (Wolf et al., 1981; Yaklich et al., 1986; Ma et al., 2004). A subepidermal layer of cells is differentiated into osteosclereids (bone-shaped cells), also termed by different authors columnar or pillar cells, hourglass cells or lagenosclereids (flask-shaped cells) depending on the shape of the cells. This layer includes conspicuous air-filled intercellular spaces (**Figure 1A**) resulting from cell shaping during testa differentiation and massive cell wall deposition in the middle part (Harris, 1983; Miller et al., 2010). Osteosclereids are the first major cell types during testa differentiation where cell death was detected, followed by parenchyma and macrosclereids (Ranathunge et al., 2010). In and around the hilum, the layer of osteosclereids merges with thick-walled, star-shaped parenchyma (**Figure 1C**). Continuity of the intercellular spaces may be related to seed desiccation and gas exchange.

THE PARENCHYMA OR NUTRIENT LAYER

The innermost part of the seed coat is composed of parenchyma cells (**Figure 1A**), which are elongated in tangential direction and result in abundant air-filled intercellular spaces. Frequently there are 5–12 cell layers of parenchyma with the inner layer in direct contact with the endosperm (Hamly, 1932, 1935). Some authors call the parenchymatous region "nutrient layer" owing to its function during embryo development (Van Dongen et al., 2003). The seed coat vascular systems are embedded into parenchyma layers and their structures vary amongst legumes; some species possess an extensive vascular systems that anastomose to form reticulated

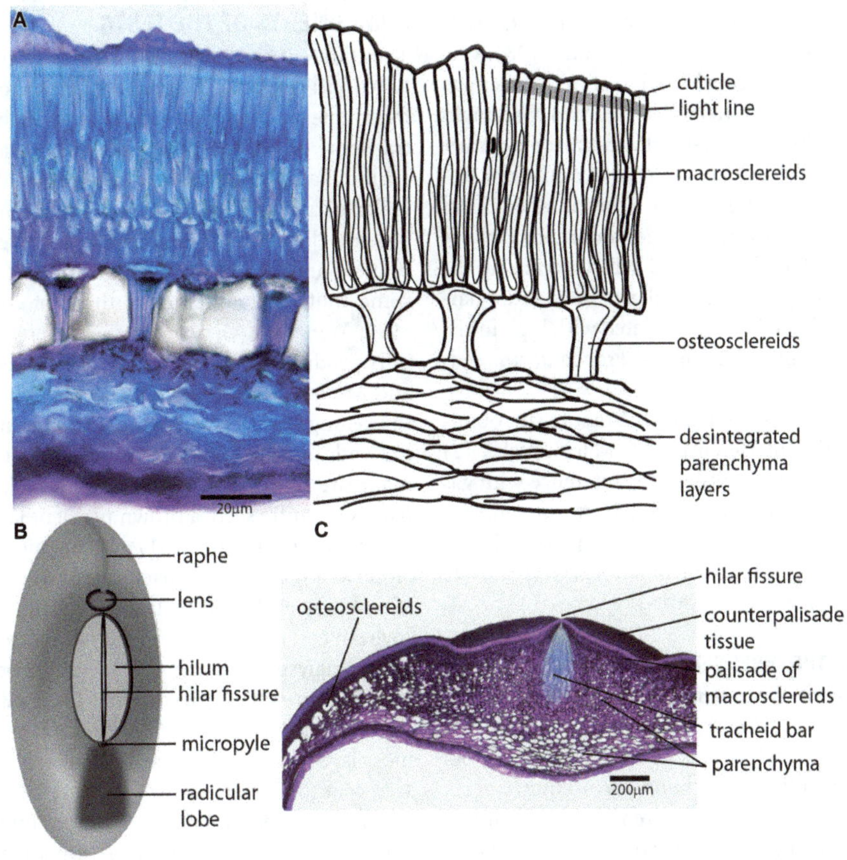

FIGURE 1 | The arrangement of Fabaceae seed coat share rather common structural features. (A) Transversal section of the seed coat of wild *Pisum sativum* subsp. *elatius* (left), with a schematic drawing (right); epidermal cells differentiate into macrosclereids, which are characterized by a cuticle-covered surface. The outer parts of the macrosclereids (sclereid caps) are frequently separated by a region of the cell wall (light line) with specific features, resulting in different optical and staining properties. The central part of the testa differentiates into osteosclereids with a specific shape caused by thickened secondary cell wall. The innermost layers of parechymatous cells frequently die during differentiation and only disintegrated fragments are left. **(B)** Generalized scheme of the seed coat morphology commonly found in Fabaceae seeds showing the most important structural features, including the hilum, lens differentiated on the raphe and micropylar pore. **(C)** Transversal section of a *Pisum sativum* seed coat in the area of the hilum. The macrosclereids of the hilar scar are covered with counter palisade tissue with a central fissure above the tracheid bar, which is surrounded by star-shaped parenchyma interconnected to intercellular spaces of a layer of osteosclereids.

networks throughout the entire seed coat (e.g., common bean, Offler and Patrick, 1984; soybean, Thorne, 1981), while other species have relatively simple vascular systems, with only a single chalazal vascular bundle and two lateral branches extending into the seed coats (e.g., pea, Spurny, 1963; Hardham, 1976; broad bean, Offler et al., 1989). During seed coat maturation, parenchyma cells lose the protoplast and the innermost layers might be crushed. Parenchyma layers are not generally related to water-impermeability. However, high callose levels in this layer might be related to low permeability of clover seed coat (Bhalla and Slattery, 1984), but a causal relationship remains to be determined.

THE MICROPYLE AND HILUM

The anatomical structure of the testa is rather homogenous except for the chalazal region (**Figure 1B**). The hilum is a distinctly oval or round abscission scar in the chalazal seed region, a relic of former connection of the seed to the maternal plant via the funiculus. There is another residual layer of palisade cells of funicular origin termed counter-palisade, which are part of the hilar scar (Lackey, 1981). There is a central fissure (hilar groove) in the hilum palisade layer overlaying the tracheid bar (**Figures 1B,C**) from the micropyle to the ovular bundle on the other side. This strip of large, pitted and lignified tracheids is commonly found in legumes although there is some variability in structure and ovular bundle position (for more detail see Lersten, 1982). A role for the hilar groove as a hygroscopically activated valve was suggested (Hyde, 1954; Lush and Evans, 1980). The fissure in the hilum opens when relative humidity is low permitting the seed to dry out whereas high relative humidity causes the fissure to close preventing the absorption of moisture.

THE LENS (STROPHIOLE)

The micropyle, an entrance pore for the pollen tube (**Figure 1B**), is inconspicuous on mimosoid and casealpinoid seeds but is discernible (often of different color) on faboid seeds (Gunn, 1981). In some species, there is a residue of the micropylar opening covered with a waxy lid (Vaughan et al., 1987). Specific outgrowth of raphe termed lens (strophiole) on the other side of the hilum (**Figure 1**) might be obvious in some species (Lersten and Gunn, 1982; Lersten et al., 1992). This structure is considered to act as a water gap (Baskin et al., 2000; Hu et al., 2009; Karaki et al., 2012) that might be open by external (e.g., heat, mechanical action, temperature variation; Baskin, 2003; Van Assche et al., 2003) or internal factors (Rolston, 1978; Argel and Paton, 1999). The palisade cells of the lens region are modified, narrower, longer and more variable than in the rest of testa. A loosely arranged cell structure on the lens-side of the hilum, a deeply grooved hilar fissure, and a narrow tracheid bar were considered to be associated with high initial water absorption in some common bean and *Psophocarpus* seeds (Deshpande and Cheryan, 1986).

Functionally, the main effects exerted by the tissues surrounding the embryo are: (1) the interference with water uptake, (2) mechanical restraint to radicle protrusion, (3) inference with gas exchange, (4) prevention of inhibitor leakage from the embryo, (5) supply of inhibitors to the embryo and 6) light penetration in species in which light plays a role in germination (Werker et al., 1979; Nowack et al., 2010; Bewley et al., 2013).

COORDINATED GROWTH OF THE THREE SEED COMPONENTS

The coordinated growth of the inner and outer integuments, which ensures that the ovule is surrounded by protective tissues, is regulated by several locally expressed transcription factors (TFs), which were first identified in *Arabidopsis* (Skinner et al., 2004). The integuments are initially undifferentiated, but rapidly undergo changes to produce a complex structure that protects the embryo and sustains its growth. In fact, differentiation of the seed coat from the ovular integuments includes some of the most dramatic cellular changes observed during seed development. Early seed coat development from the time of anthesis to seed maturation has been well described in soybean (Miller et al., 1999), pea (Van Dongen et al., 2003), *Medicago truncatula* (Wang and Grusak, 2005), and faba bean (Offler et al., 1989; Offler and Patrick, 1993; Borisjuk et al., 1995), and shows a relative homogeneity in term of ontogeny and final structure. During early legume seed development, the embryo and endosperm develop within the seed coat. The endosperm and embryo divide in parallel, with the endosperm occupying a larger volume until the beginning of seed filling or maturation, when the endosperm begins to degenerate and the embryo cells expand to accumulate storage products. Importantly, early embryo development and differentiation are controlled by the surrounding maternal tissue, and signals from the maternal plant must be transmitted through the seed coat and endosperm before they can reach the embryo. A model for the maternal control of embryo development through sugar metabolism that implicated seed coat invertases has been developed in legumes (reviewed in Weber et al., 2005). In *Arabidopsis*, mutations of the endosperm-specific LRR kinase HAIKU or the WRKY transcription regulator MINISEED3 result in limited growth of the endosperm, which in turns affects cell elongation in the seed coat resulting in smaller seeds. Conversely, mutation of the maternally expressed TF TTG2 restricts cell elongation in the seed coat, which limits endosperm growth. Whether homologous genes play a role in legume seeds remains to be shown.

DEVELOPMENT AND ACCUMULATION OF MAIN COMPONENTS OF THE TESTA

The different seed coat cell layers have three main physiological functions: (1) Production, transport, and download of metabolites for zygote development, including metabolite inter-conversions, transport of photosynthetic assimilates and photosynthesis, (2) synthesis and deposition of defense-related compounds, both phytoalexins and structural components, and (3) establishment of physical dormancy and mechanical protection. Seed coat development and structure in soybean and *Medicago truncatula* share some traits that differ from those of *Arabidopsis* (Miller et al., 1999; Wang and Grusak, 2005). Whereas *Arabidopsis* epidermal cells produce and secrete large quantities of mucilage, soybean and *Medicago* epidermal cells do not and show extensive cell wall thickening (Russi et al., 1992). The innermost seed coat cell layer, the endothelium, is a metabolically active cell layer and is the main site of synthesis of proanthocyanidins (PAs). In the *Arabidopsis* mature seed, a brown pigment layer (bpl) forms between the inner integument 1 (ii1) and the outer integument 1 (oi1) cell layers as a result of compaction of several parenchyma cell layers (Beeckman et al., 2000). The endothelium has been described in soybean (Miller et al., 2010) and was microdissected for transcriptomic analysis (Le et al., 2007). In faba bean, this layer has been termed thin-walled parenchyma (Miranda et al., 2001). Because the aleurone remains at maturity and is crushed against the innermost layer of the seed coat, it is often considered as part of the seed coat (Miller et al., 1999; Moïse et al., 2005). However, differences in their development and more importantly in their origin (maternal for the seed coat vs. zygotic for the aleurone) make these two tissues distinct entities with specific roles during seed development.

Seed coat development is tightly regulated, and several tissues present during embryogenesis and seed filling will not persist at maturity or will undergo important modifications. This is particularly true for the parenchyma layers derived from the inner ovular integuments, which proliferate and then collapse and are crushed (Miller et al., 1999; Nadeau et al., 2011). In pea, three parenchyma sub-layers can be distinguished during testa development: chlorenchyma, ground parenchyma, and branched parenchyma. Starch transiently accumulates in the plastids of these cell layers, the young seed coats being a transient storage organ for carbohydrates and proteins (Rochat and Boutin, 1992). A dramatic and transient expansion of the branched parenchyma occurs during the filling period followed by its complete compression and the formation of a boundary layer between the seed coat and the filial tissue (Van Dongen et al., 2003; Nadeau et al., 2011). In pea and common bean, branched parenchyma is the site of expression of extra-cellular invertases, and it has been suggested that the degradation of this layer initiates the storage phase through a switch from high to low ratios of hexose to sucrose in the developing seeds (Weber et al., 1997; Van Dongen et al., 2003). Important cellular changes are also observed in the testa outer cell layers and in specific cell wall thickening. Sclereids are characterized by extensive secondary cell wall formation and are usually non-living at maturity with a callose-rich wall area running parallel to the edge of the seed coat (Bhalla and Slattery, 1984; Serrato Valenti et al., 1993; Ma et al., 2004). In common bean, cell vacuoles of the thick-walled epidermal cells are often completely filled with tannins, indicating that the macrosclereid cells play a key role in hardening of the seed coat (Algan and Büyükkartal, 2000). The contrasting cellular fates above are related to different functional requirements during the course of development: nutrient transport and metabolism to sustain embryo growth during embryogenesis and seed filling and physical and structural requirements as the seed matures to insure protection.

OMICS ANALYSIS: TOWARD A GLOBAL GENE ACTIVITY PROFILE OF SEED COAT DEVELOPMENT

In general, development of the seed coat has not been characterized at the molecular level to the extent of the embryo and endosperm (Thompson et al., 2009). Transcriptomic and proteomic analyses have been used to dissect the molecular mechanisms underlying the development of the three major seed tissues, including the maternal seed coat in the legume model *Medicago truncatula* (Gallardo et al., 2007; Pang et al., 2007; Verdier et al., 2008) and in soybean (Le et al., 2007; Ranathunge et al., 2010; Miernyk and Johnston, 2013). Seed coat transcriptome and proteome were shown to be highly correlated in *Medicago* and quite distinct from that of the embryo or endosperm (Gallardo et al., 2007). Results highlighted a metabolic interdependence of these three seed components during seed filling, with certain metabolic steps or enzymes being restricted to a particular tissue. An example is several proteases specifically produced in the seed coat that may be important to provide amino acids for protein synthesis within the embryo (Gallardo et al., 2007). Similarly, seed coat specific TFs acting during early seed filling when seed coat and endosperm are active in supplying nutrients to the developing embryo were identified (Verdier et al., 2008) and could represent master regulators of seed coat development and function. More recently, a combined histology and transcriptomic analysis of the *Medicago* seed coat was performed (Verdier et al., 2013a) and a regulatory network-based analysis of transcriptome profiles during *Medicago truncatula* seed maturation was carried out (Verdier et al., 2013b). At 4 to 6 days after pollination (DAP), arrest of cell division occurs, which is compensated by cell elongation in the expanding seed coat. This cell size increase was associated with endopolyploidy and supported by transcriptomics data showing over-expression of "nucleotide metabolism" class genes. In addition, laser capture microdissection and transcriptional profiling were used to identify genes expressed in different sub-regions of soybean seeds (Le et al., 2007; http://seedgenenetwork.net/soybean). Transcriptomic data are not only available for nearly mature seed coat layers (hourglass, palisade, parenchyma) but also for the inner and outer integuments during early seed coat development, providing a source of candidate genes for further functional analyses.

POLYPHENOLIC COMPOUNDS BIOSYNTHESIS AND ACCUMULATION

The PAs, oligomers of flavan-3-ol units, have received particular attention due to their abundance in seed coats (Dixon et al., 2005; Zhao et al., 2010). PAs are also known as the chemical basis for tannins, polymeric flavonoids that comprise part of the broad and diverse group of phenolic compounds that plants produce as secondary metabolites (Winkel-Shirley, 2001). These are synthesized in the inner integument or endothelium layer. PA biosynthesis and its regulation have been dissected in *Arabidopsis* using *transparent testa* (*tt*) mutants, which regulate production, transport or storage of PAs (Lepiniec et al., 2006), and 20 genes affecting flavonoid metabolism were characterized at the molecular level (reviewed in Bradford and Nonogaki, 2009). Most notably, a set of three types [(TT2, a myeloblastosis family of transcription factors (MYB) protein family member; TT8, a basic-helix-loop-helix (bHLH) proteins; and TTG1, a WD40 protein)] of TFs was characterized that interact to regulate transcription of anthocyanidin reductase (ANR; **Figure 2**), a key enzyme producing the epicatechin building block of PAs (Baudry et al., 2004). Many of these flavonoid biosynthesis pathway genes have been found to affect dormancy of *Arabidopsis* seeds, indicating the role of pigments in this process (Debeaujon et al., 2000). Whether these genes play a similar role in legume seeds remains to be shown. Tannins are also important nutritionally because they can complex with several minerals and proteins in the gastrointestinal lumen, reducing the absorption, digestibility and availability of these nutrients (Brune et al., 1989). In *Medicago truncatula*, seed coat-expressed genes involved in PA biosynthesis and transport (**Figure 2**) have also been identified (Pang et al., 2007; Zhao and Dixon, 2009; Zhao et al., 2010), and two genes implicated in the regulation of PA biosynthesis have been isolated, a WD40-repeat TF (Pang et al., 2009) and a MYB family TF (Verdier et al., 2012).

Legumes vary in the types of PA monomers that are polymerized to form tannins (**Figure 2**). The best studied *Medicago truncatula* PAs are composed primarily of epicatechin and/or catechin units (which yield cyanidin on hydrolysis), with much lower levels of epigallocatechin or gallocatechin (yielding delphinidin) and epiafzelechin or afzelechin (yielding pelargonidin) units. In lentils, there is a balance between catechin and gallocatechin units; and the polymer fraction was more abundant than the monomer and oligomer fraction (Dueñas et al., 2003). In contrast, common beans had mainly catechin monomers (60% on average) in their seed coat tissue, with minor and variable amounts of gallocatechin and afzelechin (Díaz et al., 2010). Other genes affecting PA deposition play roles in the differentiation of the endothelium layer (Dean et al., 2011). This layer is for the most part a single cell layer, but in *Arabidopsis* limited periclinal divisions were observed in the micropylar region (Debeaujon et al., 2003). Interestingly, in *Arabidopsis*, PA accumulation begins in the micropylar region of young seeds approximately at the two-cell stage of embryo development, progressing through the seed body, and ending in the chalazal end at the heart stage of embryo development, a process reminiscent of the asynchronous differentiation of endosperm during seed development. The availability of an endothelium-specific promoter (*BANYULS* gene of *Arabidopsis*) permitted (Debeaujon et al., 2003) testing the effect of the ablation of this cell layer using a promoter:Barnase fusion. The seed, which completely lacked the pigment layer, remained viable, while embryo, and endosperm development were not being obviously affected, but testa-imposed dormancy was reduced. However, other mutants acting earlier (pre-fertilization) on the formation of the integument layers, do affect, or even prevent, seed development (Schneitz et al., 1997; Mizzotti et al., 2012). The

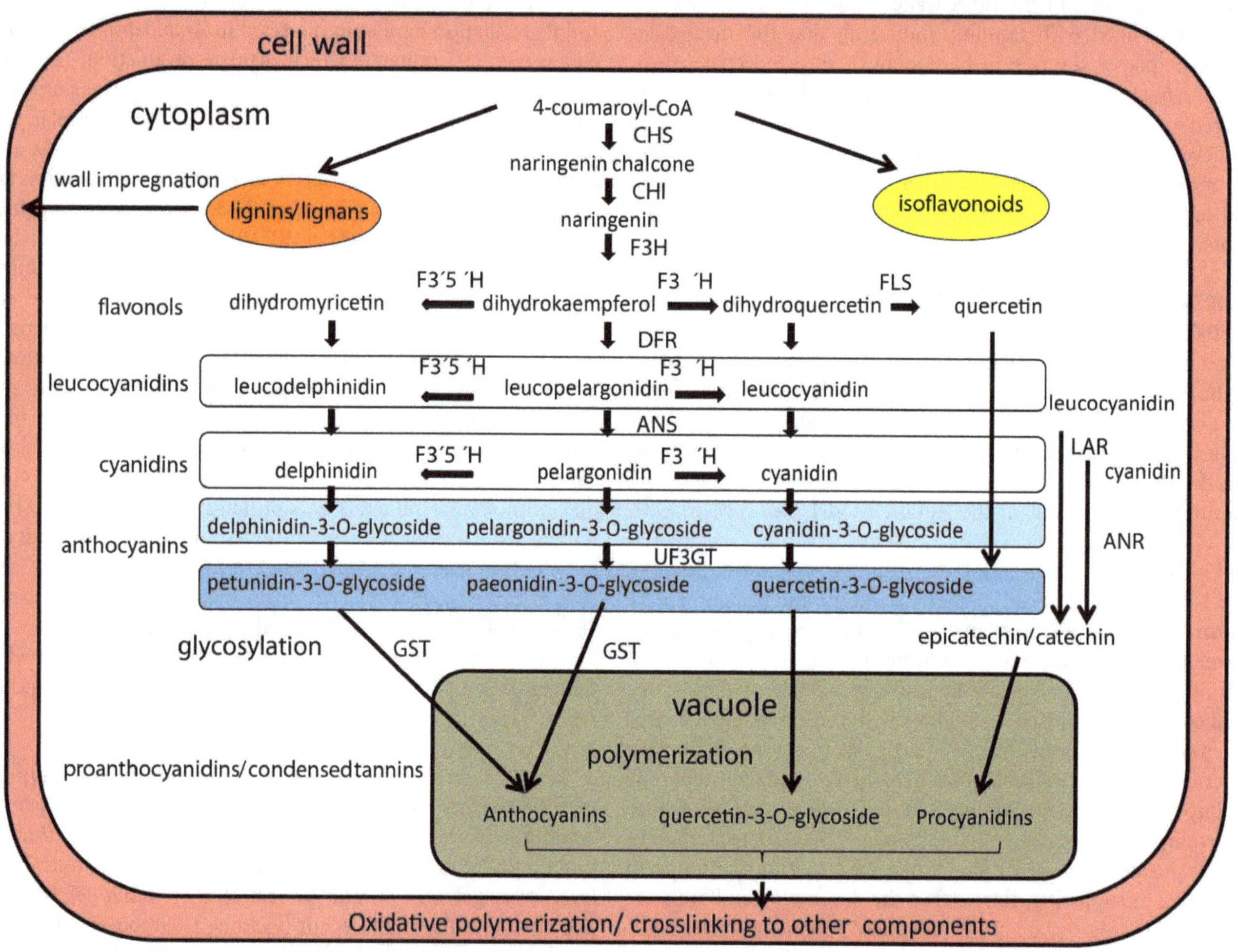

FIGURE 2 | Schematized representation of biosynthetic pathway of anthocyanins in legume seeds. Enzymes are indicated in bold italics, metabolic intermediates are in plain text. Enzymes are abbreviated as follows: anthocyanin reductase (ANR), chalcone synthase (CHS), chalcone isomerase (CHI), dihydroxyflavone reductase (DFR), flavanone 3-hydroxylase (F3H), flavonoid 3′5′hydroxylase (F3′5′H), flavonoid 3′hydroxylase (F3′H), glutathione-S-transferasse (GST), leucoanthocyanidin reductase (LAR), UDP-glucose:flavonoid 3-*O*-glycosyltransferase (UF3GT). Leucocyanidins and cyanidins are colorless, while anthocyanins and pro anthocyanins are colored. Based on Zhao et al. (2010), Gillman et al. (2011), Kovinich et al. (2011), and Reinprecht et al. (2013).

importance of pigments for early embryogenesis remains to be tested in legumes, particularly in either *Medicago truncatula* or *Lotus japonicas* models.

The high levels of lignin polymers in the seed coats support the view that they have a primary structural function, i.e., providing mechanical strength and impermeability. Polymerization of soluble phenolics to insoluble polymers has been suggested to be promoted by peroxidases (Gillikin and Graham, 1991) and catechol oxidases (Marbach and Mayer, 1974; Werker et al., 1979), which are abundant in legume seed coats. These individual compounds play important roles in defense against microbial pathogens and other biotic and abiotic stresses, as well as affecting the nutritional quality of products, and because of their health benefits, are of industrial and medicinal interest. Tannins in seed coat are particularly effective against fungal infection (Kantar et al., 1996; Winkel-Shirley, 2001).

SEED COAT COLOR PATTERN GENES

In common bean, the roles of seed coat color pattern genes such as *Bip* (*bipunctata*) and *Z* (*zonal*) genes in the accumulation of tannin also suggest that tannin first accumulate in the seed coat tissues nearest the hilum (Caldas and Blair, 2009). Patterned seeds require a *t/t* genotype for expression of a partial color, whereas a *T/-* genotype gives seeds that are totally colored. The types (form and extension) of colored patterns are then controlled by the interaction of the *t* gene with the two previously mentioned *Z* gene for patterned color expression on the testa, plus the genes *L* (*limiter*) and *J* (*joker*) which were found to be allelic to each other (Bassett, 1994, 2002, 2007). Finally, the *Z* gene is allelic with the *D* gene, which specifically determines hilum color (Bassett, 1999). Outside the legume family, 10 species were evaluated for condensed tannins in the Brassicaceae, and all of them contained tannins in the hilum including species where no tannins were found in other regions of the seed coat (Marles and Gruber,

2004). The hilum is known to be the point of attachment of the seed to the placental tissue of the ovary, but it also is the place where seed water uptake often begins, a portion of the seed that would need to be protected by tannins which have roles as antifungal metabolites. Notably, white beans, which lack any coloration, usually have no tannins or very low tannin levels and are more susceptible to root rots and other diseases (Ma and Bliss, 1978; Guzmán-Maldonado et al., 1996).

In soybean, six independent loci (*I, T, R, Wp, W1,* and *O*) control the color and distribution of pigments (Bernard and Weiss, 1973; Palmer and Kilen, 1987; Gillman et al., 2011). The best characterized *I* locus (for *inhibitor*, ecoding chalcone synthase) inhibits the production and accumulation of anthocyanins and PAs in the epidermal layer of the seed coat (Bernard and Weiss, 1973). An allelic series for the *I* gene is present in soybean, where the absence of pigmentation is controlled by the dominant allele at the *I* locus, whereas the homozygous recessive *ii* genotype produces a totally pigmented seed coat, and the alternate i^i allele results in pigmentation of the hilum (Tuteja et al., 2004). Most cultivated soybean varieties are homozygous for the *I* gene, resulting in a yellow seed coat. To date, eight chalcone synthase genes have been identified in soybean, all expressed in seed coat (reviewed in Moïse et al., 2005). Chalcone synthase is the first enzyme of the branched pathway of flavonoid biosynthesis (**Figure 2**), and it plays a role in the synthesis of secondary metabolites functioning as UV protectants, phytoalexins, insect repellents, and symbiosis initiators in various plant tissues. Another locus, *T*, affects pubescence and hilum colors, and it induces seed coat cracking. It encodes flavonoid 3′-hydroxylase, which is necessary for the formation of quercetin from kaempferol (**Figure 2**) and is responsible for the hydroxylation of the 3′position of flavonoids, leading to the production of cyanidin pigments (Zabala and Vodkin, 2003). The *W1* locus controls flower color and affects seed color only in an *iRT* background, where *W1* and *w1* alleles give imperfect black and buff seed coat colors, respectively. The *W1* allele encodes flavonoid 3′,5′-hydroxylase (Zabala and Vodkin, 2007) and causes purple flower. The *Wp* locus was suggested to code for the flavanone 3-hydroxylase (**Figure 2**) based on microarray analysis (Zabala and Vodkin, 2005), where the recessive *wp* allele resulted in change from black (*iRTWp*) to light grayish (*irtwp*) color. The *O* locus affects the color of brown seed coat (*irTO*) and has been suggested to code for ANR (Yang et al., 2010). Finally, the *R* locus controls the presence (*R*) or absence (*r*) of anthocyanins in black (*iRT*) or brown (*irT*) seed coat, respectively (Nagai, 1921). The up-regulation of anthocyanidin synthase genes suggests that the *R* locus codes for a regulatory factor (Kovinich et al., 2011). Comparably less is known about pigmentation in pea, especially in relation to the testa. Mendel's *A* gene, conferring pea flower color and testa pigmentation, has been recently identified as a bHLH TF, exerting thus pleiotropic effects on flower, leaf axils and testa pigmentation (Hellens et al., 2010). However, these traits can be genetically uncoupled, e.g., pea with colored flowers can have nearly non-pigmented seed coat (Smýkal, unpublished). The *b* gene of pea was shown to encode a defective flavonoid 3′,5′-hydroxylase, conferring pink flower color (Moreau et al., 2012). However, none of these two mutations result in alteration of seed dormancy, since they have been identified in cultivated pea lines. Also in lupin, besides the hard-seeded gene *Mollis*, the blue flower and dark seed color gene *Leucospermus* (Clements et al., 2005) confers domestication trait. Although truly wild *Vicia faba* has not identified, a locus was found to be involved in seed dormancy (i.e., *doz*) linked to a locus controlling anthocyanin and pro-anthocyanidin synthesis (i.e., *sp-v*; Ramsay, 1997). Finally, in *Trifolium subterraneum* seeds, a relationship between color, phenolic content and seed coat impermeability was found (Slattery et al., 1982).

OTHER SEED COAT COMPOUNDS INVOLVED IN PROTECTION

Besides pigments, several active enzymes have been isolated from the legume seed coat. Chitinases were isolated from soybean and are expressed late during seed development (Gijzen et al., 2001). They have been suggested to play a role in plant defense against fungal pathogens (Schlumbaum et al., 1986). Peroxidase is a major component of the protein fraction in mature soybean seed coat where it accumulates in the hourglass cells of the epidermis. Initially involved in the synthesis or modifications of extracellular polymers such as lignin or suberin, peroxidase is believed to have a function in defense when released from the hourglass cells during seed imbibition (Moïse et al., 2005). Furthermore, polysaccharides, such as galactorhamnans, are present in the innermost layer of jack bean (*Canavalia ensiformis*) seeds, and are effective against seed beetles (Oliveira et al., 2001). A well-documented barrier for preventing water entry via the seed coat is suberin accumulation (Nawrath, 2002). Suberin is composed of two distinct types of insoluble polyesters of fatty acid and glycerol. Suberin has been found in the pea testa (Spurny, 1964). Genetic evidence shows that suberin deposition controls seed permeability (Beisson et al., 2007), and glycerol-3-phosphate acyltransferase (GPAT) gene candidates for a role in suberin biosynthesis have been identified in soybean (Ranathunge et al., 2010), *Medicago truncatula* (Verdier et al., 2013a), *Arabidopsis*, and *Melilotus* (Liang et al., 2006; Beisson et al., 2007).

ROLE OF SEED COAT IN SEED DEVELOPMENT

Thorne and Rainbird (1983) developed a method, adopted by several groups, for measuring phloem unloading from seed coats by excising the immature embryo and recovering assimilates unloaded into the embryo sac. Legume seed coats were thus shown to play a critical role in the lateral transfer of assimilates and other nutrients, prior to their release to the developing embryo (Lush and Evans, 1980; Offler and Patrick, 1984, 1993). The seed coat supplies the zygote with water and oxygen, minerals, certain phytohormones including ABA and IAA, and C and N assimilates in the form of sucrose and amino acids unloaded from the phloem terminals. Further, the legume seed coat is the site of inter-conversions of the principal amino acids unloaded from the phloem, asparagine and aspartate, via asparaginase and aminotransferases, to a composition better adapted for storage protein accumulation, before unloading into t the embryo sac (Murray and Kennedy, 1980; Lanfermeijer et al., 1992). Similarly, sucrose is partly hydrolysed by extracellular invertases prior to entry into the embryo sac, contributing to the maternal control of legume seed development (reviewed in Weber et al., 2005).

Cell wall invertases promote assimilate unloading by increasing the sucrose concentration gradient in the unloading zones of the legume seed coat. The glucose released promotes embryo cell divisions, determinant of final seed size (Weber et al., 1997). Solute transfer is also facilitated by the differentiation of specialized transfer cells located in the coat parenchyma or cotyledon epidermis, i.e., at the interface between the seed coat and the embryo (Thompson et al., 2001), as found in the epidermis of faba bean and pea embryos (Bonnemain et al., 1991). Transfer cell development is accompanied by the increased expression of a sucrose-H^+ transporter gene. Both cell differentiation and gene expression have been suggested to be induced by signals coming from the maternal seed coat or to be elicited by tissue contact. Auxin, ethylene and reactive oxygen species (ROS) have been proposed as inductive signals for transfer cell differentiation (reviewed in Andriunas et al., 2013).

In addition to the production of PAs and other defense-related compounds, the seed coat is an important source of phytohormones for the developing seed, either synthesized *in situ*, or transported from the mother plant. In pea, Nadeau et al. (2011) analyzed GA content and transcript abundances of major enzymes of the GA biosynthesis pathway and studied the consequences of their inactivation in the three seed tissues during development. The results are consistent with a key role for GA in orchestrating seed coat differentiation via the production and turnover active GA forms in the seed coat during early and mid-phase seed development. In legume seeds, cytokinin (CK) concentrations are low (Slater et al., 2013), with a maximum during embryogenesis and at the beginning of the filling stage. Highest CK levels within the seed are found in the seed coat and in the liquid endosperm (Emery et al., 2000). Calculations of CK delivery rates from transport fluids to seeds suggest that CK is synthesized in situ in the seed coat and/or endosperm (Emery et al., 2000). CK was proposed to promote cell division and thus enhance sink strength, increasing solute unloading from the seed coat (Quesnelle and Emery, 2007). Auxin and its conjugated form IAA-Asp play essential roles in controlling pattern formation in the developing legume embryo, and were shown to be the major hormones present in the embryo during the early stage of development (Slater et al., 2013). In the Fabeae tribe of legumes, 4-chloro auxin is the predominant auxin species (Reinecke, 1999). Seed-produced auxin is also transported to the pericarp, where it coordinates pod development with that of the seeds (Ozga et al., 2009; Park et al., 2010).

Abscisic acid accumulates within legume seed (Liu et al., 2010; Slater et al., 2013), and genes involved in ABA biosynthesis were found to be expressed during seed filling in *Medicago truncatula* seed coat, suggesting it could be a source of the hormone (Verdier et al., 2013a). ABA is required for the seed maturation program, acting via the "master regulator" TFs to control storage and late embryogenesis accumulating (LEA) protein deposition and the acquisition of desiccation tolerance (Verdier et al., 2013b). Seed coat-mediated dormancy in *Medicago truncatula* requires ABA (Bolingue et al., 2010), probably produced by maternal seed tissues, as shown in tobacco by the reciprocal crosses of *aba* mutants, defective in ABA synthesis (Karssen et al., 1983; Frey et al., 2004). Developmental signals often result from interactions between two or more hormones. Hence fertilization-stimulated auxin production modulates synthesis of active GA (Dorcey et al., 2009), and the ratio of ABA:GA modulates seed filling (Liu et al., 2010). Liu et al. (2012) found a positive correlation between ABA and IAA concentrations and seed filling rate when comparing varieties of contrasting seed size.

ROLE OF SEED COAT IN IMPERMEABILITY TO WATER

STRUCTURAL ASPECTS

In soybean, an anatomical study showed that the only features consistently correlating with seed coat permeability to water were small cuticular cracks (Ma et al., 2004). These cracks were present in soft but not hard seeds (within the studied cultivated genotypes). The initial penetration of water into soft seeds typically occurred on the dorsal side, the location of the majority of the cracks (Ma et al., 2004). Besides anatomical differences, chemical analysis resulted in the identification of a seed coat cutin with unusual chemical composition, lacking typical mid-chain hydroxylated fatty acids but being relatively rich in other types of hydroxylated fatty acids (Shao et al., 2007). The cuticle of the impermeable soybean cultivar (again, no wild soybean was studied) contained a disproportionately high amount of hydroxylated fatty acids relative to that of the permeable ones. According to the results of Shao et al. (2007) and those of Ma et al. (2004), the difference between hard and soft soybean seeds was based on the composition and continuity of the outermost seed cuticle and the presence of small cracks in the cuticles of the latter. The differences in chemical composition were rather subtle, and the authors speculated that crucial elasticity of soybean cuticle might be related to its association with carbohydrates (Ranathunge et al., 2010).

The site of water entry into the seed coat is still a matter of debate (Meyer et al., 2007; Ranathunge et al., 2010). Various explanations were proposed to account for the different permeability of the seed coats to water in hard and soft seeds including tightly bound palisade cells (Corner, 1951; Ballard, 1973), thickened seed coat tissues (Wyatt, 1977; Miao et al., 2001), lack of pits (Chachalis and Smith, 2001), presence of endocarp deposits (Yaklich et al., 1986), dark color (Wyatt, 1977), closed hilum and/or micropyle (Hyde, 1954; Ballard, 1973; Rolston, 1978; Hu et al., 2009), modifications of the outer tangential walls of palisade cells (Werker et al., 1979), a prominent light line of the palisade cells (Harris, 1984, 1987), presence of water gaps in the lens (Dell, 1980; Hanna, 1984; Serrato Valenti et al., 1986, 1989, 1993; Serrato-Valenti et al., 1995; Van Staden et al., 1989; Shen-Miller et al., 1995; Morrison et al., 1998; Baskin et al., 2000; Burrows et al., 2009; Hu et al., 2009), as well as cracks in the cuticle of the seed coat (Morrison et al., 1998; Hu et al., 2009). In some legume species, such as *Vigna, Robinia, Acacia*, and many tropical and sub-tropical legume species (Karaki et al., 2012), water enters the seed a specialized structures such as the hilar slit and water gap (e.g., strophiole or lens; Gopinathan and Babu, 1985; Kikuchi et al., 2006) opens. The morphological characteristics of these water gaps can vary between species. It is thought that water gap structures act as an environmental sensor to fine-tune germination to coincide with an environment that provides the best chance for seedling survival and ecosystem colonization. Water gaps are usually associated with areas of the seed coat where natural openings in the ovule occurred during the seed development such as the hilum, micropyle, and chalaza. Although the soybean seed coat composition and structure are modified at the hilum and micropyle areas, these are thought not to contribute to the initial uptake of water (Ma et al., 2004). Instead, in soybeans, water mainly enters through small cracks in the seed coat to reach the embryo (Chachalis and Smith, 2000). There is some controversy on this issue, since other studies have suggested water uptake through the hilum (McDonald et al., 1988). In a related Phaseoleae tribe, the hilum, micropyle and lens have been proposed to be water entrance points for example, in *Phaseolus lunatus* (Korban et al., 1981) and *Phaseolus vulgaris* (Agbo et al., 1987).

In subfamilies Caesalpinioideae and Mimosoideae, cracks develop in the extrahilar region or in the hilum that allow water to enter the seed (Gunn, 1984, 1991; Hu et al., 2009). The hilum of faboid seed, except in flattened seeds, has a separation in palisade cells called the hilar groove (Lhotská and Chrtková, 1978), which is a diagnostic botanical feature. Hyde (1954) reported that this groove acts as hygroscopic valve, that prevents water entry from the outside but permits water to leave seed interior during maturation and drying. The seed coat becomes impermeable as drying occurs. Since the seed remains dry until released from dormancy, it has the potential to remain viable for many years. Seed dormancy not only prevents immediate germination, but it also regulates the time, conditions, and location where germination will occur.

The occurrence of precocious germination under humid conditions indicates that seeds can become germinable prior to maturation drying. There are contrasting results concerning whether some degree of drying is required for switching seeds from dormant to a germinative mode. Prior to desiccation, the seed undergoes developmental changes and mostly anabolic metabolism associated with formation of the embryo and its surrounding structures and the deposition of the major storage reserves. Following desiccation and rehydratation, seed metabolism becomes largely catabolic to support germination. Thus, for most desiccation-tolerant (orthodox) seeds, including legumes, maturation drying clearly switches the seed to a germination mode upon subsequent rehydration. Desiccation-intolerant (recalcitrant) seeds can effect this switch without dehydratation, as can orthodox seeds developing inside fleshy fruits (tomatoes, melons), having the capacity to germinate without maturation drying. However, premature removal of the seeds from the fruit inevitably severs the maternal connections. In the case of common bean and soybean seeds, it was reported that fresh developing seeds are unable to germinate without prior dehydratation. On the other hand, it was found that *Phaseolus* seeds near the end of the seed filling period are capable of germinating within the fruit with their funicular connection intact if water was injected into the pods. To survive in dry state (with less than 10% moisture content on a dry-weight basis), a seed has to avoid damage to its cellular components, both during water loss and upon subsequent rehydratation. Damage does occur, but is limited to a level that can be repaired and that involves accumulation of non-reducing sugars (sucrose, trehalose), oligosaccharides (raffinose) and other solutes such as proline, glycinebetaine (cited in Bewley et al., 2013). In addition to these metabolites, large number of LEA proteins, including dehydrins and heat shock proteins (HSPs), are expressed, both responding positively to ABA.

SEED COAT PIGMENTATION AND DORMANCY

Seed coat pigmentation was shown to correlate with imbibition ability in several legumes. Browning of seed coat during maturation was found to be associated with its impermeabilization in common bean (Caldas and Blair, 2009; Díaz et al., 2010), chickpea (Legesse and Powell, 1996), yardlong bean (Kongjaimun et al., 2012), soybean (Liu et al., 2007), faba bean (Ramsay, 1997), and pea (Marbach and Mayer, 1974; Werker et al., 1979).

The genes or quantitative trait loci (QTLs) for seed color and loss of seed dormancy in azuki bean (*Vigna angularis*) were shown to be closely linked, and there is a significant correlation between these two traits (Isemura et al., 2007). A positive correlation has also been found between phenolic content, the activity of catechol oxidase and seed dormancy in wild pea seeds (Werker et al., 1979). Recently, epicatechin, cyanidin 3-*O*-glucoside, and delphinidin 3-*O*-glucoside were specifically isolated in wild but not in cultivated soybean seed coats, with epicatechin showing a significant positive correlation with hardseededness (Zhou et al., 2010). Combined analysis of seed coats of black vs. brown isogenic lines of soybean, indicated over-accumulation of anthocyanins, altered procyanidin, and reduced flavonol, benzoic acid, and isoflavone content in black seeds, as a result of altered transcription of numerous biosynthetic pathway genes (Kovinich et al., 2011). Increased β-1,3-glucans (callose) deposition in cell walls during maturation is associated with increased dormancy in a number of species (Finch-Savage and Leubner-Metzger, 2006), while β-1,3-glucanases, which break down callose are associated with dormancy release. Other soluble phenolic compounds, such as coumarin and chlorogenic acid, and their derivatives, or ferulic, caffeic, sinapic acids occur in the coats of many seeds. These may inhibit seed germination and could be leached out into the soil where they may inhibit neighboring seeds (a form of allelopathy). Phenolic compounds of legume seeds, however, participate in nodulation by acting as chemoattractants, promoting rhizobial growth, and inducing transcription of nodulation genes in symbiotic bacteria (Mandal et al., 2010).

GENETICS OF LEGUME SEED DORMANCY

Seed dormancy is a monogenic trait in lentil (Ladizinsky, 1985), narrowleaf lupin (Forbes and Well, 1968), yardlong (Kongjaimun et al., 2012), rice bean (Isemura et al., 2010), mungbean (Isemura et al., 2012), associated with one to two loci in common bean (Koinange et al., 1996), and two to three loci in pea (Weeden, 2007). In this last case, control of seed dormancy was via testa thickness, testa impregnation and structure of the testa surface. In azuki bean (Isemura et al., 2007; Kaga et al., 2008), four to six QTLs were associated with field germination, time of germination, testa permeability, winter survival of seeds in the soil, days to germination of winter-surviving seeds in the field and water content in seeds. In recently (~100 years) domesticated lupin, *Lupinus angustifolius*, the hard-seeded gene *mollis*, and the blue flower and dark seed color gene *leucospermus* (Clements et al., 2005) are two of the key domestication traits. A molecular marker linked to the recessive *mollis* gene was discovered and applied to lupin breeding (Boersma et al., 2007; Li et al., 2012), where wild material was used to provide new diversity. Interestingly, Ladizinsky (1985) found that two different monogenic systems operated in crosses of *Lens orientalis* and *Lens ervoides*, with cultivated lens, *Lens culinaris*. In the former, the allele for dormancy was dominant, but in the latter it was recessive. The gene for dormancy in *Lens orientalis* appeared to be linked to another one controlling pod shattering. Similarly, interspecific crosses between *Vicia sativa*, and a closely related species, *Vicia cordata* suggested a two-gene system (Donnelly et al., 1972). Breeding has allowed the development of soft-seeded summer and hard-seeded winter lines of *Vicia sativa* (cited in Büyükkartal et al., 2013). As mentioned earlier, the testa is of maternal origin. The maternal control is clearly demonstrated when a soft-seeded (e.g., with no or low testa imposed seed dormancy) plant is used as a female parent and crossed with a hard-seeded wild type. The F_1 seed is soft-seeded, and all the resultant F_2 seeds are hard-seeded, including those individual seeds possessing the homozygous soft-seediness genotype, as shown in lupin (Li et al., 2012).

DORMANCY-BREAKING REQUIREMENTS OF LEGUME SEEDS

Many legume seeds are known to be long-lived, and they are frequently found in seed bank surveys. *Melilotus* seeds survived for 17 years in the soil (Bibbey, 1948), seeds of *Malva rotundifolia* (Malvaceae) lived for 120 years in burial seed experiment (Telewski and Zeevaart, 2002) and seeds of *Astragalus distortus* germinated 24 years after sowing under near-natural conditions (Baskin and Baskin, 2014). The breaking of dormancy under natural conditions is only partly understood. As mentioned above, water and temperature are two principal environmental regulators of seed germination. In many ecosystems, fire is also an important factor and there is no doubt that germination is promoted by fire-induced heat treatment in species such as *Acacia sp.* in the legumes family (Sabiiti and Wein, 1987; Hanley and Fenner, 1998). It is not understood, however, how germination is regulated in ecosystems with a temperate climate, where fire occurs very rarely and where fluctuations in daily soil temperature are rather limited to a maximum of 10–15°C. Since temperature is relatively constant in its seasonal variations, it is arguably the most important environmental factor to synchronize seed germination with conditions suitable for seedling establishment. This is certainly valid for seasonal climate types, but in arid and semi-arid regions water may be the most important factor, whereas in the humid tropics variations in temperature and water availability appear to be virtually absent. For temperate regions species, germination in summer is accompanied by a higher risk of seedling loss due to drought or shading by leaf canopies, whereas in autumn the seedlings experience a reduced length of the growing period and risk frost damage (Probert, 2000). It has been proposed that hard seeds become permeable to water after mechanical abrasion by soil particles, decomposition of the seed coat by microbial action, ingestion and passage through the digestive tracts of an animal or by cracks in the coat caused by partial seed consumption, but little evidence is available to support these views (Gogue and Emino, 1979; Baskin and Baskin, 2000; Dueberrn de Sousa and Marcos-Filho, 2001; Fenner and Thompson, 2005). Hardseededness has been shown to protect seeds during the passage through the digestive tract (Simao Neto et al., 1987), to extend seed longevity (Mohamed-Yasseen et al., 1994) and persistence in soil seed banks (Shen-Miller et al., 1995). Dalling et al. (2011) postulated that species with physical seed dormancy rely on physical defenses to exclude predators and pathogens. In this case, rapid seed germination cannot be used to escape pathogens at the emergence stage. Recently, it was proposed that hard seeds are an anti-predator trait that evolved in response to selection by small mammal seed predators (Paulsen et al., 2013). Seeds of two legume species with dimorphic seeds ("hard" and "soft"), *Robinia pseudoacacia* and *Vicia sativa*, were offered to desert hamsters. Volatile compounds released from imbibed seeds attracted the hamsters to the seeds, but the animals could not detect buried hard or dry soft seeds. Correlations between the dormancy release mechanism and ecological habitat were tested in four legume species (van Klinken and Goulier, 2013): two wetland species (*Mimosa pigra* and *Parkinsonia aculeata*), both dispersed primarily by water and two terrestrial species (*Acacia nilotica* and *Prosopis pallida*), both dispersed primarily through vertebrate herbivores. Seed viability was largely unaffected by temperature or moisture regime, although it differed with species and was lower for non-dormant seeds.

There are two steps in breaking physical dormancy with high temperatures. At first, the preconditioning phase will occur if seeds are at constant temperatures, and the rate at which this stage is completed increases with an increase in temperature. Seeds prevented from drying (by blocking the hilum) during the first stage are more likely to become water-permeable in the second stage than those that dehydrate further during stage one. The second stage (when seeds become permeable) requires fluctuating temperatures for maximum loss of dormancy (Baskin and Baskin, 2014). Taylor (1981) suggested that thermal degradation occurs during the first stage, which results in weakening of the lens. In the second stage, physical expansion and contraction associated with temperature fluctuations cause cells in the lens to open. There are also two steps in breaking dormancy by low winter temperatures, as found in temperate-zone species. Study of *Melilotus alba*, *Vicia villosa*, and *T. pratense* showed that during a first step, low winter temperatures make seeds sensitive to alternating temperatures, and during a second step, these alternating regimes occurring in early spring cause the sensitive seeds to become water-permeable. Further, cycles of germination (but not dormancy break) occur because spring temperatures might not be adequate to open the water gap of sensitive seeds. Thus, seeds may lose their sensitivity and have to go through another winter to become sensitive again. Seeds can only respond to alternating spring temperatures if they become sensitive during the winter (Baskin and Baskin, 2014).

The alternating temperatures that break the physical dormancy of a seed depend on the amplitude of the fluctuation. Clover (*T. subterraneum*) seeds become sensitive in response to temperatures that fluctuate between 30 and 60°C over a period of several weeks or months, the fluctuations occuring on open soils in Mediterranean-line climates, and subsequently their seed coats become permeable for water (Quinlivan, 1971; Taylor, 1981, 2005; Moreno-Casasola et al., 1994; Taylor and Ewing, 1996; Taylor and Revell, 1999, 2002; de Souza et al., 2012). The proportion of soft/hard seeds seems to vary from year to year (Roberts and

Boddrell, 1985), presumably as a result of climatic effects. The influence of seasonal factors on the germination of impermeable seeds of 14 herbaceous legume species was studied by Van Assche et al. (2003). Six species (*Medicago lupulina, Melilotus, Lotus, T. pratense, T. repens,* and *Vicia cracca*) showed a marked seasonal cycle with high germination rates in spring. *Medicago arabica, T. dubium,* and *Vicia sativa,* which are typical winter annuals, germinated mostly in summer and autumn, while *Lathyrus aphaca, Lathyrus nissolia,* and *Vicia hirsuta,* germinated in all seasons except summer (Van Assche et al., 2003). The importance of daily temperature fluctuations in breaking physical dormancy was shown for several legume species. The percentages of impermeable seeds of *Stylosanthes humilis* and *Stylosanthes hamata* begun to decline in northern Australian pastures in September (early spring) when mean monthly maximum and minimum temperatures were about 67 and 28°C, respectively. The number of dormant seeds decreased until December (early summer) when rains stimulate all permeable seeds to germinate. Dormancy was not broken from January to August, when daily maximum and minimum temperatures were less than 55 and 25°C, respectively (McKeon and Mott, 1982). Impermeable seeds of *Indigofera glandulosa* become permeable (Bhat, 1968) by exposure to hight temperatures (up to 60°C). Impermeable seeds of *Lupinus digitanus, Lupinus luteus, Medicago trilobus,* and *T. subterraneum* subjected to alternating temperature regimes had highest germination rates under temperature regimes of 60/15°C for 4 months (Quinlivan, 1961). Dormancy loss was determined by the maximum daily temperature, provided there was a minimum daily temperature fluctuation of at least 15°C. The maximum daily temperatures required for loss of dormancy varies with the species: *T. subterraneum,* 30°C; *T. hirtum, T. cherleri, T. cernum,* 40°C; *Medicago truncatula, Medicago littoralis, Medicago polymorpha,* 50°C; and *Lupinus varius,* 60°C (Quinlivan, 1968). Such temperature fluctuations are prevented by seeds shaded by plant litter or by soil burial. Continuous moisture of substrate could be an important clue for some seeds to become water permeable. However, this factor has received comparably less attention in the literature. Several species did not imbibe seeds for long periods, up to 3 years for *Stylobasium spathulatum* among Surianaceae (Baskin et al., 2006). The seeds of *Acacia nilotica,* which grows along rivers in Sudan and Egypt and is subjected to flooding, imbibe and germinate at best when soaked for 18 weeks (prolonged submersion decreases germination rates again), which corresponds to the average duration of annual flooding periods (Warrag and Eltigani, 2005). There is also a promoting effect of seed burial in soil which could be related to increased moisture or microbial action (Baskin and Baskin, 2000). Testing of seed germination in presence or absence (using sterilized seeds) of microbes resulted in increase of germination for the former. This was most pronounced in *Vigna minima* seeds, which imbibed within 186 h in the presence of soil suspension but failed to do so in pure water for 30 days (Gopinathan and Babu, 1985). Similarly, passage through the animal's digestive tract, often results in better germination of impermeable seeds. It is assumed that it acts via acid enzymatic digestion of testa or mechanical scarification (cited in Baskin and Baskin, 2014). Interesting interaction was found between the *Acacia sp.* seeds, bruchid beetles that lay eggs on the seeds and mammals and birds that feed on the seeds. Gazelles feed on pods of *Acacia sp.* in the Negev Desert of Israel and disperse the seeds, which have superior germination as a result. Moreover, the seeds infected with bruchid larvae germinate with higher rates (Halevy, 1974). Seeds containing a bruchid beetle larvae are more likely to germinate before the parasite destroys the embryo if seeds go through the digestive tract of an animal (Pellew and Southgate, 1984). On the other hand, these insect larvae cause substantial proportion of the otherwise water-impermeable seeds to imbibe and germinate, despite part of the seeds have destroyed embryo as shown for *Acacia sp., Vicia sativa, Ulex europaeus,* and *Gleditsia japonica* (cited in Baskin and Baskin, 2014).

Some species produce heterogeneously colored seeds with different degrees of hardness. For example, the tree legume species *Senna obtusifolia* produces 90% hard-coat seeds with 10% soft-coat seeds (in Baskin and Baskin, 1998). 82–93% of the soft-coat seeds germinated, while only 15–32% of the hard-seed-coat ones germinated. This heterogeneity may be an important ecophysiological strategy, since soft-coat seeds can germinate in the spring in temperate regions, whereas hard-coat seeds cannot germinate until late spring–summer, when high temperatures cause an increase in seed-coat permeability. The seeds of the legumes *Adenocarpus decorticans, Astragalus granatensis ssp. Granatensis,* and *Cytisus reverchonii* (all endemic to the Betic Cordillera, Spain) collected at different altitudes, required different temperature for germination (Angosto and Matilla, 1993).

Several artificial techniques are used to break physical dormancy in seeds, including mechanical, thermal and chemical scarification, enzymes, dry storage, percussion, low temperatures, radiation and high atmospheric pressures (Baskin and Baskin, 2014).

LEGUME SEED DORMANCY AND DOMESTICATION

The development of agriculture was one of the key transitions in human history, and a central part of this was the evolution of new plant forms that were selected and became domesticated crops. The domestication of wild plants into crop plants can be viewed as an accelerated evolution, the result of both human and natural selection (Abbo et al., 2012, 2014). Domestication is often described as a quality in which morphological (and genetic) changes are found amongst cultivated plants in comparison to those in wild populations (Hancock, 2012; Zohary et al., 2012). These domestication-triggered changes represent adaptations to cultivation and human harvesting, accompanied by genetic changes. A common set of traits has been recorded for domesticated, but otherwise un-related crops, which collectively have been called the "domestication syndrome" (Harlan, 1971; Hammer, 1984). These traits are linked to successful early growth of planted seeds and include loss of germination inhibition and increase of seed size (**Figure 3**). Members of the Fabaceae have been domesticated in parallel with cereal domestications (Abbo et al., 2009). One of the major differences between the wild progenitors of Near Eastern grain legumes and cereals concerns the low germination rate imposed by the hard seed coat of these legumes (Ladizinsky, 1979, 1985; Werker et al., 1979; Abbo et al., 2009, 2014; Fuller and Allaby, 2009). Timing of seed germination is thus one of the key steps both in natural and agricultural ecosystems and

FIGURE 3 | Comparison of cultivated and wild pea seeds. Macrograph of modern cultivated pea cv. Cameor with transparent testa **(A)** and wild *Pisum sativum* subsp. *elatius* JI64 with pigmented testa and visible rough (gritty) testa surface **(B)**. Scale bars = 3 mm. Transversal section of toluidine blue-stained seed coat of domesticated *Pisum sativum* cv. Cameor **(C)** and wild *Pisum sativum* subsp. *elatius* JI64 **(D)**. Scale bars = 50 mm.

is a major factor for crop production. In contrast to wild species, crops tend to germinate as soon as they are imbibed and planted making seed dormancy a potentially unwanted trait. The selection acts on loss of dormancy during cultivation (Fuller and Allaby, 2009). Seed imbibition also has a crucial role in the ability of most grain legumes to undergo cooking. Hence, reduction of seed coat thickness has led to a concurrent reduction of seed coat impermeability. Ladizinsky (1987) argued that the very low germination rates in wild pulses, in particular lentil, would have precluded their successful cultivation on the basis of very low yields from planted seeds. He therefore suggested that hunter–gatherers must have selected wild mutants with quick germination for cultivation (Ladizinsky, 1987; Weiss et al., 2006). That germplasm could have been part of a "pre-cultivation domestication" process. The experimental harvest of wild lentils by Abbo et al. (2008) provided strong support for Ladizinsky's (1985, 1987, 1998) arguments. This also holds true for peas, where intact wild seeds which have a germination rate of only 2.6–7% in a given year (Abbo et al., 2011, 2014). These results suggest that free germination was a more important trait for the domestication of wild pea (and possibly lentil and chickpea as well) than the mode of seed dispersal. However, too low seed dormancy levels reduce seed quality and may trigger preharvest sprouting, which can occur in cereals. Therefore, seeds of crop plants require a well-balanced level of seed dormancy. Preharvest sprouting is rarely a problem in legume seeds since they are produced within a pod. Similarly, high-quality mungbean and common bean seeds are difficult to produce in humid tropical regions because of susceptibility to weather damage, which, however, can be mitigated by hardseededness (Hamphry et al., 2005). Whilst seed dormancy has been largely removed in grain legume crops, it remains in a number of important fodder crops and less domesticated species. Moreover, even in highly domesticated legume crops such as soybean, which has been selected for seeds that imbibe water rapidly and uniformly, some varieties produce some hard seeds (Rolston, 1978). This constitutes a major problem during food processing where rapid and uniform hydration is important for the production of quality foods.

CONCLUSION

The seed testa plays important roles in seed development and the beginning of a new plant generation. The seed coat provides not just structural and protective functions, but as discussed in this review, has a decisive role in timing of seed germination of legumes by regulating water uptake. This control is fundamental under variable natural conditions where the establishment of young plants might influence the species evolutionary success. Such control was largely removed from domesticated crop plants, which are largely characterized by immediate seed germination. Moreover, water uptake allows us as well as early farmers, to cook and make legume seeds edible, thus providing important protein levels to the human diet. Although we have a good understanding of the genetic control of germination, we are just starting to identify the genes involved in this process. Although not formally demonstrated, the main testa pigments, PAs, are hypothesized to play a role in seed coat permeability. These pigments are also known as antioxidants with beneficial effects on human health, including cardioprotective, anticancer, and anti-inflammatory roles. These compounds also protect animals by binding to proteins in feed and slowing fermentation in the rumen, thereby reducing microbial production of methane.

Currently available analytical tools applied to a widening range of germplasm will help dissect germination in legumes more precisely. Detailed structural and chemical description of crop seed coats will provide the comparative basis for further studies and might lead to the identification of novel phenolic substances with potential health and nutritional benefits.

ACKNOWLEDGMENTS

Petr Smýkal research is funded by Grant Agency of Czech Republic, 14-11782S project. Matthew W. Blair is funded by an Evans Allen grant from the USDA.

REFERENCES

Abbo, S., Lev-Yadun, S., and Gopher, A. (2012). Plant domestication and crop evolution in the Near East: on events and processes. *Crit. Rev. Plant Sci.* 31, 241–257. doi: 10.1080/07352689.2011.645428

Abbo, S., Pinhasi van-Oss, R., Gopher, A., Saranga, Y., Ofner, I., and Peleg, Z. (2014). Plant domestication versus crop evolution: a conceptual framework for cereals and grain legumes. *Trends Plant Sci.* 19, 351–360. doi: 10.1016/j.tplants.2013.12.002

Abbo, S., Rachamim, E., Zehavi, Y., Zezak, I., Lev-Yadun, S., and Gopher, A. (2011). Experimental growing of wild pea in Israel and its bearing on Near Eastern plant domestication. *Ann. Bot.* 107, 1399–1404. doi: 10.1093/aob/mcr081

Abbo, S., Saranga, Y., Peleg, Z., Lev-Yadun, S., Kerem, Z., and Gopher, A. (2009). Reconsidering domestication of legumes versus cereals in the ancient near east. *Q. Rev. Biol.* 84, 29–50. doi: 10.1086/596462

Abbo, S., Zezak, I., Schwartz, E., Lev-Yadun, S., and Gopher, A. (2008). Experimental harvesting of wild peas in Israel: implications for the origins of near east farming. *J. Arch. Sci.* 35, 922–929. doi: 10.1016/j.jas.2007.06.016

Agbo, G. N., Hosfield, M. A., Uebersax, M. A., and Klomparens, K. (1987). Seed microstructure and its relationship to water uptake in isogenic lines and a cultivar of dry beans (*Phaseolus vulgaris* L.). *Food Microstruct.* 6, 91–102.

Algan, G., and Büyükkartal, H. N. B. (2000). Ultrastructure of seed coat development in the natural tetraploid *Trifolium pratense* L. *J. Agron. Crop Sci.* 184, 205–213. doi: 10.1046/j.1439-037x.2000.00373.x

Andriunas, F. A., Zhang, H. M., Xia, X., Patrick, J. W., and Offler, C. E. (2013). Intersection of transfer cells with phloem biology-broad evolutionary trends, function, and induction. *Front. Plant Sci.* 4:221. doi: 10.3389/fpls.2013.00221

Angosto, T., and Matilla, A. J. (1993). Variations in seeds of three endemic leguminous species at different altitudes. *Physiol. Plant.* 87, 329–334. doi: 10.1111/j.1399-3054.1993.tb01738.x

Argel, P., and Paton, C. (1999). "Overcoming legume hardseededness," in *Forage Seed Production: Tropical and Subtropical Species*, Vol. 2, eds D. S. Loch and J. E. Ferguson (Wallingford: CAB International), 247–265.

Ballard, L. A. T. (1973). Physical barriers to germination. *Seed Sci. Technol.* 1, 285–303.

Baskin, C. C. (2003). Breaking physical dormancy in seeds – focussing on the lens. *New Phytol.* 158, 229–232. doi: 10.1046/j.1469-8137.2003.00751.x

Baskin, C. C., and Baskin, J. M. (2014). *Seeds: Ecology, Biogeography and Evolution of Dormancy and Germination*, 2nd Edn. New York: Academic Press.

Baskin, C. C., Baskin, J. M., and Li, X. (2000). Taxonomy, anatomy and evolution of physical dormancy in seeds. *Plant Species Biol.* 15, 139–152. doi: 10.1046/j.1442-1984.2000.00034.x

Baskin, J. M., and Baskin, C. C. (1985). The annual dormancy cycle in buried weed seeds: a continuum. *Bioscience* 35, 492–498. doi: 10.2307/1309817

Baskin, J. M., and Baskin, C. C. (1998). *Seeds. Ecology, Biogeography and Evolution of Dormancy and Germination.* New York: Academic Press.

Baskin, J. M., and Baskin, C. C. (2000). Evolutionary considerations of claims for physical dormancy-break by microbial action and abrasion by soil particles. *Seed Sci. Res.* 10, 409–413. doi: 10.1017/S0960258500000453

Baskin, J. M., Baskin, C. C., and Dixon, K. W. (2006). Physical dormancy in the endemic Australian genus *Stylobasium*, a first report for the family Surinaceae (Fabales). *Seed Sci. Rep.* 16, 229–232. doi: 10.1079/SSR2006248

Bassett, M. J. (1994). The griseoalbus (gray–white) seed coat color is controlled by an allele (p^{gri}) a the *P* locus in common bean. *Hortscience* 29, 1178–1179.

Bassett, M. J. (1999). Allelism found between two common bean genes, hilum ring colour (D) and partly coloured seed coat pattern (Z), formerly assumed to be independent. *J. Am. Soc. Hortic. Sci.* 124, 649–653.

Bassett, M. J. (2002). Inheritance of reverse margo seed coat pattern and allelism between the genes J for seed coat colour and L for partly coloured seed coat pattern in common bean. *J. Am. Soc. Hortic. Sci.* 127, 56–61.

Bassett, M. J. (2007). Genetics of seed coat colour and pattern in common bean. *Plant Breed. Rev.* 28, 239–315. doi: 10.1002/9780470168028.ch8

Baudry, A., Heim, M. A., Dubreucq, B., Caboche, M., Weisshaar, B., and Lepiniec, L. (2004). TT2, TT8, and TTG1 synergistically specify the expression of BANYULS and proanthocyanidin biosynthesis in *Arabidopsis thaliana. Plant J.* 39, 366–380. doi: 10.1111/j.1365-313X.2004.02138.x

Beeckman, T., De Rycke, R., Viane, R., and Inzé, D. (2000). Histological study of seed coat development in *Arabidopsis thaliana. J. Plant Res.* 113, 139–148. doi: 10.1007/PL00013924

Beisson, F., Li, Y., Bonaventure, G., Pollard, M., and Ohlrogge, J. B. (2007). The acyltransferase GPAT5 is required for the synthesis of suberin in seed coat and root of *Arabidopsis. Plant Cell* 19, 351–368. doi: 10.1105/tpc.106.048033

Bernard, R. L., and Weiss, M. G. (1973). "Qualitative genetics," in *Soybeans: Improvement, Production, and Uses*, 1st Edn, ed. B. E. Caldwell (Madison, WI: American Society of Agronomy), 117–154.

Bevilacqua, L. R., Fossati, F., and Dondero, G. (1987). 'Callose' in the impermeable seed coat of *Sesbania punicea. Ann. Bot.* 59, 335–341.

Bewley, J. D., Bradford, K., Hilhorst, H., and Nonogaki, H. (2013). *Seeds: Physiology of Development, Germination and Dormancy*, 3rd Edn. New York: Springer-Verlag.

Bhalla, P. L., and Slattery, H. D. (1984). Callose deposits make clover seeds impermeable to water. *Ann. Bot.* 53, 125–128.

Bhat, J. L. (1968). Seed dormancy in *Indigofera glandulosa* Willd. *Trop. Ecol.* 9, 42–51.

Bibbey, R. O. (1948). Physiological studies of weed seed germination. *Plant Physiol.* 23, 467–484. doi: 10.1104/pp.23.4.467

Boersma, J. G., Buirchell, J. B., Sivasithamparam, K., and Yang, H. (2007). Development of a PCR marker tightly linked to mollis, the gene that controls seed dormancy in *Lupinus angustifolius* L. *Plant Breed.* 126, 612–616. doi: 10.1111/j.1439-0523.2007.01417.x

Bolingue, W., Ly Vu, B., Leprince, O., and Buitink, J. (2010). Characterization of dormancy behaviour in seeds of the model legume *Medicago truncatula. Seed Sci. Res.* 20, 97–107. doi: 10.1017/S0960258510000061

Bonnemain, J. L., Bourquin, S., Renault, S., Offler, C., and Fisher, D. G. (1991). "Transfer cells structure and physiology," in *Phloem Transport and Assimilate Compartmentation*, eds J. L. Bonnemain, S. Delrot, W. J. Lucas, and J. Dainty (Nantes: Quest Editions), 178–186.

Borisjuk, L., Weber, H., Panitz, R., Manteuffel, R., and Wobus, U. (1995). Embryogenesis of *Vicia faba* L: histodifferentiation in relation to starch and storage protein synthesis. *J. Plant Physiol.* 147, 203–218. doi: 10.1016/S0176-1617(11)81507-5

Bradford, K., and Nonogaki, H. (2009). *Seed Development, Dormancy and Germination. Annual Plant Reviews*, Vol. 27. Oxford: Blackwell.

Brune, M., Rossander, L., and Halberg, L. (1989). Iron absorption and phenolic compounds: importance of different phenolic structures. *Eur. J. Clin. Nutr.* 43, 547–558.

Burrows, G. E., Virgona, J. M., and Heady, R. D. (2009). Effect of boiling water, seed coat structure and provenance on the germination of *Acacia melanoxylon* seeds. *Aust. J. Bot.* 57, 139–147. doi: 10.1071/BT08194

Butler, E. A. (1988). "The SEM and seed identification, with particular reference to the Vicieae," in *Scanning Electron Microscopy in Archeology*, Vol. 452, ed. S. L. Olsen (Oxford: BAR International Series), 215–224.

Büyükkartal, H. N., Hatice Çölgeçen, H., Pinar, N. M., and Erdoğan, N. (2013). Seed coat ultrastructure of hard-seeded and soft-seeded varieties of *Vicia sativa. Turk. J. Bot.* 37, 270–275.

Caldas, G. V., and Blair, M. W. (2009). Inheritance of condensed tannin content and relationship with seed colour and pattern genes in common bean (*Phaseolus vulgaris* L.). *Theor. Appl. Genet.* 119, 131–142. doi: 10.1007/s00122-009-1023-4

Chachalis, D., and Smith, M. L. (2000). Imbibition behavior of soybean (*Glycine max* (L.) Merrill) accessions with different testa characteristics. *Seed Sci. Technol.* 28, 321–331.

Chachalis, D., and Smith, M. L. (2001). Seed coat regulation of water uptake during imbibition in soybeans (*Glycine max* (L.) Merr.). *Seed Sci. Technol.* 29, 401–412.

Chernoff, M., Plitmann, U., and Kislev, M. E. (1992). Seed characters and testa texture in species of Vicieae: their taxonomic significance. *Isr. J. Bot.* 41, 167–186. doi: 10.1080/0021213X.1992.10677225

Clarkson, D. T., and Robards, K. (1975). "The endodermis, its structural development and physiological role," in *Structure and Function of Roots*, eds J. G. Torrey and D. T. Clarkson (London: Academic Press), 415–436.

Clements, J. C., Buirchell, B. J., Yang, H., Smith, P. M. C., Sweetingham, M. W., and Smith, C. G. (2005). "Chapter 9, Lupin," in *Genetic Resources, Chromosome Engineering, Crop Improvement*, Vol. 1, *Grain legumes*, eds R. J Singh and P. P. Jauhar (Florida: CRC Press), 231–323.

Corner, E. J. H. (1951). The leguminous seed. *Phytomorphology* 1, 117–150.

Dalling, J. W., Davis, A. S., Schutte, B. J., and Arnold, A. E. (2011). Seed survival in soil: interacting effects of predation, dormancy and the soil microbial community. *J. Ecol.* 99, 89–95. doi: 10.1111/j.1365-2745.2010.01739.x

Dean, G., Cao, Y. G., Xiang, D., Provart, N. J., Ramsay, L., Ahada, A., et al. (2011). Analysis of gene expression patterns during seed coat development in *Arabidopsis. Mol. Plant* 4, 1074–1091. doi: 10.1093/mp/ssr040

Debeaujon, I., Léon-Kloosterziel, K. M., and Koornneef, M. (2000). Influence of the testa on seed dormancy, germination, and longevity in *Arabidopsis. Plant Physiol.* 122, 403–414. doi: 10.1104/pp.122.2.403

Debeaujon, I., Nesi, N., Perez, P., Devi, M., Grandjean, O., Caboche, M., et al. (2003). Proanthocyanidin-accumulating cells in *Arabidopsis* testa: regulation of differentiation and role in seed development. *Plant Cell* 15, 2514–2531. doi: 10.1105/tpc.014043

de Candolle, A. P. (1825). *Mémoires sur la Famille des Légumineuses.* Paris: A. Belin.

Dell, B. (1980). Structure and function of the strophiolar plug in seeds of *Albizia lophantha. Am. J. Bot.* 67, 556–563. doi: 10.2307/2442296

Deshpande, S. S., and Cheryan, M. (1986). Microstructure and water uptake of *Phaseolus* and winged beans. *J. Food Sci.* 51, 1218–1223. doi: 10.1111/j.1365-2621.1986.tb13089.x

de Souza, T. V., Voltolini, C. H., Santos, M., and Silveira Paulilo, M. T. (2012). Water absorption and dormancy-breaking requirements of physically dormant seeds of *Schizolobium parahyba* (Fabaceae – Caesalpinioideae). *Seed Sci. Res.* 22, 169–176.

Díaz, A. M., Caldas, G. V., and Blair, M. W. (2010). Concentrations of condensed tannins and anthocyanins in common bean seed coats. *Food Res. Int.* 43, 595–601. doi: 10.1016/j.foodres.2009.07.014

Dixon, R. A., Xie, D. Y., and Sharma, S. B. (2005). Proanthocyanidins – a final frontier in flavonoid research? *New Phytol.* 165, 9–28. doi: 10.1111/j.1469-8137.2004.01217.x

Donnelly, E. D., Watson, J. E., and McGuire, J. A. (1972). Inheritance of hard seed in *Vicia. J. Hered.* 63, 361–365.

Dorcey, E., Urbez, C., Blázquez, M. A., Carbonell, J., and Perez-Amador, M. A. (2009). Fertilization-dependent auxin response in ovules triggers fruit development through the modulation of gibberellin metabolism in *Arabidopsis. Plant J.* 58, 318–332. doi: 10.1111/j.1365-313X.2008.03781.x

Dueberrn de Sousa, F. H. D., and Marcos-Filho, J. (2001). The seed coat as a modulator of seed-environment relationship in Fabaceae. *Braz. J. Bot.* 24, 365–375. doi: 10.1590/S0100-84042001000400002

Dueñas, M., Sun, B., Hernández, T., Estrella, I., and Spranger, M. I. (2003). Proanthocyanidin composition in the seed coat of lentils (*Lens culinaris* L.). *J. Agric. Food Chem.* 51, 7999–8004. doi: 10.1021/jf0303215

Emery, R. J., Ma, Q., and Atkins, C. A. (2000). The forms and sources of cytokinins in developing white lupine seeds and fruits. *Plant Physiol.* 123, 1593–1604. doi: 10.1104/pp.123.4.1593

Esau, K. (1965). *Plant Anatomy*, 2nd Edn. New York: John Wiley.

Fenner, M., and Thompson, K. (2005). *The Ecology of Seeds.* Cambridge: Cambridge University Press. doi: 10.1017/CBO9780511614101

Finch-Savage, W. E., and Leubner-Metzger, G. (2006). Seed dormancy and the control of germination. *New Phytol.* 171, 501–523. doi: 10.1111/j.1469-8137.2006.01787.x

Foley, M. E. (2001). Seed dormancy: an update on terminology, physiological genetics, and quantitative trait loci regulating germinability. *Weed Sci.* 49, 305–317. doi: 10.1614/0043-1745(2001)049[0305:SDAUOT]2.0.CO;2

Forbes, I., and Well, H. D. (1968). Hard and soft seededness in blue lupine, *Lupinus angustifolius* L.: inheritance and phenotype classification. *Crop Sci.* 4, 195–197. doi: 10.2135/cropsci1968.0011183X000800020018x

Frey, A., Godin, B., Bonnet, M., Sotta, B., and Marion-Poll, A. (2004). Maternal synthesis of abscisic acid controls seed development and yield in *Nicotiana plumbaginifolia. Planta* 218, 958–964. doi: 10.1007/s00425-003-1180-7

Fuller, D. Q., and Allaby, R. (2009). Seed dispersal and crop domestication: shattering, germination and seasonality in evolution under cultivation. *Ann. Plant Rev.* 38, 238–295.

Galbiati, F., Sinha Roy, D., Simonini, S., Cucinotta, M., Ceccato, L., Cuesta, C., et al. (2013). An integrative model of the control of ovule primordia formation. *Plant J.* 76, 446–455. doi: 10.1111/tpj.12309

Gallardo, K., Firnhaber, C., Zuber, H., Héricher, D., Belghazi, M., Henry, C., et al. (2007). A combined proteome and transcriptome analysis of developing *Medicago truncatula* seeds. *Mol. Cell. Proteomics* 6, 2165–2179. doi: 10.1074/mcp.M700171-MCP200

Gijzen, M., Kuflu, K., Qutob, D., and Chernys, J. T. (2001). A class I chitinase from soybean seed coat. *J. Exp. Bot.* 52, 2283–2289. doi: 10.1093/jexbot/52.365.2283

Gijzen, M., Miller, S. S., Kuflu, K., Buzzell, R. I., and Miki, B. L. (1999). Hydrophobic protein synthesized in the pod endocarp adheres to the seed surface. *Plant Physiol.* 120, 951–960. doi: 10.1104/pp.120.4.951

Gillikin, J. W., and Graham, J. S. (1991). Purification and developmental analysis of the major anionic peroxidases from the seed coat of *Glycine max. Plant Physiol.* 96, 214–220. doi: 10.1104/pp.96.1.214

Gillman, J. D., Tetlow, A., Lee, J. D., Shannon, J. G., and Bilyeu, K. (2011). Loss-of-function mutations affecting a specific *Glycine max* R2R3 MYB transcription factor result in brown hilum and brown seed coats. *BMC Plant Biol.* 11:155. doi: 10.1186/1471-2229-11-155

Gogue, G. J., and Emino, E. R. (1979). Seed coat scarification of Albizia julibrissin Durazz by natural mechanisms. *J. Am. Soc. Hortic. Sci.* 104, 421–423.

Gopinathan, M. C., and Babu, C. R. (1985). Structural diversity and its adaptive significance in seeds of *Vigna minima* (Roxb.) Ohwi and Ohashi and allies (Leguminosae-Papilionoideae). *Ann. Bot.* 56, 723–732.

Graeber, K., Nakabayashi, K., Miatton, E., Leubner-Metzger, G., and Soppe, W. J. J. (2012). Molecular mechanisms of seed dormancy. *Plant Cell Environ.* 35, 1769–1786. doi: 10.1111/j.1365-3040.2012.02542.x

Güneş, F. (2013). Seed characteristics and testa textures of *Pratensis, Orobon, Lathyrus, Orobastrum* and *Cicercula* sections from *Lathyrus* (Fabaceae) in Turkey. *Plant Syst. Evol.* 299, 1935–1953. doi: 10.1007/s00606-013-0849-z

Gunn, C. R. (1981). Seed topography in the Fabaceae. *Seed Sci. Technol.* 9, 737–757.

Gunn, C. R. (1984). *Fruits and Seeds of Genera in Subfamily Mimosoideae (Fabaceae).* (Technical Bulletin No. 1681), (Washington, DC: United States Department of Agriculture), 1–194.

Gunn, C. R. (1991). *Fruits and seeds of genera in subfamily Caesalpinioideae (Fabaceae).* (Technical Bulletin No. 1755), (Washington, DC: United States Department of Agriculture), 1–408.

Guzmán-Maldonado, H., Castellanos, J., and Gonzalez, E. (1996). Relationship between theoretical and experimentally detected tannin content of common beans (*Phaseolus vulgaris* L.). *Food Chem.* 55, 333–335. doi: 10.1016/0308-8146(95)00106-9

Halevy, G. (1974). Effects of gazelles and seed beetles (Bruchidae) on germination and establishment of *Acacia* species. *Isr. J. Bot.* 23, 120–123.

Hamly, D. H. (1932). Softening of the seeds of *Melilotus alba*. *Bot. Gaz.* 93, 345–375. doi: 10.1086/334269

Hamly, D. H. (1935). The light line in *Melilotus alba*. *Bot. Gaz.* 96, 753–757. doi: 10.1086/334521

Hammer, K. (1984). Das Domestikationssyndrom. *Kulturpflanze* 11, 11–34. doi: 10.1007/BF02098682

Hamphry, M. E., Lambrides, C. J., Chapman, S. C., Aitken, E. A. B., Imrie, B. C., and Lawn, R. J. (2005). Relationships between hardseededness and seed weight in mungbean (*Vigna radiata*) assessed by QTL analysis. *Plant Breed.* 124, 292–298. doi: 10.1111/j.1439-0523.2005.01084.x

Hancock, J. F. (2012). *Plant Evolution and Origin of Species*, 3rd Edn. Wallingford: CABI. doi: 10.1079/9781845938017.0000

Hanley, M. E., and Fenner, M. (1998). Pre-germination temperature and the survivorship and onward growth of Mediterranean fire-following plant species. *Acta Oecol.* 19, 181–187. doi: 10.1016/S1146-609X(98)80022-2

Hanna, P. J. (1984). Anatomical features of the seed coat of *Acacia kempeana* (Mueller) which relate to increased germination rate induced by heat treatment. *New Phytol.* 96, 23–29. doi: 10.1111/j.1469-8137.1984.tb03539.x

Hardham, A. R. (1976). Structural aspects of pathways of nutrient flow to developing embryo and cotyledons of *Pisum sativum* L. *Aust. J. Bot.* 24, 711–721. doi: 10.1071/BT9760711

Harlan, J. R. (1971). Agricultural origins: centers and noncenters. *Science* 174, 468–474. doi: 10.1126/science.174.4008.468

Harper, J. L. (1957). "The ecological significance of dormancy and its importance in weed control," in *Proceedings of the Fourth International Congress of Crop Protection*, Hamburg, 415–420.

Harris, W. M. (1983). On the development of macrosclereids in seed coats of *Pisum sativum* L. *Am. J. Bot.* 70, 1528–1535. doi: 10.2307/2443351

Harris, W. M. (1984). On the development of osteosclereids in seed coats of *Pisum sativum* L. *New Phytol.* 98, 135–141. doi: 10.1111/j.1469-8137.1984.tb06103.x

Harris, W. M. (1987). Comparative ultrastructure of developing seed coats of 'hard-seeded' and 'soft-seeded' varieties of soybean, *Glycine max* (L.) Merr. *Bot. Gaz.* 148, 324–331. doi: 10.1086/337660

Hellens, R. P., Moreau, C., Lin-Wang, K., and Schwinn, K. E. (2010). Identification of Mendel's white flower character. *PLoS ONE* 5:e13230. doi: 10.1371/journal.pone.0013230

Hu, X. W., Wanga, Y. R., Wua, Y. P., and Baskin, C. C. (2009). Role of the lens in controlling water uptake in seeds of two Fabaceae (Papilionoideae) species treated with sulphuric acid and hot water. *Seed Sci. Res.* 19, 73. doi: 10.1017/S0960258509301099

Hyde, E. O. C. (1954). The function of the hilum in some Papilionaceae in relation to ripening of the seed and the permeability of the testa. *Ann. Bot.* 18, 241–256.

Isemura, T., Kaga, A., Konishi, S., Ando, T., Tomooka, H. O., and Vaughan, D. A. (2007). Genome dissection of traits related to domestication in azuki bean (*Vigna angularis*) and comparison with other warm-season legumes. *Ann. Bot.* 100, 1053–1071. doi: 10.1093/aob/mcm155

Isemura, T., Kaga, A., Tabata, S., Somta, P., Srinives, P., Shimizu, T., et al. (2012). Construction of a genetic linkage map and genetic analysis of domestication related traits in mungbean (*Vigna radiata*). *PLoS ONE* 7:e41304. doi: 10.1371/journal.pone.0041304

Isemura, T., Kaga, A., Tomooka, N., Shimizu, T., and Vaughan, D. A. (2010). The genetics of domestication of rice bean, *Vigna umbellata*. *Ann. Bot.* 106, 927–944. doi: 10.1093/aob/mcq188

Kaga, A., Isemura, T., Tomooka, N., and Vaughan, D. A. (2008). The genetics of domestication of the azuki bean (*Vigna angularis*). *Genetics* 178, 1013–1036. doi: 10.1534/genetics.107.078451

Kantar, F. P., Hebblethwaite, D., and Pilbeam, C. J. (1996). Factors influencing disease resistance in high and low tannin *Vicia faba*. *J. Agric. Sci.* 127, 83–88. doi: 10.1017/S002185960007739X

Karaki, T., Watanabe, Y., Kondo, T., and Koike, T. (2012). Strophiole of seeds of the black locust acts as a water gap. *Plant Species Biol.* 27, 226–232. doi: 10.1111/j.1442-1984.2011.00343.x

Karssen, C. M. (1982). "Seasonal patterns of dormancy in weed seeds," in *The Physiology and Biochemistry of Seed Development, Dormancy and Germination*, ed. A. A. Khan (Amsterdam: Elsevier Biomedical Press), 243–270.

Karssen, C. M., Brinkhorst van der Swan, D. L. C., Breekland, A. E., and Koornneef, M. (1983). Induction of dormancy during seed development by endogenous abscisic-acid – studies on abscisic-acid deficient genotypes of *Arabidopsis thaliana* (L.) Heynh. *Planta* 157, 158–165. doi: 10.1007/BF00393650

Kikuchi, K., Koizumi, M., Ishida, N., and Kano, H. (2006). Water uptake by dry beans observed by micro-magnetic resonance imaging. *Ann. Bot.* 98, 545–553. doi: 10.1093/aob/mcl145

Koinange, E. M. K., Singh, S. P., and Gepts, P. (1996). Genetic control of the domestication syndrome in common bean. *Crop Sci.* 36, 1037–1045. doi: 10.2135/cropsci1996.0011183X003600040037x

Kongjaimun, A., Kaga, A., Tomooka, N., Somta, P., Vaughan, D. A., and Srinives, P. (2012). The genetics of domestication of yardlong bean, *Vigna unguiculata* (L.) Walp. ssp. unguiculata cv.-gr. sesquipedalis. *Ann. Bot.* 109, 1185–1200. doi: 10.1093/aob/mcs048

Korban, S. S., Coyne, D. P., and Weihing, J. L. (1981). Evaluation, variation, and genetic control of seed coat whiteness in dry beans (*Phaseolus vulgaris* L.). *J. Am. Soc. Hortic. Sci.* 166, 575–579.

Kovinich, N., Saleem, A., Arnason, J. T., and Miki, B. (2011). Combined analysis of transcriptome and metabolite data reveals extensive differences between black and brown nearly-isogenic soybean (*Glycine max*) seed coats enabling the identification of pigment isogenes. *BMC Genomics* 12:381. doi: 10.1186/1471-2164-12-381

Kurdyukov, S., Song, Y., Sheahan, M. B., and Rose, R. J. (2014). Transcriptional regulation of early embryo development in the model legume *Medicago truncatula*. *Plant Cell Rep.* 33, 349–362. doi: 10.1007/s00299-013-1535-x

Lacerda, D. R., Lemos-Filho, J. P., Goulart, M. F., Ribeiro, R. A., and Lovato, M. B. (2004). Seed-dormancy variation in natural populations of two tropical leguminous tree species: *Senna multijuga* (Caesalpinoideae) and *Plathymenia reticulata* (Mimosoideae). *Seed Sci. Res.* 14, 127–135. doi: 10.1079/SSR2004162

Lackey, J. A. (1981). Systematic significance of the epihilum in Phaseoleae (Fabaceae, Faboideae). *Bot. Gaz.* 142, 160–164. doi: 10.1086/337207

Ladizinsky, G. (1979). The origin of lentil and its wild genepool. *Euphytica* 28, 179–187. doi: 10.1007/BF00029189

Ladizinsky, G. (1985). The genetics of hard seed coat in the genus *Lens*. *Euphytica* 34, 539–543. doi: 10.1007/BF00022952

Ladizinsky, G. (1987). Pulse domestication before cultivation. *Econ. Bot.* 41, 60–65. doi: 10.1007/BF02859349

Ladizinsky, G. (1998). *Plant Evolution Under Domestication*. Dordrecht: Kluwer Academic Publishers. doi: 10.1007/978-94-011-4429-2

Lanfermeijer, F. C., van Oene, M. A., and Borstlap, A. C. (1992). Compartmental analysis of amino-acid release from attached and detached pea seed coats. *Planta* 187, 75–82. doi: 10.1007/BF00201626

Le, B. H., Wagmaister, J. A., Kawashima, T., Bui, A. Q., Harada, J. J., and Goldberg, R. B. (2007). Using genomics to study legume seed development. *Plant Physiol.* 144, 562–574. doi: 10.1104/pp.107.100362

Legesse, N., and Powell, A. A. (1996). Relationship between the development of seed coat pigmentation, seed coat adherence to the cotyledons and the rate of imbibition during the maturation of grain legumes. *Seed Sci. Technol.* 24, 23–32.

Lepiniec, L., Debeaujon, I., Routaboul, J. M., Baudry, A., Pourcel, L., Nesi, N., et al. (2006). Genetics and biochemistry of seed flavonoids. *Annu. Rev. Plant Biol.* 57, 405–430. doi: 10.1146/annurev.arplant.57.032905.105252

Lersten, N. R. (1982). Tracheid bar and vestured pits in legume seeds (Leguminosae: Papilionoideae). *Am. J. Bot.* 69, 98–107. doi: 10.2307/2442834

Lersten N. R., and Gunn, C. R. (1981). Seed morphology and testa topography in *Cicer* (Fabaceae: Faboideae). *Syst. Bot.* 6, 223–230.

Lersten, N. R., and Gunn, C. R. (1982). *Testa Characters in Tribe Vicieae, with Notes About Tribes Abreae, Cicereae, and Trifolieae (Fabaceae)*. (Technical Bulletin No. 1667), (Washington, DC: United States Department of Agriculture), 1–40.

Lersten, N. R., Gunn, C. R., and Brubaker, C. L. (1992). *Comparative Morphology of the Lens on Legume (Fabaceae) Seeds, with Emphasis on Species in Subfamilies Caesalpinioideae and Mimosoideae*. (Technical Bulletin No. 1791), (Washington, DC: United States Department of Agriculture), 44.

Lewis, G., Schrire, B., Mackinder, B., and Lock, M. (2005). *Legumes of the World*. Kew: Royal Botanic Gardens.

Lhotská, M., and Chrtková, A. (1978). *Karpologie a diasporologie: československých zástupců čeledi Fabaceae*. Prague: Academia.

Li, X., Buirchell, B., Yan, G., and Yang, H. (2012). A molecular marker linked to the mollis gene conferring soft-seediness for marker-assisted selection applicable to a wide range of crosses in lupin (*Lupinus angustifolius* L.). *Breed. Mol. Breed.* 29, 361–370. doi: 10.1007/s11032-011-9552-3

Liang, M., Davis, E., Gardner, D., Cai, X., and Wu, Y. (2006). Involvement of AtLAC15 in lignin synthesis in seeds and in root elongation of *Arabidopsis*. *Planta* 224, 1185–1196. doi: 10.1007/s00425-006-0300-6

Liu, B., Liu, X., Wang, C., Jin, J., and Herbert, S. J. (2010). Endogenous hormones in seed, leaf, and pod wall and their relationship to seed filling in soybeans. *Crop Pasture Sci.* 61, 103–110. doi: 10.1071/CP09189

Liu, W., Peffley, E. B., Powell, R. J., Auld, D. L., and Hou, A. (2007). Association of seedcoat colour with seed water uptake, germination, and seed components in guar (*Cyamopsis tetragonoloba* (L.) Taub). *J. Arid Environ.* 70, 29–38. doi: 10.1016/j.jaridenv.2006.12.011

Liu, Y. H., Cao, J. S., Li, G. J., Wu, X. H., Wang, B. G., Xu, P., et al. (2012). Genotypic differences in pod wall and seed growth relate to invertase activities and assimilate transport pathways in asparagus bean. *Ann. Bot.* 109, 1277–1284. doi: 10.1093/aob/mcs060

Lush, W. M., and Evans, L. T. (1980). The seed coats of cowpeas and other grain legumes – structure in relation to function. *Field Crops Res.* 3, 267–286. doi: 10.1016/0378-4290(80)90034-9

Ma, F., Cholewa, E., Mohamed, T., Peterson, C. A., and Gijzen, M. (2004). Cracks in the palisade cuticle of soybean seed coats correlate with their permeability to water. *Ann. Bot.* 94, 213–228. doi: 10.1093/aob/mch133

Ma, Y., and Bliss, F. A. (1978). Tannin content and inheritance in common bean. *Crop Sci.* 18, 201–204. doi: 10.2135/cropsci1978.0011183X001800020001x

Mandal, S. M., Chakraborty, D., and Dey, S. (2010). Phenolic acids act as signalling molecules in plant-microbe symbioses. *Plant Signal. Behav.* 5, 359–368. doi: 10.4161/psb.5.4.10871

Marbach, I., and Mayer, A. M. (1974). Permeability of seed coats to water as related to drying conditions and metabolism of phenolics. *Plant Physiol.* 54, 817–820. doi: 10.1104/pp.54.6.817

Marles, S. M. A., and Gruber, M. Y. (2004). Histochemical characterization of unextractable seed coat pigments and quantification of extractable lignin in the Brassicaceae. *J. Sci. Food Agric.* 84, 251–262. doi: 10.1002/jsfa.1621

Matilla, A., Gallardo, M., and Puga-Hermida, M. I. (2005). Structural, physiological and molecular aspects of heterogeneity in seeds: a review. *Seed Sci. Res.* 15, 63–76. doi: 10.1079/SSR2005203

McDonald, M. B., Vertucci, C. W., and Roos, E. E. (1988). Soybean seed imbibition: water absorption by seed parts. *Crop Sci.* 28, 993–997. doi: 10.2135/cropsci1988.0011183X002800060026x

McKeon, G. M., and Mott, J. J. (1982). The effect of temperature on the field softening of hard seed of *Stylosanthes humilis* and *S. hamata* in dry monsoonal climate. *Aust. J. Agric. Res.* 33, 75–85. doi: 10.1071/AR9820075

Meyer, C. J., Steudle, E., and Peterson, C. A. (2007). Patterns and kinetics of water uptake by soybean seeds. *J. Exp. Bot.* 58, 717–732. doi: 10.1093/jxb/erl244

Miao, Z. H., Fortune, J. A., and Gallagher, J. (2001). Anatomical structure and nutritive value of lupin seed coats. *Aust. J. Agric. Res.* 52, 985–993. doi: 10.1071/AR00117

Miernyk, J. A., and Johnston, M. L. (2013). Proteomic analysis of the testa from developing soybean seeds. *J. Proteomics* 89, 265–272. doi: 10.1016/j.jprot.2013.05.013

Milberg, P., and Andersson, L. (1997). Seasonal variation in dormancy and light sensitivity in buried seeds of eight annual weed species. *Can. J. Bot.* 75, 1998–2004. doi: 10.1139/b97-911

Miller, S. S., Bowman, L. A., Gijzen, M., and Miki, B. L. A. (1999). Early development of the seed coat of soybean (*Glycine max*). *Ann. Bot.* 84, 297–304. doi: 10.1006/anbo.1999.0915

Miller, S. S., Jin, Z., Schnell, J. A., Romero, M. C., Brown, D. C. W., and Johnson, D. A. (2010). Hourglass cell development in the soybean seed coat. *Ann. Bot.* 106, 235–242. doi: 10.1093/aob/mcq101

Miranda, M., Borisjuk, L., Tewes, A., Heim, U., Sauer, N., Wobus, U., et al. (2001). Amino acid permeases in developing seeds of *Vicia faba* L.: expression precedes storage protein synthesis and is regulated by amino acid supply. *Plant J.* 28, 61–71. doi: 10.1046/j.1365-313X.2001.01129.x

Mizzotti, C., Mendes, M. A., Caporali, E., Schnittger, A., Kater, M. M., Battaglia, R., et al. (2012). The MADS box genes SEEDSTICK and *ARABIDOPSIS* B sister play a maternal role in fertilization and seed development. *Plant J.* 70, 409–420. doi: 10.1111/j.1365-313X.2011.04878.x

Mohamed-Yasseen, Y., Barringer, S. A., Splittstoesser, W. E., and Costanza, S. (1994). The role of seed coats in seed viability. *Bot. Rev.* 60, 426–439. doi: 10.1007/BF02857926

Moïse, J. A., Han, S., Gudynaite-Savitch, L., Johnson, D. A., and Miki, B. L. A. (2005). Seed coats: structure, development, composition, and biotechnology. *In Vitro Cell. Dev. Biol. Plant* 41, 620–644. doi: 10.1079/IVP2005686

Moreau, C., Ambrose, M. J., Turner, L., Hill, L., Ellis, T. H., and Hofer, J. M. (2012). The B gene of pea encodes a defective flavonoid 3′,5′-hydroxylase, and confers pink flower color. *Plant Physiol.* 159, 759–768. doi: 10.1104/pp.112.197517

Moreno-Casasola, P., Grime, J. P., and Martinez, L. (1994). A comparative study of the effects of fluctuations in temperature and moisture supply on hard coat dormancy in seeds of coastal tropical legumes in Mexico. *J. Trop. Ecol.* 10, 67–86. doi: 10.1017/S0266467400007720

Morrison, D. A., McClay, K., Porter, C., and Rish, S. (1998). The role of the lens in controlling heat-induced breakdown of testa-imposed dormancy and native Australian legumes. *Ann. Bot.* 82, 35–40. doi: 10.1006/anbo.1998.0640

Murray, D. R., and Kennedy, I. R. (1980). Changes in activities of enzymes of nitrogen-metabolism in seed coats and cotyledons during embryo development in pea seeds. *Plant Physiol.* 66, 782–786. doi: 10.1104/pp.66.4.782

Nadeau, C. D., Ozga, J. A., Kurepin, L. V., Jin, A., Pharis, R. P., and Reinecke, D. M. (2011). Tissue-specific regulation of gibberellin biosynthesis in developing pea seeds. *Plant Physiol.* 156, 897–912. doi: 10.1104/pp.111.172577

Nagai, I. (1921). A genetico-physiological study on the formation of anthocyanin and brown pigments in plants. *J. Coll. Agric. Imp. Univ. Tokyo* 8, 1–92.

Nawrath, C. (2002). The biopolymers cutin and suberin. *Arabidopsis Book* 1, e0021. doi: 10.1199/tab.0021

Nikolaeva, M. G. (1969). *Physiology of Deep Dormancy in Seeds. Izdatel'stvo Nauka. Leningrad* (Translated from Russian by Z. Shapiro, NSF, Washington, D.C.).

Nikolaeva, M. G. (1977). "Factors controlling the seed dormancy pattern," in *The Physiology and Biochemistry of Seed Dormancy and Germination*, ed. A. A. Khan (Amsterdam: North-Holland Publishing Co.), 51–74.

Noodén, L. D., Blakley, K. A., and Grzybowski, J. M. (1985). Control of seed coat thickness and permeability in soybean: a possible adaptation to stress. *Plant Physiol.* 79, 543–545. doi: 10.1104/pp.79.2.543

Nowack, M. K., Ungru, A., Bjerkan, K. N., Grini, P. E., and Schnittger, A. (2010). Reproductive cross-talk: seed development in flowering plants. *Biochem. Soc. Trans.* 38, 604–612. doi: 10.1042/BST0380604

Offler, C. E., Nerlich, S. M., and Patrick, J. W. (1989). Pathway of photosynthate transfer in the developing seed of *Vicia faba* L. transfer in relation to seed anatomy. *J. Exp. Bot.* 40, 769–780. doi: 10.1093/jxb/40.7.769

Offler, C. E., and Patrick, J. W. (1984). Cellular structures, plasma-membrane surface-areas and plasmodesmatal frequencies of seed coats of *Phaseolus vulgaris* L. in relation to photosynthate transfer. *Aust. J. Plant Physiol.* 11, 79–99. doi: 10.1071/PP9840079

Offler, C. E., and Patrick, J. W. (1993). Pathway of photosynthate transfer in the developing seed of *Vicia faba* L: a structural assessment of the role of transfer cells in unloading from the seed coat. *J. Exp. Bot.* 44, 711–724. doi: 10.1093/jxb/44.4.711

Oliveira, A. E. A., Sassaki, G. L., Iacomini, M., da Cunha, M., Gomes, V. M., Fernandes, B. K. V. S., et al. (2001). Isolation and characterization of a galactorhamnan polysaccharide from the seed coat of *Canavalia ensiformis* that is toxic to the cowpea weevil (*Callosobruchus maculatus*). *Entomol. Exp. Appl.* 101, 225–231. doi: 10.1046/j.1570-7458.2001.00907.x

Ozga, J. A., Reinecke, D. M., Ayele, B. T., Ngo, P., Nadeau, C., and Wickramarathna, A. D. (2009). Developmental and hormonal regulation of gibberellin biosynthesis and catabolism in pea fruit. *Plant Physiol.* 150, 448–462. doi: 10.1104/pp.108.132027

Palmer, R. G., and Kilen, T. C. (1987). "Quantitative genetics: Results relevant to soybean breeding," in *Soybean: Improvements, Production, and Uses*, 2nd Edn, ed. J. R. Wilcox (Madison, WI: Agronomy Society of America), 135–209.

Pammel, L. H. (1899). *Anatomical Characters of the Seeds of Leguminosae: Chiefly Genera of Gray's Manual, Shaw School of Botany*. Ph.D. thesis, Transactions of the Academy of Science of St. Louis, Washington University, St. Louis Bibliography, St. Louis, MI.

Pang, Y. Z., Peel, G. J., Wright, E., Wang, Z., and Dixon, R. A. (2007). Early steps in proanthocyanidin biosynthesis in the model legume *Medicago truncatula*. *Plant Physiol.* 145, 601–615. doi: 10.1104/pp.107.107326

Pang, Y. Z., Wenger, J. P., Saathoff, K., Peel, G. J., Wen, J., Huhman, D., et al. (2009). A WD40 repeat protein from *Medicago truncatula* is necessary for tissue-specific anthocyanin and proanthocyanidin biosynthesis but not for trichome development. *Plant Physiol.* 151, 1114–1129. doi: 10.1104/pp.109.144022

Park, S., Ozga, J. A., Cohen, J. D., and Reinecke, D. M. (2010). Evidence for 4-Cl-IAA and IAA bound to proteins in pea fruits and seeds. *J. Plant Growth Regul.* 29, 184–193. doi: 10.1007/s00344-009-9123-6

Paulsen, T. R., Colville, L., Kranner, I., Daws, M. I., Högstedt, G., Vandvik, V., et al. (2013). Physical dormancy in seeds: a game of hide and seek? *New Phytol.* 198, 496–503. doi: 10.1111/nph.12191

Pellew, R. A., and Southgate, B. J. (1984). The parasitism of *Acacia tortilis* seeds in the Serengeti. *Afr. J. Ecol.* 22, 73–75. doi: 10.1111/j.1365-2028.1984.tb00679.x

Probert, R. J. (2000). "The role of temperature in the regulation of seed dormancy and germination," in *Seeds. The Ecology of Regeneration in Plant Communities*, 2nd Edn, ed. M. Fenner (Wallingford: CABI publishing), 261–292. doi: 10.1079/9780851994321.0261

Quesnelle, P. E., and Emery, R. J. N. (2007). *cis*-Cytokinins that predominate in *Pisum sativum* during early embryogenesis will accelerate embryo growth *in vitro*. *Can. J. Bot.* 85, 91–103. doi: 10.1139/b06-149

Quinlivan, B. J. (1961). The effect of constant and fluctuating temperatures on the permeability of the hard seeds of some legume species. *Aust. J. Agric. Res.* 12, 1009–1022. doi: 10.1071/AR9611009

Quinlivan, B. J. (1968). The softening of hard seeds of sand-plain lupin (*Lupinus varius* L.). *Aust. J. Agric. Res.* 19, 507–515. doi: 10.1071/AR9680507

Quinlivan, B. J. (1971). Seed coat impermeability in legumes. *J. Aust. Inst. Agric. Sci.* 37, 283–295.

Ramsay, G. (1997). Inheritance and linkage of a gene for testa imposed seed dormancy in faba bean (*Vicia faba* L.). *Plant Breed.* 116, 287–289. doi: 10.1111/j.1439-0523.1997.tb00998.x

Ranathunge, K., Shao, S., Qutob, D., Gijzen, M., Peterson, C. A., and Bernards, M. A. (2010). Properties of the soybean seed coat cuticle change during development. *Planta* 231, 1171–1188. doi: 10.1007/s00425-010-1118-9

Reinecke, D. M. (1999). 4-Chloroindole-3-acetic acid and plant growth. *Plant Growth Regul.* 27, 3–13. doi: 10.1023/A:1006191917753

Reinprecht, Y., Yadegari, Z., Perry, G. E., Siddiqua, M., Wright, L. C., McClean, P. E., et al. (2013). *In silico* comparison of genomic regions containing genes coding for enzymes and transcription factors for the phenylpropanoid pathway in *Phaseolus vulgaris* L. and *Glycine max* L. Merr. *Front. Plant Sci.* 4:317. doi: 10.3389/fpls.2013.00317

Riggio Bevilacqua, L., Roti-Mihelozzi, G., and Modenesi, P. (1989). The watertight dormancy of *Melilotus alba* seeds: further observations on the palisade cell wall. *Can. J. Bot.* 67, 3453–3456. doi: 10.1139/b89-422

Roberts, H. A., and Boddrell, J. E. (1985). Seed survival and seasonal pattern of seedling emergence in some Leguminosae. *Ann. Appl. Biol.* 106, 125–132. doi: 10.1111/j.1744-7348.1985.tb03102.x

Rochat, C., and Boutin, J. P. (1992). Temporary storage compounds and sucrose-starch metabolism in seed coats during pea seed development (*Pisum sativum*). *Physiol. Plant* 85, 567–572. doi: 10.1111/j.1399-3054.1992.tb04756.x

Rolston, M. P. (1978). Water impermeable seed dormancy. *Bot. Rev.* 44, 365–396. doi: 10.1007/BF02957854

Russi, L., Cocks, P. S., and Roberts, E. H. (1992). Coat thickness and hard-seededness in some *Medicago* and *Trifolium* species. *Seed Sci. Res.* 2, 243–249. doi: 10.1017/S0960258500001434

Sabiiti, E. N., and Wein, R. W. (1987). Fire and *Acacia* seeds: a hypothesis of colonization success. *J. Ecol.* 74, 937–946. doi: 10.2307/2260305

Schlumbaum, A., Mauch, F., Vögeli, U., and Boller, T. (1986). Plant chitinases are potent inhibitors of fungal growth. *Nature* 324, 365–367. doi: 10.1038/324365a0

Schneitz, K., Hülskamp, M., Kopczak, S. D., and Pruitt, R. E. (1997). Dissection of sexual organ ontogenesis: a genetic analysis of ovule development in *Arabidopsis thaliana*. *Development* 124, 1367–1376.

Serrato Valenti, G., Cornara, L., Ferrando, M., and Modenesi, P. (1993). Structural and histochemical features of *Stylosanthes scabra* (Leguminosae: Papilionoideae) seed coat as related to water entry. *Can. J. Bot.* 71, 834–840. doi: 10.1139/b93-095

Serrato Valenti, G., Melone, L., Ferrando, M., and Bozzini, A. (1989). Comparative studies on testa structure of hard-seeded and soft-seeded varieties of *Lupinus angustifolius* L. (Leguminosae) and mechanisms of water entry. *Seed Sci. Technol.* 17, 563–581.

Serrato Valenti, G., Modenesi, P., Roti-Mihelozzi, G., and Bevilacqua, L. (1986). Structural and histochemical characters of the *Prosopis tamarugo* Phil. seed coat in relation to its hardness. *Acta Bot. Nederland* 35, 475–487.

Serrato-Valenti, G., De Vries, M., and Cornara, L. (1995). The hilar region in *Leucaena leucocephala* Lam. (De Wit) seed: structure, histochemistry and the role of the lens in germination. *Ann. Bot.* 75, 569–574. doi: 10.1006/anbo.1995.1060

Shao, S., Meyer, C. J., Ma, F., Peterson, C. A., and Bernards, M. A. (2007). The outermost cuticle of soybean seeds: chemical composition and function during imbibition. *J. Exp. Bot.* 58, 1071–1082. doi: 10.1093/jxb/erl268

Shen-Miller, J., Mudgett, M. B., Schopf, J. W., Clarke, R., and Berger, R. (1995). Exceptional seed longevity and robust growth: ancient Sacred Lotus from China. *Am. J. Bot.* 82, 1367–1380. doi: 10.2307/2445863

Simao Neto, M., Jones, R. M., and Ratcliff, D. (1987). Recovery of pasture seed ingested by ruminants. 1. Seed of six tropical pasture species fed to cattle, sheep and goats. *Aust. J. Exp. Agric.* 27, 239–246. doi: 10.1071/EA9870239

Skinner, D. J., Hill, T. A., and Gasser, C. S. (2004). Regulation of ovule development. *Plant Cell* 16, S32–S45. doi: 10.1105/tpc.015933

Slater, S. M., Yuan, H. Y., Lulsdorf, M. M., Vandenberg, A., Zaharia, L. I., Han, X., et al. (2013). Comprehensive hormone profiling of the developing seeds of four grain legumes. *Plant Cell Rep.* 32, 1939–1952. doi: 10.1007/s00299-013-1505-3

Slattery, H. D., Atwell, B. J., and Kuo, J. (1982). Relationship between colour, phenolic content and impermeability in the seed coat of various *Trifolium subterraneum* L. genotypes. *Ann. Bot.* 50, 373–378.

Smýkal, P., Coyne, C. J., Ambrose, M. J., Maxted, N., Schaefer, H., Blair, M., et al. (2014). Legume crops phylogeny and genetic diversity for science and breeding. *Crit. Rev. Plant Sci.* doi: 10.1080/07352689.2014.897904 [Epub ahead of print].

Spurny, M. (1963). Cell wall structure of epidermal cells of the pea seed coat (*Pisum sativum* L.) studied by microcinematography. *Mikroskopie* 18, 272–279.

Spurny, M. (1964). Changes in the permeability of the seed coat in connection with the development of suberin adcrustations of the macrosclereids from the seed coat of the pea (*Pisum sativum* L.). *Flora* 154, 547–567.

Stevenson, T. M. (1937). Sweet clover studies on habit of growth, seed pigmentation and permeability of the seed coat. *Sci. Agric.* 17, 627–654.

Taylor, G. B. (1981). Effect of constant temperature treatments followed by alternating temperatures on the softening of hard seeds of *Trifolium subterraneum* L. *Aust. J. Plant Physiol.* 8, 547–558. doi: 10.1071/PP9810547

Taylor, G. B. (2005). Hardseededness in Mediterranean annual pasture legumes in Australia: a review. *Aust. J. Agric. Res.* 56, 645–661. doi: 10.1071/AR04284

Taylor, G. B., and Ewing, M. A. (1996). Effects of extended (4–12 years) burial on seed softening in subterranean clover and annual medics. *Aust. J. Exp. Agric.* 36, 145–150. doi: 10.1071/EA9960145

Taylor, G. B., and Revell, C. K. (1999). Effect of pod burial, light, and temperature on seed softening in yellow serradella. *Aust. J. Agric. Res.* 50, 1203–1209. doi: 10.1071/AR98206

Taylor, G. B., and Revell, C. K. (2002). Seed softening, imbibition time, and seedling establishment in yellow serradella. *Aust. J. Agric. Res.* 53, 1011–1018. doi: 10.1071/AR01201

Telewski, F. W., and Zeevaart, J. A. (2002). The 120-yr period for Dr. Beal's seed viability experiment. *Am. J. Bot.* 89, 1285–1288. doi: 10.3732/ajb.89.8.1285

Thompson, K., and Ooi, M. K. J. (2013). Germination and dormancy breaking: two different things. *Seed Sci. Res.* 23, 1. doi: 10.1017/S0960258512000190

Thompson, R., Burstin, J., and Gallardo, K. (2009). Post-genomics studies of developmental processes in legume seeds. *Plant Physiol.* 151, 1023–1029. doi: 10.1104/pp.109.143966

Thompson, R. D., Hueros, G., Becker, H., and Maitz, M. (2001). Development and functions of seed transfer cells. *Plant Sci.* 160, 775–783. doi: 10.1016/S0168-9452(01)00345-4

Thorne, J. H. (1981). Morphology and ultrastructure of maternal seed tissues of soybean in relation to the import of photosynthate. *Plant Physiol.* 67, 1016–1025. doi: 10.1104/pp.67.5.1016

Thorne, J. H., and Rainbird, R. M. (1983). An *in vivo* technique for the study of phloem unloading in seed coats of developing soybean seeds. *Plant Physiol.* 72, 268–271. doi: 10.1104/pp.72.1.268

Tuteja, J. H., Clough, S. J., Chan, W. C., and Vlodking, L. O. (2004). Tissue specific gene silencing mediated by a naturally occurring chalcone synthase gene cluster in *Glycine max. Plant Cell* 16, 819–835. doi: 10.1105/tpc.021352

Van Assche, J. A., Debucquoy, K. L. A., and Rommens, W. A. F. (2003). Seasonal cycles in the germination capacity of buried seeds of some Leguminosae (Fabaceae). *New Phytol.* 158, 315–323. doi: 10.1046/j.1469-8137.2003.00744.x

Van Dongen, J. T., Ammerlaan, A. M., Wouterlood, M., Van Aelst, A. C., and Borstlap, A. C. (2003). Structure of the developing pea seed coat and the post-pholem transport pathway of nutrients. *Ann. Bot.* 93, 729–737. doi: 10.1093/aob/mcg066

van Klinken, K. D., and Goulier, J. B. (2013). Habitat-specific seed dormancy-release mechanisms in four legume species. *Seed Sci. Res.* 23, 181–188. doi: 10.1017/S0960258513000196

Van Staden, J., Manning, J. C., and Kelly, K. M. (1989). Legume seeds – the structure: function equation. *Monogr. Syst. Bot. Missouri Bot. Gard.* 29, 417–450.

Vaughan, D. A., Bernard, R. L., Sinclair, J. B., and Kunwar, I. K. (1987). Soybean seed coat development. *Crop Sci.* 27, 759–765. doi: 10.2135/cropsci1987.0011183X002700040031x

Vázquez-Yanes, C., and Orozco-Segovia, A. (1982). Seed germination of tropical rain forest pioneer tree (*Heliocarpus donnell-smithii*) in response to diurnal fluctuation in temperature. *Physiol. Plant.* 56, 295–298. doi: 10.1111/j.1399-3054.1982.tb00341.x

Verdier, J., Dessaint, F., Schneider, C., and Abirached-Darmency, M. (2013a). A combined histology and transcriptome analysis unravels novel questions on *Medicago truncatula* seed coat. *J. Exp. Bot.* 64, 459–470. doi: 10.1093/jxb/ers304

Verdier, J., Lalanne, D., Pelletier, S., Torres-Jerez, I., Righetti, K., Bandyopadhyay, K., et al. (2013b). A regulatory network-based approach dissects late maturation processes related to the acquisition of desiccation tolerance and longevity of *Medicago truncatula* seeds. *Plant Physiol.* 163, 757–774. doi: 10.1104/pp.113.222380

Verdier, J., Kakar, K., Gallardo, K., Le Signor, C., Aubert, G., Schlereth, A., et al. (2008). Gene expression profiling of *M. truncatula* transcription factors identifies putative regulators of grain legume seed filling. *Plant Mol. Biol.* 67, 567–580. doi: 10.1007/s11103-008-9320-x

Verdier, J., Zhao, J., Torres-Jerez, I., Ge, S., Liu, C., He, X., et al. (2012). MtPAR MYB transcription factor acts as an on switch for proanthocyanidin biosynthesis in *Medicago truncatula. Proc. Natl. Acad. Sci. U.S.A.* 109, 1766–1771. doi: 10.1073/pnas.1120916109

Vleeshouwers, L. M., Bouwmeester, H. J., and Karssen, C. M. (1995). Redefining seed dormancy: an attempt to integrate physiology and ecology. *J. Ecol.* 83, 1031–1037. doi: 10.2307/2261184

Wang, H. L., and Grusak, M. A. (2005). Structure and development of *Medicago truncatula* pod wall and seed coat. *Ann. Bot.* 95, 737–747. doi: 10.1093/aob/mci080

Warrag, E. I., and Eltigani, M. A. (2005). Breaking seed coat dormancy of *Acacia nilotica* seeds under simulated natural habit conditions in Sudan. *Trop. Ecol.* 46, 127–132.

Weber, H., Borisjuk, L., Heim, U., Sauer, N., and Wobus, U. (1997). A role for sugar transporters during seed development: molecular characterization of a hexose and a sucrose carrier in fava bean seeds. *Plant Cell* 9, 895–908. doi: 10.1105/tpc.9.6.895

Weber, H., Borisjuk, L., and Wobus, U. (2005). Molecular physiology of legume seed development. *Annu. Rev. Plant Biol.* 56, 253–279. doi: 10.1146/annurev.arplant.56.032604.144201

Weeden, N. F. (2007). Genetic changes accompanying the domestication of *Pisum sativum*: is there a common genetic basis to the 'domestication syndrome' for legumes? *Ann. Bot.* 100, 1017–1025. doi: 10.1093/aob/mcm122

Weiss, E., Kislev, M. E., and Hartmann, A. (2006). Autonomous cultivation before domestication. *Science* 312, 1608–1610. doi: 10.1126/science.1127235

Werker, E., Marbach, I., and Mayer, A. M. (1979). Relation between the anatomy of the testa, water permeability and the presence of phenolics in the genus *Pisum. Ann. Bot.* 43, 765–771.

White, J. (1908). The occurrence of an impermeable cuticle on the exterior of certain seeds. *Proc. Royal Soc. Victoria* 21, 203–210.

Winkel-Shirley, B. (2001). Flavonoid biosynthesis. a colourful model for genetics, biochemistry, cell biology, and biotechnology. *Plant Physiol.* 126, 485–493. doi: 10.1104/pp.126.2.485

Wolf, W., Baker, F. L., and Bernard, R. L. (1981). Soybean seed-coat structural features: pits, deposits and cracks. *Scan. Electron Microsc.* 3, 531–544.

Wyatt, J. E. (1977). Seed coat and water absorption properties of seed of near-isogenic snap bean lines differing in seed coat color. *J. Am. Soc. Hortic. Sci.* 102, 478–480.

Yaklich, R. W., Vigil, E. L., and Wergin, W. P. (1986). Pore development and seed coat permeability in soybean. *Crop Sci.* 26, 616–624. doi: 10.2135/cropsci1986.0011183X002600030041x

Yang, K., Jeong, N., Moon, J. K., Lee, Y. H., Lee, S. H., Kim, H. M., et al. (2010). Genetic analysis of genes controlling natural variation of seed coat and flower colours in soybean. *J. Hered.* 101, 757–768. doi: 10.1093/jhered/esq078

Zabala, G., and Vodkin, L. O. (2003). Cloning of the pleiotropic T locus in soybean and two recessive alleles that differentially affect structure and expression of the encoded flavonoid 3 hydroxylase. *Genetics* 163, 295–309.

Zabala, G., and Vodkin, L. O. (2005). The wp mutation of *Glycine max* carries a gene fragment-rich transposon of the CACTA superfamily. *Plant Cell* 17, 2619–2632. doi: 10.1105/tpc.105.033506

Zabala, G., and Vodkin, L. O. (2007). A rearrangement resulting in small tandem repeats in the F3′5′H gene of white flower genotypes is associated with the soybean W1 locus. *Crop Sci.* 47, S113–S124. doi: 10.2135/cropsci2006.12.0838tpg

Zhao, J., and Dixon, R. A. (2009). MATE transporters facilitate vacuolar uptake of epicatechin 3′-*o*-glucoside for proanthocyanidin biosynthesis in *Medicago truncatula* and *Arabidopsis. Plant Cell* 21, 2323–2340. doi: 10.1105/tpc.109.067819

Zhao, J., Pang, Y., and Dixon, R. A. (2010). The mysteries of proanthocyanidin transport and polymerization. *Plant Physiol.* 153, 437–443. doi: 10.1104/pp.110.155432

Zhou, S., Sekizaki, H., Yang, Z., Sawa, S., and Pan, J. (2010). Phenolics in the seed coat of wild soybean (*Glycine soja*) and their significance for seed hardness and seed germination. *J. Agric. Food Chem.* 58, 10972–10978. doi: 10.1021/jf102694k

Zohary, D., Hopf, M., and Weiss, E. (2012). *Domestication of Plants in the Old World*, 4th Edn. Oxford: Oxford University Press.

Conflict of Interest Statement: The authors declare that the research was conducted in the absence of any commercial or financial relationships that could be construed as a potential conflict of interest.

11

Physical, metabolic and developmental functions of the seed coat

Volodymyr Radchuk and Ljudmilla Borisjuk*

Heterosis, Molecular Genetics, Leibniz-Institut für Pflanzengenetik und Kulturpflanzenforschung, Gatersleben, Germany

Edited by:
Paolo Sabelli, University of Arizona, USA

Reviewed by:
Stewart Gillmor, Centro de Investigación y de Estudios Avanzados del Instituto Politécnico Nacional, Mexico
Philip W. Becraft, Iowa State University, USA

***Correspondence:**
Ljudmilla Borisjuk, Heterosis, Molecular Genetics, Leibniz-Institut für Pflanzengenetik und Kulturpflanzenforschung, D-06466 Gatersleben, SA, Germany
e-mail: borysyuk@ipk-gatersleben.de

The conventional understanding of the role of the seed coat is that it provides a protective layer for the developing zygote. Recent data show that the picture is more nuanced. The seed coat certainly represents a first line of defense against adverse external factors, but it also acts as channel for transmitting environmental cues to the interior of the seed. The latter function primes the seed to adjust its metabolism in response to changes in its external environment. The purpose of this review is to provide the reader with a comprehensive view of the structure and functionality of the seed coat, and to expose its hidden interaction with both the endosperm and embryo. Any breeding and/or biotechnology intervention seeking to increase seed size or modify seed features will have to consider the implications on this tripartite interaction.

Keywords: seed development, nutrients supply, seed photosynthesis, PCD, maternal–filial interface

INTRODUCTION

The evolution of sexual reproduction and the seed underlies much of the evolutionary success of the flowering plants. The most distinctive characteristic of the angiosperms is the double fertilization event, followed by the development of a seed encased in maternal tissue, referred to as the seed coat (or testa). The enclosure of the developing embryo affords it protection and thereby enhances its chances of reaching maturity and establishing the subsequent generation; this feature has not been achieved by species belonging to other clades of the plant kingdom. The progenitor structure of the angiosperm seed on the female side is the ovary, and its final form comprises an embryo, an endosperm, and the seed coat. The embryo results from the fusion between an egg cell and a sperm nucleus, while the endosperm develops from the fusion between the two central cell nuclei and a second sperm nucleus to produce (in diploid species) a triploid structure. The seed coat is entirely maternal in origin. When fertilization fails, the structure degenerates rapidly, thereby ensuring that the assimilate invested in an aborted seed is recycled (Roszak and Köhler, 2011). Post fertilization, the development of the seed relies on a coordinated interaction between the seed coat, the embryo, and the endosperm. The molecular basis of seed development has been intensively studied (Lafon-Placette and Köhler, 2014), but until now, the lack of suitable *in vivo* analytical methods has hampered systematic investigations of either the metabolism occurring or the internal structures developing within the growing seed. Here, a description is given of our current understanding of the functional role of the seed coat in the developing seed.

FROM OVULE TO SEED COAT

The seed coat originates from cell layers surrounding the ovule. The analysis of a number of *Arabidopsis thaliana* mutants has revealed its structure and function, as well as identifying many of the genes involved in its development (Haughn and Chaudhury, 2005; Figueiredo and Köhler, 2014). Seed coat development is repressed prior to fertilization by dosage-sensitive, sporophytically active polycomb-type-proteins that are expressed in the maternal tissue surrounding the female gametophyte (Roszak and Köhler, 2011). The fertilization generates a signal that relieves the polycomb type protein-mediated repression, resulting in the initiation of seed coat formation (Roszak and Köhler, 2011).

The *A. thaliana* seed coat is composed of five cell layers: the three-layered inner integument and the two-layered outer integument; each of these layers follows a distinct path during seed development. The endothelium (the innermost cell layer) synthesizes proanthocyanidins (PAs), which first condense into tannins, then oxidize to impart the brown pigmentation seen in the mature seed of many species (Lepiniec et al., 2006). The two adjacent cell layers are crushed together as the seed expands (Nakaune et al., 2005). The outer integument undergoes extensive differentiation, regulated by the *YABBY* family transcription factor *INNER NO OUTER* (Kelley and Gasser, 2009), going on to form the subepidermal and epidermal cell layers. The former of these generates a thickened wall on the side facing the epidermis (Haughn and Chaudhury, 2005), while the latter produces a pectinaceous carbohydrate referred to as mucilage (Arsovski et al., 2010; Haughn and Western, 2012). The outer integument is associated with a suberized layer, and the endothelium with a cutin-like polyester layer (Molina et al., 2008). In leguminous species, the seed coat is typically a multi-layered structure, including both macro- and osteosclerids in its outer integument and parenchyma in its inner integument (van Dongen et al., 2003; Verdier et al., 2012). In the cereal grain (strictly a caryopsis rather than a seed, since the ovary wall is fused with the seed coat), the endothelium and the outer integument each form a pair of cell layers, while the enlarged pericarp takes over some of the key functions of the seed coat (Sreenivasulu et al., 2010). The various impacts of the seed coat are illustrated for a contrasting set of species in **Figure 1**.

The development of the endothelium has been revealed by the analysis of *A. thaliana* mutants impaired in seed coat pigmentation. A number of relevant genes have been isolated, classified for the most part into either transcription factors or genes required for the synthesis and compartmentation of PA flavonoid compounds (Haughn and Chaudhury, 2005). Comprehensive transcriptomic descriptions of the developing *A. thaliana* seed coat have provided a wealth of information relevant to how the process occurs in other species (Dean et al., 2011). The *Medicago truncatula* myb transcription factor gene *MtPAR* has been shown to be a key regulator of PA synthesis, and its transcription co-localizes with the site of PA accumulation in the seed coat (Verdier et al., 2012). Key *M. truncatula* genes along with the precursor transporter *MATE1* (involved in PA synthesis) have been isolated and characterized by Zhao and Dixon (2009). Orthologs of *BANYULS*, which encodes anthocyanidin reductase (Albert et al., 1997), have been identified in oilseed rape and its close relatives *Brassica rapa* and *B. oleracea* (Auger et al., 2010). The transcriptional regulation of flavonoid metabolism is less well understood in legumes and cereals, perhaps because the genes underlying PA synthesis have been lost during domestication, with the result that white-seededness is commonplace in these taxa. Consequently, in contrast to its wild relatives, cultivated barley (similarly to rice and wheat) does not accumulate substantial amounts of PA (Sang, 2009). The relationship of secondary PA metabolism with both developmental regulation and the stress response has the potential to contribute significantly to future crop improvement and is being investigated by a number of research groups (Debeaujon et al., 2000; Bassoi and Flintham, 2005; Lepiniec et al., 2006; Gao et al., 2013).

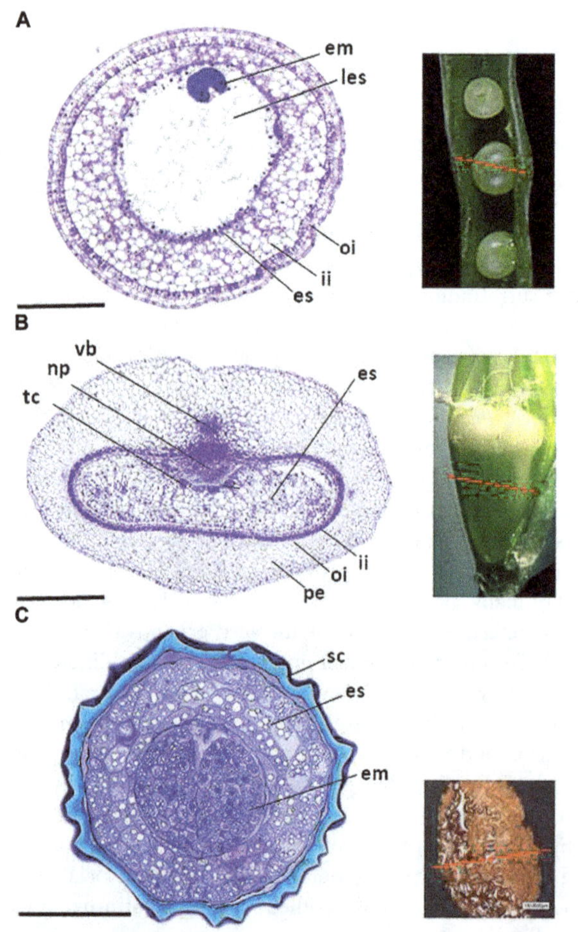

FIGURE 1 | The structure of (A) the oilseed rape seed, (B) the barley caryopsis, and (C) the tobacco seed. em, embryo; es, endosperm; tc, endosperm transfer cell; ii, endothelium; les, liquid endosperm; vb, main vascular bundle; np, nucellar projection; oi, outer integument; pe, pericarp; sc, seed coat. Bars: 0.5 mm.

Elucidation of the development of the *A. thaliana* outer integuments has relied on mutants that produce either less mucilage than the wild type or those which produce mucilage of a different composition. Both regulatory and structural genes have been recognized (Haughn and Chaudhury, 2005; Arsovski et al., 2010; Haughn and Western, 2012). The set of WD repeat, bHLH and *myb* transcription factors that regulate outer integument development partially overlaps with the factors controlling trichome initiation and development, the regulation of anthocyanin production and endothelial development, although the relevant interaction partners are distinct (Schiefelbein, 2003; Bernhardt et al., 2005; Haughn and Chaudhury, 2005; Gonzalez et al., 2008). For example, outer integument differentiation is controlled by the proteins TTG1, myb5/TT2, and TT8/EGL3, which also drive the transcription of *ABE1*, *ABE4*, *GH*, *GL2*, and *mybL2* (Gonzalez et al., 2008; Li et al., 2009). The *A. thaliana* model has been informative for understanding the molecular basis of the synthesis of cotton fibers, which arise from the epidermal cells of the outer integument and are distributed all over the seed's surface (Lee et al., 2007; Liu et al., 2012; Ruan, 2013). Several of the regulatory genes involved in fiber initiation have proven to be homologs of *A. thaliana* genes (e.g., *TTG1* and *GL2*) responsible for trichome formation and the differentiation of the outer integument. The current understanding is that a transcriptional myb/bHLH/WD repeat complex is required for this initiation process (Yang and Ye, 2013). A full understanding of the regulatory machinery operating in the epidermal cells will aid in achieving further improvement in a number of cotton seed traits (Yan et al., 2009; Efe et al., 2010) as well as the development of sustainable means of processing seeds and the fibers (Kimmel and Day, 2001; Stiff and Haigler, 2012).

NO LIFE WITHOUT PROTECTION

In many seeds, the epidermal layer of the seed coat generates a cuticle which represents a physical barrier between the seed and its external environment. Neither viruses nor bacteria are able to penetrate an intact mature seed cuticle (Singh and Mathur, 2004; Gergerich and Dolja, 2006). The only entry points into a mature seed of this type for a pathogen are the micropyle – which represents the point of entry of the pollen tube – and the funiculus, which links the maternal vascular system to the seed integument. The immature seed coat is less robust, so it offers less protection against pathogen penetration, which can occur via either the ovary wall or the stigma. Mechanically damaged cuticles offer an alternative path for pathogen invasion (Singh and Mathur, 2004). Integrity of seed coat surface is extremely important for seed quality and fitness during seed storage or germination, and diverse technologies are available for preserving and enhancing of seed surface (Black and Halmer, 2006; Brooker et al., 2007).

An additional layer of protection is provided in certain seeds by the deposition of toxic compounds such as cyanogenic glycosides, terpenoids, and flavonoids. The issue of seed coat chemistry has especial resonance in relation to the presence of glucosinolates in brassicaceous crops (Bohinc et al., 2012). The accumulation of phenolics in plant tissues is considered to be an adaptive response to adverse environmental conditions (Lattanzio et al., 2006; Vermerris and Nicholson, 2007).

Since plants lack mobile defender cells, they are forced to rely on the innate immunity of every cell and on the production of signal molecules by invaded cells and their subsequent sensing (Jones and Dangl, 2006). The small and highly stable cysteine-rich peptides referred to as defensins actively inhibit pathogen invasion in both plants and animals (Stotz et al., 2009). Defensins genes induced by pathogen infection have been identified in a number of plant species (Thomma et al., 2002; Lay and Anderson, 2005; Carvalho and Gomes, 2009). Their products are concentrated mainly in the peripheral/bordering cells, as typified in barley and rice (Kovalchuk et al., 2010), and are released following tissue damage (Thomma et al., 2002; Lay and Anderson, 2005). Defensin production can also be promoted by certain abiotic stress agents, and also by exposing plants to the phytohormones methyl jasmonate, ethylene, or salicylic acid (Lay and Anderson, 2005). The expression of defensins in response to a variety of biotic and abiotic stimuli implies the possibility of cross-talk between distinct signal transduction pathways and gene expression programs involved in cellular signaling and growth regulation (Hanks et al., 2005; Okuda et al., 2009). Plant defensins have become the focus of a considerable body of biotechnological research (Carvalho and Gomes, 2009; Kovalchuk et al., 2010).

The barrier function of the seed coat does not extend to gases, since it is in most cases at least semi-permeable (Welbaum and Bradford, 1990; Beresniewicz et al., 1995). The seed coat epidermis in the mature seed features no, or at best only scarce, functional stomata (Cochrane and Duffus, 1979; Geisler and Sack, 2002). In conjunction with the chemical composition of the cuticle, this implies a rather limited capacity for gas exchange (Nutbeam and Duffus, 1978; Sinclair et al., 1987; Sinclair, 1988). It was already demonstrated some 40 years ago that most of the gas exchange activity occurring within the pea seed is located in the micropylar region (Wager, 1974). The diffusivity of carbon dioxide through plant tissue is much higher than that of oxygen, since (unlike oxygen) carbon dioxide is readily soluble in water and so can move from cell to cell in the form of the carbonate ion. The presence of gas-filled intercellular spaces is therefore likely to be essential for translocation of oxygen within the seed. Synchrotron X-ray computer tomography has identified such spaces in the developing seeds of both *A. thaliana* (Cloetens et al., 2006) and oilseed rape *in vivo* (Verboven et al., 2013). In the latter species, both the seed coat and the hypocotyl are well supplied with void spaces, unlike the cotyledons, where the spaces are small and only poorly inter-connected (**Figure 2**). *In silico* modeling has revealed a three orders of magnitude range in oxygen diffusivity from the seed coat to particular embryonic tissues (Verboven et al., 2013). The multiple void spaces present in the seed coat suggest that gas exchange is effective within this part of the seed. There is a lack of any interconnectivity with the embryo, so the seed coat void network is likely to be autonomous. Both the seed cuticle and the lipid-containing aleurone layer of the endosperm have been identified as barriers to oxygen exchange, the former between the seed coat and external atmosphere and the latter between the seed coat and the endosperm/embryo. The oxygen pool stored in the voids of oilseed rape seed is consumed about once per minute. Since the developing seed has a high respiratory rate, it requires an additional supply of oxygen to maintain aerobic respiration.

Oxygen micro-sensor measurements made within the seed of faba bean, pea (Rolletschek et al., 2002, 2003) and soybean (Borisjuk et al., 2005), and within the grains of barley (Rolletschek et al., 2004), wheat (van Dongen et al., 2004), and maize (Rolletschek et al., 2005) have established that hypoxia is the norm. In the developing seed, it may be advantageous to keep the oxygen level low, because the bioenergetic efficiency of mitochondria is usually increased at low oxygen levels (Gnaiger et al., 2000). Thus, a low internal oxygen concentration in the seed may stimulate carbon use efficiency. Low oxygen levels help to avoid the formation of toxic concentrations of reactive oxygen species, which damage cellular structures and require the expenditure of energy for repair (Borisjuk and Rolletschek, 2009). In maize, the level of expression of detoxification genes (encoding glutathione *S*-transferase, superoxide dismutase, and ascorbate peroxidase) decreases during grain development (Méchin et al., 2007), consistent with a reduction in oxygen availability. To summarize, maintaining a low oxygen level within the seed has been proposed to provide a means for the developing seed to control the local level of metabolic activity (Borisjuk and Rolletschek, 2009). Deep within the mature seed, the inhibition of gas exchange can generate a state of near-anoxia, which may help to ensure the remarkable longevity of seeds (Shen-Miller, 2002). While the mechanistic basis of seed longevity is not fully understood, an important component is likely the control of oxidation (Hendry, 1993; Smirnoff, 2010; Bailly and Kranner, 2011). Practical methods to prolong seed viability in *ex situ* gene banks exploit this natural phenomenon by hermetically sealing the seed in order to maintain a high level of carbon dioxide within; this is combined with careful drying down and refrigeration, which help to slow seed metabolism/respiration and suppress oxidation processes (Kranner et al., 2010).

PERCEIVING ENVIRONMENTAL CUES

The seed coat's function is simultaneously to protect the embryo and to transmit information regarding the external environment. An impenetrable seed coat may help to keep the embryo safe, but at the same time it would exclude the sensing of environmental cues. The evolutionary solution to this dilemma is to combine certain structural features with appropriate levels of metabolic and photosynthetic activity in the seed coat.

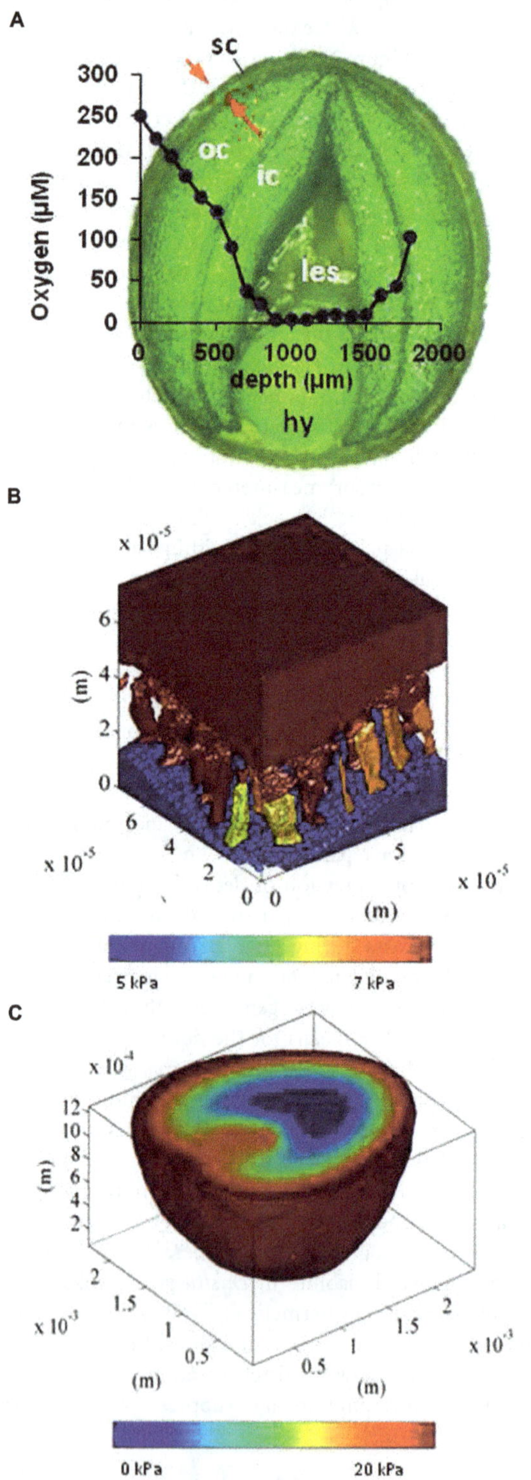

FIGURE 2 | Experimental measurements and *in silico* modeling of oxygen distribution in the developing oilseed rape seed. **(A)** Oxygen concentration (blue line) along a transect of the seed, as determined by a needle micro-sensor. The x axis plots the penetration of the sensor. **(B)** *In silico* modeling of oxygen concentration in the pore spaces of the seed coat [indicated by arrows in **(A)**]. **(C)** Oxygen concentration map (color-coded). For details see Verboven et al. (2013). hy, hypocotyl; ic, inner cotyledon; les, liquid endosperm; oc, outer cotyledon; sc, seed coat.

As plant species vary so much with respect to the distribution and amount of chlorenchyma in their developing seed, it is difficult to make meaningful generalizations regarding seed photosynthesis. However, out of 19 major crop species, only maize grains lack chlorophyll (Bewley and Black, 1994). Both the seed coat and embryo of pea (Tschiersch et al., 2012), soybean (Saito et al., 1989), oilseed rape (Borisjuk et al., 2013) and faba bean (Rolletschek et al., 2003) are photosynthetically active during seed development. The immature caryopsis of barley, wheat, rice and other grasses features a photosynthetically active pericarp (Bewley and Black, 1994; Rolletschek et al., 2004). The site of photosynthetic electron transport coincides with that of chlorophyll (Tschiersch et al., 2012), as for example in the barley pericarp (**Figures 3A,B**). When exposed to light, the chloroplastids (**Figure 3C**) produce sufficient ATP and NAPDH to meet local energy demand. Given the very short half life of both ATP and NADPH, it is likely that little long distance transport occurs from their site of synthesis. Non-photosynthetic plastids within the pericarp depend entirely on an external supply of ATP, just as is the case for other non-photosynthetic tissues (Möhlmann et al., 1994; Möhlmann and Neuhaus, 1997). The spatial separation between the endosperm and the photosynthetically active pericarp implies that seed photosynthesis does not make any direct energy contribution to assimilate storage in the endosperm.

Photosynthetic activity in the seed coat, as in the leaves, fixes carbon dioxide (Nutbeam and Duffus, 1978; Caley et al., 1990), and generates oxygen (Rolletschek et al., 2004; Tschiersch et al., 2012). This process is saturated in seeds at a light intensity some fivefold below that applicable for leaves. Oxygen production and carbon dioxide fixation combine to maintain a

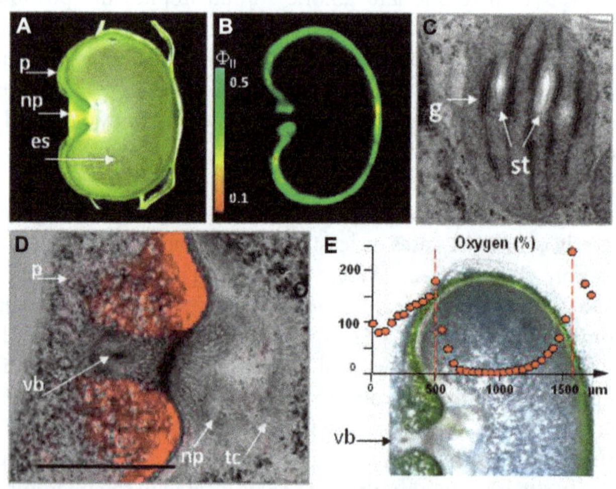

FIGURE 3 | Photosynthesis in the barley pericarp. **(A)** Cross-section of a grain. **(B)** The effective quantum yield of photosystem II (FII) across a cross-section of a 12-day-old caryopsis, measured at a light intensity of 160 µmol quanta m^{-2} s^{-1}. The scale shows the relationship between the color and FII. **(C)** A transmission electron micrograph of seed chlorenchyma plastids. **(D)** Chlorophyll auto-fluorescence within the crease region of the pericarp. **(E)** Oxygen levels within the caryopsis, as measured by a micro-sensor. For details see Tschiersch et al. (2012). es, endosperm; g, grana; np, nucellar projection; p, pericarp; st, starch grain; tc, transfer cell; vb, vascular bundle.

consistent gaseous environment within the seed. The effect of photosynthetic oxygen evolution exceeding the oxygen demand of the respiring seed coat is an increase in the internal oxygen level (Patrick et al., 1995; Rolletschek et al., 2002, 2003), which serves to relieve hypoxic stress and thereby enhances the synthetic activity of the seed (Greenway and Gibbs, 2003; Rolletschek et al., 2005). Importantly, the tissues through which nutrients are transported to the endosperm/embryo are oxygen depleted (Melkus et al., 2011). In the dark, the oxygen level can fall below 0.1% of the ambient atmospheric concentration (Rolletschek et al., 2011), but the chlorenchyma layer that surrounds the region ensures that the level of oxygen present is much higher in the light than this (**Figures 3D,E**). Both nutrient transport to, and storage activity within the endosperm rely heavily on respiratory energy and thus on a steady supply of photosynthetically derived oxygen. Experiments tracking the incorporation of labeled sucrose into starch have shown that the process is stimulated by both light and oxygen (Gifford and Bremner, 1981), underlining the dependence of storage activity on a supply of oxygen. Similarly, assimilate supply to the dicotyledonous seed is also oxygen-dependent, as shown by phloem unloading experiments (Thorne, 1982).

The seed coat's high rate of respiration, along with its low permeability with respect to carbon dioxide, contributes to elevating the seed's internal level of carbon dioxide. However, high concentrations of carbon dioxide do promote phosphoenolpyruvate carboxylase activity, which serves to encourage carbon dioxide re-fixation and so restricts its loss (Wager, 1974; Harvey et al., 1976; Flinn, 1985; Araus et al., 1993; Golombek et al., 1998). Limiting carbon dioxide loss in this way can make an important contribution to the seed's overall carbon budget (Vigeolas et al., 2003; Rolletschek et al., 2004). In addition, re-fixation of carbon dioxide is mediated by Rubisco activity during photosynthesis (Goffman et al., 2004; Ruuska et al., 2004). The seed's rate of carbon dioxide uptake from the atmosphere is much lower than the leaf's (Brar and Thies, 1977; Watson and Duffus, 1988), in line with both a lower activity of photosynthesis-associated enzymes (Duffus and Rosie, 1973) and a limited rate of metabolic turn-over (Schwender and Ohlrogge, 2002; Sriram et al., 2004). Critical factors should be considered such as (1) the low density of stomata on the surface of the developing seed, and (2) the low amount of chlorenchyma. Most relevant experiments have disregarded the re-assimilation of internally produced CO_2 and hence probably underestimated the actual sizes of the occurring fluxes (Araus et al., 1993). Nevertheless, the generally held conclusion still stands that the contribution of seed photosynthesis to dry matter production (via net CO_2 fixation) is low.

One likely hypothesis regarding the evolutionary significance of retaining photosynthetic capacity in the seed coat suggests that the interception and processing of light by the seed coat gives the seed the means to sense its external environment, which is integrated with the hormonal, metabolic, and other signals brought to the seed through the phloem. Assimilate generated in the leaf is exported into the phloem in the form of sugar. The developing seed can benefit from the capacity to anticipate a burst of sugar arriving via the phloem, since this would facilitate the seed's rapid adjustment to its synthesis of storage products. Photosynthesis has a marked effect on the entrainment and maintenance of robust circadian rhythms (Haydon et al., 2013). In this way, the retention of seed photosynthesis can provide a means of tuning the seed's metabolism to the quantity and quality of the light available to the mother plant.

HIGHWAYS AND BYWAYS TRAVELED DURING SOLUTE TRANSFER

Assimilates produced by the mother plant are delivered to the developing seed via the same conduit, the vascular system, which brings hormonal signals and the necessary protein- and RNA-based messages. Collectively, this enables the coordination of physiological and developmental processes at the whole organism level (van Bel et al., 2013). As the vascular system does not extend beyond the seed coat (Patrick et al., 1995), the embryo and the endosperm are apoplastically isolated from the mother plant and are therefore somewhat autonomous. Several pathways for nutrient flow are available, depending on seed size and structure. The smallest seeds (orchid seeds can be as small as 200 μm in diameter) have no vascular structure; rather, the zygote forms a haustorium which extends toward the terminus of the mother plant's vascular system. Slightly larger seeds form a bundle of pro-vascular elements. Medium-sized seeds develop a simple, well developed collateral bundle (van Dongen et al., 2003). Finally, in large-seeded species, the vascular bundle is bulky enough to anastomose, thereby allowing for the distribution of nutrients throughout the seed (Vinogradova and Falaleev, 2012). In the *A. thaliana* seed, the vascular tissue terminates at the junction of the funiculus and ovule, and in the maize kernel, the vascular bundle terminates at the placenta–chalazal region (**Figure 4A**; Dermastia et al., 2009; Gómez et al., 2009; Costa et al., 2012). In contrast, wheat, barley and rice grains form a vascular system that extends over the whole length of the grain (**Figure 4B**; Sreenivasulu et al., 2010). The vascular architecture of *Fabaceae* species seeds is highly variable, ranging from a single chalazal vein in the *Viciae* and *Trifolieae* to an extensive anastomosed arrangement in the *Phaseoleae* (**Figure 4C**; van Dongen et al., 2003; Weber et al., 2005; Verdier et al., 2013).

Vascular structures are typically embedded within parenchymatous tissue. Adjacent parenchyma cells are interconnected by plasmodesmata, forming a symplastic continuum (domain). The plasmodesmata within these domains are larger than elsewhere in the seed (Ruan and Patrick, 1995; Oparka et al., 1999; Stadler et al., 2005a,b), which facilitates the movement of small molecules such as sugars and peptides. The maternal symplasm represents the major route for nutrients to reach the seed (Wang and Fisher, 1994; Borisjuk et al., 2002; van Dongen et al., 2003). In *A. thaliana*, each integument forms an independent symplasm (Stadler et al., 2005a; Ingram, 2010), which acts as an extension of the phloem (Stadler et al., 2005a,b). In common bean, faba bean, and pea, assimilate unloading sites are distributed throughout the seed coat parenchyma, with the possible exception of the branched parenchyma (Patrick and Offler, 1995). In the small grain crops, the nucellar projection is the focus of a well organized transport route. The cellular architecture of the nucellar projection has been described in wheat (Wang and Fisher, 1994) and barley (Thiel

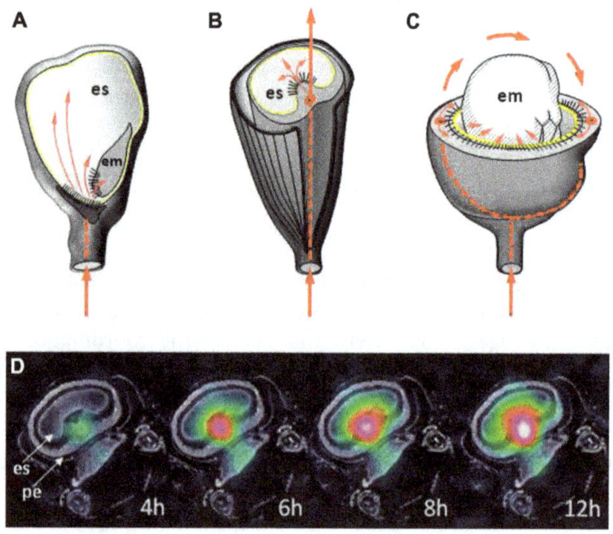

FIGURE 4 | The vascular bundle (thick red lines) and post phloem routes (thin red arrows) of solute transfer in (A) maize, (B) barley, and (C) pea. The yellow line indicates the outer surface of the endosperm and the black stripes indicate the maternal–filial interface. **(D)** The monitoring of sucrose allocation (indicated by color code) resulting from a 12-hour period of feeding with 13C sucrose to the stem at the onset of seed filling stage in barley. The time elapsed since the beginning of the feeding is shown. For details see Melkus et al. (2011). em, embryo; es, endosperm; pe, pericarp.

et al., 2008; Melkus et al., 2011). A characteristic feature of tissues adjacent to the nutrient transport route is the presence of multiple symplastic junctions, large intercellular spaces and cell wall invaginations. The cells of the nucellar projection are extended toward the endosperm, thereby directing the flow of nutrients into the seed. With the exception of the crease region, a thick cuticular layer borders the pericarp and encloses the whole endosperm. In rice, two routes have been identified for nutrients to reach the developing grain: one is analogous to the nucellar projection, while the second passes through the nucellar epidermis (Oparka and Gates, 1981; Krishnan and Dayanandan, 2003).

The structure of the seed impedes the direct visualization of the site of the interaction between maternal and filial tissue. Various dyes and fluorescence- or isotope-labeled substances have been employed to follow nutrient (mainly sucrose) transport (Fisher and Cash-Clark, 2000; Stadler et al., 2005a,b). However, this experimental approach has the major disadvantage of being destructive. Invasive methods inevitably risk inducing artifacts with respect to both metabolite distribution and enzymatic activity. Non-invasive technologies, in the form of biosensors or imaging platforms like Foerster resonance energy transfer (FRET), Positron Emissions Tomographie (PET), and nuclear magnetic resonance (NMR) provide potentially superior alternatives (Frommer et al., 2009; Jahnke et al., 2009; Borisjuk et al., 2012). Real-time information on signaling and metabolite levels with subcellular granularity can be obtained *in vivo* with the help of genetically encoded FRET nanosensors (Frommer et al., 2009). PET does appear to be an appropriate platform for *in planta* analysis (Jahnke et al., 2009). When ^{11}C is the target isotope, its spatial resolution of 1.4 mm (Phelps, 2004) suits it for the study of long distance translocation. NMR – and especially ^{13}C NMR – is less sensitive than PET, but it delivers a fivefold higher level of in-plane resolution than PET, and can be used for real time monitoring (Melkus et al., 2011). The dynamic NMR-based imaging of sucrose in barley seed was integrated with flux balance analysis (FBA), which operated with more than 250 biochemical and transport reactions occurring in the cytosol, mitochondrium, plastid, and extracellular space. This approach has helped to unravel the complex biochemical processes affecting sucrose distribution in the grain (Melkus et al., 2011; Rolletschek et al., 2011).

DELIVERING NUTRIENTS ACROSS THE MATERNAL–FILIAL INTERFACE

Some experimental evidence has been obtained to support the view that the delivery of metabolites to the embryo bypasses the endosperm (Yeung and Meinke, 1993; Weijers et al., 2003; Stadler et al., 2005a; Morley-Smith et al., 2008; Ungru et al., 2008; Pignocchi et al., 2009). However, other data suggest that the process is, in fact, mediated by the endosperm, largely because compromised endosperm development is so often associated with aberrant embryo growth (Chaudhury et al., 2001; Choi et al., 2002; Garcia et al., 2005; Ingouff et al., 2006). In either case, the intergenerational transfer of materials occurs via the apoplastic space. The specialized cellular structures developed at the tissue margins coordinate nutrient delivery from the seed coat into the seed itself. Transfer cells develop invaginated cell walls, thereby increasing the surface area of their plasma membrane and hence their capacity to transport nutrients (Andriunas et al., 2013). Nutrient transporters such as sucrose transporter 1 (Weber et al., 2005; Melkus et al., 2009) and amino acid permease 1 (AAP1; Tegeder, 2014) are typically present in both maternal and filial cells. Often they appear as tissue-specific isoforms: examples are the tonoplast intristic proteins (TIPs) for water (Gattolin et al., 2010) and Siliques Are Red 1 (SIAR1) for amino acids (Ladwig et al., 2012); some can change from efflux to influx mode in response to metabolic signals (Ladwig et al., 2012). The transfer cells positioned on either side of the apoplast (Zhang et al., 2007) act as the gateway for nutrient flow, as demonstrated *in vivo* by NMR in the barley caryopsis (**Figure 4D**; Melkus et al., 2011; Rolletschek et al., 2011).

The maternally located efflux transfer cells are responsible for the release of nutrients into the apoplast, and form cell wall ingrowths which direct the flow toward the seed (Stadler et al., 2005b; Zhang et al., 2007). The plasma membranes in these cells are enriched with respect to aquaporins, membrane transporters, and channels for sugars, amino acids and peptides, inorganic ions, and other compounds (Zhang et al., 2007; Thiel et al., 2008; Bihmidine et al., 2013). These efflux cells, like the cells of the barley and wheat nucellar projections, typically undergo programmed cell death (PCD; Zhou et al., 2009; Radchuk et al., 2011), which contributes to nutrient transfer to the filial tissue. In the maize seed coat placenta–chalazal region, PCD is coordinated with endosperm cellularization and is completed prior to the beginning of the storage phase. In this way, PCD functions as an adaptive process to facilitate the passage of solutes (Kladnik et al., 2004). In barley, the extensive vacuolization of cells in the nucellar projection allows for the transient accumulation of sucrose, which is released together with the complete cell contents to the apoplast after cell disintegration. A defective nucellar projection compromises nutrient flow into the endosperm, resulting in a reduction in final grain size (Radchuk et al., 2006; Melkus et al., 2011; Yin and Xue, 2012). Although important for the seed's fate, the identity, and mechanics of efflux constituents and transporters are only poorly understood (Braun, 2012; Patrick et al., 2013).

The influx transfer cells in monocotyledonous species lie on the surface of the endosperm, directly opposite the maternal unloading site (Thiel et al., 2008; Monjardino et al., 2013; Lopato et al., 2014). The development and function of these transfer cells have been comprehensively and recently reviewed (Lopato et al., 2014; Thiel, 2014). In dicotyledonous species, the transfer cells usually face the seed coat (Borisjuk et al., 2002; Offler et al., 2003; Olsen, 2004). A delay in the *trans*-differentiation of the embryonic epidermal cells to form transfer cells in the pea mutant *E2748* has a negative impact on embryo growth and seed viability (Borisjuk et al., 2002). Maize grains defective for the formation of basal endosperm transfer cells exhibit a shrunken kernel phenotype, as exhibited in the mutants *reduced grain filling 1* (Maitz et al., 2000), *globby1* (Costa et al., 2003), *baseless1* (Gutiérrez-Marcos et al., 2006), *empty pericarp 4* (Gutiérrez-Marcos et al., 2007), and *miniature1* (Kang et al., 2009).

The coordinated differentiation of opposing transfer cells requires a functional interaction between them, so presumably relies on an effective signaling mechanism. How this interaction operates is unclear, but a possible sequence of events has been suggested by Weber et al. (2005) and Andriunas et al. (2013). In the dicotyledonous seed, the expanding cotyledon makes contact with the seed coat, after which the innermost thin-walled parenchyma cells are gradually crushed (Offler et al., 1989; Harrington et al., 1997). The stress, akin to wounding, may induce an ethylene burst (Harrington et al., 1997; Zhou et al., 2010). In response, a secondary ethylene burst in the adjacent embryo cells could be mediated by the auto-regulated expression of 1-aminocyclopropane-1-carboxylic (ACC) synthase (Chang et al., 2008). The process initiates the *trans*-differentiation of epidermal cells into transfer cells (Zhou et al., 2010). Crushing of the seed coat is also coupled with a decrease in the activity of extracellular seed coat-specific invertase (Weber et al., 1996a,b), which leads to a local reduction in the level of intracellular glucose (Borisjuk et al., 1998). The lowered glucose level, sensed via a hexokinase-dependent pathway, removes the glucose-induced repression of ethylene-insensitive 3 (EIN3) and triggers an ethylene-signaling cascade, driving transfer cell differentiation (Dibley et al., 2009; Andriunas et al., 2011). As shown in both *in vitro* and *in vivo* experiments, transfer cell formation across a wide range of plant species involves an interaction between phytohormones, sugar, and reactive oxygen species (Dibley et al., 2009; Forestan et al., 2010; Zhou et al., 2010; Andriunas et al., 2012; Xia et al., 2012). The signals and signaling pathways responsible for the induction of transfer cell formation may be conserved across the monocotyledon/dicotyledon divide (Andriunas et al., 2013). When the promoter of the maize transfer cell-specific transcription factor *ZmMRP1* (Gómez et al., 2009) was fused to a *GUS* reporter gene and inserted into maize, *A. thaliana*, tobacco, and barley, GUS activity could be identified in regions of active transport between source and sink tissues in each of these species (Barrero et al., 2009), supporting the idea that the processes involved in transfer cell differentiation are similar across a diversity of plant species, and that differentiation is initiated by conserved induction signals.

COMMUNICATING BETWEEN ADJACENT SEED COMPARTMENTS

Coordination of seed development clearly requires communication between seed compartments, and in particular a level of feedback between the seed coat and the endosperm/embryo. Transporters localized at the embryo surface seem to be regulated by the metabolite concentrations present in the seed apoplast, but it is unclear how these transporters contribute to coordinating carbon partitioning between the maternal and filial tissues of the seed. For example, storage protein synthesis in the *A. thaliana* embryo and final seed weight depend on nitrogen availability and are mediated by AAP1, which is expressed in both the embryo and the seed coat. In both the seed coat and the endosperm of the *aap1* loss-of-function mutant, amino acid levels are higher than in the wild type, whereas in the embryo, the content of storage proteins and carbohydrate is lower (Sanders et al., 2009). Similarly, phloem amino acid concentrations regulate nitrogen loading into the oilseed rape seed (Lohaus and Moellers, 2000; Tilsner et al., 2005; Tegeder, 2014).

Nutrient release from the seed coat needs to be precisely tuned via a fast and sensitive mechanism such as, for example, cell turgor, which directly depends on the activity of vacuolar sucrose transporters (Walker et al., 2000). A turgor-homeostatic mechanism in the seed coat could sense a loss of solute from the seed apoplast and then could act to balance this by adjusting efflux activity (Patrick and Offler, 2001). The growth of the endosperm, therefore, may trigger a feedback signal to the seed coat, which is then transmitted via a calcium signaling cascade (Zhang et al., 2007) to drive cell elongation. Such a mechanism could allow the endosperm to coordinate aspects of seed development (Borisjuk et al., 2002; Melkus et al., 2009).

Sucrose, hexoses, and amino acids can all provide a regulatory signal (Koch, 2004; Weber et al., 2005; Ruan et al., 2010). Sugar responsiveness is a prominent feature of genes contributing to the sink strength of developing organs, and provides an important mechanism for sink adjustment to source delivery (Xiong et al., 2013b). A number of genes involved in sucrose metabolism are up-regulated by sugars (Smidansky et al., 2002; Kang et al., 2009). As a result, the strongest sink is the one most efficiently up-regulated by the supply of assimilate (Bihmidine et al., 2013). A balancing via down-regulation is also feasible (Kang et al., 2009). An invertase prominent in regulating sucrose unloading (Cheng et al., 1996; Weber et al., 2005; Chourey et al., 2011) has been proposed to enhance sugar signaling in the context of establishing assimilate sinks (Weber et al., 1996a, 2005; Ruan et al., 2010; Aoki et al., 2012). The expression of the myb-like transcription factor *ZmMRP-1*, a key regulator of transfer cell differentiation (Gómez et al., 2009), is modulated by various carbohydrates, with glucose being the most effective inducer (Barrero et al., 2009). ZmMRP1 transcriptionally activates a number of transfer cell-specific genes in the maize endosperm (Gómez et al., 2002); one of these is *Meg1*, which encodes a small cysteine-rich peptide localizing to the plasma membrane of differentiating endosperm transfer cells, where it regulates the expression of *cell wall invertase 2* (Costa et al., 2012). The strong maternal influence over placental-like functions is conferred by genomic imprinting, which has been attributed to maternal–filial co-adaptation (Wolf and Hager, 2006; Gehring et al., 2009). *Meg1* is one of more than a hundred imprinted genes active in the endosperm (Raissig et al., 2011; Waters et al., 2011; Wolff et al., 2011; Zhang et al., 2011), and is the first to have been identified as having a role in regulating the flow of nutrients to the embryo (Costa et al., 2012). *Meg1* also acts in tripartite (seed coat-endosperm-embryo) interaction and regulates maternal nutrient uptake, sucrose partitioning, and seed weight.

MAINTAINING A LEVEL OF CONTROL OVER SEED SIZE

The developing maternal tissue has an effect on endosperm filling and thus also final seed size. Several genes associated with seed size in *A. thaliana* are expressed in the seed coat (Haughn and Chaudhury, 2005; Roszak and Köhler, 2011). Among them are the transcription factors *ARF2/MNT* [which restricts seed size by suppressing cell proliferation in the integuments (Schruff et al., 2006; Li et al., 2008)], *AP2* (Jofuku et al., 2005; Ohto et al., 2005, 2009), *TTG2* and *EOD3/CYP78A6*, which control cell expansion in the integuments (Garcia et al., 2005; Ohto et al., 2009; Fang et al., 2012) and *KLUH/CYTOCHROME P450 78A5*, which stimulates cell proliferation in the endothelium. The up-regulation of *KLUH* increases seed size, produces larger seedlings and increases seed oil content (Adamski et al., 2009). *NARS1* and *NARS2* are expressed in the outer integument, acting redundantly to regulate seed shape and embryogenesis (Kunieda et al., 2008); the seed of the *nars1 nars2* double mutant are abnormally shaped.

At least 400 quantitative trait loci (QTL) related to grain size have been identified in rice and candidate genes have been identified for some of these (Huang et al., 2013). *Dwarf1* is strongly transcribed in the early developing pericarp and only weakly in the endosperm (Izawa et al., 2010). This gene encodes the α subunit of the heterotrimeric G protein (Ashikari et al., 1999; Fujisawa et al., 1999), which is suspected of controlling cell number, since its loss-of-function mutant displays a ubiquitous reduction in cell number (Izawa et al., 2010). A second candidate gene, *gif1*, encodes a cell wall invertase required for carbon partitioning during early grain filling (Wang et al., 2008). Its transcript is only detectable in the (maternal) vascular tissue, suggesting that its role is associated with sucrose unloading (Wang et al., 2008). The expansion of parenchymatous cell layers seen in a faba bean accession (large- versus small-seeded) may reflect the activity of cell wall invertase 1 (Weber et al., 1996a).

Several factors involved in ubiquitin-related activity have been shown to influence seed size (Li and Li, 2014). In rice, a ubiquitous RING-type protein displaying E3 ubiquitin ligase activity (encoded by *GW2*) negatively regulates grain size by restricting cell division. Its loss-of-function mutant forms an enlarged spikelet hull, which allows for a greater contact area between the endosperm and the seed coat (Song et al., 2007). An uncharacterized protein, encoded by a candidate gene for a QTL for seed width mapping to chromosome 5, interacts with polyubiquitin, and acts to limit grain size, possibly by its involvement in the ubiquitin-proteasome pathway (Weng et al., 2008). However, because ubiquitin-related genes are so widely expressed – including within the endosperm – it is unclear whether their function is exclusively under maternal control. The *A. thaliana* gene *DA2* is a homolog of *GW2*; it acts in the maternal tissue to restrict the growth of the seed. The *da1* mutant produced larger and heavier seeds then wild type ("da" means "large" in Chinese; Li et al., 2008). The growth-restricting factor DA1 is an ubiquitin receptor which determines final seed size by restricting the period of integument cell proliferation (Li et al., 2008; Xia et al., 2013). The gene underlying a major grain length QTL in rice encodes a putative phosphatase 2A-type protein harboring a Kelch-like repeat domain. Its effect is manifested by inducing a higher cell density on the outer surface of the glumes and the ovary (Zhang et al., 2012).

The influence of the maternal tissue on caryopsis size has been well documented in cereals, although quite how this is achieved at the molecular and metabolic level remains unresolved (Huang et al., 2013). The extensive synteny and conserved gene structure among cereals has allowed much of the knowledge gained from rice to be exploited in crops such as maize, wheat, and barley. In particular, orthologs of *GW2* have been identified in both wheat (Su et al., 2011) and maize (Li et al., 2010).

SMALL PROVISIONS FOR GOOD REASON

At an early developmental stage, the seed coat in dicots, and the pericarp in monocots accumulate a significant amount of starch (**Figure 5**). In *A. thaliana*, oilseed rape, and pea, starch accumulation occurs in the cells of the outer integument during the growth phase (Abirached-Darmency et al., 2005; Haughn and Chaudhury, 2005; Borisjuk et al., 2013). In the *A. thaliana* seed coat, the most abundant transcript level of the genes encoding starch synthesis enzymes are observed during the pre-globular and globular developmental stages (Khan et al., 2014). Starch granules in the chalazal part of the seed coat are smaller and less abundant than in the distal part, a trait which is mirrored by a differential starch synthesizing enzyme transcript profile (Khan et al., 2014). In *M. truncatula*, an abundance of starch granules accumulates transiently in the seed coat from the embryo heart stage all the way up to mid-maturation (Verdier et al., 2013). Based on the behavior of pea seeds lacking ADP-Glc pyrophosphorylase activity, it appears that transiently produced starch is required for sink acquisition, maximal embryo growth and final seed size (Rochat et al., 1995; Vigeolas et al., 2004). The role of this transient starch is not completely clear. Young maternal tissue may perhaps have evoloved a starch storage function to ensure a sufficient assimilate sink strength, which becomes redundant once the growing seed's own sink is established (Radchuk et al., 2009). Accumulating assimilate in the form of starch is energetically highly efficient (Schwender, 2008), so it is unsurprising that the transient maternal sink accumulates starch rather than protein or lipid.

Transient starch is utilized for the growth and development of the maternal tissue (Xiong et al., 2013a), for example, by providing a source of carbohydrate to reinforce the cell wall with

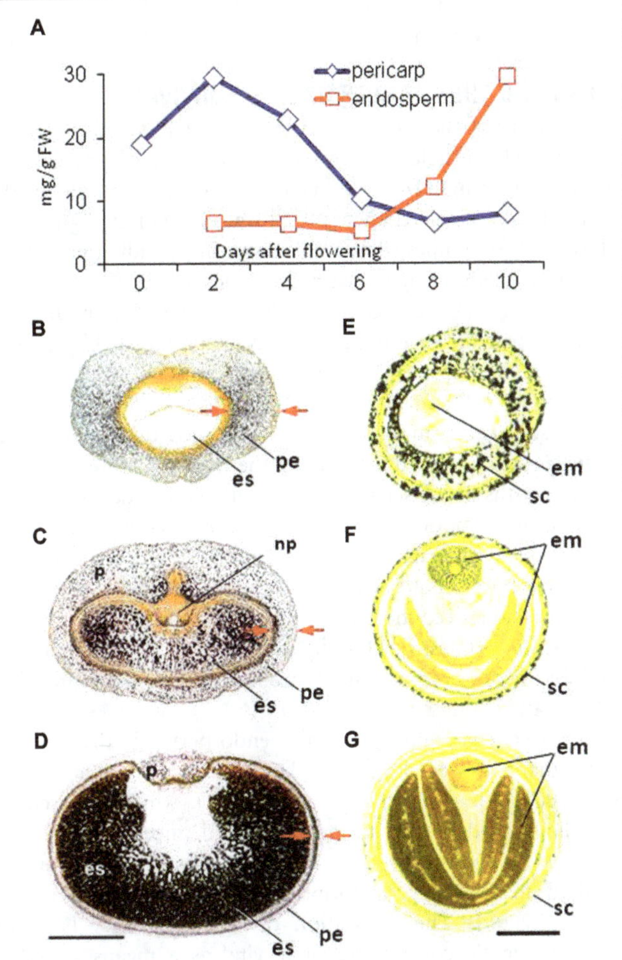

FIGURE 5 | Transient starch accumulation in the maternal tissues of the barley caryopsis and the oilseed rape seed. (A) Transient starch accumulation (visualized in black by iodine staining) in the pericarp of the developing barley caryopsis at **(B)** 2, **(C)** 8, and **(D)** 14 days after fertilization. The pericarp is indicated by arrows. For details see Radchuk et al. (2009). **(E–G)** Starch in the seed coat of young and maturing seeds of oilseed rape. For details see Borisjuk et al. (2013). em, embryo; es, endosperm; np, nucellar projection; pe, pericarp; sc, seed coat.

pectinaceous mucilage (Khan et al., 2014). It may also help support the growth of the endosperm and embryo (Radchuk et al., 2009; Verdier et al., 2013). In *Zingiberales* and *Caryophyllales* species, the nucellus, rather than degenerating, accumulates large amounts of starch (López-Fernández and Maldonado, 2013), forming a so-called perisperm which persists until the seed is mature. In quinoa and the grain amaranth, the perisperm consists of dead, thin-walled cells completely filled with starch granules, producing a structure which strongly resembles the cereal endosperm (López-Fernández and Maldonado, 2013).

The synthesis of transient starch is performed by a similar set of genes as is used in the leaf (Radchuk et al., 2009). Its mode of breakdown in the cereal pericarp is distinct between living and dying cells. In living chlorenchyma cells, it most likely occurs via a pathway similar to that followed in the photosynthesizing leaf; this involves the phosphorylation of the starch granule surface, making it accessible for the degrading enzyme β-amylase. Plastid-localized BAM5, BAM6, and BAM7 β-amylases are thought to produce maltose, acting either at the granule surface or on linear malto-oligosaccharides. The action of iso-amylase 3 on the granule, on the other hand, releases soluble malto-oligosaccharides, which can be metabolized by disproportionating enzyme 1 (DPE1), liberating glucose, and larger malto-oligosaccharides for continued degradation. After its transport to the cytosol, maltose can be further converted to glucose by DPE2. In dying cells, the mode of starch breakdown resembles that occuring in the germinating grain, which requires a combination of α- and β-amylase activity. AMY1 is active in the pericarp and nucellar tissue of a developing grain and is responsible for most of the α-amylase activity seen in germinating grains (Radchuk et al., 2009). A plausible starch degradation pathway in dying pericarp cells involves the joint activity of AMY1 and AMY4. Linear malto-oligosaccharides released by the action of these enzymes should provide an appropriate substrate for β-amylase 2. The molecular identity of the gene/enzyme responsible for the conversion of maltose to glucose remains to be identified.

Seed coat tissues may also serve as a transient depot for proteins and microelements. In faba bean, the storage protein legumin B is deposited in the seed coat at mid-embryogenesis (Panitz et al., 1995), while the wheat pericarp and nucellus accumulate significant quantities of calcium, copper, iron, molybdenum, magnesium, manganese, and phosphorus (Wu et al., 2013; Xiong et al., 2013a). The physiological significance of this accumulation has not yet been elucidated.

DYING QUIETLY

Because cell division in the maternal tissue ceases soon after fertilization, further enlargement only occurs through cell expansion (Radchuk et al., 2011; Figueiredo and Köhler, 2014). The rapid growth of the endosperm and embryo requires the triggering of PCD to remove maternal cells in order to allow seed expansion. In both dicotyledonous and monocotyledonous seeds, the early stages of endosperm expansion are at the expense of the nucellus (Domínguez et al., 2001; Greenwood et al., 2005; Lombardi et al., 2007; Zhou et al., 2009; Radchuk et al., 2011). After the cereal nucellus has degenerated, the next tissue to undergo PCD is the pericarp, starting from its innermost cell layer (Radchuk et al., 2011). The chlorenchyma is retained in a viable, functional state almost up to physiological maturity. In *A. thaliana* and the castor oil plant, PCD occurs first in the endosperm and later in the integuments (Greenwood et al., 2005; Nakaune et al., 2005).

As in animal cells, the molecular basis for PCD in plants relies on caspase-like activity. Although no caspase homologs have been identified in plants, plants do harbor proteases sharing some similarity to animal caspases. Caspase-1-like, caspase-3-like, and caspase-6-like activities have all been detected in the degenerating chayote (*Sechium edule*) nucellus (Lombardi et al., 2007). Vacuolar processing enzyme (VPE, also referred to as legumain) has caspase-1-like activities (Hara-Nishimura et al., 2005), while phytaspase possesses caspase-6-like activity (Chichkova et al., 2010). A seed specific δVPE produced by *A. thaliana* is present in the two inner cell layers of the seed coat. In a mutant defective for δ*VPE*, PCD is delayed and the seed coat remains thick throughout development. In contrast, in the wild type, the two layers undergo PCD very early during seed development, reducing their thickness by more than 50% (Nakaune et al., 2005). The spatial and temporal patterns of *HvVPE4* transcription coincides with the onset of PCD in the barley pericarp (Radchuk et al., 2011). The product of *HvVPE2a* (also called nucellain), together with those of *HvVPE2b*, *HvVPE2c*, and *HvVPE2d*, are important for the timely degeneration of the nucellus and the nucellar projection (Linnestad et al., 1998; Radchuk et al., 2011). HvVPE2b possesses caspase-1 like activity (Julián et al., 2013). The supposed role of the barley VPEs in grain development still requires experimental confirmation. Novel technologies (Tsiatsiani et al., 2012) might help to identify the target(s) of VPE.

A large number of proteases are present in degenerating maternal tissue, some of which may be active components of PCD (Sreenivasulu et al., 2006; Thiel et al., 2008). The ricinosome, a castor oil plant specific organelle, contains a large quantity of a pro-cysteine endopeptidase (CysEP), which serves to disintegrate the nucellar cells, leaving crushed and folded cell wall residues in the apoplastic space (Greenwood et al., 2005). Nuclear DNA fragmentation has been detected in the nucellus of the castor oil plant (Greenwood et al., 2005), chayote (Lombardi et al., 2007), barley (Linnestad et al., 1998; Radchuk et al., 2011) and wheat (Domínguez et al., 2001), as well as in the *A. thaliana* endothelium (Nakaune et al., 2005).

The transcription factor *OsMADS29* has been described as a regulator of PCD in the rice nucellus (Yang et al., 2012; Yin and Xue, 2012). *OsMADS29* transcripts are concentrated in the nucellus and the nucellar projection (Yin and Xue, 2012), but are also detectable in the inner layers of the pericarp, in other maternal seed tissues and in the embryo (Yang et al., 2012; Nayar et al., 2013). PCD is slowed in an *OsMADS29* knockdown line, leading to a reduction in starch accumulation in the endosperm and the production of shrunken or aborted grain (Yin and Xue, 2012). An alternative role for this transcription factor – in relation to hormone homeostasis, plastid biogenesis and starch synthesis – has been suggested by Nayar et al. (2013). OsMADS29 is thought to bind to the promoters of cysteine protease genes (Yin and Xue, 2012). The down-regulation of *OsMADS29* suppresses the transcription of *VPE* genes in the grain (Yang et al., 2012). As yet, however, there is no consensus regarding the localization of either its transcript or its gene product, its primary target(s) or its likely function during grain development. The transcription of its barley homolog, *HvMADS29*, is restricted to the nucellus and the nucellar projection and coincides with that of *Jekyll* and *HvVPE2a*. The promoter regions of *Jekyll*, *HvVPE2a*, *HvVPE2b*, *and HvVPE2d* contain the same CArG-like regions recognized by OsMADS29, which implies that they are all transcriptionally regulated by HvMADS29. Jekyll is a key player in grain development (Radchuk et al., 2006), and is also active in nurse tissues, where it mediates the gametophyte-sporophyte interaction in both the gynoecium and the androecium (Radchuk et al., 2012). Its down-regulation slows PCD in the nucellus and nucellar projection, although the mechanistic basis of this effect is unclear. The *Jekyll* product, which is unique to *Pooideae*, is a small, cysteine-rich protein deposited within the intracellular membranes (Radchuk et al., 2012). It has no significant similarity to other proteases or any protein of known function, and has no *in vitro* protease activity.

IMPROVING THE CHANCES OF SEED SURVIVAL

The capacity of the seed coat to limit water loss and to protect against mechanical damage persists beyond seed maturation. The mechanical strength of the seed coat is achieved primarily by the accumulation of sclerenchyma. Cell walls containing lignin, cellulose, and sometimes silica are effective in providing protection against attacks by fungi, insects, and herbivores (Lanning and Eleuterius, 1992). Seed coat constituents (e.g., PAs in *Brassica napus*) impair the digestibility and are being targeted by genetic and molecular approaches to improve nutritional value of seeds (Auger et al., 2010; Yu, 2013).

The seed coat can also contribute to seed dispersion: some species produce winged seed, where the wing structure is formed by outgrowths of the seed coat, while others produce hairy seed. Some of the compounds found on the seed coat have industrial applications. Under natural conditions, the cotton boll (its fibers are almost pure cellulose) will tend to increase the dispersion of the seeds, and indeed the use of cotton for fabric is known to date to prehistoric times. Genetically modified cotton has increased yield, but further improvements are needed (Molina et al., 2008; Kathage and Qaim, 2012; Ruan, 2013).

The seed coat, along with the endosperm, is the primary determinant of seed dormancy (Debeaujon et al., 2000; Bethke et al., 2007; Iglesias-Fernández et al., 2007), which represents a physiological adaptation to environmental uncertainty (Bewley et al., 2013). Dormancy dictates the environmental conditions required to trigger germination (Finch-Savage and Leubner-Metzger, 2006). The acquisition of seed dormancy is under complex genetic control, but it is vital as a means of assuring the survival of natural plant populations (Penfield and King, 2009; Graeber et al., 2012). Seed dormancy can be a critical trait for breeders (Fuller and Allaby, 2009; Gao et al., 2013; Smýkal et al., 2014) and would represent a prime target for biotechnology intervention, provided that its regulation were better understood (Flintham, 2000; Graeber et al., 2012).

A combination of external factors, such as light, temperature, water, and chemicals play an important role in breaking seed dormancy. Many small seeded species, the seeds of which contain little stored energy, are triggered to germinate by exposure to light. In some cases the minimum duration of exposure required can be measured in milliseconds (Quail, 1997). Phytochromes represent the main vehicle for seeds to sense light (Smith, 2000; Mochizuki et al., 2009). The linkage between the light-regulated trigger and the hormone-mediated induction of germination in *A. thaliana* has been been explored by Cho et al. (2012). A rapid adaptation to light fluctuation can represent a key competitive advantage in natural plant populations. A particularly striking example of how the seed coat contributes to seed survival is provided by species that have adapted to bushfires. Australian Banksia species are destroyed when burnt, but the fire stimulates the opening of their seed-bearing follicles and promotes the germination of buried seed. The smoke from bushfires contains as many as 5000 different compounds (Nelson et al., 2012), including a number of substances proven to stimulate germination (Nelson et al.,

2012). While volatiles such as ethylene and nitric oxide are not very persistent in the soil, other smoke compounds (in particular, karrikins, and cyanohydrins) are quite stable in the upper layers of the soil where dormant seeds tend to be found. The seed coat of a mature *Banksia sp.* seed is permeable to these molecules. The seeds of hundreds of plant species have been tested with many smoke compounds, and the mode of action of some has been elucidated in *A. thaliana* (Nelson et al., 2010; Flematti et al., 2013). It has even been suggested that some of these compounds have been co-opted through evolution as signals for germination (Nelson et al., 2012; Challis et al., 2013).

CONCLUDING REMARKS

During seed development, the seed coat (dicots) and pericarp (monocots) serve a number of functions, most of which have evolved to protect the seed and to promote the development of the embryo and the endosperm within it. The architecture, chemical composition and metabolism of the seed coat work together to ensure effective responses to both biotic and abiotic factors. Nutrients passing from the mother plant to the developing embryo and endosperm must traverse the seed coat, which therefore controls seed development and seed filling. Specialized tissues have developed in a coordinated fashion on either side of the apoplast to direct and facilitate nutrient flow toward the growing embryo and endosperm. The seed coat and the endosperm act together to determine final seed size. The fine-tuning of nutrient flow from the seed coat to their endosperm and embryo is controlled at the genetic, epigenetic, and metabolic level, but how the interplay is achieved *in vivo* remains to be clarified.

Photosynthesis in the seed coat provides oxygen to the hypoxic regions deep within the developing seed. The overwhelming proportion of the nutrition supplied to the seed is provided by the leaf of the mother plant, the delivery of which is tied to the circadian rhythm. Adapting the seed's metabolism to this uneven flow of nutrition is facilitated by the seed's ability to sense light. In the course of seed development, the maternal tissues undergo PCD, thereby providing both the space and nutrients for the growth of the filial tissue. Finally, an outer seed envelope is built which is important for providing protection for the mature seed, enabling the establishment of dormancy and aiding in seed dispersal.

ACKNOWLEDGMENTS

We thank Prof. T. Altmann for discussion and the German Federal Ministry of Education and Research (Deutsche Forschungsgemeinschaft, BO-1917/4-1), the German Plant Phenotyping Network (DPPN) for funding.

REFERENCES

Abirached-Darmency, M., Abdel-gawwad, M. R., Conejero, G., Verdeil, J. L., and Thompson, R. (2005). In situ expression of two storage protein genes in relation to histo-differentiation at mid-embryogenesis in *Medicago truncatula* and *Pisum sativum* seeds. *J. Exp. Bot.* 56, 2019–2028. doi: 10.1093/jxb/eri200

Adamski, N. M., Anastasiou, E., Eriksson, S., O'Neill, C. M., and Lenhard, M. (2009). Local maternal control of seed size by KLUH/CYP78A5-dependent growth signaling. *Proc. Natl. Acad. Sci. U.S.A.* 106, 20115–20120. doi: 10.1073/pnas.0907024106

Albert, S., Delseny, M., and Devic, M. (1997). BANYULS, a novel negative regulator of flavonoid biosynthesis in the *Arabidopsis* seed coat. *Plant J.* 11, 289–299. doi: 10.1046/j.1365-313X.1997.11020289.x

Andriunas, F. A., Zhang, H. M., Weber, H., McCurdy, D. W., Offler, C. E., and Patrick, J. W. (2011). Glucose and ethylene signalling pathways converge to regulate trans-differentiation of epidermal transfer cells in *Vicia* narbonensis cotyledons. *Plant J.* 68, 987–998. doi: 10.1111/j.1365-313X.2011.04749.x

Andriunas, F. A., Zhang, H. M., Xia, X., Offler, C. E., McCurdy, D. W., and Patrick, J. W. (2012). Reactive oxygen species form part of a regulatory pathway initiating trans-differentiation of epidermal transfer cells in *Vicia faba* cotyledons. *J. Exp. Bot.* 63, 3617–3630. doi: 10.1093/jxb/ers029

Andriunas, F. A., Zhang, H. M., Xia, X., Patrick, J. W., and Offler, C. E. (2013). Intersection of transfer cells with phloem biology–broad evolutionary trends, function, and induction. *Front. Plant Sci.* 4:221. doi: 10.3389/fpls.2013.00221

Aoki, N., Hirose, T., and Furbank, R.T. (2012). "Sucrose transport in higher plants: from source to sink," in *Photosynthesis: Plastid Biology, Energy Conversion, and Carbon Assimilation*, eds J. J. Eaton-Rye, B. C. Tripathy, and T. D. Sharkey (Dordrecht: Springer), 703–729.

Araus, J. L., Brown, H. R., Febrero, A., and Serret, M. D. (1993). Ear photosynthesis, carbon isotope discrimination and the contribution of respiratory CO_2 to differences in grain mass in durum wheat. *Plant Cell Environ.* 16, 383–392. doi: 10.1111/j.1365-3040.1993.tb00884.x

Arsovski, A. A., Haughn, G. W., and Western, T. L. (2010). Seed coat mucilage cells of *Arabidopsis thaliana* as a model for plant cell wall research. *Plant Signal. Behav.* 5, 796–801. doi: 10.4161/psb.5.7.11773

Ashikari, M., Wu, J., Yano, M., Sasaki, T., and Yoshimura, A. (1999). Rice gibberellin-insensive dwarf mutant gene Dwarf 1 encodes the α-subunit of GTP-binding protein. *Proc. Natl. Acad. Sci. U.S.A.* 96, 10284–10289. doi: 10.1073/pnas.96.18.10284

Auger, B., Marnet, N., Gautier, V., Maia-Grondard, A., Leprince, F., Renard, M., et al. (2010). A detailed survey of seed coat flavonoids in developing seeds of *Brassica napus* L. *J. Agric. Food Chem.* 26, 6246–6256. doi: 10.1021/jf903619v

Bailly, C., and Kranner, I. (2011). Analyses of reactive oxygen species and antioxidants in relation to seed longevity and germination. *Methods Mol. Biol.* 773, 343–367. doi: 10.1007/978-1-61779-231-1_20

Barrero, C., Royo, J., Grijota-Martinez, C., Faye, C., Paul, W., Sanz, S., et al. (2009). The promoter of ZmMRP-1, a maize transfer cell-specific transcriptional activator, is induced at solute exchange surfaces and responds to transport demands. *Planta* 229, 235–247. doi: 10.1007/s00425-008-0823-0

Bassoi, M. C., and Flintham, J. (2005). Relationship between grain colour and preharvest sprouting-resistance in wheat. *Pesqui. Agropecu. Bras.* 40, 981–988. doi: 10.1590/S0100-204X2005001000006

Beresniewicz, M. M., Taylor, A. G., Goffinet, M. C., and Terhune, B. T. (1995). Characterization and location of a semipermeable layer in seed coats of leek and onion (*Liliaceae*), tomato and pepper (*Solanaceae*). *Seed Sci. Technol.* 23, 123–134.

Bernhardt, C., Zhao, M., Gonzalez, A., Lloyd, A., and Schiefelbein, J. (2005). The bHLH genes GL3 and EGL3 participate in an intercellular regulatory circuit that controls cell patterning in the *Arabidopsis* root epidermis. *Development* 132, 291–298. doi: 10.1242/dev.01565

Bethke, P. C., Libourel, I. G. L., Aoyama, N., Chung, Y. Y., Still, D. W., and Jones, R. L. (2007). The *Arabidopsis* aleurone layer responds to nitric oxide, gibberellin, and abscisic acid and is sufficient and necessary for seed dormancy. *Plant Physiol.* 143, 1173–1188. doi: 10.1104/pp.106.093435

Bewley, J. D., and Black, M. (1994). *Seeds.* New York: Plenum Press. doi: 10.1007/978-1-4899-1002-8

Bewley, J. D., Bradford, K., Hilhorst, H., and Nonogaki, H. (2013). *Seeds: Physiology of Development, Germination and Dormancy.* New York: Springer. doi: 10.1007/978-1-4614-4693-4

Bihmidine, S., Hunter, C. T. 3rd., Johns, C. E., Koch, K. E., and Braun, D. M. (2013). Regulation of assimilate import into sink organs: update on molecular drivers of sink strength. *Front. Plant Sci.* 4:177. doi: 10.3389/fpls.2013.00177

Black, M. H., and Halmer, P. (2006). *The Encyclopedia of Seeds: Science, Technology and Uses.* Wallingford: CABI.

Bohinc, T., Ban, S. G., Ban, D., and Trdan, S. (2012). Glucosinolates in plant protection strategies: a review. *Arch. Biol. Sci.* 64, 821–828. doi: 10.2298/ABS1203821B

Borisjuk, L., Neuberger, T., Schwender, J., Heinzel, N., Sunderhaus, S., Fuchs, J., et al. (2013). Seed architecture shapes embryo metabolism in oilseed rape. *Plant Cell* 25, 1625–1640. doi: 10.1105/tpc.113.111740

Borisjuk, L., Nguyen, T. H., Neuberger, T., Rutten, T., Tschiersch, H., Claus, B., et al. (2005). Gradients of lipid storage, photosynthesis and plastid differentiation in developing soybean seeds. *New Phytol.* 167, 761–776. doi: 10.1111/j.1469-8137.2005.01474.x

Borisjuk, L., and Rolletschek, H. (2009). The oxygen status of the developing seed. *New Phytol.* 182, 17–30. doi: 10.1111/j.1469-8137.2008.02752.x

Borisjuk, L., Rolletschek, H., and Neuberger, T. (2012). Surveying the plant's world by magnetic resonance imaging. *Plant J.* 70, 129–146. doi: 10.1111/j.1365-313X.2012.04927.x

Borisjuk, L., Walenta, S., Weber, H., Mueller-Klieser, W., and Wobus, U. (1998). High-resolution histographical mapping of glucose concentrations in developing cotyledons of *Vicia faba* in relation to mitotic activity and storage processes: glucose as a possible developmental trigger. *Plant J.* 15, 583–591. doi: 10.1046/j.1365-313X.1998.00214.x

Borisjuk, L., Wang, T. L., Rolletschek, H., Wobus, U., and Weber, H. (2002). A pea seed mutant affected in the differentiation of the embryonic epidermis is impaired in embryo growth and seed maturation. *Development* 129, 1595–1607.

Brar, G., and Thies, W. (1977). Contribution of leaves, stem, siliques and seeds to dry-matter accumulation in ripening seeds of rapeseed, *Brassica napus* L. *Z. Pflanzenphysiol.* 82, 1–13. doi: 10.1016/S0044-328X(77)80128-1

Braun, D. M. (2012). SWEET! The pathway is complete. *Science* 335, 173–174. doi: 10.1126/science.1216828

Brooker, N. L., Lagalle, C. D., Zlatanic, A., Javni, I., and Petrovic, Z. (2007). Soy polyol formulations as novel seed treatments for the management of soil-borne diseases of soybean. *Commun. Agric. Appl. Biol. Sci.* 72, 35–43.

Caley, C. Y., Duffus, C. M., and Jeffcoat, B. (1990). Photosynthesis in the pericarp of developing wheat grains. *J. Exp. Bot.* 41, 303–307. doi: 10.1093/jxb/41.3.303

Carvalho, A. O., and Gomes, V. M. (2009). Plant defensins – prospects for the biological functions and biotechnological properties. *Peptides* 30, 1007–1020. doi: 10.1016/j.peptides.2009.01.018

Challis, R. J., Hepworth, J., Mouchel, C., Waites, R., and Leyser, O. (2013). A role for MORE AXILLARY GROWTH1 (MAX1) in evolutionary diversity in strigolactone signaling upstream of MAX2. *Plant Physiol.* 161, 1885–1902. doi: 10.1104/pp.112.211383

Chang, S. H., Lu, L. S., Wang, N. N., and Chang, Y. Y. (2008). Negative feedback regulation of system-1 ethylene production by the tomato 1-aminocyclopropane-1-carboxylate synthase 6 gene promoter. *Plant Sci.* 175, 149–160. doi: 10.1016/j.plantsci.2007.11.004

Chaudhury, A. M., Koltunow, A., Payne, T., Luo, M., Tucker, M. R., Ming, L., et al. (2001). Control of early seed development. *Annu. Rev. Cell Dev. Biol.* 17, 677–699. doi: 10.1146/annurev.cellbio.17.1.677

Cheng, W. H., Taliercio, E. W., and Chourey, P. S. (1996). The miniature1 seed locus of maize encodes a cell wall invertase required for normal development of endosperm and maternal cells in the pedicel. *Plant Cell* 8, 971–983. doi: 10.1105/tpc.8.6.971

Chichkova, N. V., Shaw, J., Galiullina, R. A., Drury, G. E., Tuzhikov, A. I., Kim, S. H., et al. (2010). Phytaspase, a relocalisable cell death promoting plant protease with caspase specificity. *EMBO J.* 29, 1149–1161. doi: 10.1038/emboj.2010.1

Cho, J. N., Ryu, J. Y., Jeong, Y. M., Park, J., Song, J. J., Amasino, R. M., et al. (2012). Control of seed germination by light-induced histone arginine demethylation activity. *Dev. Cell* 22, 736–748. doi: 10.1016/j.devcel.2012.01.024

Choi, Y., Gehring, M., Johnson, L., Hannon, M., Harada, J. J., Goldberg, R. B., et al. (2002). DEMETER, a DNA glycosylase domain protein, is required for endosperm gene imprinting and seed viability in *Arabidopsis*. *Cell* 110, 33–42. doi: 10.1016/S0092-8674(02)00807-3

Chourey, P. S., Li, Q., and Cevallos-Cevallos, J. (2011). Pleiotropy and its limited dissection through a metabolic gene Miniature1 (Mn1) that encodes a cell wall invertase in developing seeds of maize. *Plant Sci.* 184, 45–53. doi: 10.1016/j.plantsci.2011.12.011

Cloetens, P., Mache, R., Schlenker, M., and Lerbs-Mache, S. (2006). Quantitative phase tomography of *Arabidopsis* seeds reveals intercellular void network. *Proc. Natl. Acad. Sci. U.S.A.* 103, 14626–14630. doi: 10.1073/pnas.0603490103

Cochrane, M. P., and Duffus, C. M. (1979). Morphology and ultrastructure of immature cereal grains in relation to transport. *Ann. Bot.* 44, 67–77.

Costa, L. M., Gutierrez-Marcos, J. F., Brutnell, T. P., Greenland, A. J., and Dickinson, H. G. (2003). The globby1-1 (glo1-1) mutation disrupts nuclear and cell division in the developing maize seed causing alterations in endosperm cell fate and tissue differentiation. *Development* 130, 5009–5017. doi: 10.1242/dev.00692

Costa, L. M., Yuan, J., Rouster, J., Paul, W., Dickinson, H., and Gutierrez-Marcos, J. F. (2012). Maternal control of nutrient allocation in plant seeds by genomic imprinting. *Curr. Biol.* 22, 160–165. doi: 10.1016/j.cub.2011.11.059

Dean, G., Cao, Y., Xiang, D., Provart, N. J., Ramsay, L., Ahad, A., et al. (2011). Analysis of gene expression patterns during seed coat development in *Arabidopsis*. *Mol. Plant* 4, 1074–1091. doi: 10.1093/mp/ssr040

Debeaujon, I., Léon-Kloosterziel, K. M., and Koornneef, M. (2000). Influence of the testa on seed dormancy, germination, and longevity in *Arabidopsis*. *Plant Physiol.* 122, 403–414. doi: 10.1104/pp.122.2.403

Dermastia, M., Kladnik, A., Koce, J. D., and Chourey, P. S. (2009). A cellular study of teosinte *Zea mays* subsp. Parviglumis (*Poaceae*) caryopsis development showing several processes conserved in maize. *Am. J. Bot.* 96, 1798–1807. doi: 10.3732/ajb.0900059

Dibley, S. J., Zhou, Y., Andriunas, F. A., Talbot, M. J., Offler, C. E., Patrick, J. W., et al. (2009). Early gene expression programs accompanying trans-differentiation of epidermal cells of *Vicia faba* cotyledons into transfer cells. *New Phytol.* 182, 863–877. doi: 10.1111/j.1469-8137.2009.02822.x

Domínguez, F., Moreno, J., and Cejudo, F. J. (2001). The nucellus degenerates by a process of programmed cell death during the early stages of wheat grain development. *Planta* 213, 352–360. doi: 10.1007/s004250000517

Duffus, C. M., and Rosie, R. (1973). Some enzyme-activities associated with chlorophyll containing layers of immature barley pericarp. *Planta* 114, 219–226. doi: 10.1007/BF00389037

Efe, L., Mustafayev, A. S., and Killi, F. (2010). Agronomic, fiber and seed quality of naturally colored cotton in east Mediterranean region of Turkey. *Pak. J. Bot.* 42, 3865–3873.

Fang, W., Wang, Z., Cui, R., Li, J., and Li, Y. (2012). Maternal control of seed size by EOD3/CYP78A6 in *Arabidopsis thaliana*. *Plant J.* 70, 929–939. doi: 10.1111/j.1365-313X.2012.04907.x

Figueiredo, D. D., and Köhler, C. (2014). Signalling events regulating seed coat development. *Biochem. Soc. Trans.* 42, 358–363. doi: 10.1042/BST20130221

Finch-Savage, W. E., and Leubner-Metzger, G. (2006). Seed dormancy and the control of germination. *New Phytol.* 171, 501–523. doi: 10.1111/j.1469-8137.2006.01787.x

Fisher, D. B., and Cash-Clark, C. E. (2000). Sieve tube unloading and post-phloem transport of fluorescent tracers and proteins injected into sieve tubes via severed aphid stylets. *Plant Physiol.* 123, 125–138. doi: 10.1104/pp.123.1.125

Flematti, G. R., Waters, M. T., Scaffidi, A., Merritt, D. J., Ghisalberti, E. L., Dixon, K. W., et al. (2013). Karrikin and cyanohydrin smoke signals provide clues to new endogenous plant signalling compounds. *Mol. Plant* 6, 29–37. doi: 10.1093/mp/sss132

Flinn, A. M. (1985). "Carbon dioxide fixation in developing seeds," in *The Pea Crop: A Basis for Improvement*, eds P. D. Hebblethwaite, M. C. Heath, and T. C. K. Dawkins (London: Butterworths), 349–358.

Flintham, J. E. (2000). Different genetic components control coat-imposed and embryo-imposed dormancy in wheat. *Seed Sci. Res.* 10, 43–50. doi: 10.1017/S0960258500000052

Forestan, C., Meda, S., and Varotto, S. (2010). ZmPIN1-mediated auxin transport is related to cellular differentiation during maize embryogenesis and endosperm development. *Plant Physiol.* 152, 1373–1390. doi: 10.1104/pp.109.150193

Frommer, W. B., Davidson, M. W., and Campbell, R. E. (2009). Genetically encoded biosensors based on engineered fluorescent proteins. *Chem. Soc. Rev.* 38, 2833–2841. doi: 10.1039/b907749a

Fujisawa, Y., Kato, T., Ohki, S., Ishikawa, A., Kitano, H., Sasaki, T., et al. (1999). Suppression of the heterotrimeric G protein causes abnormal morphology, including dwarfism, in rice. *Proc. Natl. Acad. Sci. U.S.A.* 96, 7575–7580. doi: 10.1073/pnas.96.13.7575

Fuller, D. Q., and Allaby, R. (2009). Seed dispersal and crop domestication: shattering, germination and seasonality in evolution under cultivation. *Ann. Plant Rev.* 38, 238–295. doi: 10.1002/9781444314557.ch7

Gao, X., Hu, C. H., Li, H. Z., Yao, Y. J., Meng, M., Dong, J., et al. (2013). Factors affecting pre-harvest sprouting resistance in wheat (*Triticum aestivum* L.): a review. *J. Anim. Plant Sci.* 23, 556–565.

Garcia, D., Fitz Gerald, J. N., and Berger, F. (2005). Maternal control of integument cell elongation and zygotic control of endosperm growth are coordinated to determine seed size in *Arabidopsis*. *Plant Cell* 17, 52–60. doi: 10.1105/tpc.104.027136

Gattolin, S., Sorieul, M., and Frigerio, L. (2010). Mapping of tonoplast intrinsic proteins in maturing and germinating *Arabidopsis* seeds reveals dual localization of embryonic TIPs to the tonoplast and plasma membrane. *Mol. Plant* 4, 180–189. doi: 10.1093/mp/ssq051

Gehring, M., Bubb, K. L., and Henikoff, S. (2009). Extensive demethylation of repetitive elements during seed development underlies gene imprinting. *Science* 324, 1447–1451. doi: 10.1126/science.1171609

Geisler, M., and Sack, F. D. (2002). Variable timing of developmental progression in the stomatal pathway in *Arabidopsis* cotyledons. *New Phytol.* 153, 469–476. doi: 10.1046/j.0028-646X.2001.00332.x

Gergerich, R. C., and Dolja, V. V. (2006). Introduction to plant viruses, the invisible foe. *Plant Heal. Instruct.* doi: 10.1094/PHI-I-2006-0414-01

Gifford, R. M., and Bremner, P. M. (1981). Accumulation and conversion of sugars by developing wheat grains. 2. Light requirement for kernels cultured in vitro. *Aust. J. Plant Physiol.* 8, 631–640. doi: 10.1071/PP9810631

Gnaiger, E., Méndez, G., and Hand, S. C. (2000). High phosphorylation efficiency and depression of uncoupled respiration in mitochondria under hypoxia. *Proc. Natl. Acad. Sci. U.S.A.* 97, 11080–11085. doi: 10.1073/pnas.97.20.11080

Goffman, F., Ruckle, M., Ohlrogge, J., and Shachar-Hill, Y. (2004). Carbon dioxide concentrations are very high in developing oilseeds. *Plant Physiol. Biochem.* 42, 703–708. doi: 10.1016/j.plaphy.2004.07.003

Golombek, S., Heim, U., Horstmann, C., Wobus, U., and Weber, H. (1998). Phosphoenolpyruvate carboxylase in developing seeds of *Vicia faba* L.: gene expression and metabolic regulation. *Planta* 208, 66–72. doi: 10.1007/s0042500 50535

Gómez, E., Royo, J., Guo, Y., Thompson, R., and Hueros, G. (2002). Establishment of cereal endosperm expression domains: identification and properties of a maize transfer cell-specific transcription factor, ZmMRP-1. *Plant Cell* 14, 599–610. doi: 10.1105/tpc.010365

Gómez, E., Royo, J., Muñiz, L. M., Sellam, O., Paul, W., Gerentes, D., et al. (2009). The maize transcription factor myb-related protein-1 is a key regulator of the differentiation of transfer cells. *Plant Cell* 21, 2022–2035. doi: 10.1105/tpc.108.065409

Gonzalez, A., Zhao, M., Leavitt, J. M., and Lloyd, A. M. (2008). Regulation of the anthocyanin biosynthetic pathway by the TTG1/bHLH/Myb transcriptional complex in *Arabidopsis* seedlings. *Plant J.* 53, 814–827. doi: 10.1111/j.1365-313X.2007.03373.x

Graeber, K., Nakabayashi, K., Miatton, E., Leubner-Metzger, G., and Soppe, W. J. (2012). Molecular mechanisms of seed dormancy. *Plant Cell Environ.* 35, 1769–1786. doi: 10.1111/j.1365-3040.2012.02542.x

Greenway, H., and Gibbs, J. (2003). Mechanisms of anoxia tolerance in plants. II. Energy requirements for maintenance and energy distribution to essential processes. *Funct. Plant Biol.* 30, 999–1036. doi: 10.1071/PP98096

Greenwood, J. S., Helm, M., and Gietl, C. (2005). Ricinosomes and endosperm transfer cell structure in programmed cell death of the nucellus during *Ricinus* seed development. *Proc. Natl. Acad. Sci. U.S.A.* 102, 2238–2243. doi: 10.1073/pnas.0409429102

Gutiérrez-Marcos, J. F., Costa, L. M., and Evans, M. M. (2006). Maternal gametophytic baseless1 is required for development of the central cell and early endosperm patterning in maize (*Zea mays*). *Genetics* 174, 317–329. doi: 10.1534/genetics.106.059709

Gutiérrez-Marcos, J. F., Dal Prà, M., Giulini, A., Costa, L. M., Gavazzi, G., Cordelier, S., et al. (2007). Empty pericarp4 encodes a mitochondrion-targeted pentatricopeptide repeat protein necessary for seed development and plant growth in maize. *Plant Cell* 19, 196–210. doi: 10.1105/tpc.105.039594

Hanks, J. N., Snyder, A. K., Graham, M. A., Shah, R. K., Blaylock, L. A., Harrison, M. J., et al. (2005). Defensin gene family in *Medicago truncatula*: structure, expression and induction by signal molecules. *Plant Mol. Biol.* 58, 385–399. doi: 10.1007/s11103-005-5567-7

Hara-Nishimura, I., Hatsugai, N., Nakaune, S., Kuroyanagi, M., and Nishimura, M. (2005). Vacuolar processing enzyme: an executor of plant cell death. *Curr. Opin. Plant Biol.* 8, 404–408. doi: 10.1016/j.pbi.2005.05.016

Harrington, G. N., Nussbaumer, Y., Wang, X. D., Tegeder, M., Franceschi, V. R., and Frommer, W. B. (1997). Spatial and temporal expression of sucrose transport related genes in developing cotyledons of *Vicia faba*. *Protoplasma* 200, 35–50. doi: 10.1007/BF01280733

Harvey, D. M., Hedley, C. L., and Keely, R. (1976). Photosynthetic and respiratory studies during pod and seed development in *Pisum sativum* L. *Ann. Bot.* 40, 993–1001.

Haughn, G., and Chaudhury, A. (2005). Genetic analysis of seed coat development in *Arabidopsis*. *Trends Plant Sci.* 10, 472–477. doi: 10.1016/j.tplants.2005.08.005

Haughn, G. W., and Western, T. L. (2012). *Arabidopsis* seed coat mucilage is a specialized cell wall that can be used as a model for genetic analysis of plant cell wall structure and function. *Front. Plant Sci.* 3:64. doi: 10.3389/fpls.2012.00064

Haydon, M. J., Mielczarek, O., Robertson, F. C., Hubbard, K. E., and Webb, A. A. (2013). Photosynthetic entrainment of the *Arabidopsis thaliana* circadian clock. *Nature* 502, 689–692. doi: 10.1038/nature12603

Hendry, G. A. F. (1993). Oxygen, free radical processes and seed longevity. *Seed Sci. Res.* 3, 141–153. doi: 10.1017/S0960258500001720

Huang, R., Jiang, L., Zheng, J., Wang, T., Wang, H., Huang, Y., et al. (2013). Genetic bases of rice grain shape: so many genes, so little known. *Trends Plant Sci.* 18, 218–226. doi: 10.1016/j.tplants.2012.11.001

Iglesias-Fernández, R., Matilla, A. J., Pulgar, I., and Torre, F. (2007). Ripe fruits of *Sisymbrium officinale* L. contain heterogeneous endospermic seeds with different germination rates. *Seed Sci. Biotechnol.* 1, 18–24.

Ingouff, M., Jullien, P. E., and Berger, F. (2006). The female gametophyte and the endosperm control cell proliferation and differentiation of the seed coat in *Arabidopsis*. *Plant Cell* 18, 3491–3501. doi: 10.1105/tpc.106.047266

Ingram, G. C. (2010). Family life at close quarters: communication and constraint in angiosperm seed development. *Protoplasma* 247, 195–214. doi: 10.1007/s00709-010-0184-y

Izawa, Y., Takayanagi, Y., Inaba, N., Abe, Y., Minami, M., Fujisawa, Y., et al. (2010). Function and expression pattern of the alpha subunit of the heterotrimeric G protein in rice. *Plant Cell Physiol.* 51, 271–281. doi: 10.1093/pcp/pcp186

Jahnke, S., Menzel, M. I., Van Dusschoten, D., Roeb, G. W., Bühler, J., Minwuyelet, S., et al. (2009). Combined MRI–PET dissects dynamic changes in plant structures and functions. *Plant J.* 59, 634–644. doi: 10.1111/j.1365-313X.2009.03888.x

Jofuku, K. D., Omidyar, P. K., Gee, Z., and Okamuro, J. K. (2005). Control of seed mass and seed yield by the floral homeotic gene APETALA2. *Proc. Natl. Acad. Sci. U.S.A.* 102, 3117–3122. doi: 10.1073/pnas.0409893102

Jones, J. D., and Dangl, J. L. (2006). The plant immune system. *Nature* 444, 323–329. doi: 10.1038/nature05286

Julián, I., Gandullo, J., Santos-Silva, L. K., Diaz, I., and Martinez, M. (2013). Phylogenetically distant barley legumains have a role in both seed and vegetative tissues. *J. Exp. Bot.* 64, 2929–2941. doi: 10.1093/jxb/ert132

Kang, B. H., Xiong, Y., Williams, D. S., Pozueta-Romero, D., and Chourey, P. S. (2009). Miniature1-encoded cell wall invertase is essential for assembly and function of wall-ingrowth in the maize endosperm transfer cell. *Plant Physiol.* 151, 1366–1376. doi: 10.1104/pp.109.142331

Kathage, J., and Qaim, M. (2012). Economic impacts and impact dynamics of Bt (*Bacillus thuringiensis*) cotton in India. *Proc. Natl. Acad. Sci. U.S.A.* 109, 11652–11656. doi: 10.1073/pnas.1203647109

Kelley, D. R., and Gasser, C. S. (2009). Ovule development: genetic trends and evolutionary considerations. *Sex. Plant Reprod.* 22, 229–234. doi: 10.1007/s00497-009-0107-2

Khan, D., Millar, J. L., Girard, I. J., and Belmonte, M. F. (2014). Transcriptional circuitry underlying seed coat development in *Arabidopsis*. *Plant Sci.* 219–220, 51–60. doi: 10.1016/j.plantsci.2014.01.004

Kimmel, L. B., and Day, M. P. (2001). New life for an old fiber: attributes and advantages of naturally colored cotton. *AATCC Rev.* 1, 32–36.

Kladnik, A., Chamusco, K., Dermastia, M., and Chourey, P. (2004). Evidence of programmed cell death in post-phloem transport cells of the maternal pedicel tissue in developing caryopsis of maize. *Plant Physiol.* 136, 3572–3581. doi: 10.1104/pp.104.045195

Koch, K. (2004). Sucrose metabolism: regulatory mechanisms and pivotal roles in sugar sensing and plant development. *Curr. Opin. Plant Biol.* 7, 235–246. doi: 10.1016/j.pbi.2004.03.014

Kovalchuk, N., Li, M., Wittek, F., Reid, N., Singh, R., Shirley, N., et al. (2010). Defensin promoters as potential tools for engineering disease resistance in cereal grains. *Plant Biotechnol. J.* 8, 47–64. doi: 10.1111/j.1467-7652.2009.00465.x

Kranner, I., Beckett, R. P., Minibayeva, F. V., and Seal, C. E. (2010). What is stress? Concepts, definitions and applications in seed science. *New Phytol.* 188, 655–673. doi: 10.1111/j.1469-8137.2010.03461.x

Krishnan, S., and Dayanandan, P. (2003). Structural and histochemical studies on grain-filling in the caryopsis of rice (*Oryza sativa* L.). *J. Biosci.* 28, 455–469. doi: 10.1007/BF02705120

Kunieda, K., Mitsuda, N., Ohme-Takagi, M., Takeda, S., Aida, M., Tasaka, M., et al. (2008). NAC family proteins NARS1/NAC2 and NARS2/NAM in the outer integument regulate embryogenesis in *Arabidopsis. Plant Cell* 20, 2631–2642. doi: 10.1105/tpc.108.060160

Ladwig, F., Stahl, M., Ludewig, U., Hirner, A. A., Hammes, U. Z., Stadler, R., et al. (2012). Siliques are Red1 from *Arabidopsis* acts as a bidirectional amino acid transporter that is crucial for the amino acid homeostasis of siliques. *Plant Physiol.* 158, 1643–1655. doi: 10.1104/pp.111.192583

Lafon-Placette, C., and Köhler, C. (2014). Embryo and endosperm, partners in seed development. *Curr. Opin. Plant Biol.* 17, 64–69. doi: 10.1016/j.pbi.2013.11.008

Lanning, F. C., and Eleuterius, L. N. (1992). Silica and ash in seeds of cultivated grains and native plants. *Ann. Bot.* 69, 151–160.

Lattanzio, V., Lattanzio, V. M. T., and Cardinali, A. (2006). "Role of phenolics in the resistance mechanisms of plants against fungal pathogens and insects," in *Phytochemistry: Advances in Research*, ed. F. Imperato (Trivandrum: Research Signpost), 23–67.

Lay, F. T., and Anderson, M. A. (2005). Defensins – components of the innate immune system in plants. *Curr. Protein Pept. Sci.* 6, 85–101. doi: 10.2174/1389203053027575

Lee, J. J., Woodward, A. W., and Chen, Z. J. (2007). Gene expression changes and early events in cotton fibre development. *Ann. Bot.* 100, 1391–1401. doi: 10.1093/aob/mcm232

Lepiniec, L., Debeaujon, I., Routaboul, J. M., Baudry, A., Pourcel, L., Nesi, N., et al. (2006). Genetics and biochemistry of seed flavonoids. *Annu. Rev. Plant Biol.* 57, 405–430. doi: 10.1146/annurev.arplant.57.032905.105252

Li, Q., and Li, L. (2014). Uniquitin-mediated control of seed size in plant. *Front. Plant Sci.* 5:332. doi: 10.3389/fpls.2014.00332

Li, Q., Li, L., Yang, X., Warburton, M. L., Bai, G., Dai, J., et al. (2010). Relationship, evolutionary fate and function of two maize co-orthologs of rice GW2 associated with kernel size and weight. *BMC Plant Biol.* 10:143. doi: 10.1186/1471-2229-10-143

Li, S. F., Milliken, O. N., Pham, H., Seyit, R., Napoli, R., Preston, J., et al. (2009). The *Arabidopsis* MYB5 transcription factor regulates mucilage synthesis, seed coat development, and trichome morphogenesis. *Plant Cell* 21, 72–89. doi: 10.1105/tpc.108.063503

Li, Y., Zheng, L., Corke, F., Smith, C., and Bevan, M. W. (2008). Control of final seed and organ size by the DA1 gene family in *Arabidopsis thaliana. Genes Dev.* 22, 1331–1336. doi: 10.1101/gad.463608

Linnestad, C., Doan, D. N., Brown, R. C., Lemmon, B. E., Meyer, D. J., Jung, R., et al. (1998). Nucellain, a barley homolog of the dicot vacuolar-processing protease, is localized in nucellar cell walls. *Plant Physiol.* 118, 1169–1180. doi: 10.1104/pp.118.4.1169

Liu, Q., Llewellyn, D. J., Singh, S. P., and Green, A. G. (2012). "Cotton seed development: opportunities to add value to a byproduct of fiber production," in *Flowering and fruiting in cotton*, eds D. M. Oosterhuis and J. T. Cothren (Cordova: The Cotton Foundation Press), 131–162.

Lohaus, G., and Moellers, C. (2000). Phloem transport of amino acids in two *Brassica napus* L. genotypes and one B. carinata genotype in relation to their seed protein content. *Planta* 211, 833–840. doi: 10.1007/s004250000349

Lombardi, L., Casani, S., Ceccarelli, N., Galleschi, L., Picciarelli, P., and Lorenzi, R. (2007). Programmed cell death of the nucellus during *Sechium edule* Sw. seed development is associated with activation of caspase-like proteases. *J. Exp. Bot.* 58, 2949–2958. doi: 10.1093/jxb/erm137

Lopato, S., Borisjuk, N., Langridge, P., and Hrmova, M. (2014). Endosperm transfer cell-specific genes and proteins: structure, function and applications in biotechnology. *Front. Plant Sci.* 5:64. doi: 10.3389/fpls.2014.00064

López-Fernández, M. P., and Maldonado, S. (2013). Programmed cell death during quinoa perisperm development. *J. Exp. Bot.* 64, 3313–3325. doi: 10.1093/jxb/ert170

Maitz, M., Santandrea, G., Zhang, Z., Lal, S., Hannah, L. C., Salamini, F., et al. (2000). rgf1, a mutation reducing grain filling in maize through effects on basal endosperm and pedicel development. *Plant J.* 23, 29–42. doi: 10.1046/j.1365-313x.2000.00747.x

Méchin, V., Thévenot, C., Le Guilloux, M., Prioul, J. L., and Damerval, C. (2007). Developmental analysis of maize endosperm proteome suggests a pivotal role for pyruvate orthophosphate dikinase. *Plant Physiol.* 143, 1203–1219. doi: 10.1104/pp.106.092148

Melkus, G., Rolletschek, H., Fuchs, J., Radchuk, V., Grafahrend-Belau, E., Sreenivasulu, N., et al. (2011). Dynamic $^{13}C/^{1}H$ NMR imaging uncovers sugar allocation in the living seed. *Plant Biotechnol. J.* 9, 1022–1037. doi: 10.1111/j.1467-7652.2011.00618.x

Melkus, G., Rolletschek, H., Radchuk, R., Fuchs, J., Rutten, T., Wobus, U., et al. (2009). The metabolic role of the legume endosperm: a noninvasive imaging study. *Plant Physiol.* 151, 1139–1154. doi: 10.1104/pp.109.143974

Mochizuki, M., Tanaka, R., Grimm, B., Masuda, T., Moulin, M., Smith, A. G., et al. (2009). The cell biology of tetrapyrroles: a life and death struggle. *Trends Plant Sci.* 15, 488–499. doi: 10.1016/j.tplants.2010.05.012

Möhlmann, T., and Neuhaus, H. E. (1997). Precursor and effector dependency of lipid synthesis in amyloplasts isolated from developing maize and wheat endosperm. *J. Cereal Sci.* 26, 161–167. doi: 10.1006/jcrs.1997.0116

Möhlmann, T., Scheibe, R., and Neuhaus, H. E. (1994). Interaction between fatty-acid and starch synthesis in isolated amyloplasts from cauliflower floral buds. *Planta* 194, 492–497. doi: 10.1007/BF00714461

Molina, I., Li-Beisson, Y., Beisson, F., Ohlrogge, J. B., and Pollard, M. (2008). Identification of an *Arabidopsis* feruloyl-coenzyme A transferase required for suberin synthesis. *Plant Physiol.* 151, 1317–1328. doi: 10.1104/pp.109.144907

Monjardino, P., Rocha, S., Tavares, A. C., Fernandes, R., Sampaio, P., Salema, R., et al. (2013). Development of flange and reticulate wall ingrowths in maize (*Zea mays* L.) endosperm transfer cells. *Protoplasma* 250, 495–450. doi: 10.1007/s00709-012-0432-4

Morley-Smith, E. R., Pike, M. J., Findlay, K., Köckenberger, W., Hill, L. M., Smith, A. M., et al. (2008). The transport of sugars to developing embryos is not via the bulk endosperm in oilseed rape seeds. *Plant Physiol.* 147, 2121–2130. doi: 10.1104/pp.108.124644

Nakaune, S., Yamada, K., Kondo, M., Kato, T., Tabata, S., Nishimura, M., et al. (2005). A vacuolar processing enzyme, δVPE, is involved in seed coat formation at the early stage of seed development. *Plant Cell* 17, 876–887. doi: 10.1105/tpc.104.026872

Nayar, S., Sharma, R., Tyagi, A. K., and Kapoor, S. (2013). Functional delineation of rice MADS29 reveals its role in embryo and endosperm development by affecting hormone homeostasis. *J. Exp. Bot.* 64, 4239–4253. doi: 10.1093/jxb/ert231

Nelson, D. C., Flematti, G. R., Ghisalberti, E. L., Dixon, K. W., and Smith, S. M. (2012). Regulation of seed germination and seedling growth by chemical signals from burning vegetation. *Annu. Rev. Plant Biol.* 63, 107–130. doi: 10.1146/annurev-arplant-042811-105545

Nelson, D. C., Flematti, G. R., Riseborough, J. A., Ghisalberti, E. L., Dixon, K. W., and Smith, S. M. (2010). Karrikins enhance light responses during germination and seedling development in *Arabidopsis thaliana. Proc. Natl. Acad. Sci. U.S.A.* 107, 7095–7100. doi: 10.1073/pnas.0911635107

Nutbeam, A. R., and Duffus, C. M. (1978). Oxygen-exchange in pericarp green layer of immature cereal-grains. *Plant Physiol.* 62, 360–362. doi: 10.1104/pp.62.3.360

Offler, C. E., McCurdy, D. W., Patrick, J. W., and Talbot, M. J. (2003). Transfer cells: cells specialized for a special purpose. *Annu. Rev. Plant Biol.* 54, 431–454. doi: 10.1146/annurev.arplant.54.031902.134812

Offler, C. E., Nerlich, S. M., and Patrick, J. W. (1989). Pathway of photosynthate transfer in the developing seed of *Vicia faba* L. transfer in relation to seed anatomy. *J. Exp. Bot.* 40, 769–780. doi: 10.1093/jxb/40.7.769

Ohto, M. A., Fischer, R. L., Goldberg, R. B., Nakamura, K., and Harada, J. J. (2005). Control of seed mass by APETALA2. *Proc. Natl. Acad. Sci. U.S.A.* 102, 3123–3128. doi: 10.1073/pnas.0409858102

Ohto, M. A., Floyd, S. K., Fischer, R. L., Goldberg, R. B., and Harada, J. J. (2009). Effects of APETALA2 on embryo, endosperm, and seed coat development determine seed size in *Arabidopsis. Sex. Plant Reprod.* 22, 277–289. doi: 10.1007/s00497-009-0116-1

Okuda, S., Tsutsui, H., Shiina, K., Sprunck, S., Takeuchi, H., Yui, R., et al. (2009). Defensin-like polypeptide LUREs are pollen tube attractants secreted from synergid cells. *Nature* 458, 357–361. doi: 10.1038/nature07882

Olsen, O. A. (2004). Nuclear endosperm development in cereals and *Arabidopsis thaliana. Plant Cell* 16, S214–S227. doi: 10.1105/tpc.017111

Oparka, K. J., and Gates, P. (1981). Transport of assimilates in the developing caryopsis of rice (*Oryza sativa* L.). *Planta* 152, 388–396. doi: 10.1007/BF00385354

Oparka, K. J., Roberts, A. G., Boevink, P., Santa Cruz, S., Roberts, I., Pradel, K. S., et al. (1999). Simple, but not branched, plasmodesmata allow the nonspecific trafficking of proteins in developing tobacco leaves. *Cell* 97, 743–754. doi: 10.1016/S0092-8674(00)80786-2

Panitz, R., Borisjuk, L., Manteuffel, R., and Wobus, U. (1995). Transient expression of storage-protein genes during early embryogenesis of *Vicia faba*: synthesis and metabolization of vicilin and legumin in the embryo, suspensor and endosperm. *Planta* 196, 765–774. doi: 10.1007/BF01106772

Patrick, J. W., Botha, F. C., and Birch, R. G. (2013). Metabolic engineering of sugars and simple sugar derivatives in plants. *Plant Biotechnol. J.* 11, 142–156. doi: 10.1111/pbi.12002

Patrick, J. W., and Offler, C. E. (1995). Post-sieve element transport of sucrose in developing seeds. *Aust. J. Plant Physiol.* 22, 681–702. doi: 10.1071/PP9950681

Patrick, J. W., and Offler, C. E. (2001). Compartmentation of transport and transfer events in developing seeds. *J. Exp. Bot.* 52, 551–564. doi: 10.1093/jexbot/52.356.551

Patrick, J. W., Offler, C. E., and Wang, X.-D. (1995). Cellular pathway of photosynthate transport in coats of developing seed of *Vicia faba* L. and *Phaseolus vulgaris* L. I. Extent of transport through the coat symplast. *J. Exp. Bot.* 46, 35–47. doi: 10.1093/jxb/46.1.35

Penfield, S., and King, J. (2009). Towards a systems biology approach to understanding seed dormancy and germination. *Proc. Biol. Sci.* 276, 3561–3569. doi: 10.1098/rspb.2009.0592

Phelps, M. E. (2004). *PET: Molecular Imaging and Its Biological Application.* New York: Springer.

Pignocchi, C., Minns, G. E., Nesi, N., Koumproglou, R., Kitsios, G., Benning, C., et al. (2009). ENDOSPERM DEFECTIVE1 is a novel microtubule-associated protein essential for seed development in *Arabidopsis. Plant Cell* 21, 90–105. doi: 10.1105/tpc.108.061812

Quail, P. H. (1997). An emerging molecular map of the phytochrome. *Plant Cell Environ.* 20, 657–665. doi: 10.1046/j.1365-3040.1997.d01-108.x

Radchuk, V., Borisjuk, L., Radchuk, R., Steinbiss, H. H., Rolletschek, H., Broeders, S., et al. (2006). Jekyll encodes a novel protein involved in the sexual reproduction of barley. *Plant Cell* 18, 1652–1666. doi: 10.1105/tpc.106.041335

Radchuk, V., Kumlehn, J., Rutten, T., Sreenivasulu, N., Radchuk, R., Rolletschek, H., et al. (2012). Fertility in barley flowers depends on Jekyll functions in male and female sporophytes. *New Phytol.* 194, 142–157. doi: 10.1111/j.1469-8137.2011.04032.x

Radchuk, V., Weier, D., Radchuk, R., Weschke, W., and Weber, H. (2011). Development of maternal seed tissue in barley is mediated by regulated cell expansion and cell disintegration and coordinated with endosperm growth. *J. Exp. Bot.* 62, 1217–1227. doi: 10.1093/jxb/erq348

Radchuk, V. V., Borisjuk, L., Sreenivasulu, N., Merx, K., Mock, H. P., Rolletschek, H., et al. (2009). Spatio-temporal profiling of starch biosynthesis and degradation in the developing barley grain. *Plant Physiol.* 150, 190–204. doi: 10.1104/pp.108.133520

Raissig, M. T., Baroux, C., and Grossniklaus, U. (2011). Regulation and flexibility of genomic imprinting during seed development. *Plant Cell* 23, 16–26. doi: 10.1105/tpc.110.081018

Rochat, C., Wuillème, S., Boutin, J.-P., and Hedley, C. L. (1995). A mutation at the *rb* gene, lowering ADPGPPase activity, affects storage product metabolism of pea seed coats. *J. Exp. Bot.* 46, 415–421. doi: 10.1093/jxb/46.4.415

Rolletschek, H., Borisjuk, L., Koschorreck, M., Wobus, U., and Weber, H. (2002). Legume embryos develop in a hypoxic environment. *J. Exp. Bot.* 53, 1099–1107. doi: 10.1093/jexbot/53.371.1099

Rolletschek, H., Koch, K., Wobus, U., and Borisjuk, L. (2005). Positional cues for the starch/lipid balance in maize kernels and resource partitioning to the embryo. *Plant J.* 42, 69–83. doi: 10.1111/j.1365-313X.2005.02352.x

Rolletschek, H., Melkus, G., Grafahrend-Belau, E., Fuchs, J., Heinzel, N., Schreiber, F., et al. (2011). Combined noninvasive imaging and modeling approaches reveal metabolic compartmentation in the barley endosperm. *Plant Cell* 23, 3041–3054. doi: 10.1105/tpc.111.087015

Rolletschek, H., Weber, H., and Borisjuk, L. (2003). Energy status and its control on embryogenesis of legumes. Embryo photosynthesis contributes to oxygen supply and is coupled to biosynthetic fluxes. *Plant Physiol.* 132, 1196–1206. doi: 10.1104/pp.102.017376

Rolletschek, H., Weschke, W., Weber, H., Wobus, U., and Borisjuk, L. (2004). Energy state and its control on seed development: starch accumulation is associated with high ATP and steep oxygen gradients within barley grains. *J. Exp. Bot.* 55, 1351–1359. doi: 10.1093/jxb/erh130

Roszak, P., and Köhler, C. (2011). Polycomb group proteins are required to couple seed coat initiation to fertilization. *Proc. Natl. Acad. Sci. U.S.A.* 108, 20826–20831. doi: 10.1073/pnas.1117111108

Ruan, Y. L. (2013). Boosting seed development as a new strategy to increase cotton fiber yield and quality. *J. Integr. Plant Biol.* 55, 572–575. doi: 10.1111/jipb.12074

Ruan, Y. L., Jin, Y., Yang, Y. J., Li, G. J., and Boyer, J. S. (2010). Sugar input, metabolism, and signaling mediated by invertase: roles in development, yield potential, and response to drought and heat. *Mol. Plant* 3, 942–955. doi: 10.1093/mp/ssq044

Ruan, Y. L., and Patrick, J. W. (1995). The cellular pathway of post-phloem sugar-transport in developing tomato fruit. *Planta* 196, 434–444. doi: 10.1007/BF00203641

Ruuska, S. A., Schwender, J., and Ohlrogge, J. B. (2004). The capacity of green oilseeds to utilize photosynthesis to drive biosynthetic processes. *Plant Physiol.* 136, 2700–2709. doi: 10.1104/pp.104.047977

Saito, G. Y., Chang, Y. C., Walling, L. L., and Thomson, W. W. (1989). A correlation in plastid development and cytoplasmic ultrastructure with nuclear gene-expression during seed ripening in soybean. *New Phytol.* 113, 459–469. doi: 10.1111/j.1469-8137.1989.tb00357.x

Sanders, A., Collier, R., Trethewy, A., Gould, G., Sieker, R., and Tegeder, M. (2009). AAP1 regulates import of amino acids into developing *Arabidopsis* embryos. *Plant J.* 59, 540–552. doi: 10.1111/j.1365-313X.2009.03890.x

Sang, T. (2009). Genes and mutations underlying domestication transitions in grasses. *Plant Physiol.* 149, 63–70. doi: 10.1104/pp.108.128827

Schiefelbein, J. (2003). Cell-fate specification in the epidermis, a common patterning mechanism in the root and shoot. *Curr. Opin. Plant Biol.* 6, 74–78. doi: 10.1016/S136952660200002X

Schruff, M. C., Spielman, M., Tiwari, S., Adams, S., Fenby, N., and Scott, R. J. (2006). The AUXIN RESPONSE FACTOR 2 gene of *Arabidopsis* links auxin signalling, cell division, and the size of seeds and other organs. *Development* 133, 251–261. doi: 10.1242/dev.02194

Schwender, J. (2008). Metabolic flux analysis as a tool in metabolic engineering of plants. *Curr. Opin. Biotechnol.* 19, 131–137. doi: 10.1016/j.copbio.2008.02.006

Schwender, J., and Ohlrogge, J. B. (2002). Probing in vivo metabolism by stable isotope labeling of storage lipids and proteins in developing *Brassica napus* embryos. *Plant Physiol.* 130, 347–361. doi: 10.1104/pp.004275

Shen-Miller, J. (2002). Sacred lotus, the long-living fruits of China Antique. *Seed Sci. Res.* 12, 131–143. doi: 10.1079/SSR2002112

Sinclair, T. R. (1988). Oxygen and temperature effects on soybean seed coat respiration rates. *Plant Physiol.* 86, 124–128. doi: 10.1104/pp.86.1.124

Sinclair, T. R., Ward, J. P., and Randall, C. A. (1987). Soybean seed growth in response to long-term exposures to differing oxygen partial pressures. *Plant Physiol.* 83, 467–468. doi: 10.1104/pp.83.3.467

Singh, D., and Mathur, S. B. (2004). *Histopathology of Seed-Borne Infections.* Boca Raton: CRC Press.

Smidansky, E. D., Clancy, M., Meyer, F. D., Lanning, S. P., Blake, N. K., Talbert, L. E., et al. (2002). Enhanced ADP-glucose pyrophosphorylase activity in wheat endosperm increases seed yield. *Proc. Natl. Acad. Sci. U.S.A.* 99, 1724–1729. doi: 10.1073/pnas.022635299

Smirnoff, N. (2010). Tocochromanols: rancid lipids, seed longevity, and beyond. *Proc. Natl. Acad. Sci. U.S.A.* 107, 17857–17858. doi: 10.1073/pnas.1012749107

Smith, H. (2000). Phytochromes and light signal perception by plants—an emerging synthesis. *Nature* 407, 585–591. doi: 10.1038/35036500

Smýkal, P., Vernoud, V., Blair, M. W., Soukup, A., and Thompson, R. D. (2014). The role of the testa during development and in establishment of dormancy of the legume seed. *Front. Plant Sci.* 5:351. doi: 10.3389/fpls.2014.00351

Song, X. J., Huang, W., Shi, M., Zhu, M. Z., and Lin, H. X. (2007). A QTL for rice grain width and weight encodes a previously unknown RING-type E3 ubiquitin ligase. *Nat. Genet.* 39, 623–630. doi: 10.1038/ng2014

Sreenivasulu, N., Borisjuk, L., Junker, B. H., Mock, H. P., Rolletschek, H., Seiffert, U., et al. (2010). Barley grain development toward an integrative view. *Int. Rev. Cell Mol. Biol.* 281, 49–89. doi: 10.1016/S1937-6448(10)81002-0

Sreenivasulu, N., Radchuk, V., Strickert, M., Miersch, O., Weschke, W., and Wobus U. (2006). Gene expression patterns reveal tissue-specific signaling networks controlling programmed cell death and ABA-regulated maturation in developing barley seeds. *Plant J.* 47, 310–327. doi: 10.1111/j.1365-313X.2006.02789.x

Sriram, G., Fulton, D. B., Iyer, V. V., Peterson, J. M., Zhou, R., Westgate, M. E., et al. (2004). Quantification of compartmented metabolic fluxes in developing soybean embryos by employing biosynthetically directed fractional 13C labeling, two-dimensional [13C, 1H] nuclear magnetic resonance, and comprehensive isotopomer balancing. *Plant Physiol.* 136, 3043–3057. doi: 10.1104/pp.104.050625

Stadler, R., Lauterbach, C., and Sauer, N. (2005a). Cell-to-cell movement of green fluorescent protein reveals post-phloem transport in the outer integument and identifies symplastic domains in *Arabidopsis* seeds and embryos. *Plant Physiol.* 139, 701–712. doi: 10.1104/pp.105.065607

Stadler, R., Wright, K. M., Lauterbach, C., Amon, G., Gahrtz, M., Feuerstein, A., et al. (2005b). Expression of GFP-fusions in *Arabidopsis* companion cells reveals non-specific protein trafficking into sieve elements and identifies a novel post-phloem domain in roots. *Plant J.* 41, 319–331. doi: 10.1111/j.1365-313X.2004.02298.x

Stiff, M. R., and Haigler, H. C. (2012). "Recent advances in cotton development," in *Flowering and Fruiting in Cotton*, eds D. M. Oosterhuis and J. T. Cothren (Cordova: The Cotton Foundation Press), 163–190.

Stotz, H. U., Thomson, J. G., and Wang, Y. (2009). Plant defensins: defense, development and application. *Plant Signal. Behav.* 4, 1010–1012. doi: 10.4161/psb.4.11.9755

Su, Z., Hao, C., Wang, L., Dong, Y., and Zhang, X. (2011). Identification and development of a functional marker of TaGW2 associated with grain weight in bread wheat (*Triticum aestivum* L.). *Theor. Appl. Genet.* 122, 211–223. doi: 10.1007/s00122-010-1437-z

Tegeder, M. (2014). Transporters involved in source to sink partitioning of amino acids and ureides: opportunities for crop improvement. *J. Exp. Bot.* 65, 1865–1878. doi: 10.1093/jxb/eru012

Thiel, J. (2014). Development of endosperm transfer cells in barley. *Front. Plant Sci.* 5:108. doi: 10.3389/fpls.2014.00108

Thiel, J., Weier, D., Sreenivasulu, N., Strickert, M., Weichert, N., Melzer, M., et al. (2008). Different hormonal regulation of cellular differentiation and function in nucellar projection and endosperm transfer cells – a microdissection-based transcriptome study of young barley grains. *Plant Physiol.* 148, 1436–1452. doi: 10.1104/pp.108.127001

Thomma, B. P., Cammue, B. P., and Thevissen, K. (2002). Plant defensins. *Planta* 216, 193–202. doi: 10.1007/s00425-002-0902-6

Thorne, J. H. (1982). Temperature and oxygen effects on C14-labeled photosynthate unloading and accumulation in developing soybean seeds. *Plant Physiol.* 69, 48–53. doi: 10.1104/pp.69.1.48

Tilsner, J., Kassner, N., Struck, C., and Lohaus, G. (2005). Amino acid contents and transport in oilseed rape (*Brassica napus* L.) under different nitrogen conditions. *Planta* 221, 328–338. doi: 10.1007/s00425-004-1446-8

Tschiersch, H., Liebsch, G., Borisjuk, L., Stangelmayer, A., and Rolletschek, H. (2012). An imaging method for oxygen distribution, respiration and photosynthesis at a microscopic level of resolution. *New Phytol.* 196, 926–936. doi: 10.1111/j.1469-8137.2012.04295.x

Tsiatsiani, L., Gevaert, K., and Van Breusegem, F. (2012). Natural substrates of plant proteases: how can protease degradomics extend our knowledge? *Physiol. Plant.* 145, 28–40. doi: 10.1111/j.1399-3054.2011.01534.x

Ungru, A., Nowack, M. K., Reymond, M., Shirzadi, R., Kumar, M., Biewers, S., et al. (2008). Natural variation in the degree of autonomous endosperm formation reveals independence and constraints of embryo growth during seed development in *Arabidopsis thaliana. Genetics* 179, 829–841. doi: 10.1534/genetics.107.084889

van Bel, A. J., Helariutta, Y., Thompson, G. A., Ton, J., Dinant, S., Ding, B., et al. (2013). Phloem: the integrative avenue for resource distribution, signaling, and defense. *Front. Plant Sci.* 4:471. doi: 10.3389/fpls.2013.00471

van Dongen, J. T., Ammerlaan, A. M., Wouterlood, M., Van Aelst, A. C., and Borstlap, A. C. (2003). Structure of the developing pea seed coat and the post-phloem transport pathway of nutrients. *Ann. Bot.* 91, 729–737. doi: 10.1093/aob/mcg066

van Dongen, J. T., Roeb, G. W., Dautzenberg, M., Froehlich, A., Vigeolas, H., Minchin, P. E. H., et al. (2004). Phloem import and storage metabolism are highly coordinated by the low oxygen concentrations within developing wheat seeds. *Plant Physiol.* 135, 1809–1821. doi: 10.1104/pp.104.040980

Verboven, P., Herremans, E., Borisjuk, L., Helfen, L., Ho, Q. T., Tschiersch, H., et al. (2013). Void space inside the developing seed of *Brassica napus* and the modelling of its function. *New Phytol.* 199, 936–947. doi: 10.1111/nph.12342

Verdier, J., Dessaint, F., Schneider, C., and Abirached-Darmency, M. (2013). A combined histology and transcriptome analysis unravels novel questions on *Medicago truncatula* seed coat. *J. Exp. Bot.* 64, 459–470. doi: 10.1093/jxb/ers304

Verdier, J., Zhao, J., Torres-Jerez, I., Ge, S., Liu, C., He, X., et al. (2012). MtPAR MYB transcription factor acts as an on switch for proanthocyanidin biosynthesis in *Medicago truncatula. Proc. Natl. Acad. Sci. U.S.A.* 109, 1766–1771. doi: 10.1073/pnas.1120916109

Vermerris, W., and Nicholson, R. (2007). Phenolic Compound Biochemistry. *Springer e-Books* ISBN 1402051646.

Vigeolas, H., Möhlmann, T., Martini, N., Neuhaus, H. E., and Geigenberger, P. (2004). Embryo-specific reduction of ADP-Glc pyrophosphorylase leads to an inhibition of starch synthesis and a delay in oil accumulation in developing seeds of oilseed rape. *Plant Physiol.* 136, 2676–2686. doi: 10.1104/pp.104.046854

Vigeolas, H., Van Dongen, J. T., Waldeck, P., Hühn, D., and Geigenberger, P. (2003). Lipid storage metabolism is limited by the prevailing low oxygen concentrations within developing seeds of oilseed rape. *Plant Physiol.* 133, 2048–2060. doi: 10.1104/pp.103.031963

Vinogradova, I. S., and Falaleev, O. V. (2012). Formation of the vascular system of developing bean (*Phaseolus limensis* L.) seeds according to nuclear magnetic resonance microtomography. *Ontogenez* 43, 28–38. doi: 10.1134/S1062360412010079

Wager, H. G. (1974). Effect of subjecting peas to air enriched with carbon-dioxide. 1. Path of gaseous diffusion, content of CO_2 and buffering of tissue. *J. Exp. Bot.* 85, 330–337. doi: 10.1093/jxb/25.2.330

Walker, N. A., Zhang, W. H., Harrington, G., Holdaway, N., and Patrick, J. W. (2000). Effluxes of solutes from developing seed coats of *Phaseolus vulgaris* L. and *Vicia faba* L.: locating the effect of turgor in a coupled chemiosmotic system. *J. Exp. Bot.* 51, 1047–1055. doi: 10.1093/jexbot/51.347.1047

Wang, E., Wang, J., Zhu, X., Hao, W., Wang, L., Li, Q., et al. (2008). Control of rice grain-filling and yield by a gene with a potential signature of domestication. *Nat. Genet.* 40, 1370–1374. doi: 10.1038/ng.220

Wang, N., and Fisher, D. B. (1994). The use of fluorescent tracers to characterize the post-phloem transport pathway in maternal tissues of developing wheat grains. *Plant Physiol.* 104, 17–27. doi: 10.1104/pp.104.1.17

Waters, A. J., Makarevitch, I., Eichten, S. R., Swanson-Wagner, R. A., Yeh, C.-T., Xu, W., et al. (2011). Parent-of-origin effects on gene expression and DNA methylation in the maize endosperm. *Plant Cell* 23, 4221–4233. doi: 10.1105/tpc.111.092668

Watson, P. A., and Duffus, C. M. (1988). Carbon-dioxide fixation by detached cereal caryopses. *Plant Physiol.* 87, 504–509. doi: 10.1104/pp.87.2.504

Weber, H., Borisjuk, L., and Wobus, U. (1996a). Controlling seed development and seed size in *Vicia faba*: a role for seed coat-associated invertases and carbohydrate state. *Plant J.* 10, 823–834. doi: 10.1046/j.1365-313X.1996.10050823.x

Weber, H., Buchner, P., Borisjuk, L., and Wobus, U. (1996b). Sucrose metabolism during cotyledon development of *Vicia faba* L. is controlled by the concerted action of both sucrose-phosphate synthase and sucrose synthase: expression patterns, metabolic regulation and implications for seed development. *Plant J.* 9, 841–850. doi: 10.1046/j.1365-313X.1996.9060841.x

Weber, H., Borisjuk, L., and Wobus, U. (2005). Molecular physiology of legume seed development. *Annu. Rev. Plant Biol.* 56, 253–279. doi: 10.1146/annurev.arplant.56.032604.144201

Weijers, D., van Hamburg, J. P., van Rijn, E., Hooykaas, P. J. J., and Offringa, R. (2003). Diphtheria toxin-mediated cell ablation reveals interregional communication during *Arabidopsis* seed development. *Plant Physiol.* 133, 1882–1892. doi: 10.1104/pp.103.030692

Welbaum, G. E., and Bradford, K. J. (1990). Water relations of seed development and germination in muskmelon (*Cucumis melo* L.): IV. Characteristics of the perisperm during seed development. *Plant Physiol.* 92, 1038–1045. doi: 10.1104/pp.92.4.1038

Weng, J., Gu, S., Wan, X., Gao, H., Guo, T., Su, N., et al. (2008). Isolation and initial characterization of GW5, a major QTL associated with rice grain width and weight. *Cell Res.* 18, 1199–1209. doi: 10.1038/cr.2008.307

Wolf, J. B., and Hager, R. (2006). A maternal-offspring co-adaptation theory for the evolution of genomic imprinting. *PLoS Biol.* 4:e380. doi: 10.1371/journal.pbio.0040380

Wolff, P., Weinhofer, I., Seguin, J., Roszak, P., Beisel, C., Donoghue, M. T., et al. (2011). High-resolution analysis of parent-of-origin allelic expression in the *Arabidopsis* endosperm. *PLoS Genet.* 7:e1002126. doi: 10.1371/journal.pgen.1002126

Wu, B., Andersch, F., Weschke, W., Weber, H., and Becker, J. S. (2013). Diverse accumulation and distribution of nutrient elements in developing wheat grain studied by laser ablation inductively coupled plasma mass spectrometry imaging. *Metallomics* 5, 1276–1284. doi: 10.1039/c3mt00071k

Xia, T., Li, N., Dumenil, J., Li, J., Kamenski, A., Bevan, M. W., et al. (2013). The ubiquitin receptor DA1 interacts with the E3 ubiquitin ligase DA2 to regulate seed and organ size in *Arabidopsis*. *Plant Cell* 25, 3347–3359. doi: 10.1105/tpc.113.115063

Xia, X., Zhang, H. M., Andriunas, F. A., Offler, C. E., and Patrick, J. W. (2012). Extracellular hydrogen peroxide, produced through a respiratory burst oxidase/superoxide dismutase pathway, directs ingrowth wall formation in epidermal transfer cells of *Vicia faba* cotyledons. *Plant Signal. Behav.* 7, 1–4. doi: 10.4161/psb.21320

Xiong, F., Yu, X. R., Zhou, L., Wang, F., and Xiong, A. S. (2013a). Structural and physiological characterization during wheat pericarp development. *Plant Cell Rep.* 32, 1309–1320. doi: 10.1007/s00299-013-1445-y

Xiong, Y., McCormack, M., Li, L., Hall, Q., Xiang, C., and Sheen, J. (2013b). Glucose-TOR signalling reprograms the transcriptome and activates meristems. *Nature* 496, 181–186. doi: 10.1038/nature12030

Yan, H., Hua, Z., Qian, G., Wang, M., Du, G., and Chen, J. (2009). Analysis of the chemical composition of cotton seed coat by Fourier-transform infrared (FT-IR) microspectroscopy. *Cellulose* 16, 1099–1107. doi: 10.1007/s10570-009-9349-2

Yang, C., and Ye, Z. (2013). Trichomes as models for studying plant cell differentiation. *Cell. Mol. Life Sci.* 70, 1937–1948. doi: 10.1007/s00018-012-1147-6

Yang, X., Wu, F., Lin, X., Du, X., Chong, K., Gramzow, L., et al. (2012). Live and let die – the B(sister) MADS-box gene OsMADS29 controls the degeneration of cells in maternal tissues during seed development of rice (*Oryza sativa*). *PLoS ONE* 7:e51435. doi: 10.1371/journal.pone.0051435

Yeung, E. C., and Meinke, D. W. (1993). Embryogenesis in angiosperms: development of the suspensor. *Plant Cell* 5, 1371–1381. doi: 10.1105/tpc.5.10.1371

Yin, L. L., and Xue, H. W. (2012). The MADS29 transcription factor regulates the degradation of the nucellus and the nucellar projection during rice seed development. *Plant Cell* 24, 1049–1065. doi: 10.1105/tpc.111.094854

Yu, C. Y. (2013). Molecular mechanism of manipulating seed coat coloration in oilseed *Brassica* species. *J. Appl. Genet.* 54, 135–145. doi: 10.1007/s13353-012-0132-y

Zhang, M., Zhao, H., Xie, S., Chen, J., Xu, Y., Wang, K., et al. (2011). Extensive, clustered parental imprinting of protein-coding and noncoding RNAs in developing maize endosperm. *Proc. Natl. Acad. Sci. U.S.A.* 108, 20042–20047. doi: 10.1073/pnas.1112186108

Zhang, W. H., Zhou, Y., Dibley, K. E., Tyerman, S. D., Furbank, R. T., and Patrick, J. W. (2007). Nutrient loading of developing seeds. *Funct. Plant Biol.* 34, 314–331. doi: 10.1071/FP06271

Zhang, X., Wang, J., Huang, J., Lan, H., Wang, C., Yin, C., et al. (2012). Rare allele of OsPPKL1 associated with grain length causes extra-large grain and a significant yield increase in rice. *Proc. Natl. Acad. Sci. U.S.A.* 109, 21534–21539. doi: 10.1073/pnas.1219776110

Zhao, J., and Dixon, R. A. (2009). MATE transporters facilitate vacuolar uptake of epicatechin 3′-O-glucoside for proanthocyanidin biosynthesis in *Medicago truncatula* and *Arabidopsis*. *Plant Cell* 21, 2323–2340. doi: 10.1105/tpc.109.067819

Zhou, Y., Andriunas, F. A., Offler, C. E., McCurdy, D. W., and Patrick, J. W. (2010). An epidermal-specific ethylene signal cascade regulates trans-differentiation of transfer cells in *Vicia faba* cotyledons. *New Phytol.* 185, 931–943. doi: 10.1111/j.1469-8137.2009.03136.x

Zhou, Z., Wang, L., Li, J., Song, X., and Yang, C. (2009). Study on programmed cell death and dynamic changes of starch accumulation in pericarp cells of *Triticum aestivum* L. *Protoplasma* 236, 49–58. doi: 10.1007/s00709-009-0046-7

Conflict of Interest Statement: The authors declare that the research was conducted in the absence of any commercial or financial relationships that could be construed as a potential conflict of interest.

12

Genomic dissection of the seed

Michael G. Becker[1†], Ssu-Wei Hsu[2†], John J. Harada[2] and Mark F. Belmonte[1*]

[1] *Department of Biological Sciences, University of Manitoba, Winnipeg, MB, Canada*
[2] *Department of Plant Biology, University of California Davis, Davis, CA, USA*

Edited by:
Paolo Sabelli, University of Arizona, USA

Reviewed by:
David G. Oppenheimer, University of Florida, USA
Hannetz Roschzttardtz, University of Wisconsin-Madison, USA
Wilco Ligterink, Wageningen University, Netherlands

***Correspondence:**
Mark F. Belmonte, Department of Biological Sciences, University of Manitoba, 50 Sifton Road, Winnipeg, MB R3T 2N2, Canada
e-mail: mark.belmonte@umanitoba.ca

† Michael G. Becker and Ssu-Wei Hsu have contributed equally to this work.

Seeds play an integral role in the global food supply and account for more than 70% of the calories that we consume on a daily basis. To meet the demands of an increasing population, scientists are turning to seed genomics research to find new and innovative ways to increase food production. Seed genomics is evolving rapidly, and the information produced from seed genomics research has exploded over the past two decades. Advances in modern sequencing strategies that profile every molecule in every cell, tissue, and organ and the emergence of new model systems have provided the tools necessary to unravel many of the biological processes underlying seed development. Despite these advances, the analyses and mining of existing seed genomics data remain a monumental task for plant biologists. This review summarizes seed region and subregion genomic data that are currently available for existing and emerging oilseed models. We provide insight into the development of tools on how to analyze large-scale datasets.

Keywords: *Arabidopsis*, next generation sequencing, oilseed, RNA seq, seed, soybean, transcriptome

INTRODUCTION

With the world population expected to reach over 9 billion by the middle of the 21st century, one of the biggest challenges facing humanity will be the production of sustainable food supplies (Godfray et al., 2010; Cleland, 2013). To accommodate world food demands, it is estimated that crop production will need to double without increasing current agricultural land use (Foley et al., 2011; Tilman et al., 2011; Ray et al., 2013). Since the direct consumption of seeds and their use as animal feed account for more than 70% of the human diet (Sreenivasulu and Wobus, 2013), recent discussions on food security have turned to enhancing crop production through seed genomics. Seed genomics is the study of genomes and the expression of genes that are required to make a seed. This includes the spatial and temporal expression and regulation of all genes active during seed development. While classical breeding strategies have proven to be effective in producing more robust and productive plant cultivars, they can be complemented and greatly improved through the utilization of genomics-based knowledge (Tester and Langridge, 2010; Feuillet et al., 2011; Langridge and Fleury, 2011).

A seed is formed upon fertilization of the female gametophyte and early stages of development involve the deterioration of maternal gametophytic structures and the establishment of the sporophyte. Seed development is initiated by a double fertilization event that results in a seed that can be divided into three distinct regions: the embryo, the endosperm, and the seed coat (SC; Ohad et al., 1999; Le et al., 2007). In the first fertilization event, a sperm, and egg cell nucleus fuse, resulting in a zygotic embryo. The embryo is part of the next sporophytic plant generation. The endosperm results from the second fertilization event between the sperm and central cell, and it will serve to support the embryo during the early stages of seed development and/or seedling growth. Finally, the seed coat (SC) is of maternal origin and is derived from the integuments that form during ovule development. The SC transfers assimilates from the maternal plant and serves to protect the embryo throughout seed development. Further, the developmental programs that underlie seed development can be divided into two distinct phases. First, during morphogenesis, the body plan of the embryo is established and the nuclei of the endosperm proliferate. Second, during the maturation phase, large shifts in gene activity are observed across all three regions of the seed, initiating the accumulation of storage materials that help to protect the embryo in preparation for desiccation.

We can further dissect seed regions into subregions. In numerous plants including *Arabidopsis*, the zygote differentiates into the embryo proper, which will become cotyledonous and eventually form the vegetative plant, and the suspensor, which acts to facilitate communication between the embryo proper and surrounding seed regions. The endosperm develops into three distinct subregions: the micropylar endosperm (MCE, proximal to the embryo), the peripheral endosperm (PEN), and the chalazal endosperm (CZE, distal to the embryo; Brown et al., 2003). The maternally derived SC can be divided into two subregions, the chalazal seed coat (CZSC), and distal SC (**Figure 1A**). Depending on the model seed, these subregions can further be divided into tissue and cell types.

This review focuses on the genomic analysis of seed regions and subregions using established and emerging plant models. We discuss how genomics has been used successfully to study the development of the embryo, endosperm, and SC regions of the seed, and how new cutting-edge tools can be used to further dissect every cell and tissue of the seed into subregions for further interrogation. Finally, we present tools on how to analyze large-scale transcriptome datasets.

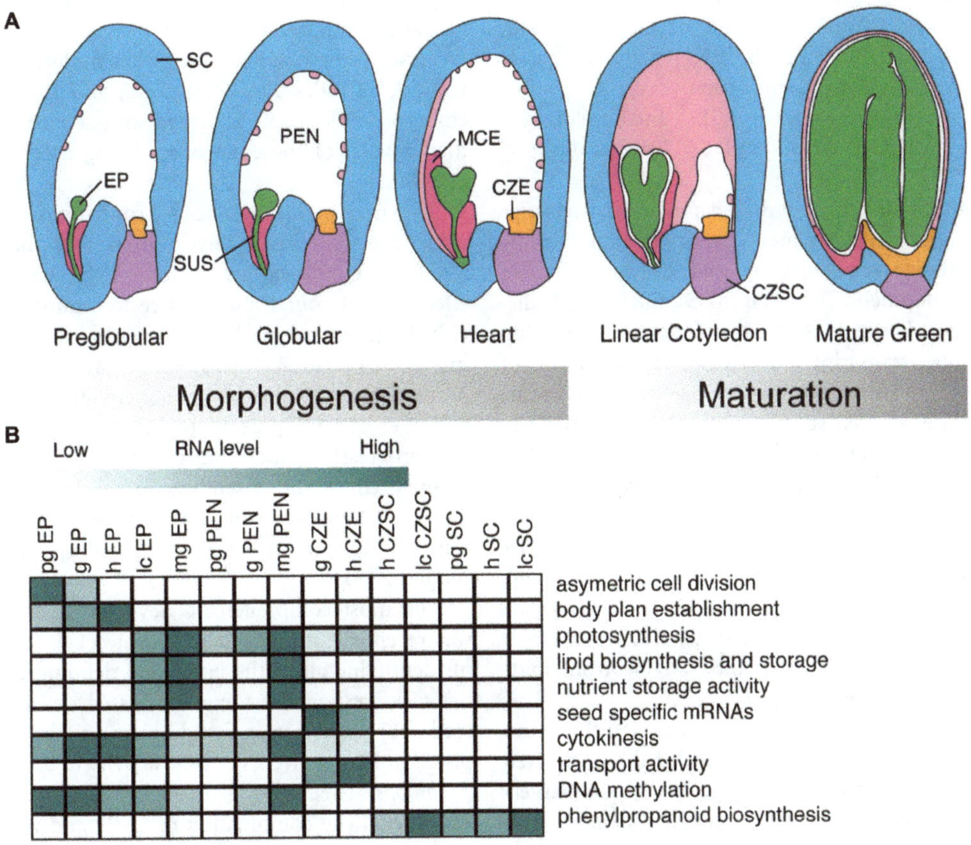

FIGURE 1 | Development and biological functions of *Arabidopsis* seed subregions. (A) Representation of seed subregions in *Arabidopsis* from the preglobular to mature green stages of development. Green, embryo proper (EP); dark pink, micropylar endosperm (MCE); light pink, peripheral endosperm (PEN); orange, chalazal endosperm (CZE); purple, chalazal seed coat(CZSC); blue, seed coat (SC). **(B)** Heat map visualization of representative Gene Ontology terms, biological processes, and metabolic pathways found in different subregions of the seed discussed in the review. Preglobular (pg); globular (g); heart (h); linear cotyledon (lc); mature green (mg). Dark green color represents activity in a particular subregion of the seed over developmental time.

CHARACTERIZING THE SEED TRANSCRIPTOME

In the current genomics era we have uncovered a number of developmental and regulatory pathways responsible for making a seed. However, we still have yet to fully understand all of the mechanisms responsible for the coordination of gene activity underlying the sophisticated development of all seed regions and subregions. Many regulatory mechanisms surrounding primary and secondary metabolism, hormone regulation, gene imprinting, transcriptional-, translational-, and post-translational regulation all operate in concert to mediate the complex processes occurring during seed development. These processes are under the regulation of 100s and 1000s of genes that are often obscured by genetic redundancy and thus difficult to identify using traditional forward genetics screens (Curtin et al., 2011). Arguably, the best way to investigate coordinated events such as cell fate specification, differentiation, and morphogenesis of the developing seed is by monitoring the expression of large gene sets with high throughput genomics-focused microarray and sequencing strategies. Bioinformatic analyses can then be used to identify transcriptional networks and key regulators of seed development.

As Next Generation Sequencing experiments such as deep genomic sequencing, RNA-, small RNA-, and DNA methylome-sequencing become commonplace in the laboratory and as sequencing technologies continue to evolve, the challenge faced by the scientific community is no longer the acquisition of data, but rather compiling and analyzing the data. Publicly available databases like NCBI (The National Center for Biotechnological Information[1]), GEO (Gene Expression Omnibus[2]), and SRA (Sequence Reads Archive[3]) contain large amounts of DNA microarray and nucleic acid sequence data that can be queried and mined to provide answers to challenging biological questions about the seed.

USING DNA MICROARRAYS TO PROFILE THE SEED

The Affymetrix ATH1 GeneChip microarray was one of the most widely used tools to profile the *Arabidopsis* transcriptome, and it was used to investigate numerous processes underlying

[1] http://www.ncbi.nlm.nih.gov
[2] http://www.ncbi.nlm.nih.gov/geo
[3] http://www.ncbi.nlm.nih.gov/sra

the seed including gibberellin response (Ogawa et al., 2003), response to abscisic acid (Nishimura et al., 2007), seed dormancy (Finch-Savage et al., 2007), seed imbibition (Nakabayashi et al., 2005; Preston et al., 2009), seed germination (Dean Rider et al., 2003; Penfield et al., 2006; Dekkers et al., 2013), and development (Day et al., 2008; Le et al., 2010; Dean et al., 2011; Belmonte et al., 2013; Khan et al., 2014).

Le et al. (2010) published the *Arabidopsis* seed transcriptome at seven stages of development from ovule to seedling and identified putative regulators of seed development. At each stage of development, between 8779 and 13,722 distinct mRNAs were detected at the level of the GeneChip with 15,563 unique transcripts detected over all stages of seed development. Of these, only 2% (289) of the transcripts were considered seed-specific with the vast majority being specific to a given stage of development (e.g., globular-cotyldeon). Of these seed-specific genes, 17% coded for transcription factors (TFs) and contained known regulators of seed development, such as *LEAFY COTYLEDON1* (*LEC1*), *LEAFY COTYLEDON2* (*LEC2*), *FUSCA3* (*FUS3*), and *MEDEA* (Le et al., 2010).

Similar analyses were conducted for developing soybean seed from five developmental time points ranging from mid-maturation through seed desiccation (Jones et al., 2010). This study noted an increase in TF activity late in seed development. TFs accumulating late in development included those involved in ethylene and auxin responses, as well as genes that were largely uncharacterized in soybean. Orthologous genes in *Arabidopsis* and rice suggest these genes are involved in processes such as abscisic acid and gibberellic acid signaling, sugar and nitrogen metabolism, and germination.

PROFILING THE SEED USING LASER MICRODISSECTION COUPLED WITH MICROARRAYS

Traditional studies that isolated seed regions like the embryo, endosperm, and SC for seed genomics used forceps or fine needles. The lack of precision of these manual techniques makes it nearly impossible to isolate individual regions without contamination from neighboring cells or tissues. These challenges limit the resolution of genomics research and dilute low abundant transcripts that may otherwise be detected using more sophisticated dissection methods. Regardless of the dissection tool used, the advancement of genomics-based seed research relies on contamination-free isolation of the cells and tissues of interest.

Currently, the most successful way to dissect regions and subregions of the seed for genomics studies without contamination of other cells types is through laser microdissection (LMD) technologies (Khan et al., 2014). Whole-seed mRNA profiling experiments provided some of the most informative seed genomic data across developmental time for *Arabidopsis* and soybean, but the application of LMD to these seeds provided higher resolution and more sensitive profiles of gene activity in developing seed. For example, Casson et al. (2005) dissected the *Arabidopsis* embryo to study mechanisms associated with apical / basal polarity. This study detected expression of ~65% of the 22,810 probe sets on the ATH1 array during the early stages of embryo development. Characterization of the spatial and temporal expression of 220 genes known to cause defects in embryo development when mutated, including *PASTICCINO1*, *PINOID*, *PIN-FORMED3*, and *PIN-FORMED4* during embryo development provided insight into their control. Further, several of these genes are being used as markers for the embryo.

The endosperm has been a difficult seed region to study using transcriptome analysis given that the endosperm subregions are not easily isolated. LMD has proven to be an effective and contamination-free technique to isolate the individual subregions of the endosperm for transcriptional profiling (Day et al., 2008). An initial study identified 800 genes, 27 encoding TFs that are preferentially expressed during early endosperm development. Biological processes associated with the progression and control of the cell cycle, DNA processing, chromatin assembly, protein synthesis, cytoskeleton- and microtubule-related processes, and cell/organelle biogenesis were all predicted to characterize endosperm proliferation and cellularization.

The most comprehensive developmental series of any seed was recently published by Belmonte et al. (2013) with the goal of identifying all of the genes and defining the gene regulatory networks responsible for guiding seed development. Thirty-six seed subregions across five developmental stages revealed complex dominant patterns of gene activity in both space and time in *Arabidopsis* (Belmonte et al., 2013; data available at seedgenenetwork.net). The combination of LMD and the ATH1 GeneChip identified at least 17,594 distinct mRNAs that are detectable during seed development and 1,316 of those mRNAs are specifically expressed in the *Arabidopsis* seed compared to vegetative and reproductive tissues. Similar data describing mRNA profiles at high spatial resolution are also available for soybean from experiments that used the Affymetrix soybean GeneChip to analyze 40 subregions across four developmental stages (Le et al., 2007; data available at seedgenenetwork.net). The reader is referred to Nelson et al. (2006) and Day et al. (2007) for reviews on methods and protocols used for LMD of plant tissues (**Figure 1B**).

USING NEXT GENERATION SEQUENCING TO PROFILE THE SEED

There are a number of advantages to NGS sequencing technology when compared to DNA microarrays: (i) the ability to detect low abundance transcripts, (ii) the identification of novel alternatively spliced isoforms of mRNAs, (iii) little requirement for *a priori* knowledge of the organism, (iv) increased sensitivity in the detection of differentially expressed genes, (v) more reproducible results, and (vi) the ability to compare expression profiles between distantly related organisms. For example, RNA sequencing facilitated the study of oil accumulation in four non-model oilseeds (or "emerging models"): castor (*Ricinus communis*), rapeseed (*Brassica napus*), burning bush (*Euonymus alatus*), and nasturtium (*Tropaeolum majus*; Troncoso-Ponce et al., 2011). These species differ in their location for oil deposition, triacylglycerol composition and content. Analysis of the data revealed a core set of well-conserved enzymes involved in triacylglycerol production that exhibit similar temporal expression

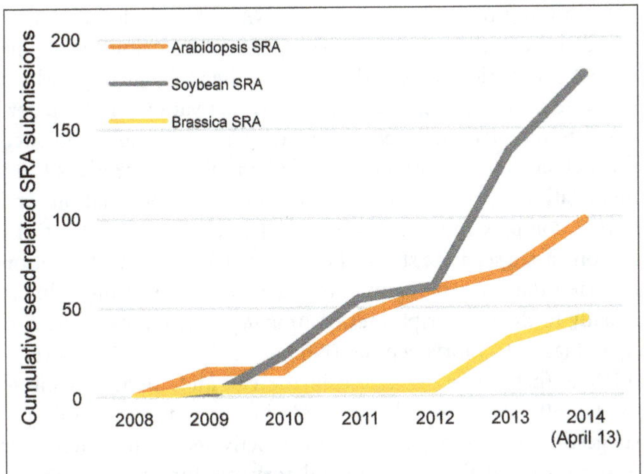

FIGURE 2 | Cumulative seed related Sequence Read Archive (SRA) submissions in *Arabidopsis* (orange), Brassica (yellow), and soybean (gray) from 2008 to 2014 (April 13) through the National Center for Biotechnology Information.

patterns in all species, suggesting a conserved evolutionary relationship in the production of seed oil. Putative regulators and mediators of oil production in *Arabidopsis* were identified and an online resource, "ARALIP[4]," was established to facilitate utilization of these data. It is important to note that while NGS has several advantages over microarray technology, the detection of low abundant transcripts as well the detection of alternative splice sites is largely dependent on the depth of sequencing and should be carefully considered during the design of the experiment.

Many other RNA sequencing studies of seed genomics have focused on soybean, largely because of its global economic importance. An indication of this emphasis is that seed-related submissions of soybean RNA sequencing data to the SRA and NCBI databases nearly double those of *Arabidopsis* (**Figure 2**). This has produced several large datasets for soybean seed development. Two particular studies stand out, one that profiled the whole soybean seed at seven time points between 10 and 42 days after fertilization (Severin et al., 2010), and an independent study focusing on whole soybean seeds at 15–65 days after fertilization (Chen et al., 2012). These studies showed that 49,151 transcripts are detected during seed development, ~12,000 mRNAs more than the 37,500 transcripts represented on the current soybean Affymetrix array. Furthermore, 9930–14,058 (Severin et al., 2010) and 11,592–16,255 (Chen et al., 2012) transcripts are differentially expressed compared to the earliest stage of seed development. Both of these studies provide examples of how RNA sequencing data can be mined using a range of bioinformatics approaches including gene ontology term enrichment and co-expression analyses.

Next Generation Sequencing also provides an effective method for the characterization of small RNA (sRNA) populations within the developing seed. Two classes of sRNAs highly expressed within seed tissues are microRNAs (miRNAs) and small interfering RNAs (siRNAs). miRNAs are ~21 nucleotide, single-stranded, non-coding RNAs that mediate the degradation or translational inhibition of target mRNAs with complementary nucleotide sequences (Chen, 2012). siRNAs, derived from double-stranded RNA, cause the degradation of target mRNAs and carry out *de novo* deposition of repressive chromatin marks and will be discussed later in this review.

Much of the recent work profiling sRNAs during seed development focus on economically important emerging models. Two independent studies examined sRNA populations in soybean, focusing on the identification of miRNAs active during development and their putative targets (Song et al., 2011; Shamimuzzaman and Vodkin, 2012). Of the miRNA targets identified, 50% (Song et al., 2011) and 82% (Shamimuzzaman and Vodkin, 2012) were TFs, including auxin response factors and growth regulating factors. Eleven annotations were found in both datasets including Argonaute Protein, Auxin Response Factor, Growth Regulating Factor, HD-ZIP TF, No Apical Meristem protein, TCP Family TF, and Nuclear Factor YA. These studies also report an increase in mRNA target diversity late in development, suggesting miRNAs have a role in the shift into maturation, which agrees with data from earlier work done with *Arabidopsis* (Tang et al., 2012).

Huang et al. (2013) characterized *B. napus* sRNA populations in whole seeds at nine time points in development and in dissected endosperm, embryo, and SC at three of those stages. Similar to *Arabidopsis* and soybean, the authors suggest that miRNAs have a role in controlling seed maturation. In addition, 279 miRNAs were identified that had been previously reported, including 182 in *Arabidopsis* and 56 in soybean. Also in *B. napus*, Zhao et al. (2012) characterized miRNA populations in high- and low-oil content seeds, and they identified putative miRNA regulators of oil metabolism.

Several databases for miRNAs and their mRNA targets are available to the researcher. Currently 427, 573, and 92 mature miRNA sequences for *Arabidopsis*, soybean, and canola, respectively, have been deposited in miRBase, an online database for published miRNA sequences ([5]Kozomara and Griffiths-Jones, 2014). Another database, MiRTarBase ([6]Hsu et al., 2014) contains experimentally confirmed miRNA-target interactions (Hsu et al., 2014). In addition, MiRFANs ([7]Liu et al., 2012) stores miRNA functional annotations specifically for *Arabidopsis*, and it includes an analysis toolbox.

The production of data from Next Generation Sequencing studies is providing the scientific community with vast amounts of genomic data that can be mined to answer many important biological questions about the seed. Dramatic improvements to seed transcriptome experiments, including enhanced sequencing chemistries and better bioinformatics tools should provide the necessary tools and data required to answer these questions. With Next Generation Sequencing, subtle changes to the transcriptome can now be detected with high confidence and exploited to identify most of the genes and gene products responsible for seed development.

[4]http://aralip.plantbiology.msu.edu/

[5]http://www.mirbase.org

[6]http://mirtarbase.mbc.nctu.edu.tw

[7]www.cassava-genome.cn/mirfans

GENOMICS OF EMBRYO DEVELOPMENT

Embryogenesis is the developmental period during which the zygote differentiates into the mature embryo. Embryo development can be divided temporally into two phases, morphogenesis and maturation (Goldberg et al., 1994). During the morphogenesis phase, the diploid zygote derived from fertilization of the egg cell by a sperm cell undergoes an asymmetric cell division, producing the apical and basal cells (Lau et al., 2012). In many plants, the apical cell gives rise to most of the embryo proper. The basal cell develops largely into the suspensor, although the uppermost suspensor cell divides to form the hypophysis that will become the quiescent center of the root apical meristem and the central root cap cells of the embryo proper. Development of the embryo proceeds along two primary axes. Along the apical-basal axis, the embryo becomes sequentially partitioned into specific pattern elements that become the cotyledons, shoot apical meristem, hypocotyl, root, and root apical meristem. The embryo proper also becomes compartmentalized along its radial axis to generate the embryonic tissue systems: procambium, ground tissue, and protoderm. The suspensor is an ephemeral structure of the embryo that serves a structural role by pushing the embryo proper into the nutrient-rich endosperm and a physiological role by transferring nutrients and growth factors to the embryo proper at early developmental stages (Kawashima and Goldberg, 2010).

As the embryo transitions from the morphogenesis to the maturation phase, morphogenetic processes, including cell division, become largely repressed (Harada, 1997; Vicente-Carbajosa and Carbonero, 2005). During the maturation phase, the embryo acquires the ability to withstand stresses imposed by desiccation that occur late in seed development and accumulates storage proteins, lipids, and/or carbohydrates to massive amounts, causing the embryo to grow as a result of cell expansion. The storage macromolecules serve as a nutrient source for the developing seedling during post-germinative development. By the end of the maturation phase, the embryo is quiescent metabolically and arrested developmentally, and it remains in this state until conditions appropriate for germination and post-germinative development are perceived.

CONTRIBUTIONS OF THE MATERNAL AND PATERNAL GENOMES TO EARLY EMBRYO DEVELOPMENT

The zygote represents the first stage of the morphogenesis phase, and two studies have addressed the question of when the zygotic genome becomes active transcriptionally following fertilization of the egg cell. In animals, early embryonic development is regulated by maternal mRNAs deposited in the egg prior to fertilization, and the zygotic genome becomes transcriptionally active several cell cycles after fertilization (Tadros and Lipshitz, 2009). The maternal-to-zygotic transition was analyzed in *Arabidopsis* by sequencing RNAs from early stage embryos derived from crosses between plants of different ecotypes and using single nucleotide polymorphisms to distinguish mRNAs derived from maternal and paternal alleles. Autran et al. (2011) reported that the majority of mRNAs in an *Arabidopsis* embryo at the two to four cell embryo proper stage are from the maternal genome, although approximately 10% of mRNAs are encoded by paternal alleles at this early stage. The paternal contribution to the mRNA population increased to 36% by the globular stage, which was interpreted to represent a gradual activation of the paternal genome. Paternal genome activity is maternally regulated through epigenetic mechanisms involving RNA-dependent DNA methylation, KRYPTONITE-mediated histone methylation, and CAF-1 complex-induced histone exchange (Autran et al., 2011). By contrast, a separate study of the maternal-to-zygotic transition reported that maternal and paternal genomes contribute almost equally to the transcriptomes of *Arabidopsis* embryos at the earliest stages of embryogenesis (Nodine and Bartel, 2012). Many mRNAs that are undetectable in the egg and sperm constitute the top 50% most abundant mRNAs in one or two-cell embryos, suggesting that the zygotic genome is activated immediately after fertilization and plays a major regulatory role during early embryogenesis. Discrepancies between the findings of these two studies may have resulted from the use of different *Arabidopsis* ecotypes by the two laboratories (Baroux et al., 2013). Alternatively, the high proportion of maternally derived mRNAs may have resulted from contamination of embryo samples by mRNAs from the SC that is entirely of maternal origin (Nodine and Bartel, 2012). Nevertheless, both studies demonstrated that the maternal-to-zygotic transition occurs at the earliest stage of embryo development in *Arabidopsis*.

ROLE OF microRNAs IN THE TRANSITION FROM THE MORPHOGENESIS TO MATURATION PHASE

The transition from the morphogenesis to the maturation phase represents a major shift in the developmental programs that occur during embryogenesis (Harada, 1997; Vicente-Carbajosa and Carbonero, 2005). The transcriptomes of *Arabidopsis* embryos that were isolated from the seed by LMD or hand dissection were profiled at several stages of development (Xiang et al., 2011; Belmonte et al., 2013), and these studies demonstrated that gene expression changes dramatically as embryos transition into the maturation phase. For example, the vast majority of mRNAs that accumulate in the embryo proper at a specific stage of development do so at the maturation phase. This gene set is enriched for those involved in maturation processes, including mRNAs encoding storage proteins, oilbody proteins, and proteins involved in lipid storage.

microRNAs play a critical role in controlling the transition from the morphogenesis to the maturation phase (Nodine and Bartel, 2010; Willmann et al., 2011). The role of miRNAs in controlling the transition from morphogenesis to maturation phase was revealed by studies of mutations affecting *DICER-LIKE1* (*DCL1*), which encodes an enzyme required for miRNA biosynthesis. Early in embryo development, loss-of-function *dcl1* mutants display abnormal cell division patterns in the hypophysis, a cell that will become incorporated into the root apical meristem, and in subprotodermal regions of the embryo. These finding were interpreted to suggest that miRNAs are required for embryo patterning events that occur during the morphogenesis phase (Nodine and Bartel, 2010; Willmann et al., 2011). Transcriptome analyses showed that mRNAs that normally accumulate specifically during the maturation phase, including those encoding storage proteins, oil body proteins, lipid biosynthesis enzymes, and several transcriptional regulators of the maturation phase, accumulate prematurely in *dcl1* mutant embryos. By contrast, two TFs, ASIL1, and HDA6/SIL1, that normally repress maturation genes after germination were downregulated in *dcl1* mutants (Willmann et al., 2011). These results, along with the finding that chloroplast maturation occurs earlier in *dcl1* mutant than wild-type embryos, were interpreted to indicate that miRNAs are required to repress maturation processes during the morphogenesis phase and that the precocious onset of the maturation phase in *dcl1* mutants causes defects in pattern formation. In particular, one set of miRNAs and their target mRNAs were implicated to mediate temporal control of the maturation phase. In *dcl1* mutants, disruption of *miR156* accumulation causes the premature upregulation of two differentiation promoting TFs, SPL10, and SPL11, and experiments analyzing the effects of altering *SPL10* and *SPL11* expression suggested that they are at least partially responsible for repressing the maturation processes early in embryogenesis. A different miRNA, *miR166*, has been shown to repress genes expressed specifically during the maturation phase in vegetatively growing plants (Tang et al., 2012). Together, these observations suggest miRNAs play critical roles in controlling embryonic processes.

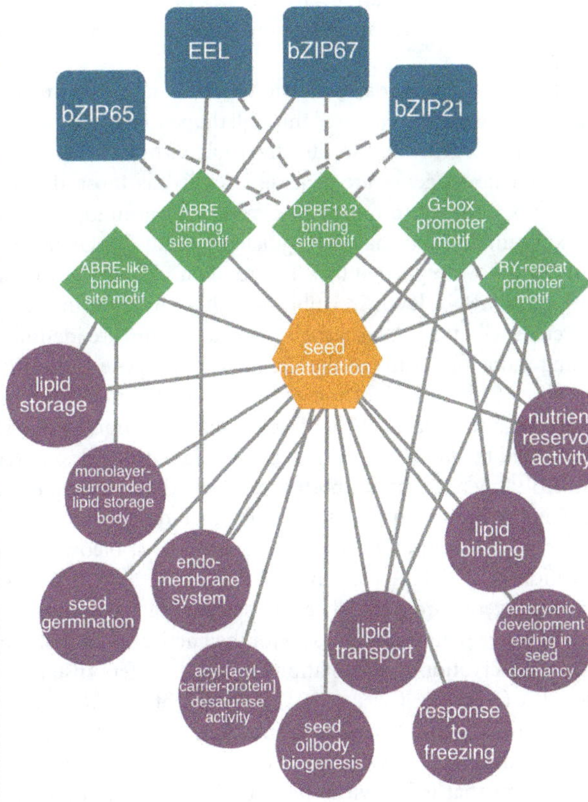

FIGURE 3 | Predicted bZIP-regulated seed maturation network. bZIP TFs (blue squircles) are predicted (dashed lines) or known (solid lines) to bind to DNA sequence motifs (green diamonds) within the 1 kb upstream region of the transcription start site in genes associated with enriched GO terms like lipid storage, nutrient reservoir activity and seed oilbody biogenesis ($P < 0.001$, hypergeometric distribution, purple circles). Genes associated with the network are co-expressed during seed maturation (orange hexagons). Modified from Belmonte et al. (2013).

MATURATION GENE REGULATORY NETWORKS

Several studies have focused on understanding the gene regulatory networks that operate during the maturation phase of seed development (reviewed by Santos Mendoza et al., 2005; Gutierrez et al., 2007; Braybrook and Harada, 2008; Holdsworth et al., 2008; Suzuki and McCarty, 2008; Junker et al., 2010). To gain insight into embryo maturation gene regulatory networks, Belmonte et al. (2013) identified DNA sequence motifs that are overrepresented in the 5′ flanking regions of a set of genes that are expressed in embryos specifically during the maturation phase. TFs that are known or predicted to bind these overrepresented DNA sequence motifs were also identified, permitting a putative gene regulatory network to be created. The network included a number of *cis*-acting DNA elements that have been shown previously to regulate genes expressed during the maturation phase, including the *ABRE*, *ABRE-like*, *DPBF1*, *DPBF2*, and *RY* motifs. Identified among the TFs known to bind these motifs were EEL and bZIP67, which are known to regulate genes during the maturation phase. An example of a maturation gene regulatory network is presented in **Figure 3** and a description of the construction of gene regulatory networks is presented in "Identifying regulatory networks required to program the *Arabidopsis* seed" below.

Studies to characterize regulators of the maturation phase have focused on the *Arabidopsis* LEC1, LEC2, FUS3, and ABI3 TFs (Koornneef et al., 1984; Meinke, 1992; Keith et al., 1994; Meinke et al., 1994; West et al., 1994). LEC1 is a HAP3 (a.k.a. NF-YB) subunit of the CCAAT-binding (NF-Y) TF (Lotan et al., 1998), whereas LEC2, FUS3, and ABI3 are B3-domain TFs (Giraudat et al., 1992; Luerssen et al., 1998; Stone et al., 2001). The central roles of these maturation TFs in controlling embryo and seed development was established initially through investigations of mutations in these genes. Loss-of-function mutations in these maturation TF genes cause embryo lethality or the ablation of embryo parts, because mutant embryos are intolerant of desiccation and storage protein and lipid accumulation is defective. Ectopic expression of these maturation TF genes induces somatic embryo development, fatty acid biosynthesis, oil body accumulation and storage protein biosynthesis in vegetative cells (Parcy et al., 1994; Lotan et al., 1998; Kagaya et al., 2005a; Santos Mendoza et al., 2005; Mu et al., 2008; Stone et al., 2008; Feeney et al., 2013).

The maturation TFs LEC1, LEC2, FUS3, and ABI3 are involved in complex and redundant regulatory interactions during embryo development (reviewed by Braybrook and Harada, 2008; Junker et al., 2010). Genetic and molecular experiments have shown that LEC1 functions upstream of LEC2, FUS3, and ABI3 and, therefore, is likely to act at or near the top of the regulatory hierarchy controlling maturation (Kagaya et al., 2005b; To et al., 2006). Redundancy is observed in interactions among the other maturation TFs that is dependent on their spatial location in the embryo (To et al., 2006). For example, the *FUS3* gene is regulated by LEC1, LEC2, and ABI3 in cotyledons, by LEC2 and ABI3 in the embryonic axis, and by LEC2 and FUS3 in the root tip. Together, the results suggest that these maturation TFs play key but complex roles in the regulatory network controlling the maturation phase of seed development.

In recent years, initial dissection of the maturation gene regulatory network has occurred through the genome-wide identification of target genes that are directly regulated by the maturation TFs. Direct target genes are generally defined as those that are bound by a TF, as determined by chromatin immunoprecipitation experiments, and that are regulated by that TF. Genes that are up- and downregulated by a TF are often identified by comparing their mRNA levels in embryos with a mutation in the TF gene versus wild type. Alternatively, regulated genes are identified by using inducible forms of the TF. Imposing a gene expression constraint on the identification of direct target genes is important, because fewer than 10% of genes that are bound by a TF are regulated by that TF (Farnham, 2009). Genome-wide analysis identified 98 genes that are both bound by ABI3 and regulated following the induction of ABI3 activity, including genes encoding 2S seed albumins, 12S seed storage globulins, oleosins, and desiccation-related LEA proteins (Monke et al., 2012). Most of these target genes are generally expressed during the maturation phase, and they require abscisic acid for their activation, consistent with the observation that mutations in *ABI3* confers insensitivity to ABA (Koornneef et al., 1984). Analysis of the ABI3 target genes identified two DNA sequence motifs that are both overrepresented in the first 250 bp upstream of the transcription start site: a RY element that is known to be bound by ABI3 and a G-box motif. The G-box is part of a well-characterized ABA-responsive element (e.g., *ABRE*) motif that interacts with bZIP TFs. These finding are consistent with previous studies showing that ABI3 interacts with a bZIP TF to regulate the transcription of genes involved in maturation processes (Nakamura et al., 2001; Lara et al., 2003).

Target genes for another B3-domain maturation TF, FUS3, were identified from embryonic culture tissue overexpressing the *AGL15* gene that expresses *FUS3* constitutively (Wang and Perry, 2013). FUS3 target genes were enriched for maturation processes, and showed a 17% overlap with ABI3 target genes. The 5′ flanking regions of the FUS3 target genes were enriched for RY and G-box motifs. These studies confirmed on a genome-wide scale that there is at least partial redundancy in the functions of FUS3 and ABI3. FUS3 also directly regulates another B3-domain TF, VAL1, which along with VAL2 and VAL3, acts as repressors of the maturation network during seedling development (Suzuki and McCarty, 2008). FUS3 was also shown to regulate miRNA genes, including *miR156*, *miR160*, *miR166*, *miR169*, *miR369*, and *miR390*. Thus, FUS3 may be involved in controlling the shift from the morphogenesis to maturation phase given the proposed role of *miRNA156* in this transition.

Genetic and molecular studies place LEC1 at or near the top of the regulatory hierarchy controlling the maturation phase (Kagaya et al., 2005a; To et al., 2006). Analysis of genes that are bound and regulated by LEC1 identified two genes, *LEC1-LIKE* and *FATTY ACID BIOSYNTHESIS2*, which suggested a potential role for LEC1 in lipid biosynthesis and other maturation processes (Junker et al., 2010). Other direct target genes regulated by LEC1 are involved in auxin and brassinosteroid biosynthesis and signaling, light responses and transcription regulation. The studies also demonstrated an interaction between LEC1 and ABA signaling. For example, although LEC1 can bind to the 5′ flanking sequences of the *YUC10* gene that encodes an auxin biosynthetic enzyme in the absence of ABA, LEC1-induced *YUC10* expression is ABA dependent. Together, these results suggest that LEC1 plays an integrative role during plant development.

GENOMICS OF ENDOSPERM DEVELOPMENT

Endosperm development is initiated with the fertilization of the central cell of the female gametophyte by a sperm cell and proceeds through three distinct stages in most angiosperms: syncytial, cellularization, and cellular (Olsen, 2004; Li and Berger, 2012). During the syncytial stage, the endosperm undergoes nuclear divisions without corresponding cell divisions, generating a syncytium of nuclei that each associates with a cytoplasmic region to form nuclear-cytoplasmic domains (Brown et al., 1999). This period of syncytial development is followed by cellularization in which cell walls form around nuclear cytoplasmic domains, beginning after the eighth nuclear divisions in *Arabidopsis*. Cellularization proceeds in a wave-like manner from the micropylar to the chalazal ends of the endosperm (**Figure 1A**). During the cellular stage, additional endosperm cells are formed through cytokinesis primarily at the periphery of the endosperm. Complex patterning of the endosperm is perhaps best exemplified by the *Brassicaceae*, including *Arabidopsis* and canola, in which three distinct endosperm subregions form corresponding to their positions within the seed: micropylar, peripheral, and chalazal (**Figure 1A**). These spatial domains are specified at the earliest stage of endosperm development in that their nuclear, cytoskeletal, and cytoplasmic characteristics and positions within the endosperm are distinguished by the fourth mitotic division (Brown et al., 2003). Depending upon the species, the endosperm remains largely intact throughout seed development as occurs in cereal grains, or it degrades as in *Arabidopsis*, canola, and soybean seeds.

ENDOSPERM DOMAINS HAVE DISTINCT AND OVERLAPPING FUNCTIONS

Transcriptome analyses of the *Arabidopsis* endosperm have provided novel insights into the relationship between the micropylar, peripheral, and CZE subregions. Previous work using LMD to profile endosperm mRNA populations provided the first characterization of gene expression genome-wide in the micropylar, peripheral, and chalazal subregions (Belmonte et al., 2013). These studies showed that a small subset is expressed specifically in each endosperm subregion at virtually all stages of development, suggesting strongly that each subregion fulfills a unique function within the seed. In particular, the CZE has the largest number of genes that are expressed specifically in a single subregion of the seed and the most seed-specific genes among all subregions. Analyses of these CZE-specific genes showed that they encoded rate-limiting enzymes involved in the biosynthesis of the hormones gibberellic acid, abscisic acid, and cytokinin (Day et al., 2008; Belmonte et al., 2013), confirming the work of others who localized these enzymes to the CZE (Miyawaki et al., 2004; Lefebvre et al., 2006; Hu et al., 2008). Chalazal endosperm-derived abscisic acid, cytokinin, and gibberellic acid, respectively, are involved in controlling seed dormancy, endosperm cellularization, and growth of maternal tissues. Thus, the CZE may serve as a hub that supplies hormones to regulate developmental processes in developing seeds.

Analyses of the transcriptome datasets uncovered dominant patterns of gene activity for mRNAs that are involved in processes critical for seed development and that occur in all three endosperm domains and in the embryo. Clustering analyses identified a number of different gene sets that are expressed at early stages of seed development in the embryo and micropylar and PEN, but their expression in the CZE is delayed until the late developmental stages (Belmonte et al., 2013). One set encodes proteins involved in cytokinesis, consistent with the observation that embryo cells undergo cytokinesis concurrently with mitosis, whereas endosperm cellularization proceeds from the micropylar to the chalazal ends of the endosperm. Another set is involved in photosynthesis and carbon metabolism, a surprising result given that these processes were known to occur in the embryo but much less was known about their role in the endosperm. Additional analyses provided strong evidence that maturation processes occur not only in the embryo but also in all endosperm subregions. Together, these results emphasize a strong degree of overlap in gene expression programs between the embryo and endosperm regions of the seed.

GENOMIC IMPRINTING AND THE CONTROL OF SEED SIZE

The endosperm has a profound influence on seed size. It has been shown or hypothesized that the size of the endosperm early in seed development, the timing of cellularization of endosperm cells, the provisioning of maternally derived nutrients from the endosperm to the embryo, and the influence of the endosperm on the proliferation and elongation of SC cells are major determinants in specifying seed size (Scott et al., 1998; Garcia et al., 2003, 2005; Melkus et al., 2009; Ohto et al., 2009). The endosperm influences seed size through parent-of-origin effects. Parent-of-origin effects are exemplified by genetic crosses between plants of different ploidy levels. Progeny from interploidy crosses that have an excess of maternal genomes (e.g., tetraploid female crossed with diploid male) produce seeds that are smaller than self-fertilized diploid plants, whereas plants with an excess of paternal genomes (e.g., diploid female by tetraploid male) produce larger seeds (Scott et al., 1998). The parental conflict theory has been proposed to explain the antagonistic influences of the mother and father. It is hypothesized that in polygamous organisms, the father will attempt to enhance the allocation of maternally derived resources specifically to his offspring to maximize their growth, whereas the mother will try to distribute resources equally to all offspring to equalize their growth (Haig and Westoby, 1989).

Parental influences on seed size are thought to be mediated by genomic imprinting. Imprinted genes are expressed following fertilization predominately from either the maternal or paternal alleles unlike the vast majority of genes that are expressed nearly equally from both alleles. Imprinted genes are thought to control resource allocation to the embryo and therefore support its growth. Consistent with this hypothesis, an imprinted gene has been shown to be involved in controlling maternal nutrient uptake and seed biomass (Costa et al., 2012). Imprinted genes have been identified using RNA sequencing experiments in which *Arabidopsis* plants of different ecotypes were crossed, and mRNAs from maternal and paternal alleles in the progeny were distinguished based on single nucleotide polymorphisms (Gehring et al., 2011; Hsieh et al., 2011; Wolff et al., 2011). These studies identified between 60 and 208 imprinted genes and showed that maternally expressed imprinted genes (MEGs) are more prevalent than paternally expressed imprinted genes (PEGs). Although these studies rarely identified any genes as being imprinted in the embryo, a recent study by Raissig et al. (2013) identified 11 MEGs and one PEG in the *Arabidopsis* embryo.

Genomic imprinting is regulated through epigenetic mechanisms involving DNA methylation and the Polycomb Repressive Complex 2 (PRC2). 5′-Methylcytosine in DNA is an epigenetic mark that is often associated with transcriptionally silenced genes, and PRC2 mediates gene silencing through the trimethylation of lysine 27 of histone H3 (H3K27me3, Kohler et al., 2012). To dissect the mechanisms regulating imprinted genes, Hsieh et al. (2011) and Wolff et al. (2011) analyzed the effects of mutations that cause defects in DNA methylation, DNA demethylation, and the PRC2 complex on gene imprinting. Collectively, their results showed that the DNA methylation status of MEGs correlated strongly with their imprinting. During female gametophyte development, the genome of the central cell, that is the maternal precursor of the endosperm, becomes hypomethylated globally due to the activity of DME, a DNA glycosylase that removes methylcytosine residues from DNA (Gehring et al., 2009; Hsieh et al., 2009). Hypomethylation of MEGs in the central cell results in the expression of maternal alleles of MEGs in the endosperm, whereas the paternal alleles retain their DNA methylation marks and remain silenced. The paternal alleles of some MEGs have also been shown to be silenced through the PRC2 pathway. By contrast, the paternal alleles of PEGs are active, but the maternal alleles are silenced predominately through the PRC2 pathway. These studies support the idea that demethylation of the maternal allele of some PEGs is required to permit the gene to be silenced by the PRC2 (Weinhofer et al., 2010). Thus, a complex set of epigenetic regulatory mechanisms underlies genomic imprinting.

A potential causal link between parent-of-origin effects and endosperm size came from studies of 24 nucleotide p4 siRNAs in developing endosperm (Lu et al., 2012). p4 siRNAs, which in endosperm are derived specifically from the maternal genome, function in RNA-dependent DNA methylation to target specific loci for methylation (Mosher et al., 2009; Law and Jacobsen, 2010). p4 siRNAs primarily target transposable elements for DNA methylation. However, a significant fraction of genes are closely associated with transposons, and methylation of some of these transposons influences the expression of the linked gene. Genome-wide profiling of sRNAs in interploidy crosses of *Arabidopsis* showed that 24 nt siRNAs corresponding to specific genomic loci were strongly overrepresented in endosperm of seeds with a maternal genome excess relative to seeds with a paternal genome excess. Several of these loci corresponded to genes encoding AGL TFs, one of which has been shown to inhibit endosperm cellularization

(Kang et al., 2008). These findings were interpreted to indicate that p4 siRNAs targeting AGL TFs are overrepresented in endosperm with a maternal genome excess, causing premature repression of the expression of *AGL* genes and precocious cellularization, resulting in a smaller seed. Together, these findings indicate a critical role for the endosperm in several aspects of seed development.

GENOMICS OF SEED COAT DEVELOPMENT

Compared to the embryo and endosperm, the SC has received little attention at the genomics level. The maternally derived SC is responsible, in part, for the evolutionary success of the seed, and it plays an integral role in filling (Verdier et al., 2013), protection, and dispersal of seeds (Haughn and Chaudhury, 2005). The SC region, like the embryo and endosperm, can further be divided into subregions based on morphological and anatomical features. For example, in *Arabidopsis* and canola, the distal SC comprised the inner and outer integuments, undergoes dramatic anatomical transformations including cell expansion, changes in cell wall deposition, and anthocyanin and mucilage accumulation followed by programed cell death, all in preparation for seed dormancy. Conversely, the CZSC, located proximal to the funiculus, is found at the junction with the maternal plant. In seeds of legumes, like soybean, a total of six subregions have been identified: (i) endothelium, (ii) hour glass, (iii) palisades, (iv) parenchyma, (v) epidermis, and (vi) hilum. The hilum in soybean is considered to be similar in function to the CZSC in *Arabidopsis* and presents the first point of entry of material destined for filial seed compartments.

While the development and anatomy of the SC in oilseeds, such as *Arabidopsis*, soybean, and canola, have been extensively studied (Beeckman et al., 2000; Western et al., 2000; Windsor et al., 2000; Macquet et al., 2007; Young et al., 2008; Dean et al., 2011) there is remarkably little information about the genes and gene regulatory networks underlying this multicellular structure. Even less information is available about the genomics of SC development in emerging model crop systems. Of the few studies that have examined the SC at the genomics level (Jiang and Deyholos, 2010; Dean et al., 2011; Belmonte et al., 2013; Khan et al., 2014), data suggest the SC is more similar to maternal tissues than to the embryo or the endosperm. Despite the vast amount of data currently being generated and the different technologies being employed to study the SC, it is still unclear how many genes are active in each subregion and how those numbers change between species.

TRANSCRIPTIONAL REGULATION IN THE SEED COAT

When comparing the SC to other seed regions, *Arabidopsis* is the best plant model studied to date. Hierarchical clustering of GeneChip data showed differences between each subregion of the SC. It is clear that global similarities and differences exist in the SC region compared to the embryo and endosperm and have likely evolved over time to protect the embryo and to adapt to environmental conditions (Debeaujon et al., 2000). Quantitative differences in gene activity within subregions of the SC provided insight into the biological processes underlying its development. Dominant patterns of gene expression were identified from comprehensive RNA profiling of *Arabidopsis* seed subregions. This analysis identified sets of genes that show spatial (between different subregions) and temporal (across seed development) differences in expression (Belmonte et al., 2013; Khan et al., 2014). Co-expressed gene sets were shown to represent biological processes associated with the development of SC color (Zhang et al., 2013), anthocyanin deposition (Debeaujon et al., 2003), and mucilage accumulation (Western et al., 2001), which have been extensively studied using forward genetic analyses. These studies revealed essential processes associated with the SC that are controlled by individual genes or small sets of genes, yet it was still unclear how all of these processes may be coordinated over the lifecycle of the seed.

Cellular processes that occur in the SC have been independently shown to be controlled by TFs belonging to MYB (Nesi et al., 2001; Penfield et al., 2001), HD-Zip (Johnson et al., 2002; Ishida et al., 2007), and MADS-Box (Nesi et al., 2002; Huang et al., 2011) families. Our comprehensive SC transcriptome analysis identified all of these TF mRNAs in a single analysis (Khan et al., 2014). Not only were all of these known regulators identified in our experiment, we also identified a number of possible gene targets responsible for cell fate specification, the accumulation of mucilage, the deposition of anthocyanin, flavonoid biosynthesis, and SC color.

TRANSCRIPTIONAL REGULATION OF SEED COAT COLOR

Seed coat color is an agroeconomically important trait and is determined by the presence or absence of flavonoids, more specifically, proanthocyanidins. Flavanoids are secondary metabolites produced in plants derived from the phenylpropanoid pathway and are thought to have a number of functional roles, including photoprotection (Agati et al., 2013) and cellular signaling (Pourcel et al., 2013). Proanthocyanidins accumulate exclusively in the SC. When cells in the SC die, the proanthocyanidins oxidize and polymerize to form brown pigments that darken the seed. Mutants that have defects in proanthocyanidin production form a lighter colored or transparent SC (yellow/green). The yellow/green SC coloration is often associated with other desired agroeconomic traits such as thinner SCs, decreased fiber, and higher protein and oil contents (Simbaya et al., 1995; Lipsa et al., 2011; Jiang et al., 2013b). Proanthocyanidin deficient mutants do not appear to have any major physiological disturbances other than SC color; however, some evidence suggests the mutants may have diminished responses to abiotic/biotic stress (Pourcel et al., 2013), longevity, and germination (Dean et al., 2011; Jiang et al., 2013a).

Seminal work in the genetics and biochemistry of SC color in *Arabidopsis* revealed complex networks of genes and gene products responsible for this trait (Yu, 2013). In canola and soybean, genes that contribute to SC color are more difficult to identify genetically due to redundancies within the genome. RNA sequencing of brown- and yellow-coated *B. juncea* revealed three dihydroflavonol reductase genes and three anthocyanin reductase genes that were highly expressed in the brown-seeded variety with almost no detectable expression in the yellow-seeded variety (Liu et al., 2013a). The expression of three phenylpropanoid biosynthetic genes, ten flavonoid biosynthetic genes and four regulatory genes were studied using qRT-PCR at seven developmental stages in yellow- and brown-seeded *B. napus*. Two propanoid biosynthetic genes (*PHENYLALANINE AMMONIA LYASE, TRANS-CINNAMATE 4-MONOOXYGENASE*), two flavonoid biosynthetic genes (*TRANSPARENT TESTA4, 6*), five anthocyandin/proanthocyandin biosynthetic genes (*3,4-DICHLOROPHENOL GLYCOSYLTRANSFERASE 2, TRANSPARENT TESTA3, 10, 12, 18*), and three TFs (*TRANSPARENT TESTA8, TRANSPARENT TESTA GLABRA1, 2*) had different expression patterns in yellow seeds (Qu et al., 2013). Further, eleven quantitative trait loci mediating SC color and fiber content were identified using high-density SNP arrays in canola (Liu et al., 2013a). Together genomics studies of SC color provide new targets for improving desirable traits, such as seed oil quality, and highlight the genetic complexity of SC color (Liu et al., 2013b). The analysis and identification of new QTLs combined with RNA sequence data should provide the information needed to design improved breeding strategies.

TRANSCRIPTIONAL REGULATION IN THE CHALAZAL SEED COAT

While the distal SC has been the primary focus of numerous functional studies, the CZSC has not been studied in the same detail. Bioinformatic analysis of CZSC mRNA populations uncovered a number of transport processes that showed dynamic programs of activity across development. These processes had not been described previously because of the inaccessibility of the CZSC within the seed for experimental analysis. For example, genes associated with phloem unloading including *SUCROSE-PROTON SYMPORTER 2* (*SUC2*) and a complement of *SWEET* genes encoding sucrose efflux transporters, amino acid transport genes including *BIDIRECTIONAL AMINO ACID TRANSPORTER 1* (*BAT1*), *AMINO ACID PERMEASE 2* (*AAP2*), water transport genes encoding tonoplast intrinsic proteins (TIP1;1), and plasma membrane intrinsic protein are all expressed in the CZSC. These findings support the hypothesis that transport processes are enriched in the CZSC. Co-expression networks generated from transcriptome data provided insight into the regulation of these transport processes. A putative G-box regulated network controlling water and sugar transport in the developing seed through bZIP TFs, including bZIP25, bZIP28, and LRL1 (Khan et al., 2014). Functional characterization of these transcriptional regulators predicted to be associated with CZSC function presents a new avenue of targeted seed improvement through modification of maternally derived subregions.

IDENTIFYING REGULATORY NETWORKS REQUIRED TO PROGRAM THE *Arabidopsis* SEED

To better understand the underlying transcriptional mechanisms required to program the seed, an integrative systems biology approach should be applied that incorporates molecular and computational biology. First, large-scale datasets are required for such an approach, and excellent sources of seed genomic data are available at databases such as GEO and NCBI as discussed previously. However, mining this data effectively requires the development of more advanced and user-friendly tools that are available to a broader scientific audience through online databases. Tools from the BioArrayResource[8], Genevestigator[9], and The *Arabidopsis* Information Resource[10] are all excellent resources for genomics-based data including but not limited to whole seed, seed region, and seed subregion datasets. In addition, the seedgenenetwork.net database houses whole seed, seed region, and seed subregion transcriptome, sRNA, and DNA methylome datasets from *Arabidopsis* and soybean. Although usability of online tools continues to improve, it remains difficult to identify genes with key roles in seed development with these online tools.

Using high-resolution seed datasets from *Arabidopsis* (Le et al., 2010; Belmonte et al., 2013; Khan et al., 2014), we developed a user-friendly bioinformatics program to identify transcriptional circuits from large-scale datasets at every stage of the seed lifecycle[11]. We identified genes, focusing our attention on TFs that are predicted to control biological processes across developmental time or that are specific to a seed subregion, including the embryo proper, micropylar endosperm, CZE, or the distal and CZSCs. The transcriptional module analysis is based on the association of a specific set of co-expressed genes with their enriched Gene Ontology terms, known DNA sequence motifs, metabolic processes, and TF families and presents the user with possible gene targets regulating biological processes within the seed.

For example, we identified a transcriptional module consisting of genes expressed specifically in the micropylar endosperm and that are enriched for the WRKY DNA sequence motif in their 5′ flanking regions. Our model predicts *MINISEED3* to control processes associated with the endomembrane system in the early stages of seed development. While *MINISEED3* has previously been shown to localize to the micropylar endosperm (Luo et al., 2005), the model allows us to predict gene targets of this TF which were previously unknown (**Figure 4A**). We also studied a putative transcriptional network underlying the CZE. Up until recently, genetic information about this understudied subregion was lacking. However, through our integrative bioinformatics approach we identified a putative *CIRCADIAN CLOCK ASSOCIATED1*-regulated transcriptional circuit controlling ubiquitin-dependent protein catabolic processes (**Figure 4B**). Within the SC, we identified a number of regulators that have been previously associated with SC development, allowing a high degree of confidence in our predictive transcriptional modules (**Figure 4C**). The TRANSPARENT TESTA GLABRA complex is implicated in the regulation of flavonoid biosynthesis, and several MYB TFs (including MYB5) are implicated in the regulation of mucilage biosynthesis and the differentiation of the outer integuments (Khan et al., 2014).

While this type of data analyses has been used successfully to identify existing transcriptional circuits, the real power of this approach lies in the identification of unknown interactions and prediction of the biological processes controlled by a TF. One of the caveats to this method is that a well-annotated genome must

[8]www.bar.utoronto.ca

[9]www.genevestigator.com

[10]www.arabidopsis.org

[11]http://seedgenenetwork.net/presentation#software

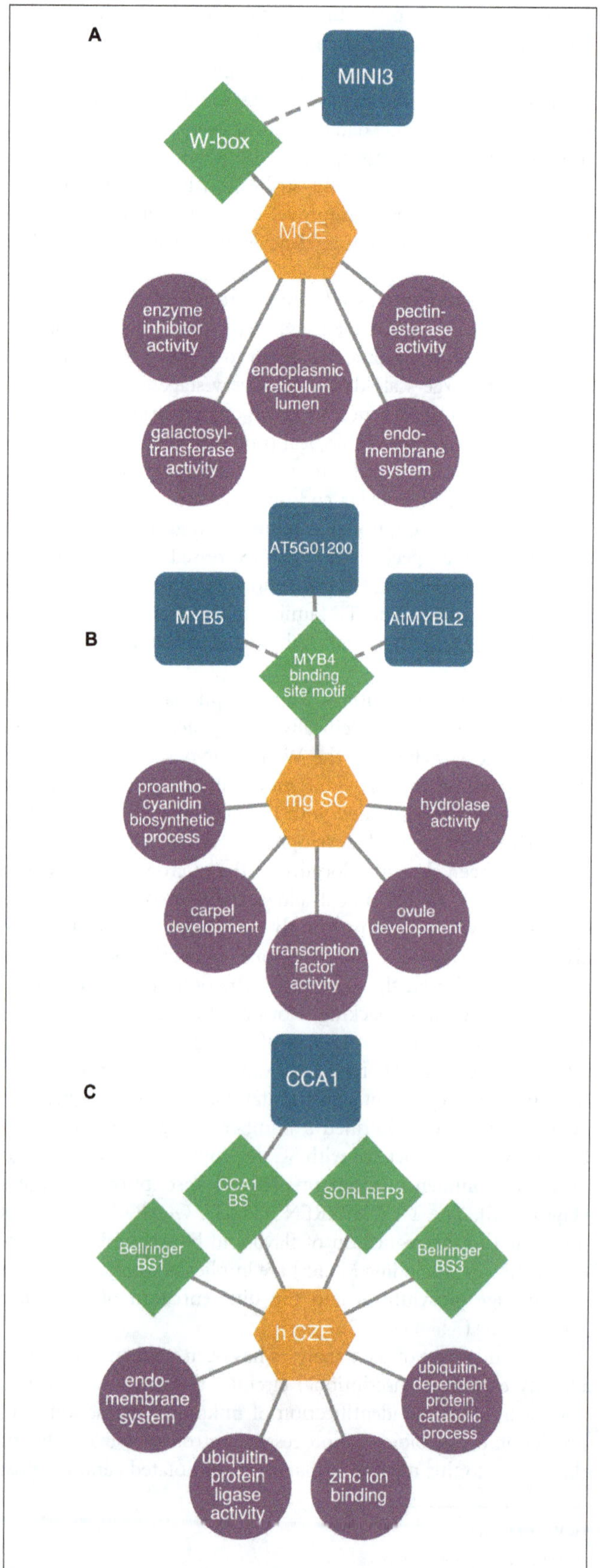

FIGURE 4 | Continued

FIGURE 4 | Continued
Predictive transcriptional circuits in subregions of the *Arabidopsis* seed. (A) *MINISEED3* (*MINI3*)-W-box transcriptional circuit in the micropylar endosperm (MCE) regulating processes like the endomembrane system. **(B)** A *CIRCADIAN CLOCK ASSOCIATED1* (*CCA1*) module in the chalazal endosperm (CZE) of heart-stage seeds. **(C)** A *MYB* transcriptional module in the mature green (mg) seed coat (SC) predicted to control processes like proanthocyanidin metabolism and ovule and carpel development. TFs (blue squircles) are predicted (dashed lines) or known (solid lines) to bind to DNA sequence motifs (green diamonds) within the 1 kb upstream region of the transcription start site in genes associated with enriched ($P < 0.001$, hypergeometric distribution) GO terms (purple circles) within patterns of co-expressed gene sets (orange hexagons). All networks are modified from Belmonte et al. (2013).

be available as a reference. Thus, one of the challenges in emerging crop systems will be the annotation of genomes for which genomics research is still in its early stages. While we are beginning to understand some of the molecular mechanisms underlying the development and properties of different seed subregions and regions, the interconnectedness of these transcriptional circuits will remain a priority in the effort to elucidate the complex regulatory pathways responsible for seed development. The spectacular increase in genomic resources applicable to the seed will enable a more comparative approach to uncover and study both conserved and unique transcriptional circuits among related seed species such as the *Brassicaceae* or the *Leguminosae*. Current efforts are directed at implementing and developing computational programs to identify gene regulatory networks for important crop species like canola and soybean. The ability to predict transcriptional circuits in cell and tissue types previously thought to be inaccessible within the seed provides unprecedented insight into the regulation of biological processes over developmental time.

IDENTIFICATION OF TFs ESSENTIAL FOR SEED DEVELOPMENT

Analysis of putative gene regulatory networks is an excellent way to identify possible regulators of seed development. However, experimental validation and functional characterization of the TFs are required to validate the network. Identification of essential seed genes is a cumbersome task yet remains a priority for those interested in studying seed biology and genomics. While research has focused on essential seed genes that when mutated cause a seed lethal phenotype, other mutant phenotypes may result in defects in metabolic pathways or biochemical processes, cellular development, morphology, or other more subtle molecular phenotypes. Through our work, we have identified a number of region- and subregion-specific TFs; however, the vast majority of mutant alleles of these regulators failed to show a seed lethal phenotype (Le et al., 2010; Belmonte et al., 2013). Thus, the function of most subregion-specific TF mRNAs discovered in our work remains unknown.

Much has been learned about the seed through the use of forward genetics. Forward genetics involves generation of random mutations within an organism through radiation-, chemical-, or insertion-induced mutagenesis followed by screening for an aberrant phenotype. Systems for phenotyping mutants are becoming increasingly automated (Fiorani and Schurr, 2013), and NGS strategies are being used to map the mutation site in what is being referred to as "fast-forward" genetics (Schneeberger and Weigel, 2011). Through forward genetics, an extensive collection of *Arabidopsis* T-DNA mutants is available through the SALK Institute (Alonso et al., 2003), and a database of essential seed genes has been established at seedgenes.org (Meinke et al., 2008) and the *Arabidopsis* Biological Resource Center.

As we continue to characterize the seed genome, forward genetics becomes increasingly ineffective as the likelihood of discovering previously uncharacterized mutants decreases. Molecular tools such as RNA interference and over-expression lines have provided researchers with important information about their genes of interest. However, new genome editing techniques utilizing the CLUSTERED REGULARLY INTERSPACED SHORT PALINDROMIC REPEATS (CRISPR)/CRISPR-Associated System (CAS; Xie and Yang, 2013), Transcription Activator-Like Effector Nucleases (TALENs; Christian et al., 2013), and Zinc Finger Nucleases (ZFNs; Zhang et al., 2010; de Pater et al., 2013), are becoming popular alternatives to classical mutagenesis. Unlike the previous approaches that relied solely on chance, emerging technologies provide an efficient means to achieve targeted mutagenesis and target multiple alleles simultaneously (Curtin et al., 2011). In addition, there is the potential for targeting non-coding regions of the genome to elucidate regulatory functions of nucleic acid sequences (Gaj et al., 2013). Of these systems, the most recent to emerge is the CRISPR/CAS system. Unlike ZFNs and TALENs that rely on complicated protein–DNA interactions, the CRISPR/CAS system uses guiding RNAs and simple base pairing between the RNA construct and target site. In addition, this technology has the ability to perform multiple genome edits by targeting more than one location simultaneously (Cong et al., 2013). This technology is also proving to have several additional practical applications, such as the modification of gene expression *in vivo* through gene fusion to transcriptional activation or repression domains (Bikard et al., 2013) or for the labeling of individual chromosomal loci (Chen et al., 2013). Taken together, the ability to manipulate transcriptional networks and fine-tune gene expression would prove valuable tools for the molecular dissection and engineering of seeds.

FUTURE DIRECTIONS

It is an exciting time to study the underlying mechanisms of seed development through genomics. The complex morphological and metabolic transformations of the seed lend themselves to intensive genomic interrogation. While seminal work dissecting cells, tissues, and organs of both *Arabidopsis* and soybean seeds has revealed an incredible abundance of information, there are still pressing questions when it comes to the coordination and regulation of seed development at the cellular and tissue levels. To answer these questions seed biologists are using modern sequencing strategies. The incredible amount of information produced by these technologies is overwhelming, and the information extracted from these analyses will only continue to improve as we perfect the chemistries and foster new collaborations with mathematics, statistics and computer science. These in-depth analyses yield significant information about the transcriptional circuitry underlying complex tissue systems responsible for the development of the seed. Moreover, identification of transcriptional regulators from large-scale datasets will provide the necessary starting point for research focusing on improving seeds.

To achieve these goals, plant biologists are coupling cutting-edge technologies that are capable of dissecting or isolating individual cells and tissues of the seed with sequencing platforms. In addition to mRNA profiling, LMD has been coupled to genomics strategies such as bisulfite sequencing to study global changes in DNA methylation marks, degradome sequencing to study miRNA cleavage sites, and ChIP sequencing to identify protein/TF DNA interactions during seed development. DNA sequencing, bisulfite sequencing, RNA and small-RNA sequencing, degradome sequencing, ChIP sequencing, and CLIP sequencing (protein–RNA interactions) each provide a piece to the developmental puzzle, and sophisticated integrative computational analyses will be required to put all of the pieces together. Thus, the development of integrative computational tools to analyze complex and possibly disparate datasets in all plants will remain a major challenge for the scientific community.

Despite the tremendous advances in genomics-focused research including NGS platforms and the continuing reduction in the cost and production of high-resolution datasets, functional characterization of genes responsible for seed development, especially in emerging model systems, remains a challenge. Functional testing and characterization of the biological information derived from the billions of data points that sample the dynamic biological processes underlying seed development will take decades using current molecular biology tools. Thus, high-throughput functional characterization of genes and gene products remains a top priority for plant biologists. There are four areas of seed genomics and its application that we suggest need to be targeted to further improve our understanding of the seed: (i) update and curate small- and large-scale genomics data in publicly available databases; (ii) implement user-friendly data analysis pipelines and educate scientists on how to use them effectively; (iii) profile and characterize the genomes of emerging models important for global crop production and development; (iv) functionally characterize every gene responsible for plant traits relevant to sustainable agriculture.

CONCLUSION

Current advancements in seed genomics are illuminating the genetic forces driving seed development. It is now possible to identify most of the genes responsible for guiding seed development in every cell, tissue, and organ throughout the seed lifecycle. Together, modern breeding strategies that include information derived from genomics-based research will provide the necessary tools to improve seeds: seeds with improved nutritional value, that can endure adverse environmental conditions, or one that can withstand biological attack. Our dependence on seeds for food, fuel, and other resources means seed improvement research through genomics will continue to have a significant impact on global biosustainability.

ACKNOWLEDGMENTS

This work was supported in part by a grant from the Plant Genome Program of the National Science Foundation to John J. Harada and a National Science and Engineering Research Council Discovery Grant to Mark F. Belmonte.

REFERENCES

Agati, G., Brunetti, C., Di Ferdinando, M., Ferrini, F., Pollastri, S., and Tattini, M. (2013). Functional roles of flavonoids in photoprotection: new evidence, lessons from the past. *Plant Physiol. Biochem.* 72, 35–45. doi: 10.1016/j.plaphy.2013.03.014

Alonso, J. M., Stepanova, A. N., Leisse, T. J., Kim, C. J., Chen, H., Shinn, P., et al. (2003). Genome-wide insertional mutagenesis of *Arabidopsis thaliana*. *Science* 301, 653–657. doi: 10.1126/science.1086391

Autran, D., Baroux, C., Raissig, M. T., Lenormand, T., Wittig, M., Grob, S., et al. (2011). Maternal epigenetic pathways control parental contributions to *Arabidopsis* early embryogenesis. *Cell* 145, 707–719. doi: 10.1016/j.cell.2011.04.014

Baroux, C., Autran, D., Raissig, M. T., Grimanelli, D., and Grossniklaus, U. (2013). Parental contributions to the transcriptome of early plant embryos. *Curr. Opin. Genet. Dev.* 23, 72–74. doi: 10.1016/j.gde.2013.01.006

Beeckman, T., De Rycke, R., Viana, R., and Inze, D. (2000). Histological study of seed coat development in *Arabidopsis thaliana*. *J. Plant Res.* 113, 139–148. doi: 10.1007/PL00013924

Belmonte, M. F., Kirkbride, R. C., Stone, S. L., Pelletier, J. M., Bui, A. Q., Yeung, E. C., et al. (2013). Comprehensive developmental profiles of gene activity in regions and subregions of the *Arabidopsis* seed. *Proc. Natl. Acad. Sci. U.S.A.* 110, E435–E444. doi: 10.1073/pnas.1222061110

Bikard, D., Jiang, W., Samai, P., Hochschild, A., Zhang, F., and Marraffini, L. (2013). Programmable repression and activation of bacterial gene expression using an engineered CRISPR-Cas system. *Nucleic Acids Res.* 41, 7429–7437. doi: 10.1093/nar/gkt520

Braybrook, S. A., and Harada, J. J. (2008). LECs go crazy in embryo development. *Trends Plant Sci.* 13, 624–630. doi: 10.1016/j.tplants.2008.09.008

Brown, R. C., Lemmon, B. E., and Nguyen, H. (2003). Events during the first four rounds of mitosis establish three developmental domains in the syncytial endosperm of *Arabidopsis thaliana*. *Protoplasma* 222, 167–174. doi: 10.1007/s00709-003-0010-x

Brown, R. C., Lemmon, B. E., Nguyen, H., and Olsen, O. A. (1999). Development of endosperm in *Arabidopsis thaliana*. *Sex. Plant Reprod.* 12, 32–42. doi: 10.1007/s004970050169

Casson, S., Spencer, M., Walker, K., and Lindsey, K. (2005). Laser capture microdissection for the analysis of gene expression during embryogenesis of *Arabidopsis*. *Plant J.* 42, 111–123. doi: 10.1111/j.1365-313X.2005.02355.x

Chen, B., Gilbert, L. A., Cimini, B. A., Schnitzbauer, J., Zhang, W., Li, G.-W., et al. (2013). Dynamic imaging of genomic loci in living human cells by an optimized CRISPR/Cas system. *Cell* 156, 1479–1491. doi: 10.1016/j.cell.2013.12.001

Chen, H., Wang, F.-W., Dong, Y.-Y., Wang, N., Sun, Y.-P., Li, X.-Y., et al. (2012). Sequence mining and transcript profiling to explore differentially expressed genes associated with lipid biosynthesis during soybean seed development. *BMC Plant Biol.* 12:122. doi: 10.1186/1471-2229-12-122

Chen, X. (2012). Small RNAs in development – insights from plants. *Curr. Opin. Genet. Dev.* 22, 361–367. doi: 10.1016/j.gde.2012.04.004

Christian, M., Qi, Y., Zhang, Y., and Voytas, D. F. (2013). Targeted mutagenesis of *Arabidopsis thaliana* using engineered TAL effector nucleases. *G3 (Bethesda)* 3, 1697–1705. doi: 10.1534/g3.113.007104

Cleland, J. (2013). World population growth: past, present and future. *Environ. Resour. Econ.* 55, 543–554. doi: 10.1007/s10640-013-9675–9676

Cong, L., Ran, F. A., Cox, D., Lin, S., Barretto, R., Habib, N., et al. (2013). Multiplex genome engineering using CRISPR/Cas systems. *Science* 339, 819–823. doi: 10.1126/science.1231143

Costa, L. M., Yuan, J., Rouster, J., Paul, W., Dickinson, H., and Gutierrez-Marcos, J. F. (2012). Maternal control of nutrient allocation in plant seeds by genomic imprinting. *Curr. Biol.* 22, 160–165. doi: 10.1016/j.cub.2011.11.059

Curtin, S. J., Zhang, F., Sander, J. D., Haun, W. J., Starker, C., Baltes, N. J., et al. (2011). Targeted mutagenesis of duplicated genes in soybean with zinc-finger nucleases. *Plant Physiol.* 156, 466–473. doi: 10.1104/pp.111.172981

Day, R. C., Herridge, R. P., Ambrose, B. A., and Macknight, R. C. (2008). Transcriptome analysis of proliferating *Arabidopsis* endosperm reveals biological implications for the control of syncytial division, cytokinin signaling, and gene expression regulation. *Plant Physiol.* 148, 1964–1984. doi: 10.1104/pp.108.128108

Day, R. C., McNoe, L., and Macknight, R. C. (2007). Evaluation of global RNA amplification and its use for high-throughput transcript analysis of laser-microdissected endosperm. *Int. J. Plant Genomics* 2007:61028. doi: 10.1155/2007/61028

de Pater, S., Pinas, J. E., Hooykaas, P. J. J., and van der Zaal, B. J. (2013). ZFN-mediated gene targeting of the *Arabidopsis* protoporphyrinogen oxidase gene through *Agrobacterium*-mediated floral dip transformation. *Plant Biotechnol. J.* 11, 510–515. doi: 10.1111/pbi.12040

Dean, G., Cao, Y., Xiang, D., Provart, N. J., Ramsay, L., Ahad, A., et al. (2011). Analysis of gene expression patterns during seed coat development in *Arabidopsis*. *Mol. Plant* 4, 1074–1091. doi: 10.1093/mp/ssr040

Dean Rider, S., Henderson, J. T., Jerome, R. E., Edenberg, H. J., Romero-Severson, J., and Ogas, J. (2003). Coordinate repression of regulators of embryonic identity by PICKLE during germination in *Arabidopsis*. *Plant J.* 35, 33–43. doi: 10.1046/j.1365-313X.2003.01783.x

Debeaujon, I., Léon-Kloosterziel, K. M., and Koornneef, M. (2000). Influence of the testa on seed dormancy, germination, and longevity in *Arabidopsis*. *Plant Physiol.* 122, 403–414. doi: 10.1104/pp.122.2.403

Debeaujon, I., Nesi, N., Perez, P., Devic, M., Grandjean, O., Caboche, M., et al. (2003). Proanthocyanidin-accumulating cells in *Arabidopsis* testa: regulation of differentiation and role in seed development. *Plant Cell* 15, 2514–2531. doi: 10.1105/tpc.014043.1

Dekkers, B. J. W., Pearce, S., van Bolderen-Veldkamp, R. P., Marshall, A., Widera, P., Gilbert, J., et al. (2013). Transcriptional dynamics of two seed compartments with opposing roles in *Arabidopsis* seed germination. *Plant Physiol.* 163, 205–215. doi: 10.1104/pp.113.223511

Farnham, P. J. (2009). Insights from genomic profiling of transcription factors. *Nat. Rev. Genet.* 10, 605–616. doi: 10.1038/nrg2636

Feeney, M., Frigerio, L., Cui, Y., and Menassa, R. (2013). Following vegetative to embryonic cellular changes in leaves of *Arabidopsis* overexpressing LEAFY COTYLEDON2. *Plant Physiol.* 162, 1881–1896. doi: 10.1104/pp.113.220996

Feuillet, C., Leach, J. E., Rogers, J., Schnable, P. S., and Eversole, K. (2011). Crop genome sequencing: lessons and rationales. *Trends Plant Sci.* 16, 77–88. doi: 10.1016/j.tplants.2010.10.005

Finch-Savage, W. E., Cadman, C. S. C., Toorop, P. E., Lynn, J. R., and Hilhorst, H. W. M. (2007). Seed dormancy release in *Arabidopsis* Cvi by dry after-ripening, low temperature, nitrate and light shows common quantitative patterns of gene expression directed by environmentally specific sensing. *Plant J.* 51, 60–78. doi: 10.1111/j.1365-313X.2007.03118.x

Fiorani, F., and Schurr, U. (2013). Future scenarios for plant phenotyping. *Annu. Rev. Plant Biol.* 64, 267–291. doi: 10.1146/annurev-arplant-050312-120137

Foley, J. A., Ramankutty, N., Brauman, K. A., Cassidy, E. S., Gerber, J. S., Johnston, M., et al. (2011). Solutions for a cultivated planet. *Nature* 478, 337–342. doi: 10.1038/nature10452

Gaj, T., Gersbach, C. A., and Barbas, C. F. (2013). ZFN, TALEN, and CRISPR/Cas-based methods for genome engineering. *Trends Biotechnol.* 31, 397–405. doi: 10.1016/j.tibtech.2013.04.004

Garcia, D., Fitz Gerald, J. N., and Berger, F. (2005). Maternal control of integument cell elongation and zygotic control of endosperm growth are coordinated to determine seed size in *Arabidopsis*. *Plant Cell* 17, 52–60. doi: 10.1105/tpc.104.027136

Garcia, D., Saingery, V., Chambrier, P., Mayer, U., Jurgens, G., and Berger, F. (2003). *Arabidopsis* haiku mutants reveal new controls of seed size by endosperm. *Plant Physiol.* 131, 1661–1670. doi: 10.1104/pp.102.018762

Gehring, M., Bubb, K. L., and Henikoff, S. (2009). Extensive demethylation of repetitive elements during seed development underlies gene imprinting. *Science* 324, 1447–1451. doi: 10.1126/science.1171609

Gehring, M., Missirian, V., and Henikoff, S. (2011). Genomic analysis of parent-of-origin allelic expression in *Arabidopsis thaliana* seeds. *PLoS ONE* 6:e23687. doi: 10.1371/journal.pone.0023687

Giraudat, J., Hauge, B. M., Valon, C., Smalle, J., Parcy, F., and Goodman, H. M. (1992). Isolation of the *Arabidopsis* ABI3 gene by positional cloning. *Plant Cell* 4, 1251–1261. doi: 10.1105/tpc.4.10.1251

Godfray, H. C. J., Beddington, J. R., Crute, I. R., Haddad, L., Lawrence, D., Muir, J. F., et al. (2010). Food security: the challenge of feeding 9 billion people. *Science* 327, 812–818. doi: 10.1126/science.1185383

Goldberg, R. B., de Paiva, G., and Yadegari, R. (1994). Plant embryogenesis: zygote to seed. *Science* 266, 605–614. doi: 10.1126/science.266.5185.605

Gutierrez, L., Van Wuytswinkel, O., Castelain, M., and Bellini, C. (2007). Combined networks regulating seed maturation. *Trends Plant Sci.* 12, 294–300. doi: 10.1016/j.tplants.2007.06.003

Haig, D., and Westoby, M. (1989). Parent-specific gene-expression and the triploid endosperm. *Am. Nat.* 134, 147–155. doi: 10.1086/284971

Harada, J. J. (1997). "Seed maturation and control of germination," in *Advances in Cellular and Molecular Biology of Plants*, Vol. 4, *Cellular and Molecular Biology of Seed Development*, eds B. A. Larkins and I. K. Vasil (Dordrecht: Springer), 545–592.

Haughn, G., and Chaudhury, A. (2005). Genetic analysis of seed coat development in *Arabidopsis. Trends Plant Sci.* 10, 472–477. doi: 10.1016/j.tplants.2005.08.005

Holdsworth, M. J., Bentsink, L., and Soppe, W. J. (2008). Molecular networks regulating *Arabidopsis* seed maturation, after-ripening, dormancy and germination. *New Phytol.* 179, 33–54. doi: 10.1111/j.1469-8137.2008.02437.x

Hsieh, T. F., Ibarra, C. A., Silva, P., Zemach, A., Eshed-Williams, L., Fischer, R. L., et al. (2009). Genome-wide demethylation of *Arabidopsis* endosperm. *Science* 324, 1451–1454. doi: 10.1126/science.1172417

Hsieh, T.-F. F., Shin, J., Uzawa, R., Silva, P., Cohen, S., Bauer, M. J., et al. (2011). Regulation of imprinted gene expression in *Arabidopsis* endosperm. *Proc. Natl. Acad. Sci. U.S.A.* 108, 1755–1762. doi: 10.1073/pnas.1019273108

Hsu, S.-D., Tseng, Y.-T., Shrestha, S., Lin, Y.-L., Khaleel, A., Chou, C.-H., et al. (2014). miRTarBase update 2014: an information resource for experimentally validated miRNA-target interactions. *Nucleic Acids Res.* 42, D78–D85. doi: 10.1093/nar/gkt1266

Huang, D., Koh, C., Feurtado, J. A., Tsang, E. W. T., and Cutler, A. J. (2013). MicroRNAs and their putative targets in *Brassica napus* seed maturation. *BMC Genomics* 14:140. doi: 10.1186/1471-2164-14-140

Huang, J., DeBowles, D., Esfandiari, E., Dean, G., Carpita, N. C., and Haughn, G. W. (2011). The *Arabidopsis* transcription factor LUH/MUM1 is required for extrusion of seed coat mucilage. *Plant Physiol.* 156, 491–502. doi: 10.1104/pp.111.172023

Hu, J. H., Mitchum, M. G., Barnaby, N., Ayele, B. T., Ogawa, M., Nam, E., et al. (2008). Potential sites of bioactive gibberellin production during reproductive growth in *Arabidopsis. Plant Cell* 20, 320–336. doi: 10.1105/tpc.107.057752

Ishida, T., Hattori, S., Sano, R., Inoue, K., Shirano, Y., Hayashi, H., et al. (2007). *Arabidopsis* TRANSPARENT TESTA GLABRA2 is directly regulated by R2R3 MYB transcription factors and is involved in regulation of GLABRA2 transcription in epidermal differentiation. *Plant Cell* 19, 2531–2543. doi: 10.1105/tpc.107.052274

Jiang, J., Shao, Y., Li, A., Lu, C., Zhang, Y., and Wang, Y. (2013a). Phenolic composition analysis and gene expression in developing seeds of yellow- and black-seeded *Brassica napus. J. Integr. Plant Biol.* 55, 537–551. doi: 10.1111/jipb.12039

Jiang, J., Shao, Y., Li, A., Zhang, Y., Wei, C., and Wang, Y. (2013b). FT-IR and NMR study of seed coat dissected from different colored progenies of *Brassica napus–Sinapis alba* hybrids. *J. Sci. Food Agric.* 93, 1898–1902. doi: 10.1002/jsfa.5986

Jiang, Y., and Deyholos, M. K. (2010). Transcriptome analysis of secondary-wall-enriched seed coat tissues of canola (*Brassica napus* L.). *Plant Cell Rep.* 29, 327–342. doi: 10.1007/s00299-010-0824-x

Johnson, C. S., Kolevski, B., and Smyth, D. R. (2002). TRANSPARENT TESTA GLABRA2, a trichome and seed coat development gene of *Arabidopsis*, encodes a WRKY transcription factor. *Plant Cell* 14, 1359–1375. doi: 10.1105/tpc.001404.covered

Jones, S. I., Gonzalez, D. O., and Vodkin, L. O. (2010). Flux of transcript patterns during soybean seed development. *BMC Genomics* 11:136. doi: 10.1186/1471-2164-11-136

Junker, A., Hartmann, A., Schreiber, F., and Baumlein, H. (2010). An engineer's view on regulation of seed development. *Trends Plant Sci.* 15, 303–307. doi: 10.1016/j.tplants.2010.03.005

Kagaya, Y., Okuda, R., Ban, A., Toyoshima, R., Tsutsumida, K., Usui, H., et al. (2005a). Indirect ABA-dependent regulation of seed storage protein genes by FUSCA3 transcription factor in *Arabidopsis. Plant Cell Physiol.* 46, 300–311. doi: 10.1093/pcp/pci031

Kagaya, Y., Toyoshima, R., Okuda, R., Usui, H., Yamamoto, A., and Hattori, T. (2005b). LEAFY COTYLEDON1 controls seed storage protein genes through its regulation of FUSCA3 and ABSCISIC ACID INSENSITIVE3. *Plant Cell Physiol.* 46, 399–406. doi: 10.1093/pcp/pci048

Kang, I.-H. H., Steffen, J. G., Portereiko, M. F., Lloyd, A., and Drews, G. N. (2008). The AGL62 MADS domain protein regulates cellularization during endosperm development in *Arabidopsis. Plant Cell* 20, 635–647. doi: 10.1105/tpc.107.055137

Kawashima, T., and Goldberg, R. B. (2010). The suspensor: not just suspending the embryo. *Trends Plant Sci.* 15, 23–30. doi: 10.1016/j.tplants.2009.11.002

Keith, K., Kraml, M., Dengler, N. G., and McCourt, P. (1994). fusca3: a heterochronic mutation affecting late embryo development in *Arabidopsis. Plant Cell* 6, 589–600. doi: 10.1105/tpc.6.5.589

Khan, D., Millar, J. L., Girard, I. J., and Belmonte, M. F. (2014). Transcriptional circuitry underlying seed coat development in *Arabidopsis. Plant Sci.* 219–220, 51–60. doi: 10.1016/j.plantsci.2014.01.004

Kohler, C., Wolff, P., and Spillane, C. (2012). Epigenetic mechanisms underlying genomic imprinting in plants. *Annu. Rev. Plant Biol.* 63, 331–352. doi: 10.1146/annurev-arplant-042811–105514

Koornneef, M., Reuling, G., and Karssen, C. M. (1984). The isolation and characterization of abscisic-acid insensitive mutants of *Arabidopsis thaliana. Physiol. Plant.* 61, 377–383. doi: 10.1111/j.1399-3054.1984.tb06343.x

Kozomara, A., and Griffiths-Jones, S. (2014). miRBase: annotating high confidence microRNAs using deep sequencing data. *Nucleic Acids Res.* 42, D68–D73. doi: 10.1093/nar/gkt1181

Langridge, P., and Fleury, D. (2011). Making the most of "omics" for crop breeding. *Trends Biotechnol.* 29, 33–40. doi: 10.1016/j.tibtech.2010.09.006

Lara, P., Onate-Sanchez, L., Abraham, Z., Ferrandiz, C., Diaz, I., Carbonero, P., et al. (2003). Synergistic activation of seed storage protein gene expression in *Arabidopsis* by ABI3 and two bZIPs related to OPAQUE2. *J. Biol. Chem.* 278, 21003–21011. doi: 10.1074/jbc.M210538200

Lau, S., Slane, D., Herud, O., Kong, J., and Jurgens, G. (2012). Early embryogenesis in flowering plants: setting up the basic body pattern. *Annu. Rev. Plant Biol.* 63, 483–506. doi: 10.1146/annurev-arplant-042811-105507

Law, J. A., and Jacobsen, S. E. (2010). Establishing, maintaining and modifying DNA methylation patterns in plants and animals. *Nat. Rev. Genet.* 11, 204–220. doi: 10.1038/nrg2719

Le, B. H., Cheng, C., Bui, A. Q., Wagmaister, J. A., Henry, K. F., Pelletier, J., et al. (2010). Global analysis of gene activity during *Arabidopsis* seed development and identification of seed-specific transcription factors. *Proc. Natl. Acad. Sci. U.S.A.* 107, 8063–8070. doi: 10.1073/pnas.1003530107

Le, B. H., Wagmaister, J. A., Kawashima, T., Bui, A. Q., Harada, J. J., and Goldberg, R. B. (2007). Using genomics to study legume seed development. *Plant Physiol.* 144, 562–574. doi: 10.1104/pp.107.100362

Lefebvre, V., North, H., Frey, A., Sotta, B., Seo, M., Okamoto, M., et al. (2006). Functional analysis of *Arabidopsis* NCED6 and NCED9 genes indicates that ABA synthesized in the endosperm is involved in the induction of seed dormancy. *Plant J.* 45, 309–319. doi: 10.1111/j.1365-313X.2005.02622.x

Li, J., and Berger, F. (2012). Endosperm: food for humankind and fodder for scientific discoveries. *New Phytol.* 195, 290–305. doi: 10.1111/j.1469-8137.2012.04182.x

Lipsa, F. D., Snowdon, R., and Friedt, W. (2011). Quantitative genetic analysis of condensed tannins in oilseed rape meal. *Euphytica* 184, 195–205. doi: 10.1007/s10681-011-0546–543

Liu, H., Jin, T., Liao, R., Wan, L., Xu, B., Zhou, S., et al. (2012). miRFANs: an integrated database for *Arabidopsis thaliana* microRNA function annotations. *BMC Plant Biol.* 12:68. doi: 10.1186/1471-2229-12-68

Liu, L., Qu, C., Wittkop, B., Yi, B., Xiao, Y., He, Y., et al. (2013a). A high-density SNP map for accurate mapping of seed fibre QTL in *Brassica napus* L. *PLoS ONE* 8:e83052. doi: 10.1371/journal.pone.0083052

Liu, X., Lu, Y., Yuan, Y., Liu, S., Guan, C., Chen, S., et al. (2013b). De novo transcriptome of *Brassica juncea* seed coat and identification of genes for the biosynthesis of flavonoids. *PLoS ONE* 8:e71110. doi: 10.1371/journal.pone.0071110

Lotan, T., Ohto, M., Yee, K. M., West, M. A., Lo, R., Kwong, R. W., et al. (1998). *Arabidopsis* LEAFY COTYLEDON1 is sufficient to induce embryo development in vegetative cells. *Cell* 93, 1195–1205. doi: 10.1016/S0092-8674(00)81463-4

Lu, J., Zhang, C., Baulcombe, D. C., and Chen, Z. J. (2012). Maternal siRNAs as regulators of parental genome imbalance and gene expression in endosperm of *Arabidopsis* seeds. *Proc. Natl. Acad. Sci. U.S.A.* 109, 5529–5534. doi: 10.1073/pnas.1203094109

Luerssen, H., Kirik, V., Herrmann, P., and Misera, S. (1998). FUSCA3 encodes a protein with a conserved VP1/AB13-like B3 domain which is of functional importance for the regulation of seed maturation in *Arabidopsis thaliana*. *Plant J.* 15, 755–764. doi: 10.1046/j.1365-313X.1998.00259.x

Luo, M., Dennis, E. S., Berger, F., Peacock, W. J., and Chaudhury, A. (2005). MINISEED3 (MINI3), a WRKY family gene, and HAIKU2 (IKU2), a leucine-rich repeat (LRR) KINASE gene, are regulators of seed size in *Arabidopsis*. *Proc. Natl. Acad. Sci. U.S.A.* 102, 17531–17536. doi: 10.1073/pnas.0508418102

Macquet, A., Ralet, M.-C., Kronenberger, J., Marion-Poll, A., and North, H. M. (2007). In situ, chemical and macromolecular study of the composition of *Arabidopsis thaliana* seed coat mucilage. *Plant Cell Physiol.* 48, 984–999. doi: 10.1093/pcp/pcm068

Meinke, D., Muralla, R., Sweeney, C., and Dickerman, A. (2008). Identifying essential genes in *Arabidopsis thaliana*. *Trends Plant Sci.* 13, 483–491. doi: 10.1016/j.tplants.2008.06.003

Meinke, D. W. (1992). A homoeotic mutant of *Arabidopsis thaliana* with leafy cotyledons. *Science* 258, 1647–1650. doi: 10.1126/science.258.5088.1647

Meinke, D. W., Franzmann, L. H., Nickle, T. C., and Yeung, E. C. (1994). Leafy cotyledon mutants of *Arabidopsis*. *Plant Cell* 6, 1049–1064. doi: 10.1105/tpc.6.8.1049

Melkus, G., Rolletschek, H., Radchuk, R., Fuchs, J., Rutten, T., Wobus, U., et al. (2009). The metabolic role of the legume endosperm: a noninvasive imaging study. *Plant Physiol.* 151, 1139–1154. doi: 10.1104/pp.109.143974

Miyawaki, K., Matsumoto-Kitano, M., and Kakimoto, T. (2004). Expression of cytokinin biosynthetic isopentenyltransferase genes in *Arabidopsis*: tissue specificity and regulation by auxin, cytokinin, and nitrate. *Plant J.* 37, 128–138. doi: 10.1046/j.1365-313X.2003.01945.x

Monke, G., Seifert, M., Keilwagen, J., Mohr, M., Grosse, I., Hahnel, U., et al. (2012). Toward the identification and regulation of the *Arabidopsis thaliana* ABI3 regulon. *Nucleic Acids Res.* 40, 8240–8254. doi: 10.1093/nar/gks594

Mosher, R. A., Melnyk, C. W., Kelly, K. A., Dunn, R. M., Studholme, D. J., and Baulcombe, D. C. (2009). Uniparental expression of PolIV-dependent siRNAs in developing endosperm of *Arabidopsis*. *Nature* 460, 283–286. doi: 10.1038/nature08084

Mu, J., Tan, H., Zheng, Q., Fu, F., Liang, Y., Zhang, J., et al. (2008). LEAFY COTYLEDON1 is a key regulator of fatty acid biosynthesis in *Arabidopsis*. *Plant Physiol.* 148, 1042–1054. doi: 10.1104/pp.108.126342

Nakabayashi, K., Okamoto, M., Koshiba, T., Kamiya, Y., and Nambara, E. (2005). Genome-wide profiling of stored mRNA in *Arabidopsis thaliana* seed germination: epigenetic and genetic regulation of transcription in seed. *Plant J.* 41, 697–709. doi: 10.1111/j.1365-313X.2005.02337.x

Nakamura, S., Lynch, T. J., and Finkelstein, R. R. (2001). Physical interactions between ABA response loci of *Arabidopsis*. *Plant J.* 26, 627–635. doi: 10.1046/j.1365-313x.2001.01069.x

Nelson, T., Tausta, S. L., Gandotra, N., and Liu, T. (2006). Laser microdissection of plant tissue: what you see is what you get. *Annu. Rev. Plant Biol.* 57, 181–201. doi: 10.1146/annurev.arplant.56.032604.144138

Nesi, N., Debeaujon, I., Jond, C., Stewart, A. J., Jenkins, G. I., Caboche, M., et al. (2002). The TRANSPARENT TESTA16 locus encodes the *ARABIDOPSIS* BSISTER MADS domain protein and is required for proper development and pigmentation of the seed coat. *Plant Cell* 14, 2463–2479. doi: 10.1105/tpc.004127

Nesi, N., Jond, C., Debeaujon, I., Caboche, M., and Lepiniec, L. (2001). The *Arabidopsis* TT2 gene encodes an R2R3 MYB domain protein that acts as a key determinant for proanthocyanidin accumulation in developing seed. *Plant Cell* 13, 2099–2114.

Nishimura, N., Yoshida, T., Kitahata, N., Asami, T., Shinozaki, K., and Hirayama, T. (2007). ABA-Hypersensitive Germination1 encodes a protein phosphatase 2C, an essential component of abscisic acid signaling in *Arabidopsis* seed. *Plant J.* 50, 935–949. doi: 10.1111/j.1365-313X.2007.03107.x

Nodine, M. D., and Bartel, D. P. (2010). MicroRNAs prevent precocious gene expression and enable pattern formation during plant embryogenesis. *Genes Dev.* 24, 2678–2692. doi: 10.1101/gad.1986710

Nodine, M. D., and Bartel, D. P. (2012). Maternal and paternal genomes contribute equally to the transcriptome of early plant embryos. *Nature* 482, 94–97. doi: 10.1038/nature10756

Ogawa, M., Hanada, A., Yamauchi, Y., Kuwahara, A., Kamiya, Y., and Yamaguchi, S. (2003). Gibberellin biosynthesis and response during *Arabidopsis* seed germination. *Plant Cell* 15, 1591–1604. doi: 10.1105/tpc.011650.ble

Ohad, N., Yadegari, R., Margossian, L., Hannon, M., Michaeli, D., Harada, J. J., et al. (1999). Mutations in FIE, a WD polycomb group gene, allow endosperm development without fertilization. *Plant Cell* 11, 407–416. doi: 10.1105/tpc.11.3.407

Ohto, M. A., Floyd, S. K., Fischer, R. L., Goldberg, R. B., and Harada, J. J. (2009). Effects of APETALA2 on embryo, endosperm, and seed coat development determine seed size in *Arabidopsis*. *Sex. Plant Reprod.* 22, 277–289. doi: 10.1007/s00497-009-0116-111

Olsen, O. A. (2004). Nuclear endosperm development in cereals and *Arabidopsis thaliana*. *Plant Cell* 16, 214–228. doi: 10.1105/tpc.017111

Parcy, F., Valon, C., Raynal, M., Gaubier-Comella, P., Delseny, M., and Giraudat, J. (1994). Regulation of gene expression programs during *Arabidopsis* seed development: roles of the ABI3 locus and of endogenous abscisic acid. *Plant Cell* 6, 1567–1582. doi: 10.1105/tpc.6.11.1567

Penfield, S., Li, Y., Gilday, A. D., Graham, S., and Graham, I. A. (2006). *Arabidopsis* ABA INSENSITIVE4 regulates lipid mobilization in the embryo and reveals repression of seed germination by the endosperm. *Plant Cell* 18, 1887–1899. doi: 10.1105/tpc.106.041277.1

Penfield, S., Meissner, R. C., Shoue, D. A., Carpita, N. C., and Bevan, M. W. (2001). MYB61 is required for mucilage deposition and extrusion in the *Arabidopsis* seed coat. *Plant Cell* 13, 2777–2791. doi: 10.1105/tpc.010265.1

Pourcel, L., Irani, N. G., Koo, A. J. K., Bohorquez-Restrepo, A., Howe, G. A., and Grotewold, E. (2013). A chemical complementation approach reveals genes and interactions of flavonoids with other pathways. *Plant J.* 74, 383–397. doi: 10.1111/tpj.12129

Preston, J., Tatematsu, K., Kanno, Y., Hobo, T., Kimura, M., Jikumaru, Y., et al. (2009). Temporal expression patterns of hormone metabolism genes during imbibition of *Arabidopsis thaliana* seeds: a comparative study on dormant and non-dormant accessions. *Plant Cell Physiol.* 50, 1786–1800. doi: 10.1093/pcp/pcp121

Qu, C., Fu, F., Lu, K., Zhang, K., Wang, R., Xu, X., et al. (2013). Differential accumulation of phenolic compounds and expression of related genes in black- and yellow-seeded *Brassica napus*. *J. Exp. Bot.* 64, 2885–2898. doi: 10.1093/jxb/ert148

Raissig, M. T., Bemer, M., Baroux, C., and Grossniklaus, U. (2013). Genomic imprinting in the *Arabidopsis* embryo is partly regulated by PRC2. *PLoS Genet.* 9:e1003862. doi: 10.1371/journal.pgen.1003862

Ray, D. K., Mueller, N. D., West, P. C., and Foley, J. A. (2013). Yield trends are insufficient to double global crop production by 2050. *PLoS ONE* 8:e66428. doi: 10.1371/journal.pone.0066428

Santos Mendoza, M., Dubreucq, B., Miquel, M., Caboche, M., and Lepiniec, L. (2005). LEAFY COTYLEDON 2 activation is sufficient to trigger the accumulation of oil and seed specific mRNAs in *Arabidopsis* leaves. *FEBS Lett.* 579, 4666–4670. doi: 10.1016/j.febslet.2005.07.037

Schneeberger, K., and Weigel, D. (2011). Fast-forward genetics enabled by new sequencing technologies. *Trends Plant Sci.* 16, 282–288. doi: 10.1016/j.tplants.2011.02.006

Scott, R. J., Spielman, M., Bailey, J., and Dickinson, H. G. (1998). Parent-of-origin effects on seed development in *Arabidopsis thaliana*. *Development* 125, 3329–3341.

Severin, A. J., Woody, J. L., Bolon, Y.-T., Joseph, B., Diers, B. W., Farmer, A. D., et al. (2010). RNA-Seq Atlas of *Glycine max*: a guide to the soybean transcriptome. *BMC Plant Biol.* 10:160. doi: 10.1186/1471-2229-10–160

Shamimuzzaman, M., and Vodkin, L. (2012). Identification of soybean seed developmental stage-specific and tissue-specific miRNA targets by degradome sequencing. *BMC Genomics* 13:310. doi: 10.1186/1471-2164-13–310

Simbaya, J., Slominski, B. A., Rakow, G., Campbell, L. D., Downey, R. K., and Bell, J. M. (1995). Quality characteristics of yellow-seeded *Brassica* seed meals: protein, carbohydrates, and dietary fiber components. *J. Agric. Food Chem.* 43, 2062–2066. doi: 10.1021/jf00056a020

Song, Q.-X. X., Liu, Y.-F. F., Hu, X.-Y. Y., Zhang, W.-K. K., Ma, B., Chen, S.-Y. Y., et al. (2011). Identification of miRNAs and their target genes in developing soybean seeds by deep sequencing. *BMC Plant Biol.* 11:5. doi: 10.1186/1471-2229-11-5

Sreenivasulu, N., and Wobus, U. (2013). Seed-development programs: a systems biology-based comparison between dicots and monocots. *Annu. Rev. Plant Biol.* 64, 189–217. doi: 10.1146/annurev-arplant-050312-120215

Stone, S. L., Braybrook, S. A., Paula, S. L., Kwong, L. W., Meuser, J., Pelletier, J., et al. (2008). *Arabidopsis* LEAFY COTYLEDON2 induces maturation traits and auxin activity: implications for somatic embryogenesis. *Proc. Natl. Acad. Sci. U.S.A.* 105, 3151–3156. doi: 10.1073/pnas.0712364105

Stone, S. L., Kwong, L. W., Yee, K. M., Pelletier, J., Lepiniec, L., Fischer, R. L., et al. (2001). LEAFY COTYLEDON2 encodes a B3 domain transcription factor that induces embryo development. *Proc. Natl. Acad. Sci. U.S.A.* 98, 11806–11811. doi: 10.1073/pnas.201413498

Suzuki, M., and McCarty, D. R. (2008). Functional symmetry of the B3 network controlling seed development. *Curr. Opin. Plant Biol.* 11, 548–553. doi: 10.1016/j.pbi.2008.06.015

Tadros, W., and Lipshitz, H. D. (2009). The maternal-to-zygotic transition: a play in two acts. *Development* 136, 3033–3042. doi: 10.1242/dev.033183

Tang, X., Bian, S., Tang, M., Lu, Q., Li, S., Liu, X., et al. (2012). MicroRNA-mediated repression of the seed maturation program during vegetative development in *Arabidopsis*. *PLoS Genet.* 8:e1003091. doi: 10.1371/journal.pgen.1003091

Tester, M., and Langridge, P. (2010). Breeding technologies to increase crop production in a changing world. *Science* 327, 818–822. doi: 10.1126/science.1183700

Tilman, D., Balzer, C., Hill, J., and Befort, B. L. (2011). Global food demand and the sustainable intensification of agriculture. *Proc. Natl. Acad. Sci. U.S.A.* 108, 20260–20264. doi: 10.1073/pnas.1116437108

To, A., Valon, C., Savino, G., Guilleminot, J., Devic, M., Giraudat, J., et al. (2006). A network of local and redundant gene regulation governs *Arabidopsis* seed maturation. *Plant Cell* 18, 1642–1651. doi: 10.1105/tpc.105.039925

Troncoso-Ponce, M., Kilaru, A., Cao, X., Durrett, T. P., Fan, J., Jensen, J. K., et al. (2011). Comparative deep transcriptional profiling of four developing oilseeds. *Plant J.* 68, 1014–1027. doi: 10.1111/j.1365-313X.2011.04751.x

Verdier, J., Dessaint, F., Schneider, C., Abirached-darmency, M., and Bitonti, B. (2013). A combined histology and transcriptome analysis unravels novel questions on *Medicago truncatula* seed coat. *J. Exp. Bot.* 64, 459–470. doi: 10.1093/jxb/err313

Vicente-Carbajosa, J., and Carbonero, P. (2005). Seed maturation: developing an intrusive phase to accomplish a quiescent state. *Int. J. Dev. Biol.* 49, 645–651. doi: 10.1387/ijdb.052046jc

Wang, F., and Perry, S. E. (2013). Identification of direct targets of FUSCA3, a key regulator of *Arabidopsis* seed development. *Plant Physiol.* 161, 1251–1264. doi: 10.1104/pp.112.212282

Weinhofer, I., Hehenberger, E., Roszak, P., Hennig, L., and Kohler, C. (2010). H3K27me3 profiling of the endosperm implies exclusion of polycomb group protein targeting by DNA methylation. *PLoS Genet.* 6:e1001152. doi: 10.1371/journal.pgen.1001152

West, M. A. L., Yee, K. M., Danao, J., Zimmerman, J. L., Fischer, R. L., Goldberg, R. B., et al. (1994). LEAFY COTYLEDON1 is an essential regulator of late embryogenesis and cotyledon identity in *Arabidopsis*. *Plant Cell* 6, 1731–1745. doi: 10.1105/tpc.6.12.1731

Western, T. L., Burn, J., Tan, W. L., Skinner, D. J., Martin-mccaffrey, L., Moffatt, B. A., et al. (2001). Isolation and characterization of mutants defective in seed coat mucilage secretory cell development in *Arabidopsis*. *Plant Physiol.* 127, 998–1011. doi: 10.1104/pp.010410.upon

Western, T. L., Skinner, D. J., and Haughn, G. W. (2000). Differentiation of mucilage secretory cells of the *Arabidopsis* seed coat. *Plant Physiol.* 122, 345–356. doi: 10.1104/pp.122.2.345

Willmann, M. R., Mehalick, A. J., Packer, R. L., and Jenik, P. D. (2011). MicroRNAs regulate the timing of embryo maturation in *Arabidopsis*. *Plant Physiol.* 155, 1871–1884. doi: 10.1104/pp.110.171355

Windsor, J. B., Symonds, V. V., Mendenhall, J., and Lloyd, A. M. (2000). *Arabidopsis* seed coat development: morphological differentiation of the outer integument. *Plant J.* 22, 483–493. doi: 10.1046/j.1365-313x.2000.00756.x

Wolff, P., Weinhofer, I., Seguin, J., Roszak, P., Beisel, C., Donoghue, M. T., et al. (2011). High-resolution analysis of parent-of-origin allelic expression in the *Arabidopsis* endosperm. *PLoS Genet.* 7:e1002126. doi: 10.1371/journal.pgen.1002126

Xiang, D., Venglat, P., Tibiche, C., Yang, H., Risseeuw, E., Cao, Y., et al. (2011). Genome-wide analysis reveals gene expression and metabolic network dynamics during embryo development in *Arabidopsis*. *Plant Physiol.* 156, 346–356. doi: 10.1104/pp.110.171702

Xie, K., and Yang, Y. (2013). RNA-guided genome editing in plants using a CRISPR-Cas system. *Mol. Plant* 6, 1975–1983. doi: 10.1093/mp/sst119

Young, R. E., McFarlane, H. E., Hahn, M. G., Western, T. L., Haughn, G. W., and Samuels, L. (2008). Analysis of the Golgi apparatus in *Arabidopsis* seed coat cells during polarized secretion of pectin-rich mucilage. *Plant Cell* 20, 1623–1638. doi: 10.1105/tpc.108.058842

Yu, C.-Y. (2013). Molecular mechanism of manipulating seed coat coloration in oilseed *Brassica* species. *J. Appl. Genet.* 54, 135–145. doi: 10.1007/s13353-012-0132-y

Zhang, F., Maeder, M. L., Unger-wallace, E., Hoshaw, J. P., Reyon, D., Christian, M., et al. (2010). High frequency targeted mutagenesis in *Arabidopsis thaliana* using zinc finger nucleases. *Proc. Natl. Acad. Sci. U.S.A.* 107, 12028–12033. doi: 10.1073/pnas.0914991107

Zhang, K., Lu, K., Qu, C., Liang, Y., Wang, R., Chai, Y., et al. (2013). Gene silencing of BnTT10 family genes causes retarded pigmentation and lignin reduction in the seed coat of *Brassica napus*. *PLoS ONE* 8:e61247. doi: 10.1371/journal.pone.0061247

Zhao, Y.-T., Wang, M., Fu, S.-X., Yang, W.-C., Qi, C.-K., and Wang, X.-J. (2012). Small RNA profiling in two *Brassica napus* cultivars identifies microRNAs with oil production- and development-correlated expression and new small RNA classes. *Plant Physiol.* 158, 813–823. doi: 10.1104/pp.111.187666

Conflict of Interest Statement: The authors declare that the research was conducted in the absence of any commercial or financial relationships that could be construed as a potential conflict of interest.

13

Soybean seed proteome rebalancing

Eliot M. Herman*

School of Plant Sciences, BIO5 Institute, University of Arizona, Tucson, AZ, USA

Edited by:
Brian A. Larkins, University of Nebraska–Lincoln, USA

Reviewed by:
Ján A. Miernyk, University of Missouri, USA
L. Curtis Hannah, University of Florida, USA

***Correspondence:**
Eliot M. Herman, School of Plant Sciences, BIO5 Institute, University of Arizona, BIO5 Institute Room 249, 1657 East Helen Street, Tucson, AZ 85721-0240, USA
e-mail: emherman@email.arizona.edu

The soybean seed's protein content and composition are regulated by both genetics and physiology. Overt seed protein content is specified by the genotype's genetic framework and is selectable as a breeding trait. Within the genotype-specified protein content phenotype soybeans have the capacity to rebalance protein composition to create differing proteomes. Soybeans possess a relatively standardized proteome, but mutation or targeted engineering can induce large-scale proteome rebalancing. Proteome rebalancing shows that the output traits of seed content and composition result from two major types of regulation: genotype and post-transcriptional control of the proteome composition. Understanding the underlying mechanisms that specifies the seed proteome can enable engineering new phenotypes for the production of a high-quality plant protein source for food, feed, and industrial proteins.

Keywords: protein, proteome, seed, storage protein, soybean

SOYBEANS ARE A GLOBAL PROTEIN COMMODITY

Among the global commodity of crops, soybean has an almost unique role, being high enough in protein content to provide the nitrogen (N) needed for efficient large-scale animal feed production. Soybeans possess economically valued oil and protein and is an archetype seed used to dissect the processes that specify seed compositional output traits. Over the past decade, considerable public and industry funds have been invested to create soybean community resources, including genomic, transcript, SNP and SSR maps, proteomics, as well as supported a broad range of bioactivity, biochemical, nutritional, and agronomic projects.

PROTEIN CONTENT AS A GENOTYPE

The genome of soybean specifies the genetic framework for seed formation and maturation, and it controls the expression, mix, and timing for synthesis of the storage metabolite traits (Wilson, 2004 for general information; Hartwig and Kilen, 1991; Wilcox and Shibles, 2001; Chung et al., 2003; Nichols et al., 2006; Bolon et al., 2010). Plant breeding has shown that different soybean genotypes specify a standardized, often line/cultivar-specific protein content (Wilson, 2004). The genetic program that produces seeds is simultaneously manifested in the embryo, endosperm, and maternal plant. Genetic marker analysis, using SNPs, has identified QTLs that demonstrate the overt protein and oil content has a strong genetic determinant (see Diers et al., 1992; Cregan et al., 1999; Zhao-Ming et al., 2011 for examples); these traits have supported generations of breeders who have enhanced soybean as a crop (Brim and Burton, 1979; Carter et al., 1986; Cober and Voldeng, 2000; Wilson, 2004).

Of the three individual components that comprise seeds, two are reproductive progeny: the endosperm and the zygotic embryo, which result from the double fertilization and are enclosed in the maternal-origin seed coat that connects the maturing seed to the maternal plant. Each reproductive-phase soybean plant consists of a coordinated network of embryos and endosperms for the common goal of maximizing reproductive output. Historically, breeding programs have selected traits for enhancement of seed productivity, storage metabolite content, and important agronomic performance traits of the maternal plant. One way to view seed production by plants is as a population that produces (maternal) and distributes (maternal and endosperm) nutrient metabolites to the embryonic sink. During soybean seed development the endosperm undergoes progressive programmed cell death that is completed prior to the accumulation of stored metabolites. By the onset of protein and oil accumulation, only a single cell layer of aleurone remains from the endosperm that encapsulates the embryo, separating it from the inner surface of the maternal seed coat. But the physiological role, if any, of the aleurone in regulating nutrient flux to the developing embryo has not been investigated. Viewed in this way, the soybean plant's progeny are a population of aleurones and embryos interacting with the metabolite flux. The non-synchronized developing population of seeds must both compete and synchronize with the common maternal nutrient source. To assure the mature seeds are nearly all equivalent in composition, independent of their particular position on the maternal plant, their developmental program and physiological regulation must be coordinated along with the capacity of the maternal plant to nourish them.

The embryo's genotype specifies maturation-stages (Fehr et al., 1971) that are controlled by transcription factors that provide the developmental framework for storage substance accumulation (Hill and Breidenbach, 1974; Goldberg et al., 1981a,b; Mienke et al., 1981; Walling et al., 1986; Naito et al., 1988; Harada et al., 1989; Perez-Grau and Goldberg, 1989; Nielsen and Nam, 1999; To et al., 2006). A large number of soybean seed-specific DNA binding proteins have been identified (http://casp.rnet.missouri.edu/soydb/), and some of these have been shown to regulate specific seed maturation specific genes. (Chen et al., 1986; Jofuku et al., 1987; Allen et al., 1989; Lessard et al., 1991; Bäumlein et al., 1992; Kwong et al., 2003; Wang et al., 2007). A key role of some transcription factors is to regulate the metabolic and developmental processes that support storage substance accumulation (Kroj et al., 2003; Gutierrez et al., 2007; Santos-Mendoza et al., 2008 for reviews). Understanding how cooperation between the embryo, endosperm, and maternal organs is integrated at the level of gene expression and cross-regulation of metabolism is important for creating models of the source-sink relationship of seed-fill.

NUTRIENT SOURCE DEFINES SEED PROTEIN ACCUMULATION

Whole plant physiological experiments demonstrate that nutrient distribution to seeds is highly regulated. The seed protein output trait is primarily regulated by controlling the composition of seeds, with the total number of seeds being a consequence of nutrient availability. The average size of seeds from small size plants compared with larger size plants, differs only a little, but the total number of seeds produced is directly related to the total biomass of the maternal plant and its mobilized metabolite source potential. In an agronomic context, this defines yield. From the perspective of the plant, the protein content genotype maximizes the potential of an individual seeds with the overall yield of seeds depending on the available biomass/growth conditions.

Although the maternal plant, endosperm, and embryo function in concert to form the seed, their metabolic interaction occurs without a direct, contiguous flow of nutrients, as each is apoplastically isolated from the other (Thorne, 1980; Thorne and Rainbird, 1983; Egli and Bruening, 2001; Patrick and Offler, 2001 for review). The metabolite flux from the maternal plant through the aleurone to the embryo results from coordinated secretion from the source and uptake by the sink, and this potentially determines the storage output trait (Borisjuk et al., 2003, 2004). In annual plants, such as soybean, the maternal plant must grow rapidly and produce nutrient-capture organs (i.e., roots and leaves).

The relationship between the maternal plant (source) and the seed (sink) has been investigated with increasingly more sophisticated tools and concepts for the past 40 years. For soybean, early studies focused on accumulation of vegetative proteins, primarily in foliage, as the nitrogen store that is mobilized to the seed and determines the accumulation of storage substances. This model of resource acquisition parallels that of most other seed plants, where metabolites in foliage are later mobilized to the seed. The amount of carbon fixed during photosynthesis is highly responsive to the environment, and the maternal plant manages carbon flux so it can distribute nutrients based on their availability and the demand of the seed (endosperm/embryo) sink (see Fellows et al., 1979; Borchers-Zampini et al., 1980 for early examples). A number of studies have shown that leaf proteins, predominantly Rubisco, accumulate over time (Schaefer et al., 1981a,b). In addition, soybean leaves accumulate a vegetative storage protein (VSP), a member of the vacuolar acid phosphatase family (DeWald et al., 1992; Staswick et al., 1994). VSP accumulation is highly responsive to nitrogen availability (Franceschi and Giaquinta, 1983; Franceschi et al., 1983; Staswick, 1989a,b), and it increases with depodding, i.e., removal of the seed sink. This observation led to proposals that VSP is a necessary adjunct that provides additional nitrogen resources for the seed. By silencing the VSP gene, it was later shown that VSP does not appear to make a difference in soybean seed protein content (Staswick et al., 2001).

Accumulated leaf proteins are mobilized by specific proteases (see Ragster and Chrispeels, 1979, 1981a,b) found in leaf cell vacuoles and plastids; the enzymes mediate the hydrolysis of Rubisco and VSP as well as other less abundant leaf proteins. Removal of all maturing seeds, except in one portion of the plant, leads to redistribution of source leaf nutrient flux (Carlson and Brun, 1984) indicating there must be (unknown) feedback-regulation between the seed and the nutrient source that is manifested through a long distance signal. Systems biology approaches could determine how the source size/composition is regulated, and how its mobilization is coordinated with the draw of the sink.

Leaf photosynthate and the products of protein hydrolysis produce a metabolite flux consisting predominantly of sucrose, glutamine and asparagine (Hsu et al., 1984; Rainbird et al., 1984; Krishnan et al., 2011); for nitrogen-fixing legumes in particular, there are also ureides from xylem fluid. With respect to the amino acid flux, the input from glutamine and asparagine has different characteristics. Skokut et al. (1982), using ^{15}N-NMR showed that there is no discrimination between the amino and amide N of glutamine, but for Asn the amino N is incorporated into protein twice as efficiently as the amide N, indicating a key role of asparagine in transamination. Asn may also be directly incorporated into proteins, with dual labeled ^{13}C and ^{15}N Asn being incorporated directly without scrambling the labels (Schaefer et al., 1981a,b). Within the seed, free Asn accounts for a larger fraction of amino acids (33–49%), with the fractional amount varying by genotype (Schaefer et al., 1981a,b). Asn dominates seed coat free amino acid efflux, assayed as apoplastic fluid, from the seed cup, i.e., the seed coat with embryo removed (Gifford and Thorne, 1985; Murray, 1987; De Jong et al., 1997; Hernández-Sebastià et al., 2005; Pandurangan et al., 2012). It is unclear whether the tissue source used for these experiments was derived from the inner surface of the aleurone or the maternal seed coat, or a combination of both, since the aleurone often adheres to the inner side of the seed coat. This shows that distinct from the embryo, the seed coat (perhaps comprising aleurone and seed coat) has an amino acid composition similar to the source (assayed as xylem sap), containing Asn as the dominant N-source, and is about 10-fold higher in abundance than Gln (Thorne and Rainbird, 1983; Lohaus et al., 1998; Zhang et al., 2010; Krishnan et al., 2011; Pandurangan et al., 2012).

PROTEIN COMPOSITION PLASTICITY AND THE SEED PROTEIN CONTENT GENOTYPE

The soybean seed protein output trait has two primary components: total protein content and the composition of individual proteins (the proteome). For soybeans, like many other seeds (Herman and Larkins, 1999 for review), the two major storage proteins, glycinin (11S legumin type) and conglycinin (7S vicilin type), dominate the proteome. The soybean seed proteome also includes many moderately abundant proteins that are bioactive and allergenic, such as the Kunitz and Bowman-Birk trypsin inhibitors, lectin, P34 allergen, sucrose binding protein, urease, oleosins (Herman and Burks, 2011) and several thousand low abundance proteins, including enzymes that mediate metabolism, synthesize storage substances, and create the structural framework of the cell. The specific mix of proteins and each protein's abundance within the proteome determines the total amino acid composition trait.

Since the development of plant transformation techniques, there have been many attempts to express genes to induce accumulation of large quantities of foreign proteins in seeds. The goals of these projects were often to alter the nutritional quality of seeds, by increasing essential amino acids such as methionine in soybean, or to use the seeds as protein bioreactors. Even with strong storage protein promoters to regulate the transgene expression, the most frequent outcome of such experiments was to produce relatively small amounts of the heterologous protein (about 1% of total) and for the protein and amino acid composition to be little altered compared to control. These minor composition changes occurred whether the protein was targeted to the vacuole or accreted in ER-derived protein bodies, suggesting the limit of protein accumulation is independent of its deposition site. The observed heterologous protein production, compared to the expectations of the engineering design, indicated there must be seed regulatory mechanisms that limit foreign protein production so as not to significantly alter the seed protein content phenotype.

The converse experiment is to silence the intrinsic major storage protein genes and assess the impact on the seed's protein content. Kinney et al. (2001) showed total soybean protein content was conserved after silencing conglycinin, which constitutes about 20% of the total seed protein. The resulting seeds accumulated more glycinin, which apparently compensated for the missing conglycinin. Of the five glycinin-genes in the soybean genome, the protein encoded by the glycinin A4 gene tends to accrete in the ER, producing protein (ER)-bodies (Herman and Schmidt, 2004; Herman, 2008), which are not normally found in soybean (Kinney et al., 2001). Mori et al. (2004) showed similar observations in a conglycinin mutant obtained from screening a collection that exhibited the same phenotype of "glycinin rebalancing," and for some of this protein to remain as proglycinin and to accumulate ER-derived protein bodies. Schmidt and Herman (2008) showed that introducing a gene encoding a foreign protein into the conglycinin-glycinin rebalancing increased the accumulation of the heterologous protein. A GFP-HDEL gene, as a glycinin-gene mimic allele with a glycinin promoter and terminator, was constructed. The addition of the HDEL ER-retention sequence was intended to promote accretion of the protein in the ER to form ER-(protein)-bodies and mimic the accreting glycinin ORF by substituting the GFP-HDEL protein. The expression of GFP-HDEL in the standard cv *Jack* soybean background resulted in about 1% of the total seed protein accumulated as GFP in protein bodies, a level typical of experiments of heterologous protein production in seeds. However, introgressing the GFP-HDEL glycinin mimic allele into the conglycinin-silenced line resulted in about eightfold increased accumulation of GFP as the glycinin mimic was utilized to compensate for the conglycinin shortfall (**Figure 1**).

To further test regulation of the protein content genotype and its capacity to allow for proteome alterations, RNAi silencing of

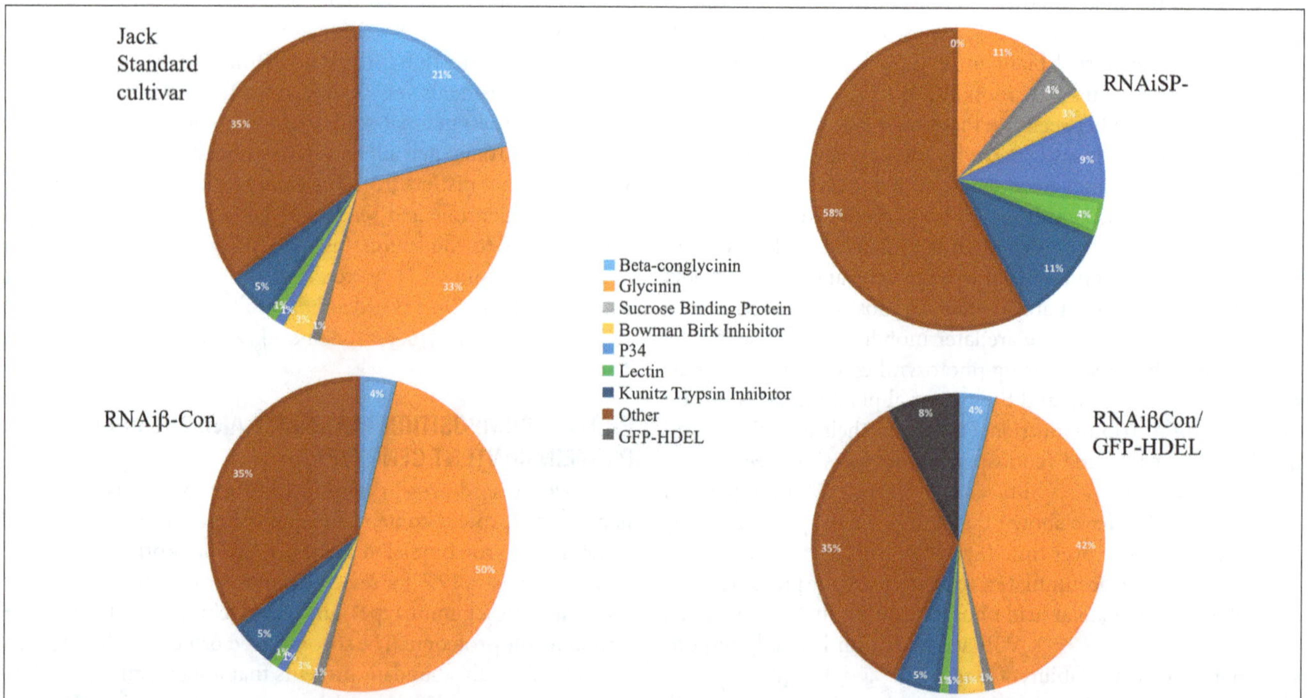

FIGURE 1 | Variations of seed composition within the context of the protein content genotype are shown. The pie charts show the relative distribution of proteins comprising the seed proteome of the standard soybean line Jack, a line that silences β-conglycinin α,α′, another line that silences both β-conglycinin and glycinin, and a line where GFP-HDEL forming ER-derived protein bodies mimic of glycinin in the background of β-conglycinin silencing. Each of these seeds contains about the same total protein content while the proteome of each of these seeds varies. This demonstrates how the genomic framework controls the protein content genotype that can be rebalanced by protein composition plasticity.

both the glycinin and conglycinin storage proteins (*SP*- or storage protein minus) lines were created that eliminated over two thirds of the protein content of standard soybean seeds (Schmidt et al., 2011). *SP*- seeds exhibited a number of different phenotypes that included conserving the same seed protein content as the parental line, due to compensatory increases of other vacuolar proteins, including Kunitz Trypsin Inhibitor, Lectin, P34, and sucrose binding protein (**Figure 1**). These compensating proteins accumulated at levels up to 11X more than normal, and each protein's increase occurred without a parallel increase in its steady-state RNA transcript abundance. This suggests that the protein content trait is determined by genotype but the abundance of proteome members occurs at the translational level.

FREE ASN IS AN INDICATOR OF ALTERED PROTEIN CONTENT AND COMPOSITION

Free Asn is correlated with soybean seed protein content and composition. One line of experimental evidence has shown that in soybean seed cultivars with a range of protein contents there is a positive correlation between high free Asn and high protein content (Hernández-Sebastià et al., 2005; Pandurangan et al., 2012). For the *SP*- storage protein silenced soybean, free Asn increased 5.8X over the standard type (Schmidt et al., 2011). Perhaps in response to the elevated free Asn, the steady-state transcript (RNAseq) level for asparaginase increased 6.5X over the conventional controls. In standard lines the asparaginase level was previously correlated with protein content in standard lines (Wan et al., 2006). Together, these observations suggest there is a correlation of protein increase, whether by genotype selection for higher protein content or by increased abundance of individual proteins within the context of the protein genotype, with changes in free Asn and asparaginase. This suggests the free Asn level is a nitrogen status indicator (Miller et al., 2008), either as a regulator of, or as a component of, the processes that specify protein content and composition.

CULTURED SOMATIC AND ZYGOTIC EMBRYOS EXHIBIT AN EXCESSIVE GROWTH TRAIT

Ex vivo zygotic embryo and somatic embryo cultures are often used as proxies for seed maturation; however, there are significant differences in metabolic behavior of embryos that form *in vivo* and *in vitro* (see Thompson et al., 1977; Obendorf et al., 1983, 1984; Raper et al., 1984; Finer, 1988; Hayati et al., 1996; Santarem et al., 1997; Chanprame et al., 1998; Pipolo et al., 2004; Iyer et al., 2008; Nishizawa and Ishimoto, 2009; Allen and Young, 2013). RNA expression profiling showed that somatic embryos produce a relatively standard set of seed-specific transcripts (Thibaud-Nissen et al., 2003). *Ex vivo* cultures exhibited fidelity with *in planta* seeds, but exhibit differences in the content of accumulated reserve substances (Pipolo et al., 2004). Gln has been shown to be an effective N-input source for these cultures (Saravitz and Raper, 1995; Schmidt et al., 2005) and is often used as the experimental N-source in nutrition-flux studies (He et al., 2011; Allen and Young, 2013; Truong et al., 2013, for recent examples), even though it is Asn that accounts for the large majority of the actual maternal source N *in planta* (Lea and Miflin, 1980; Lohaus et al., 1998; Lima and Sodek, 2003). A recent paper by Allen and Young (2013) showed in cultured zygotic soybean embryos that ^{14}C-Gln supplied 36–46% of the carbon of amino acids. In another study using somatic embryos, Truong et al. (2013) showed that increasing Gln in extrinsic culture media resulted in increased protein content, without greater oil content, showing that Gln is preferentially used to synthesize protein. This is consistent with older NMR observations on ^{13}C and ^{15}N that showed the amino and amido N for Gln as well as the carbon, is non-discriminatory when incorporated into the protein sink (Schaefer et al., 1981b; Skokut et al., 1982).

Taken together, these observations support a model where the maternal source supplies Asn (Pandurangan et al., 2012) as the N-source for zygotic embryos, but experimental *ex vivo* embryos can effectively use Gln. The difference between Asn and Gln may be important in the context of the morphological and compositional differences between *in planta* zygotic and cultured somatic and zygotic embryos. The media used for soybean culture varies, although Gln as the N-source dominates (Haga and Sodek, 1987), particularly in SHaM media (Schmidt et al., 2005), which was developed for transgenic embryo maturation (see Schmidt and Herman, 2008; Schmidt et al., 2011 for examples of its use). It is also used in some *ex vivo* nutritional studies (Truong et al., 2013). The media used for immature somatic embryo culture and transformation (Finer, 1988; Finer and McMullen, 1991; Walker and Parrott, 2001) has Asn as the N source. Tissue culture embryos used for transformation and regeneration, freed from the physical and metabolic constraints of the endosperm/aleurone and seed coat, exhibit aberrant growth (**Figures 2** and **3**), supporting

FIGURE 2 | The comparison of the size and morphology of a mature seed and a mature somatic embryo is shown. Note that the somatic embryo can exceed the seed's size, and the axis is enlarged compared to the cotyledon.

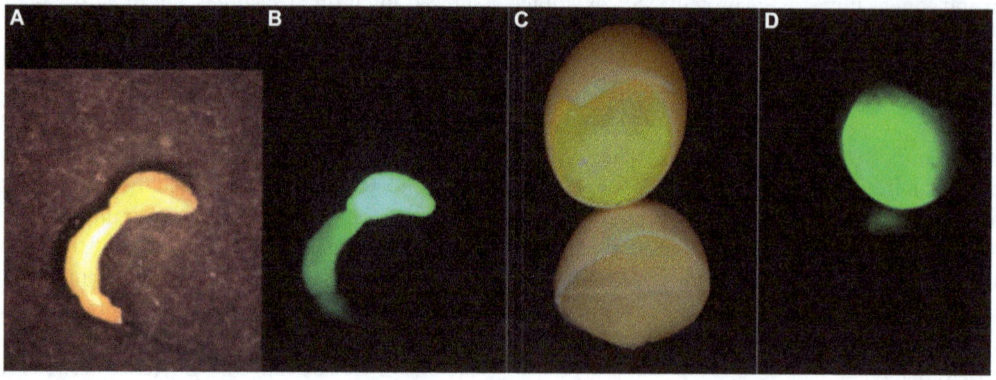

FIGURE 3 | A comparison of the expression of the storage protein trait in somatic embryos and seeds is shown as a GFP-HDEL glycinin mimic allele expressed as a storage protein proxy. The somatic embryo in white light **(A)** and UV fluorescence **(B)** shows that the primary GFP expression site is in the cotyledon that is reduced in size compared to cotyledons of seed. In seeds the cotyledons comprise the large majority of the seed's mass shown as a chipped seed illuminated with white light **(C)** and UV for GFP fluorescence **(D)**. The comparative expression the same glycinin promoter-regulated GFP in somatic embryos and seeds shows that while somatic embryos are a good proxy for seed expression there are challenges in interpreting somatic embryo results as an accurate proxy for a seed's in planta protein content and composition.

a regulatory role for the endosperm and/or seed coat in seed development (Garcia et al., 2003, 2005; Berger et al., 2006). In culture, somatic embryos form "monster" embryos with an enlarged embryonic axis and diminished, sometimes fused, cotyledons. Somatic embryos grown in the SHaM media are deemed "healthy," (i.e., large, green, well-formed), often exceeding in size a fully formed seed. Similar observations are obtained by culturing immature zygotic cotyledons that enlarge to a size that exceeds that of *in planta* seed cotyledons. This suggests the more an embryo is fed, the larger it grows, even beyond the size in a standard seed. Cultured embryos favor the accumulation of carbonaceous over nitrogenous metabolites, yielding less protein per mass than zygotic embryos. For *ex vivo* zygotic and somatic embryos, the genotype-specific protein content and its proteome phenotype appears to be less regulated, and instead the storage substance accumulation appears to have a direct relationship with nutrient input. The differences between *in planta* and *ex vivo* embryo development and storage substance accumulation indicates the significance of the *in planta* circumstance of each seed as an interactive member of a larger population.

SEED PROTEIN CONTENT AND ITS VARYING PROTEOME

For seed crops, of which soybean is a prominent example, historic and modern breeding has selected for enhanced storage substance accumulation. Generations of breeders have established that protein content is a genetically determined trait that can be selected. How protein content is regulated in relationship with protein composition appears to be a multilevel process, with the genotype establishing protein content. The rebalancing of proteome occurs in both dicots and monocots as shown by observations on soybean and maize (Wu and Messing, 2014). From the perspective of the seed that has to establish the next generation of an annual plant, its capacity to make compositional choices in response to altered metabolic circumstances has selective advantages. Understanding the processes that control proteome plasticity within the context of the protein content phenotype can enable biotechnologists to create enhanced soybeans optimized for specific end uses, such as species-specific feed or as protein bioreactors.

ACKNOWLEDGMENTS

Research in the Herman laboratory is supported by a grant from the United Soybean Board for the enhancement of seed protein content and composition. I am grateful to Dr. Monica Schmidt (University of Arizona) for her assistance with figures and comments.

REFERENCES

Allen, D. K., and Young, J. D. (2013). Carbon and nitrogen provisions alter the metabolic flux in developing soybean embryos. *Plant Physiol.* 161, 1458–1475. doi: 10.1104/pp.112.203299

Allen, R. D., Bernier, F., Lessard, P. A., and Beachy, R. N. (1989). Nuclear factors interact with a soybean beta-conglycinin enhancer. *Plant Cell* 1, 623–631.

Bäumlein, H., Nagy, I., Villarroel, R., Inzé, D., and Wobus, U. (1992). Cis-analysis of a seed protein gene promoter: the conservative RY repeat CATGCATG within the legumin box is essential for tissue-specific expression of a legumin gene. *Plant J.* 2, 233–239.

Berger, F., Grini, P. E., and Schnittger, A. (2006). Endosperm: an integrator of seed growth and development. *Curr. Opin. Plant Biol.* 9, 664–670. doi: 10.1016/j.pbi.2006.09.015

Bolon, Y. T., Joseph, B., Cannon, S. B., Granham, M. A., Diers, B. W., Farmer, A. D., et al. (2010). Complementary genetic and genomic approaches help characterize the linkage group I seed protein QTL in soybean. *BMC Plant Biol.* 10:41. doi: 10.1186/1471-2229-10-41

Borchers-Zampini, C., Glamm, A. B., Hoddinott, J., and Swanson, C. A. (1980). Alterations in source-sink patterns by modifications of source strength. *Plant Physiol.* 65, 1116–1120. doi: 10.1104/pp.65.6.1116

Borisjuk, L., Rolletschek, H., Radchuk, R., Weschke, W., Wobus, U., and Weber, H. (2004). Seed development and differentiation: a role for metabolic regulation. *Plant Biol. (Stuttg.)* 6, 375–386. doi: 10.1055/s-2004-817908

Borisjuk, L., Rolletschek, H., Wobus, U., and Weber, H. (2003). Differentiation of legume cotyledons as related to metabolic gradients and assimilate transport into seeds. *J. Exp. Bot.* 54, 503–512. doi: 10.1093/jxb/erg051

Brim, C. A., and Burton, J. W. (1979). Recurrent selection in soybeans. II. Selection for increased percent protein in seeds. *Crop Sci.* 19, 494–498. doi: 10.2135/cropsci1979.0011183X001900040016x

Carlson, D. R., and Brun, W. A. (1984). Alteration of C-assimilate partitioning in leaves of soybeans having increased reproductive loads at one node. *Plant Physiol.* 75, 887–890. doi: 10.1104/pp.75.4.887

Carter, T. E., Burton, J. W., and Brim, C. A. (1986). Registration of NC101to NC112 soybean germplasm lines contrasting in percent seed protein. *Crop Sci.* 26, 841–842. doi: 10.2135/cropsci1986.0011183X002600040057x

Chanprame, S., Kuo, T. M., and Widholm, J. M. (1998). Soluble carbohydrate content of soybean [*Glycine max* (L.) Merr.] somatic and zygotic embryos during development. *In Vitro Cell. Dev. Biol. Plant* 34, 64–68. doi: 10.1007/BF02823125

Chen, Z. L., Schuler, M. A., and Beachy, R. N. (1986). Functional analysis of regulatory elements in a plant embryo-specific gene. *Proc. Natl. Acad. Sci. U.S.A.* 83, 8560–8564. doi: 10.1073/pnas.83.22.8560

Chung, J., Babka, H. L., Graef, G. L., Staswick, P. E., Lee, D. J., Cregan, P. B., et al. (2003). The seed protein, oil, and yield QTL on soybean linkage group I. *Crop Sci.* 43, 1053–1067. doi: 10.2135/cropsci2003.1053

Cober, E. R., and Voldeng, H. D. (2000). Developing high-protein, high-yield soybean populations and lines. *Crop Sci.* 40, 39–42. doi: 10.2135/cropsci2000.40139x

Cregan, P. B., Jarvik, T., Bush, A. L., Schoemaker, R. C., Lark, K. G., Kahler, A. L., et al. (1999). An integrated genetic linkage map of the soybean genome. *Crop Sci.* 39, 1464–1490. doi: 10.2135/cropsci1999.3951464x

De Jong, A., Koerselman-Kooij, J. W., Schuurmans, J. A. M. J., and Borstlap, A. C. (1997). The mechanism of amino acid efflux from seed coats of developing pea seeds as revealed by uptake experiments. *Plant Physiol.* 114, 731–736.

DeWald, D. B., Mason, H. S., and Mullet, J. E. (1992). The soybean vegetative storage proteins VSP alpha and VSP beta are acid phosphatases active on polyphosphates. *J. Biol. Chem.* 267, 15958–15964.

Diers, B. W., Keim, P., Fehr, W. R., and Shoemaker, R. C. (1992). RFLP analysis of soybean seed protein and oil content. *Theor. Appl. Genet.* 83, 608–612. doi: 10.1007/BF00226905

Egli, D. B., and Bruening, W. P. (2001). Source–sink relationship, seed sucrose levels and seed growth rates in soybean. *Ann. Bot.* 88, 235–242. doi: 10.1006/anbo.2001.1449

Fehr, W. R., Caviness, C. E., Burmood, D. T., and Pennington, J. S. (1971). Stage of development descriptions for soybean, *Glycine max* (L.) Merril. *Crop Sci.* 11, 929–931. doi: 10.2135/cropsci1971.0011183X001100060051x

Fellows, R. J., Egli, D. B., and Leggett, J. E. (1979). Rapid changes in translocation patterns in soybeans following source-sink alterations. *Plant Physiol.* 64, 652–655. doi: 10.1104/pp.64.4.652

Finer, J. J. (1988). Apical proliferation of embryonic tissue of soybean [*Glycine max* (L.) Merrill]. *Plant Cell Rep.* 7, 238–241. doi: 10.1007/BF00272532

Finer, J. J., and McMullen, M. D. (1991). Transformation of soybean via particle bombardment of embryogenic suspension culture tissue. *In Vitro Cell. Dev. Biol. Plant* 27, 175–182. doi: 10.1007/BF02632213

Franceschi, V. R., and Giaquinta, R. T. (1983). Specialized cellular arrangements in legume leaves in relation to assimilate transport and compartmentation: comparison of the paraveinal mesophyll. *Planta* 159, 415–422. doi: 10.1007/BF00392077

Franceschi, V. R., Wittenbach, V. A., and Giaquinta, R. T. (1983). A paraveinal mesophyll of soybean leaves in relation to assimilate transfer and compartmentation III. Immunohistochemical localization of specific glycopeptides in the vacuole after depodding. *Plant Physiol.* 72, 586–589. doi: 10.1104/pp.72.2.586

Garcia, D., Fitz Gerald, J. N., and Berger, F. (2005). Maternal control of integument cell elongation and zygotic control of endosperm growth are coordinated to determine seed size in *Arabidopsis*. *Plant Cell* 17, 52–60. doi: 10.1105/tpc.104.027136

Garcia, D., Saingery, V., Chambrier, P., Mayer, U., Jurgens, G., and Berger, F. (2003). *Arabidopsis* haiku mutants reveal new controls of seed size by endosperm. *Plant Physiol.* 131, 1661–1670. doi: 10.1104/pp.102.018762

Gifford, R. M., and Thorne, J. H. (1985). Sucrose concentration at the apoplastic interface between seed coat and cotyledons of developing soybean seeds. *Plant Physiol.* 77, 863–868. doi: 10.1104/pp.77.4.863

Goldberg, R. B., Hoschek, G., Ditta, G. S., and Breidenbach, R. W. (1981a). Developmental regulation of cloned superabundant embryo mRNAs in soybean. *Dev. Biol.* 83, 218–231. doi: 10.1016/0012-1606(81)90468-1

Goldberg, R. B., Hoschek, G., Tam, S. H., Ditta, G. S., and Breidenbach, R. W. (1981b). Abundance, diversity, and regulation of mRNA sequence sets in soybean embryogenesis. *Dev. Biol.* 83, 201–217. doi: 10.1016/0012-1606(81)90467-X

Gutierrez, L., Van Wuytswinkel, O., Castelain, M., and Bellini, C. (2007). Combined networks regulating seed maturation. *Trends Plant Sci.* 12, 294–300. doi: 10.1016/j.tplants.2007.06.003

Haga, K. I., and Sodek, L. (1987). Utilization of nitrogen sources by immature soybean cotyledons in culture. *Ann. Bot.* 59, 597–601.

Harada, J. J., Barker, S. J., and Goldberg, R. B. (1989). Soybean β-conglycinin genes are clustered in several DNA regions and are regulated by transcriptional and posttranscriptional processes. *Plant Cell* 1, 415–425. doi: 10.1105/tpc.1.4.415

Hartwig, E. E., and Kilen, T. C. (1991). Yield and composition of soybean seed from parents with different protein, similar yield. *Crop Sci.* 31, 290–292. doi: 10.2135/cropsci1991.0011183X003100020011x

Hayati, R., Egli, D. B., and Crafts-Brandner, S. J. (1996). Independence of nitrogen supply and seed growth in soybean: studies using an in vitro culture system. *J. Exp. Bot.* 47, 33–40. doi: 10.1093/jxb/47.1.33

He, Y., Young, T., Clark, K., Kleppinger-Sparace, K., Bridges, W., and Sparace, S. (2011). Developmental profile of storage reserve accumulation in soybean somatic embryos. *In Vitro Cell. Dev. Biol. Plant* 47, 725–733. doi: 10.1007/s11627-011-9375-2

Herman, E. M. (2008). Endoplasmic reticulum bodies: Solving the insoluble. *Curr. Opin. Plant Biol.* 11, 672–679. doi: 10.1016/j.pbi.2008.08.004

Herman, E. M., and Burks, A. W. (2011). The impact of plant biotechnology on food allergy. *Curr. Opin. Biotechnol.* 22, 224–230. doi: 10.1016/j.copbio.2010.11.003

Herman, E. M., and Larkins, B. A. (1999). Protein storage bodies and vacuoles. *Plant Cell* 11, 601–614. doi: 10.1105/tpc.11.4.601

Herman, E. M., and Schmidt, M. A. (2004). ER to vacuole trafficking (ERvt) provides an alternate pathway for endomembrane progression for transfer to the vacuole. *Plant Physiol.* 136, 3440–3446. doi: 10.1104/pp.104.051722

Hernández-Sebastià, C., Marsolais, F., Saravitz, C., Israel, D., Dewey, R. E., and Huber, S. C. (2005). Free amino acid profiles suggest a possible role for asparagine in the control of storage-product accumulation in developing seeds of low- and high-protein soybean lines. *J. Exp. Bot.* 56, 1951–1963. doi: 10.1093/jxb/eri191

Hill, J. E., and Breidenbach, R. W. (1974). Proteins of soybean seeds. II. Accumulation of the major protein components during seed development and maturation. *Plant Physiol.* 53, 747–751. doi: 10.1104/pp.53.5.747

Hsu, F. C., Bennett, A. B., and Spanswick, R. M. (1984). Concentrations of sucrose and nitrogenous compounds in the apoplast of developing soybean seed coats and embryos. *Plant Physiol.* 75, 181–186. doi: 10.1104/pp.75.1.181

Iyer, V. V., Sriram, G., Fulton, D. B., Zhou, R., Westgate, M. E., and Shanks, J. V. (2008). Metabolic flux maps comparing the effect of temperature on protein and oil biosynthesis in developing soybean cotyledons. *Plant Cell Environ.* 31, 506–517. doi: 10.1111/j.1365-3040.2008.01781.x

Jofuku, K. D., Okamuro, J. K., and Goldberg, R. B. (1987). Interaction of an embryo DNA binding protein with a soybean lectin gene upstream region. *Nature* 328, 734–737. doi: 10.1038/328734a0

Kinney, A. J., Jung, R., and Herman, E. M. (2001). Cosuppression of the α-subunits of β-conglycinin in transgenic soybean seeds induces the formation of endoplasmic reticulum-derived protein bodies. *Plant Cell* 13, 1165–1178. doi: 10.1105/tpc.13.5.1165

Krishnan, H. B., Natarian, S. S., Bennett, J. O., and Sicher, R. C. (2011). Protein and metabolite composition of xylem sap from field-grown soybeans (*Glycine max*). *Planta* 233, 921–931. doi: 10.1007/s00425-011-1352-9

Kroj, T., Savino, G., Valon, C., Giraudat, J., and Parcy, F. (2003). Regulation of storage protein gene expression in *Arabidopsis*. *Development* 130, 6065–6073. doi: 10.1242/dev.00814

Kwong, R. W., Bui, A. Q., Lee, H., Kwong, L. W., Fischer, R. L., Goldberg, R. B., et al. (2003). LEAFY COTYLEDON1-LIKE defines a class of regulators essential for embryo development. *Plant Cell* 15, 5–18. doi: 10.1105/tpc.006973

Lea, P. J., and Miflin, B. J. (1980). "Transport and metabolism of asparagine and other nitrogen compounds within the plant," in *The Biochemistry of Plant*, eds P. K. Stumpf and E. E. Conn (New York, NY: Academic Press), 569–607.

Lessard, P. A., Allen, R. D., Bernier, F., Crispino, J. D., Fujiwara, T., and Beachy, R. N. (1991). Multiple nuclear factors interact with upstream sequences of differentially regulated beta-conglycinin genes. *Plant Mol. Biol.* 16, 397–413. doi: 10.1007/BF00023991

Lima, J. D., and Sodek, L. (2003). N-stress alters aspartate and asparagine levels of xylem sap in soybean. *Plant Sci.* 165, 649–656. doi: 10.1016/S0168-9452(03)00251-6

Lohaus, G., Buker, M., Hussmann, M., Soave, C., and Heldt, H. W. (1998). Transport of amino acids with special emphasis on the synthesis and transport of asparagine in the Illinois Low Protein and Illinois High Protein strains of maize. *Planta* 205, 181–188. doi: 10.1007/s004250050310

Mienke, D. W., Chen, J., and Beachy, R. N. (1981). Expression of storage-protein genes during soybean seed development. *Planta* 153, 130–139. doi: 10.1007/BF00384094

Miller, A. J., Fan, X., Shen, Q., and Smith, S. J. (2008). Amino acids and nitrate as signals for the regulation of nitrogen acquisition. *J. Exp. Bot.* 59, 111–119. doi: 10.1093/jxb/erm208

Mori, T., Maruyama, N., Nishizawa, K., Higasa, T., Yagasaki, K., Ishimoto, M., et al. (2004). The composition of newly synthesized proteins in the endoplasmic reticulum determines the transport pathways of soybean seed storage proteins. *Plant J.* 40, 238–249. doi: 10.1111/j.1365-313X.2004.02204.x

Murray, D. R. (1987). Nutritive role of seed coats in developing legume seeds. *Am. J. Bot.* 74, 1122–1137. doi: 10.2307/2443954

Naito, S., Dube, P. H., and Beachy, R. N. (1988). Differential expression of conglycinin alpha-subunit and beta-subunit genes in transgenic plants. *Plant Mol. Biol.* 11, 109–123. doi: 10.1007/BF00015664

Nichols, D. M., Glover, K. D., Carlson, S. R., Specht, J. E., and Diers, B. W. (2006). Fine mapping of a seed protein QTL on soybean linkage group I and its correlated effects on agronomic traits. *Crop Sci.* 46, 834–839. doi: 10.2135/cropsci2005.05-0168

Nielsen, N. C., and Nam, Y. W. (1999). "Soybean globulins," in *Seed Proteins,* eds P. R. Shewry and R. Casey (Dordrecht: Kluwer Academic Publishers), 285–313. doi: 10.1007/978-94-011-4431-5_13

Nishizawa, K., and Ishimoto, M. (2009). Maturation of somatic embryos as a model for soybean seed development. *Plant Biotechnol.* 26, 543–550. doi: 10.5511/plantbiotechnology.26.543

Obendorf, R. L., Rytko, G. T., and Byrne, M. C. (1983). Soya bean seed growth and maturation by in vitro pod culture. *Ann. Bot.* 51, 217–227.

Obendorf, R. L., Timpo, E. E., Byrne, M. C., Toai, T. V., Rytko, G. T., Hsu, F. C., et al. (1984). Soya bean seed growth and maturation in vitro without pods. *Ann. Bot.* 53, 853–863.

Pandurangan, S., Pajak, A., Molnar, S. J., Cober, E. R., Dhaubhadel, S., Hernandez-Sebastia, C., et al. (2012). Relationship between asparagine metabolism and protein concentration in soybean seed. *J. Exp. Bot.* 63, 3173–3184. doi: 10.1093/jxb/ers039

Patrick, J. W., and Offler, C. E. (2001). Compartmentation of transport and transfer events in developing seeds. *J. Exp. Bot.* 356, 551–564. doi: 10.1093/jexbot/52.356.551

Perez-Grau, L., and Goldberg, R. B. (1989). Soybean seed protein genes are regulated spatially during embryogenesis. *Plant Cell* 1, 1095–1109. doi: 10.1105/tpc.1.11.1095

Pipolo, A. E., Sinclair, T. R., and Camara, G. M. S. (2004). Protein and oil concentration of soybean seed cultured in vitro using nutrient solutions of differing glutamine concentration. *Ann. Appl. Biol.* 144, 223–227. doi: 10.1111/j.1744-7348.2004.tb00337.x

Ragster, L. E., and Chrispeels, M. J. (1981a). Autodigestion in crude extracts of soybean leaves and isolated chloroplasts as a measure of proteolytic activity. *Plant Physiol.* 67, 104–109. doi: 10.1104/pp.67.1.104

Ragster, L. V., and Chrispeels, M. J. (1981b). Hemoglobin-digesting Acid Proteinases in Soybean Leaves: characteristics and changes during leaf maturation and senescence. *Plant Physiol.* 67, 110–114. doi: 10.1104/pp.67.1.110

Ragster, L. V., and Chrispeels, M. J. (1979). Azocoll-digesting Proteinases in Soybean Leaves: characteristics and changes during Leaf Maturation and Senescence. *Plant Physiol.* 64, 857–862. doi: 10.1104/pp.64.5.857

Rainbird, R. M., Thorne, J. H., and Hardy, R. W. F. (1984). Role of amides, amino acids, and ureides in the nutrition of developing soybean seeds. *Plant Physiol.* 74, 329–334. doi: 10.1104/pp.74.2.329

Raper, C. D., Patterson, R. P., List, M. L., Obendorf, R. L., and Downs, R. J. (1984). Photoperiod effects on growth rate of in vitro cultured soybean embryos. *Bot. Gaz.* 145, 157–162. doi: 10.1086/337441

Santarem, E. R., Pelissier, B., and Finer, J. J. (1997). Effect of explant orientation, pH, solidifying agent and wounding on initiation of soybean somatic embryos. *In Vitro Cell. Dev. Biol. Plant* 33, 13–19. doi: 10.1007/s11627-997-0034-6

Santos-Mendoza, M., Dubreucq, B., Baud, S., Parcy, F., Caboche, M., and Lepiniec, L. (2008). Deciphering gene regulatory networks that control seed development and maturation in *Arabidopsis. Plant J.* 54, 608–620. doi: 10.1111/j.1365-313X.2008.03461.x

Saravitz, C. H., and Raper, C. D. (1995). Responses to sucrose and glutamine by soybean embryos grown in vitro. *Physiol. Plant.* 93, 799–805. doi: 10.1111/j.1399-3054.1995.tb05134.x

Schaefer, J., Skokut, T. A., Stejskal, E. O., McKay, R. A., and Varner, J. E. (1981a). Estimation of protein turnover in soybean leaves using magic angle double cross-polarization nitrogen 15 nuclear magnetic resonance. *J. Biol. Chem.* 256, 11574–11579.

Schaefer, J., Skokut, T. A., Stejskal, E. O., McKay, R. A., and Varner, J. E. (1981b). Asparagine amide metabolism in developing cotyledons of soybean. *Proc. Natl. Acad. Sci. U.S.A.* 78, 5978–5982. doi: 10.1073/pnas.78.10.5978

Schmidt, M. A., Barbazuk, W. B., Stanford, M., May, G., Song, Z., Hong, W., et al. (2011). Silencing of soybean seed storage proteins results in a rebalanced protein composition preserving seed protein content without major collateral changes in the metabolome and transcriptome. *Plant Physiol.* 156, 330–345. doi: 10.1104/pp.111.173807

Schmidt, M. A., and Herman, E. M. (2008). The collateral protein compensation mechanism can be exploited to enhance foreign protein accumulation in soybean seeds. *Plant Biotechnol. J.* 6, 832–842. doi: 10.1111/j.1467-7652.2008.00364.x

Schmidt, M. A., Tucker, D. M., Cahoon, E. B., and Parrott, W. A. (2005). Towards normalization of soybean somatic embryo maturation. *Plant Cell Rep.* 24, 383–391. doi: 10.1007/s00299-005-0950-z

Skokut, T. A., Varner, J. E., Schaefer, J., Stejskal, E. O., and McKay, R. A. (1982). [N]NMR determination of asparagine and glutamine nitrogen utilization for synthesis of storage proteins I developing soybeans in culture. *Plant Physiol.* 60, 308–313. doi: 10.1104/pp.69.2.308

Staswick, P. E. (1989a). Developmental regulation and the influence of plant sinks on vegetative storage protein gene expression in soybean leaves. *Plant Physiol.* 89, 309–315. doi: 10.1104/pp.89.1.309

Staswick, P. E. (1989b). Preferential loss of an abundant storage protein from soybean pods during seed development. *Plant Physiol.* 90, 1252–1255. doi: 10.1104/pp.90.4.1252

Staswick, P. E., Papa, C., Huang, J.-F., and Rhee, Y. (1994). Purification of the major soybean leaf acid phosphatase that is increased by fruit removal. *Plant Physiol.* 104, 49–57.

Staswick, P. E., Zhang, Z., Clemente, T. E., and Specht, J. E. (2001). Efficient down-regulation of the major vegetative storage protein genes in transgenic soybean does not compromise plant productivity. *Plant Physiol.* 127, 1819–1826. doi: 10.1104/pp.010638

Thibaud-Nissen, F., Shealy, R. T., Kahanna, A., and Vodkin, L. O. (2003). Clustering of microarray data reveals transcript patterns associated with somatic embryogenesis in soybean. *Plant Physiol.* 132, 118–136. doi: 10.1104/pp.103.019968

Thompson, J. F., Madison, J. T., and Muenster, A. M. E. (1977). In vitro culture of immature cotyledons of soya bean (*Glycine max* L. Merr.). *Ann. Bot.* 41, 29–39.

Thorne, J. H. (1980). Kinetics of 14C-photosynthate uptake by developing soybean fruit. *Plant Physiol.* 65, 975–979. doi: 10.1104/pp.65.5.975

Thorne, J. H., and Rainbird, R. (1983). An in vivo technique for the study of phloem unloading in seed coats of developing soybean seeds. *Plant Physiol.* 72, 268–271. doi: 10.1104/pp.72.1.268

To, A., Valon, C., Savino, G., Guilleminot, J., Devic, M., Giraudat, J., et al. (2006). A network of local and redundant gene regulation governs *Arabidopsis* seed maturation. *Plant Cell* 18, 1642–1651. doi: 10.1105/tpc.105.039925

Truong, Q., Koch, K., Yoon, J. M., Everard, J. D., and Shanks, J. V. (2013). Influence of carbon to nitrogen ratios on soybean somatic embryo (cv. Jack) growth and composition. *J. Exp. Bot.* 64, 2985–2995. doi: 10.1093/jxb/ert138

Walker, D. R., and Parrott, W. A. (2001). Effect of polyethylene glycol and sugar alcohols on soybean somatic embryo germination and conversion. *Plant Cell Tissue Organ Cult.* 64, 55–62. doi: 10.1023/A:1010601402098

Walling, L., Drews, G. N., and Goldberg, R. B. (1986). Transcriptional and post-transcriptional regulation of soybean seed protein mRNA levels. *Proc. Natl. Acad. Sci. U.S.A.* 83, 2123–2127. doi: 10.1073/pnas.83.7.2123

Wang, H. W., Zhang, B., Hao, Y. J., Huang, J., Tian, A. G., Liao, Y., et al. (2007). The soybean Dof-type transcription factor genes, GmDof4 and GmDof11, enhance lipid content in the seeds of transgenic *Arabidopsis* plants. *Plant J.* 52, 716–729. doi: 10.1111/j.1365-313X.2007.03268.x

Wan, T. F., Shao, G. H., Shan, X. C., Zeng, N. Y., and Lam, H. M. (2006). Correlation between AS1 gene expression and seed protein contents in different soybean (*Glycine max* [L.] Merr.) cultivars. *Plant Biol.* 8, 271–275. doi: 10.1055/s-2006-923876

Wilcox, J. R., and Shibles, R. M. (2001). Interrelationships among seed quality attributes in soybean. *Crop Sci.* 41, 11–14. doi: 10.2135/cropsci2001.41111x

Wilson, R. F. (2004). "Seed composition," in *Soybean: Improvement, Production and Uses*, 3rd Edn, eds H. R. Boerma and J. E. Specht (Madison, WI: American Society Of Agronomy-Crop Science Society), 621–677.

Wu, Y., and Messing, J. (2014). Proteome balancing of the maize seed for higher nutritional value. *Front. Plant Sci.* 5:240. doi: 10.3389/fpls.2014.00240

Zhang, L., Tan, Q., Lee, R., Trethewy, A., Lee, Y. H., and Tegeder, M. (2010). Altered xylem–phloem transfer of amino acids affects metabolism and leads to increased seed yield and oil content in *Arabidopsis*. *Plant Cell* 22, 3603–3620. doi: 10.1105/tpc.110.073833

Zhao-Ming, Q., Ya-nan, S., Qiong, W., Chun-yan, L., Guo-hua, H., and Qing-shan, C. (2011). A meta-analysis of seed protein concentration QTL in soybean. *Can. J. Plant Sci.* 91, 221–230. doi: 10.4141/cjps09193

Conflict of Interest Statement: The author declares that the research was conducted in the absence of any commercial or financial relationships that could be construed as a potential conflict of interest.

14

Cell cycle control and seed development

Ricardo A. Dante[1] *, Brian A. Larkins[2,3] * and Paolo A. Sabelli[3] *

[1] *Embrapa Agricultural Informatics, Campinas, Brazil*
[2] *Department of Agronomy and Horticulture, University of Nebraska, Lincoln, NE, USA*
[3] *School of Plant Sciences, University of Arizona, Tucson, AZ, USA*

Edited by:
Neelima Roy Sinha, University of California, Davis, USA

Reviewed by:
A. Mark Settles, University of Florida, USA
Pablo Daniel Jenik, Franklin & Marshall College, USA

****Correspondence:***
Ricardo A. Dante, Embrapa Agricultural Informatics, Avenida André Tosello 209, Campinas, São Paulo 13083-886, Brazil
e-mail: ricardo.dante@embrapa.br;
Brian A. Larkins, Department of Agronomy and Horticulture, University of Nebraska, 230J Whittier Research Center, 2200 Vine Street, Lincoln, NE 68583-0857, USA
e-mail: blarkins2@unl.edu;
Paolo A. Sabelli, School of Plant Sciences, University of Arizona, 303 Forbes, 1140 East South Campus Drive, Tucson, AZ 85721-0036, USA
e-mail: psabelli@ag.arizona.edu

Seed development is a complex process that requires coordinated integration of many genetic, metabolic, and physiological pathways and environmental cues. Different cell cycle types, such as asymmetric cell division, acytokinetic mitosis, mitotic cell division, and endoreduplication, frequently occur in sequential yet overlapping manner during the development of the embryo and the endosperm, seed structures that are both products of double fertilization. Asymmetric cell divisions in the embryo generate polarized daughter cells with different cell fates. While nuclear and cell division cycles play a key role in determining final seed cell numbers, endoreduplication is often associated with processes such as cell enlargement and accumulation of storage metabolites that underlie cell differentiation and growth of the different seed compartments. This review focuses on recent advances in our understanding of different cell cycle mechanisms operating during seed development and their impact on the growth, development, and function of seed tissues. Particularly, the roles of core cell cycle regulators, such as cyclin-dependent-kinases and their inhibitors, the Retinoblastoma-Related/E2F pathway and the proteasome-ubiquitin system, are discussed in the contexts of different cell cycle types that characterize seed development. The contributions of nuclear and cellular proliferative cycles and endoreduplication to cereal endosperm development are also discussed.

Keywords: cell division, cotyledon, cyclin-dependent kinase, embryo, endoreduplication, endosperm, retinoblastoma-related, seed coat

INTRODUCTION

SEED DEVELOPMENT PHASES

Angiosperms reproduce sexually via the production of seeds, which are typically derived from the fertilization of ovules, or asexually via apomixis. Mature seeds characteristically contain three major structures: a sporophyte (the embryo), nutrient storage tissues or organs (the endosperm and/or the embryonic cotyledons), which support embryogenesis and early post-embryonic sporophyte development, and a protective structure (the seed coat or testa). A prominent triploid endosperm is typically present in mature monocot seeds, while most of its cells are consumed during dicot seed development. The seed coat is a maternal structure derived from the ovule integuments that functions in seed protection, dormancy, germination, and, in some dicot species like legumes (Fabaceae or Leguminoseae family), transiently in nutrient storage. In the monocot Poaceae family (grasses), the seed coat is fused to the pericarp (the fruit coat). The maternal plant makes significant contributions to seed production by providing nutrients, conveying hormonal and environmental cues and imposing mechanical constraints on the floral structures within which seeds develop. Consequently, seed production is influenced by a range of zygotic, sporophytic, and environmental factors (Egli, 2006).

The size of a multicellular organism, its organs and tissues depend on the number and size of constituent cells. Cell number, in turn, depends on the rate of cell division, the number of dividing cells, and the duration of the cell proliferation phase during development, while the size of non-dividing cells is influenced by cell growth and cell expansion (defined as increases in cell macromolecular mass and cell volume, respectively; Sugimoto-Shirasu and Roberts, 2003). In plants, cell number generally seems to make a larger contribution than cell size to the size of comparable organs (reviewed by Mizukami, 2001). However, seed size and weight are highly influenced by cell size via the growth and expansion brought about by massive accumulation of storage compounds (proteins, lipids, and/or carbohydrates) and water intake by cotyledon or endosperm cells. In cereal and legume crops, seed growth and development typically comprise three partially overlapping phases: an initial lag phase initiated at fertilization and characterized by cell proliferation and minimal dry weight gain (phase I); seed filling, a linear phase of large dry weight gain associated with cell enlargement and accumulation of storage compounds (phase II); and a final phase of reduced dry weight gain associated with desiccation and dormancy (phase III; Egli, 2006). In phase I, various seed tissues and domains are specified and established, including the vital transfer cells, a filial conduit with the mother plant vascular tissue that nourishes the developing seed. Phase I is also characterized by uptake of sucrose, which is rapidly converted to hexoses via cell wall-bound invertase activity. Even though phase I is critical for seed development and grain yield, its contribution is indirect, as the cells generated in this phase are very small and contribute little to seed biomass. However, during phase II most of the endosperm or cotyledon cells generated in phase I accumulate storage compounds. Thus, phase II is characterized by cell enlargement due to cell growth and cell expansion, and a peak in seed water content (Egli, 2006). In storage cells of the cereal endosperm and legume cotyledons, phase II is also characterized by endoreduplication (also known as endopolyploidization, endocycling, or endoreplication), a type of cell cycle that leads to polyploidy. During phase III, water concentration decreases dramatically and physiological maturity is reached.

CELL CYCLE TYPES OCCURRING DURING SEED DEVELOPMENT

Briefly, the prototypical mitotic cycle consists of a DNA replication phase (S-phase), and a chromosome condensation and sister chromatid segregation phase (M-phase), which are preceded by G1 and G2 gap phases, respectively. Typically, this cell cycle is associated with cell division, and M-phase is generally coupled to karyokinesis and cytokinesis, which generate daughter cells with chromosome number and nuclear DNA content identical to those of their mother cell. However, in several plant tissues, cell types, and developmental stages, alternative cell cycle types can occur. In the context of seed development, frequently these are acytokinetic mitosis and endoreduplication (**Figure 1**). In acytokinetic mitosis, the mitotic cell cycle is coupled to karyokinesis in the absence of cytokinesis, thus producing a syncytium or multinucleate cell. Endoreduplication is characterized by recurrent and alternating gap and S-phase, without intervening sister chromatid segregation, karyokinesis and cytokinesis, thus resulting in polyploid cells with an unaltered number of chromosomes, but with each chromosome containing multiple chromatids (reviewed by Edgar et al., 2014). All these different cell cycle types influence the growth and development of seed structures. Early in development, asymmetric cell divisions in the embryo generate polarized daughter cells that take on diverse differentiation paths, and rapid nuclear proliferation and cellularization in the endosperm establish the initial cell populations that occupy the embryo sac. Subsequent intense cell proliferation, coupled to cell differentiation, essentially produces all the embryo and endosperm cell types and tissues.

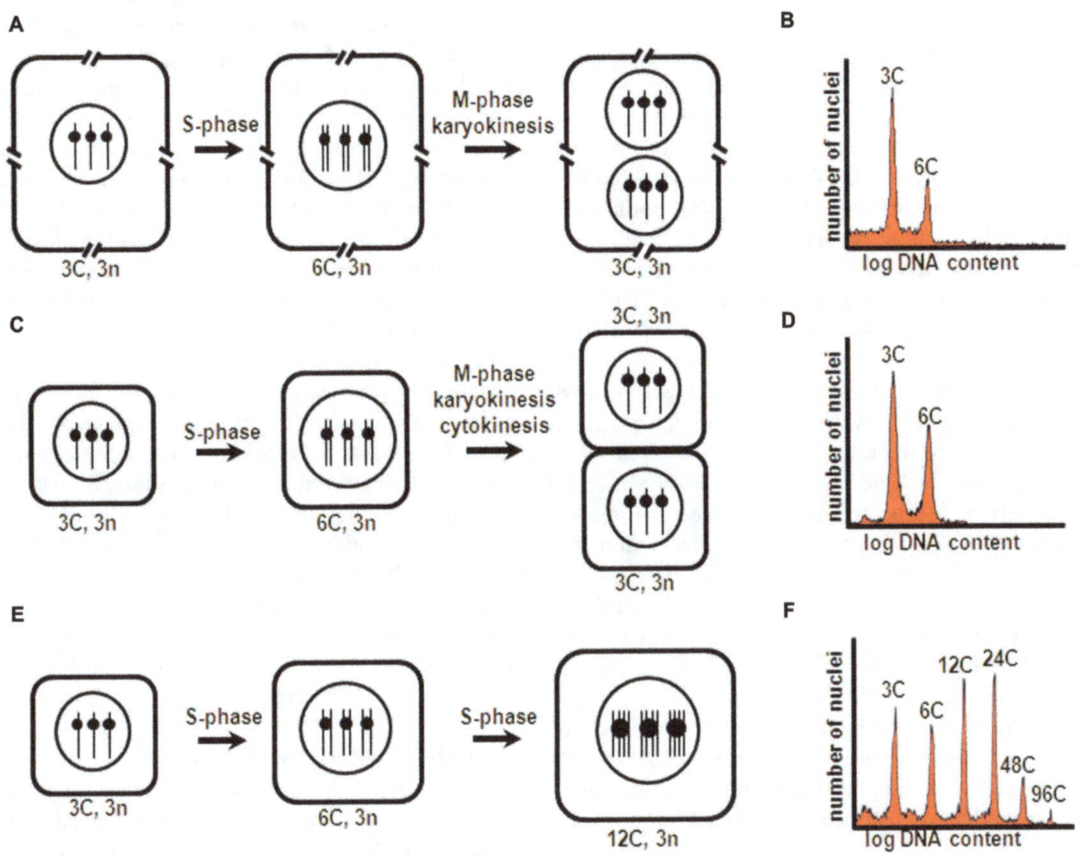

FIGURE 1 | Cell cycle types occurring during seed development. Triploid endosperm mother cells (with two maternal and one paternal chromosomal complements) are shown as an example. A hypothetical haploid number $n = 1$ is assumed for simplicity. **(A)** Acytokinetic mitosis of endosperm nuclei within the embryo sac central cell, resulting in a syncytium; **(C)** Cell proliferation through mitotic cell division following syncytium cellularization; **(E)** Endoreduplication of inner endosperm starchy cells. Cell number, size, DNA content, and chromosome number correspond to one complete cell cycle round comprising S-phase and accompanying M-phase and karyokinesis **(A,C)** and cytokinesis **(C)**, and two complete endoreduplication cycle rounds (each comprising S-phase not followed by M-phase, karyokinesis and cytokinesis) **(E)**. Interrupted cell boundaries in **(A)** indicate the large size of the embryo sac central cell. C and n, DNA content and chromosome number of a haploid cell, respectively. **(B,D,F)** show typical nuclear flow-cytometric profiles obtained for tissues undergoing asynchronous, iterative acytokinetic mitosis, mitotic cell division, and endoreduplication cycles, respectively.

Finally, endoreduplication occurs, which is inherently associated with cell enlargement and accumulation of storage compounds in specialized cotyledon or endosperm cells.

Seed biology aspects such as comparative development and anatomy of seed structures and their underlying signaling networks were reviewed in-depth recently (Sabelli and Larkins, 2009b; Nowack et al., 2010; Lau et al., 2012; Sabelli, 2012b). Likewise, the role of cell cycle regulation in plant growth and development has also been reviewed thoroughly elsewhere (De Veylder et al., 2011; Heyman and De Veylder, 2012; Edgar et al., 2014; Sabelli, 2014). Hence, we focus on recent findings that clarify the role of core cell cycle regulators and different cell cycle types in the development, growth, and function of seed structures.

CELL CYCLE CONTROL AND CORE REGULATORS IN PLANTS: AN OVERVIEW

CYCLIN-DEPENDENT KINASES AND CYCLINS

In eukaryotes, cell cycle progression is controlled by the periodic activity of various heterodimeric threonine/serine protein kinases composed of catalytic and regulatory subunits, a cyclin-dependent kinase (CDK) and a cyclin, respectively. Plants possess relatively large sets of genes encoding different CDKs and cyclins, which can interact to form a potentially large number of combinations (Van Leene et al., 2011). Plants contain eight types of CDK-like proteins (reviewed by Dudits et al., 2007). Among the major CDKs involved with cell cycle regulation are members of the A-type, which characteristically contain in their cyclin-interacting α-helix a hallmark PSTAIRE amino acid motif; these function during S-phase and at the G1/S and G2/M transitions. In the plant-specific B-type CDKs, which function primarily at the G2/M transition, the PSTAIRE motif is replaced by PPTALRE (B1-subtype) or PPTTLRE (B2-subtype). D- and F-type CDKs, also known as CDK-activating kinases (CAKs), regulate A- and B-type CDKs through phosphorylation of specific residues (reviewed by Inzé and De Veylder, 2006). Angiosperm genomes possess a cyclin complement of ~50–60 genes organized into ~10 types (Wang et al., 2004; La et al., 2006; Hu et al., 2010; Ma et al., 2013). The majority of D-type cyclins are involved in the control of the G1/S transition; A-type cyclins, S-phase, and the G2/M transition; and B-type cyclins, G2/M, and intra-mitotic transitions (Inzé and De Veylder, 2006). CDK/cyclin complexes are subjected to different levels of regulation, including binding by non-catalytic CDK-specific inhibitors (CKIs), activating or inhibitory phosphorylation of CDK subunits, and cell cycle phase-specific cyclin synthesis and proteolysis, the latter of which is mediated by the ubiquitin-proteasome system (UPS; Inzé and De Veylder, 2006). A simplified diagram depicting some major molecular mechanisms of the plant cell cycle is shown in **Figure 2**.

THE RETINOBLASTOMA-RELATED PATHWAY

In higher eukaryotes, proteins of the retinoblastoma-related (RBR) family are known as repressors of the G1/S transition for their inhibitory effect on heterodimeric E2F/DP transcription factors. These, in turn, control the expression of multiple genes required for this cell cycle transition and S-phase progression, such as those encoding the subunits of the helicase MINICHROMOSOME MAINTENANCE2-7 (MCM2-7) and the replication processivity factor proliferating cell nuclear antigen (PCNA) complexes (reviewed by Sabelli and Larkins, 2009c). RBR proteins are sequentially phosphorylated and inhibited by different CDK complexes, which relieves the block on E2F/DP-dependent gene expression and results in the transition into S-phase. In plants, CDK complexes containing D- and A-type cyclins seem to phosphorylate and thus inhibit RBR proteins. Grasses possess a more numerous and functionally diverse set of RBR proteins than most dicots, including *Arabidopsis thaliana* (Sabelli and Larkins, 2006, 2009c). The maize (*Zea mays*) genome contains at least four RBR genes that are grouped into two distinct types of duplicated genes, exemplified by *RBR1* and *RBR3* (Sabelli et al., 2005; Sabelli and Larkins, 2006). While *RBR1* functions as a repressor of cell cycle progression (Sabelli et al., 2013), *RBR3* stimulates the expression of genes encoding MCM2–7 proteins and DNA replication (Sabelli et al., 2009). *RBR3* is itself an E2F/DP target whose expression is negatively regulated by *RBR1* (Sabelli et al., 2005).

DNA REPLICATION INITIATION FACTORS

Regulation of DNA replication in plants is generally believed to follow conserved eukaryotic patterns (reviewed by Costas et al., 2011). Initiation of S-phase requires priming of chromatin via the assembly, at origins of replication, of the pre-replication complex consisting of ORIGIN OF REPLICATION COMPLEX (ORC1-6), CDC6, CDT1, and MCM2-7 proteins. ORC1-6 associate with origins of replication throughout the cell cycle but become sequentially bound by different proteins. Interaction of CDC6 and CDT1 with ORCs during G1 promotes the loading of MCM2-7, effectively licensing the origins for DNA replication (reviewed by Tuteja et al., 2011). These replication origin complexes are activated by CDKs at the G1/S transition, which leads to the recruitment of the replication machinery, the unwinding of DNA through MCM-dependent helicase activity, and DNA synthesis with the formation of replication forks. MCM protein complexes are displaced from replication origins as replication forks advance along the DNA, which prevents re-licensing of origins until S-phase and M-phase are completed. In endoreduplication cycles, DNA replication is initiated independently of M-phase, resulting in repeated rounds of DNA synthesis in the absence of mitosis. However, endoreduplication results from specific cell cycle modifications rather than merely uncontrolled activation of replication origins, which would cause over- and incomplete DNA replication (Costas et al., 2011; De Veylder et al., 2011; Sabelli, 2012a).

UBIQUITIN-DEPENDENT PROTEOLYSIS

Ubiquitin-Proteasome System-mediated proteolysis promotes the controlled destruction of several cell cycle regulators, which is critical for cell cycle phase transitions. Cyclins, CKIs, and other cell cycle regulators are targeted to the proteasome via their selective modification by various ubiquitin-protein ligases. Among the multimeric E3 ubiquitin-protein ligases functioning in plant cell cycle control are the anaphase promoting complex/cyclosome (APC/C), the Skp1/Cullin/F-box complex, and the Cullin-RING Ubiquitin Ligases (Heyman and De Veylder, 2012; Genschik et al., 2013). During late mitosis and most of G1, CDK activity is typically reduced via the targeting of A- and B-type cyclins

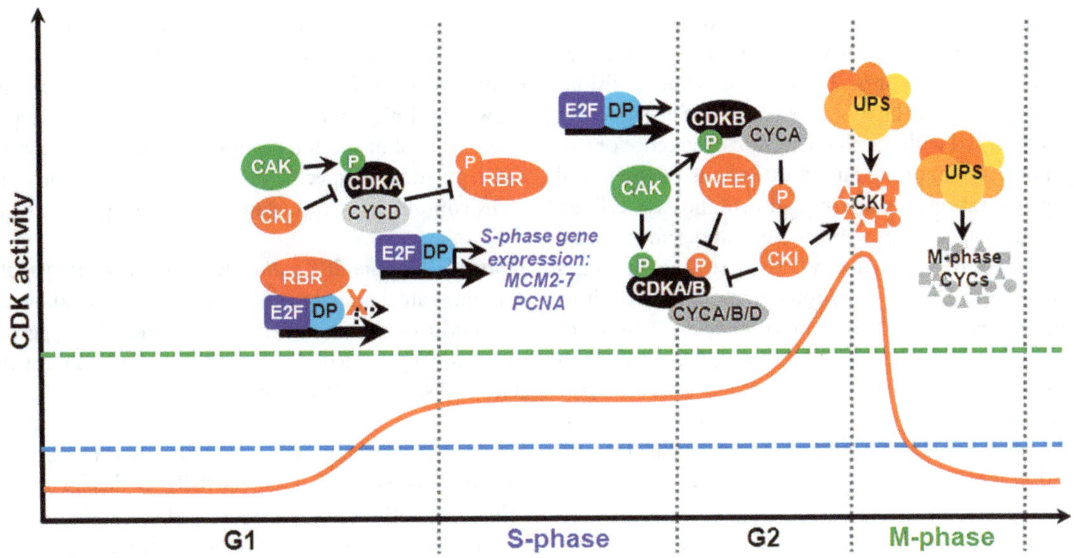

FIGURE 2 | The prototypical mitotic cell division cycle and some key molecular mechanisms that regulate its major transitions in plants. Cell cycle progression through distinct phases is driven by periodic fluctuations of cyclin-dependent kinase (CDK) activity (solid red line). Green and red circles labeled with a P letter show stimulatory and inhibitory phosphorylation, respectively. For cells to make a transition from G1 into S-phase and execute DNA replication, CDK activity must surpass an S-phase threshold (dashed blue line). A further CDK activity increase above an M-phase threshold (dashed green line) during G2 drives entry into mitosis. Both M-phase exit and origin of replication licensing require CDK activity to be reduced at the end of mitosis and maintained at low levels for most of G1. At the G1/S transition, CDKA/CYCD complexes phosphorylate and thus inactivate retinoblastoma-related (RBR), thus permitting heterodimeric E2F/DP transcription factors to stimulate an S-phase gene expression program, which results in the expression of MCM2-7 and proliferating cell nuclear antigen (PCNA) genes, among many others. CDKA/CYCD activity is positively and negatively regulated by, respectively, CDK-activating kinase (CAK)-dependent phosphorylation and binding by CKIs. Later in G2, CDK activity requires association with mitotic cyclins and is also stimulated by CAK and inhibited by CKIs. In addition, this CDK activity is inhibited by phosphorylation at specific tyrosine residues by WEE1. Certain CDKs of the B-type, whose expression is E2F/DP-dependent, promote M-phase progression by mechanisms that include interaction with A-type cyclins and stimulation of downstream CDK activity by phosphorylating and targeting certain CKIs for proteolysis by the ubiquitin-proteasome system (UPS) via different E3 ubiquitin ligases. Conversely, mitotic cyclin proteolysis by the UPS, via the APC/C, causes CDK activity to decline sharply, which is required for M-phase exit.

to the proteasome by the APC/C. In addition to its roles in the cell division cycle, E3 ligases control endoreduplication cell cycles, those occurring in trichome and root nodule cells being perhaps the best characterized such examples (Cebolla et al., 1999; Roodbarkelari et al., 2010; Heyman and De Veylder, 2012; Genschik et al., 2013).

CDK-SPECIFIC INHIBITORS

Non-catalytic inhibitors of CDK/cyclin complex activity, generally termed CKIs, have been identified as chief cell cycle regulators in all eukaryotes, and act by obstructing substrate interaction and ATP binding. Two types of CKIs have been identified and characterized in plants: INHIBITOR OF CDC2 KINASE/KIP-RELATED (ICK/KRP) and SIAMESE/SIAMESE-RELATED (Yi et al., 2014). ICK/KRPs typically bind to complexes containing A-type CDKs and D-type cyclins (Wang et al., 2008), although interaction with B-type CDKs was also reported (Nakai et al., 2006). Plant CKIs are targeted by several E3 ligases for degradation (Zhou et al., 2003; Weinl et al., 2005; Jakoby et al., 2006; Ren et al., 2008; Roodbarkelari et al., 2010). In mitotic cell cycles, ICKs/KRPs are phosphorylated by B1-type CDKs and targeted for UPS-mediated degradation, with consequent stimulation of A-type CDK activity and M-phase (Verkest et al., 2005).

The core molecular factors controlling the cell cycle in plants conform, to a large extent, to those identified in other higher eukaryotes. However, plant genomes are often characterized by a larger complement of key cell cycle genes as well as by uniquely possessing certain types of CDKs, E2Fs, and CKIs. The resulting complexity in cell cycle regulatory mechanisms, which appears to have considerable redundancy, may have evolved largely to fine-tune the cell cycle for the requirements of the sessile life style of plants.

ROLES OF DIFFERENT CELL CYCLES AND REGULATORS IN SEED DEVELOPMENT

EMBRYO

Cell proliferation, patterning, and morphogenesis during embryo development

From the first zygotic division through the early globular (8–16 cells) stages, many aspects of embryo development in monocots and dicots are conserved (Lau et al., 2012; Sabelli, 2012b). The zygote divides asymmetrically and generates an apical cell with dense cytoplasm and a large vacuolated basal cell at the chalazal and micropylar ends of the embryo sac, respectively, establishing early embryo polarity and patterning. The apical and basal cells produce, respectively, the proembryo and the suspensor, which has an embryo proper-nourishing function. Afterwards, cell proliferation and differentiation occur coordinately to produce all embryo cell types and tissues, including the cotyledon(s), which in dicots accumulate storage compounds and typically occupy a large fraction of the mature seed volume. In dicots, periclinal divisions of the octant cells result in the globular embryo. Subsequently, the heart stage is reached with the characteristic emergence of two cotyledon primordia as opposite lateral extensions of the apical end. Next, cell proliferation and differentiation of the basal cell tier lead to the torpedo stage. As growth proceeds under mechanical constraints imposed by the ovule, the embryo increasingly assumes its typical curved shape. During development, monocot and dicot embryos display increasingly different morphologies. Patterns of cell division and cell lineages usually become less organized in monocot embryos and while two prominent cotyledons emerge in dicots, the origin of the single monocot cotyledon, the scutellum, is spatially more variable. In dicots, cotyledon cells undergo endoreduplication (Dhillon and Miksche, 1983; Lemontey et al., 2000; Rewers and Sliwinska, 2012).

The importance of cell cycle control during embryogenesis extends beyond its most recognizable aspects related to cell division and endoreduplication cycles. In *Arabidopsis thaliana* and *Nicotiana tabacum*, disruption of proper developmental patterns through lengthening or impairment of cell division by interfering with CDKA;1 (Hemerly et al., 2000), DNA polymerase ε (Jenik et al., 2005), and A3-type cyclin (Yu et al., 2003) indicates the dependence of embryo patterning and morphogenesis on the correct execution of cell divisions. Also, loss-of-function of *HOBBIT (HBT)*, which encodes a homolog of the APC/C subunit CDC27 (Blilou et al., 2002), causes defects in hypophyseal cell specification and basal embryo cell division, perturbing root meristem formation (Willemsen et al., 1998). Recent investigation on *Arabidopsis* post-embryonic development indicates that moderate and high levels of CDKA;1 activity determine whether cells divide symmetrically or asymmetrically, respectively, and that CDKA;1 activity is conveyed via RBR1 and its control over cell cycle- and differentiation-related genes (Weimer et al., 2012). Similar mechanisms connecting CDKs and RBR1 appear to operate during embryogenesis (Nowack et al., 2012), and several studies have shed light on the roles of the RBR/E2F pathway, CDKs, D-type cyclins, CKIs, as well as additional APC/C components, in developing *Arabidopsis* seeds, underscoring the importance of precise cell cycle control for embryonic patterning, cell proliferation, and endoreduplication.

The role of D-type cyclins in embryogenesis

Collins et al. (2012) carried out a comprehensive investigation of D-type cyclin expression and function in developing *Arabidopsis* seeds. Various D-type cyclin genes show distinct and overlapping tissue-specific expression patterns during seed development. Developmental progression characterized in loss-of-function mutants revealed that embryo development is slower in the triple D3-type cyclin mutant, but not in single and double mutant combinations, indicating that this cyclin subtype is necessary for normal development, with individual, partly redundant components. Ectopic CYCD3;1 expression delays progression of embryonic development and causes atypical divisions in the hypophysis and suspensor. In contrast to CYCD3;1, ectopic expression of CYCD7;1, a previously uncharacterized cyclin that is not expressed in wild-type seeds, induces cell proliferation and cell enlargement in the embryo (and endosperm), causing excessive growth and higher seed lethality. These results suggest that adequate control of spatiotemporal patterns of cell division, through the regulation of specific D-type cyclins and thus possibly their CDK complexes, is important for embryo patterning and growth.

Roles of CDKs, the RBR pathway, and CKIs in embryogenesis

A combinatorial analysis of mutants recently allowed functional dissection of five members of the ICK/KRP-type of CKIs during *Arabidopsis* seed development and revealed a link between ICK/KRPs, the RBR/E2F pathway and cell proliferation (Cheng et al., 2013). CDK activity gradually increased as individual ICK/KRP T-DNA insertion mutants were combined, indicating that ICK/KRPs act at least partially as dosage-dependent CDK inhibitors. Although single-gene mutants and most multiple-gene mutants have wild-type morphological phenotypes, the quadruple *ick1ick2ick6ick7*, and the quintuple *ick1ick2ick5ick6ick7* mutants have a slightly altered leaf shape, suggesting some degree of redundancy among individual genes. The quintuple mutant has larger cotyledons, leaves, petals, and seeds than wild type. The ICK/KRP mutants generally have more numerous but smaller cells in all organs examined, and this phenotype is enhanced as the number of combined mutant genes is increased. The quintuple mutant displays extensive up-regulation of the E2F pathway via increased phosphorylation of RBR1, consistent with reduced inhibition of CDK complexes.

An additional connection between CDKs and RBR1 was provided by combinatorial analyses of their corresponding mutants (Nowack et al., 2012). Delayed development and drastically altered cell numbers and sizes are observed during embryogenesis in null *cdka;1* mutants (indicating that embryogenesis can adjust to variations of cell number and size), while loss of function of both CDKA;1 and B1-type CDKs leads to embryogenesis arrest. Because post-embryonic defects in *cdka;1* mutants can be restored by *rbr1* mutations (Nowack et al., 2012), CDKA;1 most likely acts via the RBR/E2F pathway to control embryogenic cell proliferation. *Arabidopsis DEL1* encodes an E2F-DP-like DNA binding protein that was previously shown to be mostly expressed in dividing cells and to inhibit endoreduplication (Vlieghe et al., 2005). A loss-of-function *del1* mutant exhibits a small (~11%) but significant increase in seed size (Van Daele et al., 2012), although it was not determined whether this phenotype is due to stimulated endoreduplication. Collectively, these results suggest complex interactions among plant ICK/KRPs, which can function redundantly, but also in a dosage-dependent manner, to control the activity of CDK complexes, the RBR1/E2F pathway and cell proliferation.

The influence of the APC/C in embryogenesis

Functional characterization of genes encoding APC/C subunits and activators during embryogenesis revealed roles in cell-type specification and morphogenesis, in addition to cell division and endoreduplication. Mutations in both APC4 (Wang et al., 2012) and APC1 subunits (Wang et al., 2013) cause defective gametogenesis and developmental arrest during embryogenesis, which seem to be associated with accumulation of B-type cyclin and altered auxin distribution. *SAMBA* encodes a conserved plant-specific protein that binds to and potentially regulates the APC/C (Eloy et al., 2012). *SAMBA* expression is high during embryogenesis and, to a lesser extent, early post-embryonic development. Loss-of-function *samba* mutants have defective male gametogenesis and enlarged shoot and root apical meristems that result in the production of larger seeds, embryos, leaves, and roots. Increased organ size in *samba* mutants could be attributed to a larger number of more highly endoreduplicated cells, rather than to larger cells. SAMBA binds an A2-type cyclin, which is stabilized in *samba* mutants during early development. Thus, SAMBA negatively regulates cell proliferation at least partially by targeting A2-type cyclins for proteolysis via the APC/C.

In conclusion, precise cell cycle regulation, both spatially and temporally, is critical for embryogenesis and plant reproduction. The core cell cycle regulators CDKs, cyclins, RBR1, ICK/KRPs, and the APC/C seem to play concerted roles and thus affect asymmetrical cell divisions, cell proliferation and endoreduplication during embryogenesis.

ENDOSPERM

Patterns of endosperm development

The endosperm functions in nourishing the sporophyte during embryogenesis and controlling germination and, mostly limited to monocots, also during early post-embryonic development. However, it also plays important roles in other aspects of seed development, including epigenetic regulation, coordination of cell patterning and proliferation, signaling among the major seed structures and control of seed size (Berger et al., 2006; Sabelli and Larkins, 2009b; Nowack et al., 2010; Fiume and Fletcher, 2012; Costa et al., 2014). The different cell cycle types occurring during the development of the persistent endosperm in grasses have mostly been investigated in traditional biological models and valued crop species, such as maize and rice (*Oryza sativa*).

The nuclear type of endosperm development is the most frequently encountered pattern among Angiosperms, and with regard to its early stages up to cellularization, is highly conserved in monocots and dicots (Olsen, 2004; Sabelli and Larkins, 2009b; Becraft and Gutierrez-Marcos, 2012). In this developmental type, the primary endosperm nucleus and its derivatives undergo acytokinetic mitosis iteratively, resulting in a syncytium that can comprise up to thousands of nuclei that are initially distributed around the central vacuole of the central cell (Olsen, 2004). Anticlinal cell wall deposition, forming alveoli encasing individual nuclei, creates the first endosperm cell layer, and reiteration of anticlinal cell wall deposition, alveolation, periclinal cell wall formation and periclinal cell division results in centripetal generation of additional cell layers, gradually replacing the space occupied by the central vacuole with cells. Cellularization is typically completed within three to six days after pollination (DAP) in cereals and by the torpedo stage in *Arabidopsis*. In the following developmental stage, mitotic divisions coupled to cytokinesis result in cell proliferation, thus producing most of the endosperm cells. Past the cell proliferation stage, in nonendospermic species, such as *Arabidopsis*, the endosperm is absorbed by the rapidly developing embryo and is limited to a single or few cell layers at seed maturity. In contrast, in endospermic species, the endosperm is persistent and its development progresses through relatively conserved stages, comprising a period of cell enlargement typically associated with endoreduplication, followed by maturation involving programmed cell death (PCD) of specific cell types, dehydration, and dormancy. The roles played by core cell cycle regulators in the different cell cycles of cereal endosperm development are discussed in the following sections and summarized in **Figure 3**.

Maize endosperm development: an overview

Endosperm development in maize (and related cereals) follows the general monocot developmental pattern described earlier (Sabelli and Larkins, 2009b). Starting around three DAP, as the syncytium is cellularized, endosperm growth is mostly attained by an increase in cell number through mitotic cell divisions, which peak at eight to 10 DAP (Kiesselbach, 1949; Kowles and Phillips, 1985; Lur and Setter, 1993). Initiating in the endosperm central regions, also around eight to 10 DAP and then extending centrifugally, cells gradually and asynchronously cease mitotic cell divisions and switch to endoreduplication. As a result, the nuclei of many central endosperm cells reach high DNA content levels (some in excess of 200C; C = DNA content of haploid nuclei) and contain multiple, apparently uniform, copies of chromosomes (Bauer and Birchler, 2006). In agreement with the high correlation between ploidy level and cell size observed in numerous cell types and organisms, the spatiotemporal pattern of mitosis-to-endoreduplication switch in the maize endosperm creates a gradient of nuclear ploidy and cell size. Small non- or under-endoreduplicated cells are located mostly at the peripheral aleurone and sub-aleurone layers, as opposed to the increasingly large and endoreduplicated inner starchy endosperm cells. By 16 DAP, the expanded endoreduplicated cells account for most of the endosperm volume (Vilhar et al., 2002) and as many as 75% of its cells can become endoreduplicated at later developmental stages (Dilkes et al., 2002). Starchy endosperm cells typically display concomitant accumulation of starch and proteins with endoreduplication, which has long suggested a causal relationship between these processes (Larkins et al., 2001; Kowles, 2009; Sabelli and Larkins, 2009a,b; Sabelli, 2012b). Starchy endosperm cells subsequently undergo PCD (reviewed by Young and Gallie, 2000; Sabelli, 2012a), and the peripheral aleurone layer persists as the only living tissue past seed desiccation.

The contrasting endosperm development in Brachypodium distachyon and other model cereals

In comparison with maize, rice, and other major cereals, initial analyses of seed development in the emerging grass model, *Brachypodium distachyon*, have revealed important differences with respect to cell cycle control and its possible relationship with storage compound accumulation (Guillon et al., 2012; Trafford et al., 2013). In contrast to its related species, barley (*Hordeum vulgare*), *Brachypodium* endosperm displays reduced cell proliferation and enlargement. Also, the expression of a B1-type CDK and

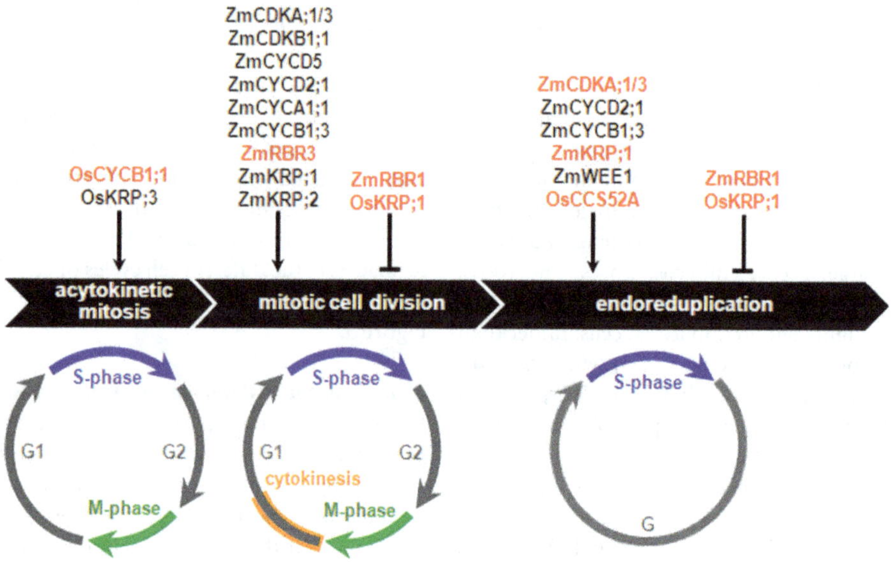

FIGURE 3 | Effects of core cell cycle regulators on the cereal endosperm cell cycles. Core cell cycle regulators that positively or negatively regulate proliferative (acytokinetic mitosis and mitotic cell division) and endoreduplication cycles are shown. Net stimulatory or inhibitory effects are assigned as a function of cell proliferation- and ploidy-related phenotypes observed in mutant or transgenic analyses carried out either in planta or plant cell cultures (red-colored fonts) or inferred from a combination of expression analyses and biochemical assays (black-colored fonts), as discussed in the text. Os, *Oryza sativa*; Zm, *Zea mays*.

an A3-type cyclin is reduced, in agreement with a withdrawal from a mitotic cell cycle program, but >6C nuclei are absent, indicating no occurrence of endoreduplication. Along with these differences, *Brachypodium* endosperm cells exhibits limited starch deposition and thickened cell walls, the cell wall polysaccharide, β-glucan, representing the main seed storage carbohydrate. Accordingly, the expression of genes involved with starch biosynthesis is reduced. Thus, an interpretation of the contrasting endosperm developmental patterns in *Brachypodium* and other cereals is that starch accumulation drives endosperm cell enlargement (Trafford et al., 2013) and, consequently, high nuclear ploidy levels are correlated with large cell sizes.

The importance of early nuclear and cell proliferation activities for endosperm development

Regulation of the syncytium-to-cellularization transition seems to be key for endosperm development and seed growth. Rice *THOUSAND-GRAIN WEIGHT 6* (*TGW6*), which encodes an indole-3-acetic acid (IAA)-glucose hydrolase, appears to stimulate, by elevating IAA levels, the expression of CYCB2;2 and E2F1 during the first three days after fertilization; it also stimulates premature cellularization of the syncytium and reduces endosperm final cell number, grain length, and weight (Ishimaru et al., 2013). In addition, suppressing the expression of rice CYCB1;1 results in delayed endosperm cellularization and seeds containing only an enlarged embryo at maturity (Guo et al., 2010). Heat stress affects rice endosperm cellularization by interfering with the expression of the epigenetic regulator FERTILIZATION-INDEPENDENT ENDOSPERM 1 of the Polycomb Repressive Complex 2 (PRC2), thus reducing seed size (Folsom et al., 2014). In *Arabidopsis*, over-expression of *SHORT HYPOCOTYL UNDER BLUE 1* (*SHB1*) promotes early endosperm nuclear proliferation, a delay in cellularization, enlarged chalazal endosperm, and enhanced proliferation and expansion of embryo cells, leading consequently to increased seed size (Zhou et al., 2009). It is possible that these effects are mediated by members of the HAIKU pathway, which function downstream of *SHB1* to promote syncytial endosperm and seed growth via epigenetic control of cytokinin signaling (Li et al., 2013).

The rate and duration of proliferative cell cycles seem to also play an important role in maize seed size. Recently, Sekhon et al. (2014) examined transcriptional and developmental changes during seed development of maize populations that were selected for large and small seed sizes (termed KLS30 and KSS30, respectively). KLS30 seeds are more than fourfold heavier and twofold larger than KSS30 seeds and contain proportionally larger endosperms. Metabolite and genome-wide expression analyses indicated that, compared to KLS30 seeds, the linear phase of the grain filling (phase II) in KSS30 seeds initiates earlier, but it also occurs at slower rates and terminates earlier. Notably, KLS30 endosperms, relative to their KSS30 counterparts, display up-regulated sucrose metabolism and expression of the cell wall invertase INCW2 (encoded by the *MINIATURE 1* gene), which is required for normal endosperm cell proliferation and expansion (Vilhar et al., 2002; Chourey et al., 2006). KLS30 endosperms also exhibit higher expression at 12–18 DAP of several genes encoding D- and B-type cyclins and APC/C subunits. Although cell number, size, and nuclear ploidy were not determined in KLS30 and KSS30 endosperms, these gene expression profiles suggest higher mitotic activity in the former. The balance between cell proliferation and endoreduplication activities as a factor influencing endosperm and seed sizes is also supported by the analyses of multiple small-seeded popcorn inbred lines in comparison to typically large-seeded dent inbred lines (Dilkes et al., 2002; Coelho et al., 2007). Popcorn lines revealed higher endosperm ploidy levels from as early as 13 DAP in comparison to dent inbred lines. This difference could be attributed to an earlier transition between the cell division to endoreduplication stages, and/or higher rates of endoreduplication in popcorn lines. The importance of correct timing for cessation of cell proliferation and commencement of cell enlargement as an underlying factor for growth has been documented (reviewed by Powell and Lenhard, 2012). Collectively, these studies suggest a causal relationship between increased periods and/or rates of cell proliferation with larger endosperms (and, consequently, larger seeds), by establishing a stronger-sink tissue, one possessing more numerous cells that subsequently enlarge and accumulate larger amounts of storage compounds.

SEED COAT DEVELOPMENT, CELL CYCLE CONTROL, AND EPIGENETIC CONTROL OF SEED DEVELOPMENT

In most Angiosperms, the two ovule integuments that enclose the nucellus differentiate into the seed coat following fertilization, and develop through stages of cell division, cell elongation, differentiation, and PCD in coordination with embryo and endosperm development (Haughn and Chaudhury, 2005). Investigation of *Arabidopsis* seed coat development revealed the existence of complex communication and interaction between these seed structures and that their underlying cell proliferation and enlargement are major determinants of seed size (reviewed by Nowack et al., 2010). Impairing the elongation of integument cells via the *transparent testa glabra 2* (*ttg2*) mutation reduces endosperm and seed growth (Garcia et al., 2005). *megaintegumental/auxin-response factor 2* (*mnt/arf2*) mutants exhibit more numerous integument cells and enlarged seeds and embryos compared to wild type (Schruff et al., 2006). Premature syncytium cellularization reduces endosperm growth and integument cell elongation in *haiku* (*iku*) mutants (Garcia et al., 2005), whereas delayed cellularization and extended endosperm cell proliferation is associated with integument cell elongation in the enlarged seeds of *apetala 2* (*ap2*) mutants (Ohto et al., 2009). The enhanced cell proliferation in *mnt/arf2* and *ap2* mutants seems to be associated with increased expression of D3- and B1-type cyclins (Schruff et al., 2006; Ohto et al., 2009), indicating that the corresponding transcription factors repress cell divisions through a pathway that involves down-regulation of these core cell cycle regulators.

Epigenetic mechanisms are particularly important for integrating growth and development of the seed coat, embryo, and endosperm. Particularly among core cell cycle regulators, RBR1 is involved in gametophyte cell differentiation and endosperm nuclear proliferation along with epigenetic regulators such as PRC2 and DNA METHYLTRANSFERASE 1 (Ebel et al., 2004; Ingouff et al., 2006; Johnston et al., 2008; Jullien et al., 2008). *rbr1* mutants display fertilization-independent endosperm development, reduced cell proliferation in the ovule integuments prior to fertilization and impaired differentiation of the seed coat (Ebel et al., 2004; Ingouff et al., 2006).

In conclusion, there appears to be extensive crosstalk and coordination, in which epigenetic control and RBR1 play significant roles, between cell cycle activity in the developing seed coat and inner seed structures, such as the embryo and endosperm. These mechanisms could modify both signaling and mechanical constrains imposed by maternal tissues on developing seed structures, and consequently could control seed size (Haughn and Chaudhury, 2005).

THE ROLE OF CORE CELL CYCLE REGULATORS IN THE CEREAL ENDOSPERM: THE MAIZE PROTOTYPE AND RELATED EXAMPLES

CONTROL OF ENDOREDUPLICATION IN ENDOSPERM

The mechanisms that control the transition from the mitotic cell cycle into endoreduplication and its progression in various cell types and species have been recently reviewed in detail (Edgar et al., 2014; Sabelli, 2014). Among plant core cell cycle regulators, certain CDK/cyclin complexes, CKIs, APC/C activators and the RBR/E2F pathway have been functionally linked to the onset and/or rates of endoreduplication cycles. Although these cell cycle regulators are widely conserved across higher eukaryotes and appear to be recurrently deployed to produce cell cycle modifications that result in endoreduplication, their individual contributions may be species- and cell-type-specific (Roodbarkelari et al., 2010; Edgar et al., 2014). In maize endosperm, induced S-phase CDK activity and inhibited M-phase CDK activity were proposed to cause endoreduplication cycles (Grafi and Larkins, 1995). In support of this model, endoreduplication is inhibited and stimulated, respectively, by over-expression of a catalytically inactive, dominant-negative form of CDKA;1 (Leiva-Neto et al., 2004) and by decreased RBR1 activity and consequent up-regulation of E2F/DP-dependent gene expression (Sabelli et al., 2013). In addition, developing maize endosperm exhibits the contrasting expression of different CKIs (Coelho et al., 2005) and functionally distinct RBR homologs of the RBR1 and RBR3 types (Grafi et al., 1996; Sabelli et al., 2005, 2009, 2013). Endosperm endoreduplication particularly correlates with the potential inhibitory phosphorylation of CDK subunits by a WEE1 homolog (Sun et al., 1999a), differential cyclin expression (Sun et al., 1999b; Dante et al., 2014) and apparent down-regulation of UPS-mediated proteolysis of members of various cyclin types, including potential mitotic cyclins (Dante et al., 2014).

CKIs

ICK/KRP-type CKIs appear to have variable roles in different cell cycle types during cereal endosperm development. KRP;1 is expressed at nearly constant levels in 7–21 DAP endosperm, while KRP;2 protein levels decline during this period, suggesting a more positive role in endoreduplicating cells for KRP;1 than KRP;2, which in contrast could be preferentially involved with regulation of the mitotic cell cycle or its transition into the endoreduplication cycle (Coelho et al., 2005). Biochemical assays showed KRP;1 activity corresponds partly to a CDK inhibitory activity existing in endoreduplicating endosperm (Grafi and Larkins, 1995). KRP;1 and KRP;2 are able to partially inhibit the complex CDK fraction that binds p13^{suc1}, and they specifically inhibit the CDK activity associated with A1- and D5-type cyclins, but not that associated with CYCB1;3 (Coelho et al., 2005). Overexpression of KRP;1 along with the wheat dwarf virus RepA protein, which antagonizes RBR1 (Grafi et al., 1996; Xie et al., 1996; Gordon-Kamm et al., 2002; Sabelli et al., 2005, 2009), causes ectopic endoreduplication in cultured, proliferating maize cells, indicating that coupling the stimulation of G1/S transition to the inhibition of certain CDK complexes is sufficient for endoreduplication onset in otherwise dividing cells (Coelho et al., 2005).

Expression and functional analyses *in planta* revealed that KRPs impact rice endosperm development. KRP;1 RNA is preferentially expressed at the mitosis-to-endoreduplication transition in wild-type rice plants, and its over-expression results in decreased kernel weight and filling rate, in addition to perturbed production and lower ploidy levels of endosperm cells (Barrôco et al., 2006). In contrast, KRP;3 RNA is most highly expressed in the syncytial endosperm, but its level declines subsequently in the cellularized endosperm, suggesting a specific function in the syncytial cell cycle or during the transition to cellularization (Mizutani et al., 2010).

THE CDK AND RBR PATHWAYS

Recently, Sabelli et al. (2013) showed that RBR1 controls multiple molecular and cellular aspects of maize endosperm development. *RBR1* down-regulation via RNAi in endosperm cells results in enhanced expression of *RBR3*-type, *MCM2–7*, and *PCNA* genes. Mitotic and endoreduplication cell cycles are both stimulated by the alleviated inhibition of RBR1 on the G1/S transition, which causes RBR1-RNAi endosperm to have 58% more cells and ~70% more DNA than its wild type counterpart by 19 DAP. However, this creates a surprising reduction in cell and nuclear sizes, in spite of increased endoreduplication, thus ruling out a causal and direct relationship between these processes, at least in the specific context of RBR1 down-regulation. Larger cell numbers and higher ploidy levels together cause a 43% increase in DNA content in mature endosperm upon RBR1 down-regulation, although no measurably altered storage protein content or kernel weight (a proxy for starch accumulation) were observed. Genetic interaction analysis of RBR1 and CDKA;1 (Leiva-Neto et al., 2004), down-regulated individually or in combination, indicated that CDKA;1 requires RBR1 for controlling endoreduplication, but conversely RBR1 represses downstream target genes independently from CDKA;1. These observations suggest distinct RBR1 activities at controlling endoreduplication, in which CDKA;1 probably participates via its inhibitory phosphorylation of RBR1, and at repressing E2F-dependent gene expression in a CDKA;1-independent manner. RBR1 down-regulated endosperm exhibits levels of p13^{suc1}-adsorbed CDK activity similar to those of wild-type endosperm even in the presence of dominant-negative CDKA;1, indicating that various CDK complexes and RBR1 are negatively and reciprocally regulated, and implying that CDKs other than CDKA;1 participate in cell cycle stimulation upon RBR1 down-regulation (Leiva-Neto et al., 2004; Sabelli et al., 2013; Dante et al., 2014). Thus, perturbing RBR1 function revealed its key roles in integrating various processes in maize endosperm, but this did not translate in altered seed size and weight, suggesting the presence of a higher order, homeostatic regulation of endosperm development (discussed in next section).

The stimulation of CDK activity in down-regulated RBR1 maize endosperm prompts the question of the identity of CDKs, besides CKDA;1, controlling cell division and endoreduplication cycles (Sabelli et al., 2013). The identification and expression of different CDKs expressed in maize endosperm were reported recently (Dante et al., 2014). Previously uncharacterized CDKs of the A-type and B1-types, termed, respectively, CDKA;3 and CDKB1;1, were found to be expressed in endosperm. Protein levels of A-type CDKs are nearly constant throughout endosperm development, whereas expression of CDKB1;1 becomes markedly reduced during the transition into the endoreduplication stage and is stimulated upon RBR1 down-regulation. These observations are in agreement with the role of A-type CDKs in both mitotic and endoreduplication cell cycles and that of B1-type CDKs specifically in the mitotic cell cycle established in other species. Similar expression patterns, cyclin binding properties, and maintenance of nearly wild-type levels of CDK activity upon combined down-regulation of RBR1 and CDKA;1 collectively indicate that CDKA;1 and CDKA;3 are partially redundant or function coordinately (Sabelli et al., 2013; Dante et al., 2014). RBR1 down-regulated endosperm possesses more numerous cells and also exhibits higher ploidy levels, indicating that both cell division and endoreduplication are stimulated in distinct spatiotemporal patterns (Sabelli et al., 2013). These results suggest some redundancy among A-type CDKs and a specialized role for CDKB1;1 in positively regulating cell division during maize endosperm development (Leiva-Neto et al., 2004; Sabelli et al., 2013; Dante et al., 2014). Consistent with this interpretation, *Arabidopsis* possesses a single A-type CDK, and its B1-type CDKs can drive progression through cell division, but not endoreduplication cycles, in the absence of CDKA;1 and RBR1 (Nowack et al., 2012).

CYCLIN/CDK COMPLEXES

The spatiotemporal expression of A-, B-, and D-type cyclins and their associated kinase activities in developing maize endosperm were recently investigated (Dante et al., 2014). Two main transcript expression patterns are apparent, one characterized by rapidly declining RNA levels with the onset of endoreduplication (A- and B-type cyclins), and the other with nearly constant RNA levels throughout endosperm development (D-type cyclins). However, these patterns are not consistent with those at the protein level, as shown by a discrepancy between declining *CYCB1;3* RNA in endoreduplicating endosperm but sustained levels of the encoded protein. While CYCB1;3 and CYCD2;1 proteins are localized to both the cytoplasm and nucleus of cells throughout the endosperm, CYCD5 protein is localized solely in the cytoplasm of peripheral cell layers. CDK activity associated with CYCA1 is tightly associated with cell division, while CYCB1;3-, CYCD2;1-, and CYCD5-associated CDK activities are highest at the transition from cell division to endoreduplication. These patterns together suggest roles for CYCA1 and CYCD5 in the cell division cycle, while CYCB1;3 and CYCD2;1 could participate in both cell division and endoreduplication. In particular, the switch to an endoreduplication program is marked by a drastic reduction in kinase activity associated with CYCA1. A-, B-, and D-type cyclins are more resistant to proteasome-dependent degradation in endoreduplicating compared to mitotic endosperm, which potentially contributes to the sustained levels of proteins, particularly CYCB1;3, in endoreduplicating cells. Consequently, the mitosis-to-endoreduplication transition and the accompanying cell enlargement typical of starchy endosperm cells are possibly associated with cell cycle modifications created by reduced proteasome-dependent proteolysis of several types of cyclins and, potentially, that of additional core cell cycle regulators (Dante et al., 2014).

UPS-MEDIATED PROTEOLYSIS

Functional and expression analyses in rice revealed roles for APC/C activators that seem in part distinct from those of their homologs in dicots. Reduced expression of CELL CYCLE SWITCH 52A (OsCCS52A), a homolog of the APC/C activating subunit, results in smaller seeds and, despite reduced nuclear and cell size of endosperm cells, only modest reduction in their ploidy levels (Su'udi et al., 2012b). Also, reduced expression of the related OsCCS52B protein negatively impacts seed and cell sizes, but has no impact on endoreduplication (Su'udi et al., 2012a). Thus, collectively, OsCCS52A and B seem to play rather minor roles in rice endosperm endoreduplication, but have important roles in controlling cell and seed sizes. Although plant CCS52 homologs are known to promote proteolysis of A- and B-type cyclins and endoreduplication (reviewed by Heyman and De Veylder, 2012), the targets and mechanisms by which OsCCS52A and B control cell and seed sizes remain unknown. Some unidentified cyclins are presumably targeted by OsCCS52A and B, but the apparent down-regulation of UPS-mediated cyclin degradation and its contribution to sustained CYCB1;3 expression in endoreduplicating maize endosperm cells (Dante et al., 2014) further suggests that CCS52 homologs have a more significant role in cyclin proteolysis in mitotic as opposed to endoreduplicating endosperm. Also, these observations are consistent with others made in dicot model species, underscoring that various plant cell cycle regulators are targeted to UPS-mediated degradation by E3 ubiquitin ligases in cell-type- and cell cycle-type-dependent manners (Roodbarkelari et al., 2010; Heyman and De Veylder, 2012). Thus, dissecting the specific roles of E3 ubiquitin ligases and the UPS at large in governing various aspects of endosperm development, including cell cycle control, merits further investigation.

Besides the apparently reduced UPS activity in endoreduplicating compared to mitotic endosperm, translational regulation may also be responsible, at least in part, for sustained CYCB1;3 protein levels despite drastically reduced amounts of its RNA, as cyclin expression is known to be regulated at this level. In addition, many levels of gene expression regulation operate extensively during seed development, as a comparative analysis of the developing maize seed transcriptome and proteome revealed large discrepancies between cognate RNA and protein levels (Walley et al., 2013). Possible underlying mechanisms include differential stability of RNA and protein pairs, transport of proteins between tissues and out-of-phase circadian accumulation of corresponding RNAs and proteins (Walley et al., 2013). Thus, the expression of CYCB1;3 and other core cell cycle regulators in different seed structures may be subject to complex regulation.

WHAT IS THE ROLE OF ENDOREDUPLICATION? EVIDENCE FROM THE MAIZE ENDOSPERM

Proliferative cell cycles are ultimately responsible for establishing the number of cells in a tissue, organ or body and, together with cell enlargement, determine their overall size. Although increased tissue/organ/body size resulting from stimulated cell proliferation has been documented in plants, typically the proliferation and enlargement of cells are inversely correlated, as more numerous cells are compensated for at the tissue/organ level by reduced cell size, essentially resulting in no overall differences (reviewed by John and Qi, 2008; Powell and Lenhard, 2012; Sabelli, 2014). A long-standing debate persists that opposes the "cell-based" or "cellular theory" (whereby cell proliferation and enlargement drive tissue/organ/body growth in a cell-autonomous manner) and the "organismal theory" (cell proliferation and enlargement follow a higher-order, supra-cellular program; Beemster et al., 2006; John and Qi, 2008; Sabelli, 2014).

While the impact of cell proliferation on tissue/organ/body size can be easily appreciated, that of endoreduplication is more controversial. Endoreduplication displays remarkable coincidence with cell enlargement in numerous cell types associated with different specialized functions. In an emerging and unifying view, endoreduplication facilitates cell expansion, growth, and accompanies differentiation in multiple cell types in which the occurrence of cell division could impair their function (Edgar et al., 2014). Among plant cells, the cellular functional specializations and attributes commonly associated with cell enlargement and endoreduplication include ability for rapid cell elongation (e.g., hypocotyl cells), branched cell morphology (e.g., trichomes), nutrient storage (e.g., cotyledons and endosperm of seeds and pericarp of fleshy fruits), hosting endosymbiotic bacteria (e.g., nodule giant cells) and interaction with pathogens and parasites (e.g., giant cells in galls and feeding sites). Some functions of endoreduplication in plant cells are only beginning to be elucidated (reviewed by John and Qi, 2008; Chevalier et al., 2011, 2013; De Veylder et al., 2011; Edgar et al., 2014; Sabelli, 2014). Recently, the contribution of endoreduplication to cell morphogenesis and cell identity maintenance (Bramsiepe et al., 2010), as well as regulation of gene expression and karyoplasmic homeostasis (Bourdon et al., 2012) were investigated. Bramsiepe et al. (2010) reported that, upon targeted reduction of endoreduplication levels by modification of core cell cycle gene expression, *Arabidopsis* trichome number is reduced due to dedifferentiation and resumption of mitosis by these cells. Conversely, promoting endoreduplication causes restoration of trichome cell identity, which revealed a role for endoreduplication in determining trichome cell fate. In endoreduplicated cells of the tomato pericarp, the nuclear surface has extensive grooves that are filled with mitochondria, thus allowing fairly constant nuclear surface/volume ratios, suggesting the existence of high ATP demand by nuclear processes

(Bourdon et al., 2012). Accordingly, rRNA and mRNA transcription in individual nuclei is positively correlated with ploidy levels.

In the terminally differentiated starchy cells of maize endosperm, endoreduplication can be viewed as a mechanism that supports increased gene expression and the enhanced metabolic activity associated with cell enlargement and massive accumulation of starch and storage protein (Larkins et al., 2001; Sabelli and Larkins, 2009a; Sabelli, 2014). However, modification of endosperm endoreduplication, by perturbing the function of core cell cycle regulators, has challenged some aspects of this paradigm (Leiva-Neto et al., 2004; Sabelli et al., 2013). Similar to maize kernels in which the expression of a dominant-negative CDKA;1 reduces endoreduplication (Leiva-Neto et al., 2004), the size, weight, and morphology of RBR1 down-regulated kernels are essentially identical to their wild-type counterparts (Sabelli et al., 2013). Thus, evidence from modification of maize endosperm endoreduplication suggests this cell cycle does not contribute to metabolic and growth processes, at least at a whole-tissue level.

Surprisingly, RBR1 down-regulation in endosperm also results in coordinated reduction in cell and nuclear sizes (Sabelli et al., 2013). This indicates a role for RBR1 in coupling DNA content to nuclear and cell sizes. Importantly, these results invoke the existence of a causal relationship between nuclear and cell sizes in endosperm cells, which is not affected by perturbing RBR1 function, in support of the general karyoplasmic ratio theory. In addition, the observation of a larger number of smaller cells in RBR1 down-regulated endosperm seems to agree with the organismal theory of development. Consequently, one interpretation of the effects of RBR1 down-regulation in maize endosperm is that suppression of RBR1 function leads to enhanced cell proliferation, which results in a larger number of cells that, nonetheless, undergo less pronounced enlargement imposed by supra-cellular control of tissue size. Nuclear sizes are adjusted to the sizes of respective cells, regardless of their higher ploidy levels, which are achieved via enhanced endoreduplication. In RBR1 down-regulated endosperm, both the uncoupling of ploidy levels from cell and nuclear size and the reduced storage protein gene expression per unit of nuclear DNA possibly arise from increased DNA methylation and chromatin condensation (Sabelli et al., 2013), in a fashion similar to *Arabidopsis* cotyledon cells (van Zanten et al., 2011). Importantly, additional chromatin produced by enhanced endoreduplication in RBR1 down-regulated endosperm seems to be less transcriptionally active (Sabelli et al., 2013), providing latent transcriptional capacity in case cell enlargement is resumed or continued. Consequently, at the individual cell level, endoreduplication in the endosperm could function in the adjustment of nuclear size to cell size (thus preserving the karyoplasmic ratio) through a process influencing chromatin condensation states and transcription. This view seems to be supported by evidence from other model systems (Wu et al., 2010; Bourdon et al., 2012).

A deeper understanding of the role endoreduplication plays in endosperm development may require genetic analyses of this cell cycle by perturbing individual genes other than those encoding core cell cycle regulators, whose possible functions at coupling endoreduplication to cellular outputs may produce confounding results. The existence of compensatory mechanisms mediating tissue homeostasis may also require approaches based on a suite of technologies such as transmission electron and fluorescence microscopy, immunohistochemistry, DNA and RNA *in situ* hybridization, *in vitro* tissue culture and flow cytometry/fluorescence-activated nuclear sorting to unmask cell-autonomous effects (Gruis et al., 2006; Bourdon et al., 2012).

CONCLUDING REMARKS

Acytokinetic mitosis, symmetric and asymmetric cell division, and cell enlargement greatly impact seed growth and development. During embryogenesis, correct execution of cell division is required for patterning and morphogenesis. Both the rates of proliferative and endoreduplication cell cycles (the latter of which being typically integral to cell enlargement and differentiation in storage compartments) and the timing of the developmental transitions between these cell cycles influence the final size of seed structures and ultimately that of the whole seed. In *Arabidopsis*, a pivotal role of its only RBR member, RBR1, in asymmetric cell divisions, gametogenesis and embryogenesis has been revealed, while in the developing cereal endosperm the RBR1 homolog is central to the control of cell proliferation and endoreduplication cycles and nuclear and cell sizes, underscoring the importance of this family of core cell cycle regulators for plant reproduction and development. Differential expansion of families of key cell cycle genes in various plant species seems to allow the establishment of both functional redundancy and specialization, creating complex cell cycle regulatory networks. Genome-wide analyses and functional gene characterization studies have recently begun to reveal potentially important differences in cell cycle control between dicots and monocots. These differences are evident from genetic analyses of members of the RBR family, whose increased complexity in grass species can allow functional diversification, as exemplified by RBR1 and RBR3. The APC/C and cyclin proteolysis appear to play less prominent roles in the control of cereal endosperm endoreduplication than in dicot root nodules, trichomes, and pericarp. More investigation is needed to unravel the functions of different cell cycle types and their underlying regulatory pathways in endosperm growth, development, and function.

ACKNOWLEDGMENTS

Research on cell cycle regulation in the Larkins laboratory was supported by grants from the Department of Energy (DE-FGO3-95ER20183 and DE–96ER20242) and Pioneer Hi-Bred International Inc. Ricardo A. Dante was supported by a graduate scholarship from the Conselho Nacional de Desenvolvimento Científico e Tecnológico of Brazil.

REFERENCES

Barrôco, R. M., Peres, A., Droual, A. M., De Veylder, L., Nguyen Le, S. L., De Wolf, J., et al. (2006). The cyclin-dependent kinase inhibitor Orysa;KRP1 plays an important role in seed development of rice. *Plant Physiol.* 142, 1053–1064. doi: 10.1104/pp.106.087056

Bauer, M. J., and Birchler, J. A. (2006). Organization of endoreduplicated chromosomes in the endosperm of *Zea mays* L. *Chromosoma* 115, 383–394. doi: 10.1007/s00412-006-0068-2

Becraft, P. W., and Gutierrez-Marcos, J. (2012). Endosperm development: dynamic processes and cellular innovations underlying sibling altruism. *Wiley Interdiscip. Rev. Dev. Biol.* 1, 579–593. doi: 10.1002/wdev.31

Beemster, G. T., Vercruysse, S., De Veylder, L., Kuiper, M., and Inze, D. (2006). The *Arabidopsis* leaf as a model system for investigating the role of cell cycle regulation in organ growth. *J. Plant Res.* 119, 43–50. doi: 10.1007/s10265-005-0234-2

Berger, F., Grini, P. E., and Schnittger, A. (2006). Endosperm: an integrator of seed growth and development. *Curr. Opin. Plant Biol.* 9, 664–670. doi: 10.1016/j.pbi.2006.09.015

Blilou, I., Frugier, F., Folmer, S., Serralbo, O., Willemsen, V., Wolkenfelt, H., et al. (2002). The *Arabidopsis* HOBBIT gene encodes a CDC27 homolog that links the plant cell cycle to progression of cell differentiation. *Genes Dev.* 16, 2566–2575. doi: 10.1101/gad.237302

Bourdon, M., Pirrello, J., Cheniclet, C., Coriton, O., Bourge, M., Brown, S., et al. (2012). Evidence for karyoplasmic homeostasis during endoreduplication and a ploidy-dependent increase in gene transcription during tomato fruit growth. *Development* 139, 3817–3826. doi: 10.1242/dev.084053

Bramsiepe, J., Wester, K., Weinl, C., Roodbarkelari, F., Kasili, R., Larkin, J. C., et al. (2010). Endoreplication controls cell fate maintenance. *PLoS Genet.* 6:e1000996. doi: 10.1371/journal.pgen.1000996

Cebolla, A., Vinardell, J. M., Kiss, E., Olah, B., Roudier, F., Kondorosi, A., et al. (1999). The mitotic inhibitor ccs52 is required for endoreduplication and ploidy-dependent cell enlargement in plants. *EMBO J.* 18, 4476–4484. doi: 10.1093/emboj/18.16.4476

Cheng, Y., Cao, L., Wang, S., Li, Y., Shi, X., Liu, H., et al. (2013). Downregulation of multiple CDK inhibitor ICK/KRP genes upregulates the E2F pathway and increases cell proliferation, and organ and seed sizes in *Arabidopsis. Plant J.* 75, 642–655. doi: 10.1111/tpj.12228

Chevalier, C., Bourdon, M., Pirrello, J., Cheniclet, C., Gévaudant, F., and Frangne, N. (2013). Endoreduplication and fruit growth in tomato: evidence in favour of the karyoplasmic ratio theory. *J. Exp. Bot.* 65, 2731–2746. doi: 10.1093/jxb/ert366

Chevalier, C., Nafati, M., Mathieu-Rivet, E., Bourdon, M., Frangne, N., Cheniclet, C., et al. (2011). Elucidating the functional role of endoreduplication in tomato fruit development. *Ann. Bot.* 107, 1159–1169. doi: 10.1093/aob/mcq257

Chourey, P. S., Jain, M., Li, Q. B., and Carlson, S. J. (2006). Genetic control of cell wall invertases in developing endosperm of maize. *Planta* 223, 159–167. doi: 10.1007/s00425-005-0039-5

Coelho, C. M., Dante, R. A., Sabelli, P. A., Sun, Y., Dilkes, B. P., Gordon-Kamm, W. J., et al. (2005). Cyclin-dependent kinase inhibitors in maize endosperm and their potential role in endoreduplication. *Plant Physiol.* 138, 2323–2336. doi: 10.1104/pp.105.063917

Coelho, C. M., Wu, S., Li, Y., Hunter, B., Dante, R. A., Cui, Y., et al. (2007). Identification of quantitative trait loci that affect endoreduplication in maize endosperm. *Theor. Appl. Genet.* 115, 1147–1162. doi: 10.1007/s00122-007-0640-z

Collins, C., Dewitte, W., and Murray, J. A. (2012). D-type cyclins control cell division and developmental rate during *Arabidopsis* seed development. *J. Exp. Bot.* 63, 3571–3586. doi: 10.1093/jxb/ers015

Costa, L. M., Marshall, E., Tesfaye, M., Silverstein, K. A. T., Mori, M., Umetsu, Y., et al. (2014). Central cell–derived peptides regulate early embryo patterning in flowering plants. *Science* 344, 168–172. doi: 10.1126/science.1243005

Costas, C., Sanchez, M. D. L., Sequeira-Mendes, J., and Gutierrez, C. (2011). Progress in understanding DNA replication control. *Plant Sci.* 181, 203–209. doi: 10.1016/j.plantsci.2011.04.020

Dante, R. A., Sabelli, P. A., Nguyen, H. N., Leiva-Neto, J. T., Tao, Y., Lowe, K. S., et al. (2014). Cyclin-dependent kinase complexes in developing maize endosperm: evidence for differential expression and functional specialization. *Planta* 239, 493–509. doi: 10.1007/s00425-013-1990-1

De Veylder, L., Larkin, J. C., and Schnittger, A. (2011). Molecular control and function of endoreplication in development and physiology. *Trends Plant Sci.* 16, 624–634. doi: 10.1016/j.tplants.2011.07.001

Dhillon, S. S., and Miksche, J. P. (1983). DNA, RNA, protein and heterochromatin changes during embryo development and germination of soybean (*Glycine max* L.). *Histochem. J.* 15, 21–37. doi: 10.1007/BF01006069

Dilkes, B. P., Dante, R. A., Coelho, C., and Larkins, B. A. (2002). Genetic analyses of endoreduplication in *Zea mays* endosperm: evidence of sporophytic and zygotic maternal control. *Genetics* 160, 1163–1177.

Dudits, D., Cserháti, M., Miskolczi, P., and Horváth, G. V. (2007). "The growing family of plant Cyclin-Dependent Kinases with multiple functions in cellular and developmental regulation," in *Annual Plant Reviews*, Vol. 32, *Cell Cycle Control and Plant Development*, ed. D. Inzé (Oxford: Blackwell Publishing Ltd), 1–30. doi: 10.1002/9780470988923.ch1

Ebel, C., Mariconti, L., and Gruissem, W. (2004). Plant retinoblastoma homologues control nuclear proliferation in the female gametophyte. *Nature* 429, 776–780. doi: 10.1038/nature02637

Edgar, B. A., Zielke, N., and Gutierrez, C. (2014). Endocycles: a recurrent evolutionary innovation for post-mitotic cell growth. *Nat. Rev. Mol. Cell Biol.* 15, 197–210. doi: 10.1038/nrm3756

Egli, D. B. (2006). The role of seed in the determination of yield of grain crops. *Aust. J. Agric. Res.* 57, 1237–1247. doi: 10.1071/AR06133

Eloy, N. B., Gonzalez, N., Van Leene, J., Maleux, K., Vanhaeren, H., De Milde, L., et al. (2012). SAMBA, a plant-specific anaphase-promoting complex/cyclosome regulator is involved in early development and A-type cyclin stabilization. *Proc. Natl. Acad. Sci. U.S.A.* 109, 13853–13858. doi: 10.1073/pnas.1211418109

Fiume, E., and Fletcher, J. C. (2012). Regulation of *Arabidopsis* embryo and endosperm development by the polypeptide signaling molecule CLE8. *Plant Cell* 24, 1000–1012. doi: 10.1105/tpc.111.094839

Folsom, J. J., Begcy, K., Hao, X., Wang, D., and Walia, H. (2014). Rice Fertilization-Independent Endosperm1 regulates seed size under heat stress by controlling early endosperm development. *Plant Physiol.* 165, 238–248. doi: 10.1104/pp.113.232413

Garcia, D., Fitz Gerald, J. N., and Berger, F. (2005). Maternal control of integument cell elongation and zygotic control of endosperm growth are coordinated to determine seed size in *Arabidopsis. Plant Cell* 17, 52–60. doi: 10.1105/tpc.104.027136

Genschik, P., Marrocco, K., Bach, L., Noir, S., and Criqui, M. C. (2013). Selective protein degradation: a rheostat to modulate cell-cycle phase transitions. *J. Exp. Bot.* 65, 2603–2615. doi: 10.1093/jxb/ert426

Gordon-Kamm, W., Dilkes, B. P., Lowe, K., Hoerster, G., Sun, X., Ross, M., et al. (2002). Stimulation of the cell cycle and maize transformation by disruption of the plant retinoblastoma pathway. *Proc. Natl. Acad. Sci. U.S.A.* 99, 11975–11980. doi: 10.1073/pnas.142409899

Grafi, G., Burnett, R. J., Helentjaris, T., Larkins, B. A., Decaprio, J. A., Sellers, W. R., et al. (1996). A maize cDNA encoding a member of the retinoblastoma protein family: involvement in endoreduplication. *Proc. Natl. Acad. Sci. U.S.A.* 93, 8962–8967. doi: 10.1073/pnas.93.17.8962

Grafi, G., and Larkins, B. A. (1995). Endoreduplication in maize endosperm: involvement of M phase-promoting factor inhibition and induction of S phase-related kinases. *Science* 269, 1262–1264. doi: 10.1126/science.269.5228.1262

Gruis, D. F., Guo, H., Selinger, D., Tian, Q., and Olsen, O. A. (2006). Surface position, not signaling from surrounding maternal tissues, specifies aleurone epidermal cell fate in maize. *Plant Physiol.* 141, 898–909. doi: 10.1104/pp.106.080945

Guillon, F., Larre, C., Petipas, F., Berger, A., Moussawi, J., Rogniaux, H., et al. (2012). A comprehensive overview of grain development in *Brachypodium* distachyon variety Bd21. *J. Exp. Bot.* 63, 739–755. doi: 10.1093/jxb/err298

Guo, J., Wang, F., Song, J., Sun, W., and Zhang, X. S. (2010). The expression of Orysa;CycB1;1 is essential for endosperm formation and causes embryo enlargement in rice. *Planta* 231, 293–303. doi: 10.1007/s00425-009-1051-y

Haughn, G., and Chaudhury, A. (2005). Genetic analysis of seed coat development in *Arabidopsis. Trends Plant Sci.* 10, 472–477. doi: 10.1016/j.tplants.2005.08.005

Hemerly, A. S., Ferreira, P. C., Van Montagu, M., Engler, G., and Inze, D. (2000). Cell division events are essential for embryo patterning and morphogenesis: studies on dominant-negative cdc2aAt mutants of *Arabidopsis. Plant J.* 23, 123–130. doi: 10.1046/j.1365-313x.2000.00800.x

Heyman, J., and De Veylder, L. (2012). The anaphase-promoting complex/cyclosome in control of plant development. *Mol. Plant* 5, 1182–1194. doi: 10.1093/mp/sss094

Hu, X., Cheng, X., Jiang, H., Zhu, S., Cheng, B., and Xiang, Y. (2010). Genome-wide analysis of cyclins in maize (*Zea mays*). *Genet. Mol. Res.* 9, 1490–1503. doi: 10.4238/vol9-3gmr861

Ingouff, M., Jullien, P. E., and Berger, F. (2006). The female gametophyte and the endosperm control cell proliferation and differentiation of the seed coat in *Arabidopsis. Plant Cell* 18, 3491–3501. doi: 10.1105/tpc.106.047266

Inzé, D., and De Veylder, L. (2006). Cell cycle regulation in plant development. *Annu. Rev. Genet.* 40, 77–105. doi: 10.1146/annurev.genet.40.110405.090431

Ishimaru, K., Hirotsu, N., Madoka, Y., Murakami, N., Hara, N., Onodera, H., et al. (2013). Loss of function of the IAA-glucose hydrolase gene TGW6 enhances rice grain weight and increases yield. *Nat. Genet.* 45, 707–711. doi: 10.1038/ng.2612

Jakoby, M. J., Weinl, C., Pusch, S., Kuijt, S. J., Merkle, T., Dissmeyer, N., et al. (2006). Analysis of the subcellular localization, function, and proteolytic control of the *Arabidopsis* cyclin-dependent kinase inhibitor ICK1/KRP1. *Plant Physiol.* 141, 1293–1305. doi: 10.1104/pp.106.081406

Jenik, P. D., Jurkuta, R. E., and Barton, M. K. (2005). Interactions between the cell cycle and embryonic patterning in *Arabidopsis* uncovered by a mutation in DNA polymerase epsilon. *Plant Cell* 17, 3362–3377. doi: 10.1105/tpc.105.036889

John, P. C., and Qi, R. (2008). Cell division and endoreduplication: doubtful engines of vegetative growth. *Trends Plant Sci.* 13, 121–127. doi: 10.1016/j.tplants.2008.01.004

Johnston, A. J., Matveeva, E., Kirioukhova, O., Grossniklaus, U., and Gruissem, W. (2008). A dynamic reciprocal RBR-PRC2 regulatory circuit controls *Arabidopsis* gametophyte development. *Curr. Biol.* 18, 1680–1686. doi: 10.1016/j.cub.2008.09.026

Jullien, P. E., Mosquna, A., Ingouff, M., Sakata, T., Ohad, N., and Berger, F. (2008). Retinoblastoma and its binding partner MSI1 control imprinting in *Arabidopsis*. *PLoS Biol.* 6:e194. doi: 10.1371/journal.pbio.0060194

Kiesselbach, T. (1949). *The Structure and Reproduction of Corn.* Lincoln: University of Nebraska.

Kowles, R. V. (2009). The importance of DNA endoreduplication in the developing endosperm of maize. *Maydica* 54, 387–399.

Kowles, R. V., and Phillips, R. L. (1985). DNA amplification patterns in maize endosperm nuclei during kernel development. *Proc. Natl. Acad. Sci. U.S.A.* 82, 7010–7014. doi: 10.1073/pnas.82.20.7010

La, H., Li, J., Ji, Z., Cheng, Y., Li, X., Jiang, S., et al. (2006). Genome-wide analysis of cyclin family in rice (*Oryza Sativa* L.). *Mol. Genet. Genomics* 275, 374–386. doi: 10.1007/s00438-005-0093-5

Larkins, B. A., Dilkes, B. P., Dante, R. A., Coelho, C. M., Woo, Y. M., and Liu, Y. (2001). Investigating the hows and whys of DNA endoreduplication. *J. Exp. Bot.* 52, 183–192. doi: 10.1093/jexbot/52.355.183

Lau, S., Slane, D., Herud, O., Kong, J., and Jurgens, G. (2012). Early embryogenesis in flowering plants: setting up the basic body pattern. *Annu. Rev. Plant Biol.* 63, 483–506. doi: 10.1146/annurev-arplant-042811-105507

Leiva-Neto, J. T., Grafi, G., Sabelli, P. A., Dante, R. A., Woo, Y. M., Maddock, S., et al. (2004). A dominant negative mutant of cyclin-dependent kinase A reduces endoreduplication but not cell size or gene expression in maize endosperm. *Plant Cell* 16, 1854–1869. doi: 10.1105/tpc.022178

Lemontey, C., Mousset-Declas, C., Munier-Jolain, N., and Boutin, J. P. (2000). Maternal genotype influences pea seed size by controlling both mitotic activity during early embryogenesis and final endoreduplication level/cotyledon cell size in mature seed. *J. Exp. Bot.* 51, 167–175. doi: 10.1093/jexbot/51.343.167

Li, J., Nie, X., Tan, J. L., and Berger, F. (2013). Integration of epigenetic and genetic controls of seed size by cytokinin in *Arabidopsis. Proc. Natl. Acad. Sci. U.S.A.* 110, 15479–15484. doi: 10.1073/pnas.1305175110

Lur, H. S., and Setter, T. L. (1993). Role of auxin in maize endosperm development (timing of nuclear DNA endoreduplication, zein expression, and cytokinin). *Plant Physiol.* 103, 273–280.

Ma, Z., Wu, Y., Jin, J., Yan, J., Kuang, S., Zhou, M., et al. (2013). Phylogenetic analysis reveals the evolution and diversification of cyclins in eukaryotes. *Mol. Phylogenet. Evol.* 66, 1002–1010. doi: 10.1016/j.ympev.2012.12.007

Mizukami, Y. (2001). A matter of size: developmental control of organ size in plants. *Curr. Opin. Plant Biol.* 4, 533–539. doi: 10.1016/S1369-5266(00)00212-0

Mizutani, M., Naganuma, T., Tsutsumi, K., and Saitoh, Y. (2010). The syncytium-specific expression of the Orysa;KRP3 CDK inhibitor: implication of its involvement in the cell cycle control in the rice (*Oryza sativa* L.) syncytial endosperm. *J. Exp. Bot.* 61, 791–798. doi: 10.1093/jxb/erp343

Nakai, T., Kato, K., Shinmyo, A., and Sekine, M. (2006). *Arabidopsis* KRPs have distinct inhibitory activity toward cyclin D2-associated kinases, including plant-specific B-type cyclin-dependent kinase. *FEBS Lett.* 580, 336–340. doi: 10.1016/j.febslet.2005.12.018

Nowack, M. K., Harashima, H., Dissmeyer, N., Zhao, X., Bouyer, D., Weimer, A. K., et al. (2012). Genetic framework of cyclin-dependent kinase function in *Arabidopsis. Dev. Cell* 22, 1030–1040. doi: 10.1016/j.devcel.2012.02.015

Nowack, M. K., Ungru, A., Bjerkan, K. N., Grini, P. E., and Schnittger, A. (2010). Reproductive cross-talk: seed development in flowering plants. *Biochem. Soc. Trans.* 38, 604–612. doi: 10.1042/bst0380604

Ohto, M. A., Floyd, S. K., Fischer, R. L., Goldberg, R. B., and Harada, J. J. (2009). Effects of APETALA2 on embryo, endosperm, and seed coat development determine seed size in *Arabidopsis. Sex. Plant Reprod.* 22, 277–289. doi: 10.1007/s00497-009-0116-1

Olsen, O. A. (2004). Nuclear endosperm development in cereals and *Arabidopsis thaliana. Plant Cell* 16(Suppl.), S214–S227. doi: 10.1105/tpc.017111

Powell, A. E., and Lenhard, M. (2012). Control of organ size in plants. *Curr. Biol.* 22, R360–R367. doi: 10.1016/j.cub.2012.02.010

Ren, H., Santner, A., Del Pozo, J. C., Murray, J. A., and Estelle, M. (2008). Degradation of the cyclin-dependent kinase inhibitor KRP1 is regulated by two different ubiquitin E3 ligases. *Plant J.* 53, 705–716. doi: 10.1111/j.1365-313X.2007.03370.x

Rewers, M., and Sliwinska, E. (2012). Endoreduplication intensity as a marker of seed developmental stage in the *Fabaceae. Cytometry A* 81, 1067–1075. doi: 10.1002/cyto.a.22202

Roodbarkelari, F., Bramsiepe, J., Weinl, C., Marquardt, S., Novak, B., Jakoby, M. J., et al. (2010). Cullin 4-ring finger-ligase plays a key role in the control of endoreplication cycles in *Arabidopsis* trichomes. *Proc. Natl. Acad. Sci. U.S.A.* 107, 15275–15280. doi: 10.1073/pnas.1006941107

Sabelli, P. A. (2012a). Replicate and die for your own good: endoreduplication and cell death in the cereal endosperm. *J. Cereal Sci.* 56, 9–20. doi: 10.1016/j.jcs.2011.09.006

Sabelli, P. A. (2012b). "Seed development: a comparative overview on biology of morphology, physiology, and biochemistry between monocot and dicot plants," in *Seed Development: OMICS Technologies toward Improvement of Seed Quality and Crop Yield*, eds G. K. Agrawal and R. Rakwal (Amsterdam: Springer), 3–25. doi: 10.1007/978-94-007-4749-4_1

Sabelli, P. A. (2014). "Cell cycle regulation and plant development," in *Handbook of Plant and Crop Physiology*, 3rd Edn, ed. M. Pessarakli (Boca Raton, FL: CRC Press), 3–32.

Sabelli, P. A., Dante, R. A., Leiva-Neto, J. T., Jung, R., Gordon-Kamm, W. J., and Larkins, B. A. (2005). RBR3, a member of the retinoblastoma-related family from maize, is regulated by the RBR1/E2F pathway. *Proc. Natl. Acad. Sci. U.S.A.* 102, 13005–13012. doi: 10.1073/pnas.0506160102

Sabelli, P. A., Hoerster, G., Lizarraga, L. E., Brown, S. W., Gordon-Kamm, W. J., and Larkins, B. A. (2009). Positive regulation of minichromosome maintenance gene expression, DNA replication, and cell transformation by a plant retinoblastoma gene. *Proc. Natl. Acad. Sci. U.S.A.* 106, 4042–4047. doi: 10.1073/pnas.0813329106

Sabelli, P. A., and Larkins, B. A. (2006). Grasses like *Mammals*? Redundancy and compensatory regulation within the retinoblastoma protein family. *Cell Cycle* 5, 352–355. doi: 10.4161/cc.5.4.2428

Sabelli, P. A., and Larkins, B. A. (2009a). The contribution of cell cycle regulation to endosperm development. *Sex. Plant Reprod.* 22, 207–219. doi: 10.1007/s00497-009-0105-4

Sabelli, P. A., and Larkins, B. A. (2009b). The development of endosperm in grasses. *Plant Physiol.* 149, 14–26. doi: 10.1104/pp.108.129437

Sabelli, P. A., and Larkins, B. A. (2009c). Regulation and function of retinoblastoma-related plant genes. *Plant Sci.* 177, 540–548. doi: 10.1016/j.plantsci.2009.09.012

Sabelli, P. A., Liu, Y., Dante, R. A., Lizarraga, L. E., Nguyen, H. N., Brown, S. W., et al. (2013). Control of cell proliferation, endoreduplication, cell size, and cell death by the retinoblastoma-related pathway in maize endosperm. *Proc. Natl. Acad. Sci. U.S.A.* 110, E1827–E1836. doi: 10.1073/pnas.1304903110

Schruff, M. C., Spielman, M., Tiwari, S., Adams, S., Fenby, N., and Scott, R. J. (2006). The AUXIN RESPONSE FACTOR 2 gene of *Arabidopsis* links auxin signalling, cell division, and the size of seeds and other organs. *Development* 133, 251–261. doi: 10.1242/dev.02194

Sekhon, R. S., Hirsch, C. N., Childs, K. L., Breitzman, M. W., Kell, P., Duvick, S., et al. (2014). Phenotypic and transcriptional analysis of divergently selected maize populations reveals the role of developmental timing in seed size determination. *Plant Physiol.* 165, 658–669. doi: 10.1104/pp.114.235424

Su'udi, M., Cha, J. Y., Ahn, I.-P., Kwak, Y.-S., Woo, Y.-M., and Son, D. (2012a). Functional characterization of a B-type cell cycle switch 52 in rice (OsCCS52B). *Plant Cell Tissue Organ Cult.* 111, 101–111. doi: 10.1007/s11240-012-0176-z

Su'udi, M., Cha, J. Y., Jung, M. H., Ermawati, N., Han, C. D., Kim, M. G., et al. (2012b). Potential role of the rice OsCCS52A gene in endoreduplication. *Planta* 235, 387–397. doi: 10.1007/s00425-011-1515-8

Sugimoto-Shirasu, K., and Roberts, K. (2003). "Big it up": endoreduplication and cell-size control in plants. *Curr. Opin. Plant Biol.* 6, 544–553. doi: 10.1016/j.pbi.2003.09.009

Sun, Y., Dilkes, B. P., Zhang, C., Dante, R. A., Carneiro, N. P., Lowe, K. S., et al. (1999a). Characterization of maize (*Zea mays* L.) Wee1 and its activity in developing endosperm. *Proc. Natl. Acad. Sci. U.S.A.* 96, 4180–4185. doi: 10.1073/pnas.96.7.4180

Sun, Y., Flannigan, B. A., and Setter, T. L. (1999b). Regulation of endoreduplication in maize (*Zea mays* L.) endosperm. Isolation of a novel B1-type cyclin and its quantitative analysis. *Plant Mol. Biol.* 41, 245–258. doi: 10.1023/A:1006315625486

Trafford, K., Haleux, P., Henderson, M., Parker, M., Shirley, N. J., Tucker, M. R., et al. (2013). Grain development in *Brachypodium* and other grasses: possible interactions between cell expansion, starch deposition, and cell-wall synthesis. *J. Exp. Bot.* 64, 5033–5047. doi: 10.1093/jxb/ert292

Tuteja, N., Tran, N. Q., Dang, H. Q., and Tuteja, R. (2011). Plant MCM proteins: role in DNA replication and beyond. *Plant Mol. Biol.* 77, 537–545. doi: 10.1007/s11103-011-9836-3

Van Daele, I., Gonzalez, N., Vercauteren, I., De Smet, L., Inzé, D., Roldan-Ruiz, I., et al. (2012). A comparative study of seed yield parameters in *Arabidopsis thaliana* mutants and transgenics. *Plant Biotechnol. J.* 10, 488–500. doi: 10.1111/j.1467-7652.2012.00687.x

Van Leene, J., Boruc, J., De Jaeger, G., Russinova, E., and De Veylder, L. (2011). A kaleidoscopic view of the *Arabidopsis* core cell cycle interactome. *Trends Plant Sci.* 16, 141–150. doi: 10.1016/j.tplants.2010.12.004

van Zanten, M., Koini, M. A., Geyer, R., Liu, Y., Brambilla, V., Bartels, D., et al. (2011). Seed maturation in *Arabidopsis thaliana* is characterized by nuclear size reduction and increased chromatin condensation. *Proc. Natl. Acad. Sci. U.S.A.* 108, 20219–20224. doi: 10.1073/pnas.1117726108

Verkest, A., Manes, C. L., Vercruysse, S., Maes, S., Van Der Schueren, E., Beeckman, T., et al. (2005). The cyclin-dependent kinase inhibitor KRP2 controls the onset of the endoreduplication cycle during *Arabidopsis* leaf development through inhibition of mitotic CDKA;1 kinase complexes. *Plant Cell* 17, 1723–1736. doi: 10.1105/tpc.105.032383

Vilhar, B., Kladnik, A., Blejec, A., Chourey, P. S., and Dermastia, M. (2002). Cytometrical evidence that the loss of seed weight in the miniature1 seed mutant of maize is associated with reduced mitotic activity in the developing endosperm. *Plant Physiol.* 129, 23–30. doi: 10.1104/pp.001826

Vlieghe, K., Boudolf, V., Beemster, G. T., Maes, S., Magyar, Z., Atanassova, A., et al. (2005). The DP-E2F-like gene DEL1 controls the endocycle in *Arabidopsis thaliana*. *Curr. Biol.* 15, 59–63. doi: 10.1016/j.cub.2004.12.038

Walley, J. W., Shen, Z., Sartor, R., Wu, K. J., Osborn, J., Smith, L. G., et al. (2013). Reconstruction of protein networks from an atlas of maize seed proteotypes. *Proc. Natl. Acad. Sci. U.S.A.* 110, E4808–E4817. doi: 10.1073/pnas.1319113110

Wang, G., Kong, H., Sun, Y., Zhang, X., Zhang, W., Altman, N., et al. (2004). Genome-wide analysis of the cyclin family in *Arabidopsis* and comparative phylogenetic analysis of plant cyclin-like proteins. *Plant Physiol.* 135, 1084–1099. doi: 10.1104/pp.104.040436

Wang, H., Zhou, Y., Bird, D. A., and Fowke, L. C. (2008). Functions, regulation and cellular localization of plant cyclin-dependent kinase inhibitors. *J. Microsc.* 231, 234–246. doi: 10.1111/j.1365-2818.2008.02039.x

Wang, Y., Hou, Y., Gu, H., Kang, D., Chen, Z., Liu, J., et al. (2012). The *Arabidopsis* APC4 subunit of the anaphase-promoting complex/cyclosome (APC/C) is critical for both female gametogenesis and embryogenesis. *Plant J.* 69, 227–240. doi: 10.1111/j.1365-313X.2011.04785.x

Wang, Y., Hou, Y., Gu, H., Kang, D., Chen, Z. L., Liu, J., et al. (2013). The *Arabidopsis* anaphase-promoting complex/cyclosome subunit 1 is critical for both female gametogenesis and embryogenesis(F). *J. Integr. Plant Biol.* 55, 64–74. doi: 10.1111/jipb.12018

Weimer, A. K., Nowack, M. K., Bouyer, D., Zhao, X., Harashima, H., Naseer, S., et al. (2012). Retinoblastoma related1 regulates asymmetric cell divisions in *Arabidopsis*. *Plant Cell* 24, 4083–4095. doi: 10.1105/tpc.112.104620

Weinl, C., Marquardt, S., Kuijt, S. J., Nowack, M. K., Jakoby, M. J., Hulskamp, M., et al. (2005). Novel functions of plant cyclin-dependent kinase inhibitors, ICK1/KRP1, can act non-cell-autonomously and inhibit entry into mitosis. *Plant Cell* 17, 1704–1722. doi: 10.1105/tpc.104.030486

Willemsen, V., Wolkenfelt, H., De Vrieze, G., Weisbeek, P., and Scheres, B. (1998). The HOBBIT gene is required for formation of the root meristem in the *Arabidopsis* embryo. *Development* 125, 521–531.

Wu, C.-Y., Rolfe, P. A., Gifford, D. K., and Fink, G. R. (2010). Control of transcription by cell size. *PLoS Biol.* 8:e1000523. doi: 10.1371/journal.pbio.1000523

Xie, Q., Sanz-Burgos, A. P., Hannon, G. J., and Gutierrez, C. (1996). Plant cells contain a novel member of the retinoblastoma family of growth regulatory proteins. *EMBO J.* 15, 4900–4908.

Yi, D., Kamei, C. L., Cools, T., Vanderauwera, S., Takahashi, N., Okushima, Y., et al. (2014). The *Arabidopsis* SIAMESE-RELATED Cyclin-Dependent Kinase Inhibitors SMR5 and SMR7 regulate the DNA damage checkpoint in response to reactive oxygen species. *Plant Cell* 26, 296–309. doi: 10.1105/tpc.113.118943

Young, T. E., and Gallie, D. R. (2000). Programmed cell death during endosperm development. *Plant Mol. Biol.* 44, 283–301. doi: 10.1023/A:1026588408152

Yu, Y., Steinmetz, A., Meyer, D., Brown, S., and Shen, W. H. (2003). The tobacco A-type cyclin, Nicta;CYCA3;2, at the nexus of cell division and differentiation. *Plant Cell* 15, 2763–2777. doi: 10.1105/tpc.015990

Zhou, Y., Li, G., Brandizzi, F., Fowke, L. C., and Wang, H. (2003). The plant cyclin-dependent kinase inhibitor ICK1 has distinct functional domains for in vivo kinase inhibition, protein instability and nuclear localization. *Plant J.* 35, 476–489. doi: 10.1046/j.1365-313X.2003.01821.x

Zhou, Y., Zhang, X., Kang, X., Zhao, X., Zhang, X., and Ni, M. (2009). SHORT HYPOCOTYL UNDER BLUE1 associates with MINISEED3 and HAIKU2 promoters in vivo to regulate *Arabidopsis* seed development. *Plant Cell* 21, 106–117. doi: 10.1105/tpc.108.064972

Conflict of Interest Statement: The authors declare that the research was conducted in the absence of any commercial or financial relationships that could be construed as a potential conflict of interest.

15

Seed dormancy and germination—emerging mechanisms and new hypotheses

Hiroyuki Nonogaki *

Department of Horticulture, Oregon State University, Corvallis, OR, USA

Edited by:
Paolo Sabelli, University of Arizona, USA

Reviewed by:
Joanna Putterill, University of Auckland, New Zealand
Pablo Daniel Jenik, Franklin and Marshall College, USA

***Correspondence:**
Hiroyuki Nonogaki, Department of Horticulture, Oregon State University, 4017 ALS Bldg., Corvallis OR 97331, USA
e-mail: hiro.nonogaki@oregonstate.edu

Seed dormancy has played a significant role in adaptation and evolution of seed plants. While its biological significance is clear, molecular mechanisms underlying seed dormancy induction, maintenance and alleviation still remain elusive. Intensive efforts have been made to investigate gibberellin and abscisic acid metabolism in seeds, which greatly contributed to the current understanding of seed dormancy mechanisms. Other mechanisms, which might be independent of hormones, or specific to the seed dormancy pathway, are also emerging from genetic analysis of "seed dormancy mutants." These studies suggest that chromatin remodeling through histone ubiquitination, methylation and acetylation, which could lead to transcription elongation or gene silencing, may play a significant role in seed dormancy regulation. Small interfering RNA and/or long non-coding RNA might be a trigger of epigenetic changes at the seed dormancy or germination loci, such as *DELAY OF GERMINATION1*. While new mechanisms are emerging from genetic studies of seed dormancy, novel hypotheses are also generated from seed germination studies with high throughput gene expression analysis. Recent studies on tissue-specific gene expression in tomato and Arabidopsis seeds, which suggested possible "mechanosensing" in the regulatory mechanisms, advanced our understanding of embryo-endosperm interaction and have potential to re-draw the traditional hypotheses or integrate them into a comprehensive scheme. The progress in basic seed science will enable knowledge translation, another frontier of research to be expanded for food and fuel production.

Keywords: chromatin remodeling, dormancy, embryo, endosperm, germination, hormone

INTRODUCTION

The ultimate role of seeds is to produce offspring and maintain species. Therefore, plants have evolved diverse strategies to ensure successful germination of this genetic delivery system. Proper distribution of seed germination, in both temporal and spatial manners, is critical for survival and proliferation of seed plants. Spatial distribution of germination is generally controlled through seed and fruit morphology, which enhances dispersal of the offspring from the maternal habitat. In contrast, temporal distribution of germination is controlled mainly by the physiological status of seeds. A variation among individual seeds in a population, in terms of physiological status, allows each seed to germinate at a different timing, which is an important strategy for seeds to avoid competition with their siblings or extinction of all individuals due to a disastrous condition. Plants have evolved seed dormancy, temporal suppression of germination under the conditions favorable to germination. Induction of seed dormancy during the maturation stage and its release at a dry state after a certain period of time, which is called "after-ripening," are widespread phenomena observed in diverse species of seed plants (Bewley et al., 2013). There may be a universal mechanism of seed dormancy as well as a species-specific variation in the regulatory mechanisms.

Hormonal regulation may be a highly conserved mechanism of seed dormancy among seed plants. Induction and maintenance of seed dormancy by abscisic acid (ABA) and dormancy release by gibberellin (GA) are observed in many species. The molecular mechanism of antagonistic function of these two hormones was unclear for many years. However, identification of the rate-limiting hormone metabolism genes, such as nine-*cis*-epoxycarotenoid dioxygenase (*NCED*), an ABA biosynthesis gene and *GA2ox*, a GA deactivation gene, and intensive analysis of their regulatory mechanisms in the last decade, have provided a comprehensive picture of ABA and GA involvement in the seed dormancy mechanisms (Seo et al., 2009). Now, we understand that seed response to light, which varies depending on species, is also controlled through hormone metabolism and signal transduction (Seo et al., 2009). Progress in seed dormancy and germination research is well summarized in recent review articles and textbooks (Graeber et al., 2012; Arc et al., 2013; Bewley et al., 2013). In this review, the main focus will be placed on the most recent discoveries from on-going research of seed dormancy and germination. Therefore, the contents of this review are not meant to be comprehensive but will highlight the "emerging" mechanisms and new hypotheses at the frontier of research.

EMERGING MECHANISMS OF SEED DORMANCY

Previously unknown seed dormancy-associated factors are emerging from on-going research, some of which enhance seed dormancy while others negatively affect it. The positive and negative regulators of seed dormancy, which will be discussed in this section, are summarized in **Table 1**. There is a risk of over-simplifying gene function with the categorization of positive and negative regulators, because there are complex regulatory mechanisms of seed dormancy, in which a single gene product could exert both positive and negative effects, including negative feedback from a positive regulator. However, to highlight the discoveries of gene function in the original research, this categorization will be used for the discussion in this section.

POSITIVE REGULATION

DOG1–CENTRAL TO SEED DORMANCY BUT UNKNOWN FOR BIOCHEMICAL FUNCTION

Quantitative trait locus (QTL) analysis using natural variation in Arabidopsis has identified the "seed dormancy-specific" loci, including the *DELAY OF GERMINATION* (*DOG*) genes (Alonso-Blanco et al., 2003; Bentsink et al., 2006, 2010), although some of them might not be strictly specific to dormancy (Chiang et al., 2013). One of them, *DOG1* has been characterized in detail. *DOG1* is expressed in seeds during the maturation stage. Loss of function of *DOG1* results in no dormancy (Bentsink et al., 2006). The genetic role of *DOG1* in seed dormancy and the significance of its expression in environment sensing and adaptation have been well documented (Kronholm et al., 2012; Footitt et al., 2013, 2014).

In contrast, the biochemical and molecular function of *DOG1* is still a mystery. *DOG1* encodes an unknown protein, for which only limited information is available. The *DOG1* cDNA shows highest similarity with a *Brassica napus* EST from an embryo library, however this gene also is not annotated. The protein with a known function that shows the highest similarity with *DOG1* is the wheat transcription factor Histone gene Binding Protein-1b (HBP-1b) (Bentsink et al., 2006). HBP-1b is a leucine zipper class transcription factor, which binds to the H3 hexamer motif ACGTCA in the promoter regions of wheat histone H3 genes (Mikami et al., 1989). This motif is required for transcription of the wheat H3 histone gene (Nakayama et al., 1989). *DOG1* has also been suggested to be a transcription factor, which is supported by its localization in the nucleus (Nakabayashi et al., 2012). However, the identity between *DOG1* and HBP-1b is not very high especially in the basic motifs and the heptad-repeat leucines in the leucine zipper structure (Tabata et al., 1991), which are conserved in HBP-1b and other H3 hexamer-binding proteins, such as tobacco Activation Sequence Factor-1 (ASF-1) (Lam et al., 1989) (**Figure 1**). Therefore, the biochemical function of *DOG1* is hardly predicted from its moderate similarity to HBP-1b. So far, direct target genes of *DOG1* that are clearly linked to the seed dormancy mechanisms have not been identified, although some dormancy up-(Dup) regulated genes [e.g., At5g43580 (*PR peptide*), At5g45540 (*unknown protein*), At5g45830 (*DOG1*), At5g47160 (*YDG/SRA domain-containing protein*)] or dormancy down-(Ddown) regulated genes [At4g19700 (*E3 ubiquitin ligase*), At5g04220 (*SYNAPTOTAGMIN3*), At5g46160 (*ribosomal protein*)] in the *DOG1* near isogenic line (NIL) have been identified (Bentsink et al., 2010).

POSSIBLE MODIFICATION AND PARTNERS OF *DOG1*

DOG1 transcript accumulates during the seed maturation stage with its peak around 14–16 days after pollination (DAP) (Bentsink et al., 2006), is reduced to about 20% in freshly harvested seeds, and disappears during imbibition (Nakabayashi et al., 2012). *DOG1* protein also accumulates during the maturation stage, however the protein level does not decrease toward the completion of seed maturation. As a consequence, freshly harvested seeds contain a relatively high level of *DOG1* protein. The protein level still remains relatively high even after 13 weeks of after-ripening when seed dormancy is already released (Nakabayashi et al., 2012). Thus, a correlation is lacking between the amount of *DOG1* protein and dormancy levels in after-ripened seeds. It has been proposed that the chemical property of *DOG1* protein, rather than its amount, is critical for *DOG1* to maintain seed dormancy and that its alteration to a non-functional form during after-ripening allows seed germination (Nakabayashi et al., 2012). In fact, there is a shift in the pI (isoelectric point) of the *DOG1* peptides prior to and following after-ripening (Nakabayashi et al., 2012).

Induction of *DOG1* in imbibed *dog1* mutant seeds with a heat-shock inducible system does not cause dormancy and allows 100% germination (Nakabayashi et al., 2012). This can be explained by the lack of protein modification discussed above. When *ABI5*, another key dormancy gene was overexpressed in Arabidopsis seeds, it was not sufficient to suppress germination. Only when the SnRK2 (Snf1-related protein kinase2), which activates ABI5, was induced in imbibed seeds, ABI5 was able to suppress seed germination (Piskurewicz et al., 2008). Therefore, it is possible that the *DOG1* protein induced by the heat-shock system was missing necessary modification in the ectopic induction experiment.

Recently, a search for possible *DOG1* partners was conducted through a yeast two-hybrid screen, which identified multiple proteins, including the PDF1 protein phosphatase 2A (Miatton, 2012). *PDF1* expression is enriched in the vascular system of the embryo (Miatton, 2012), which mimics the *DOG1* localization (Nakabayashi et al., 2012). PDF expression has its peak around 16 DAP during the maturation stage and is reduced in mature seeds, which is similar to the *DOG1* expression mentioned above. Unlike the *dog1* mutant, the *pdf1* loss of function mutant exhibits an enhanced seed dormancy phenotype (Miatton, 2012), suggesting that PDF1 is a negative regulator of seed dormancy and antagonizes *DOG1*. It is hypothesized that *DOG1* requires phosphorylation to be active, in terms of its function in seed dormancy induction and maintenance, and is dephosphorylated by PDF1, which could inactivate *DOG1* (Miatton, 2012). More analysis of PDF1 and other *DOG1*-interacting proteins will potentially provide a breakthrough in seed dormancy research.

Regardless of posttranslational modification, an alternative hypothesis to explain the lack of seed dormancy in *DOG1*-induced *dog1* seeds is that *DOG1* functions mainly during the maturation stage and the *DOG1* protein contained in mature seeds might be residual. It is possible that *DOG1* affects seed dormancy through its effects on ABA levels during maturation (Nakabayashi et al., 2012). *DOG1* has been proposed to function in a pathway independent of plant hormones. However,

Table 1 | Seed dormancy associated genes described in this article.

Symbol	Gene name	Dormancy function	Related publications
ABA1	*ABA deficient 1*	Positive	Bentsink et al., 2006
ABI3	*ABA INSENSITIVE 3, 4, 5*	Positive	Zheng et al., 2012
ABI4		Positive	Liu et al., 2007
ABI5		Positive	Piskurewicz et al., 2008
ACO1 *ACO4* *ACO5*	*1-AMINOCYCLOPROPANE-1-CARBOXYLATE OXIDASE 1, 4, 5*	Negative Negative Negative	Wang et al., 2013
AGO4	*ARGONAUTE 4*	Negative	Singh and Singh, 2012; Singh et al., 2013
ATXR7	*ARABIDOPSIS TRITHORAX-RELATED 7*	Positive	Liu et al., 2011
CYP707A	*Cytochrome P450 707A*	Negative	Wang et al., 2013
DEP	*DESPIERTO*	Positive	Barrero et al., 2010
DOG1	*DELAY OF GERMINATION 1*	Positive	Bentsink et al., 2006, 2010; Nakabayashi et al., 2012
ELF4 *ELF5*	*EARLY FLOWERING 4, 5*	Positive Positive	Liu et al., 2011
ERF9 *ERF105* *ERF112*	*ETHYLENE RESPONSE FACTOR*	Negative Negative Negative	Wang et al., 2013
GA3ox	*GA3-oxidase*	Negative	Yano et al., 2013
GA2ox	*GA2-oxidase*	Positive	
HD2B	*HISTONE DEACETYLASE 2B, 6, 19*	Negative	
HDA6 *HDA19*		Positive* Positive*	Wang et al., 2013
HDAC1	*HISTONE DEACETYLATION COMPLEX 1*	Negative*	Perrella et al., 2013
HUB1	*H2B MONOUBIQUTINATION 1*	Positive	Liu et al., 2007
KYP	*KRYPTONITE*	Negative	Zheng et al., 2012
NCED4	*NINE-CIS-EPOXYCAROTENOID DIOXYGENASE 4, 9*	Positive	Wang et al., 2013
NCED9		Positive	Liu et al., 2007
PDF1	*PDF1 protein phosphatase 2A*	Negative	Miatton, 2012
RDO2	*REDUCED DORMANCY2 (=TFIIS)*	Positive	Liu et al., 2011
RDO4	*REDUCED DORMANCY2 (=HUB1)*	Positive	Liu et al., 2007
Sdr4	*Seed dormancy 4*	Positive	Sugimoto et al., 2010
SNL1 *SNL2*	*SIN3-LIKE 1, 2*	Positive Positive	Wang et al., 2013
SnRK2	*Snf1-related protein kinase 2*	Positive	Piskurewicz et al., 2008
SUVH4	*SU(VAR)3-9 HOMOLOG4 (=KYP)*	Negative	Zheng et al., 2012
TFIIS	*Transcription elongation factor S-II*	Positive	Grasser et al., 2009
VIP7 *VIP8*	*VERNALIZATION INDEPENDENCE 7, 8*	Positive Positive	Liu et al., 2011

**HDA6 and HDA9 are known to affect ABA sensitivity negatively, which could affect seed dormancy negatively.*

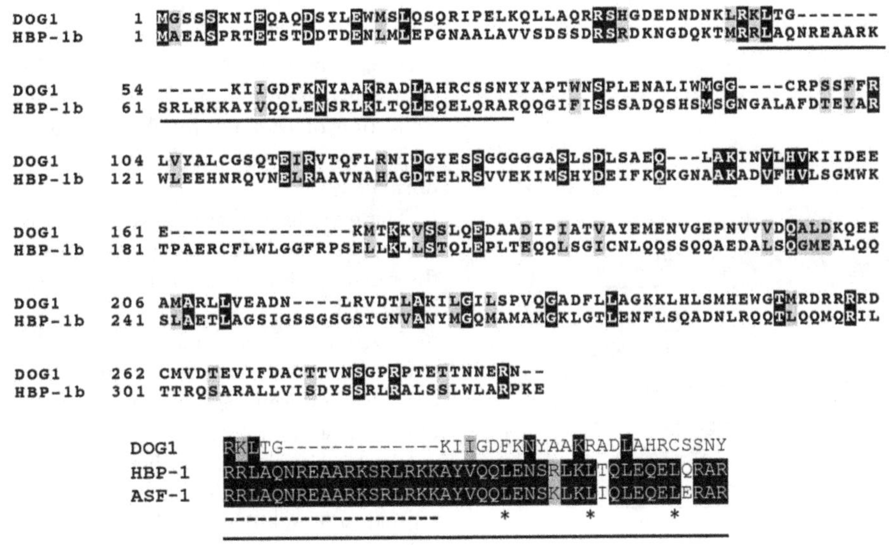

FIGURE 1 | Alignment of Arabidopsis *DOG1*, wheat HBP-1b and tobacco ASF-1. Arabidopsis *DOG1* (*DELAY OF GERMINATION1*) encodes an unknown protein, which shows some similarity to the wheat HBP-1b (Histone gene Binding Protein-1b), a leucine zipper class transcription factor (Bentsink et al., 2006). However, the *DOG1* protein does not show high identity to the leucine zipper domain in the HBP-1b (solid underline in the top panel). This region (extracted as the bottom panel) contains the basic motifs (dashed underline) and the heptad-repeat leucines (asterisks) in the leucine zipper structure (Tabata et al., 1991), which are conserved among the wheat HBP-1b, tobacco ASF-1 (Activation Sequence Factor-1) and other leucine zipper transcription factors. Sequences were aligned using the ClustralW and boxshade programs (http://www.expasy.org/genomics/sequence_alignment).

DOG1 is not able to impose seed dormancy in *aba1-1*, an ABA-deficient mutant (Bentsink et al., 2006), indicating that *DOG1* function is dependent on ABA. ABA levels are reduced in *dog1* mutants while GA levels are enhanced (Bentsink et al., 2006; Nakabayashi et al., 2012), supporting the idea of possible links between the *DOG1* and hormone pathways in seed dormancy. More information is necessary to obtain a clear picture about the hormone dependent and independent pathways of seed dormancy. To date, induction of *DOG1* specifically at the right timing during seed maturation (14–16 DAP) has not been experimentally examined. Investigation of molecular consequences upon *DOG1* induction at the right timing, including gene expression, protein phosphorylation and epigenetic changes (discussed below), will provide useful information. It should be noted that there are other dormancy(-specific) genes recently discovered, such as *Seed dormancy 4* (*Sdr4*) in rice (Sugimoto et al., 2010) and *DESPIERTO* in Arabidopsis (Barrero et al., 2010), which were not discussed here. Those genes also appear to be central to the dormancy mechanisms and are important targets of seed dormancy research.

TRANSCRIPTION ELONGATION OF SEED DORMANCY GENES

There is emerging evidence to suggest that regulation of transcriptional efficiency may be one of the core mechanisms of seed dormancy. Transcriptional efficiency is determined by recruitment of RNA polymerase II (Pol II) to the DNA template and the rate of transcription elongation after its binding to DNA. The efficiency of transcription elongation is influenced by an arrest of Pol II and its recovery from the arrest (Saunders et al., 2006). Transcription elongation factor S-II (TFIIS) assists Pol II to overcome the temporal arrest during elongation and enhances RNA synthesis (Kim et al., 2010) (**Figure 2**). A mutagenesis screen for seed dormancy in Arabidopsis yielded *reduced dormancy* (*rdo*) mutants (Leon-Kloosterziel et al., 1996; Peeters et al., 2002). *RDO2*, one of the genes identified from this screening, encoded TFIIS (Liu et al., 2011). Another independent study also found that a mutation in *TFIIS* resulted in reduced seed dormancy (Grasser et al., 2009). These results suggest that transcription elongation may be a critical part of the dormancy mechanisms.

The phenotypes of other mutants also support this contention. TFIIS and Pol II interact with the Pol II-Associated Factor 1 Complex (PAF1C) (Kim et al., 2010) (**Figure 2**). In yeast, PAF1C consists of Paf1, Rtf1, Ctr9, Leo1, and Cdc73 (Penheiter et al., 2005; Porter et al., 2005) (**Figure 2**, top-left inset). The Arabidopsis orthologs of these yeast proteins EARLY FLOWERING7 (ELF7) (= Paf1), ELF8 (= Ctr9), VERNALIZATION INDEPENDENCE4 (VIP4) (=Leo1), VIP5 (= Rtf1) and PLANT HOMOLOGOUS TO PARAFIBROMIN (PHP) (= Cdc73) have been identified (Zhang and Van Nocker, 2002; He et al., 2004; Oh et al., 2004; Yu and Michaels, 2010). Seeds of the *elf7*, *elf8*, *vip4*, and *vip5* mutants all exhibit reduced dormancy (Liu et al., 2011), suggesting the importance of PAF1C and transcription elongation for seed dormancy.

HISTONE UBIQUITINATION AND METHYLATION ASSOCIATED WITH TRANSCRIPTION ELONGATION

PAF1C interacts with Bre1, a protein involved in histone 2B (H2B) monoubiquitination (Kim et al., 2009) (**Figure 2**). Interestingly, *rdo4*, another reduced dormancy mutant in Arabidopsis, which was isolated from the same mutagenesis screening as mentioned above, has a mutation in *H2B MONOUBIQUITINATION1* (*HUB1*) gene, an Arabidopsis ortholog of *Bre1* (Liu et al., 2007).

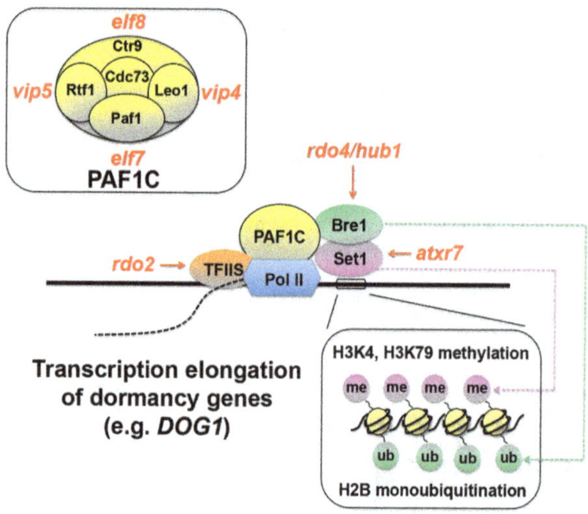

FIGURE 2 | Schematic representation of transcription elongation of seed dormancy genes. Transcription elongation factor S-II (TFIIS) assists RNA polymerase II (Pol II) and promotes transcription elongation (Saunders et al., 2006; Kim et al., 2010). Pol II-Associated Factor 1 Complex (PAF1C), which consists of Paf1, Rtf1, Ctr9, Leo1, and Cdc73 (top-left inset) in yeast (Porter et al., 2005), also functions in this process through its interaction with Bre1, which monoubiquitinates (ub) histone 2B (H2B), and Set1, which methylates (me) histone H3 lysine 4 (H3K4), and lysine 79 (H3K79) (bottom-right inset) (Sun and Allis, 2002; Zhu et al., 2005; Kim et al., 2009). These chromatin-remodeling events and their positive effects on transcription elongation are thought to be critical for induction of seed dormancy genes, because mutants in many of these components (*rdo2, rdo4, atxr7, elf 7, elf8, vip4,* and *vip5*) exhibit reduced seed dormancy (Liu et al., 2011). Red italic symbols indicate Arabidopsis mutants corresponding to the yeast protein components. *atxr7, arabidopsis trithorax-related 7; elf, early flowering; hub1, h2b monoubiquitination1; rdo, reduced dormancy; vip, vernalization independence.*

Bre1 interacts with Set1, which methylates histone 3 lysine 4 and lysine 79 (H3K4, H3K79) (Sun and Allis, 2002; Zhu et al., 2005) and promotes gene expression (**Figure 2**). A mutation in the *Set1* ortholog *ARABIDOPSIS TRITHORAX-RELATED 7* (*ATXR7*) also causes reduced dormancy in seeds (Liu et al., 2011). These results reinforce the idea that regulation of transcription elongation efficiency is an essential part of seed dormancy and suggest the significance of chromatin remodeling in the regulatory mechanisms.

H2B monoubiquitination and H3K4 and H3K79 methylation, which is dependent on H2B monoubiquitination (Nakanishi et al., 2009), are thought to activate gene expression (Henry et al., 2003). Since *hub1* (=*bre1*) seeds exhibit reduced dormancy, genes down-regulated in the *hub1* mutant are good candidates for seed dormancy-imposing genes, the expression of which is promoted through transcriptional elongation. *ABA INSENSITIIVE4* (*ABI4*), *DOG1*, *NINE-CIS-EPOXYCAROTENOID DIOXYGENASE9* (*NCED9*) and other genes have been identified as possible targets of HUB1/RDO4 (Liu et al., 2007). *RDO2* (*TFIIS*) and *RDO4* (*HUB1*), two positive regulators of transcription are induced during the same stages of seed maturation (~18–19 DAP). There is a significant overlap between *rdo2* and *rdo4*, in terms of differentially expressed genes in the mutants. These results suggest that RDO2 and RDO4 might share common targets. Intriguingly, *DOG1* is one of the genes commonly down-regulated in the two mutants (Liu et al., 2011). Activation of *DOG1* through chromatin remodeling and transcriptional elongation might be an important mechanism of seed dormancy.

The hypothesis that seed dormancy is regulated by the efficiency of transcription elongation of *DOG1* is also supported by the recent analysis of the *tfIIs* mutant, in which seed dormancy is reduced but reverted to the wild-type level by an extra copy of *DOG1* (Mortensen and Grasser, 2014). However, when the *hub1/rdo4* mutant is crossed with the NIL carrying *DOG1*-Cvi, which causes deep seed dormancy, the resulting seeds still show dormancy at a level between *hub1* and *DOG1*-Cvi NIL. Similar results are observed when the *hub1/rdo4* was transformed with the Cvi *DOG1* genomic fragment. The incomplete alleviation of dormancy from NIL *DOG1* by *hub1/rdo4* mutation in both cases suggests that *HUB1* is not epistatic to *DOG1*. In contrast, the combination of *hub1* and *DOG3*-Cvi resulted in no seed dormancy, suggesting that *HUB1* functions in the same pathway as *DOG3* to affect seed dormancy (Liu et al., 2007). More analyses of the specific targets of epigenetic modification and transcriptional elongation will be necessary to draw a clear picture about seed dormancy regulation through these processes.

REPRESSION OF SEED GERMINATION GENES THROUGH HISTONE DEACETYLATION

While activation of dormancy genes through transcription elongation appears to be critical for dormancy induction, continuous repression of seed germination-associated genes is also probably an essential part of dormancy maintenance. There is evidence that histone deacetylation is imperative for repression of genes positively affecting seed germination. In yeast and mammals, histone deacetylase (HDAC) interacts with SWI-INDEPENDENT3 (SIN3), an amphipathic helix repeat protein, removes acetyl groups from lysine in the histone tails, and creates a transcriptionally inactive state of chromatin (Kadosh and Struhl, 1998; Lai et al., 2001; Grzenda et al., 2009) (**Figure 3**). In Arabidopsis, SIN3-LIKE1 (SNL1) physically interacts with HDA19, an Arabidopsis HDAC ortholog, both *in vitro* and *in planta* (Wang et al., 2013). The Arabidopsis genome contains *SNL2*, which is partially redundant to *SNL1*. Seeds of the *snl1 snl2* double mutant exhibit reduced dormancy. A reduced dormancy phenotype is also observed in *hda19* mutant seeds (Wang et al., 2013). These results indicate that SNLs and HDA19 are positive regulators of seed dormancy. It appears that proper repression of the SNL-HDA19 targets, which are most likely germination-inducing genes, through histone deacetylation is essential for normal seed dormancy. Acetylation of H3K9/18 and H3K14 is increased in the *snl1 snl2* double mutant (Wang et al., 2013), which confirms that in wild-type seeds the SIN3-HDAC complex deacetylates histones and puts repressive marks on the chromatin (Richon and O'Brien, 2002) (**Figure 3**).

Global gene expression analysis between the *snl1 snl2* double mutant and wild-type seeds with RNA sequencing identified possible targets of SNL-HDA19. Ethylene biosynthesis

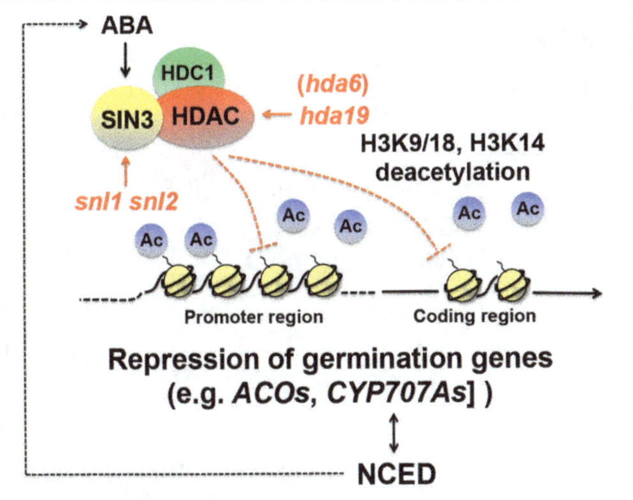

FIGURE 3 | Schematic representation of repression of seed germination genes through histone deacetylation. In yeast, a histone deacetylase (HDAC) interacts with SWI-INDEPENDENT3 (SIN3) (Kadosh and Struhl, 1998; Lai et al., 2001; Grzenda et al., 2009). HDA19, an HDAC ortholog in Arabidopsis, interacts with SIN3-LIKEs (SNLs) (Wang et al., 2013) and HDC1 (Histone Deacetylation Complex1) (Perrella et al., 2013), and removes acetyl groups (Ac) from histone 3 lysine9/18 (H3K9/18) and lysine14 (H3K14) and represses genes positively affecting germination, such as *1-AMINOCYCLOPROPANE-1-CARBOXYLATE OXIDASEs* (*ACOs*) and *CYP707As*, ABA deactivation genes (Wang et al., 2013). Deacetylation occurs in both the promoter and coding regions. Both *snl* and *hda19* mutations cause reduced dormancy (Wang et al., 2013). Expression of *NCED4*, an ABA biosynthesis gene, is reduced in the *snl1 snl 2* double mutant, suggesting that the SNL-HDA19 complex imposes seed dormancy also through the promotion of ABA biosynthesis. *SNL* expression is promoted by ABA (Wang et al., 2013), which suggests that there is a positive feedback loop to maintain high ABA levels through the SNL-HDA19 pathway.

genes, such as *1-AMINOCYCLOPROPANE-1-CARBOXYLATE OXIDASE1* (*ACO1*), *ACO4*, and *ACO5* and ethylene response genes, such as *ETHYLENE RESPONSE FACTOR9* (*ERF9*), *ERF105*, and *ERF112*, were up-regulated in the mutant (Wang et al., 2013). Quantitative PCR combined with chromatin immunoprecipitation with the H3K9/18 acetylation-specific antibodies showed that the *ACOs* and *ERFs* genes were indeed hyperacetylated in the mutant, which mainly occurred in the promoter region but were also found in the coding region (Wang et al., 2013). These results suggest that SNL-HDA19 causes seed dormancy by suppressing the ethylene pathway, which positively affects seed germination in Arabidopsis (Chiwocha et al., 2005; Arc et al., 2013).

In contrast, the same study suggests that SNL-HDA19 increases ABA levels and thereby enhances seed dormancy. *CYP707A1* and *CYP707A2*, ABA deactivation genes, which reduce ABA levels, were up-regulated in the *snl1 snl2* double mutants. Consistently, *NCED4*, an ABA biosynthesis gene, was down-regulated in the same mutant (Wang et al., 2013). These results suggest that SNL-HDA19 suppresses *CYP707As* and activates *NCED4* in wild type, both of which increase ABA levels and enhance seed dormancy. Interestingly, ABA stimulates *SNL1* and *SNL2* expression (Wang et al., 2013), which suggests that there is positive feedback regulation to maintain high levels of ABA through the histone deacetylation pathway (**Figure 3**). While this study suggests that ABA levels are positively affected by SNL-HDA19, other studies suggest that ABA sensitivity is negatively regulated by HDA19 (and HDA6). Mutations in *HDA6* and *HDA19* cause ABA hypersensitivity during germination (Chen et al., 2010; Chen and Wu, 2010). Loss of function in Histone Deacetylation Complex1 (HDC1), another component of the SNL- and HDA19-containing complex, which physically interacts with HDA6 and HDA19 (**Figure 3**), also causes ABA hypersensitivity in seedlings. *HDC1* overexpression promotes seedling emergence (Perrella et al., 2013), although detailed information about *sensu stricto* germination and a dormancy phenotype of the mutant seeds is not available. The significance of the opposite effects of the HDAC multiprotein complex to ABA levels (positive) and sensitivity (negative) in the regulatory mechanisms of seed dormancy is not known. It is possible that the seemingly counterintuitive effects are associated with negative feedback regulation.

NEGATIVE REGULATION

REPRESSION OF DORMANCY GENES AND ACTIVATION OF GERMINATION GENES THROUGH HISTONE DEACETYLATION

HISTONE DEACETYLASE 2B (*HD2B*), another *HDAC* gene, is also involved in seed dormancy. In this case, it negatively affects seed dormancy (Yano et al., 2013). This discovery was made through a combination of genome-wide association mapping (GWA) (Atwell et al., 2010) and transcriptomics. The efficiency of QTL analysis using different accessions of Arabidopsis, such as Cvi, L*er*, and Col, for seed dormancy is well exemplified by the successful identification and characterization of the *DOG* genes (Alonso-Blanco et al., 2003; Bentsink et al., 2006, 2010). Since the comparison of a few different Arabidopsis accessions is so powerful, multiplying this approach using many accessions with natural variations in seed dormancy is expected to produce fruitful outcomes in seed dormancy research, especially when it is combined with GWA, which identified a number of single nucleotide polymorphisms (SNPs) likely associated with various phenotypes (Atwell et al., 2010). Based on this concept, 113 accessions were analyzed to identify SNPs associated with natural variation in seed dormancy using GWA and transcriptomics, which identified *HD2B* as a strong candidate of a seed dormancy-associated gene. *HD2B* expression levels are significantly lower in 24 dormant accessions than 28 less-dormant accessions, although there are some exceptions. When the highly dormant Cvi line was transformed with the genomic fragment of Col *HD2B* (termed ^{Col}*HD2B*/Cvi), mature seeds of ^{Col}*HD2B*/Cvi exhibited reduced dormancy, which was not evident immediately after harvest without cold stratification but became clear when seeds were stratified or partially after-ripened (Yano et al., 2013).

Cold stratification releases seed dormancy through an increase in GA levels. *GA3ox1*, a rate-limiting GA biosynthesis gene, is induced by cold stratification (Yamauchi et al., 2004), which triggers expansion of cortex cells in the radicle/hypocotyl region and then generates growth potential of the embryo for germination (Ogawa et al., 2003). Evidence suggests that HD2B mediates this dormancy-releasing process. In ^{Col}*HD2B*/Cvi seeds, expression of *GA3ox1* and *GA3ox2* and GA_4 levels are increased, while expression of *GA2ox2*, a GA deactivation gene, is reduced compared to wild-type Cvi seeds (Yano et al., 2013). Since HDAC represses gene expression through histone deacetylation, *GA2ox2* repression could be a direct effect of HD2B. In contrast, the up-regulation of *GA3ox* genes may be through repression of their upstream regulators or some other mechanisms. It is interesting that the three separate hormone pathways (ethylene, ABA, and GA) associated with seed dormancy are regulated by histone deacetylation. These results demonstrate that epigenetic regulation through chromatin remodeling is a robust mechanism to alter hormone levels in seeds.

SILENCING OF SEED DORMANCY GENES THROUGH HISTONE AND DNA METHYLATION

The studies mentioned above showed that HDAC could affect seed dormancy either positively (HDA19) or negatively (HD2B), depending on the target genes. Histone methylation also affects seed dormancy in both ways. While H3K4 and H3K79 methylation activates gene expression and causes seed dormancy as mentioned above (Set1 or ATXR7), dimethylation of H3K9 (H3K9me2), a repressive mark, occurs on the chromatin associated with seed dormancy genes. Analysis of gene silencing at the Arabidopsis *SUPERMAN* (*SUP*) locus identified the KRYPTONITE (KYP) methyltransferase, which causes H3K9me2 (**Figure 4**). The methylated histone recruits the DNA methyltrasferase CHROMOMETHYLASE3 (CMT3) through its interaction with HETEROCHROMATIN PROTEIN1 (HP1) and triggers the methylation of cytosine nucleotides of DNA and silences the gene (Jackson et al., 2002; Johnson et al., 2007) (**Figure 4**). KYP is SU(VAR)3-9 (Rea et al., 2000) HOMOLOG 4 and is also called SUVH4. The *kyp-2* mutant seeds show enhanced dormancy, suggesting that KYP/SUVH4 suppresses seed dormancy genes. Interestingly, again, *DOG1* is one of the up-regulated genes in the mutant, as well as *ABI3* (Zheng et al., 2012). These results suggest that histone methylation caused by KYP/SUVH4 induces silencing of *DOG1* and *ABI3* through DNA methylation and negatively affects seed dormancy.

The KYP-CMT3 gene-silencing pathway mediates RNA-directed DNA methylation (RdDM), which is triggered by small interfering RNAs (siRNAs) produced by DICER-LIKE3 (DCL3) and their loading onto ARGONAUTE4 (AGO4) (Zilberman et al., 2004; Tran et al., 2005) (**Figure 4**). AGO proteins are components of the RNA-induced silencing complex (RISC) and are involved in gene silencing. While AGO1 and AGO10 proteins function mainly in posttranscriptional gene silencing (PTGS) through the MIR (microRNA) and TAS (trans-acting siRNA) pathways, the AGO4/AGO6/AGO9 clade proteins are associated with transcriptional gene silencing (TGS) through RdDM (Mallory and Vaucheret, 2010). Little information is available for silencing of seed dormancy genes through RdDM, however a possible involvement of AGO4 in seed dormancy regulation has been suggested from studies of cereal seed dormancy. *AGO1003*, an *ARGONAUTE* (*AGO*)*4_9* gene in barley, is expressed differentially in the embryos of dormant and non-dormant grains and is thought to function as a negative regulator of seed dormancy through RdDM (Singh and Singh, 2012). A separate study

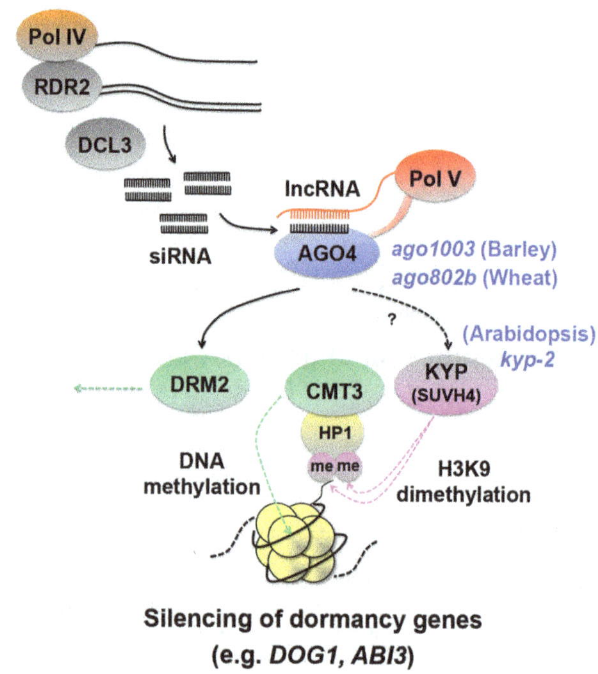

FIGURE 4 | Schematic representation of silencing of dormancy genes through histone and DNA methylation. RNA polymerase IV (Pol IV) transcripts are converted to double-stranded RNA by RNA-Dependent RNA polymerase 2 (RDR2), which are then processed into 24-nt siRNAs by DICER-LIKE3 (DCL3) (Xie et al., 2004; Herr et al., 2005; Onodera et al., 2005; Law et al., 2011). siRNAs are loaded onto ARGONAUTE4 (AGO4) (Qi et al., 2006) and interact with long non-coding RNAs (lncRNAs) produced by Pol V, which are thought to function as scaffold transcripts to guide siRNAs to specific loci to be silenced (Wierzbicki et al., 2008, 2009; Wierzbicki, 2012). In this way, the AGO4 complex containing siRNAs and lncRNAs triggers RNA-directed DNA methylation (RdDM) (Wierzbicki, 2012). A possible event downstream of AGO4 is histone 3 lysine 9 dimethylation (H3K9me2) by the KRYPTONITE (KYP), which causes HETEROCHROMATIN PROTEIN1 (HP1) to bind to the modified histone and recruit CHROMOMETHYLASE3 (CMT3), a DNA methyltransferase that induces gene silencing (Jackson et al., 2002; Zilberman et al., 2004; Tran et al., 2005; Johnson et al., 2007). A mutation in *KYP* in Arabidopsis causes enhanced dormancy and up-regulation of *DOG1* and *ABI3* (Zheng et al., 2012), suggesting that the seed dormancy genes are silenced by the KYP-CMT3 pathway. The AGO4 complex is also involved in gene silencing by DOMAINS REARRANGED METHYLTRANSFERASE2 (DRM2) (Zilberman et al., 2004; Wierzbicki, 2012), although DRM2 involvement in seed dormancy regulation is not known. Direct evidence for siRNAs and lncRNAs involvement in *DOG1* and *ABI3* regulation is lacking, however AGO4 has been shown to be a negative regulator of dormancy in barley and wheat seeds (Singh and Singh, 2012; Singh et al., 2013). The Arabidopsis, barley and wheat seed dormancy mutants corresponding to the protein components in the RdDM pathway are indicated by blue italic symbols.

in wheat supports this hypothesis. *AGO802B*, a wheat ortholog of *AGO4_9* gene is expressed during grain development (5–20 DAP). *AGO802B* expression is significantly lower in preharvest sprouting (PHS)-resistant (i.e., more dormant) varieties than in susceptible ones (Singh et al., 2013). This result also suggests that *AGO4* is a negative regulator of dormancy. It is not known whether specific coding genes are subjected to silencing through RdDM in wheat seeds. However, analysis of 5S

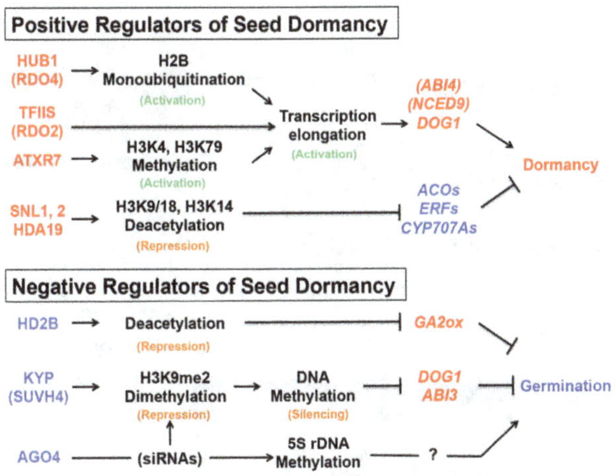

FIGURE 5 | Summary of the seed dormancy or germination pathways described in this article. The positive (red) and negative (blue) regulators of seed dormancy and their roles in the chromatin remodeling, DNA modification or siRNA pathways are indicated, together with promotive (arrows) or suppressive (blocked arrows) effects on the downstream genes (italics). Active (green) or repressive (orange) marks on histones or DNA are also indicated. See text for gene and protein symbols and references.

ribosomal DNA from PHS-resistant and susceptible varieties with the methylation-sensitive restriction enzyme Msp*I* suggested that ribosomal DNA methylation was reduced in PHS-resistant varieties (Singh et al., 2013), supporting the hypothesis that AGO4 enhances histone and DNA methylation and acts as a negative regulator of seed dormancy.

The chromatin-remodeling factors mentioned above include both positive and negative regulators of seed dormancy, which could be considered as negative and positive regulators of seed germination, respectively. The description "activation of seed dormancy genes" or "repression of seed germination genes," which could mean the same consequence (dormancy or no germination), is confusing. It is even more confusing when the description is combined with different terminology of histone modification, such as histone (de)acetylation, monoubiquitination or (de)methylation, because they could be either repressive or active marks depending on the position of residues in the histone tail. To avoid the confusion, the positive and negative regulators of seed dormancy, their roles in chromatin and DNA modification, and possible consequences in gene expression downstream are summarized in **Figure 5**.

NEW HYPOTHESES FOR GERMINATION EVENTS

REMAINING BARRIERS OF SEED GERMINATION

A quiescent state of the embryo is changed when molecular repression on seed germination genes is removed, which is probably orchestrated with silencing of dormancy genes. However, an active embryo is still unable to complete germination when the suppressive force, or mechanical resistance, of the covering tissues, such as the testa and endosperm, exceeds embryo growth potential. When the embryo is not dormant, it is the mechanical resistance of the covering tissues that mainly determines whether the embryo emerges from imbibed seeds. In fact, the embryos in dormant seeds in many species are able to grow when they are excised from seeds, which is called coat-imposed dormancy (Bewley et al., 2013). While further increase in embryo growth potential may still be necessary, alteration of the properties of covering tissues plays a significant role in germination. The testa in a mature seed is generally a non-living tissue, therefore the major reduction in the mechanical resistance of the covering tissues depends on physiological changes in the living endosperm. Changes in the properties of the endosperm significantly affect timing of radicle emergence in non-dormant seeds also. Therefore, the mechanisms of endosperm weakening have been a focal point in seed germination research.

Basic information about endosperm weakening is summarized in other literature (Linkies et al., 2010; Bewley et al., 2013). Briefly, the micropylar region of endosperm (ME) surrounds the radicle tip and provides an opposing force to it (**Figure 6**), which is reduced during germination through weakening. The mechanical resistance of ME is mainly due to the thick and rigid cell walls in this tissue. Therefore, cell wall modification is thought to play an essential role in ME weakening (Bewley et al., 2013). In fact, genes encoding cell wall-modifying proteins, such as xyloglucan endotransglycosylase/hydrolases (XTHs) and expansins (EXPs), are expressed exclusively in ME of Arabidopsis (Dekkers et al., 2013), *Lepidium sativum* (Voegele et al., 2011) and tomato (Chen and Bradford, 2000; Chen et al., 2002) seeds during germination. While distinct cell wall architecture is observed in ME of seeds depending on plant species and family (Lee et al., 2012a), ME weakening by cell wall modifying proteins seems to be a widely conserved mechanism of germination.

EMBRYO-ENDOSPERM INTERACTION IN TOMATO SEEDS

A high throughput transcriptome analysis of germinating tomato seeds showed enrichment of cell wall-associated genes in ME (Martinez-Andujar et al., 2012), supporting the hypothesis discussed above. In this study, tomato seeds were dissected into the endosperm cap (EC, equivalent to ME), lateral endosperm (LE), radicle-half embryo (R), and cotyledon-half embryo (C) (**Figure 6A**). In addition to the cell wall-associated genes, PR (pathogenesis-related) or wound-response genes were detected as another major group of ME-enriched genes. The 5′ upstream sequences of the ME-enriched PR genes contain the conserved sequences, including the DNA motifs targeted by ethylene response factors (ERFs). Interestingly, *Tomato ERF1* (*TERF1*), an experimentally validated upstream regulator of the PR genes, was also one of the ME-enriched genes in tomato seeds (Martinez-Andujar et al., 2012). These results suggest that TERF1 is a major upstream regulator in ME and induces other ME genes, such as PR- or wound response genes and possibly cell wall-associated genes also.

The degradation of cell wall in ME of tomato seeds, which is accompanied by disappearance of storage vacuoles and lipid bodies from the cells, is initiated at the inner cells adjacent to the radicle tips (**Figure 6B**), suggesting that ME activation is under the control of the embryo. A traditional view of the mechanism of ME gene induction is that diffusible signals, such as GA, or non-diffusible signals, such as peptide ligands, are secreted

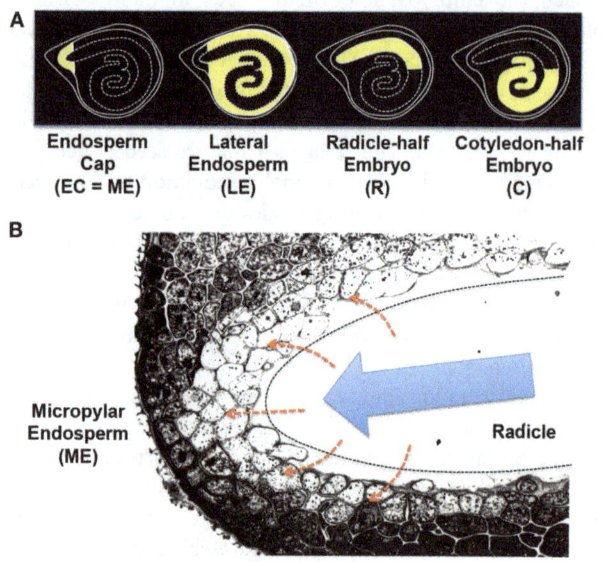

FIGURE 6 | Tomato seed tissues and regulatory mechanisms of endosperm weakening. (A) Schematic representation of the four different tissue parts of tomato seeds used for the transcriptome analysis (Martinez-Andujar et al., 2012). Endosperm cap [EC = ME (micropylar endosperm)], lateral endosperm (LE), radicle-half embryo (R), and cotyledon-half embryo (C) are highlighted by yellow filling. **(B)** Photograph of micropylar endosperm (ME) cells. The radicle tip (outlined by dashed line) has been removed. Note that cell walls are fuzzy and protein storage vacuoles and lipid bodies are disappearing in the inner layers of ME cells facing the embryo (Nonogaki et al., 1998). Possible secretion of GA or peptides from the embryo to the endosperm (red dashed arrows) and the growth potential generated by the embryo (blue arrow), which provides pressure onto the endosperm, are indicated in the scheme.

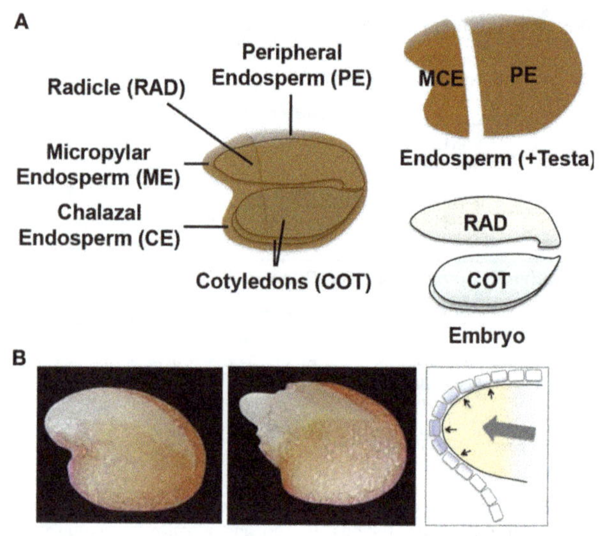

FIGURE 7 | Arabidopsis seed tissues and regulatory mechanisms of endosperm weakening. (A) Schematic representation of the four different tissue parts of Arabidopsis seeds used for the transcriptome analysis (Dekkers et al., 2013). The micropylar (ME) plus charazal (CE) endosperm (MCE), peripheral endosperm (PE, similar to LE), radicle (RAD) and cotyledons (COT) are indicated in the scheme. **(B)** Photographs of Arabidopsis seeds at testa rupture (TR, left) and endosperm rupture (ER, right). Schematic representation of ME is shown to the right with the growth potential of the embryo (gray arrow) and pressure (black arrows) placed onto the single cell layer of endosperm (with purple filling).

from the embryo to ME (**Figure 6B**, red dashed arrows) and then stimulate gene expression in this tissue. However, the new finding about the TERF1 cascade and possible involvement of a PR- or wounding response in ME gene expression generated a new hypothesis of "mechanosensing." In this hypothesis, pressure, rather than chemical molecules, which is generated by the embryo and placed onto ME cells (**Figure 6B**, blue arrow), triggers a wound response, *TERF1* expression, and then induction of the downstream genes in ME.

THE "TOUCH" GENES IN ARABIDOPSIS SEEDS

A similar but more comprehensive and dynamic transcriptomic analysis in Arabidopsis seeds provided supporting evidence for the mechanosensing hypothesis. It is technically difficult to dissect ME from Arabidopsis seeds. Therefore, in this study gene expression was compared for the micropylar and charazal endosperm (MCE), peripheral endosperm (PE, similar to LE), radicle (RAD), and cotyledons (COT) (Dekkers et al., 2013) (**Figure 7A**). The high-resolution data set included many time points including those before and after testa rupture (TR) and endosperm rupture (ER), which are the signature events during germination and at the completion of germination, respectively (**Figure 7B**). This study demonstrated that TR was marked by activation of the specific genes in MCE, such as *TOUCH3* and *TOUCH4*, which are known to be induced by touch or thigmotropism (Braam, 2005). The comparison of MCE genes at TR in Arabidopsis seeds with the genes up-regulated by touching the aerial part of Arabidopsis plants (Lee et al., 2005) showed significant overlaps. These results suggest that ME gene induction in Arabidopsis seeds is also caused by touch or mechanosensing (Dekkers et al., 2013).

No conclusive evidence has been obtained to date for the mechanosensing or touch hypothesis. However, the new findings have great potential to re-draw the traditional view of ME gene regulation, which is a core mechanism of germination. It is well known that GA stimulates ME gene expression in the GA-deficient *gib-1* tomato seeds, which absolutely require GA for radicle emergence (Groot and Karssen, 1987; Nonogaki et al., 2000). The GA requirement for ME gene expression can be substituted by co-incubation of ME with the embryonic axes, suggesting that the embryo produces GA and secretes it to the endosperm (Groot and Karssen, 1987). There seems to be no doubt that ME gene expression is under the control of GA and the embryo. However, it should be noted that exogenous GA stimulates gene expression in both ME and LE when tomato seeds are dissected, while only ME is responsive when GA is applied to intact seeds (Martinez-Andujar et al., 2012). This raises the question as to why LE in an intact seed remains unaffected by GA or why only ME is responsive to it? The new hypothesis (mechanosensing or touch) could answer these questions. If the GA-dependent embryonic effects on ME gene expression are not directly exerted through chemical secretion but are indirectly mediated by the pressure provided by the radicle tip, the highly localized gene

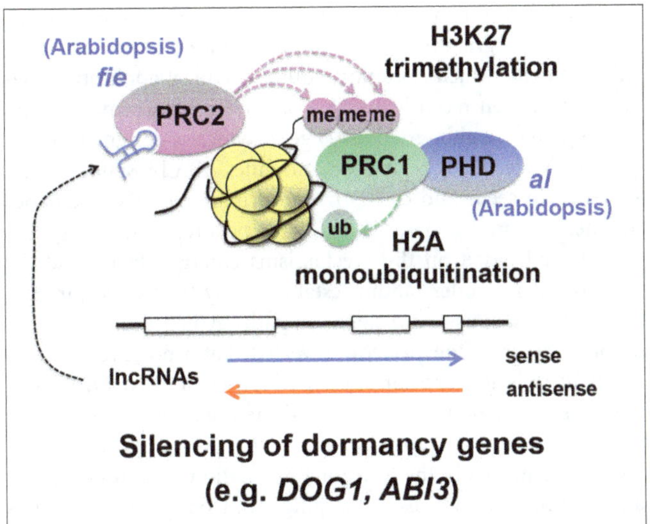

FIGURE 8 | Hypothetical integration of the known lncRNA-PRC pathway into the silencing mechanisms of seed dormancy genes. In this scheme, long non-coding RNAs (lncRNAs) (Swiezewski et al., 2009; Heo and Sung, 2011), interact with Polycomb Repressive Complex 2 (PRC2), which causes histone 3 lysine 27 trimethylation (H3K27me3) (Simon and Kingston, 2009; De Lucia and Dean, 2011). This histone modification recruits PRC1, which monoubiquitinates H2A (Simon and Kingston, 2009). While H2B monoubiquitination promotes transcription elongation (see **Figure 2**), H2A monoubiquitination is thought to be a repressive mark and silence genes (Simon and Kingston, 2009). A mutation in *FERTILIZATION INDEPENDENT ENDOSPERM(FIE)*, an essential component of PRC2, causes enhanced dormancy (Bouyer et al., 2011), supporting the idea that PRC suppresses dormancy genes and promotes germination. A mutation in ALFIN1-like (AL), a Plant Homeo Domain (PHD) finger that interacts with PRC1, also promotes dormancy (Molitor et al., 2014). Evidence has not been obtained for the involvement of specific lncRNAs in suppression of dormancy genes through PRC.

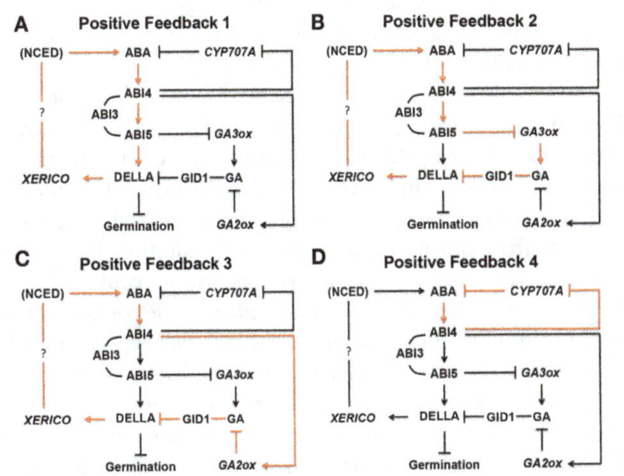

FIGURE 9 | Positive feedback loops in ABA biosynthesis in seeds. (A) In Positive Feedback 1, ABA produced by NCED, a rate-limiting ABA biosynthesis enzyme, induces ABIs. ABI3, and ABI5 interacts with each other while ABI4 induces ABI5 by binding its promoter region. ABI5 binds to the promoter region of a *DELLA* gene, such as *RGL2*, and up-regulates its expression. DELLA then promotes expression of *XERICO*, which increases ABA biosynthesis through unknown mechanism(s). In this way, the originally produced ABA in seeds enhances ABA biosynthesis through positive feedback. **(B)** In Positive Feedback 2, ABI5 down-regulates *GA3ox*, a GA biosynthesis gene, and reduces GA and GA response by GID1, a GA receptor. Reduced GA levels stabilize DELLA protein, such as RGL2, and increases ABA biosynthesis through *XERICO*, as described above. **(C)** In Positive Feedback 3, ABI4 up-regulates *GA2ox*, a GA deactivation gene, resulting in the same outcome as Positive Feedback 2. **(D)** ABI4 down regulates *CYP707A*, an ABA deactivation gene. Therefore, ABA starts to accumulate in seeds, which further enhances the same pathway through positive feedback. In these schemes, many other components, which may be participating in the pathways, and negative feedback loops are omitted. ABI, ABA INSENSITIVE; *CYP707A*, *CYTOCHROME P450 707A*; DELLA, D (aspartic acid) E (glutamic acid) L (leucine) L (leucine) A (alanine) protein; GA, gibberellin; GA2ox; GA 2-oxidase; GA3ox, GA 3-oxidase; GID1, GA INSENSITIVE DWARF; NCED, nine-*cis*-epoxycarotenoid dioxygenase; RGA, REPRESSOR OF GAI; RGL2, RGA-LIKE 2; XERICO, "XERICO" (Greek for drought tolerant). The schemes are based on Ko et al. (2006), Zentella et al. (2007), Ariizumi et al. (2008), Piskurewicz et al. (2008), Bossi et al. (2009), Lee et al. (2012b), Cantoro et al. (2013), Kong et al. (2013), Lim et al. (2013), and Shu et al. (2013).

expression in ME, which is in close contact with the radicle tip, could be explained. Since generation of embryo growth potential, which causes the pressure onto ME, is dependent on GA (Ni and Bradford, 1993; Yamaguchi et al., 2001), the concept of pressure-triggered stimulation of ME gene expression is well integrated with the traditional concept (and evidence) of GA- and embryo dependency of ME gene expression. While the possibility of direct stimulation of ME by GA or insoluble secondary messengers should not be excluded, the recent data sets provided the new concept for embryo-endosperm interaction and opened the next phase of seed germination research.

PERSPECTIVES FOR BASIC RESEARCH AND KNOWLEDGE TRANSLATION

MORE DISCOVERIES EXPECTED THROUGH EPIGENETIC STUDY

A number of discoveries were made in the recent studies of seed dormancy and germination. More significant discoveries will probably be made from epigenetic studies of seed dormancy and germination over the next few years. While bioinformatics and systems biology could generate new hypotheses, the exciting discoveries happening from characterization of seed dormancy mutants look very convincing and promising. Exploring these emerging mechanisms with forward genetics and biochemical and molecular approaches will result in more progress in seed dormancy research. The information obtained from individual mutants of chromatin remodeling was assembled into several schemes in this article to provide an overview of the frontier of this field. However, information to connect each component precisely in the schemes is still missing. For example, while histone methylation and subsequent silencing of *DOG1* by DNA methylation seems likely, contribution of DCL3, AGO4, and RdDM to the *DOG1*-dependent dormancy pathway is not clear (**Figure 4**). It is possible that siRNAs and long non-coding RNAs (lncRNAs), including antisense transcripts (Yamada et al., 2003; Liu et al., 2010; Sun et al., 2013), are involved in repression of key dormancy genes. Recent studies suggest that the Polycomb Repressive Complex (PRC), which is involved in histone methylation and gene silencing, also targets *DOG1* (Bouyer et al., 2011; Muller et al., 2012; Molitor et al., 2014). This is very interesting because

PRC is known to mediate gene silencing triggered by expression of long non-coding RNA, at least in the case of the flowering gene *FLOWERING LOCUS C* (Swiezewski et al., 2009; De Lucia and Dean, 2011; Heo and Sung, 2011). It is possible that some dormancy genes are regulated through the lncRNA-PRC pathway (**Figure 8**), which could maintain dormancy genes "dormant." Missing information in the current schemes of regulatory mechanisms of seed dormancy and germination genes might already be emerging from other epigenetic studies. In addition, the current schemes, which seem to be separate pathways, could be combined with each other and integrated into a single comprehensive scheme, through more discoveries. The crosstalk between the histone deacetylation and DNA methylation pathways is known (To et al., 2011; Kim et al., 2012), however little is known about their interaction directly linked to the seed dormancy mechanisms. This might be one of the areas in which the major discoveries could be made in the future.

KNOWLEDGE TRANSLATION OF SEED HORMONE BIOLOGY

The topic of hormonal regulation of seed dormancy, such as the regulation of ABA or GA biosynthesis and deactivation enzymes by the environmental signals (e.g., light and temperature), was minimized in the discussion above, because it is well summarized elsewhere (Finkelstein et al., 2008; Seo et al., 2009) and this article focuses on emerging mechanisms and new hypotheses. Nonetheless, this is probably the area of seed biology that has been most advanced in the last decade, and from a knowledge translation point of view, this area has the greatest potential for agricultural application. For example, identification of the rate-limiting ABA biosynthesis gene *NCED* advanced our understanding of thermoinhibition of lettuce seed germination, which is a critical issue in agriculture. Now, we understand that thermoinhibition of germination at high temperature, which could induce secondary dormancy, is caused by *NCED* expression (Argyris et al., 2008, 2011). Likewise, screening of wheat populations for mutations in ABA 8′-hydroxyase, an ABA deactivation enzyme, has successfully identified the genetic lines, which are potentially resistant to PHS, another serious issue in agriculture (Chono et al., 2013). A separate screen for a mutation in the *ENHANCED RESPONSE to ABA* (*ERA*) gene also isolated PHS-resistant wheat lines (Schramm et al., 2013). The information about *MOTHER OF FT AND TFL1* (*MFT*) gene, which is a recently identified member of the ABA and GA signaling pathways in Arabidopsis (Xi et al., 2010), has already been translated into wheat (Nakamura et al., 2011; Lei et al., 2013; Liu et al., 2013).

More progressive efforts are being made to translate seed hormone biology. It has been demonstrated that direct manipulation of the rate-limiting enzymes in the hormone metabolism pathways can successfully be used to alter seed performance. Silencing *NCED* with RNA interference can promote germination in lettuce seeds (Huo et al., 2013). In contrast, chemical induction of *NCED*, a single gene, was sufficient to suppress precocious germination in Arabidopsis, which can also be applied to PHS prevention in cereal crops (Martínez-Andújar et al., 2011). While the latter approach was tested in the model system Arabidopsis, the gene induction experiments in this study were performed with the chemical ligand that has been approved for field application by the U.S. Environmental Protection Agency, making the principle applicable to agriculture. Even more advanced system of *NECD* enhancement, which does not require chemical application, has been established recently, using a positive feedback mechanism. In this system, a chimeric *NCED* gene, which is designed to trigger positive feedback regulation, amplifies ABA biosynthesis and signaling in seeds and causes hyperdormancy in a spontaneous manner (Nonogaki et al., 2014). This positive feedback system was created based on the mechanisms emerged from, and the comprehensive understanding established by, the past research on the ABA metabolism and signaling pathway in seeds. The translational research unexpectedly revealed that a positive feedback mechanism is also present in the native system of *NCED* expression in seeds (Nonogaki et al., 2014), demonstrating the synergy between basic and translational research. Other positive feedback mechanisms in the hormonal regulation of seed dormancy and germination are also emerging from on-going discoveries (summarized in **Figure 9**). More findings and understanding of elegant pathways in nature will provide greater opportunities of knowledge translation, another frontier of research that should be expanded in the future.

ACKNOWLEDGMENTS

I am grateful to Roger Beachy, World Food Center, University of California, Davis, USA, for collaboration and continuous support in the translational biology projects described in this article, and Khadidiatou Sall and Mariko Nonogaki for critical reading of and helpful suggestions for the manuscript.

REFERENCES

Alonso-Blanco, C., Bentsink, L., Hanhart, C. J., Vries, H. B.-D., and Koornneef, M. (2003). Analysis of natural allelic variation at seed dormancy loci of *Arabidopsis thaliana*. *Genetics* 164, 711–729.

Arc, E., Sechet, J., Corbineau, F., Rajjou, L., and Marion-Poll, A. (2013). ABA crosstalk with ethylene and nitric oxide in seed dormancy and germination. *Front. Plant Sci.* 4:63. doi: 10.3389/fpls.2013.00063

Argyris, J., Dahal, P., Hayashi, E., Still, D. W., and Bradford, K. J. (2008). Genetic variation for lettuce seed thermoinhibition is associated with temperature-sensitive expression of abscisic acid, gibberellin, and ethylene biosynthesis, metabolism, and response genes. *Plant Physiol.* 148, 926–947. doi: 10.1104/pp.108.125807

Argyris, J., Truco, M., Ochoa, O., McHale, L., Dahal, P., Van Deynze, A., et al. (2011). A gene encoding an abscisic acid biosynthetic enzyme (*LsNCED4*) collocates with the high temperature germination locus *Htg6.1* in lettuce (*Lactuca sp.*). *Theor. Appl. Genet.* 122, 95–108. doi: 10.1007/s00122-010-1425-3

Ariizumi, T., Murase, K., Sun, T.-P., and Steber, C. M. (2008). Proteolysis-independent downregulation of DELLA repression in Arabidopsis by the gibberellin receptor GIBBERELLIN INSENSITIVE DWARF1. *Plant Cell* 20, 2447–2459. doi: 10.1105/tpc.108.058487

Atwell, S., Huang, Y. S., Vilhjalmsson, B. J., Willems, G., Horton, M., Li, Y., et al. (2010). Genome-wide association study of 107 phenotypes in *Arabidopsis thaliana* inbred lines. *Nature* 465, 627–631. doi: 10.1038/nature08800

Barrero, J. M., Millar, A. A., Griffiths, J., Czechowski, T., Scheible, W. R., Udvardi, M., et al. (2010). Gene expression profiling identifies two regulatory genes controlling dormancy and ABA sensitivity in Arabidopsis seeds. *Plant J.* 61, 611–622. doi: 10.1111/j.1365-313X.2009.04088.x

Bentsink, L., Hanson, J., Hanhart, C. J., Blankestijn-de Vries, H., Coltrane, C., Keizer, P., et al. (2010). Natural variation for seed dormancy in *Arabidopsis* is regulated by additive genetic and molecular pathways. *Proc. Natl. Acad. Sci. U.S.A.* 107, 4264–4269. doi: 10.1073/pnas.1000410107

Bentsink, L., Jowett, J., Hanhart, C. J., and Koornneef, M. (2006). Cloning of *DOG1*, a quantitative trait locus controlling seed dormancy in *Arabidopsis*. *Proc. Natl. Acad. Sci. U.S.A.* 103, 17042–17047. doi: 10.1073/pnas.0607877103

Bewley, J. D., Bradford, K. J., Hilhorst, H. W. M., and Nonogaki, H. (2013). *Seeds: Physiology of Development, Germination and Dormancy*. New York, NY: Springer. doi: 10.1007/978-1-4614-4693-4

Bossi, F., Cordoba, E., Dupre, P., Mendoza, M. S., Roman, C. S., and Leon, P. (2009). The Arabidopsis ABA-INSENSITIVE (ABI) 4 factor acts as a central transcription activator of the expression of its own gene, and for the induction of ABI5 and SBE2.2 genes during sugar signaling. *Plant J.* 59, 359–374. doi: 10.1111/j.1365-313X.2009.03877.x

Bouyer, D., Roudier, F., Heese, M., Andersen, E. D., Gey, D., Nowack, M. K., et al. (2011). Polycomb repressive complex 2 controls the embryo-to-seedling phase transition. *PLoS Genet.* 7:e1002014. doi: 10.1371/journal.pgen.1002014

Braam, J. (2005). In touch: plant responses to mechanical stimuli. *New Phytol.* 165, 373–389. doi: 10.1111/j.1469-8137.2004.01263.x

Cantoro, R., Crocco, C. D., Benech-Arnold, R. L., and Rodríguez, M. V. (2013). *In vitro* binding of *Sorghum bicolor* transcription factors ABI4 and ABI5 to a conserved region of a *GA 2-OXIDASE* promoter: possible role of this interaction in the expression of seed dormancy. *J. Exp. Bot.* 64, 5721–5735. doi: 10.1093/jxb/ert347

Chen, F., and Bradford, K. J. (2000). Expression of an expansin is associated with endosperm weakening during tomato seed germination. *Plant Physiol.* 124, 1265–1274. doi: 10.1104/pp.124.3.1265

Chen, F., Nonogaki, H., and Bradford, K. J. (2002). A gibberellin-regulated xyloglucan endotransglycosylase gene is expressed in the endosperm cap during tomato seed germination. *J. Exp. Bot.* 53, 215–223. doi: 10.1093/jexbot/53.367.215

Chen, L.-T., Luo, M., Wang, Y.-Y., and Wu, K. (2010). Involvement of *Arabidopsis* histone deacetylase HDA6 in ABA and salt stress response. *J. Exp. Bot.* 61, 3345–3353. doi: 10.1093/jxb/erq154

Chen, L.-T., and Wu, K. (2010). Role of histone deacetylases HDA6 and HDA19 in ABA and abiotic stress response. *Plant Signal. Behav.* 5, 1318–1320. doi: 10.4161/psb.5.10.13168

Chiang, G. C., Barua, D., Dittmar, E., Kramer, E. M., de Casas, R. R., and Donohue, K. (2013). Pleiotropy in the wild: the dormancy gene *DOG1* exerts cascading control on life cycles. *Evolution* 67, 883–893. doi: 10.1111/j.1558-5646.2012.01828.x

Chiwocha, S. D., Cutler, A. J., Abrams, S. R., Ambrose, S. J., Yang, J., Ross, A. R., et al. (2005). The *etr1-2* mutation in *Arabidopsis thaliana* affects the abscisic acid, auxin, cytokinin and gibberellin metabolic pathways during maintenance of seed dormancy, moist-chilling and germination. *Plant J.* 42, 35–48. doi: 10.1111/j.1365-313X.2005.02359.x

Chono, M., Matsunaka, H., Seki, M., Fujita, M., Kiribuchi-Otobe, C., Oda, S., et al. (2013). Isolation of a wheat (*Triticum aestivum* L.) mutant in ABA 8′-hydroxylase gene: effect of reduced ABA catabolism on germination inhibition under field condition. *Breed. Sci.* 63, 104–115. doi: 10.1270/jsbbs.63.104

Dekkers, B. J., Pearce, S., van Bolderen-Veldkamp, R. P., Marshall, A., Widera, P., Gilbert, J., et al. (2013). Transcriptional dynamics of two seed compartments with opposing roles in Arabidopsis seed germination. *Plant Physiol.* 163, 205–215. doi: 10.1104/pp.113.223511

De Lucia, F., and Dean, C. (2011). Long non-coding RNAs and chromatin regulation. *Curr. Opin. Plant Biol.* 14, 168–173. doi: 10.1016/j.pbi.2010.11.006

Finkelstein, R., Reeves, W., Ariizumi, T., and Steber, C. (2008). Molecular aspects of seed dormancy. *Annu. Rev. Plant Biol.* 59, 387–415. doi: 10.1146/annurev.arplant.59.032607.092740

Footitt, S., Clay, H. A., Dent, K., and Finch-Savage, W. E. (2014). Environment sensing in spring-dispersed seeds of a winter annual Arabidopsis influences the regulation of dormancy to align germination potential with seasonal changes. *New Phytol.* 202, 929–939. doi: 10.1111/nph.12694

Footitt, S., Huang, Z., Clay, H. A., Mead, A., and Finch-Savage, W. E. (2013). Temperature, light and nitrate sensing coordinate Arabidopsis seed dormancy cycling, resulting in winter and summer annual phenotypes. *Plant J.* 74, 1003–1015. doi: 10.1111/tpj.12186

Graeber, K. A. I., Nakabayashi, K., Miatton, E., Leubner-Metzger, G., and Soppe, W. J. J. (2012). Molecular mechanisms of seed dormancy. *Plant Cell Environ.* 35, 1769–1786. doi: 10.1111/j.1365-3040.2012.02542.x

Grasser, M., Kane, C. M., Merkle, T., Melzer, M., Emmersen, J., and Grasser, K. D. (2009). Transcript elongation factor TFIIS is involved in *Arabidopsis* seed dormancy. *J. Mol. Biol.* 386, 598–611. doi: 10.1016/j.jmb.2008.12.066

Groot, S. P. C., and Karssen, C. M. (1987). Gibberellins regulate seed germination in tomato by endosperm weakening: a study with gibberellin-deficient mutants. *Planta* 171, 525–531. doi: 10.1007/BF00392302

Grzenda, A., Lomberk, G., Zhang, J.-S., and Urrutia, R. (2009). Sin3: master scaffold and transcriptional corepressor. *Biochim. Biophys. Acta.* 1789, 443–450. doi: 10.1016/j.bbagrm.2009.05.007

He, Y., Doyle, M. R., and Amasino, R. M. (2004). PAF1-complex-mediated histone methylation of *FLOWERING LOCUS C* chromatin is required for the vernalization-responsive, winter-annual habit in *Arabidopsis*. *Genes Dev.* 18, 2774–2784. doi: 10.1101/gad.1244504

Henry, K. W., Wyce, A., Lo, W. S., Duggan, L. J., Emre, N. C., Kao, C. F., et al. (2003). Transcriptional activation via sequential histone H2B ubiquitylation and deubiquitylation, mediated by SAGA-associated Ubp8. *Genes Dev.* 17, 2648–2663. doi: 10.1101/gad.1144003

Heo, J. B., and Sung, S. (2011). Vernalization-mediated epigenetic silencing by a long intronic noncoding RNA. *Science* 331, 76–79. doi: 10.1126/science.1197349

Herr, A. J., Jensen, M. B., Dalmay, T., and Baulcombe, D. C. (2005). RNA polymerase IV directs silencing of endogenous DNA. *Science* 308, 118–120. doi: 10.1126/science.1106910

Huo, H., Dahal, P., Kunusoth, K., McCallum, C. M., and Bradford, K. J. (2013). Expression of *9-cis-EPOXYCAROTENOID DIOXYGENASE4* is essential for thermoinhibition of lettuce seed germination but not for seed development or stress tolerance. *Plant Cell* 25, 884–900. doi: 10.1105/tpc.112.108902

Jackson, J. P., Lindroth, A. M., Cao, X., and Jacobsen, S. E. (2002). Control of CpNpG DNA methylation by the KRYPTONITE histone H3 methyltransferase. *Nature* 416, 556–560. doi: 10.1038/nature731

Johnson, L. M., Bostick, M., Zhang, X., Kraft, E., Henderson, I., Callis, J., et al. (2007). The SRA methyl-cytosine-binding domain links DNA and histone methylation. *Curr. Biol.* 17, 379–384. doi: 10.1016/j.cub.2007.01.009

Kadosh, D., and Struhl, K. (1998). Histone deacetylase activity of Rpd3 is important for transcriptional repression *in vivo*. *Genes Dev.* 12, 797–805. doi: 10.1101/gad.12.6.797

Kim, J., Guermah, M., McGinty, R. K., Lee, J.-S., Tang, Z., Milne, T. A., et al. (2009). RAD6-mediated transcription-coupled H2B ubiquitylation directly stimulates H3K4 methylation in human cells. *Cell* 137, 459–471. doi: 10.1016/j.cell.2009.02.027

Kim, J., Guermah, M., and Roeder, R. G. (2010). The human PAF1 complex acts in chromatin transcription elongation both independently and cooperatively with SII/TFIIS. *Cell* 140, 491–503. doi: 10.1016/j.cell.2009.12.050

Kim, J.-M., To, T. K., and Seki, M. (2012). An epigenetic integrator: new insights into genome regulation, environmental stress responses and developmental controls by HISTONE DEACETYLASE 6. *Plant Cell Physiol.* 53, 794–800. doi: 10.1093/pcp/pcs004

Ko, J.-H., Yang, S. H., and Han, K.-H. (2006). Upregulation of an Arabidopsis RING-H2 gene, *XERICO*, confers drought tolerance through increased abscisic acid biosynthesis. *Plant J.* 47, 343–355. doi: 10.1111/j.1365-313X.2006.02782.x

Kong, Y., Chen, S., Yang, Y., and An, C. (2013). ABA-insensitive (ABI) 4 and ABI5 synergistically regulate *DGAT1* expression in *Arabidopsis* seedlings under stress. *FEBS Lett.* 587, 3076–3082. doi: 10.1016/j.febslet.2013.07.045

Kronholm, I., Pico, F. X., Alonso-Blanco, C., Goudet, J., and de Meaux, J. (2012). Genetic basis of adaptation in *Arabidopsis thaliana*: local adaptation at the seed dormancy QTL *DOG1*. *Evolution* 2287–2302. doi: 10.1111/j.1558-5646.2012.01590.x

Lai, A., Kennedy, B. K., Barbie, D. A., Bertos, N. R., Yang, X. J., Theberge, M.-C., et al. (2001). RBP1 recruits the mSIN3-histone deacetylase complex to the pocket of retinoblastoma tumor suppressor family proteins found in limited discrete regions of the nucleus at growth arrest. *Mol. Cell. Biol.* 21, 2918–2932. doi: 10.1128/MCB.21.8.2918-2932.2001

Lam, E., Benfey, P. N., Gilmartin, P. M., Fang, R. X., and Chua, N. H. (1989). Site-specific mutations alter *in vitro* factor binding and change promoter expression pattern in transgenic plants. *Proc. Natl. Acad. Sci. U.S.A.* 86, 7890–7894. doi: 10.1073/pnas.86.20.7890

Law, J. A., Vashisht, A. A., Wohlschlegel, J. A., and Jacobsen, S. E. (2011). SHH1, a homeodomain protein required for DNA methylation, as well as RDR2, RDM4, and chromatin remodeling factors, associate with RNA polymerase IV. *PLoS Genet.* 7:e1002195. doi: 10.1371/journal.pgen.1002195

Lee, D., Polisensky, D. H., and Braam, J. (2005). Genome-wide identification of touch- and darkness-regulated Arabidopsis genes: a focus on calmodulin-like and *XTH* genes. *New Phytol.* 165, 429–444. doi: 10.1111/j.1469-8137.2004.01238.x

Lee, K. J., Dekkers, B. J., Steinbrecher, T., Walsh, C. T., Bacic, A., Bentsink, L., et al. (2012a). Distinct cell wall architectures in seed endosperms in representatives of the Brassicaceae and Solanaceae. *Plant Physiol.* 160, 1551–1566. doi: 10.1104/pp.112.203661

Lee, K. P., Piskurewicz, U., Tureckova, V., Carat, S., Chappuis, R., Strnad, M., et al. (2012b). Spatially and genetically distinct control of seed germination by phytochromes A and B. *Genes Dev.* 26, 1984–1996. doi: 10.1101/gad.194266.112

Lei, L., Zhu, X., Wang, S., Zhu, M., Carver, B. F., and Yan, L. (2013). *TaMFT-A1* is associated with seed germination sensitive to temperature in winter wheat. *PLoS ONE* 8:e73330. doi: 10.1371/journal.pone.0073330

Leon-Kloosterziel, K. M., van de Bunt, G. A., Zeevaart, J. A., and Koornneef, M. (1996). Arabidopsis mutants with a reduced seed dormancy. *Plant Physiol.* 110, 233–240. doi: 10.1104/pp.110.1.233

Lim, S., Park, J., Lee, N., Jeong, J., Toh, S., Watanabe, A., et al. (2013). ABA-INSENSITIVE3, ABA-INSENSITIVE5, and DELLAs interact to activate the expression of *SOMNUS* and other high-temperature-inducible genes in imbibed seeds in *Arabidopsis. Plant Cell* 25, 4863–4878. doi: 10.1105/tpc.113.118604

Linkies, A., Graeber, K., Knight, C., and Leubner-Metzger, G. (2010). The evolution of seeds. *New Phytol.* 186, 817–831. doi: 10.1111/j.1469-8137.2010.03249.x

Liu, F., Marquardt, S., Lister, C., Swiezewski, S., and Dean, C. (2010). Targeted 3′ processing of antisense transcripts triggers *Arabidopsis FLC* chromatin silencing. *Science* 327, 94–97. doi: 10.1126/science.1180278

Liu, S., Sehgal, S. K., Li, J., Lin, M., Trick, H. N., Yu, J., et al. (2013). Cloning and characterization of a critical regulator for pre-harvest sprouting in wheat. *Genetics* 195, 263–273. doi: 10.1534/genetics.113.152330

Liu, Y., Geyer, R., van Zanten, M., Carles, A., Li, Y., Hörold, A., et al. (2011). Identification of the *Arabidopsis REDUCED DORMANCY 2* gene uncovers a role for the polymerase associated factor 1 complex in seed dormancy. *PLoS ONE* 6:e22241. doi: 10.1371/journal.pone.0022241

Liu, Y., Koornneef, M., and Soppe, W. J. J. (2007). The absence of histone H2B monoubiquitination in the *Arabidopsis hub1* (*rdo4*) mutant reveals a role for chromatin remodeling in seed dormancy. *Plant Cell* 19, 433–444. doi: 10.1105/tpc.106.049221

Mallory, A., and Vaucheret, H. (2010). Form, function, and regulation of ARGONAUTE proteins. *Plant Cell* 22, 3879–3889. doi: 10.1105/tpc.110.080671

Martinez-Andujar, C., Ordiz, M. I., Huang, Z., Nonogaki, M., Beachy, R. N., and Nonogaki, H. (2011). Induction of 9-cis-epoxycarotenoid dioxygenase in *Arabidopsis thaliana* seeds enhances seed dormancy. *Proc. Natl. Acad. Sci. U.S.A.* 108, 17225–17229. doi: 10.1073/pnas.1112151108

Martinez-Andujar, C., Pluskota, W. E., Bassel, G. W., Asahina, M., Pupel, P., Nguyen, T. T., et al. (2012). Mechanisms of hormonal regulation of endosperm cap-specific gene expression in tomato seeds. *Plant J.* 71, 575–586. doi: 10.1111/j.1365-313X.2012.05010.x

Miatton, E. (2012). *Characterization of PDF1 and Its Interaction With DELAY OF GERMINATION1 (DOG1) in the Control of Seed Dormancy in Arabidopsis thaliana.* Köln: University of Köln.

Mikami, K., Takase, H., Tabata, T., and Iwabuchi, M. (1989). Multiplicity of the DNA-binding protein HBP-1 specific to the conserved hexameric sequence ACGTCA in various plant gene promoters. *FEBS Lett.* 256, 67–70. doi: 10.1016/0014-5793(89)81719-3

Molitor, A. M., Bu, Z., Yu, Y., and Shen, W. H. (2014). *Arabidopsis* AL PHD-PRC1 complexes promote seed germination through H3K4me3-to-H3K27me3 chromatin state switch in repression of seed developmental genes. *PLoS Genet.* 10:e1004091. doi: 10.1371/journal.pgen.1004091

Mortensen, S. A., and Grasser, K. D. (2014). The seed dormancy defect of *Arabidopsis* mutants lacking the transcript elongation factor TFIIS is caused by reduced expression of the *DOG1* gene. *FEBS Lett.* 588, 47–51. doi: 10.1016/j.febslet.2013.10.047

Muller, K., Bouyer, D., Schnittger, A., and Kermode, A. R. (2012). Evolutionarily conserved histone methylation dynamics during seed life-cycle transitions. *PLoS ONE* 7:e51532. doi: 10.1371/journal.pone.0051532

Nakabayashi, K., Bartsch, M., Xiang, Y., Miatton, E., Pellengahr, S., Yano, R., et al. (2012). The time required for dormancy release in *Arabidopsis* is determined by DELAY OF GERMINATION1 protein levels in freshly harvested seeds. *Plant Cell* 24, 2826–2838. doi: 10.1105/tpc.112.100214

Nakamura, S., Abe, F., Kawahigashi, H., Nakazono, K., Tagiri, A., Matsumoto, T., et al. (2011). A wheat homolog of MOTHER OF FT AND TFL1 acts in the regulation of germination. *Plant Cell* 23, 3215–3229. doi: 10.1105/tpc.111.088492

Nakanishi, S., Lee, J. S., Gardner, K. E., Gardner, J. M., Takahashi, Y. H., Chandrasekharan, M. B., et al. (2009). Histone H2BK123 monoubiquitination is the critical determinant for H3K4 and H3K79 trimethylation by COMPASS and Dot1. *J. Cell Biol.* 186, 371–377. doi: 10.1083/jcb.200906005

Nakayama, T., Ohtsubo, N., Mikami, K., Kawata, T., Tabata, T., Kanazawa, H., et al. (1989). Cisacting sequences that modulate transcription of wheat histone H3 and 3′processing of H3 premature mRNA. *Plant Cell Physiol.* 30, 825–832.

Ni, B. R., and Bradford, K. J. (1993). Germination and dormancy of abscisic acid- and gibberellin-deficient mutant tomato (*Lycopersicon esculentum*) seeds (sensitivity of germination to abscisic acid, gibberellin, and water potential). *Plant Physiol.* 101, 607–617.

Nonogaki, H., Gee, O. H., and Bradford, K. J. (2000). A germination-specific endo-beta -mannanase gene is expressed in the micropylar endosperm cap of tomato seeds. *Plant Physiol.* 123, 1235–1246. doi: 10.1104/pp.123.4.1235

Nonogaki, H., Nomaguchi, M., Okumoto, N., Kaneko, Y., Matsushima, H., and Morohashi, M. (1998). Temporal and spatial pattern of the biochemical activation of the endosperm during and following imbibition of tomato seeds. *Physiol. Plant.* 102, 236–242. doi: 10.1034/j.1399-3054.1998.1020211.x

Nonogaki, M., Sall, K., Nambara, E., and Nonogaki, H. (2014). Amplification of ABA biosynthesis and signaling through a positive feedback mechanism in seeds. *Plant J.* 78, 527–539. doi: 10.1111/tpj.12472

Ogawa, M., Hanada, A., Yamauchi, Y., Kuwahara, A., Kamiya, Y., and Yamaguchi, S. (2003). Gibberellin biosynthesis and response during Arabidopsis seed germination. *Plant Cell* 15, 1591–1604. doi: 10.1105/tpc.011650

Oh, S., Zhang, H., Ludwig, P., and van Nocker, S. (2004). A mechanism related to the yeast transcriptional regulator Paf1c is required for expression of the Arabidopsis *FLC/MAF* MADS box gene family. *Plant Cell* 16, 2940–2953. doi: 10.1105/tpc.104.026062

Onodera, Y., Haag, J. R., Ream, T., Costa Nunes, P., Pontes, O., and Pikaard, C. S. (2005). Plant nuclear RNA polymerase IV mediates siRNA and DNA methylation-dependent heterochromatin formation. *Cell* 120, 613–622. doi: 10.1016/j.cell.2005.02.007

Peeters, A. J. M., Blankestijn-de Vries, H., Hanhart, C. J., Léon-Kloosterziel, K. M., Zeevaart, J. A. D., and Koornneef, M. (2002). Characterization of mutants with reduced seed dormancy at two novel *rdo* loci and a further characterization of *rdo1* and *rdo2* in Arabidopsis. *Physiol. Plant.* 115, 604–612. doi: 10.1034/j.1399-3054.2002.1150415.x

Penheiter, K. L., Washburn, T. M., Porter, S. E., Hoffman, M. G., and Jaehning, J. A. (2005). A posttranscriptional role for the yeast Paf1-RNA polymerase II complex is revealed by identification of primary targets. *Mol. Cell* 20, 213–223. doi: 10.1016/j.molcel.2005.08.023

Perrella, G., Lopez-Vernaza, M. A., Carr, C., Sani, E., Gossele, V., Verduyn, C., et al. (2013). Histone deacetylase complex1 expression level titrates plant growth and abscisic acid sensitivity in Arabidopsis. *Plant Cell* 25, 3491–3505. doi: 10.1105/tpc.113.114835

Piskurewicz, U., Jikumaru, Y., Kinoshita, N., Nambara, E., Kamiya, Y., and Lopez-Molina, L. (2008). The gibberellic acid signaling repressor RGL2 inhibits *Arabidopsis* seed germination by stimulating abscisic acid synthesis and ABI5 activity. *Plant Cell* 20, 2729–2745. doi: 10.1105/tpc.108.061515

Porter, S. E., Penheiter, K. L., and Jaehning, J. A. (2005). Separation of the *Saccharomyces cerevisiae* Paf1 complex from RNA polymerase II results in changes in its subnuclear localization. *Eukaryotic Cell* 4, 209–220. doi: 10.1128/EC.4.1.209-220.2005

Qi, Y., He, X., Wang, X.-J., Kohany, O., Jurka, J., and Hannon, G. J. (2006). Distinct catalytic and non-catalytic roles of ARGONAUTE4 in RNA-directed DNA methylation. *Nature* 443, 1008–1012. doi: 10.1038/nature05198

Rea, S., Eisenhaber, F., O'Carroll, D., Strahl, B. D., Sun, Z.-W., Schmid, M., et al. (2000). Regulation of chromatin structure by site-specific histone H3 methyltransferases. *Nature* 406, 593–599. doi: 10.1038/35020506

Richon, V. M., and O'Brien, J. P. (2002). Histone deacetylase inhibitors: a new class of potential therapeutic agents for cancer treatment. *Clin. Cancer Res.* 8, 718–728, 662–664.

Saunders, A., Core, L. J., and Lis, J. T. (2006). Breaking barriers to transcription elongation. *Nat. Rev. Mol. Cell Biol.* 7, 557–567. doi: 10.1038/nrm1981

Schramm, E., Nelson, S., Kidwell, K., and Steber, C. (2013). Increased ABA sensitivity results in higher seed dormancy in soft white spring wheat cultivar 'Zak.' *Theor. Appl. Genet.* 126, 791–803. doi: 10.1007/s00122-012-2018-0

Seo, M., Nambara, E., Choi, G., and Yamaguchi, S. (2009). Interaction of light and hormone signals in germinating seeds. *Plant Mol. Biol.* 69, 463–472. doi: 10.1007/s11103-008-9429-y

Shu, K., Zhang, H., Wang, S., Chen, M., Wu, Y., Tang, S., et al. (2013). ABI4 regulates primary seed dormancy by regulating the biogenesis of abscisic acid and gibberellins in Arabidopsis. *PLoS Genet.* 9:e1003577. doi: 10.1371/journal.pgen.1003577

Simon, J. A., and Kingston, R. E. (2009). Mechanisms of Polycomb gene silencing: knowns and unknowns. *Nat. Rev. Mol. Cell Biol.* 10, 697–708. doi: 10.1038/nrm2763

Singh, M., and Singh, J. (2012). Seed development-related expression of ARGONAUTE4_9 class of genes in barley: possible role in seed dormancy. *Euphytica* 188, 123–129. doi: 10.1007/s10681-012-0624-1

Singh, M., Singh, S., Randhawa, H., and Singh, J. (2013). Polymorphic homoeolog of key gene of RdDM pathway, *ARGONAUTE4_9* class is associated with pre-harvest sprouting in wheat (*Triticum aestivum* L.). *PLoS ONE* 8:e77009. doi: 10.1371/journal.pone.0077009

Sugimoto, K., Takeuchi, Y., Ebana, K., Miyao, A., Hirochika, H., Hara, N., et al. (2010). Molecular cloning of *Sdr4*, a regulator involved in seed dormancy and domestication of rice. *Proc. Natl. Acad. Sci. U.S.A.* 107, 5792–5797. doi: 10.1073/pnas.0911965107

Sun, Q., Csorba, T., Skourti-Stathaki, K., Proudfoot, N. J., and Dean, C. (2013). R-loop stabilization represses antisense transcription at the *Arabidopsis FLC* locus. *Science* 340, 619–621. doi: 10.1126/science.1234848

Sun, Z.-W., and Allis, C. D. (2002). Ubiquitination of histone H2B regulates H3 methylation and gene silencing in yeast. *Nature* 418, 104–108. doi: 10.1038/nature00883

Swiezewski, S., Liu, F., Magusin, A., and Dean, C. (2009). Cold-induced silencing by long antisense transcripts of an *Arabidopsis* Polycomb target. *Nature* 462, 799–802. doi: 10.1038/nature08618

Tabata, T., Nakayama, T., Mikami, K., and Iwabuchi, M. (1991). HBP-1a and HBP-1b: leucine zipper-type transcription factors of wheat. *EMBO J.* 10, 1459–1467.

To, T. K., Kim, J.-M., Matsui, A., Kurihara, Y., Morosawa, T., Ishida, J., et al. (2011). *Arabidopsis* HDA6 regulates locus-directed heterochromatin silencing in cooperation with MET1. *PLoS Genet.* 7:e1002055. doi: 10.1371/journal.pgen.1002055

Tran, R. K., Zilberman, D., de Bustos, C., Ditt, R. F., Henikoff, J. G., Lindroth, A. M., et al. (2005). Chromatin and siRNA pathways cooperate to maintain DNA methylation of small transposable elements in *Arabidopsis*. *Genome Biol.* 6, R90. doi: 10.1186/gb-2005-6-11-r90

Voegele, A., Linkies, A., Muller, K., and Leubner-Metzger, G. (2011). Members of the gibberellin receptor gene family *GID1* (*GIBBERELLIN INSENSITIVE DWARF1*) play distinct roles during *Lepidium sativum* and *Arabidopsis thaliana* seed germination. *J. Exp. Bot.* 62, 5131–5147. doi: 10.1093/jxb/err214

Wang, Z., Cao, H., Sun, Y., Li, X., Chen, F., Carles, A., et al. (2013). *Arabidopsis* paired amphipathic helix proteins SNL1 and SNL2 redundantly regulate primary seed dormancy via abscisic acid-ethylene antagonism mediated by histone deacetylation. *Plant Cell* 25, 149–166. doi: 10.1105/tpc.112.108191

Wierzbicki, A. T. (2012). The role of long non-coding RNA in transcriptional gene silencing. *Curr. Opin. Plant Biol.* 15, 517–522. doi: 10.1016/j.pbi.2012.08.008

Wierzbicki, A. T., Haag, J. R., and Pikaard, C. S. (2008). Noncoding transcription by RNA polymerase Pol IVb/Pol V mediates transcriptional silencing of overlapping and adjacent genes. *Cell* 135, 635–648. doi: 10.1016/j.cell.2008.09.035

Wierzbicki, A. T., Ream, T. S., Haag, J. R., and Pikaard, C. S. (2009). RNA polymerase V transcription guides ARGONAUTE4 to chromatin. *Nat. Genet.* 41, 630–634. doi: 10.1038/ng.365

Xi, W., Liu, C., Hou, X., and Yu, H. (2010). *MOTHER OF FT AND TFL1* regulates seed germination through a negative feedback loop modulating ABA signaling in *Arabidopsis*. *Plant Cell* 22, 1733–1748. doi: 10.1105/tpc.109.073072

Xie, Z., Johansen, L. K., Gustafson, A. M., Kasschau, K. D., Lellis, A. D., Zilberman, D., et al. (2004). Genetic and functional diversification of small RNA pathways in plants. *PLoS Biol.* 2:E104. doi: 10.1371/journal.pbio.0020104

Yamada, K., Lim, J., Dale, J. M., Chen, H., Shinn, P., Palm, C. J., et al. (2003). Empirical analysis of transcriptional activity in the *Arabidopsis* genome. *Science* 302, 842–846. doi: 10.1126/science.1088305

Yamaguchi, S., Kamiya, Y., and Sun, T. (2001). Distinct cell-specific expression patterns of early and late gibberellin biosynthetic genes during Arabidopsis seed germination. *Plant J.* 28, 443–453. doi: 10.1046/j.1365-313X.2001.01168.x

Yamauchi, Y., Ogawa, M., Kuwahara, A., Hanada, A., Kamiya, Y., and Yamaguchi, S. (2004). Activation of gibberellin biosynthesis and response pathways by low temperature during imbibition of *Arabidopsis thaliana* seeds. *Plant Cell* 16, 367–378. doi: 10.1105/tpc.018143

Yano, R., Takebayashi, Y., Nambara, E., Kamiya, Y., and Seo, M. (2013). Combining association mapping and transcriptomics identify *HD2B* histone deacetylase as a genetic factor associated with seed dormancy in *Arabidopsis thaliana*. *Plant J.* 74, 815–828. doi: 10.1111/tpj.12167

Yu, X., and Michaels, S. D. (2010). The Arabidopsis Paf1c complex component CDC73 participates in the modification of *FLOWERING LOCUS C* chromatin. *Plant Physiol.* 153, 1074–1084. doi: 10.1104/pp.110.158386

Zentella, R., Zhang, Z.-L., Park, M., Thomas, S. G., Endo, A., Murase, K., et al. (2007). Global analysis of DELLA direct targets in early gibberellin signaling in *Arabidopsis*. *Plant Cell* 19, 3037–3057. doi: 10.1105/tpc.107.054999

Zhang, H., and Van Nocker, S. (2002). The *VERNALIZATION INDEPENDENCE 4* gene encodes a novel regulator of *FLOWERING LOCUS C*. *Plant J.* 31, 663–673. doi: 10.1046/j.1365-313X.2002.01380.x

Zheng, J., Chen, F., Wang, Z., Cao, H., Li, X., Deng, X., et al. (2012). A novel role for histone methyltransferase KYP/SUVH4 in the control of *Arabidopsis* primary seed dormancy. *New Phytol.* 193, 605–616. doi: 10.1111/j.1469-8137.2011.03969.x

Zhu, B., Zheng, Y., Pham, A.-D., Mandal, S. S., Erdjument-Bromage, H., Tempst, P., et al. (2005). Monoubiquitination of human histone H2B: the factors involved and their roles in HOX gene regulation. *Mol. Cell* 20, 601–611. doi: 10.1016/j.molcel.2005.09.025

Zilberman, D., Cao, X., Johansen, L. K., Xie, Z., Carrington, J. C., and Jacobsen, S. E. (2004). Role of *Arabidopsis ARGONAUTE4* in RNA-directed DNA methylation triggered by inverted repeats. *Curr. Biol.* 14, 1214–1220. doi: 10.1016/j.cub.2004.06.055

Conflict of Interest Statement: A patent application has been filed for a technology described in this article. The author declares that the research was conducted in the absence of any commercial or financial relationships that could be construed as a potential conflict of interest.

16

Conserved *cis*-regulatory modules in promoters of genes encoding wheat high-molecular-weight glutenin subunits

Catherine Ravel[1,2]*, Samuel Fiquet[1,2], Julie Boudet[1,2], Mireille Dardevet[1,2], Jonathan Vincent[1,2], Marielle Merlino[1,2], Robin Michard[1,2] and Pierre Martre[1,2]

[1] Institut National de la Recherche Agronomique, UMR1095, Genetics, Diversity and Ecophysiology of Cereals, Clermont-Ferrand, France
[2] UMR1095, Genetics, Diversity and Ecophysiology of Cereals, Department of Biology, Blaise Pascal University, Aubière, France

Edited by:
Paolo A. Sabelli, University of Arizona, USA

Reviewed by:
Juan José Ripoll, University of California, San Diego, USA
Nigel G. Halford, Rothamsted Research, UK

***Correspondence:**
Catherine Ravel, Institut National de la Recherche Agronomique, UMR1095, Genetics, Diversity and Ecophysiology of Cereals, 5 chemin de Beaulieu, F-63 100 Clermont-Ferrand, France
e-mail: catherine.ravel@clermont.inra.fr

The concentration and composition of the gliadin and glutenin seed storage proteins (SSPs) in wheat flour are the most important determinants of its end-use value. In cereals, the synthesis of SSPs is predominantly regulated at the transcriptional level by a complex network involving at least five *cis*-elements in gene promoters. The high-molecular-weight glutenin subunits (HMW-GS) are encoded by two tightly linked genes located on the long arms of group 1 chromosomes. Here, we sequenced and annotated the HMW-GS gene promoters of 22 electrophoretic wheat alleles to identify putative *cis*-regulatory motifs. We focused on 24 motifs known to be involved in SSP gene regulation. Most of them were identified in at least one HMW-GS gene promoter sequence. A common regulatory framework was observed in all the HMW-GS gene promoters, as they shared conserved *cis*-regulatory modules (CCRMs) including all the five motifs known to regulate the transcription of SSP genes. This common regulatory framework comprises a composite box made of the GATA motifs and GCN4-like Motifs (GLMs) and was shown to be functional as the GLMs are able to bind a bZIP transcriptional factor SPA (Storage Protein Activator). In addition to this regulatory framework, each HMW-GS gene promoter had additional motifs organized differently. The promoters of most highly expressed x-type HMW-GS genes contain an additional box predicted to bind R2R3-MYB transcriptional factors. However, the differences in annotation between promoter alleles could not be related to their level of expression. In summary, we identified a common modular organization of HMW-GS gene promoters but the lack of correlation between the *cis*-motifs of each HMW-GS gene promoter and their level of expression suggests that other *cis*-elements or other mechanisms regulate HMW-GS gene expression.

Keywords: *cis*-elements, conserved *cis*-regulatory modules (CCRMs), high-molecular-weight glutenin subunits (HMW-GS), transcriptional regulation, seed storage proteins (SSPs), transcription factors (TFs), wheat (*Triticum aestivum* L)

INTRODUCTION

Wheat is one of the three most economically important crops in the world with maize and rice, with a global annual production of about 700 Mt in 2012 (FAOSTAT; http://faostat.fao.org/). Wheat is a broad term for crops including tetraploid species ($2n = 28$) like durum wheat (*Triticum turgidum* spp. *durum*) and hexaploid species ($2n = 42$) like bread wheat (*T. aestivum* spp. *aestivum*). Wheat is one of the most important sources of carbohydrates and vegetable proteins in human diets as it accounts for about 20% of all calories and proteins consumed. It is mostly transformed before it is consumed, and each type of transformation depends on the unique visco-elastic properties of gluten, a network formed by water and seed storage proteins (SSPs). It is mainly the SSPs that determine the technological quality of wheat flour (for instance, see reviews by Shewry et al., 2002 and Shewry, 2009). Prolamins, the major component of wheat SSPs, comprise monomeric gliadins and polymeric glutenins. The latters have both low- (LMW-GS) and high- (HMW-GS) molecular-weight subunits. Glutenins account for 30–50% of the total SSP content of grain, with HMW-GS alone representing up to 12% of the total. Glutenins strongly influence dough elasticity (Payne et al., 1987; Shewry et al., 2002), with HMW-GS more so than LMW-GS (Branlard and Dardevet, 1985; Gupta and MacRitchie, 1994; He et al., 2005).

As glutenins are so important for technological quality, the genes coding for HMW-GS have been extensively studied. The genome of the hexaploid bread wheat is divided into three subgenomes (called A, B, and D) forming three homoeologous groups. HMW-GS are encoded by the three loci *Glu-A1, -B1* and *-D1* located on the long arms of the group 1 chromosomes. As confirmed by the sequencing of these three regions (Gu et al., 2006), each locus consists of two closely linked paralogous genes, *Glu-1-1* and *Glu-1-2*, that encode x-type and y-type HMW-GS, respectively. Thus, bread wheat HMW-GS form a small multigene family of six genes with two orthologous sets of *Glu-1-1* and *Glu-1-2* genes (Allaby et al., 1999). HMW-GS genes are highly polymorphic (e.g., Payne and Lawrence, 1983). These six genes are not always all expressed. *Glu-A1-2* is silent so from three to five HMW-GS genes are usually expressed in grain. A duplication of *Glu-B1-1* is observed in lines with the overexpressed Bx7 HMW-GS giving an additional expressed gene (Ragupathy et al., 2008). SSPs are specifically expressed in the endosperm and all HMW-GS have similar patterns of expression and represent 60–65% of the total RNA from the endosperm between 10 and 30 days post anthesis (Shewry et al., 2009).

SSP synthesis is primarily controlled both spatially and temporally at the transcriptional level. Transcription factors (TFs) bind specifically to short conserved DNA sequences (5–15 nucleotides) called *cis*-regulatory elements or *cis*-elements, which are usually located in the proximal promoter of genes and characterized by a consensus motif. In barley (*Hordeum vulgare*), the regulatory mechanisms of SSP genes have been extensively studied by transient expression experiments using an hordein promoter (Mena et al., 1998; Vicente-Carbajosa et al., 1998; Oñate et al., 1999; Diaz et al., 2002, 2005; Isabel-La Moneda et al., 2003; Rubio-Somoza et al., 2006a,b; Moreno-Risueno et al., 2008) and have been described as a network of *cis*-elements and their interacting TFs (Rubio-Somoza et al., 2006a). This network is conserved in other cereals as reviewed by Verdier and Thompson (2008) and Xi and Zheng (2011). It consists of five *cis*-elements able to recognize eight TFs belonging to four families (bZIP of the Opaque-2 family, and the B3, DOF, and MYB proteins), which are all reported to be activators of SSP genes. More precisely, the GCN4 like-motif (GLM, 5′-ATGAG/CTCAT-3′) and the prolamin box (P-box, or PB, 5′-TGTAAAG-3′), also called the endosperm motif, constitute the bipartite endosperm box, which plays a key role in activating the expression of prolamin genes as also shown in wheat (Hammond-Kosack et al., 1993). GLM is recognized by bZIP TFs, like BLZ1 and BLZ2 in barley (Vicente-Carbajosa et al., 1998; Oñate et al., 1999) or SPA (Storage Protein Activator) in wheat (Albani et al., 1997), while the P-box is bound by PBF and SAD, both DOF-type TFs (Vicente-Carbajosa et al., 1997; Mena et al., 1998; Diaz et al., 2005). Two additional *cis*-elements, 5′-AACA/TA-3′ and 5′-TATC/GATA-3′ core sequences, are able to bind R2R3-MYB (notably GAMYB) and R1MYB (MCB1 and MYBS3) TFs, respectively (Diaz et al., 2002; Rubio-Somoza et al., 2006a,b). The last *cis*-regulatory sequence is the RY repeat (5′-CATGCATG-3′), which binds FUSCA3, a B3 protein (Bäumlein et al., 1992; Moreno-Risueno et al., 2008). In addition to these DNA-protein interactions, protein-protein interactions consolidate the formation of larger complexes that regulate SSP expression (Rubio-Somoza et al., 2006b).

Wheat promoters of α-gliadin classes (Van Herpen et al., 2008), LMW-GS (Hammond-Kosack et al., 1993; Conlan et al., 1999), and HMW-GS (Norre et al., 2002) have been functionally analyzed. Van Herpen et al. (2008) reported differences in regulatory-elements between promoter sequences of α-gliadin genes from A and B genomes. The LMW-GS promoter studied is characterized by a tandem repeat of two endosperm motifs known as the long endosperm box that is important for controlling endosperm-specific expression (Hammond-Kosack et al., 1993). Thomas and Flavell (1990) and Norre et al. (2002) analyzed extensively the promoters of *Glu-D1* by transient expression assay in tobacco and maize. A 38-bp enhancer element has been identified (Thomas and Flavell, 1990). In addition, the promoter of *Glu-D1-1* contains an atypical endosperm box where the P-box is associated with a G-like box of the ACGT family able to bind bZIP proteins (Norre et al., 2002). Moreover, these authors suggested that the enhancer element may act with the G-like box to increase reporter gene expression.

The exponential growth of genomic sequence databases, and the development of specialized databases of *cis*-acting elements in plants (Higo et al., 1999; Rombauts et al., 1999), coupled with the development of bioinformatics tools to discover specific motifs in DNA or protein sequences (e.g., MEME; Bailey et al., 2006), greatly facilitate the *in silico* analysis of promoters. However, the discovery of *cis*-regulatory elements is hindered by the variability within their sequences, which typically tolerate nucleotide substitutions without a loss of functionality. There are ways of taking this variability into account when predicting the presence of *cis*-regulatory elements (Stormo, 2000). Another aspect to consider is that, in higher eukaryotes, TFs often regulate gene expression by binding DNA in cooperation with other regulatory proteins. As reviewed by Armone and Davidson (1997), separate *cis*-elements of a given promoter often interact with different parts of an overall regulatory complex. This type of organization of *cis*-elements in a region of up to a few 100 bases in the vicinity of the gene being regulated is called a *cis*-regulatory module (CRM), where the relative positions of *cis*-elements and the distances between them are crucial.

Recently, the LMW-GS and HMW-GS gene promoters have been analyzed *in silico* (Juhász et al., 2011; Makai et al., 2013). The *cis*-acting elements present in published sequences of LMW-GS genes, mainly ESTs, were computationally retrieved and differences in the numbers and combinations of specific sequences were highlighted allowing the identification of conserved non-coding sequence regions (CRMs). Models for the transcriptional regulation of LMW-GS genes were then proposed (Juhász et al., 2011). The promoter profiles of HMW-GS genes are highly conserved in the Triticeae family despite differences between paralogous genes (Makai et al., 2013). Here the aim was to understand in more detail the transcriptional regulation of HMW-GS genes through a comparative promoter analysis. The promoters of the main alleles at each HMW-GS gene were analyzed *in silico* for the predicted presence of *cis*-regulatory elements. The organization of these elements within orthologous (homoeologous) and paralogous copies was compared. This work shows the presence of conserved CRMs (CCRMs). In addition, the HMW-GS gene promoters were sequenced in a set of wheat lines to determine whether their sequence variability correlates with the organization of *cis*-elements and hence the expression levels of these genes. A functional analysis of conserved regions consisting of *cis*-motifs potentially able to bind bZIP TFs was carried out by using transient expression and electrophoretic mobility shift assays (EMSA).

MATERIALS AND METHODS

DIVERSITY ANALYSIS

Forty-two lines representative of the genetic diversity (Haseneyer et al., 2008; Ravel et al., 2009) and of the main electrophoretic alleles of HMW-GS of the INRA worldwide hexaploid wheat (*Triticum aestivum* L.) core collection (Balfourier et al., 2007) were analyzed (**Table 1**). Genomic DNA was extracted from leaves as described in Ravel et al. (2009) and used for PCR amplification of the proximal promoter of HMW-GS genes. Fragments of approximately 700–1100 nucleotides were obtained (Supplementary Table 1) and sequenced. We did not amplify *Glu-A1-2* genomic DNA as it was silent in all the 42 lines. Diversity indices including nucleotide diversity (π), number of segregating sites (θ), number of haplotypes (H), haplotype diversity (Hd), and Tajima's *D*-test of neutral evolution were calculated for each sequence with SNiPlay (Dereeper et al., 2011).

EXPRESSION ANALYSIS

To quantify HMW-GS gene expression, RNA was extracted from developing grains harvested at 400°C days after anthesis from 13 lines representing the main promoter alleles (**Table 1**). Lines were cultivated in the greenhouse as described in Ravel et al. (2009). For each of the four lines 964, 1288, 2135, and 4874, four independent biological replicates were obtained. Two independent biological replicates were used for each of the nine remaining accessions. Quantitative-real-time PCR (qRT-PCR) was performed as described in Ravel et al. (2009) using a LightCycler® 480 II sequence detection system and the LightCycler 480 SYBR Green I Master (Roche) according to the manufacturer's instructions. Primer pairs used for qRT-PCR and their amplification efficiency are given in Supplementary Table 2. The specificity of each primer pairs was confirmed by a single peak in the real-time melting temperature curves for each gene.

Amplification plots and predicted threshold cycle values were obtained with LightCycler 480 SW 1.5 software (Roche). Genes coding for glyceraldehyde 3-phosphate dehydrogenase (GAPDH), elongation factor 1 alpha (eF1α), β-tubulin, and 18S RNA were used as internal controls to normalize expression results (Ravel et al., 2009). The geometric mean of control gene expression was calculated so that HMW-GS gene expression could be quantified and normalized also taking into account the efficiency of each primer pair.

PROMOTER ANNOTATIONS

Twenty motifs known to participate in the regulation of SSP and two light responsive motifs were selected from the PLACE *cis*-motif database, which contains 469 entries (**Table 2**; Higo et al., 1999). We included a light responsive (Abox) and a circadian rhythm-responsive (CIACADIANLELHC) motif because diurnal fluctuations in carbohydrate pools and Opaque 2 (O2) binding activity during seed filling may impact SSP synthesis (Ciceri et al., 1997, 1999; Carman and Bishop, 2004). We also added two additional motifs, 5′-AACNNA-3′ and 5′-TATAWA-3′, which were not in the PLACE database. The first motif is able to bind a MYB protein from rice (*Oriza sativa*) belonging to the GAMYB sub-family (Takaiwa et al., 1996). The second motif is the TATA-variant sequence of SSP genes involved in the formation of a transcription initiation complex (Fauteux and Strömvik, 2009; Bernard et al., 2010).

Both strands of the 1-kb region upstream of the start codon for the six HMW-GS genes from cv. Renan retrieved from public databases (DQ537335.1, DQ537336.1, and DQ537337.1 for *Glu-A1*, *Glu-B1*, and *Glu-D1*, respectively; Gu et al., 2006) and the promoter sequences of the five (i.e., all but *Glu-A1-2*) HMW-GS genes obtained in this study for 42 lines (including cv. Renan) of the INRA worldwide hexaploid wheat core collection were annotated using a custom-made PERL program (named PlantPAD) that extracts the name, sequence and coordinates of the motifs and produces a graphical representation of the query sequence on which the starting position of each *cis*-motif is plotted. Based on the assumption that functional *cis*-motifs are conserved among HMW-GS genes, we used PlantPAD to search for co-occurrence of *cis*-motifs in these genes. To build the consensus, the program considers each motif and its coordinates (the position of its first nucleotide relative to the start codon). Any motif that appears at the same coordinates (±5 bp) in all the sequences being annotated is considered to be conserved. As insertion-deletion events (indels) within a sequence cause motifs to shift along the gene, the program also recognizes conserved motifs which appear in all the sequences with the same coordinates plus or minus the shift size (the length of indels). The consensus is then plotted and the distances between conserved motifs corresponding to those found in more than 50% of the sequences are analyzed. Such a consensus is designed to highlight the conserved regulatory regions. This approach was used to analyze separately both sets of orthologous genes and produce a consensus plot for each of them. These consensuses were then used to generate an overall consensus annotation of HMW-GS gene promoters.

FUNCTIONAL VALIDATION

Particle bombardment was performed in developing wheat endosperm to validate *cis*-motifs potentially able to bind bZIP TFs. The promoter of *Glu-B1-1* gene (hereafter termed PrBx7) was amplified and cloned using the primers from cv. Renan given in Supplementary Table 1. A total of 747-bp upstream fragment of the start codon was used. In addition, to assess the role of the distal conserved regulatory regions of this promoter, the 597-bp fragment from the start codon (hereafter termed tPrBx7) was synthesized.

All constructs used for transient expression assay were obtained using Gateway technology (Invitrogen). Three entry clones were used (pDONRP4-P1R, pDONR221, and pDONRP2R-P3). pDONRP4-P1R contained the rice actin promoter, PrBx7 or tPrBx7, while pDONR221 and pDONRP2R-P3 contained a reporter gene (either GUS or GFP, respectively) and the 3′-terminator nopaline synthase gene (3′-NOS). Three expression pDESTR4-R3-based vectors (pAct-GFP, pPrBx7-GUS, and ptPrBx7-GUS) were created. A transient promoter activation assay based on co-bombardment with pPrBx7-GUS or ptPrBx7-GUS and pAct-GFP constructs was performed using immature endosperm from cv. Récital collected at 230°C day after anthesis from plant grown in the greenhouse under optimal growth conditions. Seeds were surface-sterilized and endosperms were carefully isolated. Endosperms were cultured on Murashige and Skoog medium supplemented with maltose (100 g L^{-1}) for 2–3 h for subsequent bombardment. Gold particles (0.6 μm in diameter; Bio-Rad) were prepared with 500 ng of a 1:1 molar ratio mixture of pAct-GFP and pPrBx7-GUS or ptPrBx7-GUS.

Table 1 | Country of origin, protein coding alleles, and haplotypes of the promoters of five HMW-GS genes for 42 accessions of the INRA worldwide hexaploid wheat core collection.

Name[a]	Country of origin[b]	*Glu-A1-1*		*Glu-B1-1*		*Glu-B1-2*		*Glu-D1-1*		*Glu-D1-2*	
		Protein[c]	Promoter[d]	Protein[c]	Promoter[d]	Protein[c]	Promoter[d]	Protein[c]	Promoter[d]	Protein[c]	Promoter[d]
A4 (748)	AFG	1	h1	7	h1	8	h1	3	h1	12	h1
Aifeng NO 4 (822)	CHN	1	h1	7	h1	HZ	h1	HZ	h2	HZ	h1
ARCHE (964)	FRA	null	h2	6	h2	8	h2	2	h2	12	h1
AURORE (1110)	AUS	2*	h3	7	h1	9	h1	2	h2	12	h1
BALKAN (1192)	YUG	2*	h3	7	h1	9	h1	5	h3	10	h2
BARBU DU FINISTERE (1323)	FRA	null	h2	20	h1	20	h3	2	h2	12	h1
BELLIEI 590 (1288)	HUN	2*	h3	7	h1	9	h1	5	h3	10	h2
CHINESE SPRING (2135)	CHN	null	h2	7	h1	8	h1	2	h2	12	h1
CHORTANDINKA (2153)	RUS	null	h1	HZ	h1	HZ	h1	HZ	h3	HZ	h2
CHYAMTANG (2171)	NPL	null	h2	7	h1	8	h1	2	h2	12	h1
COPPADRA (2330)	TUR	2*	h3	7	h1	8	h1	3	h2	12	h1
COTIPORA (2353)	BRA	2*	h3	N	h1	N	h3	2	h2	12	h1
COURTOT (2358)	FRA	2*	h3	7	h1	8	h1	2	h2	12	h1
DI7202-103 (2526)	FRA	1	h1	7	h1	8	h1	5	h3	10	h2
GLENLEA (3358)	CAN	2*	h3	7OE	h1	8	h1	5	h3	10	h2
GODOLLOI 15 (3366)	HUN	null	h2	N	h1	N	h4	5	h3	10	h2
JO3045 (3942)	FIN	2*	h3	7	h1	9	ND	2	h2	12	h1
M708//G25/N163 (4482)	ISR	2*	h4	HZ	h1	HZ	h1	2	h2	12	h1
MARS DE SUEDE ROUGE BARBU (4645)	FRA	null	h2	7	h1	8	h1	2	h2	12	h1
MISKAAGANI (4874)	LBN	2*	h3	N	h1	N	h1	2	h2	12	h1
MOCHO DE ESPIGA BRANCA (4901)	PRT	2*	h3	13	h3	16	h1	2	h2	12	h1
N46 (5088)	ISR	null	h2	7	h1	8	h1	2	h2	12	h1
NABU EPI BLANC (5102)	NPL	null	h2	7	h1	8	h1	2	h2	12	ND
NANKING 25 (5116)	CHN	null	h2	7	h1	8	h1	2	h2	12	h1
NEPAL 84 (5166)	NPL	null	h2	7	h1	8	h1	2	h2	12	h1
NP120 (5308)	IND	null	h2	17	h1	18	h1	2	h2	12	h1
NYU BAY (5399)	JPN	null	h2	7	h1	8	h4	2	h2	12	h1
OPAL (5486)	DEU	1	h1	7	h1	9	h1	5	h3	10	h2
PITIC 62 (5748)	MEX	1	h1	7	h1	8	h1	2	h2	12	h1
RECITAL (6027)	FRA	2*	h3	6	h2	8	h2	5	h3	10	h2
RENAN (6086)	FRA	2*	h3	7	h1	8	h1	5	h3	10	h2
SEU SEUN 27 (6529)	KOR	null	h2	7	h4	8	h1	4	h2	12	h1
RALET (8048)	FRA	null	h5	20	h1	20	h3	2	h2	12	h1
ZANDA (8058)	BEL	1	h2	20	h1	20	h3	2	h2	12	h1
HOPEA (9048)	FIN	1	N	6	h2	8	h2	2	h2	12	h1
FRUH-WEIZEN (13310)	DEU	null	h2	22	h5	22	h5	5	h3	10	h2
ORNICAR (13471)	FRA	2*	h3	6	h2	8	h2	5	h3	10	h2
TALDOR (13476)	FRA	2*	h3	7	h1	8	h1	4	h2	12	h1
APACHE (13481)	FRA	null	h2	7	h1	8	h1	3	h2	12	h1
OPATA 85 (13811)	MEX	2*	h3	7	h1	9	h1	5	h3	10	h2
SYNTHETIQUE-W7984 (13812)	MEX	null	h6	7	h1	8	h1	N	h4	N	ND
BLE DE REDON BLANC 1/2 LACHE 1 1 (15658)	FRA	1	h1	13	h3	16	h1	2	h5	12	h1

Accessions used for expression studies are shown in bold.

[a] *Accession no. in the INRA Triticeae Genetic Resources Collection (http://www6.clermont.inra.fr/umr1095) is given in brackets.*

[b] *Country names are given as three-letter ISO codes (http://www.unc.edu/~rowlett/units/codes/country.htm).*

[c] *Protein coding allele for the x- or y-type HMW-GS identified by SDS-PAGE. HZ, heterozygous.*

[d] *Haplotype of the promoter for HMW-GS genes. ND, no data.*

Table 2 | Characteristics of *cis*-motifs from PLACE database and bibliographic references used to annotate the promoters of HMW-GS genes.

Names	Name used in PLACE	Sequence	Binding transcription factor[a]	References
DOF				
DOF core	DOFCOREZM	AAAG	**DOF** SAD, *PBF*, BPBF WPBF	Hammond-Kosack et al., 1993; Yanagisawa and Schmidt, 1999
Pbox1	PROLAMINBOXOSGLUB1	TGCAAAG		Wu et al., 2000; Isabel-La Moneda et al., 2003
Pbox2	300CORE	TGTAAAG		Thomas and Flavell, 1990
Pbox3	−300ELEMENT	TGHAAARK		Marzábal et al., 1998
bZIP				
GLM 1	GCN4OSGLUB1	TGAGTCA	**bZIP** *O2*	Albani et al., 1997; Vicente-Carbajosa et al., 1998; Wu et al., 2000
GLM 2	−300MOTIFZMZEIN	RTGAGTCAT	BLZ1, BLZ2	Thomas and Flavell, 1990; Oñate et al., 1999
GLM 3	GLMHVCHORD	RTGASTCAT	SPA	Norre et al., 2002
ACGT core	ACGTATERD1	ACGT		
G-box motif 1	ABREATCONSENSUS	YACGTGGC		Kang et al., 2002; Choi et al., 2000
G-box motif 2	ABRELATERDI	ACGTG		
CAAT	CAATBOX1	CAAT	**bZIP**[b]	Shirsat et al., 1989
RY-REPEAT				
RY_core	RYREPEATLEGUMINBOX	CATGCAY	**AB3/VP1** FUSCA3	Fujiwara and Beachy, 1994; Moreno-Risueno et al., 2008; Van Herpen et al., 2008
AACA MYB				
AACA motif 1	AACACOREOSGLUB1	AACAAAC	**R2R3-MYB** GaMYB GaMYB	Takaiwa et al., 1996; Suzuki et al., 1998; Wu et al., 2000; Diaz et al., 2002
AACA motif 2	ANAERO1CONSENSUS	AAACAAA		
MYB1AT	MYB1AT	WAACCA		
AACA motif 3	Not referred in PLACE	AACNNA		
GATA MYB				
GATA box 1	GATABOX	GATA	**R1MYB** MCB1 MYBS3	Rubio-Somoza et al., 2006a
GATA box 2	MYBST1	GGATA	**MYB**	Baranowskij et al., 1994
OTHERS				
E box	EBOXBNNAPA	CANNTG	**bHLH**	Chaudhary and Skinner, 1999
CCAAT	CCAATBOX1	CCAAT	**HAP** *CBF*	Albani and Robert, 1995
ESP	ESPASGL01	ACATGTCATCATGT	Not identified	Vickers et al., 2006
TATA-variant	Not refered in PLACE	TATAWA	TATA-box-binding proteins	Fauteux and Strömvik, 2009
CIACADIANLELHC	CIACADIANLELHC	CAANNNNATC		Piechulla et al., 1998
ABox	PALBOXAPC	CCGTCC		

[a] *Transcription factor families are indicated in bold followed by the name of corresponding transcription factor in maize (italics), barley, wheat (underlined), or other species (italics and underlined).*

[b] *Interaction not functionally validated.*

Bombardments were conducted at a distance of 6 cm from the stopping plate using a biolistic helium gun device (PDS-1000, Bio-Rad) with a pressure of 6.21 MPa. Following bombardment, endosperms were incubated for 2 days in the dark at 24°C in a Murashige and Skoog medium supplemented with 3% (w/v) sucrose and 0.15 mM of each of the 20 proteinogenic amino acids. For GUS expression, endosperms were stained with 5-bromo-4-chloro-3-indolyl glucuronide according to Jefferson et al. (1987). Endosperms were observed using a MZ16 F stereomicroscope equipped with a DFC300 FX digital camera (Leica Microsystems) and GUS and GFP activities were determined by counting the number of blue and green cells, respectively. Expression results were normalized by dividing the number GUS foci by the number of GFP foci. For each construct, 10 independent bombardments of eight endosperms each were performed. The pAct-GFP construct was used to determine the efficiency of bombardment as proposed by Eini et al. (2013).

The DNA-binding activity of *cis*-motifs with SPA was studies by EMSA. The SPA protein was expressed in *E. coli* (BL21 AI strain) by cloning *Spa* cDNA into the pDEST17 plasmid vector

(Invitrogen) producing pHis-SPA. *Spa* expression was induced with 0.2% (w/v) arabinose for 3 h. Proteins extracts were obtained after re-suspension of the induced cells in a 10 mM Tris buffer (pH 8) containing 6 M urea and 100 mM NaH_2PO_4 (10 mL g^{-1} pellet). Purification of the recombinant protein was achieved by loading protein extracts onto a Ni^{2+}-NTA resin and bound proteins were eluted in a 10 mM Tris buffer (pH 4.5) containing 6 M urea and 100 mM NaH_2PO_4. The eluate was dialyzed against a 10 mM Tris buffer (pH 8.3) containing 2 M urea, 100 mM NaH_2PO_4, 100 mM KCl, 0.02% Tween-20, 10% glycerol, and 0.5 mM phenylmethylsulfonyl fluoride (PMSF) for 36 h to renaturate the recombinant protein and then against a 10 mM Tris buffer (pH 7.5) containing 50 mM KCl, 1 mM dithiothreitol, 0.02% Tween™ 20, 10% glycerol, and 0.5 mM PMSF for 16 h. The dialysate was then concentrated with an Amicon 10 kDa filter (Millipore).

DNA oligonucleotides able to bind bZIP TFs (GLM and G-box) used in EMSA are described in Supplementary Table 3. Each single-strand oligonucleotide was labeled using the Biotin 3′ End DNA Labeling Kit (Pierce) following the manufacturer's instructions and hybridized for 30 min at the annealing temperature of the probes. The labeled dsDNA probe (20 fmol) was incubated with 560 ng to 4 μg of recombinant His-SPA protein in 20 μL of a binding buffer containing 10 mM Tris (pH 7.5), 2 mM dithiothreitol, 100 mM KCl, 10% glycerol, 0.05% nonyl phenoxypolyethoxylethanol, 2 mM ethylenediaminetetraacetic acid, 100 ng μL^{-1} poly(dI.dC), 250 ng μL^{-1} fish sperm DNA, 0.5 mM PMSF for 30 min at room temperature. DNA-protein complexes were analyzed by non-denaturing 6% polyacrylamide gel electrophoresis in a 45 mM Tris, 45 mM Borate, and 1 mM ethylenediaminetetraacetic acid buffer (pH 8.3). After separation (100 V, 1 h at 4°C), gels were electroblotted to nylon membranes using the same buffer (380 mA, 45 min at 4°C). The biotin end-labeled DNA was detected using streptavidin, horseradish peroxidase conjugate following the manufacturer's instructions (LightShift Chemiluminescent EMSA kit, Pierce).

STATISTICAL ANALYSES

All statistical analyses were done using R 3.0 software (R Core Team, 2013). The normality of and homogeneity of variances of expression data were tested by the Shapiro–Wilk and Bartlett's tests, respectively. Depending on the results of previous analyses, expression data were submitted to non-parametric or parametric variance analysis with the Kruskal–Wallis or the general linear model procedure. Multiple comparison tests between groups after Kruskal–Wallis tests were done with the Kruskalmc function while the Student–Newman–Keuls test was used to compare means after the general linear model procedure. The Kruskal–Wallis and Student–Newman–Keuls tests used were those available in the R "agricolae" (version 1.1-8) package (De Mendiburu, 2014), all other tests were done using the R "Stats" (version 2.15.3) package. All the data were used in a first analysis based on a model with one factor (gene). In a second step, analyses were carried out gene per gene to study the promoter haplotype factor.

To analyze the differences in expression of HMW-GS genes and haplotypes one-way ANOVA were performed. First, an ANOVA with the gene as the main factor was carried out. The four lines with the null allele at *Glu-A1-1* and the line with protein allele 7 overexpressed (7OE) at *Glu-B1-1* were excluded from this analysis to avoid bias. Secondly, ANOVAs with the promoter haplotype as the main factor were performed for each gene (including the null allele at *Glu-A1-1*).

Differences in normalized expression from transient expression assays were analyzed using *t*-test. All statistically significant differences were judge at 5%.

RESULTS

THE VARIABILITY OF THE PROMOTER IS NOT SYSTEMATICALLY CONNECTED WITH PHENOTYPIC VARIABILITY

The variability in the nucleotide sequence of the promoters of the five HMW-GS genes was extensively studied by sequencing a set of 42 lines representative of the diversity present in the INRA worldwide hexaploid wheat core collection. The following results deal with the noncoding DNA region upstream of the start codon given that for HMW-GS genes the transcription start site (TSS) is about 60 bases upstream of the start codon for translation. In some cases, the hybridization sites of reverse primers were downstream of the start codons, so the sizes of the upstream fragments studied ranged from 467 to 1138 bp. A total of 36 single-base changes, 2 single-base insertion-deletions (indels) and 1 larger indel were identified in an average of 3858 bp promoter sequence per line (**Table 3**, Supplementary Table 4). These specific regions have an average of one polymorphism every 100 bases. The number of polymorphisms varied between promoters. *Glu-B1-2* promoter has one polymorphism every 58 bp, threefold more frequently than the *Glu-D1-2* promoter, which has one polymorphism every 145 bp. One large deletion of 54 bp spanning from 291 to 344 upstream of the start codon in the *Glu-B1-1* promoter was observed in two lines (accession nos. 4901 and 15658). Thus, nucleotide diversity estimated by the mean pairwise difference (π) varied from one promoter to another, ranging from 1.5×10^{-3} for *Glu-D1-1* to 3.0×10^{-3} for *Glu-B1-1*. Except for *Glu-D1-2*, we observed that the nucleotide diversity (π) and the number of segregating sites (θ) are about equal in values as confirmed by the non-significant Tajima's D statistic (**Table 3**). This suggests that there has been no particular pattern of selection in these regions.

The polymorphisms are linked by a high level of linkage disequilibrium (data not shown). Therefore, for all loci, most of the lines clustered into two main haplotypes with the remaining haplotypes being generally represented by single lines. Notably, the number of haplotypes found for each promoter fits the number of protein coding alleles for *Glu-D1-2* only (**Table 1**, **Figure 1**). For *Glu-B1-1*, we observed more protein coding alleles than promoter haplotypes. For the three other *Glu1* genes, we observed more promoter haplotypes than protein coding alleles. Each electrophoretic allele, except for *Glu-B1-2* alleles, tends to have a more-frequent promoter haplotype (**Figure 1**).

THE VARIABILITY OF THE HMW-GS GENE PROMOTER IS OFTEN CONNECTED WITH THE LEVEL OF GENE TRANSCRIPTION

To assess whether the gene transcrition level is influenced by the promoter haplotype of each HMW-GS gene, HMW-GS

Table 3 | Number of electrophoretic alleles revealed by SDS-PAGE, haplotype and diversity statistics for the promoters of five HMW-GS genes from 42 accessions of the INRA worldwide hexaploid wheat core collection.

Genes	No. of electrophoretic alleles[a]	Promoter length (bp)	No. of polymorphic sites/No. of indels[b]	No. of haplotypes[c]	Haplotype diversity, Hd	Nucleotide diversity, π	No. of segregating sites, θ	Tajima's statistic
Glu-A1-1	3	878	7 (4)/0	6 (3)	0.71	1.51×10^{-3}	1.60×10^{-3}	NS[d]
Glu-B1-1	7	747	10 (3)/1 (0) *54*	5 (2)	0.34	2.67×10^{-3}	3.16×10^{-3}	NS
Glu-B1-2	6	465	7 (2)/1 (0) *1*	5 (1)	0.45	3.03×10^{-3}	3.03×10^{-3}	NS
Glu-D1-1	4	667	5 (3)/0	5 (2)	0.49	1.48×10^{-3}	1.52×10^{-3}	NS
Glu-D1-2	2	1163	7 (0)/1 (0) *1*	2 (0)	0.43	2.93×10^{-3}	1.39×10^{-3}	$P < 0.05$

[a] *Protein coding allele for the x- or y-type HMW-GS identified by SDS-PAGE.*

[b] *The number of singletons (i.e., a polymorphism found in a single line) is given in brackets; the size of indels is indicated in italics.*

[c] *The number of haplotypes including a single line is indicated in brackets.*

[d] *NS, not significant.*

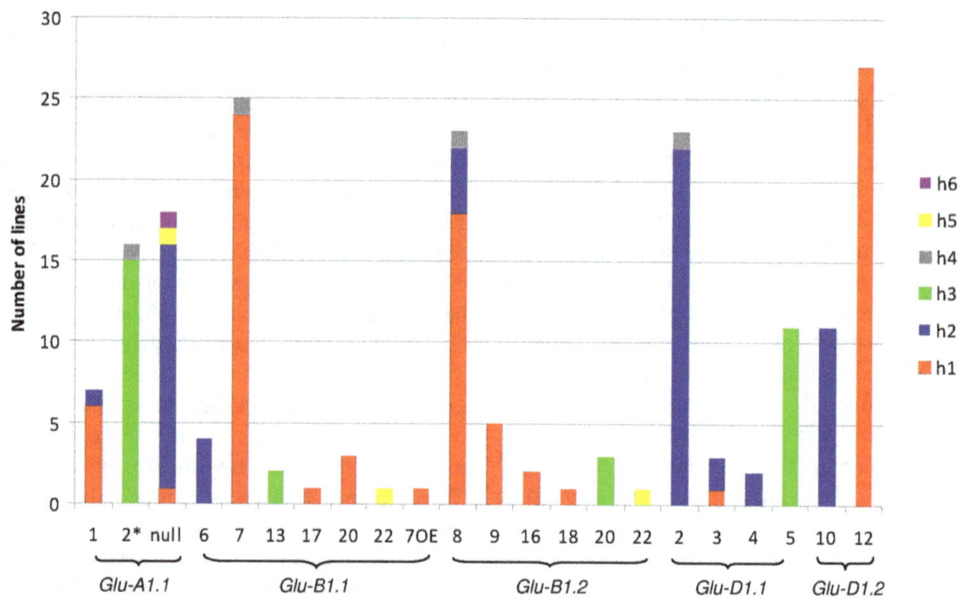

FIGURE 1 | Number of lines of each haplotype of HMW-GS gene promoter for all electrophoretic forms of HMW-GS present in the set of 42 lines studied. The promoter haplotypes are named h1 to h6. The same color is used for the same haplotype number of a given gene promoter. Although the color is identical for all h1 haplotypes, their sequences differ (e.g., the sequences of haplotype h1 at *Glu-A1-1* and *-B1-1* are different).

transcripts were quantified at 400°C days after anthesis for 13 lines by qRT-PCR (**Table 1**). The five HMW-GS genes had different levels of transcription ($P = 2 \times 10^{-16}$). On average, *Glu-B1-1* and *Glu-D1-1* showed a higher level of transcription than the remaining genes, while *Glu-D1-2* was expressed at lower levels (**Table 4**). The two x-type HMW-GS genes were expressed up to 10-fold higher than the genes coding the y-type. The transcription of *Glu-A1-1* was intermediate.

Among the four accessions with the null allele at *Glu-A1-1*, three harbor the h2 promoter haplotype and one the h5 haplotype. These two haplotypes differ by only one single nucleotide polymorphism (SNP) and their transcription was close to zero (**Table 5**). The transcription of the two other promoter haplotypes for *Glu-A1-1* were not different ($P = 0.95$). One line (accession no. 8058) harbors the h2 haplotype but has the protein allele 1 and had a transcription close to that of the h1 and h3 promoter haplotypes. For *Glu-B1-1*, once the line with the Bx7OE protein allele was discarded, the promoter haplotype effect was significant ($P = 0.014$). The transcription for the h1, h3, and h4 haplotypes was similar and, on average, 2.6-fold higher than that for haplotype h2 (**Table 5**), which only includes the Bx6 protein allele (**Figure 1**). The line with the Bx7OE protein allele has the h1 promoter haplotype, as most of the BX7 protein alleles, but it expressed *Glu-B1-1* at a level (195.33 ± 29.25, $n = 2$) twice that of Bx7 lines. For *Glu-B1-2*, the haplotype effect was significant ($P = 0.023$) and transcription from h1 was higher than from h3 (**Table 5**). For this gene, the promoter haplotypes were not linked with separate protein alleles (**Figure 1**). The RNA expression of the *Glu-D1-1* and *Glu-D1-2* alleles was not influenced by their promoter haplotypes (data not shown).

Table 4 | Comparison of the transcription levels of HMW-GS genes at 400°C days after anthesis for 13 lines of INRA worldwide hexaploid wheat core collection.

Genes	No. of lines[a]	RNA expression levels[b]
Glu-A1-1	9 (22)	56.70 ± 3.99 (B)
Glu-B1-1	12 (32)	83.64 ± 7.65 (A)
Glu-B1-2	13 (34)	17.14 ± 0.97 (C)
Glu-D1-1	13 (34)	83.69 ± 5.02 (A)
Glu-D1-2	13 (34)	8.063 ± 0.42 (D)

Data are means ± 1 SE.

[a] *The number of data points is indicated in brackets.*

[b] *Different letters in brackets indicate a significant difference (α = 5%) calculated according to a Kruskal–Wallis non-parametric test followed by the Kruskal multiple comparisons test.*

Table 5 | Multiple comparison of the mean levels of RNA expression from promoter alleles of HMW-GS genes at 400°C days after anthesis.

Genes	Promoter haplotype	No. of lines[a]	RNA expression levels[b]
Glu-A1-1	h1	1 (2)	58.25 ± 1.83 (B)
	h2 (null)[d]	3 (10)	1.40 ± 0.18 (A)
	h2 (1)[d]	1 (2)	57.38 ± 7.70 (B)
	h3	6 (16)	58.56 ± 5.19 (B)
	h5	1 (2)	1.41 ± 0.32 (A)
Glu-B1-1[c]	h1	7 (20)	102.27 ± 9.08 (A)
	h2	3 (8)	35.36 ± 3.43 (B)
	h3	1 (2)	87.23 ± 7.35 (A)
	h4	1 (2)	86.94 ± 11.50 (A)
Glu-B1-2	h1	8 (22)	18.96 ± 1.24 (A)
	h2	3 (8)	14.74 ± 1.28 (AB)
	h3	2 (4)	11.89 ± 1.45 (B)

Data are means ± 1 SE.

[a] *The number of data points is indicated in brackets.*

[b] *Different letters in brackets indicate a significant difference (α = 5%) calculated according to a Kruskal–Wallis non-parametric test followed by the Kruskal multiple comparisons test.*

[c] *The line accession no. 3358 with the 7OE allele was discarded.*

[d] *For Glu-A1-1 haplotype 2, results for the null and 1 protein alleles (indicated in brackets) were treated as two different haplotypes in the ANOVA.*

These results highlight different RNA expression levels for different HMW-GS genes and, for three HMW-GS genes, the effects of the promoter haplotype. Thus, differences in the regulation of these genes might stem from the organization of the *cis*-motifs in their promoters.

COMMON *cis*-MOTIFS ORGANIZATION OF HMW-GS GENE PROMOTERS

To analyze the organization of *cis*-motifs in HMW-GS gene promoter, we first searched for similar patterns in the 1-kb promoter region of the six HMW-GS genes of cv. Renan, as HMW-GS genes have similar expression patterns during development and in response to environmental factors. We then compared the consensus organization of *cis*-motifs found for cv. Renan with that found for the haplotypes of each gene to relate differences in *cis*-motifs organization with differences in gene expression.

In all six HMW-GS gene promoters of cv. Renan we found all the 24 *cis*-motifs we annotated but the Pbox2 and ESP motifs. Most of these motifs were annotated several times and a total of 44 (for *Glu-B1-2*) to 54 (for *Glu-D1-2*) *cis*-motifs per gene were annotated. All the *cis*-motifs able to bind all TFs known to regulate the expression of SSP genes were present, but the typical bipartite endosperm box was not found. The number of *cis*-motifs found was over-estimated as the sequences of a few motifs (**Table 2**) were nested within some others. Most of the nested *cis*-motifs bind TFs of the same family (**Table 2**). Therefore, we took into account only the longest motif where nested motifs were predicted, which reduced the number of *cis*-motifs per gene by 15–24%. Motifs able to bind MYB TFs (GAMYB, MCB1, MYBS3) were predominant, with 9–14 *cis*-motifs per gene, followed by motifs able to bind bZIP TFs, with 9–13 *cis*-motifs per gene, and DOF TFs (PBF, SAD), with 4–8 *cis*-motifs per gene. The CAAT *cis*-motif accounted for about two-thirds of the total number of *cis*-motifs able to bind bZIP TFs (**Table 6**).

The organization of orthologous promoters from cv. Renan showed few differences (**Figures 2A,B**) on the plus strand. For x-type HMW-GS genes, the organization was well conserved between 0 and −400 (nucleotide position relative to the start site). The TATA-box was at −90. A few differences were detected like an AACA motif at −144 in *Glu-A1-1* and *-D1-1*, which was absent in the orthologous B sequence. Between −400 and −1000, the organization was also well conserved but a 55-bp insertion in the *Glu-B1-1* promoter shifted the *cis*-motif upstream (i.e., at a more negative nucleotide position) of the insertion. Interestingly, we discovered a composite box named the GLM-GATA box. This box includes two repeated units, each of them made of a GATA motif and a GLM separated by a third GGATA motif. The relative positions of the constitutive *cis*-motifs in this box were conserved among the three orthologous sequences of cv. Renan (**Figure 2**). An ACGT motif was present a few bases upstream of this box in the B and D sequences. About 50 and 200 nucleotides upstream of this box, a DOF core motif (AAG) and an AACA motif (able to bind R2R3-MYB TFs), respectively, were detected in all the homoeologs. Downstream of this box, we found an AACA motif able to bind R2R3-MYB and the RY repeat.

Similar observations were made for the y-type sequences (**Figures 2B,C**). *Cis*-motifs organization presented many similarities between positions 0 and −400, although the promoter of *Glu-B1-2* includes some additional motifs at about position −150. In addition, the entire composite GLM-GATA box was lacking in the promoters of *Glu-A1-2* and *-B1-2*, the latter containing only a single copy of the GLM. None of these three sequences included the ACGT motif near the GLM-GATA found in the x-type HMW-GS gene promoters. We observed a composite motif at position −400, which was conserved in these three homoeologous sequences, composed of a G-box and three consecutive MYB motifs (two GATA and one AACA motifs). At about position −400, a deletion shortened the distances between the motifs at −400 and the adjacent ones on the *Glu-A1-2* promoter causing

Table 6 | Number of motifs in the upstream 1000-bp region of the six HMW-GS genes from the hexaploid wheat cv. Renan.

Motif sequence	Binding transcription factor	*Glu-A1-1*	*Glu-B1-1*	*Glu-D1-1*	*Glu-A1-2*	*Glu-B1-2*	*Glu-D1-2*
DOF							
AAAG	DOF	6	4	4	7	4	5
TGCAAAG	DOF		1	1	1		1
TGHAAARK	DOF	1		2			1
bZIP							
ACGT[a]	bZIP		1	1	2	3	2
ACGTG[a]	bZIP	1					
RTGAGTCAT[b]	bZIP				1	1	
TGAGTCA[b]	bZIP	2	2	2			2
YACGTGGC[b]	bZIP		1	1	1	1	1
CAAT	bZIP[c]	7	9	7	6	6	7
AACA MYB							
AAACAAA	R2R3-MYB	2	1	2	1	1	1
WAACCA	GAMYB			1	2	2	1
AACNNA	R2R3-MYB	1	1	1	1	1	1
GATA MYB							
GGATA	MYB/R1MYB	8	7	9	7	8	7
RY REPEAT							
CATGCA	AB3/VP1	2	2	2	3	1	2
OTHERS							
CANNTG	bHLH	4	4	4	7	5	4
CCAAT	HAP	2	3	2	2	2	2
TATAWA	TATA-box-Binding Proteins	1	1	1	1	1	1
CAANNNNATC		1	2	2	2	2	2
CCGTCC		1	1		1		1
	Total	39	40	42	45	38	41

[a] Related to GLM.
[b] Related to G-box motifs.
[c] Interaction not functionally validated.

a deletion of a few motifs. For the three y-type homoeologous genes, an RY repeat and an AACA motif (binding R2R3-MYB) were located between position −400 and the GLM-GATA box.

The overall consensus generated from all HMW-GS genes of cv. Renan (**Figure 2C**) consisted of 21 motifs including motifs able to bind all the TFs known to regulate SSP synthesis so far. They were organized into five CCRMs. CCRMs were numbered from 1 to 5 from the start codon and composed of two to five *cis*-elements. As expected, CCRM1, a few nucleotides upstream of the TSS, was composed of the TATA-box variant and the CAAT motif. CCRM2 included a G-box-like motif and a CAAT motif, nested into an E-box (CANNTG), while CCRM3 clustered two GATA boxes. CCRM4 was the most interesting module. It included the incomplete GLM-GATA box, an AACA motif and the RY repeat. The GLM-GATA box was incomplete because of a missing GLM in the cv. Renan allele at *Glu-B1-2*. The fifth module, CCRM5, has a DOF motif and a CAAT box nested into an E-box and is located between positions −900 and −1000 in all promoters. A few bases downstream of CCRM5, E-boxes and circadian motifs were conserved. No typical bipartite endosperm box was detected. On the minus strand, we noted an over-representation of the DOF core AAAG motif (data not shown).

For each HMW-GS gene, except *Glu-B1-1*, the annotation of haplotypes was almost identical (**Figures 3, 4**). Three groups were observed for *Glu-B1-1*. Haplotypes h2 and h5 have identical annotations, but compared to the other haplotypes, they contain an additional RY repeat at position −160. The second group contains h1 and h4, which are distinct from h3 because of an indel. Distances between motifs upstream and downstream of position −400 are therefore shorter in h3 than in the other haplotypes. In addition, a bZIP motif present in the insertion is deleted in h3. The haplotype h3 of *Glu-D1-1* promoter differs from other haplotypes as it has two additional bZIP motifs, one being a G-box.

The relative position of the GLM-box was conserved in all haplotypes of the three orthologous sequences of the x-type HMW-GS genes (**Figure 3**) and the y-type *Glu-D1-2* gene (**Figure 4**). For *Glu-B1-2*, the region sequenced in this study did not cover the GLM-GATA box (**Figure 4**), but the analysis of *Glu-B1-2* promoter sequences of cv. Chinese Spring (KC20630) and Xiaoyan

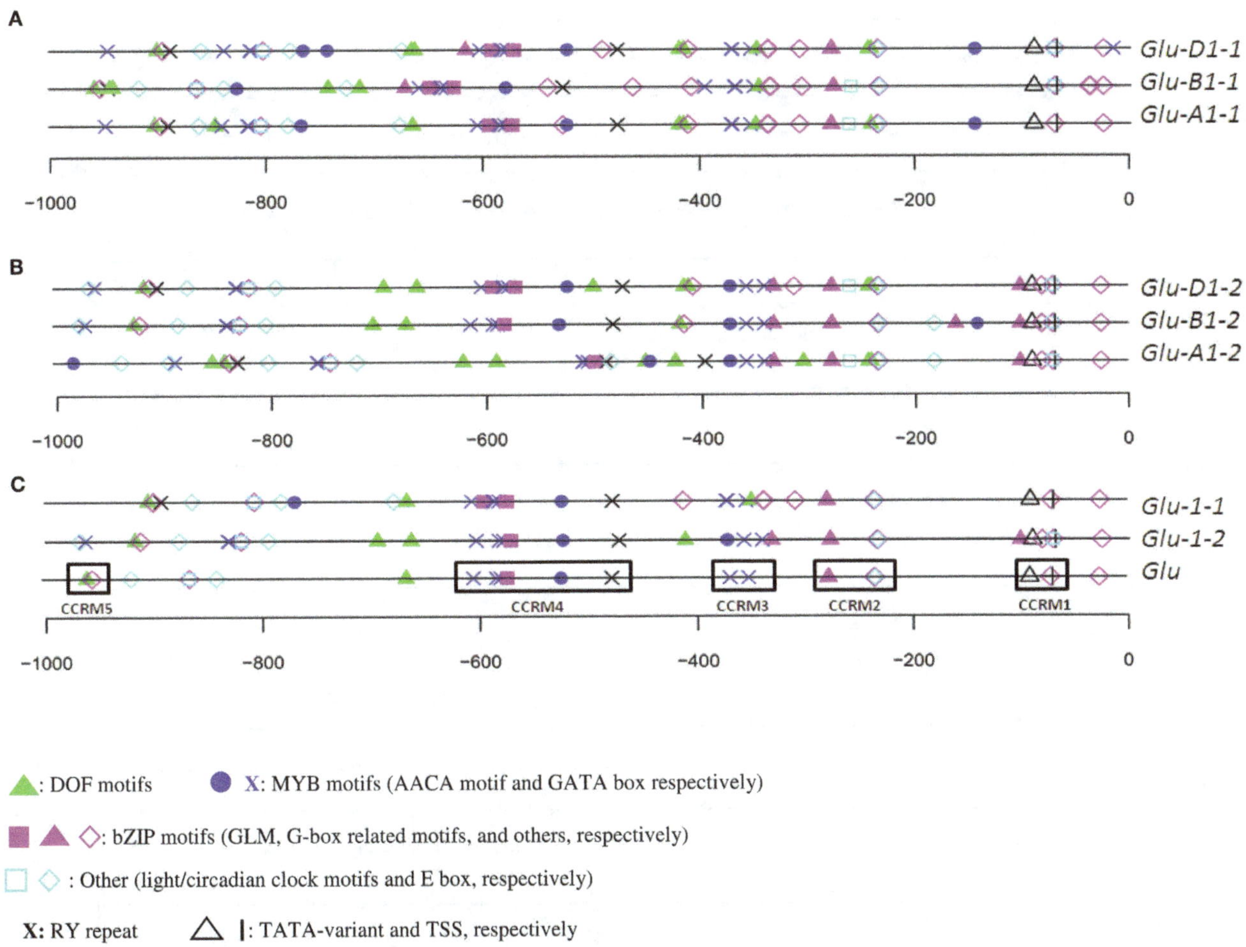

FIGURE 2 | *In silico* annotation of HMW-GS promoters of cv. Renan. Positions are indicated relatively to the start site. Sequences were obtained from public databases. **(A)** x-type *Glu-A1-1, -B1-1,* and *-D1-1* homoeologs; **(B)** y-types *Glu-A1-2, -B1-2,* and *-D1-2* homoeologs; and **(C)** consensus annotations of all orthologous set of sequences (*Glu-1-1* and *Glu-1-2* for the x- and y-type HMW-GS promoters, respectively), and of paralogous sequences (*Glu*) with its five conserved *cis*-regulatory modules CCRM1 to CCRM5.

54 (EU137874), available in public databases, shows that, in these cases, the relative position of the GLM-box is also conserved in this gene (data not shown).

THE GLM-GATA BOX IS INVOLVED IN THE REGULATION OF *Glu-B1-1* EXPRESSION

To investigate the involvement of the GLM-GATA box in the regulation of HMW-GS gene expression, we analyzed the effect of the 5′ deletion from positions −747 to −597 (fragment carrying the GLM-GATA box) by transient expression experiment (**Figures 5, 6**). The deletion of the GLM-GATA box reduced normalized GUS expression by 59%.

To verify the potential binding activity of the two GLMs (GLM1 and GLM2 at positions −647 and −626, respectively) present in the GLM-GATA box of the *Glu-B1-1* gene promoter, we performed EMSAs with synthetic oligonucleotides and a recombinant SPA protein expressed as a His fusion in *E. coli* (**Figure 7**). We also determined the *in vitro* binding of SPA to the G-box motif, which was previously shown to bind bZIP proteins (Norre et al., 2002). As shown in **Figure 7A**, arabinose treatment induced expression of a protein of 50–75 kDa that was not present in uninduced cell extracts. The apparent size of the recombinant protein determined by SDS-PAGE was larger than the expected 48 kDa molecular mass of the His-tagged SPA fusion protein. A similar apparent increase in size on SDS gels was already reported by Albani et al. (1997) in their study of SPA. The recombinant His-SPA protein was purified to near homogeneity and used for binding assays. A DNA-protein complex was clearly observed with the GLM2 motif, while the shifted band detected for the GLM1 and the G-box was considerably fainter (**Figure 7B**). No shifted band was observed when incubation was carried out with the mutated probes (*glm1*, *glm2*, and *G-box*). DNA-binding affinity of the recombinant protein seems to be greater with the GLM2 probe than the other probes tested.

DISCUSSION

Here we characterized and annotated wheat HMW-GS gene promoters. The expression of these genes in developing grain was

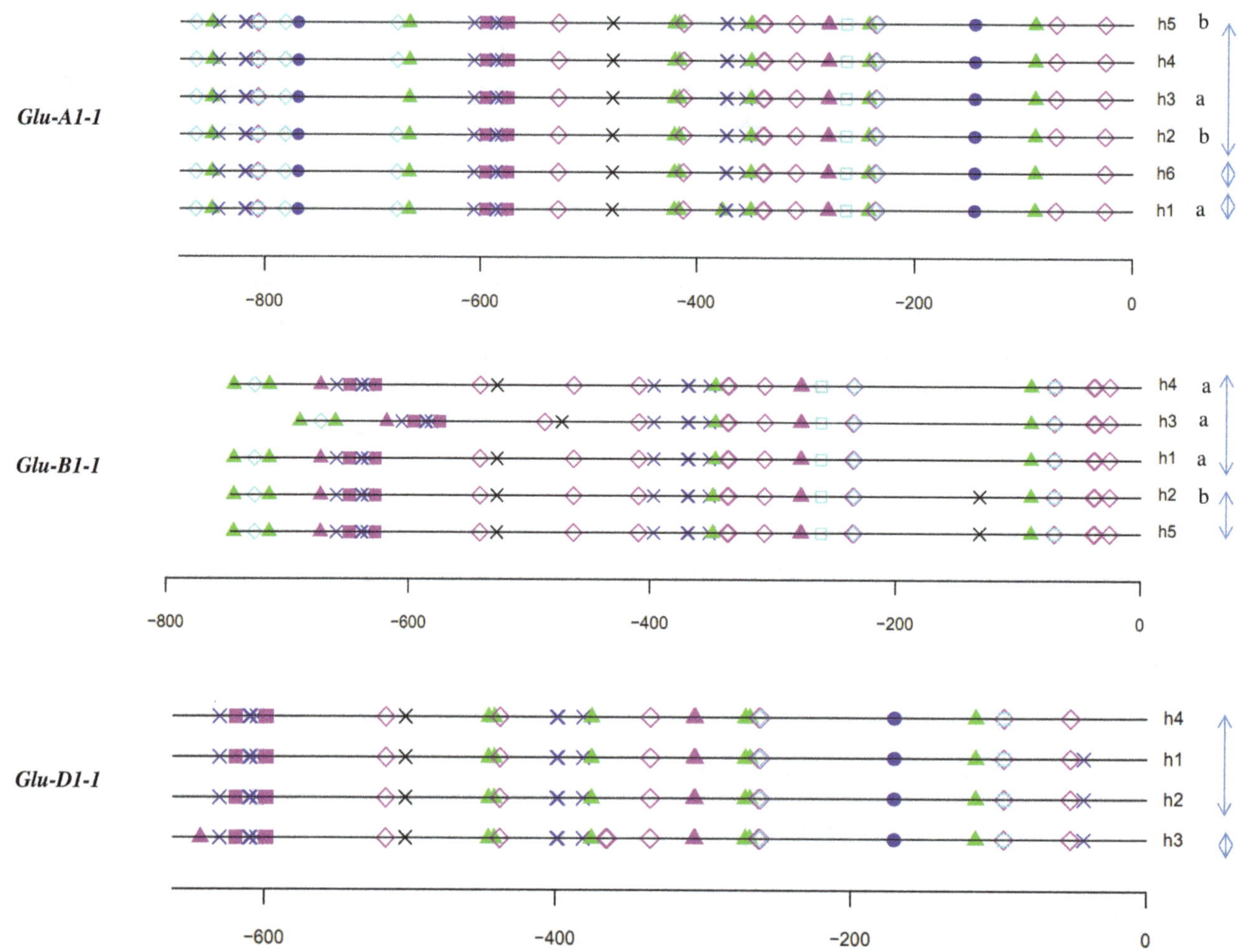

FIGURE 3 | *In silico* annotation of the x-type HMW-GS gene promoter haplotypes. Positions are indicated relative to the transcription start site. Sequences were obtained from a set of 42 lines representative of the genetic diversity of the INRA worldwide hexaploid wheat core collection. For each gene, the haplotype of the promoter is indicated by the letter h followed by the number of the haplotype. Letters a and b indicate the significantly different groups for the mean of expression for haplotypes studied by qRT-PCR. Clusters of haplotypes differing by one polymorphism are shown with gray arrows on the right. See the key to **Figure 2** for descriptions of *cis*-motif symbols.

quantified by qRT-PCR and the correlations between the variability in expression and the variability in predicted *cis*-element motifs of the corresponding promoter were also analyzed. We considered regions of 467–1138 bp upstream of the start codon. In *Arabidopsis thaliana*, based on the density of polymorphisms in gene upstream regions, functional promoters require 250–500 nucleotides upstream of the TSS (Korkuć et al., 2014). Under the assumption that promoter length is conserved, the lengths of the regions surveyed here provide a reasonable coverage of functional SSP gene promoters in wheat. Moreover, we analyzed the role of the GLM-GATA box of the *Glu-B1-1* gene promoter by transient expression assay and evaluated the functionality of the *cis*-motifs reported to bind bZIP TFs.

VARIABILITY OF HMW-GS PROMOTER HAPLOTYPES CANNOT BE USED DIRECTLY TO SCREEN FOR ELECTROPHORETIC ALLELES

In *A. thaliana*, the nucleotide variability in promoters varies depending on the function of their downstream gene (Korkuć et al., 2014). It is higher for genes involved in adaptive processes and transcriptional regulation than for genes involved in housekeeping functions. In wheat, the diversity of promoters is not widely documented so far. The range of nucleotide diversity observed for HMW-GS promoters, approximately one polymorphism every 100 bases, is comparable to that reported for the SPA promoter (Ravel et al., 2009), but is higher than the overall level of polymorphism of one SNP every 212 nucleotides reported for promoters of other genes (Ravel et al., 2006). Although upstream gene regions are somewhat constrained as they are involved in gene regulation, they are reported to show higher variability than coding regions. Constraints most likely apply to *cis*-regulatory elements (Korkuć et al., 2014). As they affect short regions, mutations could occur with little or no incidence, whereas the entire coding sequence has to withstand greater constraints. In addition, the modular organization of *cis*-elements, together with their redundancy, may buffer the effects of mutations (reviewed by Purugganan, 2000). These reasons probably explain why the

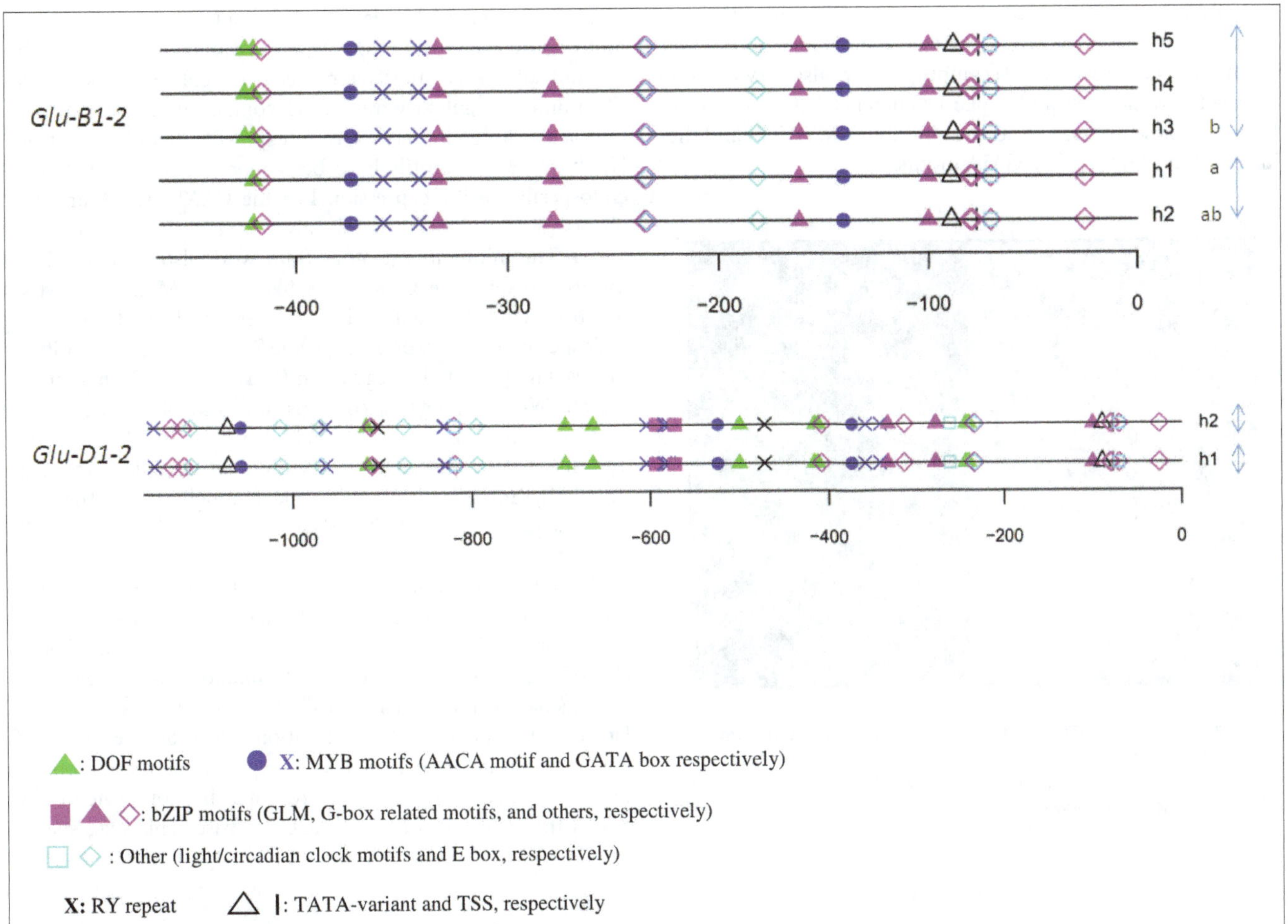

FIGURE 4 | *In silico* annotation of the y-type HMW-GS gene promoter haplotypes. Positions are indicated relative to the transcription start site. Sequences were obtained from a set of 42 lines representative of the genetic diversity of the INRA worldwide hexaploid wheat core collection. For each gene, the haplotype of the promoter is indicated by the letter h followed by the number of the haplotype. Letters a and b indicate the significantly different groups for the mean of expression for haplotypes studied by qRT-PCR. Clusters of haplotypes differing by one polymorphism are shown with gray arrows on the right.

diversity is higher in promoter regions than in coding sequences. As usually reported (e.g., Chao et al., 2009), the level of diversity was the lowest in HMW-GS sequences from the D genome with 1 polymorphism every 145 base for *Glu-D1-2*, whereas the highest level of diversity was observed for HMW-GS promoters from the B genome with, on average, one polymorphism every 60 bases.

SDS-PAGE is still routinely used for characterization of HMW-GS alleles. Developing diagnostic SNPs to identify electrophoretic forms of HMW-GS from any part of young plants would be a valuable tool to support breeding for improved flour quality. However, there are up to four haplotypes promoter sequences per electrophoretic allele or only one haplotype for several alleles. Anderson et al. (1998) already reported two different alleles for the Bx7 promoter. The promoter haplotypes perfectly match the protein alleles only for *Glu-D1-2*. Currently, the identification of a set of SNPs from the other HMW-GS promoter sequences as a shortcut to distinguish between different protein forms is not possible, so the search for diagnostic SNPs needs to continue.

A MINIMAL FRAMEWORK FOR THE TRANSCRIPTIONAL REGULATION OF HMW-GS GENES IS REVEALED

We screened for *cis*-elements known to regulate SSP synthesis among all the HMW-GS gene promoters of cv. Renan. By annotating these promoters we found that they had a few regulatory elements in common, mostly organized into five CCRMs. Since HMW-GS genes show similar patterns of spatial and temporal expression, these common *cis*-elements might be involved in their global regulation and consequently may provide a minimal regulatory framework needed for the developmental and environmental (i.e., in response to nitrogen supply) regulation of HMW-GS gene expression. Like the long endosperm box described in some LMW-GS gene promoters, which consists of two repeats of the endosperm box (Albani et al., 1997; Juhász et al., 2011), the GLM-GATA box described here for the first time is also formed by two motifs (GATA and GLM) repeated twice in most of the promoters of HMW-GS. Our results demonstrate that the GATA-GLM box has an activator effect. Its two GLMs were able to bind SPA and were thus functional *cis*-motifs. GATA and GLM motifs are reported to bind R1MYB and bZIP TFs. Modules able to bind MYB and bZIP proteins belong to the seven best-known combinations of *cis*-motifs and are also very well represented in *A. thaliana* and poplar promoters (Ding et al., 2012). However, these modules generally bind R2R3-MYB TFs and thus include AACA rather than GATA motifs.

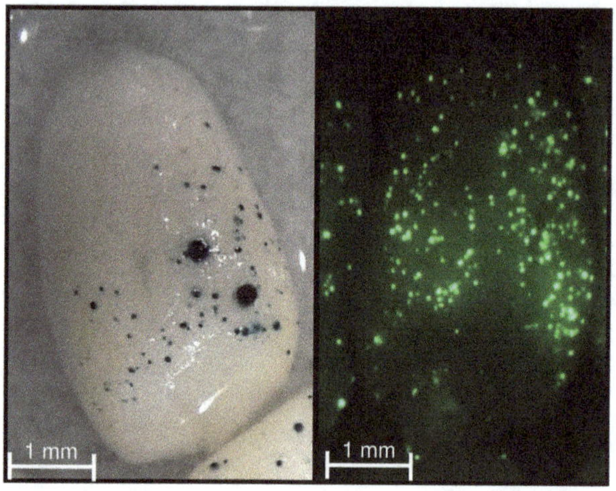

FIGURE 5 | GUS and GFP activities in wheat immature endosperm. Immature endosperm was co-bombarded with the pPrBx7-GUS and pAct-GFP constructs. Note the blue (bottom panel) and green (top panel) foci across the dorsal surface.

This GLM-GATA box is included in a CCRM with an AACA motif and a RY repeat. Notably, this conserved module is able to bind all the *cis*-motifs reported to regulate SSP synthesis. The minimal regulatory framework contains no P-box like those responsible for endosperm-specific expression of LMW-GS genes. However, several motifs have been reported to be involved in endosperm-specific expression like the CAAT, AACA and ESP motifs (Shirsat et al., 1989; Takaiwa et al., 1996; Vickers et al., 2006). The minimal regulatory framework also contains CAAT motifs. Possibly the G-box acts like the GLM in rice, which has been demonstrated to be an essential element conferring endosperm-specific expression, while P-box and AACA motifs are involved in quantitative regulation (Wu et al., 2000). In addition, the HMW-GS framework contains motifs involved in circadian rhythms. The E-box, which is able to bind bHLH and other TFs, has been reported to be involved in circadian transcriptional rhythms (Seitz et al., 2010), although exactly the same E-box sequence (5′-CATCTG-3′) was not found in the HMW-GS promoters.

Previous reports demonstrated that the 277 bp immediately upstream of the TSS are sufficient for temporal and tissue-specific regulation (Halford et al., 1989; Norre et al., 2002). There is also strong evidence indicating that mutations in this region are responsible for the silencing of *Glu-A1-2* (Halford et al., 1989). However, we did not find any mutation that could alter *cis*-motifs known to be involved in SSP gene regulation. In addition, the mutations specific to *Glu-A1-2* promoter did not create or alter any of the *cis*-motifs of the PLACE database. This suggests that

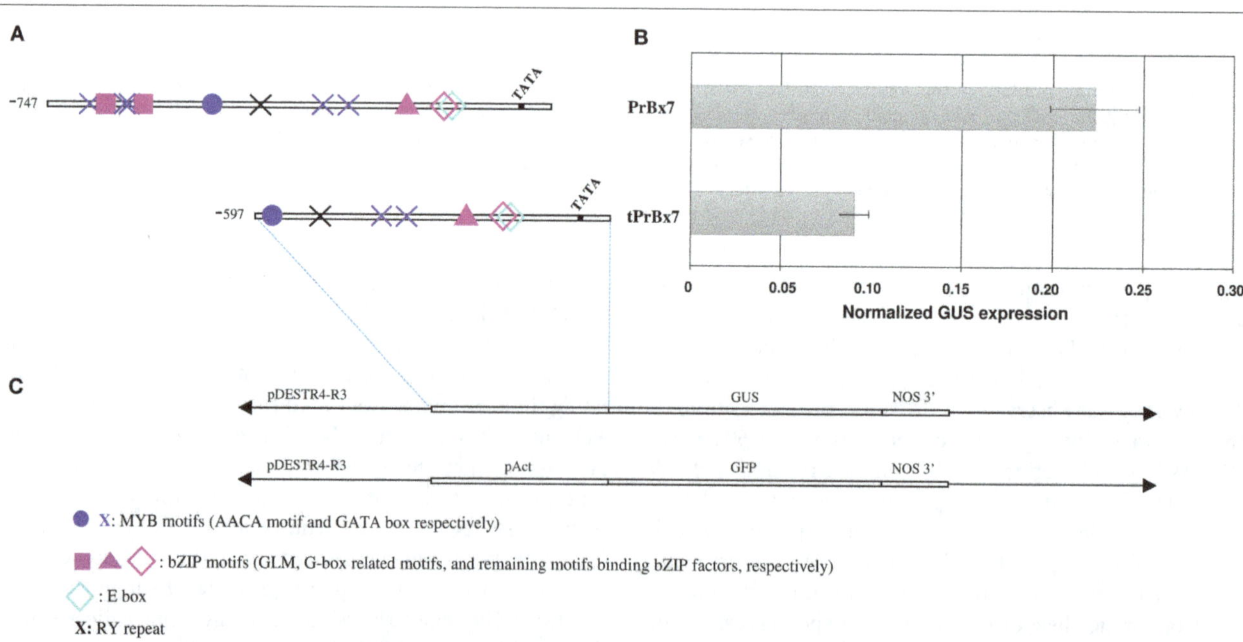

FIGURE 6 | Activity of *Glu-B1-1* gene promoter from cv. Renan (Bx7) in immature wheat endosperms using a transient expression assay. (A) Schematic representation of the constructs used. The TATA box and nucleotide positions relative to the start codon and corresponding to deleted region are indicated. Putative *cis*-regulatory elements, E-box (−259), G-box (−277), GATA box (−658, −638, −633, −368, and −350), RY motif (−525), AACA motif (−233), GLM1 and GLM2 (−647 and 626, respectively) are shown. **(B)** Normalized GUS expression of the corresponding promoters in transiently transformed endosperms. Data are the mean ± 1 SE. for $n = 10$ independent bombardments. **(C)** Schematic representation of the GUS constructs used for transformation.

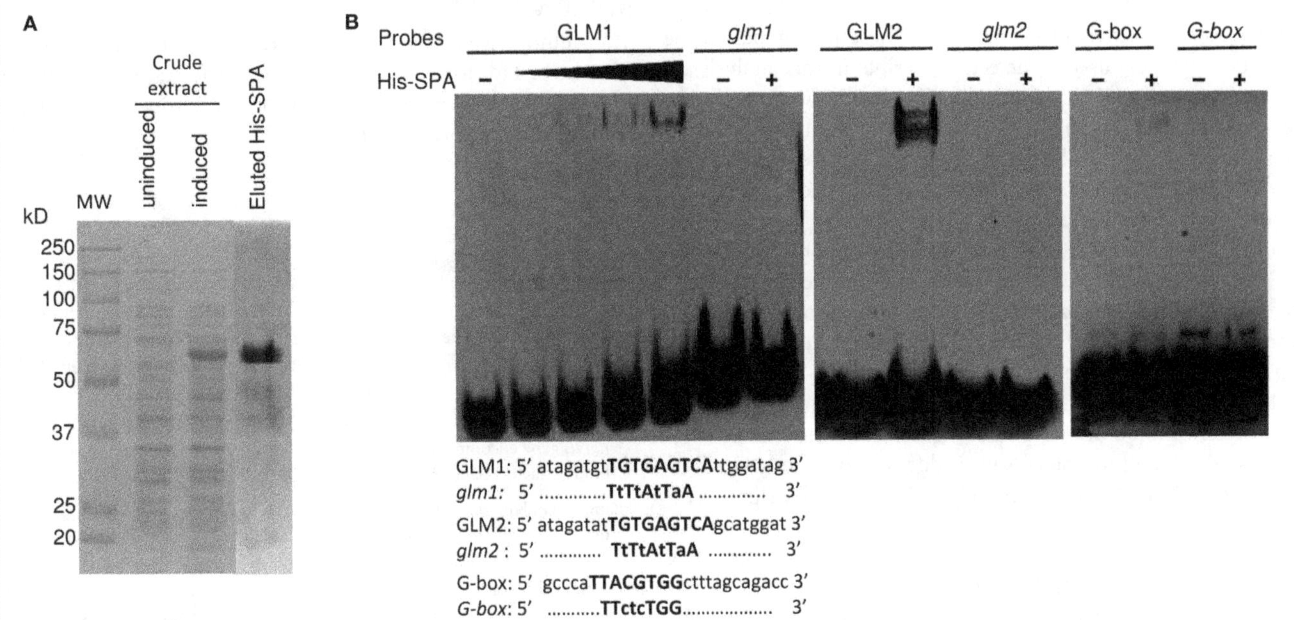

FIGURE 7 | Binding of recombinant SPA protein with the probes derived from the *Glu-B1-1* gene promoter. (A) Expression and purification of recombinant His-SPA protein. Crude extracts from uninduced and induced bacteria harboring the pHis-SPA expression vector and the eluted protein were resolved on an SDS-polyacrylamide gel. The molecular mass markers are indicated at left in kilodaltons. **(B)** EMSA of the recombinant SPA protein with the 25-bp biotin-labeled, GLM1 (−647), GLM2 (−626), and G-box (−227) probes derived from the *Glu-B1-1* gene promoter and their mutated versions *glm1*, *glm2*, and *G-box*. The sequences of the oligonucleotides used as probes are shown with the GLM1, GLM2, and G-box in bold; identical residues are represented by dots, and mutated bases are shown in lowercase.

this region may contain *cis*-motifs not yet known or that the mutations encountered in *Glu-A1-2* promoter may alter the affinity of *cis*-motifs identified for their respective TF. More precisely, this fragment contains CCRM1 and CCRM2. The latter includes the G-box found in the *Glu-D1-1* promoter and described by Norre et al. (2002) as being necessary and sufficient for expression. This box has been demonstrated to bind bZIP factors (Norre et al., 2002). CCRM2 also includes the 5′ part of the enhancer element found by Thomas and Flavell (1990), which confirms its important role. Thus, both functional validation and *in silico* analysis confirm the key role of this G-box in regulating the expression of HMW-GS genes. However, the level of expression of HMW-GS genes can be increased by adding more extensive flanking DNA (Anderson et al., 1998; Lamacchia et al., 2001), suggesting the presence of additional more distal *cis*-regulatory elements to the ones we found. This is in agreement with our results, which show a higher level of activity when the promoter of *Glu-B1-1* contained the distal GATA-GLM box. In addition, the DNA-binding affinity of SPA with one of the two GLMs of the GATA-GLM box was higher than that observed with the G-Box, suggesting a stronger role of this motif.

DIFFERENCES IN EXPRESSION ARE ONLY PARTIALLY EXPLAINED BY ANNOTATED *cis*-ELEMENTS

Our annotation strategy revealed differences at several levels: between paralogous HMW-GS genes, between orthologous HMW-GS genes and between haplotypes of a given HMW-GS gene. To investigate whether different annotated motifs induce quantitative differences in expression, we measured the level of expression from several HMW-GS promoter haplotypes. The expression of x-type gene transcripts was significantly greater than that of y-type transcripts with *Glu-B1-1* and *-D1-1* transcripts being the most expressed, *Glu-A1-1* intermediate and the two remaining genes the least abundant. This result is partially supported by GeneChip® hybridization experiments, which showed that *Glu-B1-1* is the most highly expressed HMW-GS gene in cv. Hereward (Shewry et al., 2009). However, comparing these two sources of results is not straightforward as HMW-GS probe sets cross-hybridize making it difficult to quantify the level of gene expression precisely, and only one wheat line was tested. Comparison of the consensus *cis*-motif framework of *Glu-1-1* with that of *Glu-1-2* showed several differences, which would be expected to impact their expression. Particularly, all *Glu-1-1* promoters contain an additional motif able to bind GAMYB upstream of the GLM-GATA box. Moreover, in the two most highly expressed genes, a G-box-related motif and a CAAT motif were located a few bases upstream of the GLM-GATA box and the RY repeat motif, respectively. This may enhance the activator effect of CCRM4, which contains two additional motifs.

Our results also demonstrate significant differences in the expression levels in relation to the haplotypes of the promoters for *Glu-A1-1*, *-B1-1*, and *-B1-2*. For *Glu-A1-1*, the transcription from haplotypes h2 and h5 was severely reduced for the null allele. This is in agreement with previous data on SSP synthesis in developing grains of cv. Hereward, which also has a null allele (Shewry et al., 2009). A C/T change in the coding sequence of this null allele creates a premature stop codon that could explain why this gene is inactive (De Bustos et al., 2000). However, this does not explain the low levels of expression of these haplotypes as the qRT-PCR primers used to detect transcripts in this analysis are located upstream of this mutation. The very low transcription level of this null allele may be due to sequence polymorphism in the promoter as it has been demonstrated for the null *Glu-A1-2* allele (Halford et al., 1989). There were no obvious differences in our annotation of haplotypes of the *Glu-A1-1* promoter that could explain the large differences in expression observed. This is unlike the case of *Glu-A1-2*, which is silent and shows a particular *cis*-motif organization upstream of position −370 when compared with other y-type HMW-GS genes. However, a 277-bp fragment immediately upstream of the *GluA1-2* TSS was not able to generate any transcriptional activity (Halford et al., 1989). The organization of this fragment is quite similar to that of other expressed y-type promoters, so it is difficult to hypothesize how the gene is silenced. As expected, *Glu-B1-1* in Glenlea (line accession no. 3358) strongly expresses the Bx7 subunit transcript. This over-expression is explained by a 10.3-kb duplication including a second copy of *Glu-B1-1* (Ragupathy et al., 2008). Again, our annotation of the promoter alone does not show obvious differences that could explain the different levels of expression. In agreement with the results of Halford et al. (1989), the deletion found in the h3 haplotype does not impact the level of expression, which confirms that it plays no role in transcriptional regulation.

These results suggest that other mechanisms are able to modulate HMW-GS gene expression, such as *cis*-elements located further upstream of the region studied here. This would agree with results of Wang et al. (2013), who described the presence of key regulatory sequences in the distal sequence of *Glu-B1-1*, especially a Py-rich stretch at about position -2000. This sequence has been reported to cause a high level of expression in tomato (Daraselia et al., 1996). Methylation of DNA may also be involved in HMW-GS expression regulation, as shown for hordein genes in barley (Sorensen et al., 1996; Radchuk et al., 2005), even though no CpG islands were detected in the wheat promoter regions studied here using the PlantPAN search engine (Chang et al., 2008).

In conclusion, this work reveals a minimal regulatory framework shared by all the wheat HMW-GS gene promoters. The *cis*-elements organization is conserved, including all the motifs known to be involved in the regulation of SSP genes. The conservation of this regulatory framework strongly suggests that it is involved in the regulation of this gene family. The bipartite endosperm box was not found but a CCRM with the GATA-GLM box with an RY repeat and an AACA motif is present in all the promoters. The CCRMs, which occur at similar relative positions in all the promoters of this small family, presumably have a common evolutionary origin, suggesting that they may be functional. However, validating their functional roles requires further experiments. The "*in silico* footprint" described here will help to select motifs for functional validation, as shown here by transient expression assays of *Glu-B1-1* promoter. Our annotations do not directly account for differences in expression among promoter haplotypes, suggesting that other mechanisms may be involved in regulating HMW-GS gene expression.

ACKNOWLEDGMENTS

The authors would like to thank Rachel Carol from Emendo Bioscience Ltd. for English corrections. The research leading to these results has received funding from the European Union's Seventh Framework Programme (FP7/2007–2013) under the grant agreement n° FP7-613556.

REFERENCES

Albani, D., Hammond-Kosack, M. C., Smith, C., Conlan, S., Colot, V., Holdsworth, M., et al. (1997). The wheat transcriptional activator SPA: a seed-specific bZIP protein that recognizes the GCN4-like motif in the bifactorial endosperm box of prolamin genes. *Plant Cell* 9, 171–184. doi: 10.1105/tpc.9.2.171

Albani, D., and Robert, L. S. (1995). Cloning and characterization of a *Brassica napus* gene encoding a homologue of the B subunit of a heteromeric CCAAT-binding factor. *Gene* 167, 209–213. doi: 10.1016/0378-1119(95)00680-X

Allaby, R. G., Banerjee, M., and Brown, T. A. (1999). Evolution of the high molecular weight glutenin loci of the A, B, D, and G genomes of wheat. *Genome* 42, 296–307. doi: 10.1139/g98-114

Anderson, O. D., Abraham-Pierce, F. A., and Tam, A. (1998). Conservation in wheat high-molecular-weight glutenin gene promoter sequences: comparisons among loci and among alleles of the *GLU-B1-1* locus. *Theor. Appl. Genet.* 96, 568–576. doi: 10.1007/s001220050775

Armone, M. I., and Davidson, E. H. (1997). The hardwiring of development: organization and function of genomic regulatory systems. *Development* 124, 1851–1864.

Bailey, T. L., Williams, N., Misleh, C., and Li, W. W. (2006). MEME: discovering and analysing DNA and protein sequence motifs. *Nucleic Acids Res.* 34, W369–W373. doi: 10.1093/nar/gkl198

Balfourier, F., Roussel, V., Strelchenko, P., Exbrayat-Vinson, F., Sourdille, P., Boutet, G., et al. (2007). A worldwide bread wheat core collection restricted to a full 384 deep well storage plate. *Theor. Appl. Genet.* 114, 1265–1275. doi: 10.1007/s00122-007-0517-1

Baranowskij, N., Frohberg, C., Prat, S., and Willmitzer, L. (1994). A novel DNA binding protein with homology to Myb oncoproteins containing only one repeat can function as a transcriptional activator. *EMBO J.* 13, 5383–5392.

Bäumlein, H., Nagy, I., Villarroel, R., Inze, D., and Wobus, U. (1992). *Cis*-analysis of a seed protein gene promoter: the conservation RY repeat CATGCATG within the legumin box is essential for tissue-specific expression of a legumin gene. *Plant J.* 2, 233–239.

Bernard, V., Brunaud, V., and Lecharny, A. (2010). TC-motifs at the TATA-box expected position in plant genes: a novel class of motifs involved in the transcription regulation. *BMC Genomics* 11:166. doi: 10.1186/1471-2164-11-166

Branlard, G., and Dardevet, M. (1985). Diversity of grain protein and bread wheat quality. II. Correlation between high molecular weight subunits of glutenin and flour quality characteristics. *J. Cereal Sci.* 3, 345–354. doi: 10.1016/S0733-5210(85)80007-2

Carman, J. G., and Bishop, D. L. (2004). Diurnal O2 and carbohydrate levels in wheat kernels during embryony. *J. Plant Physiol.* 161, 1003–1010. doi: 10.1016/j.jplph.2004.01.003

Chang, W. C., Lee, T. Y., Huang, H. D., Huang, H. Y., and Pan, R. L. (2008). PlantPAN: Plant Promoter Analysis Navigator, for identifying combinatorial *cis*-regulatory elements with distance constraint in plant gene group. *BMC Genomics* 9:561. doi: 10.1186/1471-2164-9-561

Chao, S., Zhang, W., Akunov, E., Sherman, J., Ma, Y., Luo, M. C., et al. (2009). Analysis of gene-derived SNP marker polymorphism in US wheat (*Triticum aestivum* L.) cultivars. *Mol. Breed.* 23, 23–33. doi: 10.1007/s11032-008-9210-6

Chaudhary, J., and Skinner, M. K. (1999). Basic helix-loop-helix proteins can act at the E-box within the serum response element of the c-fos promoter to influence hormone-induced promoter activation in Sertoli cells. *Mol. Endocrinol.* 13, 774–786. doi: 10.1210/mend.13.5.0271

Choi, H., Hong, J., Ha, J., Kang, J., and Kim, S. Y. (2000). ABFs, a family of ABA-responsive element binding factors. *J Biol. Chem.* 275, 1723–1730. doi: 10.1074/jbc.275.3.1723

Ciceri, P., Gianazza, E., Lazzari, B., Lippoli, G., Genga, A., Hoschek, G., et al. (1997). Phosphorylation of Opaque2 changes diurnally and impacts Its DNA binding activity. *Plant Cell* 9, 97–108. doi: 10.1105/tpc.9.1.97

Ciceri, P., Locatelli, F., Genga, A., Viotti, A., and Schmidt, R. J. (1999). The activity of the maize Opaque2 transcriptional activator is regulated diurnally. *Plant Physiol.* 121, 1321–1327. doi: 10.1104/pp.121.4.1321

Conlan, R. S., Hammond-Kosack, M., and Bevan, M. (1999). Transcription activation mediated by the bZIP factor SPA on the endosperm box is modulated by ESBF-1 *in vitro*. *Plant J.* 19, 173–181. doi: 10.1046/j.1365-313X.1999.00522.x

Daraselia, N. D., Tarchevskaya, S., and Narita, J. O. (1996). The promoter for tomato 3-hydroxy-3-methylglutaryl coenzyme A reductase gene 2 has unusual regulatory elements that direct high-level expression. *Plant Physiol.* 112, 727–733. doi: 10.1104/pp.112.2.727

De Bustos, A., Rubio, P., and Jouve, N. (2000). Molecular characterization of the inactive allele of the gene *Glu-A1* and the development of a set of AS-PCR markers for HMW glutenins of wheat. *Theor. Appl. Genet.* 100, 1085–1094. doi: 10.1007/s001220051390

De Mendiburu, F. (2014). *Agricolae: Statistical Procedures for Agricultural Research.* R package version 1.1-8. Available online at: http://CRAN.R-project.org/package=agricolae

Dereeper, A., Nicolas, S., Lecunff, L., Bacilieri, R., Doligez, A., Peros, J. P., et al. (2011). SNiPlay: a web-based tool for detection, management and analysis of SNPs. Application to grapevine diversity projects. *BMC Bioinformatics* 12:134. doi: 10.1186/1471-2105-12-134

Diaz, I., Martinez, M., Isabel-La Moneda, I., Rubio-Somoza, I., and Carbonero, P. (2005). The DOF protein, SAD, interacts with GAMYB in plant nuclei and activates transcription of endosperm-specific genes during barley seed development. *Plant J.* 42, 652–662. doi: 10.1111/j.1365-313X.2005.02402.x

Diaz, I., Vicente-Carbajosa, J., Abraham, Z., Martinez, M., Isabel-La Moneda, I., and Carbonero, P. (2002). The GAMYB protein from barley interacts with the DOF transcription factor BPBF and activates endosperm-specific genes during seed development. *Plant J.* 29, 453–464. doi: 10.1046/j.0960-7412.2001.01230.x

Ding, J., Hu, H., and Li, X. (2012). Thousands of *cis*-regulatory sequence combinations are shared by Arabidopsis and poplar. *Plant Physiol.* 158, 145–155. doi: 10.1104/pp.111.186080

Eini, O., Yang, N., Pyvovarenko, T., Pillman, K., Bazanova, N., Tikhomirov, N., et al. (2013). Complex regulation by Apetala2 domain-containing transcription factors revealed through analysis of the stress-responsive TdCor410b promoter from durum wheat. *PloS ONE* 8:e58713. doi: 10.1371/journal.pone.0058713

Fauteux, F., and Strömvik, M. (2009). Seed storage protein gene promoters contain conserved DNA motifs in Brassicaceae, Fabaceae and Poaceae. *BMC Plant Biol.* 9:126. doi: 10.1186/1471-2229-9-126

Fujiwara, T., and Beachy, R. N. (1994). Tissue-specific and temporal regulation of a beta-conglycinin gene: roles of the RY repeat and other *cis*-acting elements. *Plant Mol. Biol.* 24, 261–272. doi: 10.1007/BF00020166

Gu, Y. Q., Salse, J., Coleman-Derr, D., Dupin, A., Crossman, C., Lazo, G. R., et al. (2006). Types and rates of sequence evolution at the high-molecular-weight glutenin locus in hexaploid wheat and its ancestral genomes. *Genetics* 174, 1493–1504. doi: 10.1534/genetics.106.060756

Gupta, R. B., and MacRitchie, F. (1994). Allelic variation at glutenin subunit and gliadin loci, *Glu-1*, *Glu-3*, and *Gli-1* of common wheats II. Biochemical basis of the allelic effects on dough properties. *J. Cereal Sci.* 19, 19–29. doi: 10.1006/jcrs.1994.1004

Halford, N. G., Forde, J., Shewry, P. R., and Kreiss, M. (1989). Functional analysis of the upstream regions of a silent and an expressed member of a family of wheat seed protein genes in transgenic tobacco. *Plant Sci.* 62, 207–216. doi: 10.1016/0168-9452(89)90083-6

Hammond-Kosack, M. C., Holdsworth, J. M., and Bevan, W. M. (1993). *In vivo* footprinting of a low molecular weight glutenin gene (*LMWG-1Dl*) in wheat endosperm. *EMBO J.* 12, 545–554.

Haseneyer, G., Ravel, C., Dardevet, M., Balfourier, F., Sourdille, S., Charmet, G., et al. (2008). High level of conservation between genes coding for the GAMYB transcription factor in barley (*Hordeum vulgare* L.) and bread wheat (*Triticum aestivum* L.) collections. *Theor. Appl. Genet.* 117, 321–331. doi: 10.1007/s00122-008-0777-4

He, Z. H., Liu, L., Xia, X. C., Liu, J. J., and Peña, R. J. (2005). Composition of HMW and LMW glutenin subunits and their effects on dough properties, pan bread, and noodle quality of Chinese bread wheats. *Cereal Chem.* 82, 345–350.

Higo, K., Ugawa, Y., Iwamoto, M., and Korenaga, T. (1999). Plant *cis*-acting regulatory DNA elements (PLACE) database. *Nucleic Acids Res.* 27, 297–300. doi: 10.1093/nar/27.1.297

Isabel-La Moneda, I., Diaz, I., Martinez, M., Mena, M., and Carbonero, P. (2003). SAD: a new DOF protein from barley that activates transcription of a cathepsin B-like thiol protease gene in the aleurone of germinating seeds. *Plant J.* 33, 329–340. doi: 10.1046/j.1365-313X.2003.01628.x

Jefferson, R. A., Kavanagh, T. A., and Bevan, M. W. (1987). GUS fusions: beta-glucuronidase as a sensitive and versatile gene fusion marker in higher plants. *EMBO J.* 6, 3901–3907.

Juhász, A., Makai, S., Sebestyén, E., Tamás, L., and Balázs, E. (2011). Role of conserved non-coding regulatory elements in LMW glutenin gene expression. *PLoS ONE* 6:e29501. doi: 10.1371/journal.pone.0029501

Kang, J. Y., Choi, H. I., Im, M. Y., and Kim, S. Y. (2002). Arabidopsis basic leucine zipper proteins that mediated stress-responsive abscisic acid signaling. *Plant Cell.* 14, 343–357. doi: 10.1105/tpc.010362

Korkuć, P., Schippers, J. H. M., and Walther, D. (2014). Characterization and identification of *cis*-regulatory elements in Arabidopsis based on single-nucleotide polymorphism information. *Plant Physiol.* 164, 181–200. doi: 10.1104/pp.113.229716

Lamacchia, C., Shewry, P. R., Di Fonzo, N., Forsyth, J. L., Harris, N., Lazzeri, P. A., et al. (2001). Endosperm-specific activity of a storage protein gene promoter in transgenic wheat seed. *J. Exp. Bot.* 52, 243–250. doi: 10.1093/jexbot/52.355.243

Makai, S., Tamás, L., and Juhász, A. (2013). "Evolutionary differences in the transcriptional regulation of HMW glutenin genes in Triticeae," in *International Wheat Genetics Symposium* (Yokohama), 146.

Marzábal, P., Busk, P. K., Ludevid, M. D., and Torrent, M. (1998). The bifactorial endosperm box of alpha-zein gene: characterisation and function of the Pb3 and GZM *cis*-acting elements. *Plant J.* 16, 41–52. doi: 10.1046/j.1365-313x.1998.00272.x

Mena, M., Vicente-Carbajosa, J., Schmidt, R. J., and Carbonero, P. (1998). An endosperm-specific DOF protein from barley, highly conserved in wheat, binds to and activates transcription from the prolamin-box of a native B-hordein promoter in barley endosperm. *Plant J.* 16, 53–62. doi: 10.1046/j.1365-313x.1998.00275.x

Moreno-Risueno, M. A., Gonzalez, N., Diaz, I., Parcy, F., Carbonero, P., and Vicente-Carbajosa, J. (2008). FUSCA3 from barley unveils a common transcriptional regulation of seed-specific genes between cereals and Arabidopsis. *Plant J.* 53, 882–894. doi: 10.1111/j.1365-313X.2007.03382.x

Norre, F., Peyrot, C., Garcia, C., Rancé, I., Drevet, J., Theisen, M., et al. (2002). Powerful effect of an atypical bifactorial endosperm box from wheat *HMWG-Dx5* promoter in maize endosperm. *Plant Mol. Biol.* 50, 699–712. doi: 10.1023/A:1019953914467

Oñate, L., Vicente-Carbajosa, J., Lara, P., Diaz, I., and Carbonero, P. (1999). Barley BLZ2: a seed-specific bZIP protein that interacts with BLZ1 *in vivo* and activates transcription from the GCN4-like motif of B-hordein promoters in barley endosperm. *J. Biol. Chem.* 274, 9175–9182.

Payne, P. I., and Lawrence, G. J. (1983). Catalogue of alleles for the complex gene loci *GluA1*, *GluB1* and *GluD1* which code for the high-molecular-weight subunits of glutenin in hexaploid wheat. *Cereal Res. Comm.* 11, 29–35.

Payne, P. I., Nightingale, M. A., Krattiger, A. F., and Holt, L. M. (1987). The relationship between HMW glutenin subunit composition and the breadmaking quality of British grown wheat varieties. *J. Sci. Food Agric.* 40, 51–65. doi: 10.1002/jsfa.2740400108

Piechulla, B., Merforth, N., and Rudolph, B. (1998). Identification of tomato *Lhc* promoter region necessary for circadian expression. *Plant Mol. Biol.* 38, 655–662. doi: 10.1023/A:1006094015513

Purugganan, M. D. (2000). The molecular population genetics of regulatory genes. *Mol. Ecol.* 9, 1451–1461. doi: 10.1046/j.1365-294x.2000.01016.x

Radchuk, V. V., Sreenivasul, N., Radchuk, R. I., Wobus, U., and Weschke, W. (2005). The methylation cycle and its possible functions in barley endosperm development. *Plant Mol. Biol.* 59, 289–307. doi: 10.1007/s11103-005-8881-1

Ravel, C., Martre, P., Romeuf, I., Dardevet, M., El-Malki, R., Bordes, J., et al. (2009). Nucleotide polymorphism in the wheat transcriptional activator Spa influences its pattern of expression and has pleiotropic effects on grain protein composition, dough viscoelasticity, and grain hardness. *Plant Physiol.* 151, 2133–2144. doi: 10.1104/pp.109.146076

Ravel, C., Praud, S., Murigneux, A., Canaguier, A., Sapet, F., Samson, D., et al. (2006). Single-Nucleotide Polymorphisms (SNPs) frequency in a set of selected lines of bread wheat (*Triticum aestivum* L.). *Genome* 49, 1131–1139. doi: 10.1139/g06-067

R Core Team (2013). *R: A Language and Environment for Statistical Computing.* Vienna: R Foundation for Statistical Computing. Available online at: http://www.R-project.org/

Rombauts, S., Dehais, P., Van Montagu, M., and Rouze, P. (1999). PlantCARE, a plant *cis*-acting regulatory element database. *Nucleic Acids Res.* 27, 295–296. doi: 10.1093/nar/27.1.295

Rubio-Somoza, I., Martinez, M., Abraham, Z., Diaz, I., and Carbonero, P. (2006a). Ternary complex formation between HvMYBS3 and other factors involved in transcriptional control in barley seeds. *Plant J* 47, 269–281. doi: 10.1111/j.1365-313X.2006.02777.x

Rubio-Somoza, I., Martinez, M., Diaz, I., and Carbonero, P. (2006b). HvMCB1, a R1MYB transcription factor from barley with antagonistic regulatory functions during seed development and germination. *Plant J.* 45, 17–30. doi: 10.1111/j.1365-313X.2005.02596.x

Ragupathy, R., Naeem, H. A., Reimer, E., Lukow, O. M., Sapirstein, H. D., and Cloutier, S. (2008). Evolutionary origin of the segmental duplication encompassing the wheat *GLU-B1* locus encoding the overexpressed Bx7 (Bx7OE) high molecular weight glutenin subunit. *Theor. Appl. Genet.* 116, 283–296. doi: 10.1007/s00122-007-0666-2

Seitz, S. B., Weisheit, W., and Mittag, M. (2010). Multiple roles and interaction factors of an E-Box element in *Clamydomonas reinhardtii. Plant Physiol.* 152, 2243–2257. doi: 10.1104/pp.109.149195

Shewry, P. R. (2009). Wheat. *J. Exp. Bot.* 60, 1537–1553. doi: 10.1093/jxb/erp058

Shewry, P. R., Halford, N. G., Belton, P. S., and Tatham, A. S. (2002). The structure and properties of gluten: an elastic protein from wheat grain. *Philos. Trans. R. Soc. Lond. B Biol. Sci.* 357, 133–142. doi: 10.1098/rstb.2001.1024

Shewry, P. R., Underwood, C., Wan, Y., Lovegrove, A., Bhandari, D., Toole, G., et al. (2009). Storage product synthesis and accumulation in developing grains of wheat. *J. Cereal. Sci.* 50, 106–112. doi: 10.1016/j.jcs.2009.03.009

Shirsat, A., Wilford, N., Croy, R., and Boulter, D. (1989). Sequences responsible for the tissue specific promoter activity of a pea legumin gene in tobacco. *Mol. Gen. Genet.* 215, 326–331. doi: 10.1007/BF00339737

Sorensen, M. B., Muller, M., Skerritt, J., and Simpson, D. (1996). Hordein promoter methylation and transcriptional activity in wild-type and mutant barley endosperm. *Mol. Gen. Genet.* 250, 750–760. doi: 10.1007/BF02172987

Stormo, G. D. (2000). DNA binding sites: representation and discovery. *Bioinformatics* 16, 16–23. doi: 10.1093/bioinformatics/16.1.16

Suzuki, A., Wu, C. Y., Washida, H., and Takaiwa, F. (1998). Rice MYB protein OSMYB5 specifically binds to the AACA motif conserved among promoters of storage protein glutelin. *Plant Cell Physiol.* 39, 555–559. doi: 10.1093/oxfordjournals.pcp.a029404

Takaiwa, F., Yamanouchi, U., Yoshihara, T., Washida, H., Tanabe, F., Kato, A., et al. (1996). Characterization of common *cis*-regulatory elements responsible for the endosperm-specific expression of members of the rice glutelin multigene family. *Plant Mol. Biol.* 30, 1207–1221. doi: 10.1007/BF00019553

Thomas, M. S., and Flavell, R. B. (1990). Identification of an enhancer element for the endosperm-specific expression of high molecular weight glutenin. *Plant Cell* 2, 1171–1180. doi: 10.1105/tpc.2.12.1171

Van Herpen, T. W., Riley, M., Sparks, C., Jones, H. D., Gritsch, C., Dekking, E. H., et al. (2008). Detailed analysis of the expression of an alpha-gliadin promoter and the deposition of alpha-gliadin protein during wheat grain development. *Ann. Bot.* 102, 331 –342. doi: 10.1093/aob/mcn114

Verdier, J., and Thompson, R. D. (2008). Transcriptional regulation of storage protein synthesis during dicotyledon seed filling. *Plant Cell Physiol.* 49, 1263–1271. doi: 10.1093/pcp/pcn116

Vicente-Carbajosa, J., Moose, S. P., Parsons, R. L., and Schmidt, R. J. (1997). A maize zinc-finger protein binds the prolamin box in zein gene promoters and interacts with the basic leucine zipper transcriptional activator Opaque2. *Proc. Natl. Acad. Sci. U.S.A.* 94, 7685–7690. doi: 10.1073/pnas.94.14.7685

Vicente-Carbajosa, J., Oñate, L., Lara, P., Diaz, I., and Carbonero, P. (1998). Barley BLZ1: a bZIP transcriptional activator that interacts with endosperm-specific gene promoters. *Plant J.* 13, 629–640.

Vickers, C. E., Xue, G., and Gresshoff, P. M. (2006). A novel *cis*-acting element, ESP, contributes to high-level endosperm-specific expression in an oat globulin promoter. *Plant Mol. Biol.* 62, 195–214. doi: 10.1007/s11103-006-9014-1

Wang, K., Zhang, X., Zhao, Y., Chen, F., and Xia, G. (2013). Structure, variation and expression analysis of glutenin gene promoters from *Triticum aestivum* cultivar Chinese Spring shows the distal region of promoter 1Bx7 is key regulatory sequence. *Gene* 527, 484–490. doi: 10.1016/j.gene.2013.06.068

Wu, C., Washida, H., Onodera, Y., Harada, K., and Takaiwa, F. (2000). Quantitative nature of the Prolamin-box, ACGT and AACA motifs in a rice glutelin gene promoter: minimal cis-element requirements for endosperm-specific gene expression. *Plant J.* 23, 415–421. doi: 10.1046/j.1365-313x.2000.00797.x

Xi, D. M., and Zheng, C. C. (2011). Transcriptional regulation of seed storage protein genes in Arabidopsis and cereals. *Seed Sci. Res.* 21, 247–254. doi: 10.1017/S0960258511000237

Yanagisawa, S., and Schmidt, R. J. (1999). Diversity and similarity among recognition sequences of Dof transcription factors. *Plant J.* 15, 209–214. doi: 10.1046/j.1365-313X.1999.00363.x

Conflict of Interest Statement: The authors declare that the research was conducted in the absence of any commercial or financial relationships that could be construed as a potential conflict of interest.

17

Recent advances in the study of prolamin storage protein organization and function

David R. Holding*

Department of Agronomy and Horticulture, Center for Plant Science Innovation, University of Nebraska-Lincoln, Lincoln, NE, USA

Edited by:
Brian A. Larkins, University of Nebraska-Lincoln, USA

Reviewed by:
Moritz Karl Nowack, Flanders Institute for Biotechnology, Belgium
Sinead Drea, University of Leicester, UK

***Correspondence:**
David R. Holding, Department of Agronomy and Horticulture, Center for Plant Science Innovation, University of Nebraska-Lincoln, E323 Beadle Center for Biotechnology, 1901 Vine Street, Lincoln, NE, USA
e-mail: dholding2@unl.edu

Prolamin storage proteins are the main repository for nitrogen in the endosperm of cereal seeds. These stable proteins accumulate at massive levels due to the high level expression from extensively duplicated genes in endoreduplicated cells. Such abundant accumulation is achieved through efficient packaging in endoplasmic reticulum localized protein bodies in a process that is not completely understood. Prolamins are also a key determinant of hard kernel texture in the mature seed; an essential characteristic of cereal grains like maize. However, deficiencies of key essential amino acids in prolamins result in relatively poor grain protein quality. The inverse relationship between prolamin accumulation and protein quality has fueled an interest in understanding the role of prolamins and other proteins in endosperm maturation. This article reviews recent technological advances that have enabled dissection of overlapping and non-redundant roles of prolamins, particularly the maize zeins. This has come through molecular characterization of mutants first identified many decades ago, selective down-regulation of specific zein genes or entire zein gene families, and most recently through combining deletion mutagenesis with current methods in genome and transcriptome profiling. Works aimed at understanding prolamin deposition and function as well as creating novel variants with improved nutritional and digestibility characteristics, are reported.

Keywords: zein, kafirin, prolamin, storage_protein, endosperm, QPM, protein_body, deletion_mutagenesis

INTRODUCTION

Although prolamins are the dominant class of seed storage protein in many cereals, this article illustrates their function and organization in maize and sorghum, the first and fifth most globally important cereal crops. Maize and sorghum are physically distinct in terms of their vegetative and reproductive architecture with maize having separate male and female reproductive organs and sorghum having hermaphroditic flowers that produce seed less than one tenth of the size of domestic maize seed. Sorghum is also usually more water use efficient and has potential for increased cultivation in marginal lands for this reason. Despite these differences, maize and sorghum are genetically more closely related than to other grasses. This is best shown by the phylogenetic relationships between their zein and kafirin prolamin-encoding genes (Xu and Messing, 2008).

PROLAMIN GENES

Prolamins were initially distinguished as a group of proteins soluble in 70% ethanol (Osborne, 1897). However, differences in aqueous solubility and ability to form disulfide interactions, were later used to classify prolamin sub-families. The zeins are grouped into α, β, γ, and δ types based on these properties (Esen, 1987; Coleman and Larkins, 1999). Similarly, the kafirins are grouped into α, β, γ, and δ types based on their molecular weight, solubility, and gene sequence (Shull et al., 1991). Alpha zeins are encoded by four different gene sub-families (Z1A, Z1B, Z1C, and Z1D) that in the B73 reference line contain more than 40 genes in six chromosomal locations (Feng et al., 2009). There is, however, copy number and expression variation across different maize backgrounds. While substantial α-kafirin gene duplication also occurred in sorghum, all 20 α-kafirin genes are clustered at one chromosome 5 location in the BTx623 genome (Xu and Messing, 2008).

Alpha prolamins resolve at ~19- and 22-kDa on SDS-PAGE gels in both maize and sorghum. In maize, the 19-kDa α-zeins are encoded by the Z1A, Z1B, and Z1D subfamilies while the 22-kDa α-zeins are encodes by the Z1C subfamily (Song et al., 2001; Song and Messing, 2002). In sorghum, 19-kD α-kafirins are encoded by the K1α19 subfamily while the 22-kD α-kafirins are encoded by the K1α22 subfamily (Xu and Messing, 2009). Alpha prolamins cluster in a broad phylogenetic group (Group 1) as do the δ-zein genes. 10-kD and 18-kD δ-zeins are encoded by z2δ10 and z2δ18 genes in maize and k2δ2 and k2δ18 genes in sorghum (Xu and Messing, 2008, 2009). γ- and β-prolamins cluster within Group 2 (Xu and Messing, 2008, 2009). Unlike α-prolamins, Group 2 prolamin genes exist as single copies rather than highly duplicated gene families. In maize, this group consists of z2γ16 and z2γ27, encoding 16- and 27-kD γ-zeins, and z2γ50, encoding a 50-kD γ-zein. Similarly, k2γ27 and k2γ50 encode 27- and 50-kD γ-kafirins in sorghum although there is no 16-kD γ-kafirin. The maize z2γ16 gene is thought to derive from an unequal crossing-over event that occurred after allotetraploidization (Xu and Messing, 2008). Maize and sorghum also have genes encoding a 15-kD β-prolamin (z2β15 and k2β15) which are related to the γ-prolamins within Group 2 (Xu and Messing, 2008).

PROLAMIN ACCUMULATION, THEIR EFFECT ON GRAIN TEXTURE, FUNCTIONALITY, AND PROTEIN QUALITY

Sorghum kernels are usually much smaller than maize kernels and both, but especially sorghum, show considerable heterogeneity in seed size across varieties (**Figure 1**). Despite this variability, maize and sorghum seed have similar endosperm composition having a high proportion of glassy or vitreous endosperm at the periphery of the mature kernel and a central opaque region (**Figure 1**). Vitreous endosperm is important for resistance to insect and fungal damage, resilience during harvest and storage, and many end use characteristics. Although we are still learning how vitreous endosperm is formed during kernel maruration, considerable evidence suggests that accumulation and packaging of prolamins into endoplasmic reticulum (ER) protein bodies play a central role. For example, in maize the vitreous outer region of the endosperm contains much more zein than the soft, opaque interior, and environmental conditions that cause reduced zein synthesis, such as nitrogen depletion, result in kernels that are soft and starchy throughout (Tsai et al., 1978). Sorghum kernels grown in limited nitrogen conditions are smaller and lack vitreous endosperm (**Figure 1**). Duvick (1961) proposed that in the periphery of the developing starchy endosperm, there is a certain ratio of starch grains, protein bodies, and viscous cytoplasm which dries down to form a rigid glass-like structure at kernel maturity (the vitreous endosperm; **Figure 1**). Toward the center of the endosperm, where zein protein bodies are smaller and less abundant, the rigid matrix is not formed during kernel desiccation, which results in the formation of the friable, opaque kernel center (**Figure 1**; Duvick, 1961). In opaque mutants, the central opaque region extends to the periphery of the endosperm (**Figure 1**).

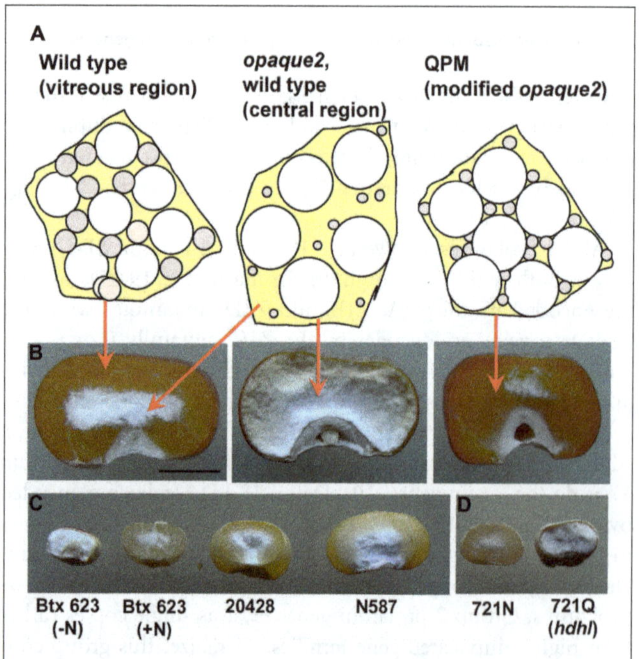

FIGURE 1 | Vitreous endosperm formation in maize and sorghum kernels. (A) Individual cells of developing endosperm are represented with the relative size and abundance of starch grains (white spheres) and zein protein bodies (gray spheres) that are thought to result in vitreous or opaque endosperm in normal as well as in *opaque2* and modified *opaque2* (QPM) kernels. **(B)** Mature kernels of wild type, *opaque-2* and QPM cracked in half to reveal extent of vitreous endosperm. **(C)** Mature sorghum kernels cracked as in **B** to reveal vitreous endosperm and size variability in sorghum grain. **(D)** High digestibility high lysine (hdhl) sorghum mutant and its wild type isoline. Scale bar in **B** is 3 mm and refers to kernels in panels **B–D**.

Protein body formation in maize is controlled at several levels, including the temporal and spatial regulation of zein gene expression, the level of transcription and interactions that occur between the different types of zein proteins (Woo et al., 2001; Kim et al., 2002). Zeins are devoid of the essential amino acids, lysine and tryptophan (Mertz et al., 1964), but account for more than 70% of maize endosperm protein. This results in an overall protein content that is especially deficient in these amino acids. The equally dominant sorghum kafirins, share this nutritional deficiency, but it is compounded by the their poor digestibility (Aboubacar et al., 2001) that results from their high degree of disufide cross-linking.

Our knowledge of how prolamins are packaged at such high levels comes largely from maize. Zeins are retained as discretely layered membrane bound accretions in the ER (Lending and Larkins, 1989; **Figure 2**). Protein bodies start as small accretions consisting entirely of γ-zein, consistent with the slightly earlier onset of γ-zein gene expression (Woo et al., 2001). As protein bodies expand, α- and δ-zeins are sequestered into the protein body core, where they become encapsulated in a shell of γ-zeins. The 19-kD α-zeins are the most abundant class and immunological evidence suggests that the 22-kD α-zeins form an intermediary layer between the central 19-kD α-zeins and the γ-zein periphery (Holding et al., 2007). Although the γ-zeins have some functional redundancy, selective down-regulation has suggested they also have specialized roles as described below (Guo et al., 2013).

MUTATIONS IN PROLAMIN GENES AND RELATED FACTORS SHED LIGHT ON PROLAMIN FUNCTIONAL ORGANIZATION

Natural and engineered mutants exhibiting reduced kernel hardness offer the opportunity to dissect the various biochemical and biophysical processes that affect vitreous endosperm formation, and consequently their study is of significant agronomic importance. Kernels of these mutants are opaque since they do not transmit light and often show defects in the accumulation of zeins or their packaging into ER-localized protein bodies. However, it is now clear that other factors are also important determinants

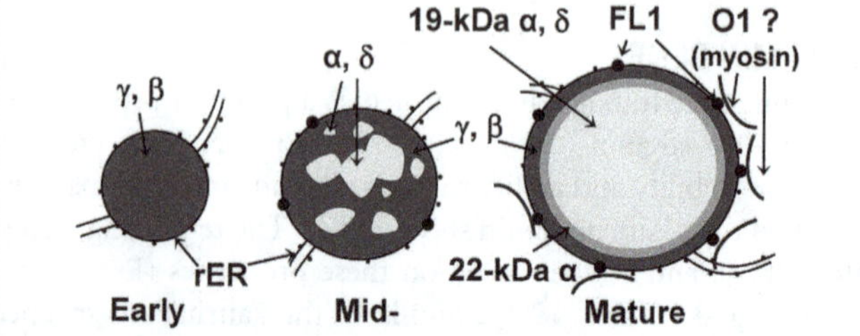

FIGURE 2 | Diagram of zein distribution in early, mid- and mature protein bodies. Small black dots in membrane represent ribosomes while large black dots represent FLOURY1 protein. Curved lines outside protein bodies represent possible direct or indirect interaction with myosin.

of kernel texture since several recent studies have shown that vitreous endosperm can be disrupted by processes that do not affect zein synthesis and protein body structure. For example, vitreous endosperm formation is abolished in the *floury1* mutant as a result of knocking out a protein body ER membrane protein which seems to be involved zein organization (Holding et al., 2007). Therefore, it is likely that further protein body-related organizational factors remain to be identified. Other opaque mutants are the result of genetic aberrations in processes unrelated to protein body formation such as amino acid biosynthesis, plastid development and cytoskeletal function (Holding et al., 2010; Myers et al., 2011; Wang et al., 2012). This indicates that further functional genomics is needed to generate a more complete understanding of the factors which control late endosperm development.

The most well-known of the maize opaque mutants is *opaque2* (*o2*) which has been widely studied because of the increased lysine and tryptophan accumulation (Mertz et al., 1964) resulting from its reduced accumulation of alpha zeins. Cloning of the *O2* gene revealed that it encodes a transcription factor that regulates α-zeins (Schmidt et al., 1990) as well as other genes such as pyruvate Pi dikinase (Maddaloni et al., 1996). Although the soft kernels and yield penalty of *o2* prevented its commercial success, subsequent breeding projects, including those in Mexico (Vasal et al., 1980) and South Africa (Geevers and Lake, 1992), led to the development of hard kernel *o2* varieties called quality protein maize (QPM). QPM kernels maintain the low levels of α-zeins and thus, retain the high levels of lysine and tryptophan but the genetic basis of *o2* endosperm modification is complicated and poorly understood. The most prominent biochemical feature of QPM endosperm is the accumulation of the 27-kD γ-zein at 2–3 fold higher levels than in wild type and *o2* (Wallace et al., 1990; Geetha et al., 1991; Lopes and Larkins, 1991). Although the genetic or epigenetic mechanism of this increase is unknown, the degree of QPM endosperm vitreousness closely correlates with the level of 27-kD γ-zein protein (Lopes and Larkins, 1991). Furthermore, the 27-kD γ-zein gene maps to the most significant QTL for endosperm modification in QPM located on chromosome 7 (Lopes and Larkins, 1995; Lopes et al., 1995; Holding et al., 2008, 2011). QPM endosperm accumulates larger numbers of small, γ-zein rich protein bodies (**Figure 1**) which are proposed to allow the formation of a rigid glassy matrix similar in texture to mature wild type endosperm (**Figure 1**). γ-zein is known to be essential for endosperm modification in QPM (Wu et al., 2010), although the extent to which it is alone sufficient is unknown.

Although the functional redundancy resulting from the multi-member α-zein gene families has prevented the identification of recessive mutants, several dominant opaque mutants have been characterized. The phenotypes in these mutants result from the accumulation of defective prolamins that interfere with normal prolamin deposition and cause ER stress responses (Coleman et al., 1997; Kim et al., 2004, 2006; Wu et al., 2013). Floury-2 (*fl2*), *Defective endosperm B30* (*De-B30*), and *Mucronate* (*Mc*) are caused by dominantly acting mutations in zein genes. *fl2*, *De-B30,* and *Mc* show pleiotropic effects and result in a general reduction of all zeins (causing increases in lysine-containing proteins as in *o2*) and lobed protein bodies (Lending and Larkins, 1992). In the case of *fl2* and *De-B30,* this result from mutations that cause the signal peptides to remain attached in the 22-kD α-zein and 19-kD α-zein respectively, resulting in aggregation of these proteins at the ER membrane (Gillikin et al., 1997; Kim et al., 2004). The *Mc* mutant (Soave and Salamini, 1984) results from a 38 bp deletion that leads to a frame-shift mutation in the 16-kD γ-zein (Kim et al., 2006). The abnormal zeins produced in these mutants result in ER stress and cause a constitutive unfolded protein response (UPR), as shown by the dramatic up-regulation of a number of UPR-associated genes (Hunter et al., 2002). In fact, elevated markers for endosperm stress is a common feature of all opaque mutants studied irrespective of discernible changes in zeins and zein protein bodies (Hunter et al., 2002). This leads to the suggestion that endosperm stress and a resulting energy crisis may be at least partially responsible for disrupting vitreous endosperm formation (Guo et al., 2012).

Zein and kafirin proteins are packaged into protein bodies that are inherently recalcitrant to digestion. This results from the disulfide cross-linked nature of the γ-prolamins themselves and the fact that that they form a shell of relatively low surface area in relation to the amount of prolamin packaged. The poor digestibility of prolamins is especially pronounced in sorghum. An opaque kernel sorghum mutant was identified in an EMS mutagenized population that had increased lysine content as a result of reduced kafirin accumulation, and most notably, a marked increase in protein digestibility (Oria et al., 2000). Called the "high digestibility high lysine" (*hdhl*) variant (**Figure 1**), this mutant has significant potential to improve the utility of sorghum as a human staple and livestock feed. The increased digestibility apparently results from increased protease accessibility caused by reticulation of kafirin protein body shape, in a manner reminiscent of *fl2*, *De-B30,* and *Mc* in maize. Furthermore, developing *hdhl* kernels also exhibit a defined UPR. These phenotypic similarities to the maize mutants prompted a directed cloning approach in which the mutation was first mapped to an alpha kafirin gene cluster and extensive genomic and cDNA sequencing identified a mutant-specific α-kafirin copy harboring point mutation (Wu et al., 2013). The mutation causes a threonine substitution of an alanine residue, that is strictly conserved at position 21 of the signal peptide of all known α-prolamins (Wu et al., 2013). This substitution causes a dominant-negative response through low level accumulation of an uncleaved α-kafirin (Wu et al., 2013).

The characterization of opaque mutants has shown that vitreous endosperm formation depends on the correct expression and processing of prolamins themselves but also on factors that may have indirect roles in prolamin protein bodies such as Floury-1 and *Opaque-1* (Holding et al., 2007; Wang et al., 2012). Floury-1 was identified as a protein body ER membrane-specific protein, through a *Mutator* knock-out line that displays normal amounts of zein proteins and normal protein body size and shape, but slightly disorganized zein organization (Holding et al., 2007). Floury-1 contains a domain of unknown function (DUF593) for which the location inside or outside the ER lumen was not determined (Holding et al., 2007). Recently, screens for endomembrane proteins that bind myosin XI proteins in *Arabidopsis* identified myosin receptor proteins that bind myosin through DUF593 (Peremyslov et al., 2013). This suggests that Fl1 may function to attach protein bodies to the cytoskeleton and may explain the absence of a severe protein body phenotype in *fl1*. Similar to *fl1, o1* also does not have reduced zein accumulation but has a reduced number of slightly smaller protein bodies (Wang et al., 2012). *O1* was identified by positional cloning and encodes a myosin XI protein that is associated with cisternal and protein body ER (Wang et al., 2012). Although it has not been demonstrated, this suggests there may be a direct or indirect functional interaction between FL1 and O1 (**Figure 2**) and may suggest that the cytoskeleton plays an essential role in endosperm maturation.

Factors unrelated to prolamins are also essential for vitreous endosperm formation as demonstrated by opaque mutants such as *mto140, o7,* and *o5* (Holding et al., 2010; Miclaus et al., 2011a; Myers et al., 2011). *MTO140* encodes a member of the maize arogenate dehydrogenase family that are involved in tyrosine biosynthesis (Holding et al., 2010), while *O7* encodes an acyl-CoA synthetase-like protein (Miclaus et al., 2011a). *O5* encodes the major biosynthetic enzyme for synthesis of chloroplast membrane lipids, monogalactosyldiacylglycerol synthase and the mutant is specifically defective in galactolipids necessary for amyloplast and chloroplast function (Myers et al., 2011). As described in the section on deletion mutagenesis below, kernel opacity is a pleiotropic characteristic of kernel mutants also displaying other phenotypes such as small kernel, rough kernel, defective kernel, viviparary, and partial empty pericarp.

TRANSGENIC EFFORTS TO OFFSET AMINO ACID DEFICIENCIES IN MAIZE

Various types of biotechnological approaches have been considered for improving the amino acid composition of maize (Holding and Larkins, 2008). In order to increase the methionine content, Lai and Messing (2002) created transgenic maize plants expressing a chimeric gene consisting of the coding region of 10-kD δ-zein and the promoter and 5′ untranslated region of the 27-kDa γ-zein. Although the effects on synthesis of endogenous high-sulfur zeins were not reported, uniformly high levels of 10-kDa δ-zein and methionine were observed and maintained over five backcross generations. Initial poultry feeding studies suggested that the transgenic grain was as effective as non-transgenic grain supplemented with free methionine.

For increasing the lysine content in maize, either non-maize, lysine-rich proteins can be expressed in an endosperm specific manner, or the zein sequences themselves can be manipulated to contain lysine. The former approach has been tried with a number of proteins which have mostly be expressed using γ- or α-zein promoters (Kriz, 2009). In order to make a significant difference to mature kernel lysine content, transgenic proteins must be driven to accumulate in very high amounts, in forms that do not interfere with the normal timing and pattern of endosperm programmed cell death and in a manner that does not induce UPR or kernel opacity. Furthermore, candidate proteins must meet stringent standards for potential allergenicity.

For the above reasons, perhaps the most promising way to elevate endosperm lysine is by modifying the coding sequences of the zein genes themselves. Lysine-containing zeins are more likely to be stored in their native ER protein body form and potentially in high enough quantities to significantly impact lysine levels. Preliminary studies in modifying a 19-kD α-zein with lysine residues, showed the transgenic protein to accumulate in protein body-like structures in *Xenopus* oocytes. Since the 19-kD α-zein is by far the most abundant zein, being packaged in the center of protein bodies (Holding et al., 2007), it may only be necessary to substitute a fraction of the native protein with a modified protein to make a significant impact on kernel protein quality. The 27-kD γ-zein, with its suspected role as the initiator of protein body formation, and as the major *o2* modifier, as well as its abundant accumulation, is also a good candidate for substituting certain amino acids with lysine. One study showed, using transient transformation of maize, that a 27-kD γ-zein in which (Pro-Lys)n sequences were inserted contiguous to or in substitution of the Pro-Xaa region, that the modified γ-zein co-localized with endogenous alpha- and gamma-zeins (Torrent et al., 1997). We are further investigating this type of approach using custom gene synthesis and the latest protein modeling programs that can assist in selection residues for substitution than are least likely to adversely affect normal zein packaging. To increase the chances of lysine-rich transgenic proteins being driven to accumulate at significant levels, it may be necessary to reduce the accumulation of native α-zeins. An effective way to do this is using RNA interference (RNAi) as described below.

Zeins are much less likely to invoke allergic reactions than the wheat prolamins, the glutens. However, when considering potential transgenic bio-fortification approaches, allergen databases must be consulted. Among several allergenic maize seed proteins, precursors of α- and β-zein have both been shown to be allergenic (Pastorello et al., 2009).

RNA INTERFERENCE LINES HAVE ALLOWED DISSECTION OF REDUNDANT AND NON-REDUNDANT PROLAMIN FUNCTION

Naturally occurring and induced opaque mutants have been extensively studied because of their potential for grain nutritional improvement and have created an understanding of prolamin packaging in ER protein bodies and its relationship to ER protein quality control and UPR. However, opaque mutants carrying nutritional benefits such as *o2* and *fl2* also carry negative pleiotropic characteristics. Furthermore, the extensive gene duplication and gene redundancy, especially in the α-prolamin classes, have resulted in a lack of recessive prolamin mutants, and an inability to infer the relative prolamin functional redundancy and non-redundancy. RNAi has been an effective tool for further addressing maize and sorghum nutritional potential as well as providing new information regarding functions of specific prolamin classes (Holding and Messing, 2013).

Initial use of RNAi to eliminate α-zeins revealed the possibility of creating dominant, non-pleiotropic low-zein lines for lysine improvement. The dominance of such transgenes circumvents one of the limitations of *o2* based varieties which is that the *o2* allele must be maintained in the homozygous mutant state, which is easily lost through wild type pollen contamination, and is especially problematic in an open pollinated QPM setting. Originally, the 22-kD α-zein was targeted and transgenic lines had considerably reduced α-zein and concomitant increase in lysine (Segal et al., 2003) despite lines accumulating substantial amounts of 19-kD α-zein. Protein bodies were of reduced size due to reduced α-zein filling and notably, exhibited distorted lobed appearances. Similarly, a study in sorghum aimed to increase lysine content and digestibility by removing α-kafirins (Kumar et al., 2012). Like the maize study, only one class of α-kafirins (22-kD) was targeted. Though a reduction in protein body size was not reported, the transgene resulted in protein body lobing similar to the 22-kD α-zein RNAi lines and dominant α-zein signal peptide mutants (Kumar et al., 2012). This suggests that 22-kD α-prolamins may be essential for correct packaging of the 19-kD α-prolamins and is consistent with the observed peripheral location of the 22-kD α-zeins relative to the 19-kD α-zeins (Holding et al., 2007). Later works targeting both the 19- and 22-kD α-zeins did not address the morphological effects on protein bodies (Huang et al., 2006; Wu and Messing, 2011). Using a chimeric α-zein RNAi cassette, comprised of ~250 bp regions of the most abundantly expressed Z1A, Z1B, Z1C, and Z1D α-zein family members in B73, and the 27-kD γ-zein promoter and the cauliflower mosaic virus 35S terminator, we suppressed both 22- and 19-kD α-zeins to low levels. This resulted in very small protein bodies but did not suppress protein body number per unit area suggesting that while α-zeins drive protein body filling, they are not involved in protein body initiation (Guo et al., 2013; **Figure 3**). In contrast to suppression of 22-kD α-zein alone, no lobing or distortion of protein bodies was observed in support of the suggestion that such phenotypes are generated by inappropriately located or unconstrained α-zeins. Similarly, in *o2* where all α-zeins are reduced, protein bodies are small but not misshapen (Geetha et al., 1991).

Apart from the *Mc* mutant, which accumulates a dominant-negative 16-kD γ-zein with a nonsense C-terminus as a result of a frame-shift mutation (Kim et al., 2006), mutants of γ-prolamins have not been described. This may be partly due to some functional redundance of different γ-prolamins but knowledge of their roles is very limited aside from their ability to form the cross-linked outer shell of ER-protein bodies. A specialized role for the 27-kD γ-zein in protein body initiation has been inferred from its increase in QPM endosperm and concomitant increase in protein body number. This is also supported by the *o15* mutant in which reduced 27-kD γ-zein leads to reduced protein body number (Dannenhoffer et al., 1995).

RNA interference lines have provided functional insight about different γ-zeins. Transgenic events that targeted the 27-/16-kD γ-zeins, whose genes share sequence similarity, as well as a separate 15-kD β-zein RNAi event caused minor morphological changes to protein bodies (Wu and Messing, 2010). Although an incidence of very small protein bodies was shown in both cases, the majority of protein bodies were of normal size and no changes in protein body number were reported (Wu and Messing, 2010). The 50-kD γ-zein was not targeted in this study since it was assumed to have a minor role in protein body formation due to its low abundance. In fact, by manipulating the ethanolic extraction procedure, it has been shown that the 50-kD γ-zein has comparable abundance to other γ-zeins except the dominantly abundant 27-kD γ-zein (Guo et al., 2013).

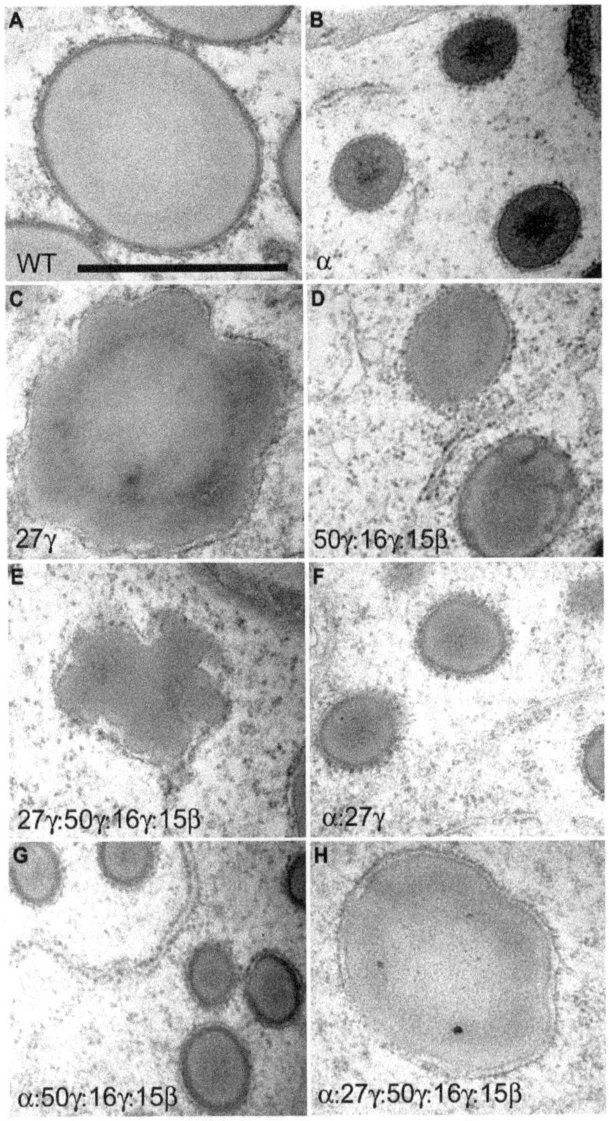

FIGURE 3 | TEM analysis showing protein body size and morphology in the fourth sub-aleurone starchy layer of 18 DAP endosperm in zein RNAi lines and their crosses. Scale bar in **A** is 1 μm and refers to all panels. RNAi transgenes present are shown in bottom left of each panel in **A–H**.

We made RNAi constructs to dissect the role of 27-kD γ-zein with respect to other γ-zeins. The first was a complete 27-kD γ-zein gene RNAi (27-γ) driven by its native promoter and terminated by the cauliflower mosaic virus 35S terminator, and this almost completely inhibited synthesis of 27-kD γ-zein and, because of sequence similarity, significantly reduced 16-kD γ-zein. A second transgene was a synthetic RNAi gene consisting of ~250 bp regions of the 16- and 50-kD γ-zein and the 15-kD β-zein (16/50/15-γ) that were selected to have the least similarity to 27-kD γ-zein. In this case, the promoter of a dominantly expressed 22-kD α-zein gene was used in this case, to avoid potential co-suppressive effects on the 27-kD γ-zein level that could have resulted from using the 27-kD γ-zein promoter. The 16/50/15-γ transgene reduced all target proteins to very low levels while leaving 27-kD at high levels (Guo et al., 2013). Protein bodies in developing endosperm of 16/50/15-γ events were of normal shape but were of reduced size, and accumulated in normal numbers (**Figure 3**; Guo et al., 2013). This illustrates that while the 16/50/15-kD γ-zeins are necessary for protein body filling and α-zein encapsulation, they are not involved in protein body initiation. Conversely, the 27-γ RNAi did not reduce protein body size and induced undulations in protein body shape (**Figure 3**), suggesting that the 27-kD γ-zein is not necessary for the bulk of protein body filling and that other γ-zeins can fulfill this role. Most notably however, protein bodies in 27-γ events were of much lower number than control, suggesting that the 27-kD γ-zein alone has the role of protein body initiation (Guo et al., 2013). Combining the γ-zein transgenes resulted in addition of the reduced number/distortion phenotypes of the 27-γ RNAi with the reduced size phenotype of the 16/50/15-γ transgene (Guo et al., 2013; **Figure 3**). Similarly, combining either of the γ-zein RNAi transgenes with the α-zein RNAi resulted in small protein bodies (**Figure 3**) with the 27-γ transgene reducing protein body number more significantly than the 16/50/15-γ transgene (Guo et al., 2013). When all three transgenes were combined to reduce all zeins proportionately, while protein body number was very low, protein bodies had a normal size and relatively normal morphology (**Figure 3**; Guo et al., 2013). This showed that maintenance of an appropriate ratio of all zeins is critical for their proper storage.

FUNCTIONAL GENOMICS OF MAIZE ENDOSPERM MATURATION USING DELETION MUTAGENESIS

While RNAi transgenes as well as different types of mutation often result in leaky expression, gene deletion mutagenesis, though random, has the advantage creating complete nulls. We investigated γ-irradiation for its potential to identify genome regions containing *o2* modifier genes in a QPM background. A small population of ~300 M3 families contained a number of recessive opaque revertant mutants (**Figure 4**) thus demonstrating its utility for identifying *o2* modifier genes as well as genes generally involved in kernel maturation (Yuan et al., 2014). This non-pleiotropic and non-lethal class of kernel mutants had varying effects on zein accumulation (**Figure 5**). The two most striking of these opaque mutants generated were unlike any previously described zein variants. The first of these, line 107, which has been thoroughly characterized at the molecular and phenotypic levels, is a null mutant of 27-kD γ-zein (Yuan et al., 2014). The second mutant, line 198, reduces 19-kD α-zeins to very low levels, in addition to the already low 22-kD α-zein caused by the *o2* mutation. Line 107 showed a complete absence of 27-kD γ-zein on SDS-PAGE gels, and the 50-kDa γ-zein was also undetectable. This was a preliminary indication that a deletion spanned both genes since they are known to be separated by only 27 kbp on chromosome 7 (Holding et al., 2008). Subsequent RT-PCR and PCR confirmed the absence transcripts and genes, while the unlinked 16-kD γ-zein was unaffected (Yuan et al., 2014). Illumina sequencing of exon-enriched genomic DNA showed that line 107 has a 1.2 Mbp deletion on chromosome 7.02 that includes both the 27- and 50-kD γ-zein genes. The increase in

FIGURE 4 | M3 ear phenotypes of selected segregating K0326Y opaque deletion mutants. K0326Y mutant line number is shown at the top center of each panel in **A–I**. Inserts in **A**,**C** show light box phenotypes. Modified from Yuan et al. (2014).

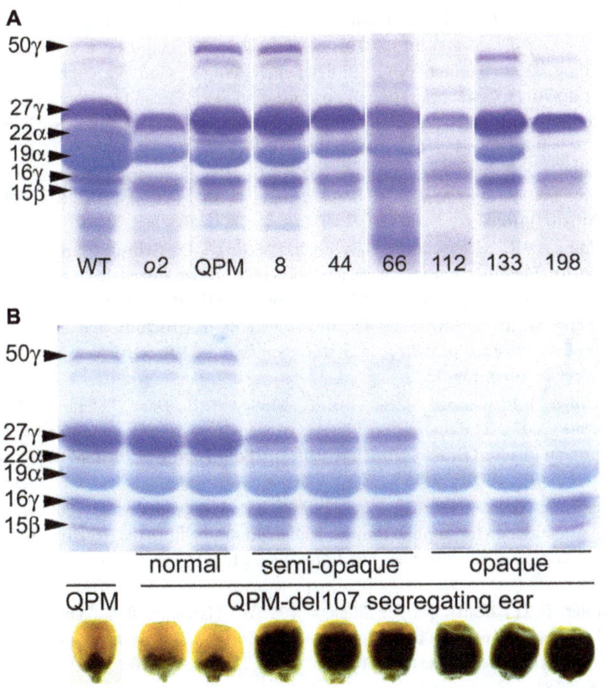

FIGURE 5 | (A) SDS-PAGE zein profiles of selected K0326Y QPM opaque reversion mutants. **(B)** SDS-PAGE zein profiles and kernel vitreousness phenotypes of QPM, and hemizygous and homozygous line 107 gamma-zein deletion mutants. Modified from Yuan et al. (2014).

27-kD γ-zein in QPM and its map position within the largest QTL for *o2* endosperm modification has long been known, but we also observed an increase in 50-kD γ-zein in QPM, possibly demonstrating a contribution to this QTL. The homozygous γ-zein deletion completely abolished endosperm modification since kernels were fully opaque. Interestingly, hemizygous kernels with a single copy of each gene accumulated intermediate amounts of both γ-zeins and were semi-modified (**Figure 5**), indicating a haploinsufficiency effect in which high level expression from both copies of these genes are necessary for vitreous endosperm formation in QPM. This unequivocally establishes 27-kD γ-zein as *o2* modifier gene. More recent work indicates that the haploinsufficiency of the 27-kD γ-zein does not apply to normal vitreous endosperm formation in a wild type (*O2/O2*) background, since kernels hemizygous for the 27-kD γ-zein deletion are fully vitreous.

The 19-kD α-zeins are encoded by genes within the Z1A1, Z1A2, Z1B, and Z1D families residing at four different loci (Miclaus et al., 2011b). Though the expression of all these genes is not equivalent within B73 and varies dramatically between genetic backgrounds (Feng et al., 2009), the very low abundance of 19-kD α-zein in QPM deletion line 198 endosperm is unlikely to result from a physical deletion at one of the four loci. Indeed, our data do not suggest a physical Z1 deletion in line 198. All Z1 classes including Z1C (22-kD α-zein) show substantially reduced transcript abundance compared with the non-mutagenized control. However, exon sequencing and RNA-seq did not reveal any physical Z1 gene deletions, cDNA sequencing did not identify missing species and bulked segregation analysis maps the mutation to a chromosome not containing Z1 loci. The more likely scenario is that the causative mutation in line 198 is within an *Opaque-2* unrelated gene with a direct or indirect role in regulating alpha-zein abundance. For other opaque mutants shown in **Figures 4** and **5**, mapping populations have been made and map positions are being used to guide prioritization of candidate gene deletions from exon-seq and RNA-seq data already generated.

Since even a small population of ~300 families in the QPM population yielded a more than 20 new opaque and small kernel mutants, we made a second mutagenized population in the B73 reference line. This population is much larger and resulted in 1793 M2 (second generation) ears. Rather than just *o2* modifier genes, the scope of this population is for more general seed functional genomics and direct manipulation of grain quality (such as by direct deletion of alpha zein sub-family loci). Since mutants are isogenic with the reference B73 genome, it will be considerably easier to assemble and utilize DNA and RNA-seq data when surveying the nature and extent of deletions within mutants.

The cells that give rise to the ear and the cells that give rise to the tassel are already specified and are physically separate in the embryonic maize shoot apical meristem (Poethig et al., 1986), and so a given mutation will not be present both ear and tassel of an M1 plant. Thus, M2 kernels will be hemizygous for such a mutation, and only show a kernel phenotype if it is dominant. Consequently, M3 families must be propagated to identify segregating recessive mutants. Among families already advanced to the M3 (10–15 plants each), there is substantial overlap between the seed phenotypes observed, since the majority of opaque mutants show some degree of reduced kernel size. Molecular genetic and biochemical characterization of these mutants will increase our understanding of the processes controlling kernel filling and its proper maturation. One priority is to identify mutants that have alterations in the ratio of zein and non-zein fractions and several mutants that have relatively increased amounts of non-zein proteins, with or without a corresponding decrease in zein proteins have been identified. F1 outcrosses are being made to Mo17 and subsequently F2 mapping populations will be used to map the causative mutations to a chromosome bin through Bulked Segregant Analysis. Map positions will be used to guide interpretation of Illumina HiSeq2500 DNA- and RNA-seg data as well as LC-MS/MS quantitative complete proteome profiling data.

CONCLUDING REMARKS

Studies of opaque mutants, especially *opaque-2* and QPM, were fueled by the prospect of maize varieties with improved protein quality. Though QPM varieties have been bred and are in use in many developing countries, their potential has not been realized in the U.S. Furthermore, characterization of other opaque mutants, especially those that do not improve protein quality, has been slow until recently. Our knowledge of the mechanisms of protein body formation and the role of zein storage proteins and other unknown factors in vitreous endosperm formation in both wild type and QPM contexts, has been limited. However, over the last decade, several technological developments have accelerated our understanding of these processes. Mutator transposon induced opaque mutants led to the identification of several non-zein factors with roles in endosperm maturation. An enhanced ability to perform map-based cloning in maize has resulted in several very old, broadly mapped opaque mutants being cloned in recent years. RNAi studies have also shed light on the redundant and non-redundant roles of zeins in protein body formation. Deletion mutagenesis is emerging as an additional way to confirm suspected *opaque-2* modifier genes and potentially lead to identification of unknown ones. We are using this approach to generate new kernel mutants that will hasten seed functional genomics when paired with current genetic mapping resources, DNA- and RNA-sequencing capacities and proteomics.

ACKNOWLEDGMENTS

David R. Holding was supported by the UNL Department of Agronomy and Horticulture and The Center for Plant Science Innovation during this work.

REFERENCES

Aboubacar, A., Axtell, J. D., Huang, C. P., and Hamaker, B. R. (2001). A rapid protein digestibility assay for identifying highly digestible sorghum lines. *Cereal Chem.* 78, 160–165. doi: 10.1094/Cchem.2001.78.2.160

Coleman, C. E., Clore, A. M., Ranch, J. P., Higgins, R., Lopes, M. A., and Larkins, B. A. (1997). Expression of a mutant alpha-zein creates the floury2 phenotype in transgenic maize. *Proc. Natl. Acad. Sci. U.S.A.* 94, 7094–7097. doi: 10.1073/pnas.94.13.7094

Coleman, C. E., and Larkins, B. A. (1999). "The prolamins of maize," in *Seed Proteins*, eds P. Shewry and R. Casey (Dordrecht: Kluwer Academic Publishers), 109–139. doi: 10.1007/978-94-011-4431-5_6

Dannenhoffer, J. M., Bostwick, D. E., Or, E., and Larkins, B. A. (1995). Opaque-15, a maize mutation with properties of a defective opaque-2 modifier. *Proc. Natl. Acad. Sci. U.S.A.* 92, 1931–1935. doi: 10.1073/pnas.92.6.1931

Duvick, D. N. (1961). Protein granules of maize endosperm cells. *Cereal Chem.* 38, 374–385.

Esen, A. (1987). A proposed nomenclature for the alcohol-soluble proteins (zeins) of maize (*Zea-mays*-L). *J. Cereal Sci.* 5, 117–128. doi: 10.1016/S0733-5210(87)80015-2

Feng, L. N., Zhu, J., Wang, G., Tang, Y. P., Chen, H. J., Jin, W. B., et al. (2009). Expressional profiling study revealed unique expressional patterns and dramatic expressional divergence of maize alpha-zein super gene family. *Plant Mol. Biol.* 69, 649–659. doi: 10.1007/s11103-008-9444-z

Geetha, K. B., Lending, C. R., Lopes, M. A., Wallace, J. C., and Larkins, B. A. (1991). Opaque-2 modifiers increase gamma-zein synthesis and alter its spatial-distribution in maize endosperm. *Plant Cell* 3, 1207–1219. doi: 10.1105/tpc.3.11.1207

Geevers, H. O., and Lake, J. K. (1992). "Development of modified opaque-2 maize in South Africa," in *Quality Protein Maize*, ed. E. T. Mertz (St. Paul, Minnesota: American Society of Cereal Chemists), 49–78.

Gillikin, J. W., Zhang, F., Coleman, C. E., Bass, H. W., Larkins, B. A., and Boston, R. S. (1997). A defective signal peptide tethers the floury-2 zein to the endoplasmic reticulum membrane. *Plant Physiol.* 114, 345–352. doi: 10.1104/pp.114.1.345

Guo, X., Ronhovde, K., Yuan, L. L., Yao, B., Soundararajan, M. P., Elthon, T., et al. (2012). Pyrophosphate-dependent fructose-6-phosphate 1-phosphotransferase induction and attenuation of hsp gene expression during endosperm modification in quality protein maize. *Plant Physiol.* 158, 917–929. doi: 10.1104/pp.111.191163

Guo, X., Yuan, L., Chen, H., Sato, S. J., Clemente, T. E., and Holding, D. R. (2013). Nonredundant function of zeins and their correct stoichiometric ratio drive protein body formation in maize endosperm. *Plant Physiol.* 162, 1359–1369. doi: 10.1104/pp.113.218941

Holding, D. R., Hunter, B. G., Chung, T., Gibbon, B. C., Ford, C. F., Bharti, A. K., et al. (2008). Genetic analysis of opaque2 modifier loci in quality protein maize. *Theor. Appl. Genet.* 117, 157–170. doi: 10.1007/s00122-008-0762-y

Holding, D. R., Hunter, B. G., Klingler, J. P., Wu, S., Guo, X., Gibbon, B. C., et al. (2011). Characterization of opaque2 modifier QTLs and candidate genes in recombinant inbred lines derived from the K0326Y quality protein maize inbred. *Theor. Appl. Gen.* 122, 783–794. doi: 10.1007/s00122-010-1486-3

Holding, D. R., and Larkins, B. A. (2008). Genetic engineering of seed storage proteins. *Bioeng. Mol. Biol. Plant Pathw.* 1, 107–133. doi: 10.1016/S1755-0408(07)01005-3

Holding, D. R., Meeley, R. B., Hazebroek, J., Selinger, D., Gruis, F., Jung, R., et al. (2010). Identification and characterization of the maize arogenate dehydrogenase gene family. *J. Exp. Bot.* 61, 3663–3673. doi: 10.1093/Jxb/Erq179

Holding, D. R., and Messing, J. (2013). "Evolution, structure, and function of prolamin storage proteins," in *Seed Genomics*, ed. P. Becraft (New York: John Wiley & Sons), 139–158.

Holding, D. R., Otegui, M. S., Li, B. L., Meeley, R. B., Dam, T., Hunter, B. G., et al. (2007). The maize floury1 gene encodes a novel endoplasmic reticulum protein involved in zein protein body formation. *Plant Cell* 19, 2569–2582. doi: 10.1105/tpc.107/053538

Huang, S., Frizzi, A., Florida, C. A., Kruger, D. E., and Luethy, M. H. (2006). High lysine and high tryptophan transgenic maize resulting from the reduction of both 19- and 22-kD alpha-zeins. *Plant Mol. Biol.* 61, 525–535. doi: 10.1007/s11103-006-0027-6

Hunter, B. G., Beatty, M. K., Singletary, G. W., Hamaker, B. R., Dilkes, B. P., Larkins, B. A., et al. (2002). Maize opaque endosperm mutations create extensive changes in patterns of gene expression. *Plant Cell* 14, 2591–2612. doi: 10.1105/tpc.003905

Kim, C. S., Gibbon, B. C., Gillikin, J. W., Larkins, B. A., Boston, R. S., and Jung, R. (2006). The maize Mucronate mutation is a deletion in the 16-kDa gamma-zein gene that induces the unfolded protein response. *Plant J.* 48, 440–451. doi: 10.1111/j.1365-313X.2006.02884.x

Kim, C. S., Hunter, B. G., Kraft, J., Boston, R. S., Yans, S., Jung, R., et al. (2004). A defective signal peptide in a 19-kD alpha-zein protein causes the unfolded protein response and an opaque endosperm phenotype in the maize De*-B30 mutant. *Plant Physiol.* 134, 380–387. doi: 10.1104/pp.103.031310

Kim, C. S., Woo, Y. M., Clore, A. M., Burnett, R. J., Carneiro, N. P., and Larkins, B. A. (2002). Zein protein interactions, rather than the asymmetric distribution of zein mRNAs on endoplasmic reticulum membranes, influence protein body formation in maize endosperm. *Plant Cell* 14, 655–672. doi: 10.1105/tpc.010431

Kriz, A. L. (2009). "Enhancement of amino acid availability in corn grain," in *Molecular Genetic Approaches to Maize Improvement*, eds A. L. Kriz and B. A. Larkins (Heidelberg: Springer), 79–89. doi: 10.1007/978-3-540-68922-5

Kumar, T., Dweikat, I., Sato, S., Ge, Z. X., Nersesian, N., Chen, H., et al. (2012). Modulation of kernel storage proteins in grain sorghum (*Sorghum bicolor* (L.) Moench). *Plant Biotechnol. J.* 10, 533–544. doi: 10.1111/j.1467-7652.2012.00685.x

Lai, J. S., and Messing, J. (2002). Increasing maize seed methionine by mRNA stability. *Plant J.* 30, 395–402. doi: 10.1046/j.1365-313X.2001.01285.x

Lending, C. R., and Larkins, B. A. (1989). Changes in the zein composition of protein bodies during maize endosperm development. *Plant Cell* 1, 1011–1023. doi: 10.1105/tpc.1.10.1011

Lending, C. R., and Larkins, B. A. (1992). Effect of the floury-2 locus on protein body formation during maize endosperm development. *Protoplasma* 171, 123–133. doi: 10.1007/BF01403727

Lopes, M. A., and Larkins, B. A. (1991). Gamma-zein content is related to endosperm modification in Quality Protein Maize. *Crop Sci.* 31, 1655–1662. doi: 10.2135/cropsci1991.0011183X003100060055x

Lopes, M. A., and Larkins, B. A. (1995). Genetic-analysis of opaque2 modifier gene activity in maize endosperm. *Theor. Appl. Gen.* 91, 274–281. doi: 10.1007/BF00220889

Lopes, M. A., Takasaki, K., Bostwick, D. E., Helentjaris, T., and Larkins, B. A. (1995). Identification of opaque 2 modifier loci in quality protein maize. *Mol. Gen. Genet.* 247, 603–613. doi: 10.1007/BF00290352

Maddaloni, M., Donini, G., Balconi, C., Rizzi, E., Gallusci, P., Forlani, F., et al. (1996). The transcriptional activator Opaque-2 controls the expression of a cytosolic form of pyruvate orthophosphate dikinase-1 in maize endosperms. *Mol. Gen. Genet.* 250, 647–654. doi: 10.1007/BF02174452

Mertz, E. T., Nelson, O. E., and Bates, L. S. (1964). Mutant gene that changes protein composition and increases lysine content of maize endosperm. *Science* 145, 279–280. doi: 10.1126/science.145.3629.279

Miclaus, M., Wu, Y. R., Xu, J. H., Dooner, H. K., and Messing, J. (2011a). The maize high-lysine mutant opaque7 Is defective in an acyl-CoA synthetase-like protein. *Genetics* 189, 1271–1280. doi: 10.1534/genetics.111.133918

Miclaus, M., Xu, J. H., and Messing, J. (2011b). Differential gene expression and epiregulation of alpha zein gene copies in maize haplotypes. *PLoS genetics* 7:e1002131. doi: 10.1371/journal.pgen.1002131

Myers, A. M., James, M. G., Lin, Q. H., Yi, G., Stinard, P. S., Hennen-Bierwagen, T. A., et al. (2011). Maize opaque5 encodes monogalactosyldiacylglycerol synthase and specifically affects galactolipids necessary for amyloplast and chloroplast function. *Plant Cell* 23, 2331–2347. doi: 10.1105/tpc.111.087205

Oria, M. P., Hamaker, B. R., Axtell, J. D., and Huang, C. P. (2000). A highly digestible sorghum mutant cultivar exhibits a unique folded structure of endosperm protein bodies. *Proc. Natl. Acad. Sci. U.S.A.* 97, 5065–5070. doi: 10.1073/pnas.0800 76297

Osborne, T. B. (1897). Amount and properties of the proteins of the maize kernel. *J. Am. Chem. Soc.* 19, 525–532. doi: 10.1021/ja02081a002

Pastorello, E. A., Farioli, L., Pravettoni, V., Scibilia, J., Conti, A., Fortunato, D., et al. (2009). Maize food allergy: lipid-transfer proteins, endochitinases, and alpha-zein precursor are relevant maize allergens in double-blind placebo-controlled maize-challenge-positive patients. *Anal. Bioanal. Chem.* 395, 93–102. doi: 10.1007/s00216-009-2945-z

Peremyslov, V. V., Morgun, E. A., Kurth, E. G., Makarova, K. S., Koonin, E. V., and Dolja, V. V. (2013). Identification of myosin XI receptors in arabidopsis defines a distinct class of transport vesicles. *Plant Cell* 25, 3022–3038. doi: 10.1105/tpc.113.113704

Poethig, R. S., Coe, E. H., and Johri, M. M. (1986). Cell lineage patterns in maize embryogenesis – a clonal analysis. *Dev. Biol.* 117, 392–404. doi: 10.1016/0012-1606(86)90308–90308

Schmidt, R. J., Burr, F. A., Aukerman, M. J., and Burr, B. (1990). Maize regulatory gene opaque-2 encodes a protein with a leucine-zipper motif that binds to zein DNA. *Proc. Natl. Acad. Sci. U.S.A* 87, 46–50. doi: 10.1073/pnas.87.1.46

Segal, G., Song, R. T., and Messing, J. (2003). A new opaque variant of maize by a single dominant RNA-interference-inducing transgene. *Genetics* 165, 387–397.

Shull, J. M., Watterson, J. J., and Kirleis, A. W. (1991). Proposed nomenclature for the alcohol-soluble proteins (Kafirins) of *Sorghum bicolor* (L Moench) based on molecular-weight, solubility, and structure. *J. Agric. Food Chem.* 39, 83–87. doi: 10.1021/Jf00001a015

Soave, C., and Salamini, F. (1984). Organization and regulation of zein genes in maize endosperm. *Philos. Trans. R. Soc. Lond. Series B Biol. Sci.* 304, 341–343. doi: 10.1098/rstb.1984.0029

Song, R., Llaca, V., Linton, E., and Messing, J. (2001). Sequence, regulation, and evolution of the maize 22-kD alpha zein gene family. *Genome Res.* 11, 1817–1825. doi: 10.1101/gr.197301

Song, R., and Messing, J. (2002). Contiguous genomic DNA sequence comprising the 19-kD zein gene family from maize. *Plant Physiol.* 130, 1626–1635. doi: 10.1104/pp.012179

Torrent, M., Alvarez, I., Geli, M. I., Dalcol, I., and Ludevid, D. (1997). Lysine-rich modified gamma-zeins accumulate in protein bodies of transiently transformed maize endosperms. *Plant Mol. Biol.* 34, 139–149. doi: 10.1023/A:1005889 314967

Tsai, C. Y., Huber, D. M., and Warren, H. L. (1978). Relationship of kernel sink for N to maize productivity. *Crop Sci.* 18, 399–404. doi: 10.2135/cropsci1978.0011183X001800030011x

Vasal, S. K., Villegas, E., Bjarnason, M., Gela, W., and Goertz, P. (1980). "Genetic modifiers and breeding strategies in developing hard endosperm opaque2 materials," in *Quality Traits of Maize for Grain and Silage Use*, eds W. G. Pollmer and R. H. Phillips (London: Martinus Nijhoff), 37–73.

Wallace, J. C., Lopes, M. A., Paiva, E., and Larkins, B. A. (1990). New methods for extraction and quantitation of zeins reveal a high content of gamma-zein in modified ppaque-2 maize. *Plant Physiol.* 92, 191–196. doi: 10.1104/pp.92.1.191

Wang, G. F., Wang, F., Wang, G., Wang, F., Zhang, X. W., Zhong, M. Y., et al. (2012). Opaque1 encodes a myosin XI motor protein that is required for endoplasmic reticulum motility and protein body formation in maize endosperm. *Plant Cell* 24, 3447–3462. doi: 10.1105/tpc.112.101360

Woo, Y. M., Hu, D. W. N., Larkins, B. A., and Jung, R. (2001). Genomics analysis of genes expressed in maize endosperm identifies novel seed proteins and clarifies patterns of zein gene expression. *Plant Cell* 13, 2297–2317. doi: 10.1105/tpc.13.10.2297

Wu, Y., Holding, D. R., and Messing, J. (2010). Gamma-zein is essential for endosperm modification in quality protein maize. *Proc. Natl. Acad. Sci. U.S.A.* 107, 12810–12815. doi: 10.1073/pnas.1004721107

Wu, Y., and Messing, J. (2010). RNA interference-mediated change in protein body morphology and seed opacity through loss of different zein proteins. *Plant Physiol.* 153, 337–347. doi: 10.1104/pp.110.154690

Wu, Y. R., and Messing, J. (2011). Novel genetic selection system for quantitative trait loci of quality protein maize. *Genetics* 188, 1019–1022. doi: 10.1534/genetics.111.131078

Wu, Y., Yuan, L., Guo, X., Holding, D. R., and Messing, J. (2013). Mutation in the seed storage protein kafirin creates a high-value food trait in sorghum. *Nat. Commun.* 4:2217. doi: 10.1038/ncomms3217

Xu, J. H., and Messing, J. (2008). Organization of the prolamin gene family provides insight into the evolution of the maize genome and gene duplications in grass species. *Proc. Natl. Acad. Sci. U.S.A* 105, 14330–14335. doi: 10.1073/pnas.0807026105

Xu, J. H., and Messing, J. (2009). Amplification of prolamin storage protein genes in different subfamilies of the Poaceae. *Theor. Appl. Genet.* 119, 1397–1412. doi: 10.1007/s00122-009-1143-x

Yuan, L., Dou, Y., Kianian, S. F., Zhang, C., and Holding, D. R. (2014). Deletion mutagenesis identifies a haploinsufficient role for γ-zein in opaque2 endosperm modification. *Plant Physiol.* 164, 119–130. doi: 10.1104/pp.113.230961

Conflict of Interest Statement: The Guest Associate Editor Brian A. Larkins declares that, despite being affiliated to the same institution as the author David R. Holding, the review process was handled objectively and no conflict of interest exists. The author declares that the research was conducted in the absence of any commercial or financial relationships that could be construed as a potential conflict of interest.

18

Expression, regulation and activity of a B2-type cyclin in mitotic and endoreduplicating maize endosperm

Paolo A. Sabelli[1]*, Ricardo A. Dante[1†], Hong N. Nguyen[1], William J. Gordon-Kamm[2] and Brian A. Larkins[1†]

[1] School of Plant Sciences, University of Arizona, Tucson, AZ, USA
[2] Pioneer Hi-Bred International Inc., Johnston, IO, USA

Edited by:
Michael J. Scanlon, Cornell University, USA

Reviewed by:
Michelle Facette, University of California, San Diego, USA
Sharon Ann Kessler, University of Oklahoma, USA

***Correspondence:**
Paolo A. Sabelli, School of Plant Sciences, University of Arizona, 303 Forbes Building, Tucson, AZ 85721, USA
e-mail: psabelli@ag.arizona.edu

†Present address:
Ricardo A. Dante, Embrapa Agricultural Informatics, Campinas, SP, Brazil;
Brian A. Larkins, Beadle Center, University of Nebraska, Lincoln, NE, USA

Cyclin-dependent kinases, the master regulators of the eukaryotic cell cycle, are complexes comprised of a catalytic serine/threonine protein kinase and an essential regulatory cyclin. The maize genome encodes over 50 cyclins grouped in different types, but they have been little investigated. We characterized a type B2 cyclin (CYCB2;2) during maize endosperm development, which comprises a cell proliferation phase based on the standard mitotic cell cycle, followed by an endoreduplication phase in which DNA replication is reiterated in the absence of mitosis or cytokinesis. *CYCB2;2* RNA was present throughout the period of endosperm development studied, but its level declined as the endosperm transitioned from a mitotic to an endoreduplication cell cycle. However, the level of CYCB2;2 protein remained relatively constant during both stages of endosperm development. CYCB2;2 was recalcitrant to degradation by the 26S proteasome in endoreduplicating endosperm extracts, which could explain its sustained accumulation during endosperm development. In addition, although CYCB2;2 was generally localized to the nucleus of endosperm cells, a lower molecular weight form of the protein accumulated specifically in the cytosol of endoreduplicating endosperm cells. In dividing cells, CYCB2;2 appeared to be localized to the phragmoplast and may be involved in cytokinesis and cell wall formation. Kinase activity was associated with CYCB2;2 in mitotic endosperm, but was absent or greatly reduced in immature ear and endoreduplicating endosperm. CYCB2;2-associated kinase phosphorylated maize E2F1 and the "pocket" domains of RBR1 and RBR3. CYCB2;2 interacted with both maize CDKA;1 and CDKA;3 in insect cells. These results suggest CYCB2;2 functions primarily during the mitotic cell cycle, and they are discussed in the context of the roles of cyclins, CDKs and proteasome activity in the regulation of the cell cycle during endosperm development.

Keywords: cell cycle, cyclin, cyclin-dependent kinase, endoreduplication, endosperm, maize, proteasome

INTRODUCTION

The eukaryotic cell division cycle is driven by the activity of complexes between a serine/threonine cyclin-dependent kinase (CDK) and a regulatory cyclin (CYC) protein. Periodic activation of different CDK/CYC complexes ensures the typically unidirectional execution of cell cycle steps, such as DNA replication and chromosome segregation (mitosis), and progress through their intervening transitions. A number of mechanisms are known to regulate CDK/CYC activity, including gene transcription, phosphorylation, and dephosphorylation of key amino acids, proteolysis by the 26S proteasome, and binding by specific inhibitors (De Veylder et al., 2007; Harashima et al., 2013). Compared to yeast and mammals, plant genomes encode a larger number of these two components of CDK complexes, and it has been challenging to dissect their function, especially because of the genetic redundancy within the two protein families. Understanding the mechanisms governing the cell cycle, coupled with the ability to manipulate it, offer the potential to enhance crop performance and yield, and several important examples are known in which alteration of cell cycle control has significantly impacted crop evolution and breeding (Sabelli, 2014). Although the cell cycle is relatively well characterized in the model species *Arabidopsis thaliana*, our understanding of cell cycle control in economically important crops, and particularly in major grasses such as maize, rice and wheat, is still in its infancy, and there is a pressing need to unravel the roles of key cell cycle-controlling genes in these crops. Because yield in maize and related cereals is in part determined by grain size and weight, it is important to understand how different cell cycle genes are regulated and how they function during seed development.

The most economically important seed tissue is the endosperm, which originates at fertilization by the fusion of one sperm cell nucleus with two polar nuclei within the central cell of the female gametophyte (reviewed in Becraft et al., 2001; Olsen, 2004; Sabelli and Larkins, 2009c). In maize, the primary triploid endosperm nucleus goes on to proliferate by acytokinetic mitoses to generate a syncytium. Deposition of cell wall material between nuclear domains followed by cell proliferation through mitotic cell division give rise to virtually all the endosperm cells by approximately

12 days after pollination (DAP). However, starting from around 8–10 DAP, inner cells of the endosperm switch to an endoreduplication type of cell cycle, which is characterized by reiterated rounds of DNA synthesis in the absence of mitosis or cell division; typically, this results in large, polyploid nuclei and cells. Endoreduplication proceeds toward the periphery of the endosperm and is roughly coincident with the massive accumulation of storage compounds, such as starch and zein proteins. From around 16 DAP, central endosperm cells begin to undergo programmed cell death, and this continues so that at seed maturity all endosperm cells are dead, except for the peripheral aleurone layer. Because the mitotic phase of endosperm development produces virtually all the cells involved in storage compound accumulation and because the endoreduplication phase is associated with rapid cell growth and endosperm expansion and filling, it is important to understand how the cell cycle is regulated during these two phases (Sabelli and Larkins, 2009b). In this context, three maize CDKs: CDKA;1, CDKA;3, and CDKB1;1; several cyclins belonging to the A1, B1, D2, and D5 types (Grafi and Larkins, 1995; Sun et al., 1999b; Leiva-Neto et al., 2004; Dante et al., 2014); two CDK inhibitors (Coelho et al., 2005); a Wee1 homolog (Sun et al., 1999a); and members of the retinoblastoma-related (RBR) protein family (Grafi et al., 1996; Sabelli et al., 2005, 2013) have been studied in some detail.

Recent genome-wide analyses revealed the cyclin family in maize comprises over 50 members (Hu et al., 2010), and for the most part these genes have not been investigated. B-type cyclins are believed to be primarily involved in the regulation of M-phase and appear to play important roles in plant growth and seed development. CYCBs are known to stimulate cell division and tissue growth in ectopic or over-expression studies (Doerner et al., 1996; Lee et al., 2003). Knockdown of CYCB1;1 in rice caused aborted seed endosperm due to abnormal cellularization of the syncytium (Guo et al., 2010) and resulted in large, triploid embryo cells (Guo et al., 2014). In addition, CYCB2;2 was implicated in the timing of rice endosperm cellularization, cell number, and grain size and yield (Ishimaru et al., 2013). Here we describe CYCB2;2 from maize with respect to its spatiotemporal expression patterns, interaction with CDKs, associated kinase activity, and proteolysis patterns during endosperm development. These data suggest this cyclin likely functions during cell division and cell wall formation in mitotic cells and becomes stabilized in endoreduplicating cells with the accumulation of a lower molecular weight form in the cytoplasm.

MATERIALS AND METHODS

PLANT MATERIALS

Maize (*Zea mays* L.) B73 plants were grown in the field or a greenhouse and hand-pollinated. Endosperms were dissected from kernels harvested at different stages of development and processed for molecular and immunohistochemical analyses as described in previous publications (Leiva-Neto et al., 2004; Sabelli et al., 2005, 2013; Dante et al., 2014).

DATABASE SEARCHES AND SEQUENCE ANALYSES

A nucleotide sequence encoding CYCB2;2 was initially obtained by querying Pioneer Hi-Bred's maize EST database. Additional searches were made in the Maize Genome Sequencing Project[1], MaizeGDB[2] (Lawrence et al., 2004), Phytozome v10[3], Rice Genome Annotation Project v7[4] (Kawahara et al., 2013), Gramene v42[5] (Ware et al., 2002), and Pfam v27[6] (Finn et al., 2014) databases and the Maize eFP Browser[7] (Sekhon et al., 2011). Functional motifs were predicted with the Eukaryotic Linear Motif Resource[8] (Dinkel et al., 2014). Subcellular localization was predicted using LocTree 3[9] (Goldberg et al., 2012), Plant-mPLoc[10] (Chou and Shen, 2010), and BaCelLo[11] software. Multiple sequence alignments were carried out with M-Coffee[12] (Di Tommaso et al., 2011) or MUSCLE (Edgar, 2004). An un-rooted Neighbor-Joining tree of a set of 35 plant B-type cyclin amino acid sequences, spanning the conserved Cyc_N and Cyc_C domains (Nugent et al., 1991), was constructed using MEGA6 software package (Tamura et al., 2013). Amino acid sequences were selected based on previous analyses (La et al., 2006; Guo et al., 2007; Hu et al., 2010; Jia et al., 2014) and novel database searches. Only one amino acid sequence per locus was selected in the case of multiple predicted transcripts. Several shorter amino acid sequences (GRMZM2G025200_P01, Loc_Os02g41720, AT1G34460, Sb07g003015, Potri.006G035200.2) were not included in the analysis. The evolutionary distances were computed using the Poisson correction method. All positions containing gaps and missing data were eliminated. There were a total of 235 positions in the final dataset.

ANALYSES OF ENDOSPERM RNA AND PROTEINS

Detailed procedures for purification of endosperm RNA and protein and their analyses by RT-PCR and immunoblotting, respectively, are given in previous publications (Sabelli et al., 2005, 2013; Dante et al., 2014). The following RT-PCR primers were used for CYCB2;2: CYCB2;2F (GAAAATGAGGCTAAGAGTTGTG-TAAG) and CYCB2;2R (GAGCTCCAGCATGAAAAATGACGCT) and actin: ACT1-F (ATTCAGGTGATGGTGTGAGCCACAC) and ACT1-R (GCCACCGATCCAGACACTGTACTTCC). Each developmental stage comprised a pool of 5–13 endosperm RNA samples. Two analysis replicates were carried out and the RNA levels averaged, normalized to those of actin control, and displayed relative to those at 7-DAP.

Analysis of *CYCB2;2* RNA accumulation patterns in 14 different tissues/developmental stages was carried out by compiling Nimblegen-derived RNA expression data from Sekhon et al. (2011), available at the Maize eFP Browser[7].

---

[1] http://www.maizesequence.org

[2] http://www.maizegdb.org/

[3] http://phytozome.jgi.doe.gov

[4] http://rice.plantbiology.msu.edu/

[5] http://www.gramene.org/

[6] http://pfam.xfam.org/

[7] http://bar.utoronto.ca/efp_maize/cgi-bin/efpWeb.cgi?dataSource = Sekhon_et_al

[8] http://elm.eu.org/

[9] https://rostlab.org/services/loctree3/

[10] http://www.csbio.sjtu.edu.cn/bioinf/plant-multi/

[11] http://gpcr.biocomp.unibo.it/bacello/

[12] http://tcoffee.crg.cat/apps/tcoffee/index.html

Immunohistochemical localization assays were carried out essentially as described by Dante et al. (2014), except for a monoclonal anti-tubulin antibody (YOL 1/34, Accurate Chemical and Scientific Corp., Westbury, NY, USA), which was used to stain microtubules. Antibodies were utilized at a concentration of 0.5–1 μg/ml. Immunoprecipitation and kinase activity assays were carried out as described previously (Dante et al., 2014).

PROTEIN EXPRESSION IN HETEROLOGOUS SYSTEMS AND ANTIBODIES

Sequences encoding selected domains of maize cyclins (amino acid residues 1–243 of CYCA1;1, 1–206 of CYCB1;3, and 4–143 of CYCB2;2) were expressed as GST fusions in *Escherichia coli* and purified according to procedures described previously (Dante et al., 2014). For CYCB2;2, a cDNA fragment encoding a 139-aa long, *N*-terminal domain was amplified with primers B2;2 4–143F (ATCGCGGATCCCGCGCGGCGGATGAAAACCGCAGACC) and B2;2 1–62R (CCGCTCGAGTCAGAGCAATTCATCTTCGTTCATAATGTCC), and cloned at the *Bam*HI and *Xho*I sites of the pGEX4T-3 vector (GE Heathcare, Life Sciences). Antibodies were generated in rabbits (anti-cyclin) and affinity purified or purchased (anti-actin) as previously reported (Dante et al., 2014).

For expression in *Drosophila* S2 cells, CYCB2;2 was PCR-amplified with primers E2-KpnI (ATCCGCGGTACCAAACATGGCGGCGCGGGCGGCTGACGAGAAC) and E2HAS2 (ATCGCGAATTCCTATTAGGCGTAATCGGGCACATCGTAGGGGTAGTTTGCACCTGAAGGAGGCGG), cloned into the *Kpn*I and *Eco*RI sites of the pHSKSMCS vector (kindly provided by Dr. Thomas Bunch, University of Arizona), expressed in *Drosophila* S2 cells either alone or co-expressed with maize CDKA;1 or CDKA;3, and analyzed essentially as described previously for other cyclins (Dante et al., 2014). CYCB2;2 was immunoprecipitated from cell lysates and assayed for interaction with co-expressed maize CDKs using an anti-PSTAIR (Sigma, catalog No. P7962) antibody (Dante et al., 2014). Immunoprecipitates were also tested for kinase activity on histone H1 substrate in *in vitro*-assays as described (Dante et al., 2014).

PROTEIN STABILITY ASSAYS

^{35}S-radiolabeled CYCB2;2, and a T417A mutant (CYCB2;2T/A) polypeptides were synthesized *in vitro* using TNT T7 Quick for PCR DNA kit (Promega, Madison, WI, USA) as described (Dante et al., 2014). These proteins were incubated with cell extracts obtained from prevalently mitotic (i.e., 7-DAP) or endoreduplicating (i.e., 15-DAP) endosperm for 90 min before being subjected to SDS-PAGE and authoradiography. The 26S proteasome-specific inhibitor, carboben-zoxyl-leucinyl-leucinyl-leucinal (MG-132, Sigma), was utilized to determine whether protein degradation was proteasome-dependent. Procedures for these assays were as previously reported (Dante et al., 2014).

RESULTS

IDENTIFICATION OF CYCB2;2 FROM MAIZE

An EST encoding a B-type cyclin from maize, hereafter termed CYCB2;2, was identified by searching Pioneer Hi-Bred Inc.'s database. Subsequently, a corresponding genomic sequence located on chromosome 2, with accession number GRMZM2G138886, was also identified. The deduced CYCB2;2 protein sequence is 424 amino acids long, with a calculated molecular weight of 47.5 kDa. It possesses two all-alpha fold domains embedded in the so-called and highly conserved "cyclin core," which are characteristic of cyclin proteins: Cyc_N (or cyclin box), which contains the CDK-binding site and is located between residues 163 and 289, and Cyc_C, which however, is less conserved among cyclins, and is located between residues 291 and 407 (Nugent et al., 1991; **Figure 1A**). A potential PRKK nuclear localization signal (NLS) is located between residues 231–234. A protein destruction box (SRRALTDIK, D-Box), targeted by the anaphase-promoting complex/cylosome (APC/C) and which is required for anaphase-specific proteolysis, is predicted at residues 26–34. A MRAIL motif, which is involved in cyclin binding to substrates containing the RxL motif, such as CDK-specific inhibitors and RBRs (Schulman et al., 1998), is located between residues 192–196.

Thr-417 represents a potential CDK phosphorylation site, which is conserved with similar phosphorylation sites in animal CycE proteins (**Figures 1A,B**). Phosphorylation of this site contributes to CycE proteolysis via the SCF-Fbw7 ubiquitin ligase pathway (Welker et al., 2003; Hwang and Clurman, 2005; Hao et al., 2007). Through database searches, a closely related cyclin from sorghum (Sb06g025380.1) was identified that also possesses this putative phosphorylation site (**Figures 1B** and **2**), though it does not appear that this motif is conserved in other plant cyclins.

Phylogenetic analyses indicated that the maize CYCB2;2 cyclin described in this study clusters with other B2-type cyclins from monocots (maize, sorghum, and rice) and dicots (*Arabidopsis* and poplar). These results are in general agreement with previous analyses (Hu et al., 2010). During the course of this study, a gene was identified on chromosome 10 (GRMZM2G061287) that encodes a closely related and previously un-reported maize cyclin, which was termed CYCB2;3. The closest cyclins from rice (CYCB2;1) and sorghum (Sb06g025380.1; **Figure 2**) also lack functional characterization to date, which makes it difficult to predict a precise function for maize CYCB2;2 based on sequence similarity.

An *N*-terminal region of CYCB2;2 (amino acid residues 4–143) that has little or no sequence identity with other maize cyclins identified so far (~39% identity with the corresponding regions in CYCB2;1, the closest maize homolog) was selected to raise specific antibodies (**Figure 1A**). This region of the protein lies outside the Cyc_N domain, and thus these antibodies were not likely to interfere with the catalytic or CDK-binding activity of CYCB2;2. The corresponding *CYCB2;2* cDNA sequence was expressed in *E. coli* as a GST fusion and polyclonal antibodies were raised in rabbits against the purified protein, and were affinity purified. These antibodies were tested against the recombinant CYCB2;2 antigen and comparable *N*-terminal regions of maize CYCA1;1 and CYCB1;3, which were also expressed as GST-fusions in *E. coli* as described previously (Dante et al., 2014; **Figure 3**). While these antibodies effectively recognized the CYCB2;2 *N*-terminal polypeptide, no cross-reactivity was observed with either CYCA1;1 or CYCB1;3.

A

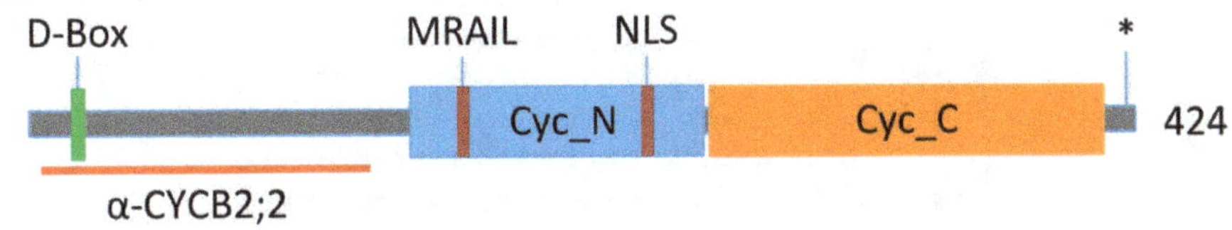

B

				*
Fruit fly	CycE	FBgn0010382	650	ESLLTPPASSHKPDE
Human	CycE1	ENSG00000105173	391	SGLLTPPQSGKKQSS
Rat	CycE1	ENSRNOG0000014786	392	SGVLTPPHSSKKQSS
Mouse	CycE1	ENSMUSG00000002068	389	SVVLTPPHSSKKQSS
Chicken	CycE1	ENSGALG00000004494	340	TGVLTPPQSSKKQPA
Zebrafish	CycE1	ENSDARG00000011133	393	TGVLTPPPSSEKPES
Human	CycE2	ENSG00000175305	388	GGIMTPPKSTEKPPG
Rat	CycE2	ENSRNOG00000008055	389	GGIMTPPKSTEKPPV
Mouse	CycE2	ENSMUSG00000028212	387	GGIMTPPKSTEKPPG
Chicken	CycE2	ENSGALG00000015985	393	AGIITPPKSTEKK–
Zebrafish	CycE2	ENSDARG00000027918	377	AGILTPPKSTEKPAN
Maize	CYCB2;2	GRMZM2G138886_P01	415	GG--TPPPSGAN---
Sorghum	CYCB	Sb06g025380.1	423	GG--TPPPSGAI---

FIGURE 1 | Key features of the maize CYCB2;2 polypeptide. (A) Schematic diagram illustrating the conserved Cyc_N and Cyc_C domains, the positions of the destruction box (D-Box), the NLS, the MRAIL motif, and the putative CDK phosphorylation site at position 417 (asterisk), as well as the *N*-terminal region selected for raising specific antibodies (red line). **(B)** Multiple sequence alignment showing the Thr-417 putative phosphorylation site in maize CYCB2;2, which is conserved among animal E-type cyclins and in a closely related protein from sorghum (Sb06g025380.1). The Thr-395 residue in the human CycE1 sequence is often reported as Thr-380 in the literature due to a 15-amino acid shorter isoform of the protein. Conserved residues are shaded. Ensembl database (http://uswest.ensembl.org/) and plant gene IDs are given.

EXPRESSION OF CYCB2;2 DURING ENDOSPERM DEVELOPMENT

The expression patterns of CYCB2;2 RNA and protein during endosperm development were analyzed by RT-PCR and western blotting, respectively (**Figure 4**). These experiments showed that *CYCB2;2* RNA levels are relatively high in 7-DAP endosperm and decline steadily thereafter, reaching by 13-DAP a minimum of less than 20% the 7-DAP reference levels, and remaining low up to 21-DAP (**Figures 4A,B**). This result is in agreement with the expression pattern of *CYCB2;2* RNA obtained through global transcriptome analyses of developing maize endosperm (12–24 DAP) and available through the Maize eFP Browser (Sekhon et al., 2011; **Figure 4C**). Additionally, the data shown in **Figure 4C** indicate that *CYCB2;2* RNA is widely expressed in maize tissues, and particularly at high levels in those known to contain an elevated proportion of mitotic cells, such as the primary root, shoot tip, leaf base, immature ear, embryo, and young endosperm.

The anti-CYCB2;2 antibody recognized a polypeptide of the expected molecular weight in endosperm extracts (**Figure 4D**). However, longer autoradiograph exposure times revealed the presence of an additional band of slightly lower molecular weight, which tended to accumulate specifically during the endoreduplication phase of endosperm development and was detectable from around 13-15 DAP (**Figure 4E**). This low-molecular-weight (LMW) CYCB2;2-related polypeptide was never detected in 7- or 9-DAP extracts; its accumulation appeared to be associated with the endoreduplication phase of endosperm development, which suggests it may play a specific role in this specialized type of cell cycle.

Comparison of the CYCB2;2 RNA and protein accumulation patterns revealed a discrepancy between the marked decline of RNA levels and the relatively constant protein levels observed between 7 and 21 DAP. This observation raises the possibility that CYCB2;2 protein may be subject to different turnover regulation in early (i.e., 7-DAP) *versus* more advanced (i.e., 13–21 DAP) endosperm developmental stages, which primarily comprise cells that are mitotic or endoreduplicating, respectively. Specifically, CYCB2;2 may not be as efficiently degraded in endoreduplicating endosperm as in mitotic endosperm.

SUBCELLULAR LOCALIZATION OF CYCB2;2 PROTEIN

Sequence analyses by subcellular localization prediction software and the presence of a putative NLS in the CYCB2;2 amino acid sequence suggested this protein could be targeted to the nucleus. We investigated the localization pattern of CYCB2;2 in endosperm, embryo, and root tip cells by immunofluorescence using anti-CYCB2;2 antibody (**Figure 5**). DNA and tubulin were stained as markers for nuclei and the microtubule cytoskeleton, respectively, in the same tissue sections. In 7-DAP endosperm, CYCB2;2 was clearly localized to the nucleus, though a weak signal was also detected in the cytoplasm (**Figures 5A–D**). Although most endosperm nuclei stained positively for CYCB2;2, the signal among nuclei differed notably, with a rather diffused CYCB2;2 accumulation pattern in some nuclei and a sharply punctuate one in others. These differences in localization patterns within a population of cells that are asynchronously engaged in the cell cycle are typical of cell cycle-regulated proteins. In 13-DAP endosperm cells, CYCB2;2 was

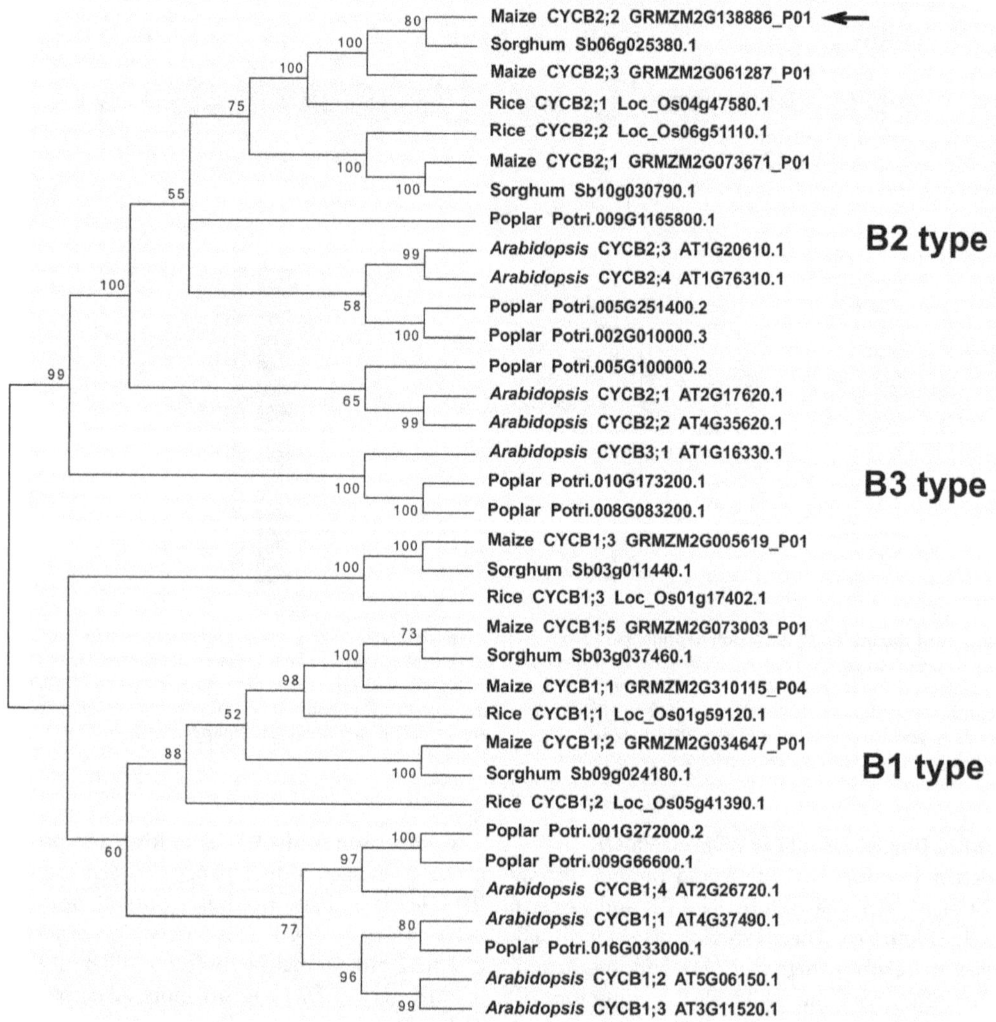

FIGURE 2 | Phylogenetic tree of B-type plant cyclins. Selected plant cyclin B amino acid sequences from *Arabidopsis*, poplar, maize, sorghum, and rice were aligned and an un-rooted tree was generated with the Neighbor-Joining method. Bootstraps values (>50) are shown next to the branches. Branches corresponding to partitions reproduced in less than 50% of bootstrap replicates are collapsed. Protein accession numbers and cyclin designations (when available) are given next to the plant species names. Three main cyclin B groups can be identified: B1-, B2-, and B3-type. The maize CYCB2;2 polypeptides described in this study is shown at the top of the B2-type group and is indicated by an arrow. The closely related GRMZM2G061287_P01 polypeptide is a newly identified B2-type cyclin from maize, which was designated as CYCB2;3. Evolutionary analyses were conducted in MEGA6 (Tamura et al., 2013).

clearly localized to the nucleus but it also showed extensive and diffuse localization in the cytoplasm, which was clearly more pronounced than in 7-DAP endosperm cells (**Figures 5E–H**). However, peripheral cells gave a relatively stronger CYCB2;2 signal than inner endosperm cells in both 7- (**Figures 5A,D**) and 13-DAP (**Figures 5E,H**) sections, but it is not clear whether this might be due to physiological differences between the two cell types or to the highly dense cytoplasm of peripheral cells at both developmental stages. Analysis of 7-DAP endosperm (**Figure 5A**, arrow indicates a telophase cell), embryo (**Figures 5I–L**, arrow indicates a late-phragmoplast cell in which the residual phragmoplast is being eroded beginning from its central zone) and meristematic root tip cells (**Figure 5M**), which typically proliferate through the mitotic cell cycle and do not undergo endoreduplication, revealed CYCB2;2 was localized to the phragmoplast between two daughter nuclei, suggesting a potential role for CYCB2;2 in late mitosis-early cytokinesis and the deposition of the new cell wall separating daughter cells. In root tip cells (**Figure 5M**), CYCB2;2 was localized to the cytoplasm but also to the mitotic spindle's mid-zone in anaphase (center arrow), the phragmoplast mid-zone in telophase (right arrow) and at the site of cell wall formation at cytokinesis (left arrow). Both in mitotic and endoreduplicating cells, the nuclear fraction of CYCB2;2 did not appear to coincide with the DNA, and thus it appeared to be generally excluded from chromatin, similarly to CYCB2;1 (Mews et al., 1997), but there was some overlap with the microtubule cytoskeleton.

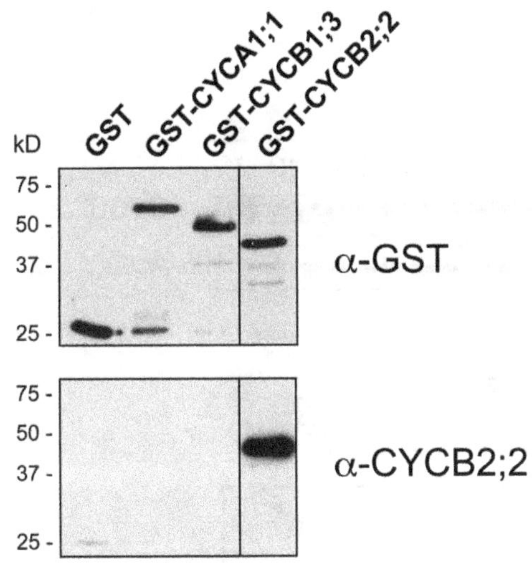

FIGURE 3 | Specificity of anti-CYCB2;2 antibodies. Affinity-purified antibodies against CYCB2;2 were tested against the CYCB2;2 antigen and selected domains of maize cyclins CYCA1;1 and CYCB1;3, expressed in *E. coli* as GST fusions. **Top**, control western blot with anti-GST antibodies; **Bottom**, western blot with anti-CYCB2;2 antibodies.

We compared the subcellular distribution patterns of CYCB2;2 between 9-DAP endosperm cells, which are mostly mitotic, and 15-DAP cells, which are primarily endoreduplicating. Subcellular fractions enriched for nuclear and cytosolic proteins were prepared from these two endosperm developmental stages and assayed for CYCB2;2 accumulation by western blotting (**Figure 6**). CYCB2;2 appeared prevalently localized to the nuclear fraction in 9-DAP endosperm, but a lower molecular weight polypeptide, corresponding to the additional LMW band shown in **Figures 4D,E**, accumulated specifically in the cytosolic fraction in 15-DAP endosperm. The presence of intact actin protein as well as tubulin and other cyclins as recently reported (Dante et al., 2014) indicates that the LMW, CYCB2;2-related band was not due to general protein degradation in the extract. These results are in agreement with the immunohistochemistry data shown in **Figure 5**, which does not discriminate signal based on molecular weight differences. In prevalently mitotic endosperm (i.e., 7–9 DAP), CYCB2;2 is mostly nuclear with relatively little signal from the cytoplasm. In endoreduplicating cells (i.e., 13–15 DAP), however, CYCB2;2 becomes more extensively localized to the cytoplasm, and the cell fractionation analysis suggests that this shift in localization is due to the specific accumulation of an extra-nuclear, LMW form of the protein (**Figure 6**). The presence of similar amounts of full-length CYCB2;2 in mitotic and endoreduplicating endosperm cells suggests this protein maybe involved in regulating cell cycle processes that are common to these two cell types, such as DNA synthesis. Alternatively, partial CYCB2;2 proteolysis and exclusion from the nucleus may be associated with the endocycle and may underscore a role for the intact protein in mitosis and/or cytokinesis. Thus, in contrast to mitotic cells, endoreduplicating endosperm cells are characterized by specific nuclear-to-cytosol redistributions of CYCB2;2, specifically suggesting that accumulation of the LMW component in the cytoplasm may be critical for the transition from the mitotic to endoreduplication phase of endosperm development.

CYCB2;2-ASSOCIATED KINASE ACTIVITY

We investigated whether CYCB2;2 is part of active kinase complexes, is capable of phosphorylating various substrates, and its activity varies between immature ear and endosperm at different stages of development. CYCB2;2 was immunoprecipitated from extracts prepared with either immature ear and 9-DAP endosperm, two tissues comprising prevalently mitotic cells, and tested for phosphorylation of histone H1 substrate by *in vitro* assays (**Figure 7A**). Whereas virtually no CYCB2;2-associated kinase activity was detected in extracts from immature ear, a strong signal was obtained from 9-DAP endosperm extracts. These patterns were similar to those of CYCB1;3, and antithetic to those of CYCD2;1 or CYCD5, which displayed much strong associated kinase activities in immature ear relative to endosperm. These experiments indicate there are specific differences with regard to the kinase activity associated with CYCB2;2 between immature ears and 9-DAP endosperm, as well as those associated with other types of cyclins. Although both tissues are known to exhibit high mitotic activity, these data underscore the presence of potentially important differences in the regulation of the cell cycle between these two tissues, and suggest a more prominent role in endosperm for B-type cyclins. We next asked whether CYCB2;2-associated kinase activity from 9-DAP endosperm could phosphorylate different recombinant polypeptides expressed as GST fusions (**Figure 7B**). The pocket domain of RBR3 was effectively phosphorylated, although RBR1 pocket was only weakly phosphorylated in this assay. Neither *N*-terminal domain of these proteins was phosphorylated, consistent with the typical predominance of conserved CDK targets specifically within the pocket domain of RBR proteins from many organisms (Sabelli et al., 2005; Sabelli and Larkins, 2009a; Burke et al., 2012). The E2F1 substrate was also phosphorylated by CYCB2;2-associated kinase activity, similarly to recent results for CYCB1;3-associated kinase activity (Dante et al., 2014). By 9-DAP, some endosperm cells could have already initiated the transition to endoreduplication, particularly those located in the center of the tissue, and thus analyses of kinase activity at this stage may not exclusively reflect mitotic cells. We thus measured CYCB2;2-associated kinase activity in 7-DAP endosperm extracts, a stage characterized exclusively by mitotic cells as shown by flow-cytometric profiles (Dante et al., 2014), in 15-DAP extracts, a stage in which cells are actively endoreduplicating and mitotic figures are rare, as well as at 11-DAP, which represents a transition stage with high rates of mitotic and endoreduplication cell cycle activities (Dante et al., 2014). As shown in **Figure 7C**, while CYCB2;2-associated kinase activity was high and virtually at the same levels in 7- and 11-DAP endosperm, it declined dramatically in 15-DAP endosperm. Collectively, these results indicate that CYCB2;2/CDK complexes are more active in mitotic than in endoreduplicating endosperm cells.

We next investigated whether CYCB2;2 can form complexes with A-type CDKs when co-expressed in a heterologous cell system, and whether such complexes are active (**Figure 8**).

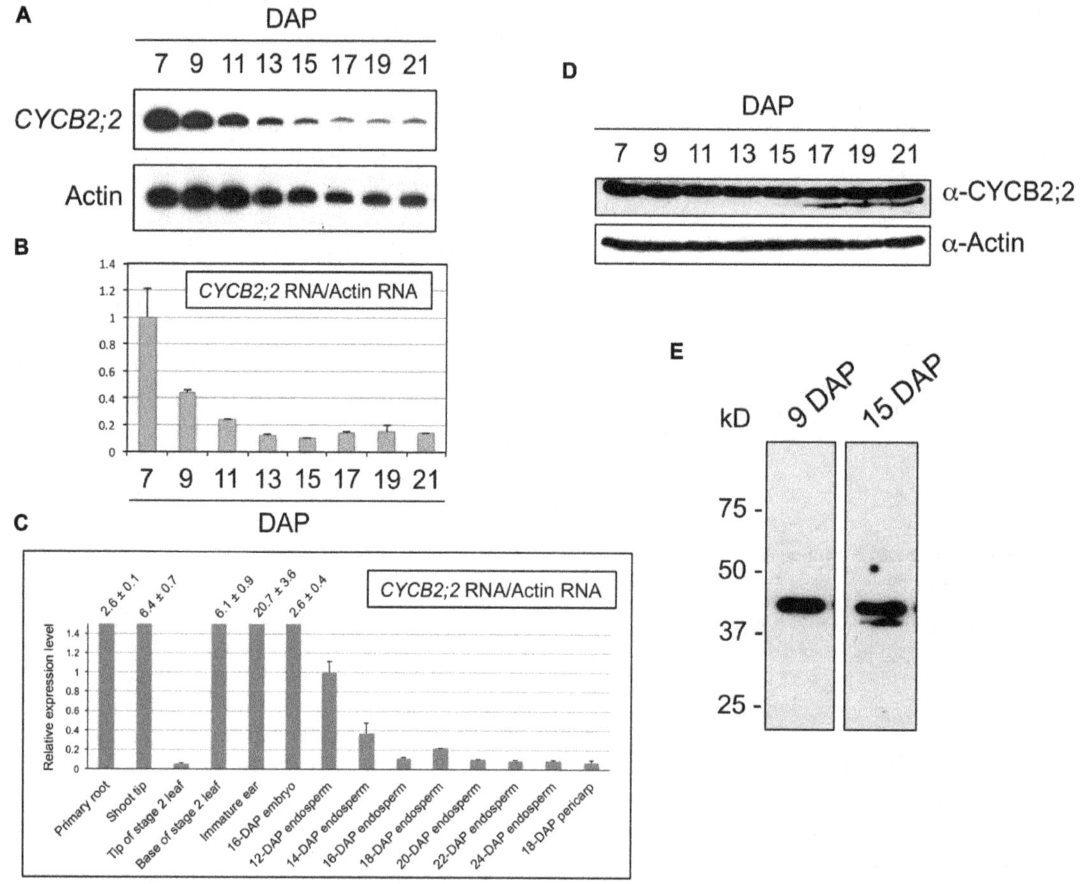

FIGURE 4 | CYCB2;2 expression patterns during maize endosperm development. (A–C) Analyses of RNA accumulations patterns. **(A,B)** Analysis of *CYCB2;2* RNA accumulation pattern during endosperm development from 7 to 21 DAP by RT-PCR. Actin RNA was used as a control. Error bars indicate standard deviations. **(C)** Analysis of CYCB2;2 RNA accumulation, relative to that of actin, in 14 maize tissues and developmental stages. Error bars indicate standard deviations. This analysis was based on data from a previous study by Sekhon et al. (2011). **(D,E)** Western blots. Actin expression pattern was used as a control. Accumulation of CYCB2;2 protein is roughly constant, despite decreasing levels of RNA. A lower molecular weight CYCB2;2-related polypeptide accumulated specifically in endoreduplicating (e.g., 15-DAP) endosperm cell extracts **(E)**. The actin blot was part of a set of common experiments carried out during a larger study and was published previously (Sabelli et al., 2005).

CYCB2;2 was co-expressed with either maize CDKA;1 or CDKA;3 in *Drosophila* S2 cells. Protein extracts were immunoprecipitated with anti-CYCB2;2 antibody and tested for the presence of either CDKA using anti-PSTAIR antibodies. In both cases, a clear interaction between CYCB2;2 and CDKA;1 or CDKA;3 was detected. However, these CDKA/CYCB2;2 complexes did not phosphorylate histone H1 substrate (**Figure 8B**), suggesting either that the CYCB2;2-associated kinase activity in endosperm extracts is due to a CDK other than CDKA;1 or CDKA;3 or that the expression system utilized for this assay was deficient for some other necessary factors, such as, for example, specific CDK-activating kinases typically acting upstream of CDKAs (Umeda et al., 2000; Shimotohno et al., 2004).

PROTEASOME-DEPENDENT DEGRADATION OF CYCB2;2 IN ENDOSPERM

The sustained accumulation of CYCB2;2 protein during endosperm development, in spite of rapidly declining levels of its RNA, suggested that perhaps CYCB2;2 was becoming resistant to degradation by the 26S proteasome during the endoreduplication phase. Indeed, we recently showed that endoreduplicating endosperm can be differentiated from mitotic endosperm by virtue of sustained expression of certain A-, B-, and D-type cyclins associated with their reduced ubiquitin-mediated proteolysis (Dante et al., 2014). A similar outcome was obtained for CYCB2;2 in the present investigation (**Figure 9**). *In vitro*-synthesized CYCB2;2 was almost entirely degraded by a 90-min incubation with mitotic (i.e., 7-DAP) endosperm extracts, whereas it was unaffected by endoreduplicating (i.e., 15-DAP) extracts (**Figure 9A**). Degradation by mitotic extracts at 7-DAP was primarily due to the 26S proteasome, as it could largely be prevented by the addition of the proteasome-specific inhibitor, MG-132 (**Figure 9B**). In addition, we tested whether Thr-417 is required for CYCB2;2 proteolysis. This residue, in fact, is a potential CDK phosphorylation site that is conserved in animal CycE proteins and is necessary for protein degradation via the ubiquitin-dependent pathway (Welker et al., 2003; Hwang and

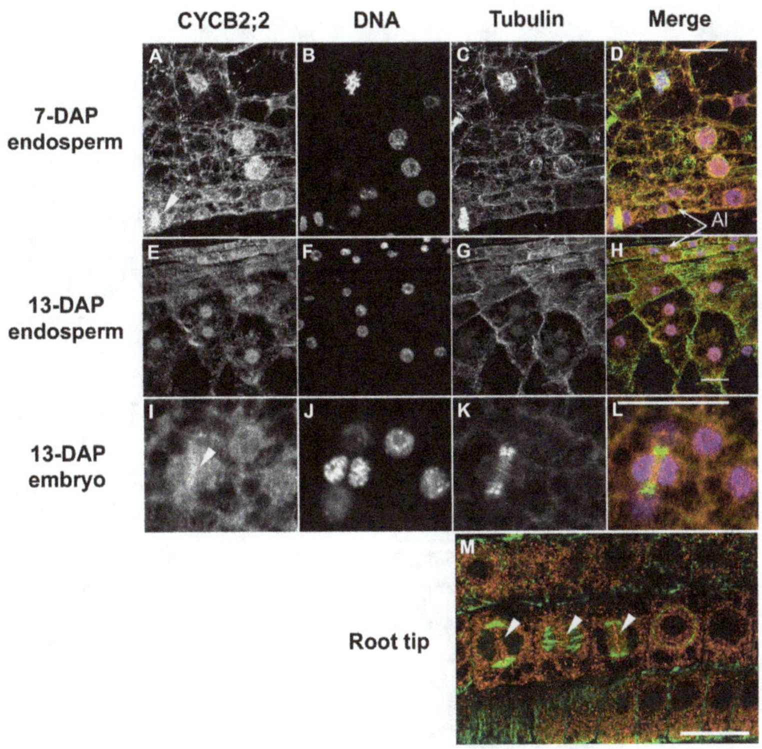

FIGURE 5 | Immunolocalization of CYCB2;2 protein in maize endosperm, embryo and root tip cells. Treatments as follows: **(A,E,I)**, anti-CYCB2;2 antibody; **(B,F,J)** DNA staining; **(C,G,K)** anti-tubulin antibody; **(D,H,L)** merge (CYCB2;2, red; DNA, blue; tubulin, green). **(A–D)** 7-DAP endosperm; **(E–H)** 13-DAP endosperm; **(I–L)** 13-DAP embryo. **(M)** Merge image of CYCB2;2 (red) and tubulin (green) staining in root tip cells. Panels **(I–L)** show CYCB2;2 localization (arrow in **I**) in a late-phragmoplast cell. Arrows in **(A,I,M)** indicate CYCB2;2 localization to the phragmoplast or cell division site. Arrows in **(M)** indicate CYCB2;2 localization in cells during cytokinesis (left arrow), anaphase (center arrow), and telophase (right arrow). The aleurone layer (Al) is indicated in **(D,H)**. Scale bars = 25 μm.

Clurman, 2005; Hao et al., 2007). Thr-417 was mutagenized to Ala-417 (**Figure 9C**), which would destroy the putative phosphorylation site in CYCB2;2, but the mutant CYCB2;2T/A protein was degraded by incubation with mitotic extracts just as effectively as the wild-type protein (**Figure 9A**), suggesting that Thr-417, at least *in vitro*, is not required for CYCB2;2 proteolysis.

DISCUSSION

Cyclin proteins of the A-, B, and D-type are essential components of CDK complexes that drive the plant cell cycle. Specifically, they contribute to the regulation of the rhythmic activation of CDK activity during the cell cycle and help determine substrate specificity. Here we report the characterization of a B2;2-type cyclin from maize during the development of the seed endosperm, which, similarly to other cereals, typically comprises two successive phases characterized by cell proliferation through the mitotic cell cycle and endoreduplication, which is associated with cell expansion and the growth of the whole tissue.

The temporal expression pattern of CYCB2;2 is characterized by relatively high RNA levels during the mitotic phase of endosperm development (i.e., 7–9 DAP) and a dramatic decline during the endoreduplication phase (i.e., 13–21 DAP). By 13-DAP, the *CYCB2;2* RNA level is less than 20% that of the 7-DAP level, and remains below this figure throughout the period of endosperm development studied up to 21-DAP. This pattern is similar to those of CYCA1;1, CYCA1;2, and CYCB1;3 and clearly distinct from those of D-type cyclins, which display less marked differences in RNA levels between mitotic and endoreduplicating

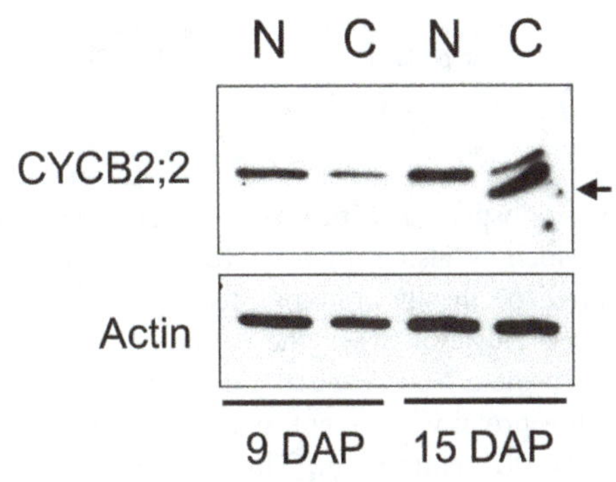

FIGURE 6 | A low-molecular-weight form of CYCB2;2 accumulates in the cytosol of endoreduplicating endosperm cells. Extracts were prepared from nuclei- (N) and cytosolic- (C) enriched fractions from 9- and 15-DAP endosperms. Western analysis with anti-CYCB2;2 antibody revealed this protein was prevalently nuclear at 9 and 15 DAP, but a lower molecular weight polypeptide accumulated in the cytosolic fraction at 15-DAP. Anti-actin antibody was used as a control.

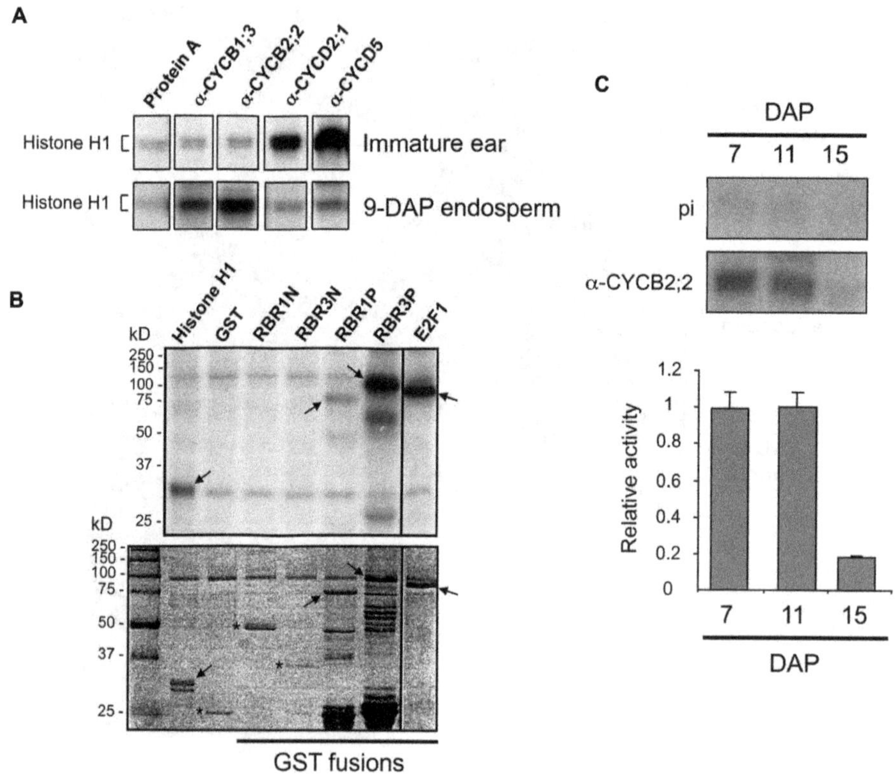

FIGURE 7 | Analysis of CYCB2;2-associated kinase activity. **(A)** Phosphorylation of histone H1 substrate by kinase activities immunoprecipitated from immature ear and 9-DAP endosperm with antibodies against CYCB2;2, and various other cyclins as indicated. Protein A indicates a control reaction carried out without antibody showing background signal due to agarose-Protein A beads. **(B)** Phosphorylation of various recombinant substrates by CYCB2;2-associated kinase activity immunoprecipitated from 9-DAP endosperm (top). Substrates include histone H1, GST, as well as GST fusions of the *N*-terminal domains of RBR1 (RBR1N) and RBR3 (RBR3N), the pocket domains of RBR1 (RBR1P) and RBR3 (RBR3P), and full-length E2F1 (shown in the stained gel, bottom). Arrows indicate specific phosphorylation of substrates. Asterisks indicate substrates that were not phosphorylated (bottom panel). **(C)** Phosphorylation of histone H1 by CYCB2;2-associated kinase immunoprecipitated from 7-, 11-, and 15-DAP endosperm. Top: Authoradiographic detection. Control assays with IgG from pre-immune serum are shown (pi). Bottom: quantitation of kinase activities. Values are averages of two independent experiments, from which background phosphorylation with pre-immune IgG was subtracted. Relative values were normalized by the highest values in each experiment. Error bars represent standard deviations.

endosperm (Dante et al., 2014). This observation suggests that the primary endosperm role of CYCB2;2 may be in the mitotic cell cycle, which is also supported by the high levels of its RNA observed specifically in mitotic tissues in the Sekhon et al. (2011) dataset. However, when the pattern of accumulation of the encoded protein was analyzed using specific antibodies, it was found the protein is present at roughly constant levels in 7- through 21-DAP endosperm. This pattern is reminiscent of that recently described for CYCB1;3 (Dante et al., 2014), and suggests that the activity of these B-type cyclins is sustained in endoreduplicating cells. Interestingly, a CYCB2;2 immunologically related polypeptide of slightly lower molecular weight accumulated specifically in endoreduplicating but not mitotic endosperm cells, suggesting it may have a specific function during endoreduplication. It is not known whether this LMW protein represents an incomplete form of CYCB2;2 due, for example, to alternative RNA splicing (though there is no evidence of such an occurrence in genomics datasets), the specific presence/absence of phosphate groups, or partial proteolysis, although the latter is unlikely since we found no evidence of CYCB2;2 degradation in endoreduplicating endosperm extracts. An alternative possibility is that this LMW polypeptide reflects the accumulation of another immunologically cross-reactive cyclin. The anti-CYCB2;2 antibodies used in this study do not cross-react with CYCB1;3 or CYCA1;1. The two most closely related maize proteins to CYCB2;2 are CYCB2;1 and the CYCB2;3 polypeptide identified in this study. CYCB2;1 has a calculated MW of 50 kDa, which is larger than that of CYCB2;2 and shares only 39% sequence identity with CYCB2;2 over the *N*-terminal region selected to raise antibodies. The *N*-terminal region of CYCB2;3, on the other hand, is 78% identical to that of CYCB2;2, which suggests potential antibody cross-reactivity between these two cyclins, but CYCB2;3 is slightly larger than CYCB2;2 (426 amino acid-long with a calculated MW of 47.7 kDa). Thus, the above features of CYCB2;1 and CYCB2;3 make it unlikely for either of them, at least as a full-length protein, to account for the CYCB2;2 LMW polypeptide detected in endoreduplicating extracts. Additionally, data on RNA levels available at the Maize eFP Browser (Sekhon et al., 2011) indicate that *CYCB2;1* and the newly identified GRMZM2G061287_T01 (encoding CYCB2;3) transcripts

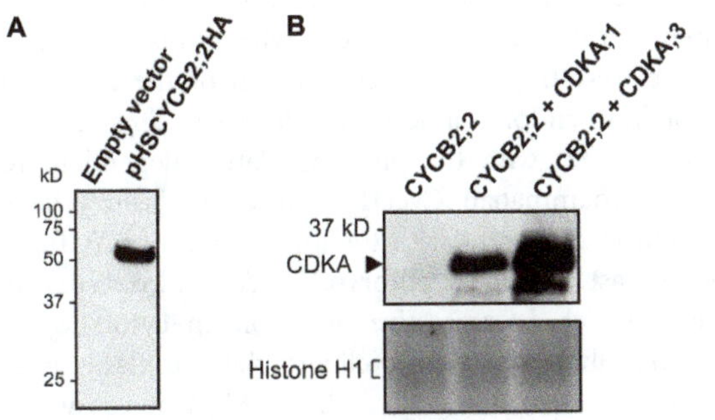

FIGURE 8 | Interaction of CYCB2;2 with maize CDKs in *Drosophila* S2 cells and analysis of associated kinase activities. (A) *Drosophila* S2 cells were transfected either with empty pHSKSMCS vector or pHSCYB2;2HA construct, and CYCB2;2HA expression was induced by a 37 °C heat-shock. CYCB2;2HA expression was assessed by western blotting with anti-CYCB2;2 antibody. **(B)** CYCB2;2;HA (CYCB2;2) was expressed in S2 cells either alone or co-expressed with maize CDKA;1 or CDKA;3, and immunoprecipitated with anti-CYCB2;2 antibody. **Top**, Western blot analysis performed on immunoprecipitates with anti-PSTAIR antibody to reveal CYCB2;2-CDKA interaction. **Bottom**, histone H1 kinase activity of immunoprecipitates. Complexes between CYCB2;2 and A-type CDKs show no activity.

decrease dramatically in the endosperm after 12-DAP, which does not support the idea they could encode the LMW CYCB2;2 band. Thus, unraveling the identity of this LMW CYCB2;2 polypeptide will require further investigation.

Analysis of sub-cellular fractions revealed CYCB2;2 is primarily a nuclear protein in both mitotic and endoreduplicating endosperm. The LMW polypeptide, however, accumulated specifically in endoreduplicating endosperm extracts, confirming data from whole cell extracts, but was solely detected in the cytosolic fraction. Clearly, whatever the identity of this protein, its accumulation pattern suggests it may play a role in determining whether endosperm cells engage in the mitotic or endoreduplication cycle. However, a possible role for this polypeptide in other cytoplasm-associated aspects of endosperm development linked to endoreduplication, such as cell expansion, storage compound deposition and grain filling, cannot be ruled out.

Based on immunohistochemical analyses, CYCB2;2 appears to be localized primarily to the nucleus of mitotic endosperm cells, whereas by 13-DAP it is also distributed extensively throughout the cytoplasm. These data are in agreement with the results from the subcellular fractionation analysis, and presumably a large proportion of the extra-nuclear signal in cells of endoreduplicating endosperm sections is due to the accumulation of the LMW form of the protein. Considerable overlap between CYCB2;2 localization and the microtubule cytoskeleton was consistently observed, which suggests that this cyclin is involved in regulating aspects of cytoskeleton structure and organization, in agreement with previous observations (Mews et al., 1997; John et al., 2001; Weingartner et al., 2004). In dividing cells, whether in the endosperm, embryo or the root tip, CYCB2;2 appeared to accumulate preferentially at

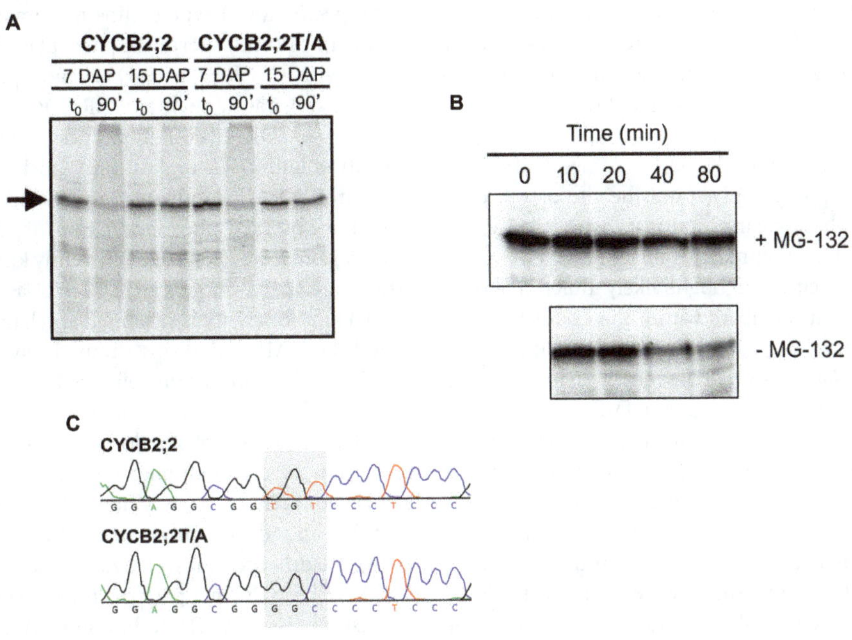

FIGURE 9 | Analysis of CYCB2;2 stability in mitotic and endoreduplicating endosperm cell extracts. (A) ^{35}S-labeled, *in vitro* synthesized CYCB2;2 and T417A mutant (CYCB2;2T/A) proteins (indicated by an arrow) were incubated for 90 min with mitotic (7-DAP) and endoreduplicating (15-DAP) endosperm extracts, separated by SDS-PAGE and detected by autoradiography. Control reactions, with added endosperm extracts but without incubation, are indicated as t_0. CYCB2;2 appears to undergo degradation specifically in 7-DAP endosperm extracts, regardless of the T417A mutation. **(B)** Time-course analysis of CYCB2;2 degradation in 7-DAP endosperm extracts in the presence (+) or absence (–) of the proteasome-specific inhibitor MG-132. **(C)** Chromatograms showing the TGT to GGC conversion (indicated by a shaded box) responsible for mutagenizing the potential Thr-417 phosphorylation site in CYCB2;2 protein to Ala-417 in the CYCB2;2T/A protein.

the intervening domains between daughter chromatin and nuclei, and it was closely associated with the phragmoplast and with the sites of deposition of the cell plate and nascent cell walls. CYCB2;2 is clearly localized to the residual phragmoplast as it is gradually eroded at cytokinesis (**Figures 5I–L**), further suggesting a role in cell wall formation and cell division. These patterns resemble those of KNOLLE (a cytokinesis-specific syntaxin required for cell plate formation) and NACK1 (a kinesin-like protein involved in the vesicular traffic required for phragmoplast outgrowth; Weingartner et al., 2004), although a large number of proteins are also known to be associated with the phragmoplast and/or the cell plate (McMichael and Bednarek, 2013).

These observations suggest a role for CYCB2;2 in cytokinesis and cell wall formation. Previous data on the expression patterns and accumulation of plant B-type cyclins indicate a mitotic role, but they are contrasting with regard to an involvement in cytokinesis. Typically, the expression profiles of B-type cyclins display a peak in early mitosis, due to transcriptional up-regulation at the G2/M-phase transition coupled to proteolysis during anaphase, and their destruction in late M-phase is essential for proper microtubule cytoskeleton re-organization in late mitosis and cell cycle progression. Over- or ectopic expression of at least some B1- and B2-type cyclins stimulate M-phase entry and cell division (Doerner et al., 1996; Schnittger et al., 2002; Lee et al., 2003; Weingartner, 2003). CYCB1;1 from tobacco and CYCB2;2 from rice associate with metaphase chromosomes and are degraded thereafter (Criqui et al., 2000; Lee et al., 2003; Weingartner, 2003). Tobacco CYCB2;2 is also degraded after prophase (Weingartner, 2003). In maize, CYCB1;1 is prevalently nuclear prior to mitosis, whereas CYCB1;2 is associated with the pre-prophase band, condensed chromosomes and the mitotic spindle and is degraded at anaphase, while CYCB2;1 is predominantly nuclear and becomes associated with the spindle and the phragmoplast at telophase (Mews et al., 1997). The persistence of maize CYCB2;2 (and CYCB2;1 – Mews et al., 1997; John et al., 2001) at the phragmoplast and the site of cell plate formation during telophase clearly contrasts with the typical degradation of most B-type cyclins after metaphase (Bulankova et al., 2013). These differences highlight likely functional specialization among different members within the cyclin B family and its subfamilies (John et al., 2001), as recently confirmed by a large study in *Arabidopsis* (Bulankova et al., 2013). Indeed, we found that maize CYCB2;2 possesses a D-Box and undergoes proteasome-dependent degradation in mitotic endosperm extracts, suggesting it may be destroyed, like other related cyclins, late in mitosis. However, our immunolocalization results indicate that the protein remains present during anaphase and telophase, suggesting that the observed proteasome-dependent degradation is part of the normal turnover regimen for this protein and may not be related to M-phase exit. Interestingly, the apparent ability of maize CYCB2;2 to elude degradation by the proteasome in endoreduplicating endosperm is reminiscent of the outcome of expressing a non-degradable form of tobacco CYCB1,1 (with a non-functional D-Box), which resulted in microtubule disorganization, inhibited phragmoplast formation and caused endomitosis and increased DNA content (Weingartner et al., 2004). It is tempting to postulate that partial proteolysis of CYCB2;2 is responsible for its exclusion from the nucleus in endoreduplicating endosperm cells, which thus would be unable to divide, resulting in an accumulation of the LMW form of the protein in endoreduplicating endosperm. This interpretation is in agreement with the phosphorylation-dependent redistribution of mammalian CycD1 to the cytoplasm (Diehl et al., 1998) and is consistent with the localization of CYCB2;2 to the phragmoplast in mitotic embryo, 7-DAP endosperm, and root tip cells, which is suggestive of a role in cytokinesis. However, redistribution of D1 cyclin to the cytoplasm is coupled with its degradation (Diehl et al., 1998), whereas we did not observe any evidence for proteolysis of CYCB2;2 in endoreduplicating cells. On the contrary, it appears that, like other cyclins, CYCB2;2 is specifically stabilized in endocycling cells (Dante et al., 2014). However, although our analyses of proteolytic activities revealed a clear difference between mitotic and endoreduplicating endosperm in carrying out CYCB2;2 degradation, they are based on a substrate synthesized *in vitro*, which could lack some co- or post-translational modification essential for targeting the protein for proteolysis in a physiological context.

CYCB2;2-associated kinase activity is relatively high in 9-DAP endosperm but only slightly above background levels in immature ear, which is similar to the situation for another maize B-type cyclin, CYCB1;3. In contrast, the kinase activities associated with CYCD2;1 and CYCD5 are high in immature ear and relatively low, though appreciably higher than background, in endosperm. Although both tissues are known to be prevalently mitotic, these differences point to some additional developmental roles specific for D-type cyclins in immature ear and for B-type cyclins in the endosperm. In particular, the kinase activity associated with CYCB2;2 in 9-DAP endosperm can phosphorylate *in vitro* substrates, such as RBR1, RBR3 and E2F1, which are part of the CDK–RBR–E2F pathway controlling the G1/S-phase transition and DNA replication (Sabelli et al., 2013). The presence of a MRAIL motif in the CYCB2;2 sequence is consistent with the observed phosphorylation of RBRs by CYCB2;2/CDK complexes. However, it is not currently known whether these substrates are actual targets of CYCB2;2-associated kinase *in vivo*. Such a kinase activity is sustained at relatively high levels between 7 and 11 DAP, and drops dramatically by 15-DAP, a stage at which starchy endosperm cells are almost exclusively endoreduplicated. The much lower kinase activity at 15-DAP suggests that CYCB2;2/CDK complexes do not significantly participate in the regulation of endoreduplication cycles. In contrast, CYCB1;3-associated activity is very low at 7-DAP, reaches a peak at 11-DAP, and drops only marginally by 15-DAP (Dante et al., 2014). Clearly the kinase activities associated with CYCB1;3 and CYCB2;2 have distinct patterns during endosperm development, suggesting that CYCB1;3, but not CYCB2;2, may also play a role in the endocycle. In insect cells, CYCB2;2 forms complexes, albeit inactive, with CDKA;1 and CDKA;3. CDKA;1 protein has not been found associated with the phragmoplast or the cell plate in maize root tip cells (Colasanti et al., 1993), and thus it is unlikely that it forms active complexes with CYCB2;2 to regulate microtubule dynamics during late mitosis and cytokinesis. Interaction between CYCB2;2 and a B-type CDK (CDKBs specifically regulate the G2/M-phase transition and suppress the endocycle in mitotic *Arabidopsis* cells – Boudolf et al., 2004, 2009), and formation of active kinase were indeed shown in rice (Lee et al., 2003). It remains to be seen whether these interactions occur in a physiological context in maize (i.e., endosperm cells) and whether they result in phosphorylation of downstream targets. However, it is tempting to speculate that inactivation of CYCB2;2/CDK complexes could contribute to down-regulating M-phase-specific kinase activity, thereby licensing endosperm cells for endoreduplication.

Together, these results suggest that CYCB2;2 may possess some specific function particularly associated with mitotic endosperm. However, it is intriguing that this protein and the closely related LMW polypeptide accumulate during the endoreduplication phase of endosperm development, although they apparently possess little, if any, associated kinase activity at this developmental phase. Evidence for a general decrease in proteasome-dependent proteolysis of cyclins that contributes to their stabilization during endosperm development has recently been obtained (Dante et al., 2014). Furthermore, comprehensive analyses have shown there is generally a weak correlation between RNA and proteins levels in maize seeds, and that proteins can disproportionally accumulate in relation to low RNA levels (Walley et al., 2013). The above paradigms seem to hold true also for CYCB2;2, and whether this protein has any role in post-mitotic endosperm development remains to be established.

ACKNOWLEDGMENTS

This work was funded in part by a grant from the U.S. Department of Energy (Grant DE-FG02-96ER20242), and by a grant from Pioneer Hi-Bred Inc. Ricardo A. Dante was supported by a scholarship from the Conselho Nacional de Desenvolvimento Cientifico e Tecnologico of Brazil.

REFERENCES

Becraft, P., Brown, R., Lemmon, B., Olsen, O.-A., and Opsahl-Ferstad, H.-G. (2001). "Endosperm development," in *Current Trends in the Embryology of Angiosperms*, eds S. Bhojwani and W. Soh (Dordrecht: Kluwer Academic Publishers), 353–374.

Boudolf, V., Lammens, T., Boruc, J., Van Leene, J., Van Den Daele, H., Maes, S., et al. (2009). CDKB1;1 forms a functional complex with CYCA2;3 to suppress endocycle onset. *Plant Physiol.* 150, 1482–1493. doi: 10.1104/pp.109.140269

Boudolf, V., Vlieghe, K., Beemster, G. T. S., Magyar, Z., Acosta, J. A. T., Maes, S., et al. (2004). The plant-specific cyclin-dependent kinase CDKB1;1 and transcription factor E2Fa-DPa control the balance of mitotically dividing and endoreduplicating cells in *Arabidopsis*. *Plant Cell* 16, 2683–2692. doi: 10.1105/tpc.104.024398

Bulankova, P., Akimcheva, S., Fellner, N., and Riha, K. (2013). Identification of *Arabidopsis* meiotic cyclins reveals functional diversification among plant cyclin genes. *PLoS Genet.* 9:e1003508. doi: 10.1371/journal.pgen.1003508

Burke, J. R., Hura, G. L., and Rubin, S. M. (2012). Structures of inactive retinoblastoma protein reveal multiple mechanisms for cell cycle control. *Genes Dev.* 26, 1156–1166. doi: 10.1101/gad.189837.112

Chou, K.-C., and Shen, H.-B. (2010). Plant-mPLoc: a top-down strategy to augment the power for predicting plant protein subcellular localization. *PLoS ONE* 5:e11335. doi: 10.1371/journal.pone.0011335

Coelho, C. M., Dante, R. A., Sabelli, P. A., Sun, Y., Dilkes, B. P., Gordon-Kamm, W. J., et al. (2005). Cyclin-dependent kinase inhibitors in maize endosperm and their potential role in endoreduplication. *Plant Physiol.* 138, 2323–2336. doi: 10.1104/pp.105.063917

Colasanti, J., Cho, S. O., Wick, S., and Sundaresan, V. (1993). Localization of the functional p34cdc2 homolog of maize in root tip and stomatal complex cells: association with predicted division sites. *Plant Cell* 5, 1101–1111. doi: 10.1105/tpc.5.9.1101

Criqui, M. C., Parmentier, Y., Derevier, A., Shen, W. H., Dong, A., and Genschik, P. (2000). Cell cycle-dependent proteolysis and ectopic overexpression of cyclin B1 in tobacco BY2 cells. *Plant J.* 24, 763–773. doi: 10.1111/j.1365-313X.2000.t01-1-.x

Dante, R. A., Sabelli, P. A., Nguyen, H. N., Leiva-Neto, J. T., Tao, Y., Lowe, K. S., et al. (2014). Cyclin-dependent kinase complexes in developing maize endosperm: evidence for differential expression and functional specialization. *Planta* 239, 493–509. doi: 10.1007/s00425-013-1990-1

De Veylder, L., Beeckman, T., and Inzé, D. (2007). The ins and outs of the plant cell cycle. *Nat. Rev. Mol. Cell Biol.* 8, 655–665. doi: 10.1038/nrm2227

Di Tommaso, P., Moretti, S., Xenarios, I., Orobitg, M., Montanyola, A., Chang, J.-M., et al. (2011). T-Coffee: a web server for the multiple sequence alignment of protein and RNA sequences using structural information and homology extension. *Nucleic Acids Res.* 39, W13–W17. doi: 10.1093/nar/gkr245

Diehl, J. A., Cheng, M., Roussel, M. F., and Sherr, C. J. (1998). Glycogen synthase kinase-3β regulates cyclin D1 proteolysis and subcellular localization. *Genes Dev.* 12, 3499–3511. doi: 10.1101/gad.12.22.3499

Dinkel, H., Van Roey, K., Michael, S., Davey, N. E., Weatheritt, R. J., Born, D., et al. (2014). The eukaryotic linear motif resource ELM: 10 years and counting. *Nucleic Acids Res.* 42, D259–D266. doi: 10.1093/nar/gkt1047

Doerner, P., Jorgensen, J. E., You, R., Steppuhn, J., and Lamb, C. (1996). Control of root growth and development by cyclin expression. *Nature* 380, 520–523. doi: 10.1038/380520a0

Edgar, R. C. (2004). MUSCLE: multiple sequence alignment with high accuracy and high throughput. *Nucleic Acids Res.* 32, 1792–1797. doi: 10.1093/nar/gkh340

Finn, R. D., Bateman, A., Clements, J., Coggill, P., Eberhardt, R. Y., Eddy, S. R., et al. (2014). The Pfam protein families database. *Nucleic Acids Res.* 42, D222–D230. doi: 10.1093/nar/gkt1223

Goldberg, T., Hamp, T., and Rost, B. (2012). LocTree2 predicts localization for all domains of life. *Bioinformatics* 28, i458–i465. doi: 10.1093/bioinformatics/bts390

Grafi, G., Burnett, R. J., Helentjaris, T., Larkins, B. A. L., DeCaprio, J. A., Sellers, W. J., et al. (1996). A maize cDNA encoding a member of the retinoblastoma protein family: involvement in endoreduplication. *Proc. Natl. Acad. Sci. U.S.A.* 93, 8962–8967. doi: 10.1073/pnas.93.17.8962

Grafi, G., and Larkins, B. A. L. (1995). Endoreduplication in maize endosperm: involvement of M phase-promoting factor inhibition and induction of S phase-related kinases. *Science* 269, 1262–1264. doi: 10.1126/science.269.5228.1262

Guo, J., Song, J., Wang, F., and Zhang, X. S. (2007). Genome-wide identification and expression analysis of rice cell cycle genes. *Plant Mol. Biol.* 64, 349–360. doi: 10.1007/s11103-007-9154-y

Guo, J., Wang, F., Song, J., Sun, W., and Zhang, X. S. (2010). The expression of Orysa;CycB1;1 is essential for endosperm formation and causes embryo enlargement in rice. *Planta* 231, 293–303. doi: 10.1007/s00425-009-1051-y

Guo, J., Wang, F., and Zhang, X. S. (2014). Knockdown expression of the B-type cyclin gene Orysa;CycB1;1 leads to triploid rice. *J. Plant Biol.* 57, 43–47. doi: 10.1007/s12374-013-0405-y

Hao, B., Oehlmann, S., Sowa, M. E., Harper, J. W., and Pavletich, N. P. (2007). Structure of a Fbw7-Skp1-Cyclin E complex: multisite-phosphorylated substrate recognition by SCF ubiquitin ligases. *Mol. Cell* 26, 131–143. doi: 10.1016/j.molcel.2007.02.022

Harashima, H., Dissmeyer, N., and Schnittger, A. (2013). Cell cycle control across the eukaryotic kingdom. *Trends Cell Biol.* 23, 345–356. doi: 10.1016/j.tcb.2013.03.002

Hu, X., Cheng, X., Jiang, H., Zhu, S., Cheng, B., and Xiang, Y. (2010). Genome-wide analysis of cyclins in maize (*Zea mays*). *Genet. Mol. Res.* 9, 1490–1503. doi: 10.4238/vol9-3gmr861

Hwang, H. C., and Clurman, B. E. (2005). Cyclin E in normal and neoplastic cell cycles. *Oncogene* 24, 2776–2786. doi: 10.1038/sj.onc.1208613

Ishimaru, K., Hirotsu, N., Madoka, Y., Murakami, N., Hara, N., Onodera, H., et al. (2013). Loss of function of the IAA-glucose hydrolase gene TGW6 enhances rice grain weight and increases yield. *Nat. Genet.* 45, 707–711. doi: 10.1038/ng.2612

Jia, R.-D., Guo, C.-C., Xu, G.-X., Shan, H.-Y., and Kong, H.-Z. (2014). Evolution of the cyclin gene family in plants. *J. Syst. Evol.* 52, 651–659. doi: 10.1111/jse.12112

John, P. C., Mews, M., and Moore, R. (2001). Cyclin/Cdk complexes: their involvement in cell cycle progression and mitotic division. *Protoplasma* 216, 119–142. doi: 10.1007/BF02673865

Kawahara, Y., de la Bastide, M., Hamilton, J. P., Kanamori, H., McCombie, W. R., Ouyang, S., et al. (2013). Improvement of the *Oryza sativa* Nipponbare reference genome using next generation sequence and optical map data. *Rice* 6:4. doi: 10.1186/1939-8433-6-4

La, H., Li, J., Ji, Z., Cheng, Y., Li, X., Jiang, S., et al. (2006). Genome-wide analysis of cyclin family in rice (*Oryza sativa* L.). *Mol. Gen. Genom.* 275, 374–386. doi: 10.1007/s00438-005-0093-5

Lawrence, C. J., Dong, Q., Polacco, M. L., Seigfried, T. E., and Brendel, V. (2004). MaizeGDB, the community database for maize genetics and genomics. *Nucleic Acids Res.* 32, D393–D397. doi: 10.1093/nar/gkh011

Lee, J., Das, A., Yamaguchi, M., Hashimoto, J., Tsutsumi, N., Uchimiya, H., et al. (2003). Cell cycle function of a rice B2-type cyclin interacting with a B-type cyclin-dependent kinase. *Plant J.* 34, 417–425. doi: 10.1046/j.1365-313X.2003.01736.x

Leiva-Neto, J. T., Grafi, G., Sabelli, P. A., Dante, R. A., Woo, Y. M., Maddock, S., et al. (2004). A dominant negative mutant of cyclin-dependent kinase A reduces endoreduplication but not cell size or gene expression in maize endosperm. *Plant Cell* 16, 1854–1869. doi: 10.1105/tpc.022178

McMichael, C. M., and Bednarek, S. Y. (2013). Cytoskeletal and membrane dynamics during higher plant cytokinesis. *New Phytol.* 197, 1039–1057. doi: 10.1111/nph.12122

Mews, M., Sek, F. J., Moore, R., Volkmann, D., Gunning, B. E. S., and John, P. C. L. (1997). Mitotic cyclin distribution during maize cell division: implications for the sequence diversity and function of cyclins in plants. *Protoplasma* 200, 128–145. doi: 10.1007/BF01283289

Nugent, J. H., Alfa, C. E., Young, T., and Hyams, J. S. (1991). Conserved structural motifs in cyclins identified by sequence analysis. *J. Cell Sci.* 25, 669–674.

Olsen, O. A. (2004). Nuclear endosperm development in cereals and *Arabidopsis thaliana*. *Plant Cell* 16, 214–227. doi: 10.1105/tpc.017111

Sabelli, P. A. (2014). "Cell cycle regulation and plant development: a crop production perspective," in *Handbook of Plant and Crop Physiology*, ed. M. Pessarakli (Boca Raton, FL: CRC Press), 3–32.

Sabelli, P. A., Dante, R. A., Leiva-Neto, J. T., Jung, R., Gordon-Kamm, W. J., and Larkins, B. A. (2005). RBR3, a member of the retinoblastoma-related family from maize, is regulated by the RBR1/E2F pathway. *Proc. Natl. Acad. Sci. U.S.A.* 102, 13005–13012. doi: 10.1073/pnas.0506160102

Sabelli, P. A., and Larkins, B. A. (2009a). Regulation and function of retinoblastoma-related plant genes. *Plant Sci.* 177, 540–548. doi: 10.1016/j.plantsci.2009.09.012

Sabelli, P. A., and Larkins, B. A. (2009b). The contribution of cell cycle regulation to endosperm development. *Sex. Plant Reprod.* 22, 207–219. doi: 10.1007/s00497-009-0105-4

Sabelli, P. A., and Larkins, B. A. (2009c). The development of endosperm in grasses. *Plant Physiol.* 149, 14–26. doi: 10.1104/pp.108.129437

Sabelli, P. A., Liu, Y., Dante, R. A., Lizarraga, L. E., Nguyen, H. N., Brown, S. W., et al. (2013). Control of cell proliferation, endoreduplication, cell size, and cell death by the retinoblastoma-related pathway in maize endosperm. *Proc. Natl. Acad. Sci. U.S.A.* 110, E1827–E1836. doi: 10.1073/pnas.1304903110

Schnittger, A., Schöbinger, U., Stierhof, Y.-D., and Hülskamp, M. (2002). Ectopic B-type cyclin expression induces mitotic cycles in endoreduplicating *Arabidopsis* trichomes. *Curr. Biol.* 12, 415–420. doi: 10.1016/S0960-9822(02)00693-0

Schulman, B. A., Lindstrom, D. L., and Harlow, E. (1998). Substrate recruitment to cyclin-dependent kinase 2 by a multipurpose docking site on cyclin A. *Proc. Natl. Acad. Sci. U.S.A.* 95, 10453–10458. doi: 10.1073/pnas.95.18.10453

Sekhon, R. S., Lin, H., Childs, K. L., Hansey, C. N., Buell, C. R., de Leon, N., et al. (2011). Genome-wide atlas of transcription during maize development. *Plant J.* 66, 553–563. doi: 10.1111/j.1365-313X.2011.04527.x

Shimotohno, A., Umeda-Hara, C., Bisova, K., Uchimiya, H., and Umeda, M. (2004). The plant-specific kinase CDKF;1 is involved in activating phosphorylation of cyclin-dependent kinase-activating kinases in *Arabidopsis*. *Plant Cell* 16, 2954–2966. doi: 10.1105/tpc.104.025601

Sun, Y. J., Dilkes, B. P., Zhang, C. S., Dante, R. A., Carneiro, N. P., Lowe, K. S., et al. (1999a). Characterization of maize (*Zea mays* L.) *Wee1* and its activity in developing endosperm. *Proc. Natl. Acad. Sci. U.S.A.* 96, 4180–4185. doi: 10.1073/pnas.96.7.4180

Sun, Y. J., Flannigan, B. A., and Setter, T. L. (1999b). Regulation of endoreduplication in maize (*Zea mays* L.) endosperm. Isolation of a novel B1-type cyclin and its quantitative analysis. *Plant Mol. Biol.* 41, 245–258. doi: 10.1023/A:1006315625486

Tamura, K., Stecher, G., Peterson, D., Filipski, A., and Kumar, S. (2013). MEGA6: molecular evolutionary genetics analysis version 6.0. *Mol. Biol. Evol.* 30, 2725–2729. doi: 10.1093/molbev/mst197

Umeda, M., Umeda-Hara, C., and Uchimiya, H. (2000). A cyclin-dependent kinase-activating kinase regulates differentiation of root initial cells in *Arabidopsis*. *Proc. Natl. Acad. Sci. U.S.A.* 97, 13396–13400. doi: 10.1073/pnas.240458997

Walley, J. W., Shen, Z., Sartor, R., Wu, K. J., Osborn, J., Smith, L. G., et al. (2013). Reconstruction of protein networks from an atlas of maize seed proteotypes. *Proc. Natl. Acad. Sci. U.S.A.* 3, E4808–E4817. doi: 10.1073/pnas.1319113110

Ware, D. H., Jaiswal, P., Ni, J., Yap, I. V., Pan, X., Clark, K. Y., et al. (2002). Gramene, a tool for grass genomics. *Plant Physiol.* 130, 1606–1613. doi: 10.1104/pp.015248

Weingartner, M. (2003). A plant cyclin B2 is degraded early in mitosis and its ectopic expression shortens G2-phase and alleviates the DNA-damage checkpoint. *J. Cell Sci.* 116, 487–498. doi: 10.1242/jcs.00250

Weingartner, M., Criqui, M.-C., Tamas, M., Binarova, P., Schmit, A.-C., Helfer, A., et al. (2004). Expression of a nondegradable cyclin B1 affects plant development and leads to endomitosis by inhibiting the formation of a phragmoplast. *Plant Cell* 16, 643–657. doi: 10.1105/tpc.020057

Welker, M., Singer, J., Loeb, K. R., Grim, J., Bloechler, A., Curien-West, M., et al. (2003). Multisite phosphorylation by Cdk2 and GSK3 controls Cyclin E degradation. *Mol. Cell* 12, 381–392. doi: 10.1016/S1097-2765(03)00287-9

Conflict of Interest Statement: A patent application has been filed on the maize CYCB2;2 nucleotide and protein sequences.

19

Analogous reserve distribution and tissue characteristics in quinoa and grass seeds suggest convergent evolution

Hernán P. Burrieza[1,2†], María P. López-Fernández[1,2†] and Sara Maldonado[1,2*]

[1] *Instituto de Biodiversidad y Biologia Experimental y Aplicada – Consejo Nacional de Investigaciones Científicas y Técnicas, Ciudad Autónoma de Buenos Aires, Argentina*

[2] *Departamento de Biodiversidad y Biología Experimental, Facultad de Ciencias Exactas y Naturales, Universidad de Buenos Aires, Ciudad Autónoma de Buenos Aires, Argentina*

Edited by:
Paolo A. Sabelli, University of Arizona, USA

Reviewed by:
Philip W. Becraft, Iowa State University, USA
Paolo A. Sabelli, University of Arizona, USA

****Correspondence:***
Sara Maldonado, Departamento de Biodiversidad y Biología Experimental, Facultad de Ciencias Exactas y Naturales, Universidad de Buenos Aires, Intendente Güiraldes 2160, Ciudad Autónoma de Buenos Aires C1428EGA, Argentina
e-mail: saram@bg.fcen.uba.ar

† Hernán P. Burrieza and María P. López-Fernández have contributed equally to this work.

Quinoa seeds are highly nutritious due to the quality of their proteins and lipids and the wide range of minerals and vitamins they store. Three compartments can be distinguished within the mature seed: embryo, endosperm, and perisperm. The distribution of main storage reserves is clearly different in those areas: the embryo and endosperm store proteins, lipids, and minerals, and the perisperm stores starch. Tissues equivalent (but not homologous) to those found in grasses can be identified in quinoa, suggesting the effectiveness of this seed reserve distribution strategy; as in cells of grass starchy endosperm, the cells of the quinoa perisperm endoreduplicate, increase in size, synthesize starch, and die during development. In addition, both systems present an extra-embryonic tissue that stores proteins, lipids and minerals: in gramineae, the aleurone layer(s) of the endosperm; in quinoa, the micropylar endosperm; in both cases, the tissues are living. Moreover, the quinoa micropylar endosperm and the coleorhiza in grasses play similar roles, protecting the root in the quiescent seed and controlling dormancy during germination. This investigation is just the beginning of a broader and comparative study of the development of quinoa and grass seeds. Several questions arise from this study, such as: how are synthesis and activation of seed proteins and enzymes regulated during development and germination, what are the genes involved in these processes, and lastly, what is the genetic foundation justifying the analogy to grasses.

Keywords: coleorhiza, endosperm, grass seed, micropylar endosperm, perisperm, quinoa seed

INTRODUCTION

Cereals, e.g., rice (*Oryza sativa* L.), wheat (*Triticum aestivum* L.), maize (*Zea mays* L.), barley (*Hordeum vulgare* L.), oat (*Avena sativa* L.), rye (*Secale cereale* L.) are members of the monocot family Poaceae, which are cultivated for the edible components of their grains or cariopses and consist of a single seed enclosed by dry and indehiscent pericarp, firmly adhered to the rest of the integuments.

Pseudocereals are dicots (thus not cereals) and include species of the Amaranthaceae (e.g., quinoa, *Chenopodium quinoa* Willd., and different species of the genus *Amaranthus*) and Polygonaceae (e.g., buckwheat, *Fagopyrum esculentum* Moench) families, also cultivated for the edible components of their grains. The dispersal unit in pseudocereals is the grain botanically called achene, which consists of a single seed enclosed in a dry and indehiscent pericarp (**Figures 1** and **2**). According to Prego et al. (1998), in quinoa the pericarp is very thin; as a result, the achene is also referred to as utricle.

Quinoa seeds are highly nutritious due to the quality of their proteins and lipids and the wide range of minerals and vitamins they store. The ability of quinoa to produce high-quality proteins under extreme environmental conditions makes it an important crop not only for Andean communities but also for the diversification of future agricultural systems. Cereals and quinoa are grain crops and thus essentially full of starch, but they also contain significant quantities of proteins, oil, and minerals. In quinoa and grasses, the distribution of main storage reserves is clearly divided and three main storage compartments can be distinguished in the mature seed: (i) a tissue that stores mainly starch, which is the perisperm in quinoa and the starchy endosperm in grasses, with (ii) an embryo that stores principally proteins and lipids, and (iii) a non-embryonic tissue that stores proteins, lipids, and minerals, which is the aleurone layer in gramineae and the micropylar cone in quinoa. In this study we demonstrate that tissues equivalent (but not homologous) to those found in grasses can be identified in quinoa (**Figure 3**), thus suggesting the effectiveness of this seed reserve compounds distribution strategy and justifying the advantages that led evolutionary distant plants to develop seeds with analogous nutrients distribution.

This investigation is part of a broad comparative study of the development of quinoa and grass seeds. Several questions arise from this investigation, such as: how are synthesis of seed proteins and enzymes regulated during development and germination, what are the genes involved in these processes, and lastly, what is the genetic foundation for the analogy with grasses.

THE FRUIT

According to Prego et al. (1998), in quinoa the pericarp is made up of papillose cells derived from the outer epidermis of the ovary

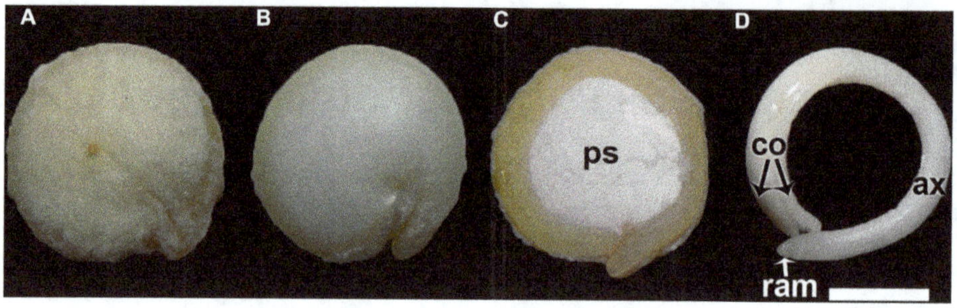

FIGURE 1 | (A) Quinoa grain; **(B)** Quinoa seed (without pericarp); **(C)** Longitudinal midsection of a quinoa seed; **(D)** Excised embryo. *ax*, hypocotyl-radicle axis; *co*, cotyledon; *ps*, perisperm; ram, root apical meristem. Bar: 1 mm.

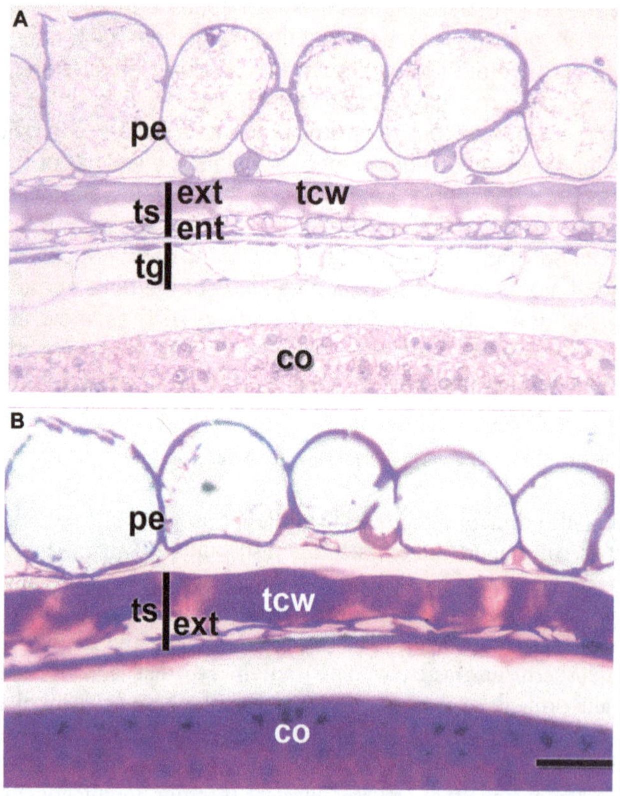

FIGURE 2 | Pericarp and integuments at two stages of quinoa seed development. (A) Young seed (torpedo stage). **(B)** Mature seed. Grains were fixed using a mixture of 2% paraformaldehyde and 1% glutaraldehyde in 0.1 M phosphate buffer, pH 7.2, and embedded in London Resin White resin, according to López-Fernández and Maldonado (2013b). Semithin sections (2 μm thick) were stained with 0.5 % Toluidine Blue. *co*, cotyledon; *ent*, endotesta; *ext*, exotesta; *pe*, pericarp; *tcw*, tangential cell wall of the cells from the outer layer of the outer integument; *tg*, tegmen; *ts*, testa.

and an inner discontinuous layer with tangentially stretched cells (**Figure 2**). The seed coat derives from ovule integuments, each one constituted by two to three layers. The seed coat consists of a testa and a tegmen, each one two layers thick. During seed development, the endotesta layer and both layers of the tegmen are almost completely consumed (**Figure 2B**); on the contrary, cells of the exotesta enlarge and develop thick tangential cell walls; these cells remain intact and are dismantled just after germination (**Figure 2**).

In grasses, the pericarp is adhered to the remains derived from the ovule integuments, which are, at maturity, usually reduced to a thin layer (Morrison, 1976). The anatomy of the pericarp is remarkably similar in all the grasses studied to date (Rost, 1973). It is composed of an outer epidermis with a thick cuticle layer and one to several subjacent cell layers (Bessey, 1894; Narayanaswami, 1953, 1955a,b,c, 1956; Rost, 1973). Even though cells of the subjacent layers are crushed, two types of cell layers, both typically containing cells with thick walls can be identified: a cross-cell layer, adjacent to the inner epidermis, whose cells are elongated transversely to the long axis of the caryopsis; and a tube cell layer, derived from the inner epidermis, whose cells are elongated parallel to the caryopsis axis (Krauss, 1933; Kent and Evers, 1994).

THE SEED

THE QUINOA SEED

The ovule is amphitropous (i.e., ovule is bent by the formation of a basal body, then both micropylar, and chalazal ends are near each other), bitegmic (with two integuments), and crassinucellate (archesporial cell cuts off a parietal cell, and parietal cell derivatives make the megasporocyte deep-seated in the ovule; Davis, 1966). The seed contains a peripheral, curved embryo surrounding a perisperm or basal body and is covered by integuments and pericarp (**Figure 1**). A micropylar endosperm forming a cone surrounds the root apical meristem of the embryo (**Figure 3**). The embryo consists of a hypocotyl–radicle axis and two cotyledons (**Figure 1D**). In the axis, both the root apical meristem, with the root cap, and the shoot apical meristem are differentiated. The shoot apical meristem forms a conical structure between the two cotyledons lacking leaf primordia (Prego et al., 1998; Burrieza et al., 2012). All embryo cells, including those of the apical meristems, store abundant proteins and lipids in the form of protein and lipid bodies, respectively. During germination, the radicle grows through the center of the cone, in the space previously occupied by the suspensor.

The suspensor connects the embryo to the nucellus during early seed development, holding the growing embryo in a fixed position within the seed and allowing nutrients to be transported to the embryo (Bozhkov et al., 2005). It is a short-lived tissue, only active

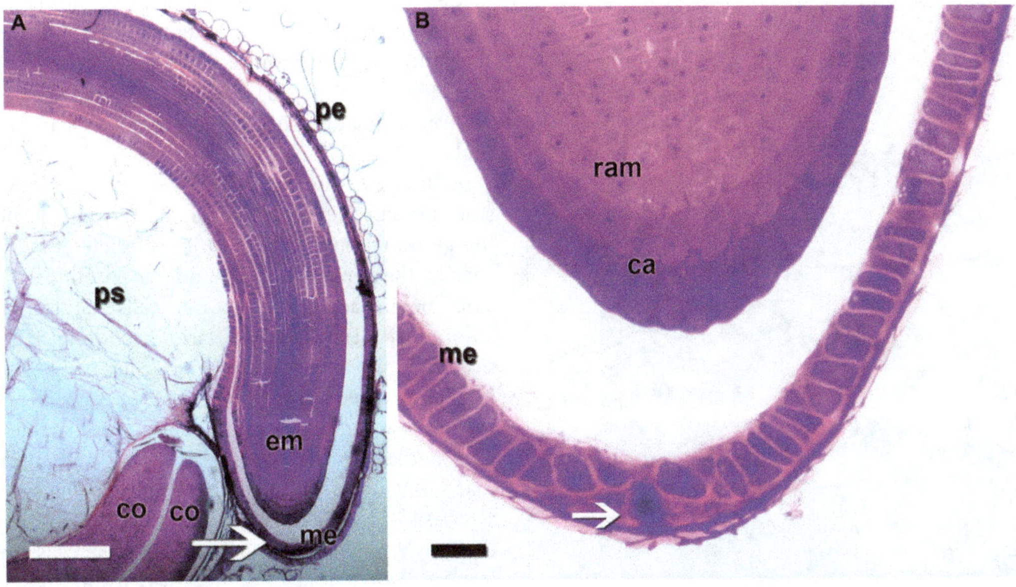

FIGURE 3 | Quinoa micropylar endosperm (in a mature seed). (A) The white arrow indicates the center of the micropylar cone; *co*, cotyledon; *em*, embryo; *me*, micropylar endosperm; *pe*, pericarp; *ps*, perisperm. **(B)** Detail of **(A)**. The white arrow indicates the central channel of the micropylar cone, which is occupied by the remains of the suspensor. *ca*, caliptra; *me*, micropylar endosperm; *ram*, root apical meristem. Semithin section (2 μm thick) was obtained from a fixed, resin-embedded, sectioned, and stained seed, as described in **Figure 2**.

in early embryogenesis. In quinoa, it is constituted by (1) a neck that connects the suspensor to the embryo proper, which is made up of cells with small vacuoles; and (2) a knob, which is composed of a set of larger basal cells that protrude into the micropylar endosperm (**Figures 4** and **5**). The knob is formed by transfer cells on the outside, with ingrowths in their outer cell walls (**Figure 4**), dense cytoplasm, and numerous small vacuoles (López-Fernández and Maldonado, 2013b). When the embryo finishes accumulating reserves its cells degenerate leaving, however, visible remains in the center of the micropylar cone (**Figure 3**). During germination, this remnant of the suspensor presumably offers less mechanical resistance and might help facilitate radicle protrusion during germination.

The perisperm is derived from the nucellus of the ovule; this portion of the nucellar tissue is not consumed by the development of the embryo sac and persists after fertilization, becoming the main nutritive tissue of the seed (**Figure 5A**). During quinoa perisperm development, three major developmental phases can be distinguished: (1) early development of the nucellus, including mitotic activity, the last stage of which takes place before anthesis, establishing the final cell number and tissue configuration; (2) cellular differentiation, which can be broken down into the partly overlapping processes of cellular expansion, endoreduplication, accumulation of starch reserves, and programmed cell death (PCD); and (3) maturation, which comprises the shutdown of biosynthetic processes, desiccation induction, and quiescence (López-Fernández and Maldonado, 2013a). Through endoreduplication, DNA content usually peaks at 8°C but some nuclei can reach up to 16 and 32°C (López-Fernández and Maldonado, 2013a). At maturity, the perisperm consists of uniform, non-living, thin-walled cells. Nuclei and other cytoplasmic organelles are absent at this stage. Cells are full of compound starch grains and simple starch grains occupy the space between compound grains (Prego et al., 1998).

Endosperm development in quinoa, as in cereals, is of the nuclear type. According to Olsen (2004), syncytial and cellularization phases of nuclear endosperm development are conserved among all groups of angiosperms. Endosperm development has only recently begun to be studied in quinoa (López-Fernández and Maldonado, 2013b). In the endosperm mother cell, nuclear divisions occur freely in the parietal cytoplasm surrounding the central vacuole. Cellularization is associated with the initiation of periclinal divisions, which occur in a centripetal direction, thereby shrinking the vacuole. While divisions are ongoing, nuclei grow in size and endoreduplicate, with the DNA content peaking at 6°C (López-Fernández and Maldonado, 2013b). When the endosperm reaches its final size, three domains are differentiated: a micropylar domain, comprising six to eight cell layers; a peripheral domain of two cell layers; and a chalazal domain of six or seven cell layers. The embryo grows at the expense of the endosperm, i.e., the chalazal and peripheral endosperms, as well as the inner layers of the micropylar endosperm are progressively dismantled (**Figure 5A**). Cells destined to be consumed during embryo development do not accumulate storage reserves. In mature seeds, the remaining endosperm forms a micropylar cone, crossed in the center by the suspensor (**Figure 4**). According to López-Fernández and Maldonado (2013b), throughout development, two major cell types make up the quinoa endosperm: the micropylar endosperm, which is constituted by a cone covering the radicle, and the ephemeral endosperm: the tissue located on either side of the growing embryo (**Figure 4**). During seed development, cells of the ephemeral endosperm

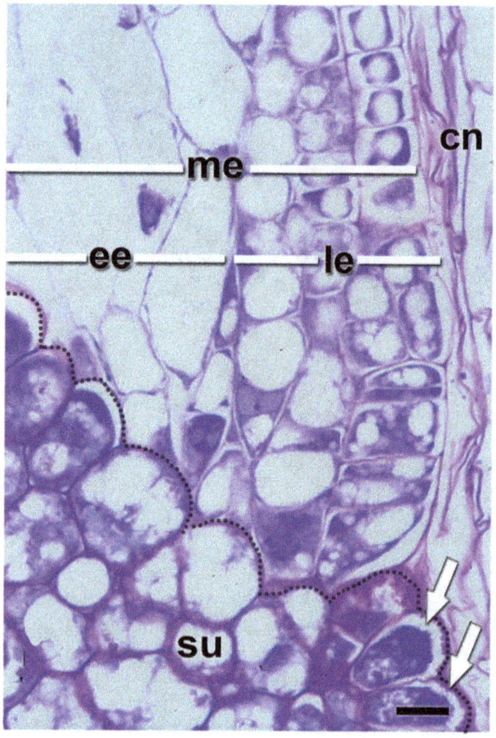

FIGURE 4 | Quinoa micropylar endosperm (at the torpedo stage). *cn*, crushed nucellus; *ee*, ephemeral endosperm; *le*, lasting endosperm; *me*, micropylar endosperm; *su*, supensor. Semithin section (2 μm thick) was obtained from a fixed, resin-embedded, sectioned, and stained seed as described in **Figure 2**. The border between the suspensor and endosperm has been highlighted with a dotted line. White arrows indicate the "festooned" outer cell walls in the transfer cells of the suspensor. Bar: 10 μm.

are crushed and finally disintegrated. In contrast, cells of the micropylar cone (**Figure 3**), which is one or two cell layers thick, store lipids and proteins that are used by the embryo during germination.

THE GRASS SEED

The ovule is hemianatropous (the hemianatropous condition results from the curvature of the ovule such that the micropyle comes to lie at right angles to the funiculus), bitegmic (with two integuments), and tenuinucellate (i.e., the archesporial cell enlarges to form the megasporocyte, then megasporocyte is subdermal; Davis, 1966). The seed contains the embryo on the adaxial face of the caryopsis, which is unilaterally covered by the endosperm (**Figure 5B**).

The endosperm is almost completely preserved as reserve tissue in mature seeds. According to Olsen (2004), throughout development, four major cell types constitute the grass endosperm, i.e., aleurone, starchy endosperm, basal endosperm transfer cells, and embryo-surrounding region (**Figure 5B**). Similar to quinoa endosperm, after fertilization, the initial endosperm nucleus divides repeatedly without cell wall formation in the parietal cytoplasm surrounding the central vacuole (Brown et al., 1994; Olsen, 2004).

Cellularization is associated with the initiation of periclinal divisions, which occur in a centripetal direction, progressively shrinking the vacuole until its disappearance (Brown et al., 1994, 1996; Olsen, 2001, 2004). The first periclinal division round is formative, originating both the aleurone and starchy initials (Brown et al., 1996).

Aleurone cells form a sheet generally composed of cells that store proteins, lipids, and mineral nutrients to be used up during germination (Jones, 1969a; Lonsdale et al., 1999). In some species, the peripheral layer undergoes rounds of periclinal divisions before assuming aleurone cell characteristics. Thus, the amount of aleurone layers varies according to the species, e.g., maize and wheat have one layer, rice has one to several layers, and barley has three layers (Buttrose, 1963; Hoshikawa, 1993). Aleurone cells accumulate proteins and lipids in protein vacuoles and lipid bodies, respectively. Upon seed germination, it is assumed that storage proteins provides the amino acids necessary for the synthesis of hydrolytic enzymes required for starch mobilizing in the starchy endosperm and that aleurone cells die by autophagy (Jones, 1969b,c; Jones and Price, 1970; Taiz and Jones, 1970; Taiz and Honigman, 1976; for reviews see Bethke et al., 1998; Becraft and Gibum, 2011).

At maturity the starchy endosperm cells consist of uniform, non-living, thin-walled cells full of starch grains but they also contain protein bodies (Reyes et al., 2011; for review see Olsen et al., 1992). During development, these cells simultaneously accumulate storage reserves and degenerate, both processes mediated by a program of developmentally controlled cell death. Starchy endosperm cells accumulate starch in plastids and prolamins in protein bodies.

Endoreduplication occurs in starchy endosperm. In maize, DNA content usually peaks at 6 and 12°C, but some nuclei reach 24, 48, 96, and 192°C (Kowles and Phillips, 1985; Kowles et al., 1990; Larkins et al., 2001; Sabelli and Larkins, 2009; Sabelli, 2012). Endoreduplication precedes and accompanies two programs that occur simultaneously in this tissue: (i) accumulation of storage reserves, and (ii) PCD (Young et al., 1997; Young and Gallie, 2000; Domínguez and Cejudo, 2014).

The region of the transfer cells is located in the basal endosperm close to the vascular tissues of the placenta (**Figure 5B**), facilitating solute transfer from the vascular bundle of the pedicel toward the endosperm (Thompson et al., 2001). Two or three cells derived from the outer layer during the process of cellularization assume transfer cell identity, developing cell wall ingrowths (Thompson et al., 2001).

The embryo-surrounding region comprises several cell layers that completely envelop the young embryo (**Figure 5B**) and are characterized by their dense cytoplasm contents, abundance of small vacuoles, and a complex membrane system but they do not accumulate storage reserves and their death generates a space filled with crushed endosperm cells surrounding the embryo (Olsen, 2004; Sabelli and Larkins, 2009). The embryo grows at the expense of these cells, causing their degradation.

The single cotyledon of grasses is transformed into the absorptive scutellum which lies between the endosperm and the embryo axis. Many grasses possess a small scale-like appendage opposite the scutellum, the so-called epilates (Saha, 1957). The shoot apical

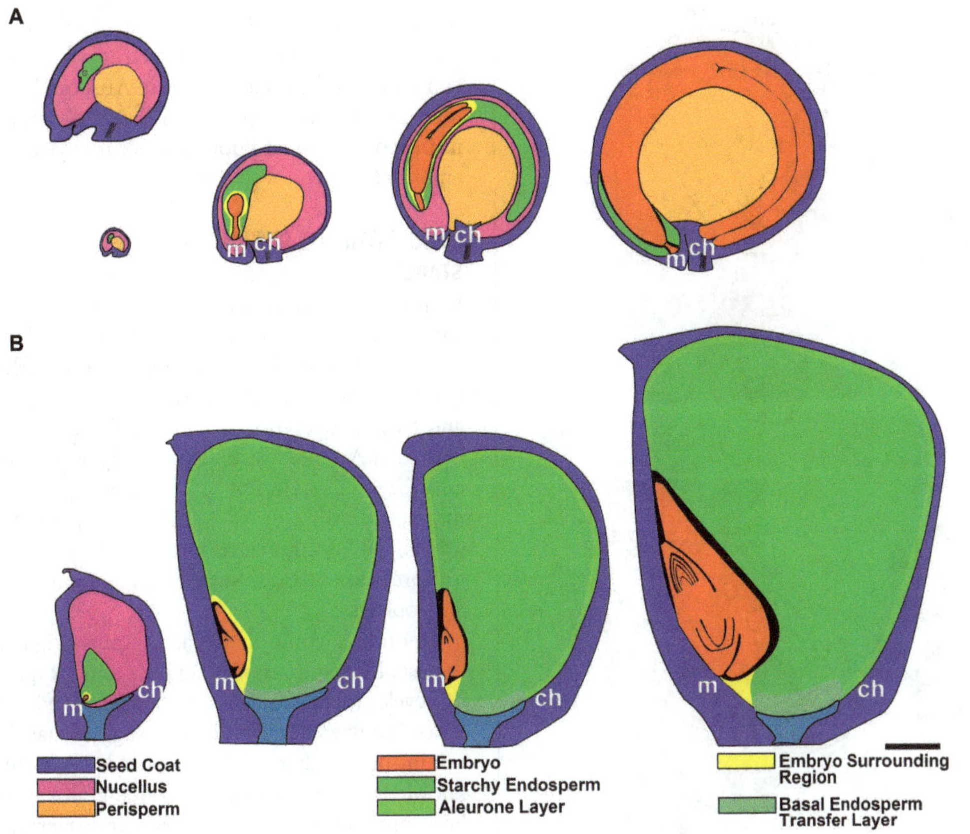

FIGURE 5 | Grain development in quinoa (A) and maize (B), from anthesis to maturity. In quinoa, the first image (ovule at anthesis) has been enlarged in the top image. *ch*, chalazal pole; *m*, micropylar pole. Bar: 0.5 mm.

meristem is covered by the coleoptile and contains several leaf primordia.

The coleorhiza is a non-vascularized, multicellular embryonic tissue that covers the cereal root apical meristem. It is an embryonic tissue located between the root apical meristem and the suspensor, characteristic of the grass embryo and absent in quinoa embryo (**Figure 6**). The coleorhiza originates together with the root cap and the suspensor (Johansen, 1950), but later in development it is separated from the radicle by a cleft (**Figure 6B**).

Although coleorhiza is involved in germination and successful establishment of all grass seedlings, information on its morphology, anatomy, and function is very sparse. In rye, Sargent and Osborne (1980) describe coleorhiza in the mature rye seed as made up of parenchymatic quiescent cells lacking vacuoles and containing a cytoplasm densely packed with ribosomes, lipid bodies largely confined to a peripheral position, a greatly reduced endomembrane system, mitochondria with a few cristae, and nuclei in which the heterochromatin is condensed. During germination of rye (Sargent and Osborne, 1980) and barley (Barrero et al., 2009) seeds, coleorhiza cells elongate and separate from each other, thereby forming intercellular spaces after consuming their own storage reserves. Eventually, the growth of embryonic roots (primary and adventitious) dismantles the tissue (Millar et al., 2006).

MICROPYLAR ENDOSPERM VS. COLEORHIZA

In quinoa, the micropylar endosperm is the only part of the tissue that persists in the mature seed, forming a cone covering the root apical meristem (**Figure 5A**). In other species (e.g., tomato, *Datura ferox*), the micropylar endosperm is just the micropylar portion of a larger tissue covering the root apical meristem. The micropylar endosperm has been studied in the species *Arabidopsis thaliana* (Millar et al., 2006; Okamoto et al., 2006), cress (Müller et al., 2006), *Datura ferox* (Mella et al., 1995; Arana et al., 2007), tomato (Toorop et al., 2000; Wu et al., 2000), coffee (Da Silva et al., 2004), cucumber (Amritphale et al., 2005; Salanenka et al., 2009), and lettuce (Nascimento et al., 2000).

In the mature seed, cells of the micropylar endosperm store hemicelluloses in cell walls, and proteins, lipids and minerals in the cytoplasm. Hemicelluloses strengthen and harden the tissue, and germination is only possible after this tissue has been sufficiently weakened (i.e., hydrolyzed) and the radicle can overcome its resistance (Psaras et al., 1981; Psaras and Georghiou, 1983; Watkins and Cantliffe, 1983; Groot and Karssen, 1987; Sánchez et al., 1990). In quinoa, control of radicle protrusion during germination is mediated, at least in part, by micropylar endosperm weakening, but its emergence occurs via the channel occupied by suspensor remains (unpublished observations).

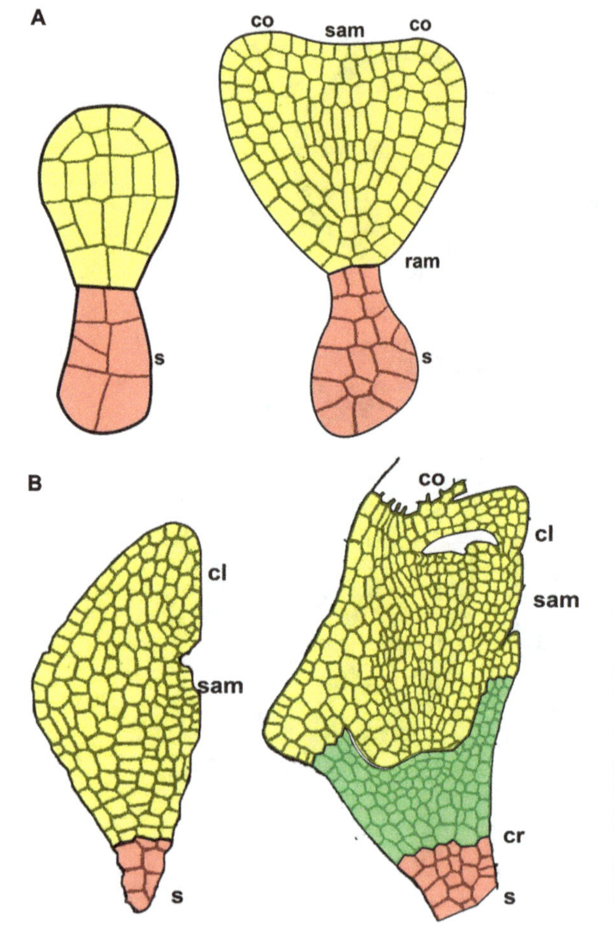

FIGURE 6 | The suspensor in two subsequent early stages of embryo development. **(A)** Quinoa seed. **(B)** Grass seed. *cl*, coleoptile; *co*, cotyledon primordium; *cr*, coleorhiza; *ram*, root apical meristem; *sam*, shoot apical meristem; *s*, suspensor.

On the other hand, endosperm weakening, as well as radicle growth potential, are known to be regulated by plant hormones (Linkies et al., 2010): abscisic acid (ABA) inhibits endosperm weakening while gibberellins (GA) act as its antagonists in a complex network integrating environmental signals such as light, temperature, water availability, and nutrient status (Kucera et al., 2005); ethylene and brassinosteroids also counteract ABA, but their effects on endosperm weakening are unknown.

In grasses, there is no micropylar endosperm, and the coleorhiza plays a role in protecting emerging roots during germination (Sargent and Osborne, 1980). According to Nishimura (1922), Howarth (1927), Walne et al. (1975), and Debaene-Gill et al. (1994), the coleorhizae also act in water and nutrient uptake, as a water reserve during dehydration and as a storage tissue. Barrero et al. (2009) propose a new role for the coleorhiza, i.e., to regulate germination in dormant seeds. Recent studies have shown that ABA 8-hydroxylase gene expression is strong and uniform in barley coleorhizae, thus suggesting a key role for this tissue in dormancy control, equivalent to that of the micropylar endosperm (Millar et al., 2006). In fact, several alternative catabolic pathways exist for the inactivation of ABA (Zhou et al., 2004; Nambara and Marion-Poll, 2005), but the reaction catalyzed by ABA 8′-hydroxlyase is considered to be predominant in ABA catabolism (Nambara and Marion-Poll, 2005). ABA 8′-hydroxlyase gene expression occurs in the coleorhiza and does not occur anywhere else in the barley embryo (Millar et al., 2006).

SEED STORAGE RESERVES

STARCH

In quinoa perisperm, starch accumulates forming compound and simple grains (Prego et al., 1998). In cereal starchy endosperm, compound grains have been reported in barley, rice, and wheat (Matsushima et al., 2010; Yun and Kawagoe, 2010). In wheat and barley, two types of starch grains are present: the large and lenticular A-type, which contains higher amylose concentrations, and the small spherical or ovoide B-type (Peng et al., 1999). In maize, according to Katz et al. (1993), starch grains are simple grains, but during development, Shannon et al. (1998) report the presence of both simple (larger) and compound (smaller) amyloplasts.

In quinoa, compound starch grains originate inside the amyloplasts by aggregation of single grains (**Figure 7A**). At the end of development, single grains are deposited in the extraplastidial space (**Figure 7B**). TEM images suggest that they are cytosolic in origin, but further studies are needed, considering that, to our knowledge, the formation of starch grains in the cytosol has not been previously reported. In rice endosperm, compound grains are generated by divisions of amyloplasts, which occur simultaneously at multiples constriction sites, and small amyloplasts bud from their surface (Yun and Kawagoe, 2009). More recently, Yun and Kawagoe (2010) demonstrated that a septum-like structure containing inner envelope membrane divides granules in the

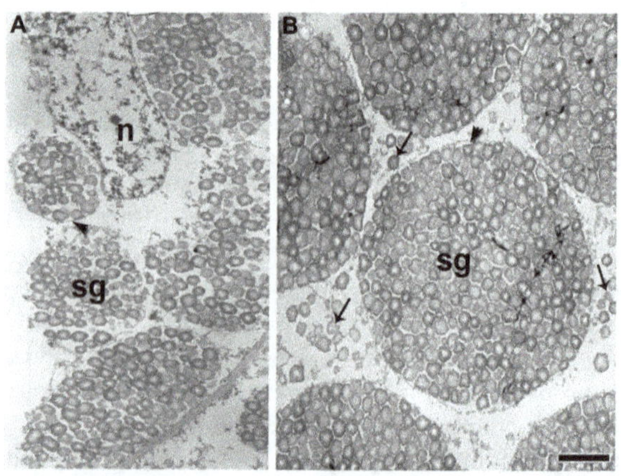

FIGURE 7 | Ultrathin sections of the quinoa perisperm in two subsequent developmental stages. Grains were fixed and embedded as described in **Figure 2**. Ultra-thin sections were mounted on grids coated with Formvar and stained with uranyl acetate followed by lead citrate. **(A)** Numerous simple starch grains are being packed inside amyloplasts (arrowhead) originating compound grains; *n*, nucleus; *sg*, starch grains. **(B)** Simple starch grains (arrows) and compound starch grains (arrowhead). Bar: 1 μm.

amyloplast and that proteins of organelle division, (including FtZ, Min, ARC5, and PDV2) play roles not only in amyloplast division but also in compound granule synthesis (Yun and Kawagoe, 2009, 2010). These results strongly suggest that amyloplast division and compound granule synthesis in rice are closely related. This needs to be investigated further, since starch biosynthesis is an essential function in plant metabolism and is highly conserved throughout the plant kingdom.

Starch is synthesized from sucrose as a result of the combined action of four distinct enzymes: ADP-Glc pyrophosphorylase (AGPase), starch synthase (SS), starch-branching (BE), and starch-debranching (DBE) enzymes (Hannah, 2007). The complement of starch biosynthetic enzymes is well conserved between plastids of tissues that make different types of starches, i.e., transitory starch (made in chloroplasts) and storage starch (made in amyloplasts; Tetlow, 2010; for review see Geigenberger, 2011). In dicot storage tissue, the synthesis of ADP-glucose by the enzyme ADP-glucose AGPase occurs entirely within plastids, as reported in potato tuber (Sweetlove et al., 1996), pea embryo, and root (Denyer and Smith, 1988; Smith, 1988). In contrast, in monocots/cereals, i.e., in maize (Denyer et al., 1996; Huang et al., 2014), barley (Thorbjørnsen et al., 1996), rice (Sikka et al., 2001), and wheat (Tetlow et al., 2003), there is evidence indicating the presence of AGPase enzymes corresponding to plastidial and cytosolic isoforms (for reviews see Hannah and James, 2008; Comparot-Moss and Denyer, 2009; Geigenberger, 2011). Furthermore, whereas SS, BE, and DBE enzymes are found within amyloplasts, AGPase activity (which represents the rate-limiting step in starch biosynthesis) is confined almost exclusively to the cytosol. According to Beckles et al. (2001), the cytosolic localization of AGP in cereal endosperm may have functional significance for partitioning large amounts of carbon into starch when sucrose is plentiful.

In quinoa, it has yet to be determined if the synthesis of ADP-glucose by the enzyme ADP-glucose AGPase occurs entirely in plastids, as in the rest of dicotyledons studied to date. However, in advanced stages of quinoa seed development, once compound starch grain formation is complete, simple starch grains do form filling the space between amyloplasts (López-Fernández and Maldonado, 2013a; **Figure 7B**).

In cereal starchy endosperm dead cells, protein bodies, which mainly store prolamins, are present in the extraplastidial space. Conversely, in dead cells of quinoa perisperm, starch is the only storage reserve.

PROTEINS, MINERALS, AND LIPIDS

In quinoa, embryo and micropylar endosperm cells store proteins and lipids in the form of protein storage vacuoles (PSVs) and lipid bodies (Prego et al., 1998; Carjuzaá et al., 2008; López-Fernández and Maldonado, 2013b), respectively. PSVs contain one or more phytin crystals in the proteinaceous matrix and proplastids contain clusters of particles of phytoferritin (Prego et al., 1998). Brinegar and Goundan (1993) describe the protein composition of quinoa seeds and report an 11S-type globulin.

A 2S cysteine-rich globulin (8–9 kDa) is also described for quinoa seeds by Brinegar et al. (1996). More recently, Castellión et al. (2008) confirm these results. Likewise, Balzotti et al. (2008) report the genomic and cDNA sequences for two 11S genes from the quinoa genome; in addition, on the basis of comparison with orthologous 11S sequences from other species, they describe the characteristics of the genes and their encoded proteins.

During development, proteins are transported from the lumen of the rough endoplasmic reticulum (RER) to the vacuole through the Golgi apparatus and PSVs are formed when vacuoles fragment (**Figure 8**). For each PSV, several electron-dense globoid crystals are formed (**Figure 8**). Before globoid formation, vesicles of phytic acid can be observed as bubbles concentrated on the outside of the PSV (**Figures 8A,B**); there, phytic acid associates with ions to form electron-dense globoids of phytin. To our knowledge, the formation of globoid crystals in the PSV had not been previously photo-documented in angiosperm seeds. Energy dispersive X-ray (EDX) analysis of globoid crystals reveals the presence of P, K, and Mg (Prego et al., 1998).

In cereals, globulins are synthesized in both the aleurone layer and starchy endosperm, and prolamins synthesized within the starchy endosperm rather than in the aleurone layer (Kriz and Schwartz, 1986; Kriz, 1989; Kriz and Wallace, 1991; for review see Shewry and Halford, 2002). On the other hand, protein body formation is tissue specific: in the aleurone layer, the pathways for all globulins and some albumins are similar to that described for quinoa, i.e., protein vesicles are transported from the lumen of the RER to the vacuole by way of the Golgi apparatus, and PSVs are formed by the subsequent fragmentation of the vacuole. Prolamins accumulate in protein bodies inside the endoplasmic reticulum (ER) of starchy endosperm cells, excluding the Golgi and vacuoles (Levanony et al., 1992; Rechinger et al., 1993; Reyes et al., 2011). Crystal globoids, which are included in the PSVs, are detected in the scutellum and aleurone layer but not in the starchy endosperm (Tanaka et al., 1973; Ogawa et al., 1977).

Lipid bodies are found in the quinoa embryo and micropylar endosperm, as well as in grass aleurone and embryo tissues. Lipid bodies originate in the ER (Schwarzenbach, 1971; Wanner et al., 1981; Hsieh and Huang, 2004). Lipids (triacylglycerides) accumulate between the bilayer leaflets at ER specific sites and, when they reach a certain size, bud off from the ER. Oleosin proteins are small (~15–30 kDa) and abundant proteins in the seeds of plants that bind to the surface of lipid bodies (Chapman et al., 2012). The synthesis and incorporation of oleosins occurs as triacylglycerides are deposited within the lipid bodies (Hsieh and Huang, 2007). During quinoa seed development, lipid bodies originate from the ER at the same time as the PSVs (**Figure 8**).

ENDOPOLIPLODY AND PROGRAMMED CELL DEATH IN STORAGE SEED TISSUES: QUINOA PERISPERM VS. CEREAL STARCHY ENDOSPERM

Quinoa perisperm consists of uniform, non-living, thin-walled cells full of starch grains. In grass starchy endosperm cells are full of starch grains, but also contain prolamin protein bodies. Thus, the two storage tissues are similar in terms of general characteristics and function, although genetically different.

During perisperm and endosperm development, two important aspects of both quinoa perisperm and cereal starchy endosperm are endoreduplication (for a review see Larkins et al.,

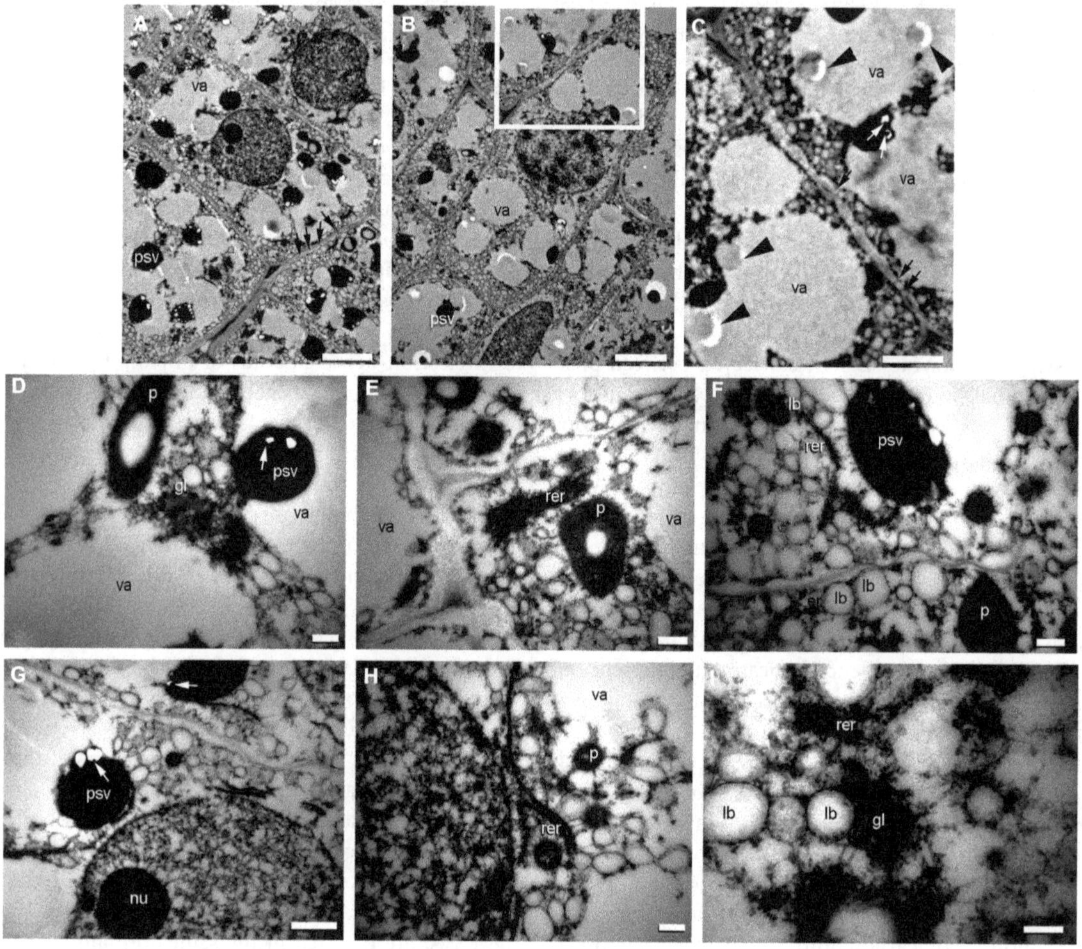

FIGURE 8 | Origin of protein storage vacuoles (PSVs) and lipid bodies in quinoa embryo. Excised embryos (torpedo stage) were processed as indicated in **Figure 7**. **(A)** Cells from the ground meristem. **(B)** Cells from the procambium. **(C)** detail of **(B)**. **(D–I)** Details of cytoplasm and organelles in cells from the ground meristem. In the different images, intense biosynthetic activity in cytoplasm can be inferred by the presence of abundant cisternae of rough endoplasmic reticulum (*rer*), numerous *PSVs* originating from the vacuoles (*va*); globoids (white arrows) can be seen inside PSV; the empty areas inside PSV contained globoid crystals before they were dissolved during treatment of tissue fixation; numerous lipid bodies (*lb*) associated with endoplasmic reticulum; frequent dictiosomes, or Golgi (*gl*) and circular vesicles with electronically dense content in proximity to the dictiosomes (see **D**); nuclei with one or more nucleoli (*nu*). Arrowheads indicate the presence in the vacuole of the precursor salts of the globoids; white arrows indicate globoids; black arrows indicate plasmodesmata. *p*, plastid. Bars: **A–B**, 2 μm; **C**, 1.5 μm; **D–I**, 0.5 μm.

2001) and PCD (Young et al., 1997). In both tissues, cell death and dismantling are temporally separated, a process that can take years depending on when germination takes place (Young et al., 1997; Young and Gallie, 1999, 2000; López-Fernández and Maldonado, 2013a). In both cases, cell death occurs during seed formation and is not due to a complete process of autophagy since the cell remains intact until germination, when the starch reserves are mobilized and the dead tissue is finally dismantled. According to van Doorn and Woltering (2005), this type of cellular death does not seem to be autophagic in the strict sense of the term. In a more recent classification, based on morphological criteria, van Doorn et al. (2011) recognize two major classes of cell death that occur in plant tissues: vacuolar cell death and necrosis. However, PCD in grass starchy endosperm (Van Doorn, 2011; van Doorn et al., 2011) and quinoa perisperm, does not strictly fall into these two categories and are classified as separate modalities (van Doorn et al., 2011).

CONCLUDING REMARKS

In spite of their different origins, quinoa perisperm and grass starchy endosperm exhibit similar developmental programs and functional fates, i.e., endoreduplication, starch accumulation, and programmed cell death. Given the conservation of this seed developmental trajectory in quinoa and cereals, we infer the existence of convergent evolution in these two phylogenetically distant taxa.

The micropylar endosperm of quinoa and the coleorhiza of cereals serve the same role during germination: both tissues store reserves, protecting the root apical meristem in the quiescent seed and control dormancy during germination.

Because of the supposedly independent origin of monocotyledons and dicotyledons, efforts to solve questions related to seed evolution and particularly that of what determines the fate of storage tissues should be furthered for both grasses and quinoa. Hence, the present study may constitute a contribution toward a more complete understanding of seed biology and thus may provide support for broader phylogenetic studies.

ACKNOWLEDGMENTS

This work was supported by the Universidad de Buenos Aires (UBACYT 20020100100232 to Sara Maldonado), the Consejo Nacional de Investigaciones Científicas y Técnicas (CONICET. Res. 810/13. PIP 0465 to Sara Maldonado) and the Fundación Juan Bautista Sauberan (to Hernán P. Burrieza and Sara Maldonado).

REFERENCES

Amritphale, D., Yoneyama, K., Takeuchi, Y., Ramakrishna, P., and Kusumoto, D. (2005). The modulating effect of the perisperm-endosperm envelope on ABA-inhibition of seed germination in cucumber. *J. Exp. Bot.* 56, 2173–2181. doi: 10.1093/jxb/eri217

Arana, M. V., Burgin, M. J., de Miguel, L. C., and Sánchez, R. A. (2007). The very-low-fluence and high-irradiance responses of the phytochromes have antagonistic effects on germination, mannan-degrading activities, and DfGA3ox transcript levels in *Datura ferox* seeds. *J. Exp. Bot.* 58, 3997–4004. doi: 10.1093/jxb/erm256

Balzotti, M. R. B., Thornton, J. N., Maughan, P. J., McClellan, D. A., Stevens, M. R., Jellen, E. N., et al. (2008). Expression and evolutionary relationship of the *Chenopodium quinoa* 11S seed storage protein gene. *Int. J. Plant Sci.* 169, 281–291. doi: 10.1086/523874

Barrero, J. M., Talbot, M. J., White, R. G., Jacobsen, J. V., and Gubler, F. (2009). Anatomical and transcriptomic studies of the coleorhiza reveal the importance of this tissue in regulating dormancy in barley. *Plant Physiol.* 150, 1006–1021. doi: 10.1104/pp.109.137901

Beckles, D. M., Smith, A. M., and Rees, T. (2001). A cytosolic ADP-glucose pyrophosphorylase is a feature of graminaceous endosperms, but not of other starch-storing organs. *Plant Physiol.* 125, 818–827. doi: 10.1104/pp.125.2.818

Becraft, P. W., and Gibum, Y. (2011). Regulation of aleurone development in cereal grains. *J. Exp. Bot.* 62, 1669–1675. doi: 10.1093/jxb/erq372

Bessey, C. E. (1894). The structure of the wheat grain. *Nebr. Agric. Exp. Stat. Bull.* 32, 100–114.

Bethke, P. C., Swanson, S. J., Hillmer, S., and Jones, R. L. (1998). From storage compartment to lytic organelle: the metamorphosis of the aleurone protein storage vacuole. *Ann. Bot.* 82, 399–412. doi: 10.1006/anbo.1998.0702

Bozhkov, P. V., Filonova, L. H., and Suárez, M. F. (2005). Programmed cell death in plant embryogenesis. *Curr. Top. Dev. Biol.* 67, 135–179. doi: 10.1016/S0070-2153(05)67004-4

Brinegar, C., and Goundan, S. (1993). Isolation and characterization of chenopodin, the 11S seed storage protein of quinoa (*Chenopodium quinoa*). *J. Agric. Food Chem.* 41, 182–185. doi: 10.1021/jf00026a006

Brinegar, C., Sine, B., and Nwokocha, L. (1996). High-cysteine 2S seed storage proteins from quinoa (*Chenopodium quinoa*). *J. Agric. Food Chem.* 44, 1621–1623. doi: 10.1021/jf950830+

Brown, R. C., Lemmon, B. E., and Olsen, O.-A. (1994). Endosperm development in barley: microtubule involvement in the morphogenetic pathway. *Plant Cell* 6, 1241–1252. doi: 10.1105/tpc.6.9.1241

Brown, R. C., Lemmon, B. E., and Olsen, O.-A. (1996). Development of the endosperm in rice (*Oryza sativa* L.): cellularization. *J. Plant Res.* 109, 301–313. doi: 10.1007/BF02344477

Burrieza, H. P., Koyro, H.-W., Martínez Tosar, L., Kobayashi, K., and Maldonado, S. (2012). High salinity induces dehydrin accumulation in *Chenopodium quinoa* cv. Hualhuas embryos. *Plant Soil* 354, 69–79. doi: 10.1007/s11104-011-1045-y

Buttrose, M. (1963). Ultrastructure of the developing aleurone cells of wheat grain. *Aust. J. Biol. Sci.* 16, 768–774.

Carjuzaá, P., Castellión, M., Distéfano, A. J., del Vas, M., and Maldonado, S. (2008). Detection and subcellular localization of dehydrin-like proteins in quinoa (*Chenopodium quinoa* Willd.) embryos. *Protoplasma* 233, 149–156. doi: 10.1007/s00709-008-0300-4

Castellión, M., Matiacevich, S., Buera, M. P., and Maldonado, S. (2008). Protein deterioration and longevity of quinoa seeds during long-term storage. *Food Chem.* 121, 952–958. doi: 10.1016/j.foodchem.2010.01.025

Chapman, K. D., Dyer, J. M., and Mullen, R. T. (2012). Biogenesis and functions of lipid droplets in plants. *J. Lipid Res.* 53, 215–226. doi: 10.1194/jlr.R021436

Comparot-Moss, S., and Denyer, K. (2009). The evolution of the starch biosynthetic pathway in cereals and other grasses. *J. Exp. Bot.* 60, 2481–2492. doi: 10.1093/jxb/erp141

Da Silva, E. A. A., Toorop, P. E., van Aelst, A. C., and Hilhorst, H. W. M. (2004). Abscisic acid controls embryo growth potential and endosperm cap weakening during coffee (*Coffea arabica* cv. Rubi) seed germination. *Planta* 220, 251–261. doi: 10.1007/s00425-004-1344-0

Davis, G. (1966). *Systematic Embryology of the Angiosperms.* New York, NY: Wiley.

Debaene-Gill, S. B., Allen, P. S., and Gardner, J. S. (1994). Morphology of the perennial ryegrass (*Lolium perenne* Poaceae) coleorhiza and emerging radicle with continuous or discontinuous hydration. *Am. J. Bot.* 81, 739–744. doi: 10.2307/2445652

Denyer, K., Dunlap, F., Thorbjørnsen, T., Keeling, P., and Smith, A. M. (1996). The major form of ADP-glucose pyrophosphorylase in maize endosperm is extra-plastidial. *Plant Physiol.* 112, 779–783. doi: 10.1104/pp.112.2.779

Denyer, K., and Smith, A. M. (1988). The capacity of plastids from developing pea cotyledons to synthesis acetyl CoA. *Planta* 173, 172–182. doi: 10.1007/BF00403008

Domínguez, F., and Cejudo, F. G. (2014). Programmed cell death (PCD), an essential process of cereal seed development and germination. *Front. Plant Sci.* 5:366. doi: 10.3389/fpls.2014.00366

Geigenberger, P. (2011). Regulation of starch biosynthesis in response to a fluctuating environment. *Plant Physiol.* 155, 1566–1577. doi: 10.1104/pp.110.170399

Groot, S. P. C., and Karssen, C. M. (1987). Gibberellins regulate seed germination in tomato by endosperm weakening: a study with gibberellin-deficient mutants. *Planta* 171, 525–531. doi: 10.1007/BF00392302

Hannah, L. C. (2007). "Starch formation in the cereal endosperm," in *Endosperm*, ed. O.-A. Olsen (Berlin: Springer-Verlag), 179–193.

Hannah, L. C., and James, M. (2008). The complexities of starch biosynthesis in cereal endosperms. *Curr. Opin. Biotechnol.* 19, 160–165. doi: 10.1016/j.copbio.2008.02.013

Hoshikawa, K. (1993). "Anthesis, fertilization and development of caryopsis," in *Science of the Rice Plant*, Vol. I, eds T. Matsuo and K. Hoshikawa (Tokyo: Morphology Food and Agriculture Policy Research Center), 339–376.

Howarth, W. O. (1927). The seedling development of *Festuca rubra* L. *var tenuifolia* MIHI, and its bearing on the morphology of the grass embryo. *New Phytol.* 26, 46–57. doi: 10.1111/j.1469-8137.1927.tb06705.x

Hsieh, K., and Huang, A. H. C. (2004). Endoplasmic reticulum, oleosins, and oils in seeds and tapetum cells. *Plant Physiol.* 136, 3427–3434. doi: 10.1104/pp.104.051060

Hsieh, K., and Huang, A. H. C. (2007). Tapetosomes in *Brassica* tapetum accumulate endoplasmic reticulum–derived flavonoids and alkanes for delivery to the pollen surface. *Plant Cell* 19, 582–596. doi: 10.1105/tpc.106.049049

Huang, B., Hennen-Bierwagen, T. A., and Myers, A. M. (2014). Functions of multiple genes encoding ADP-glucose pyrophosphorylase subunits in maize endosperm, embryo, and leaf. *Plant Physiol.* 164, 596–611. doi: 10.1104/pp.113.231605

Johansen, D. A. (1950). *Plant Embryology. Embryogeny of the Spermatophyta.* Middlesex, MA: Waltham.

Jones, R. L. (1969a). The fine structure of barley aleurone cells. *Planta* 85, 359–375. doi: 10.1007/BF00381284

Jones, R. L. (1969b). Gibberellic acid and the fine structure of barley aleurone cells. I. Changes during the lag phase of α-amylase synthesis. *Planta* 87, 119–133. doi: 10.1007/BF00386971

Jones, R. L. (1969c). Gibberellic acid and the fine structure of barley aleurone cells. II. Changes during the synthesis and secretion of α-amylase. *Planta* 88, 73–86. doi: 10.1007/BF00396117

Jones, R. L., and Price, J. M. (1970). Gibberellic acid and the fine structure of barley aleurone cells. Ill. Vacuolation of the aleurone cell during the phase of ribonuclease release. *Planta* 94, 191–202. doi: 10.1007/BF00386129

Katz, F. R., Furcsik, S. L., Tenbarge, F. L., Hauber, R. J., and Friedman, R. B. (1993). Behavior of starches derived from varieties of maize containing different genetic mutations. Effects of starch genotype on granular morphology. *Carbohyd. Polym.* 21, 133–136. doi: 10.1016/0144-8617(93)90007-Q

Kent, N. L., and Evers, A. D. (1994). *Kent's Technology of Cereals.* Oxford: Elsevier Science Ltd. doi: 10.1533/9781855736603

Kowles, R. V., and Phillips, R. L. (1985). DNA amplification patterns in maize endosperm nuclei during kernel development. *Proc. Natl. Acad. Sci. U. S. A.* 82, 7010–7014. doi: 10.1073/pnas.82.20.7010

Kowles, R. V., Srienc, F., and Phillips, R. L. (1990). Endoreduplication of nuclear DNA in the developing maize endosperm. *Dev. Genet.* 11, 125–132. doi: 10.1002/dvg.1020110202

Krauss, L. (1933). Entwicklungsgeschichte der Früchte von *Hordeum, Triticum, Bromus* und *Poa* mit besonderer Berücksichtigung ihrer Samenschalen. *Jahrb. f. Wiss. Bot.* 77, 733–808.

Kriz, A. L. (1989). Characterization of embryo globulins encoded by the maize Glb genes. *Biochem. Genet.* 27, 239–251. doi: 10.1007/BF02401804

Kriz, A. L., and Schwartz, D. (1986). Synthesis of globulins in maize embryos. *Plant Physiol.* 82, 1069–1075. doi: 10.1104/pp.82.4.1069

Kriz, A. L., and Wallace, N. H. (1991). Characterization of the maize Globulin-2 gene and analysis of two null alleles. *Biochem. Genet.* 29, 241–254. doi: 10.1007/BF00590104

Kucera, B., Cohn, M. A., and Leubner-Metzger, G. (2005). Plant hormone interactions during seed dormancy release and germination. *Seed Sci. Res.* 15, 281–307. doi: 10.1079/SSR2005218

Larkins, B. A., Dilkes, B. P., Dante, R. A., Coelho, C. M., Woo, Y. M., and Liu, Y. (2001). Investigating the hows and whys of DNA endoreduplication. *J. Exp. Bot.* 52, 183–192. doi: 10.1093/jexbot/52.355.183

Levanony, H., Rubin, R., Altschuler, Y., and Galili, G. (1992). Evidence for a novel route of wheat storage proteins to vacuoles. *J. Cell Biol.* 119, 1117–1128. doi: 10.1083/jcb.119.5.1117

Linkies, A., Schuster-Sherpa, U., Tintelnot, S., Leubner-Metzger, G., and Müller, K. (2010). Peroxidases identified in a subtractive cDNA library approach show tissue-specific transcript abundance and enzyme activity during seed germination of *Lepidium sativum*. *J. Exp. Bot.* 61, 491–502. doi: 10.1093/jxb/erp318

Lonsdale, J. E., McDonald, K. L., and Jones, R. R. L. (1999). High pressure freezing and freeze substitution reveal new aspects of fine structure and maintain protein antigenicity in barley aleurone cells. *Plant J.* 17, 221–229. doi: 10.1046/j.1365-313X.1999.00362.x

López-Fernández, M. P., and Maldonado, S. (2013a). Programmed cell death during quinoa perisperm development. *J. Exp. Bot.* 57, 3747–3753.

López-Fernández, M. P., and Maldonado, S. (2013b). Quinoa ricinosomes early mark suspensor and endosperm cells destined to die during late seed development. *Ann. Bot.* 112, 1253–1262. doi: 10.1093/aob/mct184

Matsushima, R., Maekawa, M., Fujita, N., and Sakamoto, W. (2010). A rapid, direct observation method to isolate mutants with defects in starch grain morphology in rice. *Plant Cell Physiol.* 51, 728–741. doi: 10.1093/pcp/pcq040

Mella, R., Maldonado, S., and Sánchez, R. A. (1995). Phytochrome induced structural changes and protein degradation prior to radicle protrusion in *Datura ferox* seeds. *Can. J. Bot.* 73, 1371–1378. doi: 10.1139/b95-149

Millar, A. A., Jacobsen, J. V., Ross, J. J., Helliwell, C. A., Poole, A. T., Scofield, G., et al. (2006). Seed dormancy and ABA metabolism in *Arabidopsis* and barley: the role of ABA 8'-hydroxylase. *Plant J.* 45, 942–954. doi: 10.1111/j.1365-313X.2006.02659.x

Morrison, I. N. (1976). The structure of the chlorophyll-containing cross cells and tube cells of the inner pericarp of wheat during grain development. *Bot. Gaz.* 137, 85–93. doi: 10.1086/336845

Müller, K., Tintelnot, S., and Leubner-Metzger, G. (2006). Endosperm-limited Brassicaceae seed germination: abscisic acid inhibits embryo-induced endosperm weakening of *Lepidium sativum* (cress) and endosperm rupture of cress and *Arabidopsis thaliana*. *Plant Cell Physiol.* 47, 864–877. doi: 10.1093/pcp/pcj059

Nambara, E., and Marion-Poll, A. (2005). Abscisic acid biosynthesis and catabolism. *Annu. Rev. Plant Biol.* 56, 165–185. doi: 10.1146/annurev.arplant.56.032604.144046

Narayanaswami, S. (1953). The structure and development of the cayopsis in some Indian millets. I. *Pennisetum typhoideum* Rich. *Phytomorphology* 3, 98–112.

Narayanaswami, S. (1955a). The structure and development of the caryopsis in some Indian millets. II. *Eleusine coreana* Gaerte. *Pap. Mich. Acad. Sci.* 40, 33–46.

Narayanaswami, S. (1955b). The structure and development of the cayopsis in some Indian millets. III. *Panicum miliare* Lamk. and *P. miliaceum* Linn. *Lloydia* 18, 51–73.

Narayanaswami, S. (1955c). The structure and development of the cayopsis in some Indian millets. IV. *Echinochloa frumentacea* Linl. *Phytomorphology* 5, 161–171.

Narayanaswami, S. (1956). The structure and development of the cayopsis in some Indian millets VI. *Setaria italica*. *Bot. Gaz.* 118, 112–122. doi: 10.1086/335933

Nascimento, W. M., Cantliffe, D. J., and Huber, D. J. (2000). Thermotolerance in lettuce seeds: association with ethylene and endo-β-mannanase. *J. Am. Soc. Hort. Sci.* 125, 518–524.

Nishimura, M. (1922). Comparative morphology and development of *Poa pratensis, Phleum pratense* and *Setaria italica*. *Jpn. J. Bot.* 1, 55–85.

Ogawa, M., Tanaka, K., and Kasai, Z. (1977). Note on the phytin-containing particles isolated from rice scutellum. *Cereal Chem.* 54, 1029–1034.

Okamoto, M., Kuwahara, A., Seo, M., Kushiro, T., Asami, T, Hirai, N., Kamiya, Y., et al. (2006). CYP707A1 and CYP707A2, which encode abscisic acid 8'-hydroxylases, are indispensable for proper control of seed dormancy and germination in *Arabidopsis*. *Plant Physiol.* 141, 97–107. doi: 10.1104/pp.106.079475

Olsen, O.-A. (2001). Endosperm development: cellularization and cell fate specification. *Annu. Rev. Plant Physiol. Plant Mol. Biol.* 52, 233–267. doi: 10.1146/annurev.arplant.52.1.233

Olsen, O. -A. (2004). Nuclear endosperm development in cereals and *Arabidopsis thaliana*. *Plant Cell* 16, S214–S227. doi: 10.1105/tpc.017111

Olsen, O.-A., Potter, R. H., and Kalla, R. (1992). Histo-differentiation and molecular biology of developing cereal endosperm. *Seed Sci. Res.* 2, 117–131. doi: 10.1017/S0960258500001240

Peng, M., Gao, M., Abdel-Aal, E.-S. M., Hucl, P., and Chibbar, R. N. (1999). Separation and characterization of A-and B-type starch granules in wheat endosperm. *Cereal Chem.* 76, 375–379. doi: 10.1094/CCHEM.1999.76.3.375

Prego, I., Maldonado, S., and Otegui, M. (1998). Seed structure and localization of reserves in *Chenopodium quinoa*. *Ann. Bot.* 82, 481–488. doi: 10.1006/anbo.1998.0704

Psaras, G., and Georghiou, K. (1983). Gibberellic acid-induced structural alterations in the endosperm of germinating *Lactuca sativa* L. *achenes*. *Z. Pflanzenphysiol.* 2, 15–19. doi: 10.1016/S0044-328X(83)80058-0

Psaras, G., Georghiou, K., and Mitrakos, K. (1981). Red-light induced endosperm preparation for radicle protrusion of lettuce embryos. *Bot. Gaz.* 142, 13–18. doi: 10.1086/337190

Rechinger, K. B., Simpson, D. J., Svendsen, I., and Cameron-Mills, V. (1993). A role for γ3 hordein in the transport and targeting of prolamin polypeptides to the vacuole of developing barley endosperm. *Plant J.* 4, 841–853. doi: 10.1046/j.1365-313X.1993.04050841.x

Reyes, F. C., Chung, T., Holding, D., Jung, R., Vierstra, R., and Otegui, M. S. (2011). Delivery of prolamins to the protein storage vacuole in maize aleurone cells. *Plant Cell* 23, 769–784. doi: 10.1105/tpc.110.082156

Rost, T. L. (1973). The anatomy of the caryopsis coat in mature caryopses of the yellow foxtail grass (*Setaria lutescens*). *Bot. Gaz.* 134, 32–39. doi: 10.1086/336676

Sabelli, P. A. (2012). Replicate and die for your own good: endoreduplication and cell death in the cereal endosperm. *J. Cereal Sci.* 56, 9–20. doi: 10.1016/j.jcs.2011.09.006

Sabelli, P. A., and Larkins, B. A. (2009). The development of endosperm in grasses. *Plant Physiol.* 149, 14–26. doi: 10.1104/pp.108.129437

Saha, B. (1957). Studies on the development of the embryo of *Oryza sativa* L. and the homologies of its parts. *Proc. Natl. Acad. Sci. India* 22, 86–101.

Salanenka, Y. A., Goffinet, M. C., and Taylor, A. G. (2009). Structure and histochemistry of the micropylar and chalazal regions of the perisperm–endosperm envelope of cucumber seeds associated with solute permeability and germination. *J. Am. Soc. Hort. Sci.* 134, 479–487.

Sánchez, R. A., Sunell, L., Labavitch, J. M., and Bonner, B. A. (1990). Changes in the endosperm cell walls of two *Datura* species before radical protrusion. *Plant Physiol.* 93, 89–97. doi: 10.1104/pp.93.1.89

Sargent, J. A., and Osborne, D. J. (1980). A comparative study of the fine structure of coleorhiza and root cells during the early hours of germination of rye embryos. *Protoplasma* 104, 91–103. doi: 10.1007/BF01279372

Schwarzenbach, A. M. (1971). Observations on spherosomal membranes. *Cytobiologie* 4, 145–147.

Shannon, J. C., Pien, F. M., Cao, H., and Liu, K. C. (1998). Brittle-1, an adenylate translocator, facilitates transfer of extraplastidial synthesized ADP-glucose into amyloplasts of maize endosperms. *Plant Physiol.* 117, 1235–1252. doi: 10.1104/pp.117.4.1235

Shewry, P. R., and Halford, N. G. (2002). Cereal seed storage proteins: structures, properties and role in grain utilization. *J. Exp. Bot.* 53, 947–958. doi: 10.1093/jexbot/53.370.947

Sikka, V. K., Choi, S., Kavakli, I. H., Sakulsingharoj, C., Gupta, S., Ito, H., et al. (2001). Subcellular compartmentation and allosteric regulation of the rice endosperm

ADPglucose pyrophosphorylase. *Plant Sci.* 161, 461–468. doi: 10.1016/S0168-9452(01)00431-9

Smith, A. M. (1988). Major differences in isoforms of starch branching enzyme between developing embryos of round- and wrinkled-seeded peas. *Planta* 175, 270–279. doi: 10.1007/BF00392437

Sweetlove, L. J., Burrell, M. M., and Rees, T. (1996). Characterization of transgenic potato (*Solanum tuberosum*) tubers with increased ADPglucose pyrophosphorylase. *Biochem. J.* 320, 487–492.

Taiz, L., and Honigman, W. A. (1976). Production of cell wall hydrolyzing enzymes by barley aleurone layers in response to gibberellic acid. *Plant Physiol.* 58, 380–386. doi: 10.1104/pp.58.3.380

Taiz, L., and Jones, R. L. (1970). Gibberellic acid, 8-1,3-glucanase, and the cell walls of barley aleurone layers. *Planta* 92, 73–84. doi: 10.1007/BF00385565

Tanaka, K., Yoshida, T., Asada, K., and Kasai, Z. (1973). Subcellular particles isolated from aleurone layer of rice seeds. *Arch. Biochem. Biophys.* 155, 136–143. doi: 10.1016/S0003-9861(73)80016-5

Tetlow, I. J. (2010). Starch biosynthesis in developing seeds. *Seed Sci. Res.* 21, 5–32. doi: 10.1017/S0960258510000292

Tetlow, I. J., Davies, E. J., Vardy, K. A., Bowsher, C. G., Burrell, M. M., and Emes, M. J. (2003). Subcellular localization of ADPglucose pyrophosphorylase in developing wheat endosperm and analysis of a plastidial isoform. *J. Exp. Bot.* 54, 715–725. doi: 10.1093/jxb/erg088

Thompson, R. D., Hueros, G., Becker, H. A., and Maitz, M. (2001). Development and functions of seed transfer cells. *Plant Sci.* 160, 775–783. doi: 10.1016/S0168-9452(01)00345-4

Thorbjørnsen, T., Villand, P., Denyer, K., Olsen, O.-A., and Smith, A. M. (1996). Distinct isoforms of ADPglucose pyrophosphorylase occur inside and outside the amyloplasts in barley endosperm. *Plant J.* 10, 243–250. doi: 10.1046/j.1365-313X.1996.10020243.x

Toorop, P. E., van Aelst, A. C., and Hilhorst, H. W. M. (2000). The second step of the biphasic endosperm cap weakening that mediates tomato (*Lycopersicon esculentum*) seed germination is under control of ABA. *J. Exp. Bot.* 51, 1371–1379. doi: 10.1093/jexbot/51.349.1371

Van Doorn, W. G. (2011). Classes of programmed cell death in plants, compared with those in animals. *J. Exp. Bot.* 62, 4749–4761. doi: 10.1093/jxb/err196

van Doorn, W. G., Beers, E. P., Dangl, J. L., Franklin-Tong, V. E., Gallois, P., Hara-Nishimura. I., et al. (2011). Morphological classification of plant cell deaths. *Cell Death Differ.* 18, 1241–1246. doi: 10.1038/cdd.2011.36

van Doorn, W. G., and Woltering, E. J. (2005). Many ways to exit? Cell death categories in plants. *Trends Plant Sci.* 10, 117-122. doi: 10.1016/j.tplants.2005.01.006

Walne, P. L., Haber, A. H., and Triplett, L. L. (1975). Ultrastructure of auxin-induced tumors of the coleorhiza-epiblast of wheat. *Am. J. Bot.* 62, 58–66. doi: 10.2307/2442078

Wanner, G., Formanek, H., and Theimer, R. R. (1981). The ontogeny of lipid bodies (spherosomes) in plant cells. *Planta* 151, 109–123. doi: 10.1007/BF00387812

Watkins, J. T., and Cantliffe, D. J. (1983). Mechanical resistance of the seed coat and endosperm during germination of *Capsicum annuum* at low temperature. *Plant Physiol.* 72, 146–150. doi: 10.1104/pp.72.1.146

Wu, C.-T., Leubner-Metzger, G., Meins, F. Jr., and Bradford, K. J. (2000). Class I β-1,3-glucanase and chitinase are expressed in the micropylar endosperm of tomato seeds prior to radicle emergence. *Plant Physiol.* 126, 1299–1313. doi: 10.1104/pp.126.3.1299

Young, T. E., and Gallie, D. R. (1999). Analysis of programmed cell death in wheat endosperm reveals differences in endosperm development between cereals. *Plant Mol. Biol.* 39, 915–926. doi: 10.1023/A:1006134027834

Young, T. E., Gallie, D. R., and DeMason, D. A. (1997). Ethylene mediated programmed cell death during maize endosperm development of Su and sh2 genotypes. *Plant Physiol.* 115, 737–751. doi: 10.1104/pp.115.2.737

Young, T. E., and Gallie, D. R. (2000). Regulation of programmed cell death in maize endosperm by abscisic acid. *Plant Mol. Biol.* 42, 397–414. doi: 10.1023/A:1006333103342

Yun, M.-S., and Kawagoe, Y. (2009). Amyloplast division progresses simultaneously at multiple sites in the endosperm of rice. *Plant Cell Physiol.* 50, 1617-1626. doi: 10.1093/pcp/pcp104

Yun, M.-S., and Kawagoe, Y. (2010). Septum formation in amyloplasts produces compound granules in the rice endosperm and is regulated by plastid division proteins. *Plant Cell Physiol.* 51, 1469–1479. doi: 10.1093/pcp/pcq116

Zhou, R., Cutler, A. J., Ambrose, S. J., Galka, M. M., Nelson, K. M., Squires, T. M., et al. (2004). A new abscisic acid catabolic pathway. *Plant Physiol.* 134, 361–369. doi: 10.1104/pp.103.030734

Conflict of Interest Statement: The authors declare that the research was conducted in the absence of any commercial or financial relationships that could be construed as a potential conflict of interest.

20

Evolution and development of cell walls in cereal grains

Rachel A. Burton and Geoffrey B. Fincher*

Australian Research Council Centre of Excellence in Plant Cell Walls – School of Agriculture, Food and Wine, University of Adelaide, Glen Osmond, SA, Australia

Edited by:
Paolo Sabelli, University of Arizona, USA

Reviewed by:
Sinead Drea, University of Leicester, UK
Rowan Mitchell, Rothamsted Research, UK

****Correspondence:***
Geoffrey B. Fincher, Australian Research Council Centre of Excellence in Plant Cell Walls – School of Agriculture, Food and Wine, University of Adelaide, Waite Campus, Glen Osmond, SA 5064, Australia
e-mail: geoff.fincher@adelaide.edu.au

The composition of cell walls in cereal grains and other grass species differs markedly from walls in seeds of other plants. In the maternal tissues that surround the embryo and endosperm of the grain, walls contain higher levels of cellulose and in many cases are heavily lignified. This may be contrasted with walls of the endosperm, where the amount of cellulose is relatively low, and the walls are generally not lignified. The low cellulose and lignin contents are possible because the walls of the endosperm perform no load-bearing function in the mature grain and indeed the low levels of these relatively intractable wall components are necessary because they allow rapid degradation of the walls following germination of the grain. The major non-cellulosic components of endosperm walls are usually heteroxylans and (1,3;1,4)-β-glucans, with lower levels of xyloglucans, glucomannans, and pectic polysaccharides. Pectic polysaccharides and xyloglucans are the major non-cellulosic wall constituents in most dicot species, in which (1,3;1,4)-β-glucans are usually absent and heteroxylans are found at relatively low levels. Thus, the "core" non-cellulosic wall polysaccharides in grain of the cereals and other grasses are the heteroxylans and, more specifically, arabinoxylans. The (1,3;1,4)-β-glucans appear in the endosperm of some grass species but are essentially absent from others; they may constitute from zero to more than 45% of the cell walls of the endosperm, depending on the species. It is clear that in some cases these (1,3;1,4)-β-glucans function as a major store of metabolizable glucose in the grain. Cereal grains and their constituent cell wall polysaccharides are centrally important as a source of dietary fiber in human societies and breeders have started to select for high levels of non-cellulosic wall polysaccharides in grain. To meet end-user requirements, it is important that we understand cell wall biology in the grain both during development and following germination.

Keywords: arabinoxylans, biosynthesis, cellulose, evolution, (1,3;1,4)-β-glucan, non-cellulosic polysaccharides

INTRODUCTION

Two major differences distinguish the cell walls of cereal grains from those found in seeds of other higher plant species. Firstly, the cell walls of the Poaceae family, which includes the grasses as well as the economically important cereals, are fundamentally different in composition, compared with walls in dicotyledons and in most other monocotyledons. Secondly, walls in the grain of the Poaceae are usually quite different than those found in vegetative tissues. Here we will examine emerging evolutionary evidence and potential selection pressures that might account for these two levels of differences in wall composition in cereal grains.

Studies on the evolution and development of cell walls in cereal grains have been greatly accelerated through emerging technologies and genetic resources. In examining cell wall composition during grain development, it is clear that walls vary greatly in various parts of the grain and even between adjacent cells (Burton et al., 2010). It is therefore crucial to deploy new, high resolution *in situ* methods to define the heterogeneity of wall composition in plant material that contains different cell types. Thus, sophisticated methods for determining polysaccharides present in walls during grain development are under development. For example, there has been a recent surge in the availability of reliable antibodies and carbohydrate binding modules that detect specific epitopes on wall polysaccharides (Verhertbruggen et al., 2009; Pattathil et al., 2010) and can therefore be used to distinguish different wall compositions in immunocytochemical labeling at both the light and electron microscopy levels (Wilson et al., 2012). In addition, there are new imaging methods with improved resolution, such as Fourier-transform infra-red (FT-IR), Raman and nuclear magnetic resonance (NMR) spectroscopy, and matrix-assisted laser desorption/ionization mass spectrometry imaging (MALDI–MSI). The use of these spectroscopic and immunocytochemical methods have confirmed that there is no such thing as a "standard" homogeneous cell wall in any tissue and this is no less true in the various cell types of cereal grains.

Evolutionary studies on cell wall polysaccharides have been greatly assisted by the identification of genes that encode polysaccharide synthases that are responsible for wall synthesis (Pear et al., 1996; Dhugga et al., 2004; Burton et al., 2006; Sterling et al., 2006; Doblin et al., 2009) and the recognition that the synthases are encoded by families of genes (Richmond and Somerville, 2000; Hazen et al., 2002). Our knowledge of the genes that mediate wall polysaccharide biosynthesis is increasingly assisted by the availability of genome sequences of important cereal and grass species, high throughput transcript profiling, and by the availability of rapidly expanding genetic resources for cereal species, including mutant libraries. Further exploration of non-crop grass species and the increasing use of grain development mutants, coupled with the emerging imaging and transcript analysis capabilities, will surely throw up more surprises and help us unravel the complex process of grain development. Here, we briefly review the current knowledge of wall composition in cereal grain and consider the evolutionary origins of diverse grain compositions.

MORPHOLOGY OF WALLS IN THE GRAIN

Large variations are observed in cell wall compositions between different species of grasses. Until recently most attention was focused on walls of the cereals, including wheat (Mares and Stone, 1973), barley (Fincher, 1975), and rice (Shibuya and Iwasaki, 1985). More recently, information has been published on endosperm walls from the grass *Brachypodium distachyon* (Guillon et al., 2011). Significant differences are observed in the polysaccharide compositions of the walls in these species and in the morphology of the endosperm although only a relatively narrow range of forms have been described. Indeed, Terrell (1971) surveyed 169 grass genera and found that a significant proportion of these had persistent liquid, soft, or semi-solid endosperm, the investigation of which surely has implications for grain quality and for the field of cell wall biology in general. The values in **Table 1** illustrate the differences in wall compositions between grains of selected grass species and between vegetative tissues and fruit of grass and dicotyledonous species.

The starchy endosperms of most economically important cereals display a range of morphological forms (**Figure 1A**) and a range of cell shapes and sizes across the grain. In barley there are wings of irregularly shaped starchy endosperm cells that flank a central core of prismatic cells overlying the transfer cells (TC; Becraft and Asuncion-Crabb, 2000). The outer endosperm cells in wheat are prismatic whilst the inner cells are rounded (Toole et al., 2007). In rice grain the endosperm cells are radially symmetrical and so appear to be tube-like (Srinivas, 1975). In sorghum, hard or translucent endosperm tissue surrounds a softer, opaque core (Waniska, 2000). In the former there are no air spaces and the starch granules are packed in tightly. In the softer core region there are large intergranular air spaces that affect both the properties of the tissue and the way that it reflects light. Maize kernels possess the same features (**Figure 1B**) and sorghum and maize grain can be dominated by one particular type of endosperm and thus can be predominantly soft or hard (Evers and Millar, 2002). In the same way, barley varieties can be described as mealy or steely (Ferrari et al., 2010). Grain hardness and strength, for example in sorghum and maize, is related to the packing of the starch granules within their protein matrix, rather than to the cell walls (Chandrashekar and Mazhar, 1999).

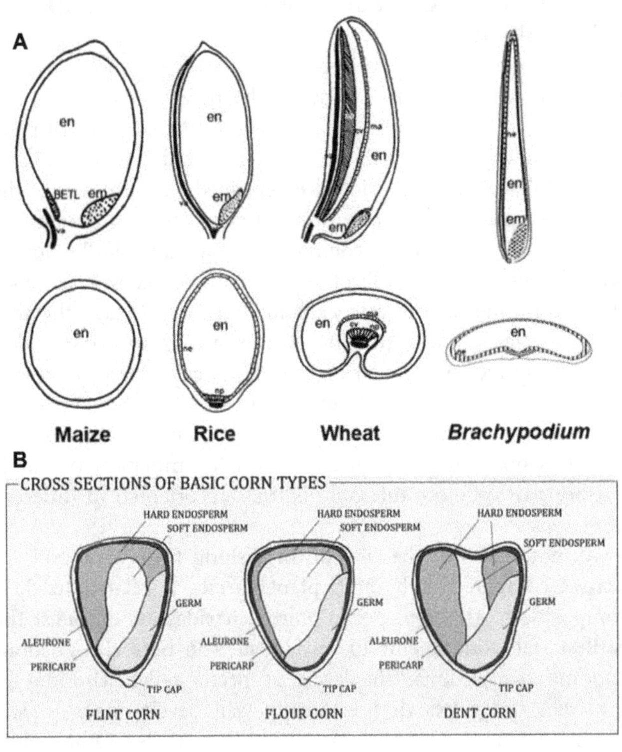

FIGURE 1 | (A) Examples of different grain morphologies. **(B)** Hard and soft endosperm proportions vary in maize kernels. Reproduced with permission from Hands and Drea (2012) and http://www.deductiveseasoning.com/2014/03/planting-and-growing-corn-for-nutrition.html. In panel **(A)**, em, embryo; en, endosperm; ma, modified aleurone; cav, cavity; va, vasculature; ne, nucellar epidermis; np, nucellar projection; BETL, basal endosperm transfer layer.

Table 1 | Selected comparisons of polysaccharide compositions in walls of vegetative tissues, fruit, and grains/seeds (% w/w).

Tissue	Hetero-xylan	(1,3;1,4)-β-Glucan	Cellulose	Hetero-mannan	Pectin	Xyloglucan	Reference
Barley coleoptiles (4 days)	32	10	35	nr	12	10	Gibeaut et al. (2005)
Barley aleurone	71	26	2	2	nd	nd	Bacic and Stone (1981)
Barley starchy endosperm	20	70	3	2	nd	nd	Fincher (1975)
Maize internodes	46	3	35	2	trace	6	Zhang et al. (2014)
Brachypodium whole grain	4.7	42.4	6	trace	nr	nr	Guillon et al. (2011)
Rice endosperm	32	nr	36.3	nr	7.3	nr	Shibuya and Iwasaki (1978)
Arabidopsis leaves	4	0	14	nr	42	20	Zablackis et al. (1995)
Grape berries (114 DPA)	7	0	31	3	45	8	Nunan et al. (1998)

nd, not detected; nr, not reported.

WALL COMPOSITION IN GRAIN DIFFERS FROM THAT IN VEGETATIVE TISSUES

In most dividing cells of vegetative tissues of higher plants, a callosic cell plate forms between the newly separated nuclei (Waterkeyn, 1967; Morrison and O'Brien, 1976). The cell plate acts as a scaffold on which the new wall is built. Cellulosic and non-cellulosic polysaccharides are deposited on both sides of the cell plate until the nascent wall eventually separates the daughter cells. The cell plate is compressed to a thin middle lamella layer that lies between walls of the two daughter cells. Wall deposition continues as the cells expand, but at this stage the wall remains relatively thin to allow this expansion to occur and is usually referred to as the primary wall. As cell expansion ceases, wall deposition continues in many cells to form a much thicker and stronger secondary wall, which can be further strengthened by the deposition of lignin and through lamination of parallel sheets of cellulose microfibrils that are oriented in different directions.

As noted above, the first distinguishing feature of walls in grasses compared with other plant species is related to their composition. Although pectic polysaccharides are amongst the earliest wall components to be deposited in both dicotyledons and monocotyledons, the levels of pectic polysaccharides in the walls of grasses decline during wall development to low levels relative to those observed in dicot walls. Other non-cellulosic polysaccharides are deposited during primary wall formation, including xyloglucans, heteromannans, and heteroxylans. In primary walls of the grasses, pectic polysaccharides and xyloglucans are found at relatively low levels, while the heteroxylans appear to form the core non-cellulosic polysaccharides of most walls (Burton and Fincher, 2009). An additional wall polysaccharide is often deposited, namely the (1,3;1,4)-β-glucans. This polysaccharide is not widely distributed outside the Poaceae and the genes that mediate its biosynthesis are believed to have evolved relatively recently. The wall composition of the grasses can be contrasted with the walls of *Arabidopsis*, where xyloglucans appear to be the core non-cellulosic polysaccharides, pectic polysaccharides remain relatively high, and the levels of heteroxylans are low (Zablackis et al., 1995).

The differences are exemplified in developing coleoptiles of barley (Gibeaut et al., 2005), where pectic polysaccharides decrease from about 30% w/w to just a few percent of walls over 6 days. Heteroxylan levels remain at about 30% w/w throughout coleoptile development, while xyloglucan levels are generally 10% w/w or less (**Figure 2**; Gibeaut et al., 2005). Similar results were reported for the composition of walls in elongating maize internodes, which can also be viewed as a useful system for monitoring developmental changes in wall composition in vegetative tissues of the Poaceae (Zhang et al., 2014).

The second distinguishing feature of wall composition in the Poaceae is seen in comparisons between vegetative tissues and grain and, more particularly, the starchy endosperm. Botanically, grains are one-seeded fruits, or caryopses (Esau, 1977). Formation of cell walls in the developing endosperm proceeds via a completely different developmental program from other tissues. Fusion of a sperm cell with two haploid central cell nuclei gives rise to a triploid endosperm nucleus. Repeated nuclear division produces many nuclei in a syncytium, which is essentially a cavity in the caryopsis. In most cases, cellularization follows, where callosic cell walls are laid down from the outside in, simultaneously separating the nuclei and apportioning them evenly into cells until the newly formed endosperm walls eventually meet at a central point to fill the coenocyte, as exemplified by rice (Brown et al., 1994), sorghum (Paulson, 1969), and barley (Wilson et al., 2006; **Figure 3**). In both cellularizing barley and rice endosperm callose is believed to be the major component of the cell walls that grow around the nuclei in the syncytium. In barley callose is found along the central cell wall at 3 days after pollination (DAP); it is present in the first and subsequent anticlinal walls from 4 DAP, in the periclinal walls at 5 DAP and disappears at 6 DAP, except in the vicinity of plasmodesmata (Wilson et al., 2006).

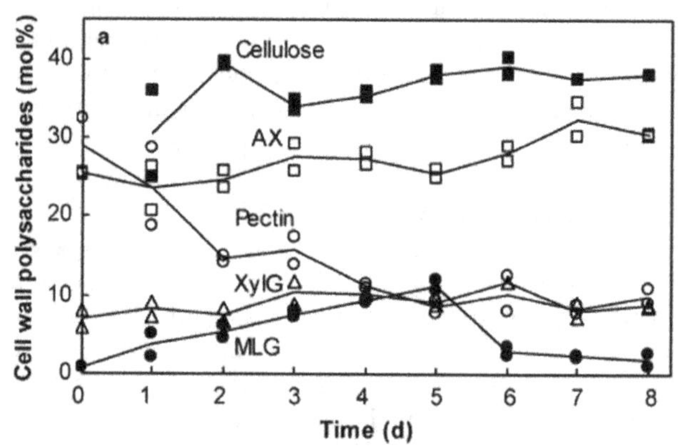

FIGURE 2 | Changes in cell wall composition during the development of barley coleoptiles. Compositions were deduced from data obtained by alditol acetate, methylation and acetic–nitric acid analyses. Changes in cellulose (filled square), arabinoxylan (AX, open square), pectic polysaccharides (pectin, open circle), xyloglucan (XylG, triangle), and (1,3;1,4)-β-glucan (MLG, filled circle) are shown. Reproduced with permission from Gibeaut et al. (2005).

Callose often re-appears much later during barley and wheat grain development (Fulcher et al., 1977; Bacic and Stone, 1981). At 28 DAP, newly deposited patches of callose are detected at irregular spacings along the aleurone–subaleurone interface of barley grain (Wilson et al., 2012). The function of these deposits is unclear but they may represent a wound response to the osmotic stresses imposed by desiccation of the maturing grain or by periods of water stress during grain maturation (Fincher, 1989).

Despite the different cellular developmental patterns in the grain, the walls of the mature grain are still composed of the polysaccharides observed in vegetative walls. However, in the endosperm of many grass species, the amount of cellulose is reduced to just a few percent on a weight basis, which can be contrasted with cellulose contents of 30% (w/w) or more of primary walls in vegetative tissues (Fincher, 2009). The low cellulose content in the endosperm is consistent with the fact that these cells have no load-bearing function, as distinct from walls in barley coleoptiles or maize stalk internodes, and because it is

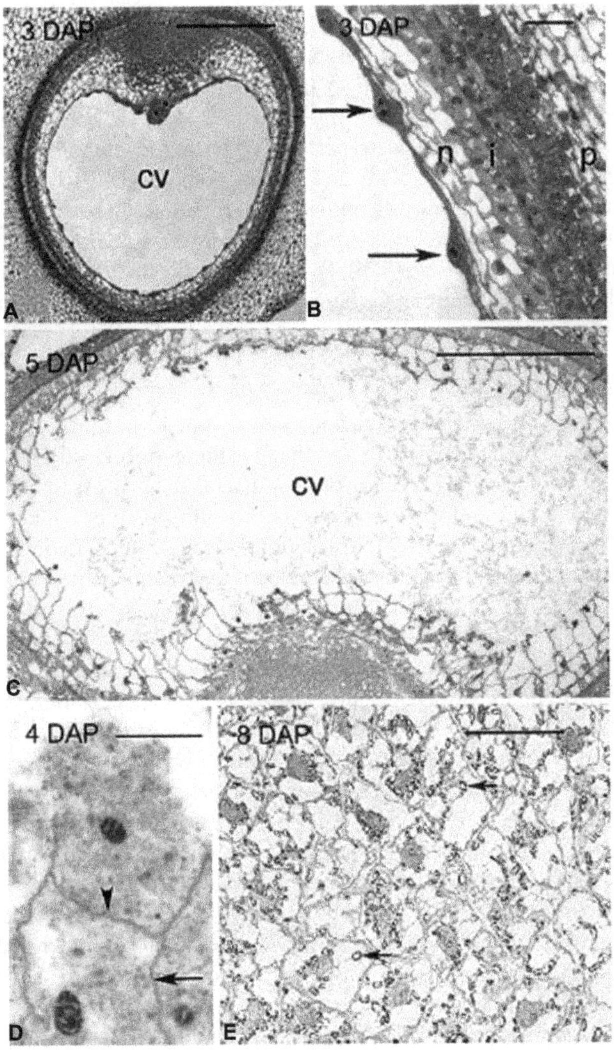

FIGURE 3 | Different stages of endosperm development in barley. Light micrographs of sections through barley grains showing stages of endosperm development from 3 to 8 DAP. **(A)** 3 DAP: a thin layer of syncytial cytoplasm surrounds a large central vacuole. **(B)** Details of the syncytium in **(A)**. Arrows indicate the position of nuclei along the perimeter of the central cell, all enclosed within discrete layers of maternal tissue. **(C)** 5 DAP: cellularization occurs centripetally with repeated cycles of anticlinal wall formation, mitosis and periclinal wall formation. **(D)** 4 DAP: shows the wavy appearance of anticlinal walls (arrow) and a periclinal wall (arrowhead) separating two recently divided daughter nuclei. **(E)** 8 DAP: the endosperm was fully cellularized and starch granules (arrows) had accumulated within each cell. cv, central vacuole; i, integuments; n, nucellus; p, pericarp. Scale bars = 300 μm **(A,C)**, 50 μm **(B,E)**, 20 μm **(D)**. Reproduced with permission from Wilson et al. (2006).

important for walls of endosperm cells to be quickly degraded in the germinated grain. High levels of cellulose in these walls would almost certainly slow their rate of degradation following germination. However, it must be noted that walls in the starchy endosperm of grain do have to withstand pressures exerted by grain expansion and later by dehydration as the grain matures, and such stresses may trigger changes in the matrix polysaccharides of the wall.

WALL COMPOSITION IN DIFFERENT TISSUES OF THE GRAIN

Most of the discussion above has been focused on the development of cell walls in the starchy endosperm of grains of the Poaceae. However, as the grain develops several other specific cell types can be distinguished. These include, in addition to the starchy endosperm cells, which are the main repository for starch and storage protein, TC, which are clustered around the vascular network that feeds the growing grain, aleurone cells, which envelop the starchy endosperm and are rich in oil and protein bodies, sub-aleurone cells that arise through periclinal division of the aleurone cells, and finally the embryo itself, which is comprised of many organ-specific vegetative tissues. Information on the cell walls of these tissues is not extensive, but some interesting data are emerging.

ALEURONE LAYER

Aleurone cells form a layer around the starchy endosperm that varies from one to three or four cells in thickness, depending on the species, and are indeed components of the endosperm as a whole. Aleurone cells are typically cuboid in shape with much thicker cell walls, usually at least twice the thickness of those in the central starchy endosperm. Aleurone cells contain a dense granular cytoplasm comprised of aleurone grains and small vacuoles containing inclusion bodies (Olsen, 2004). They are rich in proteins and oil but contain no starch and, unlike the cells of the starchy endosperm which undergo programmed cell death (Young and Gallie, 2000), they remain living in the mature grain. This is essential if they are to perform their key role in grain germination, where they synthesize and release a range of hydrolyzing enzymes that are responsible for mobilizing the storage polymers of the starchy endosperm. Aleurone cells usually remain triploid, unlike the starchy endosperm cells, which undergo endoreduplication and become polyploidy in nature (Olsen, 2001).

The walls of aleurone cells in mature barley and wheat grain have two quite distinct layers (Taiz and Jones, 1973; Bacic and Stone, 1981). The inner layer is thinner and may have higher concentrations of (1,3;1,4)-β-glucans (Wood et al., 1983). The thicker outer layer of the aleurone wall may be enriched in arabinoxylans, although ferulic acid residues were believed to be evenly distributed across the two wall layers (Fincher, 1989). The two layered structure of aleurone walls might be important during grain germination, when the thick outer layer is rapidly dissolved, while the thin, inner layer remains intact. The outer layer might be removed to facilitate the secretion of newly synthesized hydrolytic enzymes into the starchy endosperm (Van der Eb and Nieuwdorp, 1967; Gubler et al., 1987), while the retention of the thin inner layer might be necessary to maintain the physical integrity of the aleurone cells until their role in enzyme secretion is complete (Fincher, 1989). Walls of the scutellar epithelium layer, which is important in the secretion of hydrolytic enzymes into the starchy endosperm early after germination (McFadden et al., 1988), have morphological features that are similar to those of the aleurone and it is likely that the walls of the scutellar epithelium have a similar composition to those of the aleurone layer (Fincher, 1989).

The developmental cues for aleurone cells are complex and not yet fully understood. In wheat, they have a specific molecular signature by 6 days post anthesis, conferred by their position in the "surface layer" (Gillies et al., 2012). However, aleurone cell fate remains plastic up to the last cell division and specific signals are necessary to maintain cell identity (Becraft and Yi, 2011). In barley grain, aleurone cells are present at 10 DAP and their walls continue to thicken until 22 DAP when grain maturation begins (Wilson et al., 2012). Many cereals also have a zone of cells that separate the true aleurone from the starchy endosperm cells. These subaleurone layers arise from periclinal division of the aleurone cells (Becraft and Asuncion-Crabb, 2000) and in barley they are present by 14 DAP (Wilson et al., 2012). Subaleurone cells are larger than aleurone cells but smaller than starchy endosperm cells, and contain small starch granules and protein bodies.

The developmental signals that dictate the number of cell layers and hence the thickness of the aleurone layer overall are gradually being unraveled (Sabelli and Larkins, 2009). Aleurone layer thickness, the number of cell layers therein and the regularity of thickness has been examined in a range of cereals by Hands et al. (2012). Barley was found to be the only grain to consistently possess a layer more than one cell in thickness. The non-cultivated species *B. distachyon* and *Festuca pratensis* have markedly more disorganized and irregular aleurone layers, which may imply that there is a correlation between regularity of shape and domestication, since this trait may have been selected to meet certain grain quality parameters, such as speed of germination and endosperm mobilization (Hands et al., 2012). However, our knowledge of grain ultrastructure in non-crop species of the Poaceae is generally poor but increasing the number of cell layers in aleurone layers could be beneficial. Approximately half the volume of cereal bran is comprised of aleurone tissue and since this is the most dietary beneficial part of the bran, rich in proteins, oils, and other phytonutrients, increasing the amount further is desirable in human health and animal nutrition (Okarter and Liu, 2010). However, there are also milling considerations, since aleurone cell walls are so thick the cells may remain intact and their contents unobtainable (Minifie and Stone, 1988).

The core polysaccharides found in aleurone cell walls are also arabinoxylans, although relatively high levels of (1,3;1,4)-β-glucan are found in wheat and barley grain. Early work in which aleurone cells were isolated and analyzed showed that aleurone walls from wheat and barley contained about 65% arabinoxylan and about 28% (1,3;1,4)-β-glucan; cellulose and glucomannan levels were again very low (Bacic and Stone, 1981). Several groups have used immunolabeling, Raman spectroscopy, and IR microspectroscopy to monitor changes in aleurone call walls, *in situ*, during the development of wheat grain. Aleurone walls are more heterogeneous early in grain development compared with those at maturity (Jamme et al., 2008). Antibody labeling indicated the presence of the pectic polysaccharide epitopes RGI, (1,5)-α-arabinan and (1,4,)-β-galactan in the aleurone, particularly on the inner surface of the cell wall, and in the pericarp in mature grain (Jamme et al., 2008; Chateigner-Boutin et al., 2014).

Strong autofluorescence has long been known in aleurone and is attributable to high levels of the phenolic acids, ferulic acid, and *p*-coumaric acid in mature aleurone walls in wheat (Fulcher et al., 1972; Bacic and Stone, 1981; Robert et al., 2011) and other cereals. These phenolic compounds have been examined more closely by Jaaskelainen et al. (2013) using *in situ* optical and Raman microscopy. In the aleurone cells of both barley and wheat, the anticlinal walls contain high amounts of phenolic acids compounds, with much less in the inner periclinal walls. In barley, phenolic compounds were particularly strong in the outer periclinal walls. Ferulic acid, and indeed arabinoxylan, were first detected in the newly differentiated aleurone walls in barley grain at 12 DAP (Wilson et al., 2012). Jaaskelainen et al. (2013) confirmed that there is no (1,3;1,4)-β-glucan in the middle lamella of aleurone walls but that arabinoxylan is enriched here and in the outer cell wall layers.

TRANSFER CELLS

Transfer cells provide the major route for nutrient acquisition by the developing endosperm and they are therefore a key determinant of grain filling. TCs are present in a range of tissues in many plant species and they can be classified into two types, namely reticulate and flange-like. Through the deposition of secondary cell wall material, both types develop a massively expanded surface area to facilitate the transfer of nutrients. Wang et al. (1994) estimated that the plasma membrane surface area increases up to 22-fold. Reticulate types are exemplified by TCs found in *Vicia faba* cotyledons whereas flange-like types are typically found in cereals (McCurdy et al., 2008; **Figure 4**). Reticulate TCs arise from re-differentiation of epidermal cells (Offler et al., 1997), which is a very different pathway from the direct differentiation of flange-like TCs from endosperm cells in developing cereal grains. The latter occurs opposite the nucellar projection as early as 5 DAP in barley, when the first wall ingrowths appear in the syncytium (Thiel et al., 2012b). By 7 DAP the TC walls are enlarged with net-like and branched strands on the inner wall and TCs represent 6.7% of the total endosperm volume; they increase in area ninefold between 5 and 10 DAP. By 10 DAP the walls are thicker with rib-shaped projections and cells are flattened in parallel with the long axis of the grain; by 12 DAP the wall thickenings are asymmetric and irregularly spaced and the flanges have fused; and by 14 DAP TCs represent a much lower proportion of the total endosperm volume at just 0.9% (Thiel et al., 2012b). Wheat TCs develop in a similar fashion to those in barley (Zheng and Wang, 2011), whilst maize TCs present a dense network of flanges and are found in the basal endosperm (Zheng and Wang, 2010), and rice TCs are found in the aleurone layers in the dorsal region of the grain adjacent to the major vascular bundle in the pericarp. Development of TCs in rice is uneven but they also show wall in-growths (Hoshikawa and Wang, 1990).

The deposition of layers of material onto the original wall in TCs has been defined as secondary wall thickening. This occurs widely in many vegetative parts of the plant as cell expansion ceases and wall deposition continues to form a much thicker and stronger secondary wall, which can be further strengthened through the deposition of lignin and *via* lamination. We know that the major polysaccharides laid down through secondary thickening are cellulose and heteroxylans, with the deposition of lignins to further strengthen and, in some cases, to waterproof the wall. Although we know much less about the composition of TC walls, it would

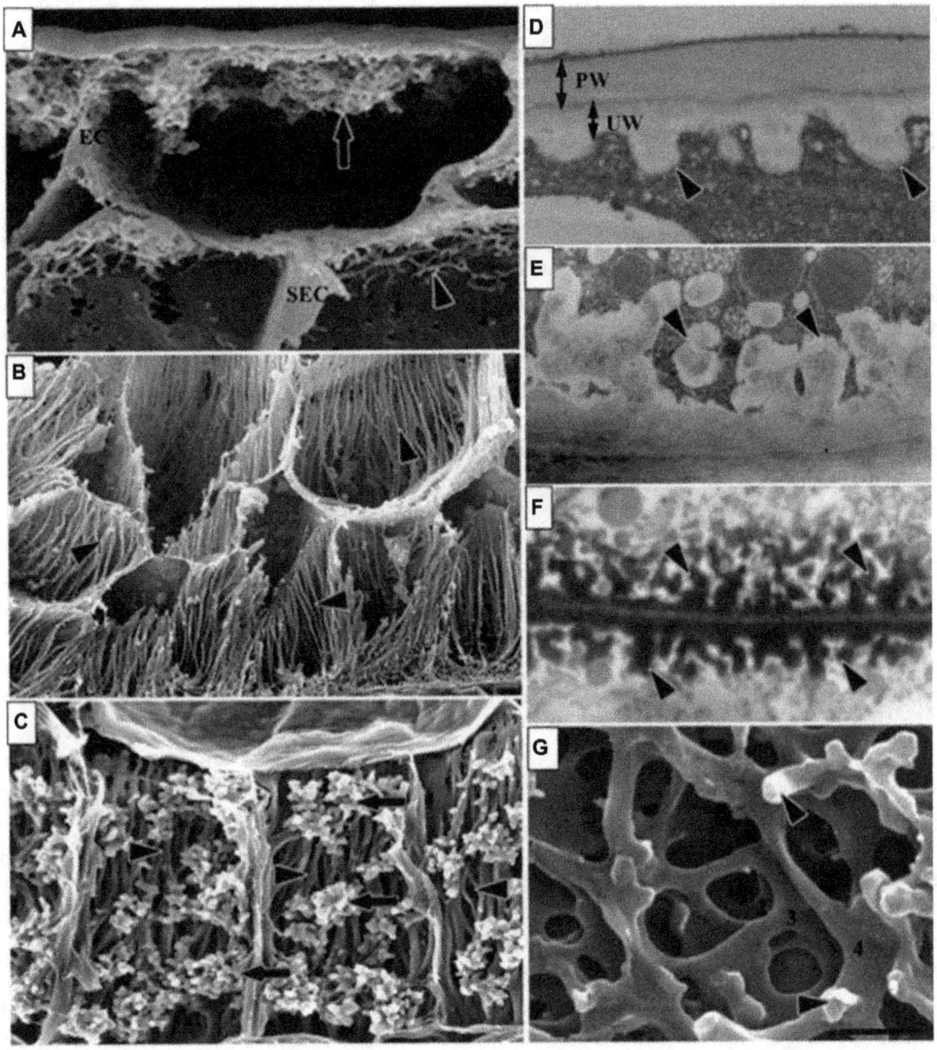

FIGURE 4 | Different types of transfer cells (TC) in cereals and other seeds. These images of TC of developing seeds illustrate various ingrowth wall morphologies. **(A)** Epidermal transfer cells (ETC) of a *Vicia faba* cotyledon with an extensive reticulate ingrowth wall labyrinth including clumps of ingrowth material (arrow) and smaller wall ingrowths in the subepidermal cells (SEC; arrowhead). **(B)** Basal endosperm TC of *Zea mays* exhibiting flange wall ingrowth morphology; arrowheads indicate small lateral protrusions from the linear ribs (modified after Talbot et al., 2002). **(C)** Thin-walled parenchyma TC located at the inner surface of the inner seed coat of *Gossypium hirsutum* with wall ingrowth flanges (darts) extending the length of each cell on which are deposited groups of reticulate wall ingrowths (arrows; modified after Pugh et al., 2010). **(D–F)** Transmission electron microscope images of portions of transverse sections of TC: **(D)** the outer periclinal wall of an adaxial epidermal cell of a *V. faba* cotyledon induced to trans-differentiate to a transfer cell morphology displaying primary wall (PW) and uniform walls (UW). **(E)** Small papillate ingrowths (darts) of a seed coat transfer cell of *V. faba* exhibiting reticulate architecture. **(F)** Antler-shaped reticulate wall ingrowths (darts) of a nucellar projection transfer cell of a developing *Triticum turgidum* var. durum seed (modified after Wang et al., 1994). **(G)** Field emission scanning electron microscope image of the cytoplasmic face of the reticulate ingrowth wall labyrinth of an abaxial epidermal transfer cell of a *V. faba* cotyledon following removal of the cytoplasm and dry cleaving (for method see Talbot et al. (2001), image modified after Talbot et al. (2001)) where the darts indicate ingrowth papillae on the most recently deposited wall layer. Single scale bar for **(A,B)** = 2.5 μm; for **(C)** = 5 μm; for **(D,E)** = 1 μm; for **(F)** = 0.25 μm; for **(G)** = 0.5 μm. Figure legend and images reproduced with permission from Andriunas et al. (2013).

seem likely that they do not resemble a typical secondary wall. Significantly, lignin is absent and in wheat, arabinoxylan is the predominant component from 5 to 23 DAP (Robert et al., 2011), and is more highly substituted than the arabinoxylan in the walls of the aleurone layer. After 23 DAP, the TC walls become enriched in (1,3;1,4)-β-glucan, which also occurs in the aleurone, and again this is not typical of secondary cell walls in other parts of the plant.

Recently, laser-microdissection methods have been used successfully to define tissue-specific transcripts and allow metabolite profiling of TCs in barley (Thiel et al., 2012a; Thiel, 2014).

MINOR WALL POLYSACCHARIDES IN THE GRAIN

The core non-cellulosic wall polysaccharides of the grain are the heteroxylans and, in some cases, (1,3;1,4)-β-glucans, while cellulose contents are usually low, as noted above (Fincher and Stone,

2004). However, there is one notable variant when it comes to wall composition in the starchy endosperm of grasses. The endosperm walls of mature rice grain are comprised of significant amounts of cellulose, up to 30% as reported by Shibuya and Nakane (1984). Cellulose is also present at higher levels during the very early stages of barley grain development (Wilson et al., 2006).

Although arabinoxylan and (1,3;1,4)-β-glucan predominate in cereal grain cell walls, we are starting to discover the presence of other polysaccharides which, although only minor components of the walls, may represent key determinants of wall plasticity and other properties. Thus, levels of pectic polysaccharides, heteromannans, and xyloglucans are low in many grains, including wheat and barley (Mares and Stone, 1973; Fincher, 1975). Again an exception here appears to be rice, which contains relatively high levels of pectin (Shibuya and Nakane, 1984) and xyloglucan (Shibuya and Misaki, 1978). Xyloglucan can also be detected in barley grain during early grain development, but appears to be transitory in nature. It first appears at 3 DAP in the central cell wall but is undetectable by 6 DAP (Wilson et al., 2012). Mannans first appear in barley endosperm walls at 5–6 DAP, after cellularization is complete and, based on the accumulation of mannose, mannans, or glucomannans continue to be deposited at low levels up to 20 DAP (Wilson et al., 2012); the final levels of mannans or glucomannans in mature wheat and barley grain are about 2–3% w/w (Mares and Stone, 1973; Fincher, 1975).

Small but significant pectic deposits have recently been reported in wheat grain (Chateigner-Boutin et al., 2014). Pectins have previously been reported in rice endosperm cell walls (Shibuya and Nakane, 1984) and in *B. distachyon* (Guillon et al., 2011) but little is known about their presence or otherwise in the majority of cereal grains. Pectins are complex, multi-domain polysaccharides that bear many different epitopes (Caffall and Mohnen, 2009). Chateigner-Boutin et al. (2014) used antibodies that recognize specific pectic epitopes on sections of developing and mature wheat grains. The inclusion of pre-labeling enzymatic digests with lichenase and xylanase to remove a portion of the major polysaccharides (1,3;1,4)-β-glucan and arabinoxylan proved to be a key step in rendering the pectic epitopes accessible. In the developing grain LM20, which recognizes methyl-esterified homogalacturonan (HG; Verhertbruggen et al., 2009), labeled the pericarp and early endosperm walls, where elasticity would be required. In older grain, large bodies containing unesterified HG, as detected by LM19, were found located in the subcuticle layer, and the reason for their presence here is currently unclear (Chateigner-Boutin et al., 2014).

EVOLUTIONARY DIFFERENCES IN HETEROXYLANS IN THE GRAIN

Consistent with the low cellulose content of endosperm walls, the levels of the core wall polysaccharide in the Poaceae, the heteroxylans, are relatively higher in the starchy endosperm, while the levels of the core polysaccharides of dicotyledonous plants, pectic polysaccharides, and xyloglucans, are generally much lower. Indeed, heteroxylans are found in all walls of the grasses and are the major non-cellulosic polysaccharide in most walls. However, there is evidence of evolutionary forces at work on the heteroxylans of the Poaceae. In dicotyledonous plants, glucuronoarabinoxylans are abundant and in some cases glucuronyl residues predominate. In the grasses, two types of heteroxylans can be distinguished. Glucuronoarabinoxylans are relatively abundant in the outer, pericarp-testa layers of the grain and in bran, while arabinoxylans are the major non-cellulosic polysaccharides of the aleurone and starchy endosperm cell walls (Fincher and Stone, 2004).

The species best characterized for arabinoxylan is wheat, where isolated endosperm walls comprise about 70% of this polysaccharide (Mares and Stone, 1973). The (1,4)-β-xylan backbone of the polysaccharide displays both structural and spatial heterogeneity with regard to its degree of substitution and this heterogeneity varies throughout endosperm development, as assessed by enzyme mapping, FT-IR, and Raman microscopy and NMR spectroscopy (Toole et al., 2007, 2009, 2010, 2012). Early in endosperm development more of the backbone (1,4)-β-linked xylosyl residues are di-substituted with arabinofuranosyl residues at the *O-2* and *O-3* positions, but as the grain matures, a higher degree of mono-substitution at the *O-3* position is observed, possibly to allow more inter-chain interactions to occur to withstand mechanical stresses as the grain dries out. Ferulic acid and to a lesser extent *p*-coumaric acid residues are ester-linked at *O-5* of some of the *O-3* mono-substituted arabinosyl groups and it has been reported that these can form covalent cross-links between arabinoxylan chains through oxidative dimerization (Iiyama et al., 1990). There is a gradient of arabinoxylan substitution patterns across the grain as prismatic cells give way to round cells (Toole et al., 2010). Barley endosperm cell walls also contain about 20% arabinoxylan (Fincher, 1975) and show subtle inter-species variation in the types and amounts of backbone substitutions (Izydorczyk, 2014). This is also evident in rye grain, which has a much higher ratio of mono- to di-substitutions than wheat (Rantanen et al., 2007).

The substitution of the extended (1,4)-β-xylan backbone with arabinofuranosyl residues sterically hinders the aggregation of the (1,4)-β-xylan chains into insoluble microfibrils and results in the formation of a long, asymmetrical polysaccharide that is partly soluble in water and can form gel-like structures in the cell wall matrix (Fincher and Stone, 2004). As expected, the degree of substitution of the (1,4)-β-xylan backbone will affect the physical properties of the polysaccharide and, in particular, its solubility. Highly substituted, soluble arabinoxylans, which have a characteristically high arabinose:xylose ratio, are found in the endosperm cells of the grain, while arabinoxylans with lower degrees of substitution are less soluble and are located in the outer layers of the grain (Fincher and Stone, 2004; Izydorczyk, 2014).

EVOLUTION OF (1,3;1,4)-β-GLUCANS IN THE GRASSES

Another key difference in walls of cereal grains compared with other seeds is the presence of (1,3;1,4)-β-glucan. This polysaccharide has an interesting distribution in the plant kingdom (Harris and Fincher, 2009). It is found in many species of the Poaceae but is also occasionally found in other Poales, and in lower plants such as the *Equisetum* spp. horsetail ferns (Trethewey et al., 2005; Fry et al., 2008; Sørensen et al., 2008), bryophytes (Popper and Fry, 2003), some fungi (Pettolino et al., 2009), brown, green and red algae (Lechat et al., 2000; Eder et al., 2008; Popper and Tuohy, 2010), and lichens (Stone and Clarke, 1992). This distribution pattern of (1,3;1,4)-β-glucans in higher and lower plants is suggestive of convergent evolution. The (1,3;1,4)-β-glucans seem to have been widely adopted only in the Poaceae, where one might conclude there is positive selection pressure to retain the polysaccharide in the walls.

The (1,3;1,4)-β-glucans of the grasses are comprised of an unsubstituted chain of glucosyl residues linked either through (1,4)-β- or (1,3)-β-linkages. About 90% of the polysaccharide chain is comprised of cellotriosyl (DP3) and cellotetraosyl (DP4) units that are linked through (1,3)-β-linkages; adjacent β-linkages are rare or absent (Buliga et al., 1986). Approximately 10% of the polysaccharide is comprised of longer chains of adjacent (1,4)-β-linkages (Woodward et al., 1983). The DP3 and DP4 units are arranged randomly along the chain (Staudte et al., 1983). The combination of the single (1,3)-β-linkages and the random arrangement of the cellotriosyl (DP3) and cellotetraosyl (DP4) units, and hence the (1,3)-β-linkages, result in an extended polysaccharide chain that has a limited capacity to align with other (1,3;1,4)-β-glucan chains. The (1,3;1,4)-β-glucans from many cereal grains are therefore at least partly soluble in water, they adopt an asymmetrical conformation and can form gel-like structures that are believed to be functionally advantageous for non-cellulosic cell wall polysaccharides in the matrix phase of the wall (Fincher and Stone, 2004).

The ratio of the DP3:DP4 units can be used to predict the solubility of the molecule and its rheological behavior (Papageorgiou et al., 2005). High and low ratios indicate a predominance of cellotriosyl and cellotetraosyl residues, respectively, and in both cases the conformation of the polysaccharide becomes more uniform and hence more capable of aligning into insoluble aggregates (Burton et al., 2010). High and low ratios are characteristic of the insoluble (1,3;1,4)-β-glucans from lower plants such as horsetail ferns and fungi (Burton et al., 2010). The DP3:DP4 ratio in (1,3;1,4)-β-glucans from the Poaceae have intermediate values, usually around 2–3:1 (Trafford and Fincher, 2014). It would appear that (1,3;1,4)-β-glucans with these structures and physical properties have evolved and are retained by the grasses for functional reasons. Nevertheless, the ratios vary considerably across cereal species (**Table 1**; Burton and Fincher, 2012) and grains in which (1,3;1,4)-β-glucans are particularly abundant often have a lower DP3:DP4 ratio and are more soluble (Trafford and Fincher, 2014). The exception here is the relatively insoluble (1,3;1,4)-β-glucan in the grain of *B. distachyon*, where this polysaccharide has a ratio of 5.8:1 and clearly has evolved to perform a storage function (Guillon et al., 2011).

Although the chemical structures of the arabinoxylans and the (1,3;1,4)-β-glucans are quite different (**Figure 5**), their physical properties are similar and well adapted to a structural role in cell walls. This is therefore an example of convergent evolution to the extant state. Arabinoxylans are extended asymmetrical molecules by virtue of their linear (1,4)-β-xylan backbone and are partly soluble because of the steric hindrance of intermolecular aggregation afforded by their arabinofuranosyl substituents. Solubility is further influenced by acetylation and feruloylation which participate in cross-link formation between arabinoxylan and other wall components. This is exemplified in wheat endosperm walls where the degree of acetylation declines affecting solubility as the grain matures (Veličković et al., 2014) and where arabinoxylan in older walls is rendered less soluble by significant ferulate cross-linking (Saulnier et al., 2009). In contrast, the (1,3;1,4)-β-glucans are extended asymmetrical molecules by virtue of the predominance of "cellulosic" (1,4)-β-glucosyl linkages along their linear backbone and are partly soluble because of the steric hindrance of aggregation caused by the random disposition of (1,3)-β-glucosyl residues that result in randomly distributed molecular kinks in the macromolecule. Just as the solubility of arabinoxylans can be predicted from the degree of substitution and cross-linking, so too can the physical properties of (1,3;1,4)-β-glucans be predicted from their DP3:DP4 ratio. Different chemical strategies have evolved to produce the same physicochemical properties in heteroxylans and (1,3;1,4)-β-glucans.

(1,3;1,4)-β-Glucan is the predominant polysaccharide in the starchy endosperm cell walls of barley and oats and comprises about 15% of starchy endosperm cell walls in wheat grain (Mares and Stone, 1973). Recently, Veličković et al. (2014) used MALDI-MS to examine the spatial distribution of both (1,3;1,4)-β-glucan and arabinoxylan across the wheat grain. They reported higher amounts of (1,3;1,4)-β-glucan and arabinoxylan in outer endosperm regions of young grain and showed that this distribution became more even in mature grain, although cells close to the embryo had walls rich in (1,3;1,4)-β-glucan at all stages of grain development (Saulnier et al., 2009). In barley,

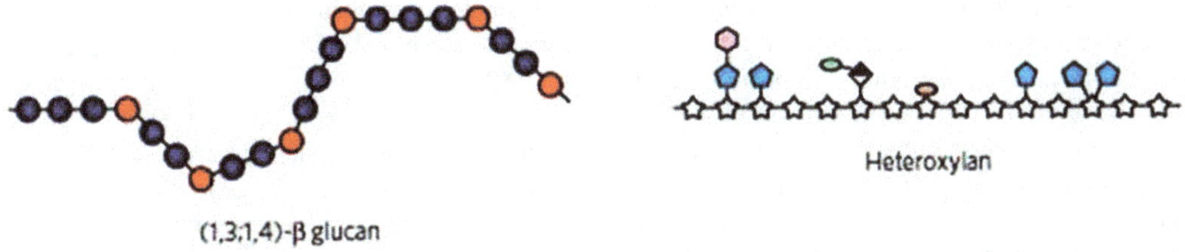

FIGURE 5 | Diagrammatical representations of the major non-cellulosic wall polysaccharides from cereal grains. The (1,3;1,4)-β-glucan **(left)** has relatively extended regions of adjacent (1,4)-β-glucosyl residues (blue) with irregularly spaced, single (1,3)-β-glucosyl residues. The latter residues form molecular "kinks" in the polysaccharide chain and limit intermolecular alignment and microfibril formation. In the heteroxylan **(right)**, intermolecular alignment of the xylan backbone (stars) and microfibril formation is limited by steric hindrance afforded by the substituents (blue, pink, etc.). Reproduced with permission from Burton et al. (2010).

the (1,3;1,4)-β-glucan is reported to be evenly distributed in endosperm walls by 10 DAP (Wilson et al., 2012), but little (1,3;1,4)-β-glucan was detected between 12 and 16 DAP in the peripheral starchy endosperm cells closest to the differentiating aleurone. This has also been noted in wheat (Philippe et al., 2006) but while this situation persists in wheat, in barley by 16 DAP (1,3;1,4)-β-glucan deposition has occurred in the peripheral starchy endosperm. There are clearly microdomains present across the endosperm where cell wall composition varies but the requirement for these subtle variations is presently unclear. There is also currently little information on spatial differences in the DP3:DP4 ratio of (1,3;1,4)-β-glucans across developing grain of any species, which is undoubtedly related to the lack of high resolution detection methods. However, the MALDI-MS method shows promise for these kinds of analyses. Veličković et al. (2014) were able to quantify oligosaccharides released by *in situ* digestion of (1,3;1,4)-β-glucans with lichenase and reported that the DP3:DP4 ratio was elevated to 7:1 in younger endosperm, compared with around 4:1 in mature tissue.

EVOLUTION OF POLYSACCHARIDE SYNTHASE GENES

Many of the enzymes that catalyze the polymerization of the backbone chains of wall polysaccharides are encoded by genes that belong to the "cellulose synthase gene superfamily." This gene family has close to 50 members in most higher plants (Richmond and Somerville, 2000; Hazen et al., 2002) and it has proved difficult to unequivocally assign functions to individual genes and some clades. The *CesA* clade encodes cellulose synthases (Pear et al., 1996; Arioli et al., 1998), but it is clear that several CesA enzymes and a number of other enzymes and/or proteins are required for an active cellulose synthesis complex (Doblin et al., 2002; Burton and Fincher, 2014). Several of the cellulose synthase-like (*Csl*) clades of the gene superfamily have been implicated in the synthesis of different wall polysaccharides. The *CslA* group of genes is likely to encode mannan and glucomannan synthases (Dhugga et al., 2004; Liepman et al., 2005). Cocuron et al. (2007) have presented evidence for a role of the *CslC* group of genes in the synthesis of the (1,4)-β-glucan backbone of xyloglucans and the genes in the *CslD* clade may be involved in cellulose synthesis, particular in cells that exhibit tip growth (Doblin et al., 2001; Favery et al., 2001; Wang et al., 2001).

A good deal of effort has been focused on the identification of genes that mediate the synthesis of the cereal grain arabinoxylans and (1,3;1,4)-β-glucans. In the case of the arabinoxylan enzymes, much of the initial work on the identification of genes involved was focused on analyses of *Arabidopsis* mutant lines and transcript profiling. These studies implicated genes from the *GT8*, *GT43*, *GT47*, and *GT61* families (Brown et al., 2007, 2009; Mitchell et al., 2007; Pena et al., 2007; Persson et al., 2007; Oikawa et al., 2010). However, these approaches are plagued with interpretative difficulties imposed by the large gene families, compensation, and pleiotropic effects in transgenic lines during proof-of-function tests, and the difficulties associated with developing reliable biochemical assays for expressed enzymes. Mitchell et al. (2007) and Pellny et al. (2012) used comparative bioinformatics analyses to predict the functions of candidate genes and concluded that genes in the GT43 and GT47 families might encode backbone (1,4)-β-xylan synthases in wheat, genes in the GT61 family might encode xylan (1,2)-α- or (1,3)-α-L-arabinosyl transferases, and that BAHD genes encode feruloyl-arabinoxylan transferases. This group recently provided additional and compelling evidence for wheat *GT61* genes, which they designated *TaXAT* for wheat, as xylan (1,3)-α-L-arabinosyl transferases (Anders et al., 2012), whilst another member of the *GT61* family in rice, called *XAX1*, was shown to be responsible for adding the xylose residues in Xylp-(1 ⟶ 2)-α-Araf-(1 ⟶ 3) substitutions (Chiniquy et al., 2012). Zeng et al. (2010) used GT43-specific antibodies to co-immunoprecipitate a complex from wheat microsomes that contained GT43, GT47, and GT75 proteins, and Lovegrove et al. (2013) used RNA interference suppression of *GT43* and *GT47* genes to reduce the total amount of arabinoxylan in wheat endosperm walls by 40–50%. Analysis of the glucuronoarabinoxylan polymer synthesized by the complex suggested a regular structure containing Xyl, Ara, and GluA in a ratio of 45:12:1. The authors suggested that this may represent a core complex in the biosynthetic process of xylans but to date we have no definitive evidence for the involvement of specific genes or proteins in the synthesis of the backbone or in the addition of certain substituents. Mortimer et al. (2010) reported that the products of two *GT8* genes mediate the addition of α-GluA and α-4-*O*-methylglucuronic acid residues to the heteroxylan of *Arabidopsis*, and Rennie et al. (2012) later established that the *GT8* gene *GUX1* performs substitution of the xylan backbone with GlcA. α-Galacturonosyl transferases that are involved in HG synthesis are also members of the *GT8* family (Yin et al., 2009). Double mutant plants for these genes (*gux1gux2*) contain xylan that is almost completely unsubstituted, but still contain wild-type amounts of the xylan backbone. This indicates that the synthesis of the backbone and its substitution can be uncoupled; a somewhat surprising observation when the behavior of such an unsubstituted and hence possibly insoluble polysaccharide in an aqueous environment is considered, although potential insolubility may be ameliorated by extensive acetylation. The domain of unknown function protein, DUF579, which was reported by Jensen et al. (2011) to be involved in xylan biosynthesis, has since been shown to encode a glucuronoxylan 4-*O*-methyl transferase that catalyzes the methyl etherification of C(O)4 of glucuronyl residues in heteroxylans of *Arabidopsis* (Urbanowicz et al., 2012).

The genes involved in the biosynthesis of (1,3;1,4)-β-glucans are reasonably well defined and include members of the *CslF* and *CslH* clades of the cellulose synthase gene superfamily. These genes are found only in the Poaceae (Hazen et al., 2002) and when transformed into *Arabidopsis thaliana* mediate the biosynthesis of (1,3;1,4)-β-glucans in the walls of transgenic plants (Burton et al., 2006; Doblin et al., 2009). As a dicotyledon, *Arabidopsis* does not normally have (1,3;1,4)-β-glucans in its walls and does not have *CslF* or *CslH* genes. These genes are members of smaller gene sub-families that contain about 10 *CslF* genes and just a few *CslH* genes (Burton and Fincher, 2012). It has not yet been demonstrated that all genes in these two clades encode (1,3;1,4)-β-glucan synthases. Additional evidence for the involvement of these genes in (1,3;1,4)-β-glucan synthesis was obtained through over-expression in barley of the *CslF6* gene driven by an endosperm-specific promoter. This resulted in increases of more than 80% in (1,3;1,4)-β-glucan content in the transgenic barley grain (Burton et al., 2010). Similarly, a mutant barley line in which there is a lesion in the *CslF6* gene has no (1,3;1,4)-β-glucan in its grain (Taketa et al., 2012). It is worth noting that the *CslF6* gene might act in concert with other proteins or enzymes during (1,3;1,4)-β-glucan synthesis and to investigate this possibility genome-wide association mapping has been used in attempts to identify other genes that might contribute to the biosynthesis or regulation of (1,3;1,4)-β-glucan synthesis (Rasmussen and Shu, 2014).

Given that the Poaceae evolved relatively recently (Feuillet et al., 2008) and that (1,3;1,4)-β-glucans are largely restricted to the Poaceae in higher plants (Harris and Fincher, 2009), it seems likely that the *CslF* and *CslH* clades evolved from other clades in the cellulose synthase gene superfamily. The *CslF* and *CslH* clades are not particularly close on the phylogenetic tree (Farrokhi et al., 2006) and this suggests that genes involved in (1,3;1,4)-β-glucan synthesis might have evolved independently on at least two occasions (Fincher, 2009). Whether these evolutionary events were based on duplication and ensuing steady changes in other *Csl* genes or whether recombination caused domain swapping in enzymes that resulted in genes encoding the (1,3;1,4)-β-glucan synthases is not known. However, it is clear that some competitive advantage must be associated with the presence of (1,3;1,4)-β-glucans in walls of the Poaceae and that selection pressure has retained the capacity of enzymes encoded by *CslF* and *CslH* genes to synthesize (1,3;1,4)-β-glucans. Detailed phylogenetic analyses indicate that the *CslF* genes shared a common ancestor with *CslD* genes and are now under a stationary selection barrier (Yin et al., 2009). A stationary selection barrier would suggest that the evolution of (1,3;1,4)-β-glucans has provided functional advantages for the Poaceae.

The recent availability of the three-dimensional structure of a bacterial cellulose synthase (Morgan et al., 2013) and a molecular model of a cellulose synthase from cotton (Sethaphong et al., 2013), provide new opportunities to link evolution at the gene level with the evolution of a new enzyme with the capacity for (1,3;1,4)-β-glucan synthesis. For example, the nascent (1,3;1,4)-β-glucan synthase enzymes might have evolved by virtue of subtle changes in the three-dimensional dispositions of active site residues or through changes in surface amino acid residues that are involved in protein–protein interactions. We are now in a position to test these possibilities.

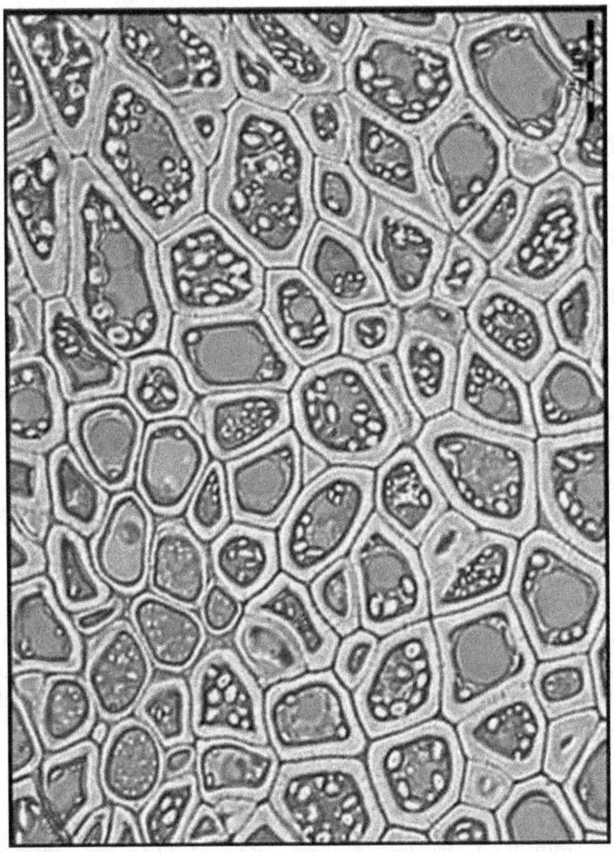

FIGURE 6 | Thick endosperm cell walls in *Brachypodium distachyon* grain. Reproduced with permission from Trafford et al. (2013).

HAVE CELL WALL POLYSACCHARIDES EVOLVED A STORAGE FUNCTION?

A striking feature of some cereal grains is the highly variable amounts of (1,3;1,4)-β-glucan that they contain; this can vary from close to zero in rice to 45% w/w in the starchy endosperm of *B. distachyon* Bd21 (Guillon et al., 2011). The starchy endosperm walls of *B. distachyon* are enormously thick compared with other cereals (**Figure 6**). In the Bd21 line there is a concomitant drop in grain starch content from values of 60–65% that are typical for grains of the Triticeae to 6% w/w (Guillon et al., 2011). Trafford et al. (2013) specifically compared grains of *B. distachyon* Bd21 and barley in terms of cell division, cell expansion, and endoreduplication during grain development. All of these processes were markedly reduced in Bd21, as were transcript levels of certain cell-cycle and starch biosynthesis genes. However, transcript levels of the (1,3;1,4)-β-glucan synthase genes, notably *BdCslF6*, were not affected. This lead to the hypothesis that the thick walls in *B. distachyon* grain are the result of continued accretion of (1,3;1,4)-β-glucan onto walls of cells that are not expanding (Trafford et al., 2013). Even though the endosperm walls of Bd21 are thicker, they contain a similar amount of (1,3;1,4)-β-glucan on a weight percentage of walls basis; the values are 80% w/w for Bd21 and about 70% w/w for barley endosperm walls. Trafford et al. (2013) suggested that if starch accumulation is a driver for cell expansion, as may occur in cereals such as wheat and barley, then the much lower level of starch synthesis in Bd21 may be primarily responsible for the reduced cell size and the concomitant re-direction of carbon into cell wall (1,3;1,4)-β-glucans.

The reasons for the variability of (1,3;1,4)-β-glucan content in cereal grains is not known, but it has been suggested that this polysaccharide acts as a secondary store of metabolizable glucose and that this function might be the key to the adoption of (1,3;1,4)-β-glucans during the evolution of the grasses (Burton and Fincher, 2012). It is clear that (1,3;1,4)-β-glucans are not essential structural components of cell walls in the Poaceae, because their levels are very low in some species and in many tissues of species that have high levels in their grain. It is equally clear that *B. distachyon* uses (1,3;1,4)-β-glucans as a storage polysaccharide in its grain, where small amounts of starch are present (Guillon et al., 2011). There is also indirect evidence that (1,3;1,4)-β-glucans are used as an alternative source of metabolizable energy in the leaves of barley seedlings (Roulin et al., 2002). During daylight hours, the (1,3;1,4)-β-glucan content of the leaves is about 10% w/w, but this rapidly decreases to close to zero when the plants are placed in the dark. Levels of the degradative enzymes, (1,3;1,4)-β-glucan endohydrolase isoenzyme EI and a broad-specificity β-glucan exohydrolase, increase when the plants are placed in the dark. When the lights are turned on again, the (1,3;1,4)-β-glucans rise to initial levels (Roulin et al., 2002). It has been argued that (1,3;1,4)-β-glucans would be a better short term form of stored glucose than starch, because they require a relatively simple enzyme system for both synthesis and subsequent depolymerization, and they can be deposited in the wall without the complexities of plastidial starch granule synthesis (Burton and Fincher, 2012). Might this be the reason that (1,3;1,4)-β-glucans are concentrated near the vasculature of young barley leaves (**Figure 7**)?

In the majority of cereal grains that have been examined, starch is the major storage carbohydrate. In *B. distachyon* storage metabolism seems to have shifted to (1,3;1,4)-β-glucan for unknown reasons, although disadvantageous mutations in starch synthase and cell cycle genes are possible explanations (Trafford et al., 2013). Grain anatomy, morphology, and development of *B. distachyon* has been well described (Guillon et al., 2011; Opanowicz et al., 2011; Hands and Drea, 2012; Hands et al., 2012) and grain characteristics have been compared with other domesticated cereals and non-crop species (Hands et al., 2012; Trafford et al., 2013). It is clear that the use of cell wall polysaccharides as a major source of storage carbohydrate in grains and seeds is not confined to *B. distachyon*. Of the grasses, species of *Bromus*, notably *Bromus mollis* (Hands and Drea, 2012), also possess thickened endosperm walls and a reduced starch content, suggesting that such a shift from starch to cell wall polysaccharides occurs elsewhere in the grasses but in general terms is unusual. However, a significant number of dicotyledonous seeds use cell wall polysaccharides rather than starch as the main storage medium in the endosperm (Buckeridge, 2010). These include mannans in coffee, lettuce, and tomato, glucomannan in orchids and galactomannans in legumes such as guar, fenugreek, and carob (Campbell and Reid, 1982; McCleary et al., 1985). In other examples, cell wall arabinogalactans are found as storage reserves in lupins, and xyloglucans are found in tamarind (Kumar and Bhattacharya, 2008) and nasturtium cotyledons (Edwards et al., 1985). Again, the shift from starch to wall polysaccharides is not widespread, but is seen in isolated species or genera.

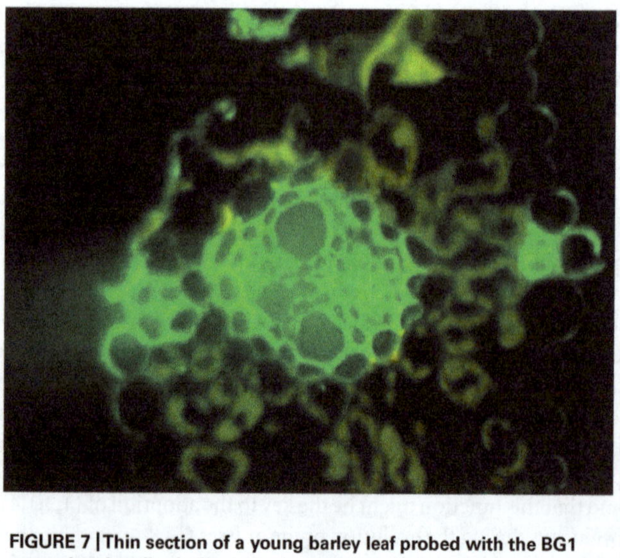

FIGURE 7 | Thin section of a young barley leaf probed with the BG1 monoclonal antibody. The high concentration of (1,3;1,4) β glucans can be seen around the vasculature and the polysaccharide appears to be associated with secondary cell walls of the vasculature and other cells. Reproduced with permission from Burton et al. (2011).

Where an alternative storage carbohydrate is found in storage tissues, one would expect to see a battery of corresponding hydrolytic enzymes expressed in the germinated grain or seed, to catalyze the efficient breakdown of the polysaccharide into component sugars for use by the growing embryo and young seedling. There is little information available on germination processes in *B. distachyon* grain but it would be interesting to see if the balance of hydrolytic enzymes has also adjusted to the paucity of starch and the dominance of (1,3;1,4)-β-glucan.

CONCLUDING REMARKS

The distinguishing features of cell walls in the grasses include the adoption of heteroxylans as the "core" non-cellulosic polysaccharide and the corresponding lower levels of xyloglucans and pectic polysaccharides. The widespread adoption of (1,3;1,4)-β-glucans in the Poaceae family also distinguishes the grasses from other monocotyledonous and dicotyledonous plants, although it is intriguing that (1,3;1,4)-β-glucans do not appear to be an essential structural component of walls in these species. One might question whether these distinguishing characteristics of cell walls of the grasses and their grain might in any way contribute to the obvious evolutionary success of the Poaceae family. It has been estimated that the grasses, which have appeared relatively recently in evolutionary history (Feuillet et al., 2008), now dominate plant ecosystems of about 20% of terrestrial land on the planet (Gaut, 2002). Does cell wall composition contribute to the ecological dominance of the grasses? Is the widespread adoption of (1,3;1,4)-β-glucans in walls of grass species and their possible function as a secondary source of metabolizable glucose important for the evolutionary success of grasses? It is clear that the (1,3;1,4)-β-glucans have appeared in other plant species, including in the walls of primitive dicots and in fungi, but they are widespread only in the Poaceae.

The non-cellulosic wall polysaccharides of plants are important components of dietary fiber and increased intake of dietary fiber has been advocated to reduce the risk of contracting serious human diseases, including colorectal cancer, type II diabetes, and cardiovascular disease (Collins et al., 2010; Burton and Fincher, 2014). Over-expression of the *HvCslF6* gene in barley led to more than 50% increases in dietary fiber in the transgenic grain (Burton et al., 2011). Given that cereal species are probably the most important source of caloric intake on Earth, one can predict that the non-cellulosic polysaccharides of cereal grains will play an important future role in human health. Conversely, the non-cellulosic wall polysaccharides present a number of technical problems in the malting and brewing industries and in animal feedstock formulations. It is therefore likely that the genetics, cell biology, and biochemistry of these polysaccharides and the enzymes that are responsible for their synthesis will remain the subjects of research interest in the immediate future. It can also be anticipated that advances in high-throughput genomics technologies, the increasing availability of complete genome sequences, and the continuing development of *in situ* methods will greatly facilitate that research.

ACKNOWLEDGMENTS

We thank the Australian Research Council for its long term support for this work. We also thank Natalie Kibble for her skilled technical assistance in the preparation of the manuscript.

REFERENCES

Anders, N., Wilkinson, M. D., Lovegrove, A., Freeman, J., Tryfona, T., Pellny, T. K., et al. (2012). Glycosyl transferases in family 61 mediate arabinofuranosyl transfer onto xylan in grasses. *Proc. Natl. Acad. Sci. U.S.A.* 109, 989–993. doi: 10.1073/pnas.1115858109

Andriunas, F. A., Zhang, H-M., Xia, X., Patrick, J. W., and Offler, C. E. (2013). Intersection of transfer cells with phloem biology: broad evolutionary trends, function, and induction. *Front. Plant Sci.* 4:221. doi: 10.3389/fpls.2013.00221

Arioli, T., Peng, L., Betzner, A. S., Burn, J., Wittke, W., Herth, W., et al. (1998). Molecular analysis of cellulose biosynthesis in *Arabidopsis. Science* 279, 717–720. doi: 10.1126/science.279.5351.717

Bacic, A., and Stone, B. A. (1981). Chemistry and organization of aleurone cell wall components from wheat and barley. *Funct. Plant Biol.* 8, 475–495. doi: 10.1071/PP9810475

Becraft, P. W., and Asuncion-Crabb, Y. (2000). Positional cues specify and maintain aleurone cell fate in maize endosperm development. *Development* 127, 4039–4048.

Becraft, P. W., and Yi, G. B. (2011). Regulation of aleurone development in cereal grains. *J. Exp. Bot.* 62, 1669–1675. doi: 10.1093/Jxb/Erq372

Brown, D. M., Goubet, F., Vicky, W. W. A., Goodacre, R., Stephens, E., Dupree, P., et al. (2007). Comparison of five xylan synthesis mutants reveals new insight into the mechanisms of xylan synthesis. *Plant J.* 52, 1154–1168. doi: 10.1111/j.1365-313X.2007.03307.x

Brown, D. M., Zhang, Z., Stephens, E., Dupree, P., and Turner, S. R. (2009). Characterization of IRX10 and IRX10-like reveals an essential role in glucuronoxylan biosynthesis in *Arabidopsis. Plant J.* 57, 732–746. doi: 10.1111/j.1365-313X.2008.03729.x

Brown, R. M., Li, L. K., Okuda, K., Kuga, S., Kudlicka, K., Drake, R., et al. (1994). In vitro cellulose synthesis in plants. *Plant Physiol.* 105, 1–2.

Buckeridge, M. S. (2010). Seed cell wall storage polysaccharides: models to understand cell wall biosynthesis and degradation. *Plant Physiol.* 154, 1017–1023. doi: 10.1104/pp.110.158642

Buliga, G. S., Brant, D. A., and Fincher, G. B. (1986). The sequence statistics and solution configuration of barley (1,3;1,4)-β-D-glucan. *Carbohyd. Res.* 157, 139–156. doi: 10.1016/0008-6215(86)85065-0

Burton, R. A., Collins, H. M., Kibble, N. A. J., Smith, J. A., Shirley, N. J., Jobling, S. A., et al. (2011). Over-expression of specific HvCslF cellulose synthase-like genes in transgenic barley increases the levels of cell wall (1,3;1,4)-β-D-glucans and alters their fine structure. *Plant Biotechnol. J.* 9, 117–135. doi: 10.1111/j.1467-7652.2010.00532.x

Burton, R. A., and Fincher, G. B. (2009). (1,3;1,4)-β-D-glucans in cell walls of the Poaceae, lower plants, and fungi: a tale of two linkages. *Mol. Plant* 2, 873–882. doi: 10.1093/mp/ssp063

Burton, R. A., and Fincher, G. B. (2012). Current challenges in cell wall biology in the cereals and grasses. *Front. Plant Sci.* 3:130. doi: 10.3389/fpls.2012.00130

Burton, R. A., and Fincher, G. B. (2014). Plant cell wall engineering: applications in biofuel production and improved human health. *Curr. Opin. Biotechnol.* 26, 79–84. doi: 10.1016/j.copbio.2013.10.007

Burton, R. A., Gidley, M. J., and Fincher, G. B. (2010). Heterogeneity in the chemistry, structure and function of plant cell walls. *Nat. Chem. Biol.* 6, 724–732. doi: 10.1038/nchembio.439

Burton, R. A., Wilson, S. M., Hrmova, M., Harvey, A. J., Shirley, N. J., Medhurst, A., et al. (2006). Cellulose synthase-like CslF genes mediate the synthesis of cell wall (1,3;1,4)-β-D-glucans. *Science* 311, 1940–1942. doi: 10.1126/science.1122975

Caffall, K. H., and Mohnen, D. (2009). The structure, function, and biosynthesis of plant cell wall pectic polysaccharides. *Carbohyd. Res.* 344, 1879–1900. doi: 10.1016/j.carres.2009.05.021

Campbell, J. M., and Reid, J. G. (1982). Galactomannan formation and guanosine 5′-diphosphate-mannose: galactomannan mannosyltransferase in developing seeds of fenugreek (*Trigonella foenum-graecum* L., *Leguminosae*). *Planta* 155, 105–111. doi: 10.1007/BF00392539

Chandrashekar, A., and Mazhar, H. (1999). The biochemical basis and implications of grain strength in sorghum and maize. *J. Cereal Sci.* 30, 193–207. doi: 10.1006/jcrs.1999.0264

Chateigner-Boutin, A. L., Bouchet, B., Alvarado, C., Bakan, B., and Guillon, F. (2014). The wheat grain contains pectic domains exhibiting specific spatial and development-associated distribution. *PLoS ONE* 9:e89620. doi: 10.1371/journal.pone.0089620

Chiniquy, D., Sharma, V., Schultink, A., Baidoo, E. E., Rautengarten, C., Cheng, K., et al. (2012). XAX1 from glycosyltransferase family 61 mediates xylosyltransfer to rice xylan. *Proc. Natl. Acad. Sci. U.S.A.* 109, 17117–17122. doi: 10.1073/pnas.1202079109

Cocuron, J. C., Lerouxel, O., Drakakaki, G., Alonso, A. P., Liepman, A. H., Keegstra, K., et al. (2007). A gene from the cellulose synthase-like C family encodes a beta-1,4 glucan synthase. *Proc. Natl. Acad. Sci. U.S.A.* 104, 8550–8555. doi: 10.1073/pnas.0703133104

Collins, H. M., Burton, R. A., Topping, D. L., Liao, M.-L., Bacic, A., and Fincher, G. B. (2010). Variability in fine structures of noncellulosic cell wall polysaccharides from cereal grains: potential importance in human health and nutrition. *Cereal Chem.* 87, 272–282. doi: 10.1094/CCHEM-87-4-0272

Dhugga, K. S., Barreiro, R., Whitten, B., Stecca, K., Hazebroek, J., Randhawa, G. S., et al. (2004). Guar seed beta-mannan synthase is a member of the cellulose synthase super gene family. *Science* 303, 363–366. doi: 10.1126/science.1090908

Doblin, M. S., De Melis, L., Newbigin, E., Bacic, A., and Read, S. M. (2001). Pollen tubes of *Nicotiana alata* express two genes from different beta-glucan synthase families. *Plant Physiol.* 125, 2040–2052. doi: 10.1104/pp.125.4.2040

Doblin, M. S., Kurek, I., Jacob-Wilk, D., and Delmer, D. P. (2002). Cellulose biosynthesis in plants: from genes to rosettes. *Plant Cell Physiol.* 43, 1407–1420. doi: 10.1093/pcp/pcf164

Doblin, M. S., Pettolino, F. A., Wilson, S. M., Campbell, R., Burton, R. A., Fincher, G. B., et al. (2009). A barley cellulose synthase-like CSLH gene mediates (1,3;1,4)-β-D-glucan synthesis in transgenic *Arabidopsis. Proc. Natl. Acad. Sci. U.S.A.* 106, 5996–6001. doi: 10.1073/pnas.0902019106

Eder, M., Tenhaken, R., Driouich, A., and Lütz-Meindl, U. (2008). Occurrence and characterization of arabinogalactan-like proteins and hemicelluloses in micrasterias (streptophyta) 1. *J. Phycol.* 44, 1221–1234. doi: 10.1111/j.1529-8817.2008.00576.x

Edwards, M., Dea, I. C., Bulpin, P. V., and Reid, J. G. (1985). Xyloglucan (amyloid) mobilisation in the cotyledons of *Tropaeolum majus* L. seeds following germination. *Planta* 163, 133–140. doi: 10.1007/BF00395907

Esau, K. (1977). *Anatomy of Seed Plants.* New York: John Wiley and Sons Inc.

Evers, T., and Millar, S. (2002). Cereal grain structure and development: some implications for quality. *J. Cereal Sci.* 36, 261–284. doi: 10.1006/jcrs.2002.0435

Farrokhi, N., Burton, R. A., Brownfield, L., Hrmova, M., Wilson, S. M., Bacic, A., et al. (2006). Plant cell wall biosynthesis: genetic, biochemical and functional genomics approaches to the identification of key genes. *Plant Biotechnol. J.* 4, 145–167. doi: 10.1111/j.1467-7652.2005.00169.x

Favery, B., Ryan, E., Foreman, J., Linstead, P., Boudonck, K., Steer, M., et al. (2001). KOJAK encodes a cellulose synthase-like protein required for root hair cell morphogenesis in *Arabidopsis. Genes Dev.* 15, 79–89. doi: 10.1101/gad.188801

Ferrari, B., Baronchelli, M., Stanca, A. M., and Gianinetti, A. (2010). Constitutive differences between steely and mealy barley samples associated with endosperm modification. *J. Sci. Food Agr.* 90, 2105–2113. doi: 10.1002/Jsfa.4058

Feuillet, C., Langridge, P., and Waugh, R. (2008). Cereal breeding takes a walk on the wild side. *Trends Genet.* 24, 24–32. doi: 10.1016/j.tig.2007.11.001

Fincher, G. B. (1975). Morphology and chemical composition of barley endosperm cell walls. *J. I. Brew.* 81, 116–122. doi: 10.1002/j.2050-0416.1975.tb03672.x

Fincher, G. B. (1989). Molecular and cellular biology associated with endosperm mobilization in germinating cereal-grains. *Annu. Rev. Plant Physiol.* 40, 305–346. doi: 10.1146/annurev.pp.40.060189.001513

Fincher, G. B. (2009). Revolutionary times in our understanding of cell wall biosynthesis and remodeling in the grasses. *Plant Physiol.* 149, 27–37. doi: 10.1104/pp.108.130096

Fincher, G. B., and Stone, B. A. (2004). *Chemistry of Nonstarch Polysaccharides.* Oxford: Elsevier.

Fry, S. C., Nesselrode, B. H. W. A., Miller, J. G., and Mewburn, B. R. (2008). Mixed-linkage (1 ⟶ 3,1 ⟶ 4)-β-D-glucan is a major hemicellulose of Equisetum (horsetail) cell walls. *New Phytol.* 179, 104–115. doi: 10.1111/j.1469-8137.2008.02435.x

Fulcher, R., Setterfield, G., Mccully, M., and Wood, P. (1977). Observations on the aleurone layer. II. Fluorescence microscopy of the aleurone–sub-aleurone junction with emphasis on possible β-1, 3-glucan deposits in barley. *Funct. Plant Biol.* 4, 917–928. doi: 10.1071/PP9770917

Fulcher, R. G., O'Brien, T. P., and Lee, J. W. (1972). Studies on the aluerone layer I. Conventional and fluorescence microscopy of the cell wall with emphasis on phenol-carbohydrate complexes in wheat. *Aust. J. Biol. Sci.* 25, 487–497.

Gaut, B. S. (2002). Evolutionary dynamics of grass genomes. *New Phytol.* 154, 15–28. doi: 10.1046/j.1469-8137.2002.00352.x

Gibeaut, D. M., Pauly, M., Bacic, A., and Fincher, G. B. (2005). Changes in cell wall polysaccharides in developing barley (*Hordeum vulgare*) coleoptiles. *Planta* 221, 729–738. doi: 10.1007/s00425-005-1481-0

Gillies, S. A., Futardo, A., and Henry, R. J. (2012). Gene expression in the developing aleurone and starchy endosperm of wheat. *Plant Biotechnol. J.* 10, 668–679. doi: 10.1111/j.1467-7652.2012.00705.x

Gubler, F., Ashford, A. E., and Jacobsen, J. V. (1987). The release of alpha-amylase through gibberellin-treated barley aleurone cell-walls: an immunocytochemical study with Lowi Cryl K4M. *Planta* 172, 155–161. doi: 10.1007/BF00394583

Guillon, F., Bouchet, B., Jamme, F., Robert, P., Quemener, B., Barron, C., et al. (2011). *Brachypodium distachyon* grain: characterization of endosperm cell walls. *J. Exp. Bot.* 62, 1001–1015. doi: 10.1093/Jxb/Erq332

Hands, P., and Drea, S. (2012). A comparative view of grain development in *Brachypodium distachyon. J. Cereal Sci.* 56, 2–8. doi: 10.1016/j.jcs.2011.12.010

Hands, P., Kourmpetli, S., Sharples, D., Harris, R. G., and Drea, S. (2012). Analysis of grain characters in temperate grasses reveals distinctive patterns of endosperm organization associated with grain shape. *J. Exp. Bot.* 63, 6253–6266. doi: 10.1093/Jxb/Ers281

Harris, P., and Fincher, G. B. (2009). "Distribution, fine structure and function of (1,3;1,4)-β-glucans in the grasses and other taxa," in *Chemistry, Biochemistry, and Biology of 1–3 Beta Glucans and Related Polysaccharides*, eds A. Bacic, G. B. Fincher, and B. A. Stone (Burlington, MA: Elsevier Academic Press), 621–654.

Hazen, S. P., Scott-Craig, J. S., and Walton, J. D. (2002). Cellulose synthase-like genes of rice. *Plant Physiol.* 128, 336–340. doi: 10.1104/pp.010875

Hoshikawa, K., and Wang, S. B. (1990). Studies on lodging in rice plants. 1. A general observation on lodged rice culms. *Jpn. J. Crop Sci.* 59, 809–814. doi: 10.1626/jcs.59.809

Iiyama, K., Lam, T. B. T., and Stone, B. A. (1990). Phenolic acid bridges between polysaccharides and lignin in wheat internodes. *Phytochemistry* 29, 733–737. doi: 10.1016/0031-9422(90)80009-6

Izydorczyk, M. S. (2014). "Barley arabinoxylans: molecular, physicochemical and functional properties," in *Barley: Chemistry and Technology*, 2nd Edn, eds P. R. Shewry and S. E. Ullrich (Minnesota, MN: AACC International Press), 97–122.

Jaaskelainen, A. S., Holopainen-Mantila, U., Tamminen, T., and Vuorinen, T. (2013). Endosperm and aleurone cell structure in barley and wheat as studied by optical and Raman microscopy. *J. Cereal Sci.* 57, 543–550. doi: 10.1016/j.jcs.2013.02.007

Jamme, F., Robert, P., Bouchet, B., Saulnier, L., Dumas, P., and Guillon, F. (2008). Aleurone cell walls of wheat grain: high spatial resolution investigation using synchrotron infrared microspectroscopy. *Appl. Spectros.* 62, 895–900. doi: 10.1366/000370208785284448

Jensen, J. K., Kim, H., Cocuron, J. C., Orler, R., Ralph, J., and Wilkerson, C. G. (2011). The DUF579 domain containing proteins IRX15 and IRX15: L affect xylan synthesis in *Arabidopsis. Plant J.* 66, 387–400. doi: 10.1111/j.1365-313X.2010.04475.x

Kumar, C. S., and Bhattacharya, S. (2008). Tamarind seed: properties, processing and utilization. *Crit. Rev. Food Sci. Nutr.* 48, 1–20. doi: 10.1080/10408390600948600

Lechat, H., Amat, M., Mazoyer, J., Buleon, A., and Lahaye, M. (2000). Structure and distribution of glucomannan and sulfated glucan in the cell walls of the red alga *Kappaphycus alvarezii* (gigartinales, rhodophyta). *J. Phycol.* 36, 891–902. doi: 10.1046/j.1529-8817.2000.00056.x

Liepman, A. H., Wilkerson, C. G., and Keegstra, K. (2005). Expression of cellulose synthase-like (Csl) genes in insect cells reveals that CslA family members encode mannan synthases. *Proc. Natl. Acad. Sci. U.S.A.* 102, 2221–2226. doi: 10.1073/pnas.0409179102

Lovegrove, A., Wilkinson, M. D., Freeman, J., Pellny, T. K., Tosi, P., Saulnier, L., et al. (2013). RNA interference suppression of genes in glycosyl transferase families 43 and 47 in wheat starchy endosperm causes large decreases in arabinoxylan content. *Plant Physiol.* 163, 95–107. doi: 10.1104/pp.113.222653

Mares, D. J., and Stone, B. A. (1973). Studies on wheat endosperm. 1. Chemical composition and ultrastructure of cell-walls. *Aust. J. Biol. Sci.* 26, 793–812.

McCleary, B. V., Clark, A. H., Dea, I., and Rees, D. A. (1985). The fine structures of carob and guar galactomannans. *Carbohyd. Res.* 139, 237–260. doi: 10.1016/0008-6215(85)90024-2

McCurdy, D. W., Patrick, J. W., and Offler, C. E. (2008). Wall ingrowth formation in transfer cells: novel examples of localized wall deposition in plant cells. *Curr. Opin. Plant Biol.* 11, 653–661. doi: 10.1016/j.pbi.2008.08.005

McFadden, G., Ahluwalia, B., Clarke, A. E., and Fincher, G. B. (1988). Expression sites and developmental regulation of genes encoding (1;3,1;4)-β-glucanase in germinated barley. *Planta* 173, 500–508. doi: 10.1007/BF00958963

Minifie, J., and Stone, B. A. (1988). Recovery of aleurone cells from wheat bran. Patent US4746073.

Mitchell, R. A. C., Dupree, P., and Shewry, P. R. (2007). A novel bioinformatics approach identifies candidate genes for the synthesis and feruloylation of arabinoxylan. *Plant Physiol.* 144, 43–53. doi: 10.1104/pp.106.094995

Morgan, J. L., Strumillo, J., and Zimmer, J. (2013). Crystallographic snapshot of cellulose synthesis and membrane translocation. *Nature* 493, 181–186. doi: 10.1038/nature11744

Morrison, I. N., and O'Brien, T. P. (1976). Cytokinesis in the developing wheat grain: division with and without a phragmoplast. *Planta* 130, 57–67. doi: 10.1007/BF00390845

Mortimer, J. C., Miles, G. P., Brown, D. M., Zhang, Z., Segura, M. P., Weimar, T., et al. (2010). Absence of branches from xylan in *Arabidopsis gux* mutants reveals potential for simplification of lignocellulosic biomass. *Proc. Natl. Acad. Sci. U.S.A.* 107, 17409–17414. doi: 10.1073/pnas.1005456107

Nunan, K. J., Sims, I. M., Bacic, A., Robinson, S. P., and Fincher, G. B. (1998). Changes in cell wall composition during ripening of grape berries. *Plant Physiol.* 118, 783–792. doi: 10.1104/pp.118.3.783

Offler, C. E., Liet, E., and Sutton, E. G. (1997). Transfer cell induction in cotyledons of *Vicia faba* L. *Protoplasma* 200, 51–64. doi: 10.1007/Bf01280734

Oikawa, A., Joshi, H. J., Rennie, E. A., Ebert, B., Manisseri, C., Heazlewood, J. L., et al. (2010). An integrative approach to the identification of *Arabidopsis* and rice genes involved in xylan and secondary wall development. *PLoS ONE* 5:e15481. doi: 10.1371/journal.pone.0015481

Okarter, N., and Liu, R. H. (2010). Health benefits of whole grain phytochemicals. *Crit. Rev. Food Sci.* 50, 193–208. doi: 10.1080/10408390802248734

Olsen, O. A. (2001). Endosperm development: cellularization and cell fate specification. *Annu. Rev. Plant Physiol. Plant Mol. Biol.* 52, 233–267. doi: 10.1146/annurev.arplant.52.1.233

Olsen, O. A. (2004). Nuclear endosperm development in cereals and *Arabidopsis thaliana. Plant Cell* 16, S214–S227. doi: 10.1105/tpc.017111

Opanowicz, M., Hands, P., Betts, D., Parker, M. L., Toole, G. A., Mills, E. N. C., et al. (2011). Endosperm development in *Brachypodium distachyon. J. Exp. Bot.* 62, 735–748. doi: 10.1093/Jxb/Erq309

Papageorgiou, M., Lakhdara, N., Lazaridou, A., Biliaderis, C., and Izydorczyk, M. (2005). Water extractable (1 ⟶ 3, 1 ⟶ 4)-β-D-glucans from barley and oats: an intervarietal study on their structural features and rheological behaviour. *J. Cereal Sci.* 42, 213–224. doi: 10.1016/j.jcs.2005.03.002

Pattathil, S., Avci, U., Baldwin, D., Swennes, A. G., Mcgill, J. A., Popper, Z., et al. (2010). A comprehensive toolkit of plant cell wall glycan-directed monoclonal antibodies. *Plant Physiol.* 153, 514–525. doi: 10.1104/pp.109.151985

Paulson, I. W. (1969). Embryogeny and caryopsis development of *Sorghum bicolor* (L.) Moench. *Crop Sci.* 9, 97–102. doi: 10.2135/cropsci1969.0011183X000900010034x

Pear, J. R., Kawagoe, Y., Schreckengost, W. E., Delmer, D. P., and Stalker, D. M. (1996). Higher plants contain homologs of the bacterial Cela genes encoding the catalytic subunit of cellulose synthase. *Proc. Natl. Acad. Sci. U.S.A.* 93, 12637–12642. doi: 10.1073/pnas.93.22.12637

Pellny, T. K., Lovegrove, A., Freeman, J., Tosi, P., Love, C. G., Knox, J. P., et al. (2012). Cell walls of developing wheat starchy endosperm: comparison of composition and RNA-Seq transcriptome. *Plant Physiol.* 158, 612–627. doi: 10.1104/pp.111.189191

Pena, M. J., Zhong, R., Zhou, G-K., Richardson, E. A., O'Neill, M. A., Darvill, A. G., et al. (2007). *Arabidopsis* irregular xylem8 and irregular xylem9: implications for the complexity of glucuronoxylan biosynthesis. *Plant Cell* 19, 549–563. doi: 10.1105/tpc.106.049320

Persson, S., Caffall, K. H., Freshour, G., Hilley, M. T., Bauer, S., Poindexter, P., et al. (2007). The *Arabidopsis* irregular xylem8 mutant is deficient in glucuronoxylan and homogalacturonan, which are essential for secondary cell wall integrity. *Plant Cell* 19, 237–255. doi: 10.1105/tpc.106.047720

Pettolino, F., Sasaki, I., Turbic, A., Wilson, S. M., Bacic, A., Hrmova, M., et al. (2009). Hyphal cell walls from the plant pathogen *Rhynchosporium secalis* contain (1,3;1,6)-beta-D-glucans, galacto- and rhamnomannans, (1,3;1,4)-beta-D-glucans and chitin. *FEBS J.* 276, 3698–3709. doi: 10.1111/j.1742-4658.2009.07086.x

Philippe, S., Saulnier, L., and Guillon, F. (2006). Arabinoxylan and (1,3;1,4)-beta-glucan deposition in cell walls during wheat endosperm development. *Planta* 224, 449–461. doi: 10.1007/s00425-005-0209-5

Popper, Z. A., and Fry, S. C. (2003). Primary cell wall composition of bryophytes and charophytes. *Ann. Bot.* 91, 1–12. doi: 10.1093/aob/mcg013

Popper, Z. A., and Tuohy, M. G. (2010). Beyond the green: understanding the evolutionary puzzle of plant and algal cell walls. *Plant Physiol.* 153, 373–383. doi: 10.1104/pp.110.158055

Pugh, D. A., Offler, C. E., Talbot, M. J., and Ruan, Y.-L. (2010). Evidence for the role of transfer cells in the evolutionary increase in seed and fiber biomass yield in cotton. *Mol. Plant* 3, 1075–1086. doi: 10.1093/mp/ssq054

Rantanen, H., Virkki, L., Tuomainen, P., Kabel, M., Schols, H., and Tenkanen, M. (2007). Preparation of arabinoxylobiose from rye xylan using family 10 *Aspergillus aculeatus* endo-1,4-beta-D-xylanase. *Carbohyd. Polym.* 68, 350–359. doi: 10.1016/j.carbpol.2006.11.022

Rasmussen, S. K., and Shu, X. (2014). Quantification of amylose, amylopectin, and β-glucan in search for genes controlling the three major quality traits in barley by genome-wide association studies. *Front. Plant Sci.* 5:197. doi: 10.3389/fpls.2014.00197

Richmond, T., and Somerville, C. (2000). The cellulose synthase superfamily. *Plant Physiol.* 124, 495–498. doi: 10.1104/pp.124.2.495

Rennie, E. A., Hansen, S. F., Baidoo, E. E., Hadi, M. Z., Keasling, J. D., and Scheller, H. V. (2012). Three members of the *Arabidopsis* glycosyltransferase family 8 are xylan glucuronosyltransferases. *Plant Physiol.* 159, 1408–1417. doi: 10.1104/pp.112.200964

Robert, P., Jamme, F., Barron, C., Bouchet, B., Saulnier, L., Dumas, P., et al. (2011). Change in wall composition of transfer and aleurone cells during wheat grain development. *Planta* 233, 393–406. doi: 10.1007/s00425-010-1306-1307

Roulin, S., Buchala, A. J., and Fincher, G. B. (2002). Induction of (1-3,1-4)-β-D-glucan hydrolases in leaves of dark-incubated barley seedlings. *Planta* 215, 51–59. doi: 10.1007/s00425-001-0721-1

Sabelli, P. A., and Larkins, B. A. (2009). The development of endosperm in grasses. *Plant Physiol.* 149, 14–26. doi: 10.1104/pp.108.129437

Saulnier, L., Robert, P., Grintchenko, M., Jamme, F., Bouchet, B., and Guillon, F. (2009). Wheat endosperm cell walls: spatial heterogeneity of polysaccharide structure and composition using micro-scale enzymatic fingerprinting and FT-IR microspectroscopy. *J. Cereal Sci.* 50, 312–317. doi: 10.1016/j.jcs.2009.05.003

Sethaphong, L., Haigler, C. H., Kubicki, J. D., Zimmer, J., Bonetta, D., Debolt, S., et al. (2013). Tertiary model of a plant cellulose synthase. *Proc. Natl. Acad. Sci. U.S.A.* 110, 7512–7517. doi: 10.1073/pnas.1301027110

Shibuya, N., and Iwasaki, T. (1978). Polysaccharides and glycoproteins in rice endosperm cell-wall. *Agric. Biol. Chem. Tokyo* 42, 2259–2266. doi: 10.1271/bbb1961.42.2259

Shibuya, N., and Iwasaki, T. (1985). Structural features of rice bran hemicellulose. *Phytochemistry* 24, 285–289. doi: 10.1016/S0031-9422(00)83538-4

Shibuya, N., and Misaki, A. (1978). Structure of hemicellulose isolated from rice endosperm cell wall: mode of linkages and sequences in xyloglucan, beta-glucan and arabinoxylan. *Agric. Biol. Chem. Tokyo* 42, 2267–2274. doi: 10.1271/bbb1961.42.2267

Shibuya, N., and Nakane, R. (1984). Pectic polysaccharides of rice *Oryza sativa* cultivar norin-29 endosperm cell walls. *Phytochemistry (Oxford)* 23, 1425–1430. doi: 10.1016/S0031-9422(00)80479-3

Sørensen, I., Pettolino, F. A., Wilson, S. M., Doblin, M. S., Johansen, B., Bacic, A., et al. (2008). Mixed-linkage (1,3;1,4)-beta-D-glucan is not unique to the poales and is an abundant component of *Equisetum arvense* cell walls. *Plant J.* 54, 510–521. doi: 10.1111/j.1365-313X.2008.03453.x

Srinivas, T. (1975). Pattern of crack formation in rice grain as influenced by shape and orientation of cells. *J. Sci. Food Agric.* 26, 1479–1482. doi: 10.1002/jsfa.2740261006

Staudte, R., Woodward, J., Fincher, G., and Stone, B. (1983). Water-soluble (1,3;1,4)-β-D-glucans in barley (*Hordeum vulgare*) endosperm. III. Distribution of celltriosyl and cellotetraosyl residues. *Carbohyd. Polym.* 3, 299–312. doi: 10.1016/0144-8617(83)90027-9

Sterling, J. D., Atmodjo, M. A., Inwood, S. E., Kolli, V. S. K., Quigley, H. F., Hahn, M. G., et al. (2006). Functional identification of an *Arabidopsis* pectin biosynthetic homogalacturonan galacturonosyltransferase. *Proc. Natl. Acad. Sci. U.S.A.* 103, 5236–5241. doi: 10.1073/pnas.0600120103

Stone, B. A., and Clarke, A. E. (1992). *Chemistry and Biology of (1,3)-β-D-Glucan.* Victoria, Australia: La Trobe University Press

Taiz, L., and Jones, R. L. (1973). Plasmodesmata and an associated cell wall component in barley aleurone tissue. *Am. J. Bot.* 60, 67–75. doi: 10.2307/2441325

Taketa, S., Yuo, T., Tonooka, T., Tsumuraya, Y., Inagaki, Y., Haruyama, N., et al. (2012). Functional characterization of barley betaglucanless mutants demonstrates a unique role for CslF6 in (1,3;1,4)-β-D-glucan biosynthesis. *J. Exp. Biol.* 63, 381–392. doi: 10.1093/jxb/err285

Talbot, M. J., Franceschi, V. R., McCurdy, D. W., and Offler, C. E. (2001). Wall ingrowth architecture in epidermal transfer cells of *Vicia faba* cotyledons. *Protoplasma* 215, 191–203. doi: 10.1007/BF01280314

Talbot, M. J., Offler, C. E., and McCurdy, D. W. (2002). Transfer cell wall architecture: a contribution towards understanding localized wall deposition. *Protoplasma* 219, 197–209. doi: 10.1007/s007090200021

Terrell, E. E. (1971). Survey of occurrences of liquid or soft endosperm in grass genera. *Bull. Torrey Bot. Club* 264–268. doi: 10.2307/2483625

Thiel, J. (2014). Development of endosperm transfer cells in barley. *Front. Plant Sci.* 5:108. doi: 10.3389/Fpls.2014.00108

Thiel, J., Hollmann, J., Rutten, T., Weber, H., Scholz, U., and Weschke, W. (2012a). 454 Transcriptome sequencing suggests a role for two-component signalling in cellularization and differentiation of barley endosperm transfer cells. *PLoS ONE* 7:e41867. doi: 10.1371/journal.pone.0041867

Thiel, J., Riewe, D., Rutten, T., Melzer, M., Friedel, S., Bollenbeck, F., et al. (2012b). Differentiation of endosperm transfer cells of barley: a comprehensive analysis at the micro-scale. *Plant J.* 71, 639–655. doi: 10.1111/j.1365-313X.2012.05018.x

Toole, G. A., Barron, C., Le Gall, G., Colquhoun, I. J., Shewry, P. R., and Mills, E. N. C. (2009). Remodelling of arabinoxylan in wheat (*Triticum aestivum*) endosperm cell walls during grain filling. *Planta* 229, 667–680. doi: 10.1007/s00425-008-0865-863

Toole, G. A., Le Gall, G., Colquhoun, I. J., Drea, S., Opanowicz, M., Bedo, Z., et al. (2012). Spectroscopic analysis of diversity in the spatial distribution of arabinoxylan structures in endosperm cell walls of cereal species in the HEALTHGRAIN diversity collection. *J. Cereal Sci.* 56, 134–141. doi: 10.1016/j.jcs.2012.02.016

Toole, G. A., Le Gall, G., Colquhoun, I. J., Nemeth, C., Saulnier, L., Lovegrove, A., et al. (2010). Temporal and spatial changes in cell wall composition in developing grains of wheat cv. *Hereward. Planta* 232, 677–689. doi: 10.1007/s00425-010-1199-1195

Toole, G. A., Wilson, R. H., Parker, M. L., Wellner, N. K., Wheeler, T. R., Shewry, P. R., et al. (2007). The effect of environment on endosperm cell-wall development in *Triticum aestivum* during grain filling: an infrared spectroscopic imaging study. *Planta* 225, 1393–1403. doi: 10.1007/s00425-006-0448-440

Trafford, K., and Fincher, G. B. (2014). "Barley grain carbohydrates: starch and cell walls," in *Barley: Chemistry and Technology*, eds P. R. Shewry and S. E. Ullrich (Minnesota, MN: AACC International Press), 71–95.

Trafford, K., Haleux, P., Henderson, M., Parker, M., Shirley, N. J., Tucker, M. R., et al. (2013). Grain development in *Brachypodium* and other grasses: possible interactions between cell expansion, starch deposition, and cell-wall synthesis. *J. Exp. Bot.* 64, 5033–5047. doi: 10.1093/Jxb/Ert292

Trethewey, J. A. K., Campbell, L. M., and Harris, P. J. (2005). (1,3;1,4)-β-D-Glucans in the cell walls of the poales (sensu lato): an immunogold labeling study using a monoclonal antibody. *Am. J. Bot.* 92, 1660–1674. doi: 10.3732/ajb.92.10.1660

Urbanowicz, B. R., Peña, M. J., Ratnaparkhe, S., Avci, U., Backe, J., Steet, H. F., et al. (2012). 4-*O*-methylation of glucuronic acid in *Arabidopsis* glucuronoxylan is catalyzed by a domain of unknown function family 579 protein. *Proc. Natl. Acad. Sci. U.S.A.* 109, 14253–14258. doi: 10.1073/pnas.1208097109

Van der Eb, A., and Nieuwdorp, P. (1967). Electron microscopic structure of the aleuron cells of barley during germination. *Acta Bot. Neerl.* 15, 690–699. doi: 10.1111/j.1438-8677.1966.tb00263.x

Veličković, D., Ropartz, D., Guillon, F., Saulnier, L., and Rogniaux, H. (2014). New insights into the structural and spatial variability of cell-wall polysaccharides during wheat grain development, as revealed through MALDI mass spectrometry imaging. *J. Exp. Bot.* 65, 2079–2091. doi: 10.1093/jxb/eru065

Verhertbruggen, Y., Marcus, S. E., Haeger, A., Ordaz-Ortiz, J. J., and Knox, J. P. (2009). An extended set of monoclonal antibodies to pectic homogalacturonan. *Carbohyd. Res.* 344, 1858–1862. doi: 10.1016/j.carres.2008.11.010

Wang, H. L., Offler, C. E., and Patrick, J. W. (1994). Nucellar projection transfer cells in the developing wheat grain. *Protoplasma* 182, 39–52. doi: 10.1007/Bf01403687

Wang, X., Cnops, G., Vanderhaeghen, R., De Block, S., Van Montagu, M., and Van Lijsebettens, M. (2001). AtCSLD3, a cellulose synthase-like gene important for root hair growth in *Arabidopsis*. *Plant Physiol.* 126, 575–586. doi: 10.1104/pp.126.2.575

Waniska, R. D. (2000). "Structure, phenolic compounds, and antifungal proteins of sorghum caryopses," in *Technical and Institutional Options for Sorghum Grain Mold Management. Proceedings of an International Consultation, 18–19 May 2000*, eds A. Chandrashekar, R. Bandyopadhyay, and A. J. Hall (Patancheru: ICRISAT), 7–33.

Waterkeyn, L. (1967). Sur l'existance d'un 'stade callosique' presente par la paroi cellulaire, au cours de la cytocinese. *C. R. Acad. Sci. D* 265, 1792–1794.

Wilson, S. M., Burton, R. A., Collins, H. M., Doblin, M. S., Pettolino, F. A., Shirley, N., et al. (2012). Pattern of deposition of cell wall polysaccharides and transcript abundance of related cell wall synthesis genes during differentiation in barley endosperm. *Plant Physiol.* 159, 655–670. doi: 10.1104/pp.111.192682

Wilson, S. M., Burton, R. A., Doblin, M. S., Stone, B. A., Newbigin, E. J., Fincher, G. B., et al. (2006). Temporal and spatial appearance of wall polysaccharides during cellularization of barley (*Hordeum vulgare*) endosperm. *Planta* 224, 655–667. doi: 10.1007/s00425-006-0244-x

Wood, P. J., Fulcher, R. G., and Stone, B. A. (1983). Studies on the specificity of interaction of cereal cell-wall components with Congo red and calcofluor: specific detection and histochemistry of (1-3),(1-4) -β-D-glucan. *J. Cereal Sci.* 1, 95–110. doi: 10.1016/S0733-5210(83)80027-7

Woodward, J. R., Fincher, G. B., and Stone, B. (1983). Water-soluble (1-3,1-4)-β-D-glucans from barley (*Hordeum vulgare*) endosperm. II. Fine structure. *Carbohyd. Polym.* 3, 207–225. doi: 10.1016/0144-8617(83)90019-X

Yin, Y., Huang, J., and Xu, Y. (2009). The cellulose synthase superfamily in fully sequenced plants and algae. *BMC Plant Biol.* 9:99. doi: 10.1186/1471-22 29-9-99

Young, T. E., and Gallie, D. R. (2000). "Programmed cell death during endosperm development," in *Programmed Cell Death in Higher Plants*, eds E. Lam, H. Fukuda, and J. Greenberg (Amsterdam: Springer), 39–57.

Zablackis, E., Huang, J., Muller, B., Darvill, A. G., and Albersheim, P. (1995). Characterization of the cell-wall polysaccharides of *Arabidopsis thaliana* leaves. *Plant Physiol.* 107, 1129–1138. doi: 10.1104/pp.107.4.1129

Zeng, W., Jiang, N., Nadella, R., Killen, T. L., Nadella, V., and Faik, A. (2010). A glucurono(arabino)xylan synthase complex from wheat contains members of the GT43, GT47, and GT75 families and functions cooperatively. *Plant Physiol.* 154, 78–97. doi: 10.1104/pp.110.159749

Zhang, Q., Cheetamun, R., Dhugga, K. S., Rafalski, J. A., Tingey, S. V., Shirley, N. J., et al. (2014). Spatial gradients in cell wall composition and transcriptional profiles along elongating maize internodes. *BMC Plant Biol.* 14:27. doi: 10.1186/1471-2229-14-27

Zheng, Y. K., and Wang, Z. (2010). Current opinions on endosperm transfer cells in maize. *Plant Cell Rep.* 29, 935–942. doi: 10.1007/s00299-010-0891-z

Zheng, Y. K., and Wang, Z. (2011). Contrast observation and investigation of wheat endosperm transfer cells and nucellar projection transfer cells. *Plant Cell Rep.* 30, 1281–1288. doi: 10.1007/s00299-011-1039-1035

Conflict of Interest Statement: The authors declare that the research was conducted in the absence of any commercial or financial relationships that could be construed as a potential conflict of interest.

Permissions

All chapters in this book were first published in ASB, by Frontiers Media SA; hereby published with permission under the Creative Commons Attribution License or equivalent. Every chapter published in this book has been scrutinized by our experts. Their significance has been extensively debated. The topics covered herein carry significant findings which will fuel the growth of the discipline. They may even be implemented as practical applications or may be referred to as a beginning point for another development.

The contributors of this book come from diverse backgrounds, making this book a truly international effort. This book will bring forth new frontiers with its revolutionizing research information and detailed analysis of the nascent developments around the world.

We would like to thank all the contributing authors for lending their expertise to make the book truly unique. They have played a crucial role in the development of this book. Without their invaluable contributions this book wouldn't have been possible. They have made vital efforts to compile up to date information on the varied aspects of this subject to make this book a valuable addition to the collection of many professionals and students.

This book was conceptualized with the vision of imparting up-to-date information and advanced data in this field. To ensure the same, a matchless editorial board was set up. Every individual on the board went through rigorous rounds of assessment to prove their worth. After which they invested a large part of their time researching and compiling the most relevant data for our readers.

The editorial board has been involved in producing this book since its inception. They have spent rigorous hours researching and exploring the diverse topics which have resulted in the successful publishing of this book. They have passed on their knowledge of decades through this book. To expedite this challenging task, the publisher supported the team at every step. A small team of assistant editors was also appointed to further simplify the editing procedure and attain best results for the readers.

Apart from the editorial board, the designing team has also invested a significant amount of their time in understanding the subject and creating the most relevant covers. They scrutinized every image to scout for the most suitable representation of the subject and create an appropriate cover for the book.

The publishing team has been an ardent support to the editorial, designing and production team. Their endless efforts to recruit the best for this project, has resulted in the accomplishment of this book. They are a veteran in the field of academics and their pool of knowledge is as vast as their experience in printing. Their expertise and guidance has proved useful at every step. Their uncompromising quality standards have made this book an exceptional effort. Their encouragement from time to time has been an inspiration for everyone.

The publisher and the editorial board hope that this book will prove to be a valuable piece of knowledge for researchers, students, practitioners and scholars across the globe.

List of Contributors

James A. Birchler
Division of Biological Science, University of Missouri-Columbia, Columbia, MO, USA

YongruiWu
Institute of Plant Physiology and Ecology, Shanghai Institutes for Biological Sciences, Chinese Academy of Sciences, Shanghai, China

Joachim Messing
Waksman Institute of Microbiology, Rutgers University, Piscataway, NJ, USA

Na Li and Yunhai Li
State Key Laboratory of Plant Cell and Chromosome Engineering, Institute of Genetics and Developmental Biology, Chinese Academy of Sciences, Beijing, China

Gad Galili
Department of Plant Sciences, TheWeizmann Institute of Science, Rehovot, Israel

Tamar Avin-Wittenberg and Alisdair R. Fernie
Max-Planck-Institut für Molekulare Pflanzenphysiologie, Potsdam-Golm, Germany

Ruthie Angelovici
Department of Biochemistry and Molecular Biology, Michigan State University, East Lansing, MI, USA

Fernando Domínguez and Francisco J. Cejudo
Instituto de Bioquímica Vegetal y Fotosíntesis, Universidad de Sevilla – Consejo Superior de Investigaciones Científicas, Sevilla, Spain

Elsa Arcalis, Verena Ibl, Jenny Peters, Stanislav Melnik and Eva Stoger
Department of Applied Genetics and Cell Biology, University of Natural Resources and Life Sciences, Vienna, Austria

Fang Bai and A. M. Settles
Horticultural Sciences Department and Plant Molecular and Cellular Biology Program, University of Florida, Gainesville, FL, USA

Davide Mainieri, Francesca Morandini, Marie Maîtrejean, Andrea Saccani, Emanuela Pedrazzini and Alessandro Vitale
Istituto di Biologia e Biotecnologia Agraria, Consiglio Nazionale delle Ricerche, Milano, Italy

Andrea Bleckmann and Thomas Dresselhaus
Cell Biology and Plant Biochemistry, Biochemie-Zentrum Regensburg, University of Regensburg, Regensburg, Germany

Svenja Alter
Plant Breeding, Center of Life and Food Sciences Weihenstephan, Technische Universität München, Freising, Germany

Petr Smýkal
Department of Botany, Faculty of Sciences, Palacký University in Olomouc, Olomouc, Czech Republic

Vanessa Vernoud and Richard D. Thompson
INRA, UMR 1347 Agroécologie, Dijon, France

Matthew W. Blair
Department of Agricultural and Environmental Sciences, Tennessee State University, Nashville, TN, USA

Aleš Soukup
Department of Experimental Plant Biology, Charles University, Prague, Czech Republic

Volodymyr Radchuk and Ljudmilla Borisjuk
Heterosis, Molecular Genetics, Leibniz-Institut für Pflanzengenetik und Kulturpflanzenforschung, Gatersleben, Germany

Michael G. Becker and Mark F. Belmonte
Department of Biological Sciences, University of Manitoba,Winnipeg, MB, Canada

Ssu-Wei Hsu and John J. Harada
Department of Plant Biology, University of California Davis, Davis, CA, USA

Eliot M. Herman
School of Plant Sciences, BIO5 Institute, University of Arizona, Tucson, AZ, USA

Ricardo A. Dante
Embrapa Agricultural Informatics, Campinas, Brazil

Paolo A. Sabelli
School of Plant Sciences, University of Arizona, Tucson, AZ, USA

Brian A. Larkins
Department of Agronomy and Horticulture, University of Nebraska, Lincoln, NE, USA
School of Plant Sciences, University of Arizona, Tucson, AZ, USA

Hiroyuki Nonogaki
Department of Horticulture, Oregon State University, Corvallis, OR, USA

Catherine Ravel, Samuel Fiquet, Julie Boudet, Mireille Dardevet, Jonathan Vincent, Marielle Merlino, Robin Michard and Pierre Martre
Institut National de la Recherche Agronomique, UMR1095, Genetics, Diversity and Ecophysiology of Cereals, Clermont-Ferrand, France
UMR1095, Genetics, Diversity and Ecophysiology of Cereals, Department of Biology, Blaise Pascal University, Aubière, France

David R. Holding
Department of Agronomy and Horticulture, Center for Plant Science Innovation, University of Nebraska-Lincoln, Lincoln, NE, USA

Paolo A. Sabelli, Ricardo A. Dante, Hong N. Nguyen, and Brian A. Larkins
School of Plant Sciences, University of Arizona, Tucson, AZ, USA

William J. Gordon-Kamm
Pioneer Hi-Bred International Inc., Johnston, IO, USA

Hernán P. Burrieza, María P. López-Fernández and Sara Maldonado
Instituto de Biodiversidad y Biologia Experimental y Aplicada - Consejo Nacional de Investigaciones Científicas y Técnicas, Ciudad Autónoma de Buenos Aires, Argentina
Departamento de Biodiversidad y Biología Experimental, Facultad de Ciencias Exactas y Naturales, Universidad de Buenos Aires, Ciudad Autónoma de Buenos Aires, Argentina

Rachel A. Burton and Geoffrey B. Fincher
Australian Research Council Centre of Excellence in Plant Cell Walls - School of Agriculture, Food and Wine, University of Adelaide, Glen Osmond, SA, Australia

Index

www.ingramcontent.com/pod-product-compliance
Lightning Source LLC
LaVergne TN
LVHW060536210926
848485LV00005B/1208